W9-DDG-064

# ENVIRONMENTAL SCIENCE

## Foundations and Applications

# ENVIRONMENTAL SCIENCE

## Foundations and Applications

**Andrew Friedland**
*Dartmouth College*

**Rick Relyea**
*University of Pittsburgh*

**David Courard-Hauri**
*Drake University*

(With Ross Jones and Susan Weisberg)

W. H. Freeman and Company

Publisher: Peter Marshall
Senior Acquisitions Editor: Jerry Correa
Executive Marketing Manager: Debbie Clare
Developmental Editor: Rebecca Kohn
Art Development: Lee Wilcox
Project Manager: Karen Misler
Media Editor: Patrick Shriner
Supplements Editor: Amanda Nietzel
Assistant Editor: Marni Rolfes
Photo Editor: Ted Szczepanski
Photo Researcher: Debbie Goodsite
Cover Designer: Diana Blume
Text Designer: Lissi Sigillo
Senior Project Editor: Mary Louise Byrd
Illustrations: Precision Graphics
Illustration Coordinator: Bill Page
Production Coordinator: Julia DeRosa
Composition: Sheridan Sellers
Printing and Binding: QuadGraphics

**Photo Credits for Opening Pages for Science Applied Essays**

*solar cells* Andreas Weber/iStockphoto.com
*faucet* Jozsef Szasz-Fabian/Dreamstime.com
*people* Phil Degginger/Alamy
*polar bear* Jan Will/iStockphoto.com
*farmer's market* Dan Moore/iStockphoto.com

Library of Congress Control Number: 2010940423
ISBN-13: 978-1-4292-4029-1
ISBN-10: 0-4292-4029-6

Printed in the United States of America

First printing

W. H. Freeman and Company
41 Madison Avenue
New York, NY 10010
Houndmills, Basingstoke RG21 6XS, England
www.whfreeman.com

To Katie, Jared, and Ethan for their interest and enthusiasm

—AJF

To Christine, Isabelle, and Wyatt for their patience and inspiration

—RAR

# Brief Contents

# Contents

**Andrew Friedland** is Richard and Jane Pearl Professor in Environmental Studies and chair of the Environmental Studies Program at Dartmouth College. Professor Friedland is known internationally for his work on the biogeochemistry of lead cycling in forests of the northeastern United States and for describing and documenting forest decline. For more than two decades, Professor Friedland has been investigating the effects of air pollution on the cycling of carbon, nitrogen, and lead in high-elevation forests of New England and the Northeast. He participated in documenting and searching for causes of red spruce decline in the montane regions of New England and New York in the 1980s and 1990s. Recently, he has been examining the impact of increased demand for wood as a fuel and the subsequent effect on carbon stored deep in forest soils.

Professor Friedland has served on panels for the National Science Foundation, the USDA Forest Service, and the Science Advisory Board of the Environmental Protection Agency. He has authored or coauthored 60 peer-reviewed publications and one book, *Writing Successful Science Proposals* (Yale University Press). He received BA degrees in biology and environmental studies and a PhD in Earth and environmental science from the University of Pennsylvania. He currently teaches introductory environmental science and energy courses. He has taught courses in forest biogeochemistry, global change, and soil science, as well as foreign study courses in Kenya.

Professor Friedland is passionate about saving energy and can be seen wandering the halls of the Environmental Studies Program at Dartmouth with a Kill A Watt meter, determining the electricity load of vending machines, data projectors, and computers. On weekends, he likes to hike in the woods and track wildlife in the snow with his two sons.

**Rick Relyea** is a professor of biology at the University of Pittsburgh and director of the Pymatuning Laboratory of Ecology. He is recognized throughout the world for his work in the fields of ecology and toxicology. Professor Relyea has served on multiple scientific panels for the National Science Foundation and is an associate editor for the journals of the Ecological Society of America. For two decades, he has conducted research on a wide range of topics including community ecology, evolution, animal behavior, and ecotoxicology. He has authored more than 80 scientific articles and book chapters and has presented research seminars throughout the world. In 2005, he was named the Chancellor's Distinguished Researcher at the University of Pittsburgh.

Professor Relyea teaches courses in ecology, evolution, and animal behavior at the undergraduate and graduate levels. He received a BS in environmental forest biology from the State University of New York College of Environmental Science and Forestry, an MS in wildlife management from Texas Tech University, and a PhD in ecology and evolution from the University of Michigan.

When not writing about the environment, Professor Relyea enjoys walking in the woods with his family. During the cold winters in Pennsylvania, he likes to build furniture from locally logged trees.

**David Courard-Hauri** is associate professor of environmental science and policy at Drake University. He teaches courses on environmental science, climate change science and policy, quantitative methods in environmental decision making, and ecological economics. He received an MA in public affairs from Princeton's Woodrow Wilson School and a PhD in chemistry from Stanford University. His work focuses on complex dynamical systems and the sources of nonproductive consumption.

# Content Advisory Board

 **Art Samel** is an associate professor at the Center of Environmental Programs at Bowling Green University. He has a joint appointment with the department of geography, where he is the chair.

 **Teri C. Balser** is an associate professor of soil and ecosystem science at the University of Wisconsin, Madison. She teaches undergraduate and graduate-level courses in soil biology, ecosystem microbiology, honors introductory biology, and environmental studies.

 **Dean Goodwin** is an adjunct faculty member at Plymouth State University, the University of New Hampshire, and Rappahannock Community College, Virginia.

 **Michael L. Denniston** is an associate professor of chemistry at Georgia Perimeter College. He teaches general chemistry and environmental science.

 **Jeffery A. Schneider** is an assistant professor of environmental chemistry at the State University of New York in Oswego, New York. He teaches general chemistry, environmental science, and environmental chemistry.

We are delighted to introduce our new textbook: *Environmental Science: Foundations and Applications*. Our mission has been to create a book that provides in-depth coverage of the core topics in the environmental science course with a contemporary, holistic approach to learning about Earth and its inhabitants. The book not only engages the fundamentals of environmental science but also shows students how environmental science informs sustainability, environmental policies, economics, and personal choices.

This book has taken shape over the course of a decade. Subject to a rigorous development and review process to make sure that the material is as accurate, clear, and engaging as possible, we wrote and rewrote until we got it right. College instructors and specialists in specific topics have checked to make sure we are current and pedagogically sound. The art development team worked with us on every graphic and photo researchers sifted through thousands of possibilities until we found the best choice for each concept we wished to illustrate. The end-of-chapter problems and solutions were also subject to review by both instructors and students. Here's what we think is special.

## A Balanced Approach with Emphasis on the Science

Daily life is filled with decisions large and small that affect our environment. From the food we eat, to the cars we drive or choose not to drive, to the chemicals we put into the water, soil, and air, the impact of human activity is wide-ranging and deep. And yet decisions about the environment are not often easy or straightforward. Is it better for the environment to purchase a new, energy-efficient hybrid car or to continue using the car you already own, or to ride a bicycle or take public transportation? Can we find ways to encourage development without creating urban sprawl? Should a dam that provides electricity for 70,000 homes be removed because it interferes with the migration of salmon?

As educators, scientists, and people concerned about sustainability, our goal is to help today's students prepare for the challenges they will face in the future. *Environmental Science: Foundations and Applications* does not preach or tell students how to conduct their lives. Rather, we focus on the science and show students how to make decisions based on their own assessments.

## Focus on Critical Thinking Through Quantitative Reasoning

The book nurtures critical thinking skills through the use of quantitative reasoning in an accessible and friendly manner. By learning how to analyze graphs, measure environmental impact on various scales, and use simple calculations to understand key concepts, students will gain the skills to interpret the wealth of data they will encounter as citizens, professionals, and consumers. Optional mathematical applications boxes (Crunch the Numbers) help students practice calculations relevant to everyday problems related to environmental science. The use of many U.S. examples will help students feel connected to the material and see the everyday relevance of the environmental science course.

# Comprehensive Foundation for the First Course in Environmental Science

We understand that students come to the environmental science course with a variety of backgrounds. Many instructors face a classroom with both majors and non-majors. *Environmental Science: Foundations and Applications* provides a thorough presentation of the core content for students who will go on to major in the environmental sciences and, at the same time, seeks to stimulate and inspire students who may never take another environmental science course. As you can see from the chapter-by-chapter description, this book provides an in-depth exploration of all the topics covered in the foundational course.

## Foundations

**Chapter 1 Environmental Science: Studying the State of Our Earth** and **Chapter 2 Environmental Systems: Matter, Energy, and Change** introduce the study of environmental science by defining the field and explaining some of the difficulties of studying a planet that has been altered by humans. They identify a number of environmental indicators for the state of the planet. These chapters introduce systems and how scientists use physical and chemical parameters to observe and measure the impact of human activity on the environment. **Chapter 3 Ecosystem Ecology: Interactions Between the Living and Nonliving World** introduces the student to the concepts of food webs, ecosystems, and the major biogeochemical cycles of the world. **Chapter 4 Global Climates and Biomes: Geographic Variations in Temperature and Precipitation** explains global climate patterns and tours the major terrestrial and aquatic biomes. **Chapter 5 Evolution and Biodiversity: Origin and Diversification of Organisms** describes the process of evolution by natural selection and traces the evolution of biodiversity from the fossil record into the present, including the modern evolution of pesticide and drug resistance. **Chapter 6 Population and Community Ecology: Distribution and Abundance of Species** explores how populations of species interact with each other in ecological communities and the factors that determine the growth of biological and human populations. **Chapter 7 Human Populations: Patterns and Processes of Human Population Growth** looks at how human population growth is affected by natural factors, as well as by human choices and behavior.

## Applications

**Chapter 8 Earth's Resources: Geologic Processes, Soil, and Minerals** looks at the physical Earth and the various systems and resources that affect humans including plate tectonics, rock weathering, and the distribution of Earth's resources. **Chapter 9 Water Resources: Supply, Distribution, and Use** explores water resources and the processes that influence water distribution and access. **Chapter 10 Land: Public and Private** discusses land classifications and the variety of uses for public and private lands. **Chapter 11 Agriculture: Feeding the World** examines agriculture and aquaculture and how the environment is affected by them. **Chapter 12 Nonrenewable Energy: Coal, Oil, Natural Gas, and Nuclear Fuels** considers the nonrenewable fossil fuels and radioactive fuels, and explores the reasons for energy demand as well as the conventional ways that we supply it. **Chapter 13 Renewable Energy: Innovative Uses of Earth, Sun, Wind, and Water** examines the various renewable energy sources that are currently or may become part of our energy portfolio. Both energy chapters consider the positive and negative impacts of all energy choices. **Chapter 14 Water Pollution: Causes and Effects** and **Chapter 15 Air Pollution: Causes, Effects, and Stratospheric Ozone Depletion** explore how the major pollutants and wastes adversely affect our water and air. **Chapter 16 Waste: Solid Waste Generation and Disposal Systems** looks at solid waste generation and the choices for solid waste disposal. **Chapter 17 Human Health and Toxicology: Environmental Sources of Health Risk** looks at the leading human health risks stemming from common diseases, newly emerging diseases, and pollutants. **Chapter 18 Conservation of Biodiversity: Protection of Earth's**

**Species and Ecosystems** examines the conservation of biodiversity and builds on the biological foundations presented in previous chapters. **Chapter 19 Global Change: Climate Alteration and Global Warming** examines how human activities have produced greenhouse gases that have warmed Earth and affected a wide variety of organisms. **Chapter 20 Working Toward Sustainability: Environmental Economics, Equity, and Policy** examines the subject of sustainability as it relates to global change, economics, and equity.

# A Pedagogical Framework to Reinforce Classroom Learning

We have built each chapter on a framework of learning tools that will help students get the most out of their first course in environmental science. Pedagogical features include:

- **Chapter opening cases:** Each chapter opens with a detailed case study that motivates the student by showing the subject of the chapter in a real-world context.

- **Understand the Key Ideas:** A list of key concepts follows the opening case. This tool helps students organize and focus their study.

- **Crunch the Numbers:** Optional boxes throughout the text demonstrate how to do simple mathematical calculations relevant to chapter concepts.

- **Gauge Your Progress:** After each major chapter section, these review questions ask students to test their understanding of the material.

- **Photos and line art:** Developed in conjunction with the text by specialists in the field of science illustration, figures have been selected and rendered for maximum visual impact.

- **Revisit the Key Ideas:** Chapter summaries are built around the Key Ideas list to reinforce chapter concepts.

- **Working Toward Sustainability:** Chapters conclude with an inspiring story of people or organizations that are making a difference to the environment.

- **Check Your Understanding:** At the end of each chapter, Check Your Understanding questions, in multiple-choice format, test student comprehension.

- **Apply the Concepts:** Two multilevel response questions at the end of each chapter help students solidify their understanding of key concepts by applying what they have learned in the chapter to relevant situations.

- **Measure Your Impact:** In the Measure Your Impact question at the end of each chapter, students are asked to calculate and answer everyday problem scenarios to assess their environmental impact and make informed decisions.

- **Engage Your Environment:** In the Engage Your Environment exercises, students learn how environmental science applies to them personally and are asked to make informed choices about the environment based on real data they collect from their home, campus, or community.

- **Science Applied:** In the Science Applied essays throughout the book, students have the opportunity to see science in action and to determine whether the science supports laws, public policy, and everyday decisions.

## We'd Love to Hear from You

Our goal—to create a balanced, holistic approach to the study of environmental science—has brought us in contact with hundreds of professionals and students. We hope this book inspires you as you have inspired us. Let us know how we're doing! Feel free to get in touch with Andy at Env.Science .Friedland@gmail.com and Rick at Env.Science.Relyea@gmail.com.

# Supplements

**Instructor's Resource CD** provides information on pacing and coverage, teaching tips, and chapter-by-chapter advice on the most important and challenging topics, as well as sample quizzes and complete solutions to the end-of-chapter questions. Instructors will also find labs that correspond to each chapter, and suggestions for additional lab resources. PowerPoint and high-resolution JPEGs of images and tables from the book are also available for classroom lecture use.

**Printed Test Bank** includes approximately 100 multiple-choice, free-response, and footprint calculation questions per chapter.

**Computerized Test Bank** includes all of the printed test bank questions in an easy-to-use computerized format. The software allows instructors to add and edit questions and prepare quizzes and tests quickly and easily.

**The eBook** fully integrates the text with the student media. The eBook also offers a range of customization tools including bookmarking, highlighting, note-taking, and a convenient glossary.

**Math Videos** review basic math concepts important to the course.

**Book Companion Site** provides quizzes, online data sets, and links to useful resources for students and instructors free of charge. The instructor's side features a gradebook and all of the resources from the Instructor's Resource CD.

**Environmental Science Portal** includes a wide array of study and review resources designed to help students master the chapter material. The portal features extensive assessment resources, prebuilt homework assignments, and a powerful adaptive quizzing engine.

**Faculty Lounge** for Environmental Science offers a free forum for instructors to share and review teaching resources.

**Course Management Cartridges** include the student and instructor materials in Blackboard, WebCT, Angel, and other selected platforms.

# Acknowledgments

## From Andy Friedland . . .

A large number of people have contributed to this book in a variety of ways. I would like to thank all of my teachers, students, and colleagues. Professors Robert Giegengack and Arthur Johnson introduced me to environmental science as an undergraduate and graduate student. My colleagues in the Environmental Studies Program at Dartmouth have contributed in numerous ways. I thank Doug Bolger, Michael Dorsey, Karen Fisher-Vanden, Coleen Fox, Jim Hornig, Rich Howarth, Ross Jones, Anne Kapuscinski, Karol Kawiaka, David Mbora, Jill Mikucki, Terry Osborne, Darren Ranco, Bill Roebuck, Jack Shepherd, Chris Sneddon, Scott Stokoe, Ross Virginia, and D.G. Webster for all sorts of contributions to my teaching in general and to this book.

In the final draft, four Dartmouth undergraduates who have taken courses from me, Matt Nichols, Travis Price, Chris Whitehead, and Elizabeth Wilkerson, provided excellent editorial, proofreading, and writing assistance. Many other colleagues have had discussions with me or evaluated sections of text including Ben Carton, Jon Kull, Jeff Schneider, Jimmy Wu, Colin Calloway, Leslie Sonder, Carl Renshaw, Xiahong Feng, Bob Hawley, Meredith Kelly, Rosi Kerr, Jay Lawrence, Jim Labelle, Tim Smith, Charlie Sullivan, Jenna Pollock, Jim Kaste, Carol Folt, Celia Chen, Matt Ayres, Kathy Cottingham, Mark McPeek, David Peart, Lisa Adams, and Richard Waddell. Graduate students and recent graduate students Andrew Schroth, Lynne Zummo, Rachel Neurath, and Chelsea Vario also contributed.

Four friends helped me develop the foundation for this textbook and shared their knowledge of environmental science and writing. I wish to acknowledge Dana Meadows and Ned Perrin, both of whom have since passed away, for all sorts of contributions during the early stages of this work. Terry Tempest Williams has been a tremendous source of advice and wisdom about topics environmental, scientific, and practical. Jack Shepherd contributed a great deal of wisdom about writing and publishing.

John Winn, Paul Matsudeiro, and Neil Campbell offered guidance with my introduction to the world of publishing. Beth Nichols and Tom Corley helped me learn about the wide variety of environmental science courses that are being taught in the United States.

A great many people worked with me at or through W. H. Freeman and provided all kinds of assistance. I particularly would like to acknowledge Jerry Correa, Ann Heath, Becky Kohn, Lee Wilcox, Karen Misler, Cathy Murphy, Hélène de Portu, Beth Howe, and Debbie Clare. I especially want to thank Lee Wilcox for art assistance, and much more, including numerous phone conversations.

Susan Weisberg, Susan Milord, Carrie Larabee, Kim Wind, and Lauren Gifford provided editorial, administrative, logistical, and other support.

I'd also like to acknowledge Dick and Janie Pearl for friendship, and support through the Richard and Jane Pearl Professorship in Environmental Studies.

Finally, I'd like to thank Katie, Jared, and Ethan Friedland, and my mother, Selma, for everything. And especially Ethan, who was willing to see so many of our read-a-thons start late or end early by my having to work on some aspect of a chapter.

## From Rick Relyea . . .

First and foremost I would like to thank my family—my wife Christine and my children Isabelle and Wyatt. Too many nights and weekends were taken from them and given to this textbook and they never complained. Their presence and patience continually inspired me to push forward and complete the project.

Much of the writing coincided with a sabbatical that I spent in Montpellier, France. I am indebted to Philippe Jarne and Patrice David for supporting and funding my time at the Centre d'Ecologie Fonctionnelle et Evolutive. I am also indebted to many individuals at my home institution for supporting my sabbatical, including Graham Hatfull and James Knapp.

Finally, I would like to thank the many people at W. H. Freeman who helped guide me through the publication process and taught me a great deal. As with any book, a tremendous number of people were responsible, including many whom I have never even met. I would especially like to thank Jerry Correa for convincing me to join this project. I thank Becky Kohn, Karen Misler, Cathy Murphy, and Lee Wilcox for translating my words and art ideas into a beautiful final product. Additional credit goes to Norma Roche and Fred Burns for their copyediting, and to Debbie Goodsite and Ted Szczepanski for finding great photos no matter how odd my request. Finally, I thank Ann Heath and Beth Howe for ensuring a high-quality product and the dozens of reviewers who constantly challenged Andy and me to write a clear, correct, and philosophically balanced textbook.

We would like to extend our deep appreciation to the following instructors who reviewed the book manuscript at various stages of development. The content experts who carefully reviewed chapters in their area of epertise are designated with an asterisk (★).

M. Stephen Ailstock, *Anne Arundel Community College*

Deniz Z. Altin-Ballero, *Georgia Perimeter College*

Daphne Babcock, *Collin County Community College District*

Jay L. Banner, *University of Texas at San Antonio*

James W. Bartolome, *University of California, Berkeley*

Brad Basehore, *Harrisburg Area Community College*

Ray Beiersdorfer, *Youngstown State University*

Grady Price Blount, *Texas A&M University, Corpus Christi*

Edward M. Brecker, *Palm Beach Community College, Boca Raton*

Anne E. Bunnell, *East Carolina University*

Ingrid C. Burke, *Colorado State University*

Anya Butt, *Central Alabama Community College*

John Callewaert, *University of Michigan*★

Kelly Cartwright, *College of Lake County*

Mary Kay Cassani, *Florida Gulf Coast University*

Young D. Choi, *Purdue University Calumet*

John C. Clausen, *University of Connecticut*★

Richard K. Clements, *Chattanooga State Technical Community College*

Jennifer Cole, *Northeastern University*

Stephen D. Conrad, *Indiana Wesleyan University*

Terence H. Cooper, *University of Minnesota*

Douglas Crawford-Brown, *University of North Carolina at Chapel Hill*

Wynn W. Cudmore, *Chemeketa Community College*

Katherine Kao Cushing, *San Jose State University*

Maxine Dakins, *University of Idaho*

Robert Dennison, *Heartland Community College*

Michael Denniston, *Georgia Perimeter College*

Roman Dial, *Alaska Pacific University*

Robert Dill, *Bergen Community College*

Michael L. Draney, *University of Wisconsin, Green Bay*

Anita I. Drever, *University of Wyoming*★

James Eames, *Loyola University New Orleans*

Kathy Evans, *Reading Area Community College*

Mark Finley, *Heartland Community College*

Eric J. Fitch, *Marietta College*

Karen F. Gaines, *Northeastern Illinois University*

James E. Gawel, *University of Washington, Tacoma*

Carri Gerber, *Ohio State University Agricultural Technical Institute*

Julie Grossman, *Saint Mary's University, Winona Campus*

Lonnie J. Guralnick, *Roger Williams University*

Sue Habeck, *Tacoma Community College*

Hilary Hamann, *Colorado College*

Sally R. Harms, *Wayne State College*

Barbara Harvey, *Kirkwood Community College*

Floyd Hayes, *Pacific Union College*

Keith R. Hench, *Kirkwood Community College*

William Hopkins, *Virginia Tech*★

Richard Jensen, *Hofstra University*

Sheryll Jerez, *Stephen F. Austin State University*

Shane Jones, *College of Lake County*

Caroline A. Karp, *Brown University*

Erica Kipp, *Pace University, Pleasantville/Briarcliff*

Christopher McGrory Klyza, *Middlebury College*★

Frank T. Kuserk, *Moravian College*

Matthew Landis, *Middlebury College*★

Kimberly Largen, *George Mason University*

Larry L. Lehr, *Baylor University*

Zhaohui Li, *University of Wisconsin, Parkside*

Thomas R. MacDonald, *University of San Francisco*

Robert Stephen Mahoney, *Johnson & Wales University*

Bryan Mark, *Ohio State University, Columbus Campus*

Paula J.S. Martin, *Juniata College*

Robert J. Mason, *Tennessee Temple University*

Michael R. Mayfield, *Ball State University*

Alan W. McIntosh, *University of Vermont*

Kendra K. McLauchlan, *Kansas State University*★

Patricia R. Menchaca, *Mount San Jacinto Community College*

Dorothy Merritts, *Franklin and Marshall College*★

Bram Middeldorp, *Minneapolis Community and Technical College*

Tamera Minnick, *Mesa State College*

Mark Mitch, *New England College*

Ronald Mossman, *Miami Dade College, North*

William Nieter, *St. John's University*

Mark Oemke, *Alma College*

Victor Okereke, *Morrisville State College*

Duke U. Ophori, *Montclair State University*

Chris Paradise, *Davidson College*

Clayton A. Penniman, *Central Connecticut State University*

Christopher G. Peterson, *Loyola University Chicago*

Craig D. Phelps, *Rutgers, The State University of New Jersey, New Brunswick*

F. X. Phillips, *McNeese State University*

Rich Poirot, *Vermont Department of Environmental Conservation*★

Bradley R. Reynolds, *University of Tennessee, Chattanooga*

Amy Rhodes, *Smith College*★

Marsha Richmond, *Wayne State University*

Sam Riffell, *Mississippi State University*

Jennifer S. Rivers, *Northeastern Illinois University*

Ellison Robinson, *Midlands Technical College*

Bill D. Roebuck, *Dartmouth Medical School*★

William J. Rogers, *West Texas A&M University*

Thomas Rohrer, *Central Michigan University*

Aldemaro Romero, *Arkansas State University*

William R. Roy, *University of Illinois at Urbana-Champaign*

Steven Rudnick, *University of Massachusetts, Boston*

Heather Rueth, *Grand Valley State University*

Eleanor M. Saboski, *University of New England*

Seema Sah, *Florida International University*

Shamili Ajgaonkar Sandiford, *College of DuPage*

Robert M. Sanford, *University of Southern Maine*

Nan Schmidt, *Pima Community College*

Jeffery A. Schneider, *State University of New York at Oswego*

Bruce A. Schulte, *Georgia Southern University*

Eric Shulenberger, *University of Washington*

Michael Simpson, *Antioch University New England*★

Annelle Soponis, *Reading Area Community College*

Douglas J. Spieles, *Denison University*

David Steffy, *Jacksonville State University*

Christiane Stidham, *State University of New York at Stony Brook*

Peter F. Strom, *Rutgers, The State University of New Jersey, New Brunswick*

Kathryn P. Sutherland, *University of Georgia*

Christopher M. Swan, *University of Maryland, Baltimore County*★

Karen Swanson, *William Paterson University of New Jersey*

Melanie Szulczewski, *University of Mary Washington*

Donald Thieme, *Valdosta State University*

Jamey Thompson, *Hudson Valley Community College*

Tim Tibbets, *Monmouth College*

John A. Tiedemann, *Monmouth University*

Conrad Toepfer, *Brescia University*

Todd Tracy, *Northwestern College*

Steve Trombulak, *Middlebury College*

Zhi Wang, *California State University, Fresno*

Jim White, *University of Colorado, Boulder*

Rich Wolfson, *Middlebury College*★

C. Wesley Wood, *Auburn University*

David T. Wyatt, *Sacramento City College*

# Chapter Highlights

## Students Develop Quantitative Reasoning Skills

### Crunch the Numbers

Students solidify their understanding of key concepts by practicing simple math calculations (including dimensional analysis, unit conversion, working with data, and reading and interpreting graphs) in "Crunch the Numbers" boxes throughout the text.

---

**CRUNCH THE NUMBERS**

**Calculating Energy Use**

Your electricity bill shows that you use 600 kWh of electricity each month. Your refrigerator, which is 15 years old, could be responsible for up to 25 percent of this electricity consumption. Newer refrigerators are more efficient, meaning that they use less energy to do the same amount of work. If you wish to conserve electrical energy and save money, should you replace your refrigerator? How can you compare the energy efficiency of your old refrigerator with that of the more efficient new models?

Your refrigerator uses 500 watts when the motor is running. The motor runs for about 30 minutes per hour (or a total of 12 hours per day).

**1.** How much energy does your current refrigerator use?

$$0.5 \text{ kW} \times 12 \text{ hours/day} = 6 \text{ kWh/day}$$

$$6 \text{ kWh/day} \times 365 \text{ days/year} = 2,190 \text{ kWh/year}$$

**2.** How much more efficient is the best new refrigerator compared with your current model?

The best new model uses 400 kWh per year. Your refrigerator uses 2,190 kWh per year.

$$2,190 \text{ kWh/year} - 400 \text{ kWh/year} = 1,790 \text{ kWh/year}$$

**3.** Assume that you are paying, on average, $0.10 per kilowatt-hour for electricity. A new refrigerator would cost you $550. You will receive a rebate of $50 from your electric company for purchasing an energy-efficient refrigerator. If you replace your refrigerator, how long will it be before your energy savings compensate you for the cost of the new appliance?

You will save

$$1,790 \text{ kWh/year} \times \$0.10/\text{kWh} = \$179/\text{year}$$

Dividing $500 by $179 indicates that in less than 3 years, you will recover the cost of the new appliance.

**Your Turn:** Environmental scientists must often convert energy units in order to compare various types of energy. For instance, you might want to compare the energy you would save by purchasing an energy-efficient refrigerator with the energy you would save by driving a more fuel-efficient car. Assume that for the amount you would spend on the new refrigerator ($500), you can make repairs to your car engine that would save you 20 gallons (76 liters) of gasoline per month. (1 liter of gasoline contains the energy equivalent of about 10 kWh.) Use this information and Table 2.1 to convert the quantities of both gasoline and electricity into joules. Compare the energy savings. Which decision would save the most energy?

---

**MEASURE YOUR IMPACT**

**Choosing a Car: Conventional or Hybrid?** One person buys a compact sedan that costs $15,000 and gets 20 miles per gallon. Another person pays $22,000 for the hybrid version of the same compact sedan, which gets 50 miles per gallon. Each owner drives 12,000 miles per year and plans on keeping the vehicle for 10 years.

(a) A gallon of gas emits 20 pounds of $CO_2$ when burned in an internal combustion engine. The average cost of a gallon of gas over the 10-year ownership period is $3.00.
  (i) Calculate how many gallons of gas each vehicle uses per year.
  (ii) Calculate the cost of the gas that each vehicle uses per year.
  (iii) Calculate the amount of $CO_2$ that each vehicle emits per year.

(b) Based on your answers to questions i–iii, complete the data table below.

| Year of operation | Sedan: total costs—purchase and gas ($) | Sedan: cumulative $CO_2$ emissions (pounds) | Hybrid: total costs—purchase and gas ($) | Hybrid: cumulative $CO_2$ emissions (pounds) |
|---|---|---|---|---|
| 1 | | | | |
| 2 | | | | |
| 3 | | | | |
| 4 | | | | |
| 5 | | | | |
| 6 | | | | |
| 7 | | | | |
| 8 | | | | |
| 9 | | | | |
| 10 | | | | |

(c) Use the data in the table to answer the following questions:
  (i) Estimate how many years it would take for the hybrid owner to recoup the extra cost of purchasing the vehicle based on savings in gas consumption.
  (ii) After the amount of time determined in (i), compare and comment on the total costs (purchase and gas) for each vehicle at that time.

  (iii) Over the 10-year ownership period, which vehicle is the more economically and environmentally costly to operate (in terms of dollars and $CO_2$ emissions), and by how much?

(d) Suggest ways that the owner of the conventional car could reduce the overall yearly $CO_2$ emissions from the vehicle.

(e) Suggest ways that the hybrid owner could become carbon-neutral in terms of operating the vehicle.

### Measure Your Impact

In the end-of-chapter "Measure Your Impact" exercises, students calculate and answer problem scenarios to assess their environmental impact and make informed decisions.

# Students Evaluate Information and Draw Their Own Conclusions

## ENGAGE YOUR ENVIRONMENT

Does species richness differ in areas of human disturbance? In this activity you'll survey two habitats, one disturbed and one undisturbed, and you will compare the species richness of both. *See Engage Your Environment Activity 5 at the end of this book.*

**Engage Your Environment**

Students make informed choices on their own based on the real data they collect in each "Engage Your Environment" activity.

# science applied

## How Should We Prioritize the Protection of Species Diversity?

As a result of human activities, we have seen a widespread decline in biodiversity across the globe. Many people agree that we should try to slow or even stop this loss. But how do we proceed? Ideally, we might want to preserve all biodiversity. In reality, preserving biodiversity requires compromises. For example, in order to preserve the biodiversity of an area, we might have to set aside land that would otherwise be used for housing developments, shopping malls, or strip mines. If we cannot preserve all biodiversity, how do we decide which species receive our attention?

In 1988, Oxford University professor Norman Myers noted that much of the world's biodiversity is concentrated in areas that make up a relatively small fraction of the globe. Part of the reason for this uneven pattern of biodiversity is that so many species are *endemic species*. **Endemic species** are species that live in a very small area of the world and nowhere else, often in isolated locations such as the Hawaiian Islands. Because they are home to so many endemic species, these isolated areas end up containing a high proportion of all the species found on Earth. Myers called these areas **biodiversity hotspots.**

Scientists originally identified 10 biodiversity hotspots, including Madagascar, western Ecuador, and the Philippines. Myers argued that these 10 areas were in need of immediate conservation attention because human activities there could have disproportionately large negative effects on the world's biodiversity. A year later, the group Conservation International adopted Myers's concept of biodiversity hotspots to guide its conservation priorities. As of 2010, Conservation International had identified the 34 biodiversity hotspots shown in FIGURE SA2.1. Although these hotspots collectively represent only 2.3 percent of the world's land area, 50 percent of all plant species and 42 percent of all vertebrate species are confined to these areas. As a result of this categorization, major conservation organizations have adjusted their funding priorities and are spending hundreds of millions of dollars to conserve these areas. What does environmental science tell us about the hotspot approach to conserving biodiversity?

### What makes a hotspot hot?

Since Norman Myers initiated the idea of biodiversity hotspots, scientists have debated which factors should be considered most important when deciding where to focus conservation efforts. For example, most scientists agree that species richness is an important factor. There are more than 1,300 bird species in the small nation of Ecuador—more than twice the number of bird species living in the United States and Canada. For this reason, protecting a habitat in Ecuador has the potential to save many more bird species than protecting the same amount of habitat in the United States. From this point of view, the choice to protect areas with a lot of species makes sense.

Identifying biodiversity hotspots is challenging, however, because scientists have not yet discovered and identified all the species on Earth. Because the distribution of plant diversity is typically much better known than that of animal diversity, the most practical way to

**Science Applied**

Students see science in action and determine for themselves whether the science supports the policy decisions described in each essay.

# Human Populations: Patterns and Processes of Human Population Growth

## The Environmental Implications of China's Growing Population

Human population size, affluence, and resource consumption all have interrelated impacts on the environment. The example of China is striking. With 1.3 billion people—20 percent of the world's population—China is the world's most populous nation. Because of its rapid economic development, it is expected to soon become the world's largest economy. Once-scarce consumer goods such as automobiles and refrigerators are becoming increasingly commonplace in China. Although the United States, with 307 million

of 1.6 births per woman. If its current population dynamics continue, China's population may begin to decline by 2040.

Population control is only one part of the picture, however. Even if China's population were to stop growing today, the country's resource consumption would continue to increase as standards of living improve. Greater numbers of Chinese people are purchasing cars, home appliances, and other material goods that are common in Western nations. All of these products require resources to produce and use. Manufacturing a refrigerator requires mining and processing raw

> The Chinese are facing considerable environmental challenges as their affluence increases.

people, is currently the world's largest consumer of resources and the greatest producer of many pollutants, China may soon surpass the United States in both consumption and pollution. It is already the largest emitter of carbon dioxide and sulfur dioxide, and it consumes one-third of commercial fish and seafood. The Chinese are facing considerable environmental challenges as their affluence increases.

Managing the presence of humans on Earth sustainably requires addressing both population growth and resource consumption. China has already taken dramatic steps to limit its population growth. Since the 1970s, China has had a "one-child" policy. Couples that restrict themselves to a single child are rewarded financially, while those with three or more children face sanctions, such as a 10 percent salary reduction. Chinese officials use numerous tools—many controversial—to meet population targets, including abortions, sterilizations, and the designation of certain pregnancies as "illegal."

China is one of only a few countries where government-mandated population control measures have significantly reduced population growth. After decades of having one of the world's highest fertility rates, China now has a fertility rate

materials such as steel and copper, producing plastic from oil, and using large quantities of electricity. Having a refrigerator in the home increases daily electricity demand. All of these processes generate carbon dioxide, air and water pollution, and other waste products.

A look at a typical Chinese city street is evidence of the country's growing affluence. Between 1985 and 2002, China's population increased 30 percent, but the number of vehicles used in China grew by over 500 percent, from 3 million to 20 million. Today, China has about 25 million private cars on the road. By 2020, China will have 140 million vehicles, according to Chinese government estimates. China is already the second largest consumer of petroleum (after the United States), and concentrations of urban air pollutants, such as carbon monoxide and photochemical smog, are on the rise. ▶

A traffic jam in Beijing, China.

◀ There are more than 250 million children under the age of 15 in China today.

179

## Chapter Opening Case Studies
An intriguing case study launches each chapter and prompts students to think about how environmental challenges relate to them.

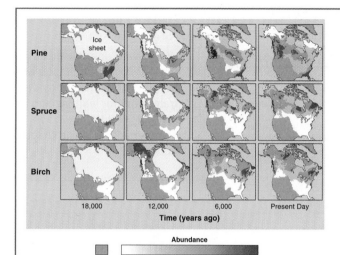

**FIGURE 5.20** **Changes in tree species distributions over time.** Pollen recovered from lake sediments indicates that plant species moved north as temperatures warmed following the retreat of the glaciers, beginning about 12,000 years ago. Areas shown in color or white were sampled for pollen, whereas areas shown in gray were not sampled. [After http://veimages.gsfc.nasa.gov//3453/boreal_model.gif.]

## Numerous U.S. Examples

Local and regional examples make the material relevant.

## Working Toward Sustainability

At the end of each chapter, students are inspired by a success story that focuses on how environmental problems are being addressed by individual action.

## ♻ WORKING TOWARD SUSTAINABILITY

### Is the Water in Your Toilet Too Clean?

In certain parts of the world, such as the United States, sanitation regulations impose such high standards on household wastewater that we classify relatively clean water from bathtubs and washing machines as contaminated. This water must then be treated as sewage. We also use clean, drinkable water to flush our toilets and water our lawns. Can we combine these two observations to come up with a way to save water? One idea that is gaining popularity throughout the developed world is to reuse some of the water we normally discard as waste.

This idea has led creative homeowners and plumbers to identify two categories of wastewater in the home:

gray water and contaminated water. **Gray water** is defined as the wastewater from baths, showers, bathroom sinks, and washing machines. Although no one would want to drink it, gray water is perfectly suitable for watering lawns and plants, washing cars, and flushing toilets. In contrast, water from toilets, kitchen sinks, and dishwashers contains a good deal of waste and contaminants and should therefore be disposed of in the usual fashion.

Around the world, there are a growing number of commercial and homemade systems in use for storing gray water to flush toilets and water lawns or gardens. For example, a Turkish inventor has designed a house-

## APPLY THE CONCEPTS

1. Look at the photograph below and answer the following questions.

Forest A                              Forest B

(a) Explain how this human impact on a forest ecosystem might affect the ability of some species to move to more suitable habitats as Earth's climate changes.
(b) Propose and explain *one* alternative plan that could have preserved this forest ecosystem.
(c) Distinguish between the terms *microevolution* and *macroevolution*. Explain how the organisms in forest A could evolve into species different from those in forest B.

obtained in hospital and non-hospital clinical settings between 2000 and 2006, has identified drug-resistant strains of *E. coli* and *Klebsiella* bacteria in more than 50 blood, urine and respiratory samples. These resistant strains, which resemble bacteria reported in Latin America, Asia and Europe, were thought to be rare in the U.S.

"This antibiotic resistance problem is likely to become widespread," said paper co-author Jan Evans Patterson, M.D., professor of medicine, infectious diseases and pathology at the UT Health Science Center. "It affects the way we will treat infections in the future. In the past, we were concerned with antibiotic resistance in the hospital primarily, but in this review many of the strains we detected were from the community. This tells us antibiotic resistance is spreading in the community, as well, and will affect how we choose antibiotics for outpatient infections."

If the trend continues, it may become difficult to select appropriate antibiotic therapy for urinary tract infections, for example. "The trend over the last decade has been to treat urinary infections empirically, to pick the drug that has worked," said James Jorgensen, Ph.D., professor of pathology, medicine, microbiology and clinical laboratory sciences at the Health Science Center. "Now it is important for physicians to culture the patient's urine to be sure they have selected the right antibiotic. The top three drugs that are often prescribed may not be effective with these

## Apply the Concepts

Multilevel response questions at the end of each chapter encourage students to apply chapter concepts to everyday situations.

# Students Identify and Master Key Ideas
## Using In-Chapter Pedagogy

### Understand the Key Ideas/ Revisit the Key Ideas

"Key Ideas," introduced at the beginning of each chapter and revisited at the end, provide a framework for learning and help students test their comprehension of the chapter material.

---

### UNDERSTAND THE KEY IDEAS

Water is a key resource for life on Earth. All organisms, including humans, require water to live, but growing human populations combined with industrialization have led to the contamination of water supplies. Water contamination has a wide variety of causes. The consequences for people and ecosystems can be severe.

After reading this chapter you should be able to

- distinguish between point and nonpoint sources of pollution.
- identify the ways in which human wastewater can cause water pollution.
- evaluate the different technologies that humans have developed for treating wastewater.
- identify the major types of heavy metals and other substances that pose serious hazards to humans and the environment.
- discuss the impacts of oil spills and how such spills can be remediated.
- identify contaminants that are nonchemical pollutants.

---

### REVISIT THE KEY IDEAS

- **Distinguish between point and nonpoint sources of pollution.**

Point sources of pollution have distinct locations, such as a pipe from a factory that discharges toxic chemicals into a stream. In contrast, nonpoint sources of pollution are more diffuse and cover very large areas, such as agricultural fields that leach fertilizer into a nearby stream.

- **Identify the ways in which human wastewater can cause water pollution.**

Human wastewater can have a number of effects on natural water bodies. Wastewater adds organic matter that increases the biochemical oxygen demand, nutrients that cause eutrophication and algal blooms, and disease-causing pathogens that can harm both humans and wildlife.

- **Evaluate the different technologies that humans have developed for treating wastewater.**

Single residences in rural areas with sufficient land space use simple septic systems that consist of a holding tank and leach field. In large communities with denser human populations and less open land, wastewater treatment plants are needed. These systems are much more complex and can be constructed either in a traditional way or by using newer, more environmentally friendly technologies.

- **Identify the major types of heavy metals and other substances that pose serious hazards to humans and the environment.**

The major inorganic compounds that are of concern for water pollution are mercury, arsenic, and acids. Most arsenic occurs in well water through natural processes, but pollution from mercury, acid precipitation, and acid mine drainage largely occur as a result of human industrial activities. The major organic compounds composing water pollution are pesticides and their inert ingredients; pharmaceuticals, including hormones; and industrial compounds, including PCBs.

- **Discuss the impacts of oil spills and how such spills can be remediated.**

Oil spills occur both from tankers that transport oil as well as from offshore drilling platforms that leak during oil extraction. There is general agreement about containing and removing the oil slicks that float on the surface of the water. However, scientists still debate whether oil spills that hit the coastline should be remediated by washing the coastline with hot water or leaving it to recover without human intervention.

- **Identify contaminants that are nonchemical pollutants.**

Though nonchemical pollutants receive much less attention, they can be very harmful. These pollutants include sediments, heat, noise, and solid waste such as garbage.

- **Explain the connections among industrialization, affluence, and water-pollution legislation.**

Most modern nations have experienced periods of industrialization and widespread pollution followed by greater affluence that allows an improvement in the quality of their waterways. Developed countries have the resources to address pollution issues. Many developing countries are still in the phase of rapid industrial growth and consequently have poor water quality.

---

### GAUGE YOUR PROGRESS

✓ What is water pollution? Why is it important to learn about water pollution?

✓ What are point and nonpoint sources? How do they differ?

✓ What are the most common types of pollutants in the water?

### Gauge Your Progress

The questions in the "Gauge Your Progress" feature, found at the end of each major section in the chapter, help students master one set of concepts before moving on to the next.

# Students Visualize the Concepts
# Using Art as a Learning Tool

**(a)** Random distribution

**(b)** Uniform distribution

**(c)** Clumped distribution

**FIGURE 6.3** **Population distributions.** Populations in nature distribute themselves in three ways. (a) Many of the tree species in this New England forest are randomly distributed, with no apparent pattern in the locations of individuals. (b) Territorial nesting birds, such as these Australasian gannets (*Morus serrator*), exhibit a uniform distribution, in which all individuals maintain a similar distance from one another. (c) Many pairs of eyes are better than one at detecting approaching predators. The clumped distribution of these meerkats (*Suricata suricatta*) provides them with extra protection.

## Instructive Art and Photo Program

The text uses visuals to make complex ideas accessible. The illustration program includes fully integrated teaching captions to help students understand and remember important concepts.

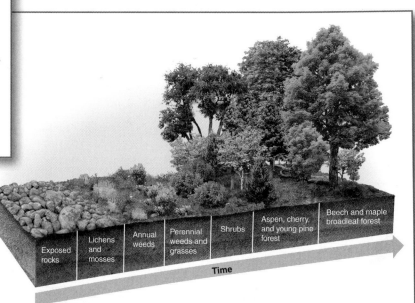

**FIGURE 6.23** **Primary succession.** Primary succession occurs in areas devoid of soil. Early-arriving plants and algae can colonize bare rock and begin to form soil, making the site more hospitable for other species to colonize later. Over time, a series of distinct communities develops. In this illustration, representing an area in New England, bare rock is initially colonized by lichens and mosses and later by grasses, shrubs, and trees.

# ENVIRONMENTAL SCIENCE

## Foundations and Applications

## The Mysterious Neuse River Fish Killer

Over the course of a few days in 1991, roughly a billion fish died in North Carolina's Neuse River. Researchers at North Carolina State University (NCSU), led by Professor JoAnn Burkholder, identified the cause of this disaster as a microscopic free-living aquatic organism in the river water. This particular organism, of the genus *Pfiesteria* (fis-TEER-ee-uh), emits a potent toxin that rapidly kills fish. When members of the research team working with the organism began to develop skin sores and experience nausea, vomiting, memory impairment, and confusion, they became concerned that people using the river for fishing, crabbing, or recreation could also be in danger.

killer. But where did these nutrients come from, and how did they get into the river? The answer probably lies in human activities along the river's banks. The Neuse flows through a region dominated by large industrial-scale hog farms, agricultural fields, and rapidly growing suburban areas, all of which contribute fertilizer runoff and nutrient-rich waste to the river water. A sudden increase in nutrient concentrations caused by these various human activities apparently started a "bloom," or rapid proliferation, of *Pfiesteria*.

The discovery of *Pfiesteria* in North Carolina rivers created panic among the area's recreation and fishing industries. The organism was subsequently found in many other locations from Delaware to Florida, where it infected fisheries and

> The discovery of *Pfiesteria* in North Carolina rivers created panic among the area's recreation and fishing industries.

As researchers continued to study *Pfiesteria,* they found that, depending on environmental conditions, the organism could have up to 24 different life stages—an incredibly large number for any organism. They found that under most conditions, swimming *Pfiesteria* fed harmlessly on algae. However, in the presence of high concentrations of nutrients and large populations of fish, *Pfiesteria* rapidly changed into a carnivore. During this carnivorous life stage, *Pfiesteria* emitted a toxin that stunned fish, then burrowed into a fish's body to feed. Once the fish died, *Pfiesteria* transformed into yet another life stage, a free-floating amoeba that engulfed the tissue sloughed off from fish corpses. Finally, when food became scarce, it could develop a protective casing and sink to the river bottom as a cyst, able to remain dormant for decades awaiting a new influx of nutrients.

Burkholder's group deduced that large influxes of nutrients into the Neuse River had triggered *Pfiesteria's* metamorphosis from harmless algae eater into carnivorous fish

discouraged tourism. Concern over *Pfiesteria* led to a $40 million loss in seafood sales in the Chesapeake Bay region alone.

While the NCSU researchers proceeded with their investigations, other investigators suggested that the "*Pfiesteria* hysteria" was overblown. Studies of humans exposed to *Pfiesteria* along rivers were inconclusive, despite additional anecdotal evidence of the symptoms that the initial researchers had experienced. Some investigators were unable to replicate the findings of Burkholder's team regarding certain *Pfiesteria* life stages. A few researchers even argued that *Pfiesteria* did not produce toxins at all. It wasn't until 2007—16 years after the fish kill that drew so much attention—that other investigators confirmed the identity of the toxin released by *Pfiesteria*. ▶

*Pfiesteria* cell.

---

◀ Despite the beautiful appearance of North Carolina's Neuse River, shown here, runoff from agriculture and housing development contributed to an environmental catastrophe in 1991.

The *Pfiesteria* story is a particularly good introduction to the study of environmental science. It shows us that human activities—for example, releasing waste material into a river—can affect the environment in complex and unexpected ways. Such unintended consequences of human activities are a key concern for environmental scientists.

The case of *Pfiesteria* also tells us that environmental science can be controversial. Following a new discovery, individuals, commercial interests, and the media may overstate the problem, understate it, or disagree with the initial report. Many years may pass before scientists understand the true nature and extent of the problem. Because the findings of environmental science often have an impact on industry, tourism, or recreation, they can create conflicts between scientific study and economic interests.

Finally, the story shows us that findings in environmental science are not always as clear-cut as they first appear. As we begin our study of environmental science, it's important to recognize that the process of scientific inquiry always builds on the work of previous investigators. In this way we accumulate a body of knowledge that eventually resolves important questions—such as what killed the fish in the Neuse River. Only with this knowledge in hand can we begin to make informed decisions on questions of appropriate policy. ■

*Sources:* P. D. R. Moeller et al., Metal complexes and free radical toxins produced by *Pfiesteria piscicida*, *Environmental Science and Technology* 41 (2006): 1166–1172; Nicholas Wade, Deadly or dull? Uproar over a microbe, *New York Times*, August 6, 2000.

## UNDERSTAND THE KEY IDEAS

Humans are dependent on Earth's air, water, and soil for our existence. However, we have altered the planet in many ways, large and small. The study of environmental science can help us understand how humans have changed the planet and identify ways of responding to those changes.

After reading this chapter you should be able to

- define the field of environmental science and discuss its importance.
- identify ways in which humans have altered and continue to alter our environment.
- describe key environmental indicators that help us evaluate the health of the planet.
- define sustainability and explain how it can be measured using the ecological footprint.
- explain the scientific method and its application to the study of environmental problems.
- describe some of the unique challenges and limitations of environmental science.

# Environmental science offers important insights into our world and how we influence it

Stop reading for a moment and look up to observe your surroundings. Consider the air you breathe, the heating or cooling system that keeps you at a comfortable temperature, and the natural or artificial light that helps you see. Our **environment** is the sum of all the conditions surrounding us that influence life. These conditions include living organisms as well as nonliving components such as soil, temperature, and the availability of water. The influence of humans is an important part of the environment as well. The environment we live in determines how healthy we are, how fast we grow, how easy it is to move around, and even how much food we can obtain. One environment may be strikingly different from another—a hot, dry desert versus a cool, humid tropical rainforest, or a coral reef teeming with marine life versus a crowded city street.

We are about to begin a study of **environmental science,** the field that looks at interactions among human *systems* and those found in nature. By **system** we mean any set of interacting components that influence one another by exchanging energy or materials. We have already seen that a change in one part of a system—for example, nutrients released into the Neuse River—can cause changes throughout the entire system.

An environmental system may be completely human-made, like a subway system, or it may be natural, like weather. The scope of an environmental scientist's work can vary from looking at a small population of individuals, to multiple populations that make up a species, to a community of interacting species, or even larger systems, such as the global climate system. Some environmental scientists are interested in regional problems. The specific case of *Pfiesteria* in the Neuse River, for example, was a regional problem. Other environmental scientists work on global issues, such as species extinction and climate change.

Many environmental scientists study a specific type of natural system known as an *ecosystem*. An **ecosystem** is a particular location on Earth whose interacting components include living, or **biotic,** components and nonliving, or **abiotic,** components.

**FIGURE 1.1** Environmental studies. The study of environmental science uses knowledge from many disciplines.

It is important for students of environmental science to recognize that environmental science is different from *environmentalism,* which is a social movement that seeks to protect the environment through lobbying, activism, and education. An **environmentalist** is a person who participates in environmentalism. In contrast, an environmental scientist, like any scientist, follows the process of observation, hypothesis testing, and field and laboratory research. We'll learn more about the scientific method later in this chapter.

So what does the study of environmental science actually include? As **FIGURE 1.1** shows, environmental science encompasses topics from many scientific disciplines, such as chemistry, biology, and Earth science. And environmental science is itself a subset of the broader field known as **environmental studies,** which includes additional subjects such as environmental policy, economics, literature, and ethics. Throughout the course of this book you will become familiar with these and many other disciplines.

We have seen that environmental science is a deeply interdisciplinary field. It is also a rapidly growing area of study. As human activities continue to affect the environment, environmental science can help us understand the consequences of our interactions with our planet and help us make better decisions about our actions.

**GAUGE YOUR PROGRESS**

✓ What factors make up an organism's environment?

✓ In what ways is the field of environmental studies interdisciplinary?

✓ Why is environmental science research important?

# Humans alter natural systems

Think of the last time you walked in a wooded area. Did you notice any dead or fallen trees? Chances are that even if you did, you were not aware that living and nonliving components were interacting all around you. Perhaps an insect pest killed the tree you saw and many others of the same species. Over time, dead trees in a forest lose moisture. The increase in dry wood makes the forest more vulnerable to intense wildfires. But the process doesn't stop there. Wildfires trigger the germination of certain tree seeds, some of which lie dormant until after a fire. And so what began with the activity of insects leads to a transformation of the forest. In this way, *biotic,* or living, factors interact with *abiotic,* or nonliving, factors to influence the future of the forest.

The global environment is composed of small-scale and large-scale systems. Within a given system, biotic and abiotic components can interact in surprisingly complex ways. In the forest example, the species of trees that are present in the forest, the insect pests, and the wildfires interact with one another: they form a system. This small forest system is part of many larger systems and, ultimately, one global system that generates, circulates, and utilizes oxygen and carbon dioxide, among other things.

Humans manipulate their environment more than any other species. We convert land from its natural state into urban, suburban, and agricultural areas (**FIGURE 1.2**). We change the chemistry of our air, water, and soil, both intentionally—for example, by adding fertilizers—and unintentionally, as a consequence of activities that generate pollution. Even where we don't manipulate the environment directly, the simple fact that we are so abundant affects our surroundings.

**FIGURE 1.2** The impact of humans on Earth. Housing development is one example of the many ways in which humans convert land from its natural state.

(a)

(b)

**FIGURE 1.3** It is impossible for millions of people to inhabit an area without altering it. (a) In 1880, fewer than 6,000 people lived in Los Angeles. (b) In 2009, Los Angeles had a population of 3.8 million people, and the greater Los Angeles metropolitan area was home to nearly 13 million people.

Humans and their direct ancestors (other members of the genus *Homo*) have lived on Earth for about 2.5 million years. During this time, and especially during the last 10,000 to 20,000 years, we have shaped and influenced our environment. As tool-using, social animals, we have continued to develop a capacity to directly alter our environment in substantial ways. *Homo sapiens*—genetically modern humans—evolved to be successful hunters: when they entered a new environment, they often hunted large animal species to extinction. In fact, early humans are thought to be responsible for the extinction of mammoths, mastodons, giant ground sloths, and many types of birds. More recently, hunting in North America led to the extinction of the passenger pigeon (*Ectopistes migratorius*) and nearly caused the loss of the American bison (*Bison bison*).

But the picture isn't all bleak. Human activities have also created opportunities for certain species to thrive. For example, for thousands of years Native Americans on the Great Plains used fire to capture animals for food. The fires they set kept trees from encroaching on the plains, which in turn created a window for an entire ecosystem to develop. Because of human activity, this ecosystem—the tallgrass prairie—is now home to numerous unique species.

During the last two centuries, the rapid and widespread development of technology, coupled with dramatic human population growth, has increased both the rate and the scale of our global environmental impact substantially. Modern cities with electricity, running water, sewer systems, Internet connections, and public transportation systems have improved human well-being, but they have come at a cost. Cities cover land that was once natural habitat. Species relying on that habitat must adapt, relocate, or go extinct. Human-induced changes

in climate—for example, in patterns of temperature and precipitation—affect the health of natural systems on a global scale. Current changes in land use and climate are rapidly outpacing the rate at which natural systems can evolve. Some species have not "kept up" and can no longer compete in the human-modified environment.

Moreover, as the number of people on the planet has grown, their effect has multiplied. Six thousand people can live in a relatively small area with only minimal environmental effects. But when 4 million people live in a modern city like Los Angeles, their combined activity will cause greater environmental damage that will inevitably pollute the water, air, and soil and introduce other consequences as well (FIGURE 1.3).

**GAUGE YOUR PROGRESS**

✓ In what ways do humans change the environment?

✓ What is the relationship between the development of technology and environmental impacts?

✓ How does human development have an impact on natural systems?

# Environmental scientists monitor natural systems for signs of stress

One of the critical questions that environmental scientists investigate is whether the planet's natural

| TABLE 1.1 | Some common environmental indicators | |
|---|---|---|
| Environmental indicator | Unit of measure | Chapter where indicator is discussed |
| Human population | Individuals | 7 |
| Ecological footprint | Hectares of land | 1 |
| Total food production | Metric tons of grain | 11 |
| Food production per unit area | Kilograms of grain per hectare of land | 11 |
| Per capita food production | Kilograms of grain per person | 11 |
| Carbon dioxide | Concentration in air (parts per million) | 19 |
| Average global surface temperature | Degrees centigrade | 19 |
| Sea level change | Millimeters | 19 |
| Annual precipitation | Millimeters | 4 |
| Species diversity | Number of species | 5, 18 |
| Fish consumption advisories | Present or absent; number of fish allowed per week | 17 |
| Water quality (toxic chemicals) | Concentration | 14 |
| Water quality (conventional pollutants) | Concentration; presence or absence of bacteria | 14 |
| Deposition rates of atmospheric compounds | Milligrams per square meter per year | 15 |
| Fish catch or harvest | Kilograms of fish per year or weight of fish per effort expended | 11 |
| Extinction rate | Number of species per year | 5 |
| Habitat loss rate | Hectares of land cleared or "lost" per year | 18 |
| Infant mortality rate | Number of deaths of infants under age 1 per 1,000 live births | 7 |
| Life expectancy | Average number of years a newborn infant can be expected to live under current conditions | 7 |

life-support systems are being degraded by human-induced changes. Natural environments provide what we refer to as **ecosystem services**—the processes by which life-supporting resources such as clean water, timber, fisheries, and agricultural crops are produced. We often take a healthy ecosystem for granted, but we notice when an ecosystem is degraded or stressed because it is unable to provide the same services or produce the same goods. To understand the extent of our effect on the environment, we need to be able to measure the health of Earth's ecosystems.

To describe the health and quality of natural systems, environmental scientists use *environmental indicators.* Just as body temperature and heart rate can indicate whether a person is healthy or sick, **environmental indicators** describe the current state of an environmental system. These indicators do not always tell us what is causing a change, but they do tell us when we might need to look more deeply into a particular issue. Environmental indicators provide valuable information about natural systems on both small and large scales. Some of these indicators are listed in Table 1.1.

In this book we will focus on the five global-scale environmental indicators listed in Table 1.2: biological diversity, food production, average global surface temperature and carbon dioxide concentrations in the

atmosphere, human population, and resource depletion. These key environmental indicators help us analyze the health of the planet. We can use this information to guide us toward **sustainability,** by which we mean living on Earth in a way that allows us to use its resources without depriving future generations of those resources. Many scientists maintain that achieving sustainability is the single most important goal for the human species. It is also one of the most challenging tasks we face.

## Biological Diversity

Biological diversity, or **biodiversity,** is the diversity of life forms in an environment. It exists on three scales: *genetic, species,* and *ecosystem* diversity. Each of these is an important indicator of environmental health and quality.

**GENETIC DIVERSITY** Genetic diversity is a measure of the genetic variation among individuals in a population. Populations with high genetic diversity are better able to respond to environmental change than populations with lower genetic diversity. For example, if a population of fish possesses high genetic diversity for disease resistance, at least some individuals are likely to survive

## TABLE 1.2    Five key global environmental indicators

| Indicator | Recent trend | Outlook for future | Overall impact on environmental quality |
|---|---|---|---|
| Biological diversity | Large number of extinctions, extinction rate increasing | Extinctions will continue | Negative |
| Food production | Per capita production possibly leveling off | Unclear | May affect the number of people Earth can support |
| Average global surface temperature and $CO_2$ concentrations | $CO_2$ concentrations and temperatures increasing | Probably will continue to increase, at least in the short term | Effects are uncertain and varied, but probably detrimental |
| Human population | Still increasing, but growth rate slowing | Population leveling off Resource consumption rates are also a factor | Negative |
| Resource depletion | Many resources are being depleted at rapid rates. But human ingenuity frequently develops "new" resources, and efficiency of resource use is increasing in many cases | Unknown | Increased use of most resources has negative effects |

whatever diseases move through the population. If the population declines in number, however, the amount of genetic diversity it can possess is also reduced, and this reduction increases the likelihood that the population will decline further when exposed to a disease.

**SPECIES DIVERSITY** Species diversity indicates the number of *species* in a region or in a particular type of habitat. A **species** is defined as a group of organisms that is distinct from other groups in its morphology (body form and structure), behavior, or biochemical properties. Individuals within a species can breed and produce fertile offspring. Scientists have identified and cataloged approximately 2 million species on Earth. Estimates of the total number of species on Earth range between 5 million and 100 million, with the most common estimate at 10 million. This number includes a large array of organisms with a multitude of sizes, shapes, colors, and roles (**FIGURE 1.4**). Scientists have observed that ecosystems with more species, that is, higher species diversity, are more resilient and productive. For example, a tropical forest with a large number of plant species growing in the understory is likely to be more productive, and more resilient to change, than a nearby tropical forest plantation with one crop species growing in the understory.

Environmental scientists often focus on species diversity as a critical environmental indicator. The number of frog species, for example, is used as an indicator of regional environmental health because frogs are exposed to both the water and the air in their ecosystem. A decrease in the number of frog species in a particular ecosystem may be an indicator of environmental problems there. Species losses in several ecosystems can indicate larger-scale environmental problems.

Not all species losses are indicators of environmental problems, however. Species arise and others go extinct as part of the natural evolutionary process. The evolution of new species, known as **speciation,** typically happens very slowly—perhaps on the order of one to three new species per year worldwide. The average rate at which species go extinct over the long term, referred to as the **background extinction rate,** is also very slow: about one species in a million every year. So with 2 million identified species on Earth, the background extinction rate should be about two species per year.

Under conditions of environmental change or biological stress, species may go extinct faster than new ones evolve. Some scientists estimate that more than 10,000 species are currently going extinct each year—5,000 times the background rate of extinction. Habitat destruction and habitat degradation are the major causes of species extinction today, although climate change, overharvesting, and pressure from introduced species also contribute to species loss. Human intervention has saved certain species, including the American bison, peregrine falcon (*Falco peregrinus*), bald eagle (*Haliaeetus leucocephalus*), and American alligator (*Alligator mississippiensis*). But other large animal species, such as the Bengal tiger (*Panthera tigris*), snow leopard (*Panthera uncia*), and West Indian manatee (*Trichechus manatus*), remain endangered and may go extinct if present trends are not reversed. Overall, the number of species has been declining (**FIGURE 1.5**).

**ECOSYSTEM DIVERSITY** Ecosystem diversity is a measure of the diversity of ecosystems or habitats that exist in a given region. A greater number of healthy and productive ecosystems means a healthier environment overall.

As an environmental indicator, the current loss of biodiversity tells us that natural systems are facing strains

**FIGURE 1.4** **Species diversity.** The variety of organisms on Earth is evidence of biological diversity.

Labels: Bacteria · Diatom · British soldier lichen · Queen angelfish · Atlas moth · Fly agaric mushroom · Giant panda · Monument plant · Colorado blue spruce

unlike any in the recent past. It is clearly an important topic in the study of environmental science, and we will look at it in greater detail in Chapters 5 and 18 of this book.

Some measures of biodiversity are given in terms of land area, so becoming familiar with measurements of land area is important to understanding them. As Crunch the Numbers "Converting to Hectares" describes,

## CRUNCH THE NUMBERS

### Converting to Hectares

In the United States we measure land area in terms of square miles and acres. However, the rest of the world measures land in terms of hectares (ha). One ha represents an area of 100 meters by 100 meters. Let's see how the two systems compare:

$$1 \text{ mile}^2 = 640 \text{ acres}$$

Given that there are 5,280 feet in a mile:

$$1 \text{ mi}^2 = (5{,}280 \text{ ft})^2 = 27{,}878{,}400 \text{ ft}^2$$

Using this information, we can determine the number of square feet in 1 acre, as follows:

$$\left(\frac{1 \text{ mi}^2}{640 \text{ acres}}\right) \times \left(\frac{27{,}878{,}400 \text{ ft}^2}{1 \text{ mi}^2}\right) = 43{,}560 \text{ ft}^2/\text{acre}$$

So—what is a hectare?

$$1 \text{ ha} = 10{,}000 \text{ m}^2$$

—that is, a square that is 100 m on each side—and

$$1 \text{ kilometer (km)} = 1{,}000 \text{ m}$$

Thus,

$$1 \text{ km}^2 = (1{,}000 \text{ m})^2 = 1{,}000{,}000 \text{ m}^2$$

Using this information, we can determine the number of hectares in 1 square kilometer.

$$\left(\frac{1{,}000{,}000 \text{ m}^2}{1 \text{ km}^2}\right) \times \left(\frac{1 \text{ ha}}{10{,}000 \text{ m}^2}\right) = 100 \text{ ha/km}^2$$

How can we compare hectares to acres? First, we need to use common units—1 km = 0.6214 mi. Converting square kilometers to square feet:

$$1 \text{ km}^2 = \left(0.6214 \text{ mi}\right)^2 \times \left(\frac{27{,}878{,}400 \text{ ft}^2}{1 \text{ mi}^2}\right) = 10{,}764{,}908 \text{ ft}^2$$

To determine the number of acres in 1 hectare:

$$\left(10{,}764{,}908 \text{ ft}^2/\text{km}^2\right) \times \left(\frac{1 \text{ km}^2}{100 \text{ ha}}\right) \times \left(\frac{1 \text{ acre}}{43{,}560 \text{ ft}^2}\right) = 2.47 \text{ acres/ha}$$

**Your Turn:** How many hectares are there in a 10,000 acre forest preserve?

**FIGURE 1.5** **Species on the brink.** Humans have saved some species from the brink of extinction, such as (a) the American bison and (b) the peregrine falcon. Other species, such as (c) the snow leopard and (d) the West Indian manatee, continue to decline toward extinction.

a hectare is a unit of area used primarily in the measurement of land.

## Food Production

The second of our five global indicators is food production: our ability to grow food to nourish the human population. Just as a healthy ecosystem supports a wide range of species, a healthy soil supports abundant and continuous food production. Food grains such as wheat, corn, and rice provide more than half the calories and protein humans consume. Still, the growth of the human population is straining our ability to grow and distribute adequate amounts of food.

In the past we have used science and technology to increase the amount of food we can produce on a given area of land. World grain production has increased fairly steadily since 1950 as a result of expanded irrigation, fertilization, new crop varieties, and other innovations.

At the same time, worldwide production of grain *per person,* also called *per capita* world grain production, has leveled off. FIGURE 1.6 shows a downward trend in wheat production since about 1985.

In 2008, food shortages around the world led to higher food prices and even riots in some places. Why did this happen? The amount of grain produced worldwide is influenced by many factors. These factors include climatic conditions, the amount and quality of land under cultivation, irrigation, and the human labor and energy required to plant, harvest, and bring the grain to market. Why is grain production not keeping up with population growth? In some areas, the productivity of agricultural ecosystems has declined because of soil degradation, crop diseases, and unfavorable weather conditions such as drought or flooding. In addition, demand is outpacing supply. The rate of human population growth has outpaced increases in food production. Furthermore, humans currently use

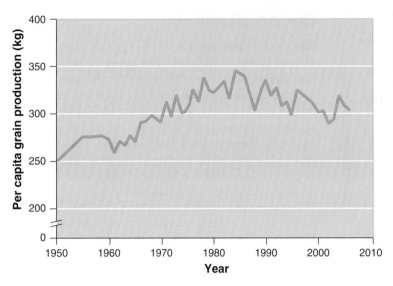

FIGURE 1.6 **World grain production per person.** Grain production has increased since the 1950s, but it has recently begun to level off. [After http://www.earth-policy.org/index.php?/indicators/C54.]

more grain to feed livestock than they consume themselves. Finally, some government policies discourage food production by making it more profitable to allow land to remain uncultivated, or by encouraging farmers to grow crops for fuels such as ethanol and biodiesel instead of food.

Will there be sufficient grain to feed the world's population in the future? In the past, whenever a shortage of food loomed, humans have discovered and employed technological or biological innovations to increase production. However, these innovations often put a strain on the productivity of the soil. Unfortunately, if we continue to overexploit the soil, its ability to sustain food production may decline dramatically. We will take a closer look at soil quality in Chapter 8 and food production in Chapter 11.

## Average Global Surface Temperature and Carbon Dioxide Concentrations

We have seen that biodiversity and abundant food production are necessary for life. One of the things that makes them possible is a stable climate. Earth's temperature has been relatively constant since the earliest forms of life began, about 3.5 billion years ago. The temperature of Earth allows the presence of liquid water, which is necessary for life.

What keeps Earth's temperature so constant? As FIGURE 1.7 shows, our thick planetary atmosphere contains many gases, some of which act like a blanket trapping heat near Earth's surface. The most important of these heat-trapping gases, called **greenhouse gases,** is carbon dioxide ($CO_2$). During most of the history of life on Earth, greenhouse gases have been present in the atmosphere at fairly constant concentrations for relatively long periods. They help keep Earth's surface within the range of temperatures at which life can flourish.

In the past two centuries, however, the concentrations of $CO_2$ and other greenhouse gases in the atmosphere have risen. During roughly the same period, as the graph in FIGURE 1.8 shows, global temperatures have fluctuated considerably, but have shown an overall increase. Many scientists believe that the increase in atmospheric $CO_2$ during the last two centuries is **anthropogenic**—derived from human activities. The two major sources of anthropogenic $CO_2$ are the combustion of fossil fuels and the net loss of forests and other habitat types that would otherwise take up and store $CO_2$ from the atmosphere. We will discuss climate in Chapter 4 and global climate change in Chapter 19.

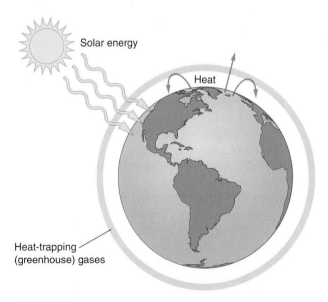

FIGURE 1.7 **The greenhouse effect.** As Earth's surface is warmed by the Sun, it radiates heat outward. Heat-trapping gases absorb the outgoing heat and reradiate some of it back to Earth. Without these greenhouse gases, Earth would be much cooler.

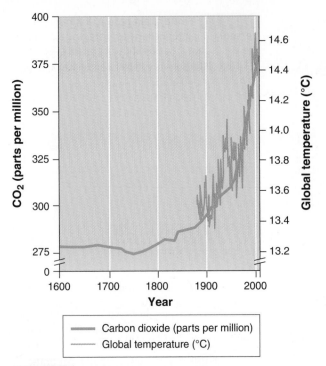

**FIGURE 1.8** Changes in average global surface temperature and in atmospheric CO₂ concentrations. Earth's average global surface temperature has increased steadily for at least the past 100 years. Carbon dioxide concentrations in the atmosphere have varied over geologic time, but have risen steadily since 1960. [After http://data.giss.nasa.gov/gistemp /2008/. http://mb-soft.com/public3/co2hist.gif.]

## Human Population

In addition to biodiversity, food production, and global surface temperature, the size of the human population can tell us a great deal about the health of our global environment. The human population is currently 6.8 billion and growing. The increasing world population places additional demands on natural systems, since each new person requires food, water, and other resources. In any given 24-hour period, 364,000 infants are born and 152,000 people die. The net result is 212,000 new inhabitants on Earth each day, or *over a million additional people every 5 days*. The rate of population growth has been slowing since the 1960s, but world population size will continue to increase for at least 50 to 100 years. Most population scientists project that the human population will be somewhere between 8.1 billion and 9.6 billion in 2050 and will stabilize between 6.8 billion and 10.5 billion by 2100.

Can the planet sustain so many people (**FIGURE 1.9**)? Even if the human population eventually stops growing, the billions of additional people will create a greater demand on Earth's finite resources, including food, energy, and land. Unless humans work to reduce these pressures, the human population will put a rapidly growing strain on natural systems for at least the first half of this century. We discuss human population issues in Chapter 7.

## Resource Depletion

Natural resources provide the energy and materials that support human civilization. But as the human population grows, the resources necessary for our survival become increasingly depleted. In addition, extracting these natural resources can affect the health of our environment in many ways. Pollution and land degradation caused by mining, waste from discarded manufactured products, and air pollution caused by fossil fuel combustion are just a few of the negative environmental consequences of resource extraction and use.

Some natural resources, such as coal, oil, and uranium, are finite and cannot be renewed or reused. Others, such as aluminum or copper, also exist in finite quantities, but can be used multiple times through reuse or recycling. Renewable resources, such as timber, can be grown and harvested indefinitely, but in some locations they are being used faster than they are naturally replenished. Crunch the Numbers "Rates of Forest Clearing" provides an opportunity to calculate rates of one type of resource depletion.

Sustaining the global human population requires vast quantities of resources. However, in addition to the total amounts of resources used by humans, we must consider resource use per capita.

Patterns of resource consumption vary enormously among nations depending on their level of development. What exactly do we mean by *development*? **Development** is defined as improvement in human well-being

**FIGURE 1.9** **Kolkata, India.** The human population will continue to grow for at least 50 years. Unless humans can devise ways to live more sustainably, these population increases will put additional strains on natural systems.

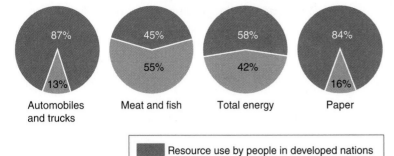

| 87% | 45% | 58% | 84% |
| 13% | 55% | 42% | 16% |
| Automobiles and trucks | Meat and fish | Total energy | Paper |

**FIGURE 1.10** **Resource use in developed and developing countries.** Only 20 percent of the world's population lives in developed countries, but that 20 percent uses most of the world's resources. The remaining 80 percent of the population lives in developing countries and uses far fewer resources per capita.

through economic advancement. Development influences personal and collective human lifestyles—things such as automobile use, the amount of meat in the diet, and the availability and use of technologies such as cell phones and personal computers. As economies develop, resource consumption also increases: people drive more automobiles, live in larger homes, and purchase more goods. These increases can often have implications for the natural environment.

According to the United Nations Development Programme, people in developed nations—including the United States, Canada, Australia, most European countries, and Japan—use most of the world's resources. FIGURE 1.10 shows that the 20 percent of the global population that lives in developed nations owns 87 percent of the world's automobiles and consumes 58 percent of all energy, 84 percent of all paper, and 45 percent of all fish and meat. The poorest 20 percent of the world's people consume 5 percent or less of these resources. Thus, even though the number of people in the developing countries is much larger than the num-

ber in the developed countries, their total consumption of natural resources is relatively small.

So while it is true that a larger human population has greater environmental impacts, a full evaluation requires that we look at economic development and consumption patterns as well. We will take a closer look at resource depletion and consumption patterns in Chapters 7, 12, and 13.

## GAUGE YOUR PROGRESS

✓ What is an environmental indicator and what does it tell us?

✓ What are the five global-scale environmental indicators we focus on in this book, and how do they help us monitor the health of the environment?

✓ How do human activities contribute to changes in the five global-scale environmental indicators?

## CRUNCH THE NUMBERS

### Rates of Forest Clearing

A Web search of environmental organizations yields several different estimates of the amount of worldwide forest clearing:

    Estimate 1: 1 acre per second

    Estimate 2: 80,000 acres per day

    Estimate 3: 32,000 ha per day

In order to compare these estimates, we need to convert them into hectares per year. There are 2.47 acres per hectare, so 1 acre = 0.40 ha.

### Estimate 1

1.0 acre/second × 0.40 ha/acre = 0.40 ha/second

0.40 ha/second × 60 seconds/minute ×
60 minutes/hour × 24 hours/day × 365 days/year =
12,614,400 ha cleared per year

### Estimate 2

80,000 acres/day × 0.40 ha/acre =
32,000 ha cleared per day

**Your Turn:** Notice that Estimate 2, when converted to hectares, is identical to Estimate 3. Now convert the estimate of 32,000 ha/day into the amount cleared per year. How does this number compare to the first estimate? Why might environmental organizations, or anyone else, choose to present similar information in different ways?

# Human well-being depends on sustainable practices

We have seen that people living in developed nations consume a far greater share of the world's resources than do people in developing countries. What effect does this consumption have on our environment? It is easy to imagine a very small human population living on Earth without degrading its environment: there simply would not be enough people to do significant damage. Today, however, Earth's population is 6.8 billion people and growing. Many environmental scientists ask how we will be able to continue to produce sufficient food, build needed infrastructure, and process pollution and waste. Our current attempts to sustain the human population have already modified many environmental systems. Can we continue our current level of resource consumption without jeopardizing the well-being of future generations?

Easter Island, in the South Pacific, provides a cautionary tale (**FIGURE 1.11**). This island, also called Rapa Nui, was once covered with trees and grasses. When humans settled the island hundreds of years ago, they quickly multiplied in its hospitable environment. They cut down trees to build homes and canoes for fishing, and they overused the island's soil and water resources. By the 1870s, almost all of the trees were gone. Without the trees to hold the soil in place, massive erosion occurred, and the loss of soil caused food production to decrease. While other forces, including diseases introduced by European visitors, were also involved in the

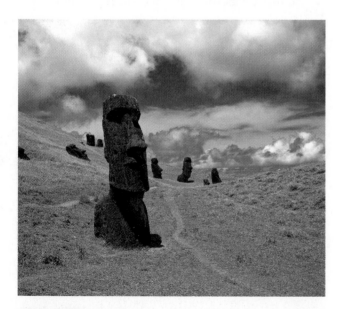

**FIGURE 1.11** **Easter Island.** The overuse of resources by the people of Easter Island is probably the primary cause for the demise of that civilization.

destruction of the population, the unsustainable use of natural resources on Easter Island appears to be the primary cause for the collapse of its civilization.

Most environmental scientists believe that there are limits to the supply of clean air and water, nutritious foods, and other life-sustaining resources our environment can provide, as well as a point at which Earth will no longer be able to maintain a stable climate. We must meet several requirements in order to live sustainably:

- Environmental systems must not be damaged beyond their ability to recover.

- Renewable resources must not be depleted faster than they can regenerate.

- Nonrenewable resources must be used sparingly.

**Sustainable development** is development that balances current human well-being and economic advancement with resource management for the benefit of future generations. This is not as easy as it sounds. The issues involved in evaluating sustainability are complex, in part because sustainability depends not only on the number of people using a resource, but also on how that resource is being used. For example, eating chicken is sustainable when people raise their own chickens and allow them to forage for food on the land. However, if all people, including city dwellers, wanted to eat chicken six times a week, the amount of resources needed to raise that many chickens would probably make the practice of eating chicken unsustainable.

Living sustainably means *acting in a way such that activities that are crucial to human society can continue*. It includes practices such as conserving and finding alternatives to nonrenewable resources as well as protecting the capacity of the environment to continue to supply renewable resources (**FIGURE 1.12**).

Iron, for example, is a nonrenewable resource derived from ore removed from the ground. It is the major constituent of steel, which we use to make many things, including automobiles, bicycles, and strong frames for tall buildings. Historically, our ability to smelt iron for steel limited our use of that resource. But as we have improved steel manufacturing technology, steel has become more readily available, and the demand for it has grown. Because of this, our current use of iron is unsustainable. What would happen if we ran out of iron? Not too long ago the depletion of iron ore might have been a catastrophe. But today we have developed materials that can substitute for certain uses of steel—for example, carbon fiber—and we also know how to recycle steel. Developing substitutes and recycling materials are two ways to address the problem of resource depletion and increase sustainability.

The example of iron leads us to a question that environmental scientists often ask: How do we determine

**FIGURE 1.12** **Living sustainably.** Sustainable choices such as bicycling to work or school can help protect the environment and conserve resources for future generations.

the importance of a given resource? If we use up a resource such as iron for which substitutes exist, it is possible that the consequences will not be severe. However, if we are unable to find an alternative to the resource—for example, something to replace fossil fuels—people in the developed nations may have to make significant changes in their consumption habits.

## Defining Human Needs

We have seen that sustainable development requires us to determine how we can meet our current needs without compromising the ability of future generations to meet their own needs. Let's look at how environmental science can help us achieve that goal. We will begin by defining *needs*.

If you have ever experienced an interruption of electricity to your home or school, you know how frustrating it can be. Without the use of lights, computers, televisions, air-conditioning, heating, and refrigeration, many people feel disconnected and uncomfortable. Almost everyone in the developed world would insist that they need—cannot live without—electricity. But in other parts of the world, people have never had these modern conveniences. When we speak of *basic needs,* we are referring to the essentials that sustain human life, including air, water, food, and shelter.

But humans also have more complex needs. Many psychologists have argued that we require meaningful human interactions in order to live a satisfying life; therefore, a community of some sort might be considered a human need. Biologist Edward O. Wilson wrote that humans exhibit **biophilia**—that is, love of life—which is a *need* to make "the connections that humans subconsciously seek with the rest of life." Thus our needs for access to natural areas, for beauty, and for social connections can be considered as vital to our well-being as our basic physical needs and must be considered as part of our long-term goal of global sustainability (**FIGURE 1.13**).

## The Ecological Footprint

We have begun to see the multitude of ways in which human activities affect the environment. As countries prosper, their populations use more resources. But economic development can sometimes improve environmental conditions. For instance, wealthier countries may be able to afford to implement pollution controls and invest money to protect native species. So although people in developing countries do not consume the same quantity of resources as those in developed nations, they may be less likely to use environmentally friendly technologies or to have the financial resources to implement environmental protections.

How do we determine what lifestyles have the greatest environmental impact? This is an important question for environmental scientists if we are to understand the effects of human activities on the planet and develop sustainable practices. Calculating sustainability, however, is more difficult than one might think. We have to consider the impacts of our activities and lifestyles on different aspects of our environment. We use land to grow food, to build on, and for parks and recreation. We require

**FIGURE 1.13** **Central Park, New York City.** New Yorkers have set aside 2,082 ha (843 acres) in the center of the largest city in the United States—a testament to the compelling human need for interactions with nature.

water for drinking, for cleaning, and for manufacturing products such as paper. We need clean air to breathe. Yet these goods and services are all interdependent: using or protecting one has an effect on the others. For example, using land for conventional agriculture may require water for irrigation, fertilizer to promote plant growth, and pesticides to reduce crop damage. This use of land reduces the amount of water available for human use: the plants consume it and the pesticides pollute it.

One method used to assess whether we are living sustainably is to measure the impact of a person or country on world resources. The tool many environmental scientists use for this purpose, the *ecological footprint,* was developed in 1995 by Professor William Rees and his graduate student Mathis Wackernagel. An individual's **ecological footprint** is a measure of how much that person consumes, expressed in area of land. That is, the output from the total amount of land required to support a person's lifestyle represents that person's ecological footprint (FIGURE 1.14).

Rees and Wackernagel maintained that if our lifestyle demands more land than is available, then we must be living unsustainably—using up resources more quickly than they can be produced, or producing wastes more quickly than they can be processed. For example, each person requires a certain number of food calories each day. We know the number of calories in a given amount

of grain or meat. We also know how much farmland or rangeland is needed to grow the grain to feed people or livestock such as sheep, chickens, or cows. If a person eats only grains or plants, the amount of land needed to provide that person with food is simply the amount of land needed to grow the plants they eat. If that person eats meat, however, the amount of land required to feed that person is greater, because we must also consider the land required to raise and feed the livestock that ultimately become meat. Thus one factor in the size of a person's ecological footprint is the amount of meat in the diet. Meat consumption is a lifestyle choice, and per capita meat consumption is much greater in developed countries.

We can calculate the ecological footprint of the food we eat, the water and energy we use, and even the activities we perform that contribute to climate change. All of these impacts determine our ecological footprint on the planet as individuals, cities, states, or nations. Calculating the ecological footprint is complex, and the details are subject to debate, but it has at least given scientists a concrete measure to discuss and refine.

Scientists at the Global Footprint Network, where Wackernagel is now president, have calculated that the human ecological footprint has reached 14 billion hectares (34.6 billion acres), or 125 percent of Earth's total usable land area. Furthermore, they have calculated that if every person on Earth lived the average lifestyle of people in the United States, we would require the equivalent of five Earths (FIGURE 1.15). Even to support the entire human population with the lifestyles we have now, we would need more than one Earth. Clearly, this level of resource consumption is not sustainable.

According to Wackernagel and Rees, if we are to sustain human life, we must ensure that our total consumption leads to an ecological footprint of no more than 11 billion hectares (27.2 billion acres). This number will need to be significantly less if we wish to preserve land for species other than humans. In order to achieve this goal, humans will have some important choices to make.

Energy   Settlements   Timber and paper   Food and fibers   Seafood

Carbon footprint   Built-up land   Forests   Fisheries

Cropland and pastures

FIGURE 1.14 **The ecological footprint.** An individual's ecological footprint is a measure of how much land is needed to supply the goods and services that individual uses. Only some of the many factors that go into the calculation of the footprint are shown here. (The actual amount of land used for each resource is not drawn to scale.)

**GAUGE YOUR PROGRESS**

✓ What is meant by basic human needs?

✓ What does it mean to live sustainably?

✓ What does an ecological footprint tell us? Why is it important to calculate?

**The human footprint.** If all people worldwide lived the lifestyle of the average U.S. citizen, the human population would need five Earths to support its resource use.

# Science is a process

In the past century humans have learned a lot about the impact of their activities on the natural world. Scientific inquiry has provided great insights into the challenges we are facing and has suggested ways to address those challenges. For example, a hundred years ago, we did not know how significantly or rapidly we could alter the chemistry of the atmosphere by burning fossil fuels. Nor did we understand the effects of many common materials, such as lead and mercury, on human health. Much of our knowledge comes from the work of researchers who study a particular problem or situation to understand why it occurs and how we can fix or prevent it. We will now look at the process scientists use to ask and answer questions about the environment.

## The Scientific Method

To investigate the natural world, scientists like JoAnn Burkholder and her colleagues, who examined the large-scale fish kill in the Neuse River, have to be as objective and methodical as possible. They must conduct their research in such a way that other researchers can understand how their data were collected and agree on the validity of their findings. To do this, scientists

follow a process known as the *scientific method*. The **scientific method** is an objective way to explore the natural world, draw inferences from it, and predict the outcome of certain events, processes, or changes. It is used in some form by scientists in all parts of the world and is a generally accepted way to conduct science.

As we can see in **FIGURE 1.16**, the scientific method has a number of steps, including *observations and questions, forming hypotheses, collecting data, interpreting results,* and *disseminating findings.*

**OBSERVATIONS AND QUESTIONS** JoAnn Burkholder and her team observed a mass die-off of fish in the Neuse River and wanted to know why it happened. Such observing and questioning is where the process of scientific research begins.

**FORMING HYPOTHESES** Observation and questioning lead a scientist to formulate a *hypothesis*. A **hypothesis** is a *testable* conjecture about how something works. It may be an idea, a proposition, a possible mechanism of interaction, or a statement about an effect. For example, we might hypothesize that when the air temperature rises, certain plant species will be more likely, and others less likely, to persist.

What makes a hypothesis testable? We can test the idea about the relationship between air temperature and plant species by growing plants in a greenhouse at different temperatures. "Fish kills are caused by something

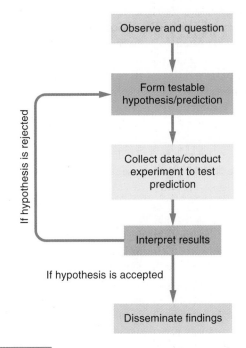

**The scientific method has a number of steps.** In an actual investigation, a researcher might reject a hypothesis and investigate further with a new hypothesis, several times if necessary, depending on the results of the experiment.

in the water" is a testable hypothesis: it speculates that there is an interaction between something in the water and the observed dead fish.

Sometimes it is easier to prove something wrong than to prove it is true beyond doubt. In this case, scientists use a *null hypothesis.* A **null hypothesi**s is a statement or idea that can be falsified, or proved wrong. The statement "Fish deaths have no relationship to something in the water" is an example of a null hypothesis.

**COLLECTING DATA** Scientists typically take several sets of measurements—a procedure called **replication.** The number of times a measurement is replicated is the **sample size** (sometimes referred to as *n*). A sample size that is too small can cause misleading results. For example, if a scientist chose three men out of a crowd at random and found that they all had size 10 shoes, she might conclude that all men have a shoe size of 10. If, however, she chose a larger sample size—100 men—it is very unlikely that all 100 individuals would happen to have the same shoe size.

Proper procedures yield results that are accurate and precise. They also help us determine the possible relationship between our measurements or calculations and the true value. **Accuracy** refers to how close a measured value is to the actual or true value. For example, an environmental scientist might estimate how many songbirds of a particular species there are in an area of 1,000 ha by randomly sampling 10 ha and then projecting or extrapolating the result up to 1,000 ha. If the extrapolation is close to the true value, it is an accurate extrapolation. **Precision** is how close to one another the repeated measurements of the same sample are. In the same example, if the scientist counted birds five times on five different days and obtained five results that were similar to one another, the estimates would be precise. **Uncertainty** is an estimate of how much a measured or calculated value differs from a true value. In some cases, it represents the likelihood that additional repeated measurements will fall within a certain range. Looking at **FIGURE 1.17**, we see that high accuracy and high precision is the most desirable result.

**INTERPRETING RESULTS** We have followed the steps in the scientific method from making observations and asking questions, to forming a hypothesis, to collecting data. What happens next? Once results have been obtained, analysis of data begins. A scientist may use a variety of techniques to assist with data analysis, including summaries, graphs, charts, and diagrams.

As data analysis proceeds, scientists begin to interpret their results. This process normally involves two types of reasoning: *inductive* and *deductive.* **Inductive reasoning** is the process of making general statements from specific facts or examples. If the scientist who sampled a songbird species in the preceding example

| Low accuracy | High accuracy | High accuracy |
| High precision | Low precision | High precision |

**FIGURE 1.17** **Accuracy and precision.** Accuracy refers to how close a measured value is to the actual or true value. Precision is how close repeated measurements of the same sample are to one another.

made a statement about all birds of that species, she would be using inductive reasoning. It might be reasonable to make such a statement if the songbirds that she sampled were representative of the whole population. **Deductive reasoning** is the process of applying a general statement to specific facts or situations. For example, if we know that, in general, air pollution kills trees, and we see a single, dead tree, we may attribute that death to air pollution. But a conclusion based on a single tree might be incorrect, since the tree could have been killed by something else, such as a parasite or fungus. Without additional observations or measurements, and possibly experimentation, the observer would have no way of knowing the cause of death with any degree of certainty.

The most careful scientists always maintain multiple working hypotheses; that is, they entertain many possible explanations for their results. They accept or reject certain hypotheses based on what the data show and do not show. Eventually, they determine that certain explanations are the most likely, and they begin to generate conclusions based on their results.

**DISSEMINATING FINDINGS** A hypothesis is never confirmed by a single experiment. That is why scientists not only repeat their experiments themselves, but also present papers at conferences and publish the results of their investigations. This dissemination of scientific findings allows other scientists to repeat the original experiment and verify or challenge the results. The process of science involves ongoing discussion among scientists, who frequently disagree about hypotheses, experimental conditions, results, and the interpretation of results. Two investigators may even obtain different results from similar measurements and experiments, as happened in the *Pfiesteria* case. Only when the same results are obtained over and over by different investigators can we begin to trust that those results are valid. In the meantime, the disagreements and discussion about contradictory findings are a valuable part of the scientific process. They help scientists refine their research to arrive at more consistent, reliable conclusions.

Like any scientist, you should always read reports of "exciting new findings" with a critical eye. Question the source of the information, consider the methods or processes that were used to obtain the information, and draw your own conclusions. This process, essential to all scientific endeavor, is known as **critical thinking.**

A hypothesis that has been repeatedly tested and confirmed by multiple groups of researchers and has reached wide acceptance becomes a **theory.** Current theories about how plant species distributions change with air temperature, for example, are derived from decades of research and evidence. Notice that this sense of *theory* is different from the way we might use the term in everyday conversation ("But that's just a theory!"). To be considered a theory, a hypothesis must be consistent with a large body of experimental results. A theory can not be contradicted by any replicable tests.

Scientists work under the assumption that the world operates according to fixed, knowable laws. We accept this assumption because it has been successful in explaining a vast array of natural phenomena and continues to lead to new discoveries. When the scientific process has generated a theory that has been tested multiple times, we can call that theory a *natural law.* A **natural law** is a theory to which there are no known exceptions and which has withstood rigorous testing. Familiar examples include the law of gravity and the laws of thermodynamics, which we will look at in the next chapter. These theories are accepted as fact by the scientific community, but they remain subject to revision if contradictory data are found.

## Case Study: The Chlorpyrifos Investigation

Let's look at what we have learned about the scientific method in the context of an actual scientific investigation. In the 1990s, scientists suspected that organophosphates—a group of chemicals commonly used in insecticides—might have serious effects on the human central nervous system. By the early part of the decade, scientists suspected that organophosphates might be linked to such problems as neurological disorders, birth defects, ADHD, and palsy. One of these chemicals, chlorpyrifos (klor-PEER-i-fos), was of particular concern because it is among the most widely used pesticides in the world, with large amounts applied in homes in the United States and elsewhere.

The researchers investigating the effects of chlorpyrifos on human health formulated a hypothesis: *chlorpyrifos causes neurological disorders and negatively affects human health.* Because this hypothesis would be hard to prove conclusively, the researchers also proposed a null hypothesis: *chlorpyrifos has no observable negative effects on the central*

*nervous system.* We can follow the process of their investigation in FIGURE 1.18.

To test the null hypothesis, the scientists designed experiments using rats. One experiment used two groups of rats, with 10 individuals per group. The first group—the *experimental group*—was fed small doses of chlorpyrifos for each of the first 4 days of life. No chlorpyrifos was fed to the second group. That second group was a **control group:** a group that experiences exactly the same conditions as the experimental group, except for the single variable under study. In this experiment, the only difference between the control group and the

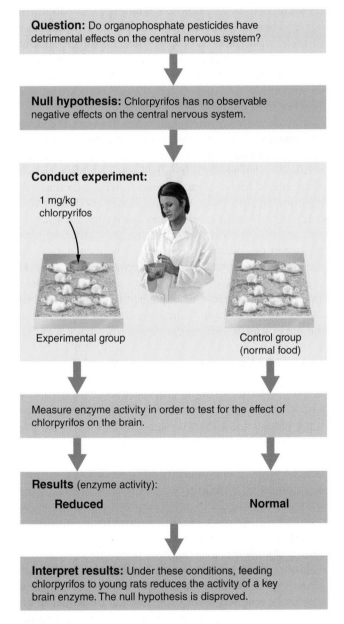

**FIGURE 1.18** **A typical experimental process.** An investigation of the effects of chlorpyrifos on the central nervous system illustrates how the scientific method is used.

experimental group was that the control group was not fed any chlorpyrifos. By designating a control group, scientists can determine whether an observed effect is the result of the experimental treatment or of something else in the environment to which all the subjects are exposed. For example, if the control rats—those that were not fed chlorpyrifos—and the experimental rats—that were exposed to chlorpyrifos—showed no differences in their brain chemistry, researchers could conclude that the chlorpyrifos had no effect. If the control group and experimental group had very different brain chemistry after the experiment, the scientists could conclude that the difference must have been due to the chlorpyrifos. At the end of the experiment, the researchers found that the rats exposed to chlorpyrifos had much lower levels of the enzyme choline acetyltransferase in their brains than the rats in the control group. But without a control group for comparison, the researchers would never have known whether the chlorpyrifos or something else caused the change observed in the experimental group.

The discovery of the relationship between ingesting chlorpyrifos and a single change in brain chemistry might seem relatively small. But that is how most scientific research works: very small steps establish that an effect occurs and, eventually, how it occurs. In this way, we progress toward a more thorough understanding of how the world works. This particular research on chlorpyrifos, combined with numerous other experiments testing specific aspects of the chemical's effect on rat brains, demonstrated that chlorpyrifos was capable of damaging developing rat brains at fairly low doses. The results of this research have been important for our understanding of human health and toxic substances in the environment.

## Controlled Experiments and Natural Experiments

The chlorpyrifos experiment we have just described was conducted in the controlled conditions of a laboratory. However, not all experiments can be done under such controlled conditions. For example, it would be difficult to study the interactions of wolves and caribou in a controlled setting because both species need large amounts of land and because their behavior changes in captivity. Other reasons that a controlled laboratory experiment may not be possible include prohibitive costs and ethical concerns.

Under these circumstances, investigators look for a *natural experiment*. A **natural experiment** occurs when a natural event acts as an experimental treatment in an ecosystem. For example, a volcano that destroys thousands of hectares of forest provides a natural experiment for understanding large-scale forest regrowth (**FIGURE 1.19**). We would never destroy that much forest just to study regrowth, but we can study such natural disasters

(a)

(b)

(c)

**FIGURE 1.19** **A natural experiment.** The Mount St. Helens eruption in 1980 created a natural experiment for understanding large-scale forest regrowth. (a) A pre-eruption forest near Mount St. Helens in 1979; (b) the same location, post-eruption, in 1982; (c) the same location in 2009 begins to show forest regrowth.

when they occur. Still other cases of natural experiments do not involve disasters. For example, we can study the process of ecological succession by looking at areas where forests have been growing for different amounts of time and comparing them. We can study the effects of species invasions by comparing uninvaded ecosystems with invaded ones.

Because a natural experiment is not controlled, many variables can change at once, and results can be difficult to interpret. Ideally, researchers compare multiple examples of similar systems in order to exclude the influences of different variables. For example, after a forest fire, researchers might not only observe how a burned forest responds to the disturbance, but also compare it with a nearby forest that did not burn. In this case, the researchers are comparing similar forests that differ in only one variable: fire. If, however, they tried to compare the burned forest with a different type of forest, perhaps one at a different elevation, it would be difficult to separate the effects of the fire from the effects of elevation. Still, because they may be the only way to obtain vital information, natural experiments are indispensable.

Returning to the study of chlorpyrifos, researchers wanted to know if human brains that were exposed to the chemical would react in the same way as rat brains. For obvious ethical reasons, researchers would never feed pesticides to humans to study their effects. Instead, they conducted a natural experiment. They looked for groups of people in similar circumstances (income, age, level of education) that varied mostly in their exposure to chlorpyrifos. That variation came from their use of pesticides containing chlorpyrifos, the frequency and location of that use, and the brand used. Researchers found that tissue concentrations of chlorpyrifos were highest in groups that worked with the chemical and among poor urban families whose exposure to residential pesticides was high. Among these populations, a number of studies connected exposure to chlorpyrifos with low birth weight and other developmental abnormalities.

## Science and Progress

The chlorpyrifos experiment is a good example of the process of science. Based on observations, the scientists proposed a hypothesis and null hypothesis. The null hypothesis was tested and rejected. Multiple rounds of additional testing gave researchers confidence in their understanding of the problem. Moreover, as the research progressed, the scientists informed the public, as well as the scientific community, about their results. Finally, in 2000, as a result of the step-by-step scientific investigation of chlorpyrifos, the U.S. Environmental Protection Agency (EPA) decided to prohibit its use for most residential applications. It also prohibited agricultural use on fruits that are eaten without peeling, such as apples and pears, or those that are especially popular with children, such as grapes.

### GAUGE YOUR PROGRESS

✓ What is the scientific method, and how do scientists use it to address environmental problems?

✓ What is a hypothesis? What is a null hypothesis?

✓ How are controlled and natural experiments different? Why do we need each type?

# Environmental science presents unique challenges

Environmental science has many things in common with other scientific disciplines. However, it presents a number of challenges and limitations that are not usually found in most other scientific fields. These challenges and limitations are a result of the nature of environmental science and the way research in the field is conducted.

## Lack of Baseline Data

The greatest challenge to environmental science is the fact that there is no undisturbed baseline—no "control planet"—with which to compare the contemporary Earth. Virtually every part of the globe has been altered by humans in some way (FIGURE 1.20). Even though some remote regions appear to be undisturbed, we can still find quantities of lead in the Greenland ice sheet, traces of the anthropogenic compound PCB in the fatty tissue of penguins in Antarctica, and invasive species from many locations carried by ship to remote tropical

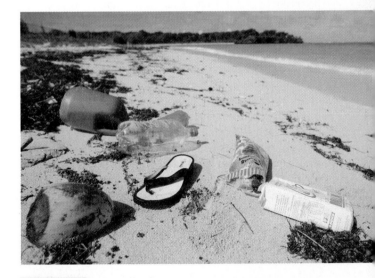

**FIGURE 1.20** **Human impacts are global.** The trash washed up onto the beach of this remote Pacific island vividly demonstrates the difficulty of finding any part of Earth unaffected by human activities.

islands. This situation makes it difficult to know the original levels of contaminants or numbers of species that existed before humans began to alter the planet. Consequently, we can only speculate about how the current conditions deviate from those of pre-human activity.

## Subjectivity

A second challenge unique to environmental science lies in the dilemmas raised by subjectivity. For example, when you go to the grocery store, the bagger may ask, "Paper or plastic?" How can we know for certain which type of bag has the least environmental impact? There are techniques for determining what harm may come from using the petrochemical benzene to make a plastic bag and from using chlorine to make a paper bag. However, different substances tend to affect the environment differently: benzene may pose more of a risk to people, whereas chlorine may pose a greater risk to organisms in a stream. It is difficult, if not impossible, to decide which is better or worse for the environment overall. There is no single measure of environmental quality. Ultimately, our assessments and our choices involve value judgments and personal opinions.

## Interactions

A third challenge is the complexity of natural and human-dominated systems. All scientific fields examine interacting systems, but those systems are rarely as complex and as intertwined as they are in environmental science. Because environmental systems have so many interacting parts, the results of a study of one system cannot always be easily applied to similar systems elsewhere.

There are also many examples in which human preferences and behaviors have as much of an effect on environmental systems as the natural laws that describe them. For example, many people assume that if we built more efficient automobiles, the overall consumption of gasoline in the United States would decrease. To decrease gas consumption, however, it is necessary not only to build more efficient automobiles, but also to get people to purchase those vehicles and use them in place of less efficient ones. During the 1990s and early 2000s, even though there were many fuel-efficient cars available, the majority of buyers in the United States continued to purchase larger, heavier, and less fuel-efficient cars, minivans, light trucks, and sport-utility vehicles. Environmental scientists thought they knew how to reduce gasoline consumption, but they neglected to account for consumer behavior. Science is the search for natural laws that govern the world around us, whereas environmental science may involve politics, law, and economics as well as the traditional natural sciences. This complexity often makes environmental science challenging and its findings the subject of vigorous and lively debate.

## Human Well-Being

As we continue our study of environmental science, we will see that many of its topics touch on human well-being. In environmental science, we study how humans impact the biological systems and natural resources of the planet. We also study how changes in natural systems and the supply of natural resources affect humans.

We know that people who are unable to meet their basic needs are less likely to be interested in or able to be concerned about the state of the natural environment. The principle of *environmental equity*—the fair distribution of Earth's resources—adds a moral issue to questions raised by environmental science. Pollution and environmental degradation are inequitably distributed, with the poor receiving much more than an equal share. Is this a situation that we, as fellow humans, can tolerate? The ecological footprint and other environmental indicators show that it would be unsustainable for all people on the planet to live like the typical North American. But as more and more people develop an ability to improve their living conditions, how do we think about apportioning limited resources? Who has the right and the responsibility to make such decisions? **Environmental justice** is a social movement and field of study that works toward equal enforcement of environmental laws and the elimination of disparities, whether intended or unintended, in how pollutants and other environmental harms are distributed among the various ethnic and socioeconomic groups within a society (FIGURE 1.21).

Our society faces many environmental challenges. The loss of biodiversity, the growing human demand for

FIGURE 1.21 **A village on the outskirts of New Delhi, India.** The poor are exposed to a disproportionate amount of pollutants and other hazards. The people shown here are recycling circuit boards from discarded electronics products.

resources, and climate change are all complex problems. To solve them, we will need to apply thoughtful analysis, scientific innovation, and strategies that consider human behavior. Around the globe today, we can find people who are changing the way their governments work, changing the way they do business, and changing the way they live their lives, all with a common goal: they are working toward sustainability. Here, and at the end of each chapter of this book, we will tell a few of their stories.

 # WORKING TOWARD SUSTAINABILITY

## Using Environmental Indicators to Make a Better City

We have seen that environmental indicators can be used to monitor conditions across a range of scales, from local to global. They are also being used by people looking for ways to apply environmental science to the urban planning process in countries as diverse as China, Brazil, and the United States.

San Francisco, California, is one example. In 1997, the city adopted a sustainability plan to go along with its newly formed Department of the Environment. The San Francisco Sustainability Plan focuses on 10 environmental concerns:

- Air quality
- Biodiversity
- Energy, climate change, and ozone depletion
- Food and agriculture
- Hazardous materials
- Human health
- Parks, open spaces, and streetscapes
- Solid waste
- Transportation
- Water and wastewater

Although some of these topics may not seem like components of urban planning, the drafters of the plan recognized that the everyday choices of city dwellers can have wide-ranging environmental impacts, both in and beyond the city. For example, purchasing local produce or organic food affects the environments and economies of both San Francisco and the agricultural areas that serve it.

For each of the 10 environmental concerns, the sustainability plan sets out a series of 5-year and long-term objectives as well as specific actions required to achieve them. These actions include public education through information sources such as Web sites and newsletters and hands-on activities such as replacing non-native plants with native trees and shrubs.

To monitor the effectiveness of the various actions, San Francisco chose specific environmental indicators for each of the 10 environmental concerns. These indicators had to indicate a clear trend toward or away from environmental sustainability, demonstrate cost-effectiveness, be understandable to the nonscientist, and be easily presented to the media. For example, to evaluate biodiversity, San Francisco uses four indicators:

| Environmental indicator | Desired trend |
| --- | --- |
| Number of volunteer hours dedicated to managing, monitoring, and conserving San Francisco's biodiversity | INCREASING |
| Number of square feet of the worst non-native species removed from natural areas | INCREASING |
| Number of surviving native plant species planted in developed parks, private landscapes, and natural areas | INCREASING |
| Abundance and species diversity of birds, as indicated by the Golden Gate Audubon Society's Christmas bird counts | INCREASING |

Together, these indicators provide a relatively inexpensive and simple way to summarize the level of biodiversity, the threat to native biodiversity from non-native species, and the amount of effort going into biodiversity protection.

More than 13 years later, what do the indicators show? In general, there has been a surprising amount of improvement. For example, in the category of solid waste, San Francisco has increased the amount of waste

**FIGURE 1.22** A "green" city. San Francisco's adoption of environmental indicators has helped it achieve many of its sustainability goals.

recycled from 30 to 70 percent, with a goal of 75 percent by 2020, and it now has the largest urban composting program in the country. San Francisco has also improved its air quality, reducing the number of days in which fine particulate matter exceeded the EPA air quality safe level, from 27 days in 2000 to 10 days in 2006. These and other successes have won the city numerous accolades: it has been selected as one of "America's Top Five Cleanest Cities" by *Reader's Digest* and as one of the "Top 10 Green Cities" by *The Green Guide*. In 2005, San Francisco was named the most sustainable city in the United States by SustainLane (**FIGURE 1.22**).

**Reference**

www.sustainlane.com.

## REVISIT THE KEY IDEAS

■ **Define the field of environmental science and discuss its importance.**

Environmental science is the study of the interactions among human-dominated systems and natural systems and how those interactions affect environments. Studying environmental science helps us identify, understand, and respond to anthropogenic changes.

■ **Identify ways in which humans have altered and continue to alter our environment.**

The impact of humans on natural systems has been significant since early humans hunted some large animal species to extinction. However, technology and population growth have dramatically increased both the rate and the scale of human-induced change.

■ **Describe key environmental indicators that help us evaluate the health of the planet.**

Five important global-scale environmental indicators are biological diversity, food production, average global surface temperature and atmospheric $CO_2$ concentrations, human population, and resource depletion.

■ **Define sustainability and explain how it can be measured using the ecological footprint.**

Sustainability is the use of Earth's resources to meet our current needs without jeopardizing the ability of future generations to meet their own needs. The ecological footprint is the land area required to support a person's (or a country's) lifestyle. We can use that information to say something about how sustainable that lifestyle would be if it were adopted globally.

■ **Explain the scientific method and its application to the study of environmental problems.**

The scientific method is a process of observation, hypothesis generation, data collection, analysis of results, and dissemination of findings. Repetition of measurements or experiments is critical if one is to determine the validity of findings. Hypotheses are tested and often modified before being accepted.

■ **Describe some of the unique challenges and limitations of environmental science.**

We lack an undisturbed "control planet" with which to compare conditions on Earth today. Assessments and choices are often subjective because there is no single measure of environmental quality. Environmental systems are so complex that they are poorly understood, and human preferences and policies may have as much of an effect on them as natural laws.

# CHECK YOUR UNDERSTANDING

1. Which of the following events has increased the impact of humans on the environment?

      I Advances in technology
      II Reduced human population growth
      III Use of tools for hunting

   (a) I only
   (b) I and II only
   (c) II and III only
   (d) I and III only
   (e) I, II, and III

2. As described in this chapter, environmental indicators
   (a) always tell us what is causing an environmental change.
   (b) can be used to analyze the health of natural systems.
   (c) are useful only when studying large-scale changes.
   (d) do not provide information regarding sustainability.
   (e) take into account only the living components of ecosystems.

3. Which statement regarding a global environmental indicator is *not* correct?
   (a) Concentrations of atmospheric carbon dioxide have been rising quite steadily since the Industrial Revolution.
   (b) World grain production has increased fairly steadily since 1950, but worldwide production of grain per capita has decreased dramatically over the same period.
   (c) For the past 130 years, average global surface temperatures have shown an overall increase that seems likely to continue.
   (d) World population is expected to be between 8.1 billion and 9.6 billion by 2050.
   (e) Some natural resources are available in finite amounts and are consumed during a one-time use, whereas other finite resources can be used multiple times through recycling.

4. Figure 1.8 (on page 10) shows atmospheric carbon dioxide concentrations over time. The measured concentration of $CO_2$ in the atmosphere is an example of
   (a) a sample of air from over the Antarctic.
   (b) an environmental indicator.
   (c) replicate sampling.
   (d) calculating an ecological footprint.
   (e) how to study seasonal variation in Earth's temperatures.

5. In science, which of the following is the most certain?
   (a) Hypothesis
   (b) Idea
   (c) Natural law
   (d) Observation
   (e) Theory

6. All of the following would be exclusively caused by anthropogenic activities *except*
   (a) combustion of fossil fuels.
   (b) overuse of resources such as uranium.
   (c) forest clearing for crops.
   (d air pollution from burning oil.
   (e) forest fires.

7. Use Figure 1.6 (on page 9) to calculate the approximate percentage change in world grain production per person between 1950 and 2000.
   (a) 10 percent
   (b) 20 percent
   (c) 30 percent
   (d) 40 percent
   (e) 50 percent

8. The populations of some endangered animal species have stabilized or increased in numbers after human intervention. An example of a species that is still endangered and needs further assistance to recover is the
   (a) American bison.
   (b) peregrine falcon.
   (c) bald eagle.
   (d) American alligator.
   (e) snow leopard.

*Questions 9 and 10 refer to the following experimental scenario:*

An experiment was performed to determine the effect of caffeine on the pulse rate of five healthy 18-year-old males. Each was given 250 mL of a beverage with or without caffeine. The men had their pulse rates measured before they had the drink (time 0 minutes) and again after they had been sitting at rest for 30 minutes after consuming the drink. The results are shown in the following table.

| Subject | Beverage | Caffeine content (mg/serving) | Pulse rate at time 0 minutes | Pulse rate at time 30 minutes |
|---|---|---|---|---|
| 1 | Water | 0 | 60 | 59 |
| 2 | Caffeine-free soda | 0 | 55 | 56 |
| 3 | Caffeinated soda | 10 | 58 | 68 |
| 4 | Coffee, decaffeinated | 3 | 62 | 67 |
| 5 | Coffee, regular | 45 | 58 | 81 |

9. Before the researchers began the experiment, they formulated a null hypothesis. The best null hypothesis for the experiment would be that caffeine
   (a) has no observable effect on the pulse rate of an individual.
   (b) will increase the pulse rates of all test subjects.
   (c) will decrease the pulse rates of all test subjects.
   (d) has no observable effects on the pulse rates of 18-year-old males.
   (e) from a soda will have a greater effect on pulse rates than caffeine from coffee.

10. After analyzing the results of the experiment, the most appropriate conclusion would be that caffeine
   (a) increased the pulse rates of the 18-year-old males tested.
   (b) decreased the pulse rates of the 18-year-old males tested.
   (c) will increase the pulse rate of any individual that is tested.
   (d) increases the pulse rate and is safe to consume.
   (e) makes drinks better than decaffeinated beverages.

## APPLY THE CONCEPTS

1. Your neighbor has fertilized her lawn. Several weeks later, she is alarmed to see that the surface of her ornamental pond, which sits at the bottom of the sloping lawn, is covered with a green layer of algae.
   (a) Suggest a feasible explanation for the algal bloom in the pond.
   (b) Design an experiment that would enable you to validate your explanation. Include and label in your answer:
      (i) a testable hypothesis.
      (ii) the variable that you will be testing.
      (iii) the data to be collected.
      (iv) a description of the experimental procedure.
      (v) a description of the results that would validate your hypothesis.
   (c) Based on the data from your experiment and your explanation of the problem, think of, and suggest, one action that your neighbor could take to help the pond recover.

2. The study of environmental science sometimes involves examining the overuse of environmental resources.
   (a) Identify one general effect of overuse of an environmental resource.
   (b) For the effect you listed above, describe a more sustainable strategy for resource utilization.
   (c) Describe how the events from Easter Island can be indicative of environmental issues on Earth today.

## MEASURE YOUR IMPACT

**Exploring Your Footprint** Make a list of the activities you did today and attempt to describe their impact on the five global environmental indicators described in this chapter. For each activity, such as eating lunch or traveling to school, make as complete a list as you can of the resources and fuels that went into the activity and try to determine the impacts of using those resources.

After completing your inventory, visit the Web site of the Global Footprint Network (www.footprintnetwork .org) and complete the personal footprint calculator. Compare the impacts you described with the impacts you are asked to identify in the personal footprint calculator.

## ENGAGE YOUR ENVIRONMENT

What are the most critical environmental issues today? In this Engage Your Environment exercise you will identify pressing global and local environmental issues and suggest ways to address these problems. *See Engage Your Environment Activity 1 at the end of this book.*

# Environmental Systems: Matter, Energy, and Change

## A Lake of Salt Water, Dust Storms, and Endangered Species

Located between the deserts of the Great Basin and the mountains of the Sierra Nevada, California's Mono Lake is an unusual site. It is characterized by eerie tufa towers of limestone rock, unique animal species, glassy waters, and frequent dust storms. Mono Lake is a *terminal lake,* which means that water flows into it, but does not flow out. As water moves through the mountains and desert soil, it picks up salt and other minerals, which it deposits in the lake. As the water evaporates, these minerals are left behind. Over time, evaporation has caused a buildup of salt concentrations so high that the lake is actually saltier than the ocean, and no fish can survive in its water.

The Mono brine shrimp (*Artemia monica*) and the larvae of the Mono Lake alkali fly (*Ephydra hians*) are two of only a few animal species that can tolerate the conditions of the lake. The brine shrimp and the fly larvae consume microscopic algae, millions of tons of which grow in the lake each year. In turn, large flocks of migrating birds, such as sandpipers, gulls, and flycatchers, use the lake as a stopover, feeding on the brine shrimp and fly larvae to replenish their energy stores. The lake is an oasis on the migration route for these birds. They have come to depend on its food and water resources. The health of Mono Lake is therefore critical for many species.

In 1913, the city of Los Angeles drew up a controversial plan to redirect water away from Mono Lake and its neighbor, the larger and shallower Owens Lake. Owens Lake was diverted first, via a 359 km (223-mile) aqueduct that drew water away from the springs and streams that kept Owens Lake full. Soon, the lake began to dry up, and by the 1930s, only an empty salt flat remained. Today the dry lake bed covers roughly 440 km² (109,000 acres). It is one of the nation's largest sources of windblown dust, which lowers visibility in the area's national parks. Even worse, because of the local geology, the dust contains high concentrations of arsenic — a major threat to human health.

In 1941, despite the environmental degradation at Owens Lake, Los Angeles extended the aqueduct to draw water from the streams feeding Mono Lake. By 1982, with less fresh water feeding the lake, its depth had decreased by half, to an average of 14 m (45 feet), and the salinity of the water had doubled to more than twice that of the ocean. The salt killed the lake's algae. Without algae to eat, the Mono brine shrimp also died. Most birds stayed away, and newly exposed land bridges allowed coyotes from the desert to prey on those colonies of nesting birds that remained.

However, just when it appeared that Mono Lake would not recover, circumstances changed. In 1994, after years of litigation led by the National Audubon Society and tireless work by environmentalists, the Los Angeles Department of Water and Power agreed to reduce the amount of water it diverted and allow the lake to refill to about two-thirds of its historical depth. By summer 2009, lake levels had risen to just short of that goal, and the ecosystem was slowly recovering. Today brine shrimp are thriving, and many birds are returning to Mono Lake. ▶

> ## Just when it appeared that Mono Lake would not recover, circumstances changed.

A California gull feeding on alkali flies.

◀ Tufa towers rise out of the salty water of Mono Lake.

Water is a scarce resource in the Los Angeles area, and demand there is particularly high. To decrease the amount of water diverted from Mono Lake, the city of Los Angeles had to reduce its water consumption. The city converted water-demanding grass lawns to drought-tolerant native shrubs, and it imposed new rules requiring low-flow shower heads and water-saving toilets. Through these seemingly small, but effective, measures, Los Angeles inhabitants were able to cut their water consumption and, in turn, protect nesting birds, Mono brine shrimp, and algae populations, as well as the rest of the Mono Lake ecosystem. ■

Sources: J. Kay, It's rising and healthy, *San Francisco Chronicle*, July 29, 2006; Mono Lake Committee, Mono Lake (2010), http://www.monolake.org/.

## UNDERSTAND THE KEY IDEAS

Most problems of interest to environmental scientists involve more than one organism and more than one physical factor. Organisms, nonliving matter, and energy all interact in natural systems. Taking a systems approach to an environmental issue, rather than focusing on only one piece of the puzzle, decreases the chance of overlooking important components of that issue.

After reading this chapter you should be able to

■ define *systems* within the context of environmental science.

■ explain the components and states of matter.

■ distinguish between various forms of energy and discuss the first and second laws of thermodynamics.

■ describe the ways in which ecological systems depend on energy inputs.

■ explain how scientists keep track of inputs, outputs, and changes to complex systems.

■ describe how natural systems change over time and space.

# Earth is a single interconnected system

The story of Mono Lake shows us that the activities of humans, the lives of other organisms, and processes in the environment are interconnected. Humans, water, animals, plants, and the desert environment all interact at Mono Lake to create a complex environmental system. The story also demonstrates a key principle of environmental science: that a change in any one factor can have other, often unexpected, effects.

In Chapter 1, we learned that a system is a set of interacting components connected in such a way that a change in one part of the system affects one or more other parts of the system. The Mono Lake system is relatively small. Other complex systems exist on a much larger scale.

A large system may contain many smaller systems within it. FIGURE 2.1 shows an example of complex, interconnecting systems that operate at multiple space and time scales: the fisheries of the North Atlantic. A physiologist who wants to study how codfish survive in the North Atlantic's freezing waters must consider all the biological adaptations of the cod to be part of one system. In this case, the fish and its internal organs are the system being studied. In the same environment, a marine biologist might study the predator–prey relationship between cod and herring. That relationship constitutes another system, which includes two fish species and the environment they live in. At an even larger scale, an oceanographer might focus on how ocean currents in a particular area affect the dispersal of cod and other fish species. A fisheries management official might study a system that includes all of the systems above as well as people, fishing technology, policy, and law.

The largest system that environmental science considers is Earth. Many of our most important current environmental issues—including human population growth and climate change—exist at the global scale. Throughout this book we will define a given system in terms of the environmental issue we are studying and the scale in which we are interested.

Whether we are investigating ways to reduce pollution, increase food supplies, or find alternatives to fossil fuels, environmental scientists must have a thorough understanding of matter and energy and their interactions within and across systems. We will begin this chapter by exploring the properties of matter. We will then discuss the various types of energy and how they influence and limit systems.

### GAUGE YOUR PROGRESS

✓ What is an environmental system? Name some examples.

✓ How do systems vary in scale, and how does a large system include a smaller system?

✓ What are the largest systems in the Mono Lake ecosystem? What are some examples of smaller systems within that system?

To a physiologist, a cod is a system.

Cod    Herring

To a marine biologist, the predator-prey relationship between two fish species forms a system.

For an oceanographer, the system might consist of ocean currents and their effects on fish populations.

Current

A fisheries manager is interested in a larger system, consisting of fish populations as well as human activities and laws.

**FIGURE 2.1** **Systems within systems.** The boundaries of an environmental system may be defined by the researcher's point of view. Physiologists, marine biologists, oceanographers, and fisheries managers would describe the North Atlantic Ocean fisheries system differently.

# All environmental systems consist of matter

What do rocks, water, air, the book in your hands, and the cells in your body have in common? They are all forms of *matter.* **Matter** is anything that occupies space and has *mass.* The **mass** of an object is defined as a measure of the amount of matter it contains. Note that the words *mass* and *weight* are often used interchangeably, but they are not the same thing. Weight is the force that results from the action of gravity on mass. Your own weight, for example, is determined by the amount of gravity pulling you toward the planet's center. Whatever your weight is on Earth, you would have a lesser weight on the Moon, where the action of gravity is less. In contrast, mass stays the same no matter what gravitational influence is acting on an object. So although your weight would change on the Moon, your mass would remain the same because the amount of matter you are made of would be the same.

## Atoms and Molecules

All matter is composed of tiny particles that cannot be broken down into smaller pieces. The basic building blocks of matter are known as atoms. An **atom** is the smallest particle that can contain the chemical properties of an element. An **element** is a substance composed of atoms that cannot be broken down into smaller, simpler components. At Earth's surface temperatures, elements can occur as solids (such as gold), liquids (such as bromine), or gases (such as helium). Atoms are so small that a single human hair measures about a few hundred thousand carbon atoms across.

Ninety-four elements occur naturally on Earth, and another 24 have been produced in laboratories. The **periodic table** lists all of the elements currently known. (For a copy of the periodic table, turn to Appendix A.) Each element is identified by a one- or two-letter symbol; for example, the symbol for carbon is C, and the symbol for oxygen is O. These symbols are used to describe the atomic makeup of **molecules,** which are particles containing more than one atom. Molecules that contain more than one element are called **compounds.** For example, a carbon dioxide molecule ($CO_2$) is a compound composed of one carbon atom (C) and two oxygen atoms ($O_2$).

As we can see in **FIGURE 2.2a**, every atom has a *nucleus,* or core, which contains protons and neutrons. Protons and neutrons have roughly the same mass—both minutely small. Protons have a positive electrical charge, like the "plus" side of a battery. The number of protons in the nucleus of a particular element—called the **atomic number**—is unique to that element. Neutrons have no electrical charge, but they are critical to the stability of nuclei because they keep the positively charged protons together. Without them, the protons would repel one another and separate.

As **FIGURE 2.2b** shows, the space around the nucleus of the atom is not empty. In this space, electrons exist in *orbitals,* which are electron clouds that extend different distances from the nucleus. Electrons are negatively

**(a)** Nitrogen atom with electrons shown in shells

Electrons

Electron shells

Nucleus

Neutron

Proton

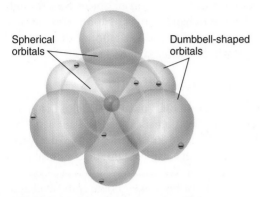

**(b)** Nitrogen atom with electrons in orbitals

Spherical orbitals

Dumbbell-shaped orbitals

**FIGURE 2.2** **Structure of the atom.** An atom is composed of protons, neutrons, and electrons. Neutrons and positively charged protons make up the nucleus. Negatively charged electrons surround the nucleus. (a) Moving electrons are commonly represented in shells. (b) In reality, however, they exist in complex orbitals.

charged, like the "minus" side of a battery, and have a much smaller mass than protons or neutrons. In the molecular world, opposites always attract, so negatively charged electrons are attracted to positively charged protons. This attraction binds the electrons to the nucleus. In a neutral atom, the numbers of protons and electrons are equal. The distribution of electrons in an orbital, particularly the outermost part of the orbital, greatly contributes to the atom's chemical characteristics. In any electron orbital, there can be only a certain number of electrons.

The total number of protons and neutrons in an element is known as its **mass number.** Because the mass of an electron is insignificant compared with the mass of a proton or neutron, we do not include electrons in mass number calculations.

Although the number of protons in a chemical element is constant, atoms of the same element may have different numbers of neutrons, and therefore different mass numbers. These various kinds of atoms are called **isotopes.** Isotopes of the element carbon, for example, all have six protons, but can occur with six, seven, or eight neutrons, yielding mass numbers of 12, 13, or 14, respectively. In the natural environment, carbon occurs as a mixture of carbon isotopes. All carbon isotopes behave the same chemically. However, biological processes sometimes favor one isotope over another. Thus certain isotopic "signatures" (that is, different ratios of isotopes) can be left behind by different biological processes. These signatures allow environmental scientists to learn about certain processes by determining the proportions of different isotopes in soil, air, water, or ice.

## Radioactivity

The nuclei of isotopes can be stable or unstable, depending on the mass number of the isotope and the number of neutrons it contains. Unstable isotopes are *radioactive.*

Radioactive isotopes undergo **radioactive decay,** the spontaneous release of material from the nucleus. Radioactive decay changes the radioactive element into a different element. For example, uranium-235 ($^{235}$U) decays to form thorium-231 ($^{231}$Th). The original atom (uranium) is called the *parent* and the resulting decay product (thorium) is called the *daughter.* The radioactive decay of $^{235}$U and certain other elements emits a great deal of energy that can be captured as heat. Nuclear power plants use this heat to produce steam that turns turbines to generate electricity.

We measure radioactive decay by recording the average rate of decay of a quantity of a radioactive element. This measurement is commonly stated in terms of the element's **half-life:** the time it takes for one-half of the original radioactive parent atoms to decay. An element's half-life is a useful parameter to know because some elements that undergo radioactive decay emit harmful radiation. Knowledge of the half-life allows scientists to determine the length of time that a particular radioactive element may be dangerous. For example, using the half-life allows scientists to calculate the period of time that people and the environment must be protected from depleted nuclear fuel, like that generated by a nuclear power plant. As it turns out, many of the elements produced during the decay of $^{235}$U have half-lives of tens of thousands of years and more. From this we can see why long-term storage of radioactive nuclear waste is so important.

The measurement of isotopes has many applications in environmental science as well as in other scientific fields. For example, carbon in the atmosphere exists in a known ratio of the isotopes carbon-12 (99 percent), carbon-13 (1 percent), and carbon-14 (which occurs in trace amounts, on the order of one part per trillion). Carbon-14 is radioactive and has a half-life of 5,730 years. Carbon-13 and carbon-12 are stable isotopes. Living organisms incorporate carbon into their tissues at roughly the known atmospheric ratio. But after an organism dies, it stops incorporating new carbon into its tissues. Over time, the radioactive carbon-14 in the organism decays to nitrogen-14. By calculating the proportion of carbon-14 in dead biological material—a technique called *carbon dating*—researchers can determine how many years ago an organism died.

## Chemical Bonds

We have seen that matter is composed of atoms, which form molecules or compounds. In order to form molecules or compounds, atoms must be able to interact or join together. This happens by means of chemical bonds of various types. Chemical bonds fall into three categories: *covalent bonds, ionic bonds,* and *hydrogen bonds.*

**COVALENT BONDS** Elements that do not readily gain or lose electrons form compounds by sharing electrons. These compounds are said to be held together by **covalent bonds.** FIGURE 2.3 illustrates the covalent bonds in a molecule of methane ($CH_4$, also called *natural gas*). A methane molecule is made up of one carbon (C) atom surrounded by four hydrogen (H) atoms. Covalent bonds form between the single carbon atom and each hydrogen atom. Covalent bonds also hold the two hydrogen atoms and the oxygen atom in a water molecule together.

**IONIC BONDS** In a covalent bond, atoms share electrons. Another kind of bond between two atoms involves the transfer of electrons. When such a transfer happens, one atom becomes electron deficient (positively charged), and the other becomes electron rich (negatively charged). This charge imbalance holds the two atoms together. The charged atoms are called *ions,* and the attraction between oppositely charged ions forms a chemical bond called an **ionic bond.** FIGURE 2.4 shows an example of this process. Sodium (Na) donates one electron to chlorine (Cl), which gains one electron, to form sodium chloride (NaCl), or table salt.

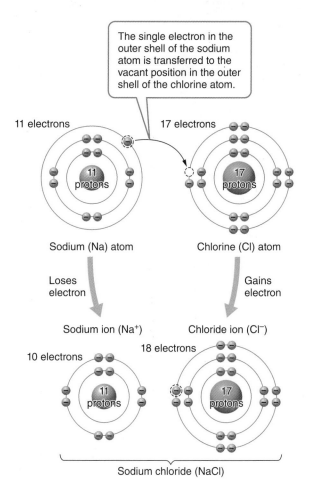

The single electron in the outer shell of the sodium atom is transferred to the vacant position in the outer shell of the chlorine atom.

Sodium chloride (NaCl)

**FIGURE 2.4** **Ions and ionic bonds.** A sodium atom and a chlorine atom can readily form an ionic bond. The sodium atom loses an electron, and the chlorine atom gains one. As a result, the sodium atom becomes a positively charged ion ($Na^+$) and the chlorine atom becomes a negatively charged ion ($Cl^-$, known in ionic form as chloride). The attraction between the oppositely charged ions—an ionic bond—forms sodium chloride (NaCl), or table salt.

An ionic bond is not usually as strong as a covalent bond. This means that the compound can readily dissolve. As long as sodium chloride remains in a salt shaker, it remains in solid form. But if you shake some into water, the salt dissolves into sodium and chloride ions ($Na^+$ and $Cl^-$).

**HYDROGEN BONDS** The third type of chemical bond is weaker than either covalent or ionic bonds. A **hydrogen bond** is a weak chemical bond that forms when hydrogen atoms that are covalently bonded to one atom are attracted to another atom on another molecule. When atoms of different elements form bonds, their electrons may be shared unequally; that is, shared electrons may be pulled closer to one atom than to the other. In some cases, the strong attraction of the hydrogen electron to other atoms creates a charge imbalance within the covalently bonded molecule.

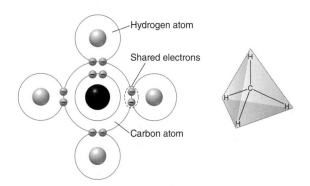

**FIGURE 2.3** **Covalent bonds.** Molecules such as methane ($CH_4$) are associations of atoms held together by covalent bonds, in which electrons are shared between the atoms. As a result of the four hydrogen atoms sharing electrons with a carbon atom, each atom has a complete set of electrons in its outer shell—two for the hydrogen atoms and eight for the carbon atom.

**(a)** Water molecule

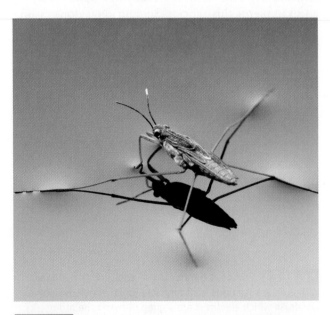

**(b)** Hydrogen bonds between water molecules

**FIGURE 2.5** **The polarity of the water molecule allows it to form hydrogen bonds.** (a) Water ($H_2O$) consists of two hydrogen atoms covalently bonded to one oxygen atom. Water is a polar molecule because its shared electrons spend more time near the oxygen atom than near the hydrogen atoms. The hydrogen atoms thus have a slightly positive charge, and the oxygen atom has a slightly negative charge. (b) The slightly positive hydrogen atoms are attracted to the slightly negative oxygen atom of another water molecule. The result is a hydrogen bond between the two molecules.

Looking at **FIGURE 2.5a**, we see that water is an excellent example of this type of asymmetric electron distribution. Each water molecule as a whole is neutral; that is, it carries neither a positive nor a negative charge. But water has unequal covalent bonds between its two hydrogen atoms and one oxygen atom. Because of these unequal bonds and the angle formed by the H-O-H bonds, water is known as a *polar* molecule. In a **polar molecule,** one side is more positive and the other side is more negative. We can see the result in **FIGURE 2.5b**: a hydrogen atom in a water molecule is attracted to the oxygen atom in a nearby water molecule. That attraction forms a hydrogen bond between the two molecules.

By allowing water molecules to link together, hydrogen bonding gives water a number of unusual properties. Hydrogen bonds also occur in nucleic acids such as DNA, the biological molecule that carries the genetic code for all organisms.

## Properties of Water

The molecular structure of water gives it unique properties that support the conditions necessary for life on Earth. Among these properties are surface tension, capillary action, a high boiling point, and the ability to dissolve many different substances—all essential to physiological functioning.

**SURFACE TENSION AND CAPILLARY ACTION** We don't generally think of water as being sticky, but hydrogen bonding makes water molecules stick strongly to one another (*cohesion*) and to certain other substances (*adhesion*). The ability to cohere or adhere underlies two unusual properties of water: *surface tension and capillary action.*

**Surface tension,** which results from the cohesion of water molecules at the surface of a body of water, creates a sort of skin on the water's surface. Have you ever seen an aquatic insect, such as a water strider, walk across the surface of the water? This is possible because of surface tension (**FIGURE 2.6**). Surface tension also makes water droplets smooth and more or less spherical as they cling to a water faucet before dropping.

**Capillary action** happens when adhesion of water molecules to a surface is stronger than cohesion between the molecules. The absorption of water by a paper towel or a sponge is the result of capillary action. This property is important in thin tubes, such as the water-conducting vessels in tree trunks, and in small pores in soil. It is also important in the transport of underground water, as well as dissolved pollutants, from one location to another.

**BOILING AND FREEZING** At the atmospheric pressures found at Earth's surface, water boils (becomes a gas) at

**FIGURE 2.6** **Surface tension.** Hydrogen bonding between water molecules creates the surface tension necessary to support this water strider. Where else in nature can you witness surface tension?

**FIGURE 2.7** **Ice floats on water.** Below 4°C, water molecules realign into a crystal lattice structure. With its molecules farther apart, solid water (ice) is less dense than liquid water. This property allows ice to float on liquid water.

100°C (212°F) and freezes (becomes a solid) at 0°C (32°F). If water behaved like structurally similar compounds such as hydrogen sulfide ($H_2S$), which boils at –60°C (–76°F), it would be a gas at typical Earth temperatures, and life as we know it could not exist. Because of its cohesion, however, water can be a solid, a gas, or—most importantly for living organisms—a liquid at Earth's surface temperatures. In addition, the hydrogen bonding between water molecules means that it takes a great deal of energy to change the temperature of water. Thus the water in organisms protects them from wide temperature swings. Hydrogen bonding also explains why geographic areas near large lakes or oceans have moderate climates. The water body holds summer heat, slowly releasing it as the atmosphere cools in the fall, and warms only slowly in spring, thereby preventing the adjacent land area from heating up quickly.

Water has another unique property: it takes up a larger volume in solid form than it does in liquid form. **FIGURE 2.7** illustrates the difference in structure between liquid water and ice. As liquid water cools, it becomes denser, until it reaches 4°C (39°F), the temperature at which it reaches maximum density. As it cools from 4°C down to freezing at 0°C, however, its molecules realign into a crystal lattice structure, and its volume expands. You can see the result any time you add an ice cube to a drink: ice floats on liquid water.

What does this unique property of water mean for life on Earth? Imagine what would happen if water acted like most other liquids. As it cooled, it would continue to become denser. Its solid form (ice) would sink, and lakes and ponds would freeze from the bottom up. As a result, very few aquatic organisms could survive in temperate and cold climates.

**WATER AS A SOLVENT** In our table salt example, we saw that water makes a good solvent. Many substances, such as table salt, dissolve well in water because their polar molecules bond easily with other polar molecules. This explains the high concentrations of dissolved ions in seawater as well as the capacity of living organisms to store many types of molecules in solution in their cells. Unfortunately, many toxic substances also dissolve well in water, which makes them easy to transport through the environment.

## Acids, Bases, and pH

Another important property of water is its ability to dissolve hydrogen- or hydroxide-containing compounds known as *acids* and *bases*. An **acid** is a substance that contributes hydrogen ions to a solution. A **base** is a substance that contributes hydroxide ions to a solution. Both acids and bases typically dissolve in water.

When an acid is dissolved in water, it dissociates into positively charged hydrogen ions ($H^+$) and negatively charged ions. Two important acids we will discuss in this book are nitric acid ($HNO_3$) and sulfuric acid ($H_2SO_4$), the primary constituents of acidic deposition, one form of which is acid rain.

Bases, on the other hand, dissociate into negatively charged hydroxide ions ($OH^-$) and positively charged ions. Some examples of bases are sodium hydroxide ($NaOH$) and calcium hydroxide ($Ca(OH)_2$), which can be used to neutralize acidic emissions from power plants.

The **pH** scale is a way to indicate the strength of acids and bases. In FIGURE 2.8, the pH of many familiar substances is indicated on the pH scale, which ranges from 0 to 14. A pH value of 7 on this scale—the pH of pure water—is neutral, meaning that the number of hydrogen ions is equal to the number of hydroxide ions. Anything above 7 is basic, or alkaline, and anything below 7 is acidic. The lower the number, the stronger the acid, and the higher the number, the more basic the substance is. The pH scale is *logarithmic,* meaning that there is a factor of 10 difference between each number on the scale. For example, a substance with a pH of 5 has 10 times the hydrogen ion concentration of a substance with a pH of 6 (it is 10 times more acidic). Water in equilibrium with Earth's atmosphere typically has a pH of 5.65 because carbon dioxide from the atmosphere dissolves in that water, making it weakly acidic.

## Chemical Reactions and the Conservation of Matter

A **chemical reaction** occurs when atoms separate from the molecules they are a part of or recombine with other molecules. In a chemical reaction, no atoms are ever destroyed or created. The bonds between particular atoms may change, however. For example, when methane ($CH_4$) is burned in air, it reacts with two molecules of oxygen ($2 O_2$) to create one molecule of carbon dioxide ($CO_2$) and two molecules of water ($2 H_2O$):

$$CH_4 + 2 O_2 \rightarrow CO_2 + 2 H_2O$$

Notice that the number of atoms of each chemical element is the same on each side of the reaction.

Chemical reactions can occur in either direction. For example, during the combustion of fuels, nitrogen gas ($N_2$) combines with oxygen gas ($O_2$) from the atmosphere to form two molecules of nitrogen oxide (NO), which is an air pollutant:

$$N_2 + O_2 \rightarrow 2 NO$$

This reaction can also proceed in the opposite direction:

$$2 NO \rightarrow N_2 + O_2$$

The observation that no atoms are created or destroyed in a chemical reaction leads us to the **law of conservation of matter,** which states that matter cannot be created or destroyed; it can only change form. For example, when paper burns, it may seem to vanish, but no atoms are lost; the carbon and hydrogen that make up the paper combine with oxygen in the air to produce carbon dioxide, water vapor, and other materials, which either enter the atmosphere or form ash. Combustion converts most of the solid paper into gases, but all of the original atoms remain. The same process occurs in a forest fire, but on a much larger scale (FIGURE 2.9). The only known exception to the law of conservation of matter occurs in nuclear reactions, in which small amounts of matter change into energy.

The law of conservation of matter tells us that we cannot easily dispose of hazardous materials. For example, when we burn material that contains heavy metals, such as an automotive battery, the atoms of the metals in the battery do not disappear. They turn up elsewhere in the environment, where they may cause a hazard to humans and other organisms. For this and other reasons, understanding the law of conservation of matter is crucial to environmental science.

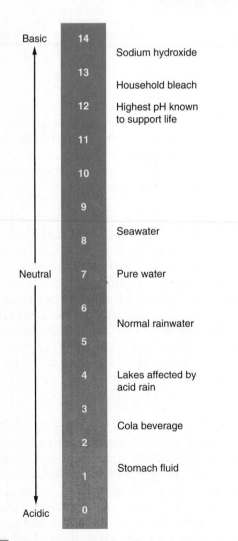

FIGURE 2.8 **The pH scale.** The pH scale is a way of expressing how acidic or how basic a solution is.

## Biological Molecules and Cells

We now have a sense of how chemical compounds form and how they respond to various processes such as burning and freezing. To further our understanding of chemical compounds, we will divide them into two basic types: *inorganic* and *organic*. **Inorganic compounds** are compounds that either (a) do not contain the element carbon or (b) do contain carbon, but only carbon bound to elements other than hydrogen. Examples include ammonia ($NH_3$), sodium chloride (NaCl), water ($H_2O$), and carbon dioxide ($CO_2$). **Organic compounds** are compounds that have carbon–carbon and carbon–hydrogen bonds. Examples of organic compounds include glucose ($C_6H_{12}O_6$) and fossil fuels, such as natural gas ($CH_4$).

Organic compounds are the basis of the biological molecules that are important to life: *carbohydrates, proteins, nucleic acids,* and *lipids.* Because these four types of molecules are relatively large, they are also known as *macromolecules.*

**CARBOHYDRATES** **Carbohydrates** are compounds composed of carbon, hydrogen, and oxygen atoms. Glucose ($C_6H_{12}O_6$) is a simple sugar (a *monosaccharide,* or single sugar) easily used by plants and animals for quick energy. Sugars can link together in long chains called *complex carbohydrates,* or *polysaccharides* ("many sugars"). For example, plants store energy as starch, which is made up of long chains of covalently bonded glucose molecules. The starch can also be used by animals that eat the plants. Cellulose, a component of plant leaves and stems, is another polysaccharide consisting of long chains of glucose molecules. Cellulose is the raw material for cellulosic ethanol, a type of fuel that has the potential to replace or supplement gasoline.

**PROTEINS** **Proteins** are made up of long chains of nitrogen-containing organic molecules called *amino acids.* Proteins are critical components of living organisms, playing roles in structural support, energy storage, internal transport, and defense against foreign substances. *Enzymes* are proteins that help control the rates of chemical reactions. The antibodies that protect us from infections are also proteins.

**NUCLEIC ACIDS** **Nucleic acids** are organic compounds found in all living cells. Long chains of nucleic acids form *DNA* and *RNA*. **DNA (deoxyribonucleic acid)** is the genetic material organisms pass on to their offspring that contains the code for reproducing the components of the next generation. **RNA (ribonucleic acid)** translates the code stored in the DNA and allows for the synthesis of proteins.

**LIPIDS** **Lipids** are smaller biological molecules that do not mix with water. Fats, waxes, and steroids are all lipids. Lipids form a major part of the membranes that surround cells.

**CELLS** We have looked at the four types of macromolecules required for life. But how do they work as part of a living organism? The smallest structural and functional component of organisms is known as a *cell*.

(a)

(b)

**FIGURE 2.10** **Organisms are composed of cells.** (a) Some organisms, such as these green algae, consist of a single cell. (b) More complex organisms, such as the Mono brine shrimp, are made up of millions of cells.

A **cell** is a highly organized living entity that consists of the four types of macromolecules and other substances in a watery solution, surrounded by a *membrane.* Some organisms, such as most bacteria and some algae, consist of a single cell. That one cell contains all of the functional structures, or *organelles,* needed to keep the cell alive and allow it to reproduce (**FIGURE 2.10a**). Larger and more complex organisms, such as Mono Lake's brine shrimp, are multicellular (**FIGURE 2.10b**).

> **GAUGE YOUR PROGRESS**
>
> ✓ What are the three types of chemical bonds?
>
> ✓ What are the unique properties of water? In what ways do those properties make life possible on Earth?
>
> ✓ What are the four types of biological molecules, and how do they differ from one another?

# Energy is a fundamental component of environmental systems

Earth's systems cannot function, and organisms cannot survive, without *energy.* **Energy** is the ability to do work, or transfer heat. Water flowing into a lake has energy because it moves and can move other objects in its path. All living systems absorb energy from their surroundings and use it to organize and reorganize molecules within their cells and to power movement. Plants absorb solar energy and use it in photosynthesis to convert carbon dioxide and water into sugars, which they then use to survive, grow, and reproduce. The sugars in plants are also an important energy source for many animals. Humans, like other animals, absorb the energy they need for cellular respiration from food. This provides the energy for our daily activities, from waking to sleeping and everything in between.

Ultimately, most energy on Earth derives from the Sun. The Sun emits **electromagnetic radiation,** a form of energy that includes, but is not limited to, visible light, ultraviolet light, and infrared energy, which we perceive as heat. The scale at the top of **FIGURE 2.11** shows these and other types of electromagnetic radiation.

Electromagnetic radiation is carried by **photons,** massless packets of energy that travel at the speed of light and can move even through the vacuum of space. The amount of energy contained in a photon depends on its *wavelength*—the distance between two peaks or troughs in a wave, as shown in the inset in Figure 2.11. Photons with long wavelengths, such as radio waves, have very low energy, while those with short wavelengths, such as X-rays, have high energy. Photons of different wavelengths are used by humans for different purposes.

## Forms of Energy

The basic unit of energy in the metric system is the *joule* (abbreviated J). A **joule** is the amount of energy used when a 1-watt light bulb is turned on for 1 second—a very small amount. Although the joule is the preferred energy unit in scientific study, many other energy units are commonly used. Conversions between these units and joules are given in Table 2.1.

**ENERGY AND POWER** Energy and *power* are not the same thing, even though we often use the words inter-

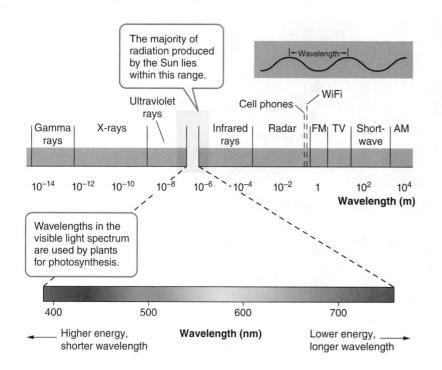

changeably. Energy is the ability to do work, whereas **power** is the rate at which work is done:

$$\text{energy} = \text{power} \times \text{time}$$
$$\text{power} = \text{energy} \div \text{time}$$

When we talk about generating electricity, we often hear about kilowatts and kilowatt-hours. The kilowatt (kW) is a unit of power. The kilowatt-hour (kWh) is a unit of energy. Therefore, the capacity of a turbine is given in kW because that measurement refers to the turbine's power. Your monthly home electricity bill reports energy use—the amount of energy from electricity that you have used in your home—in kWh. Crunch the Numbers "Calculating Energy Use" gives you an opportunity to practice working with these units.

**KINETIC AND POTENTIAL ENERGY** Energy takes a variety of forms. Many stationary objects possess a large amount of **potential energy**—energy that is stored but has not yet been released. Water impounded behind a dam contains a great deal of potential energy. When the water is released and flows downstream, that potential energy becomes **kinetic energy,** the energy of motion (**FIGURE 2.12**). The kinetic energy of moving water can be captured at a dam and transferred to a turbine and generator, and ultimately to the energy in electricity. Can you think of other common examples of kinetic energy? A car moving down the street, a flying honeybee, and a football travelling through the air all have kinetic energy. Sound also has kinetic energy because it travels in waves through the coordinated motion of atoms. Systems can contain potential energy, kinetic energy, or some of each.

| TABLE 2.1 | Common units of energy and their conversion into joules | | |
|---|---|---|---|
| **Unit** | **Definition** | **Relationship to joules** | **Common uses** |
| calorie | Amount of energy it takes to heat 1 gram of water 1°C | 1 calorie = 4.184 J | Energy expenditure and transfer in ecosystems; human food consumption |
| Calorie | Food calorie; always shown with a capital C | 1 Calorie = 1,000 calories = 1 kilocalorie (kcal) | Food labels; human food consumption |
| British thermal unit (Btu) | Amount of energy it takes to heat 1 pound of water 1°F | 1 Btu = 1,055 J | Energy transfer in air conditioners and home and water heaters |
| kilowatt-hour (kWh) | Amount of energy expended by using 1 kilowatt of electricity for 1 hour | 1 kWh = 3,600,000 J = 3.6 megajoules (MJ) | Energy use by electrical appliances, often given in kWh per year |

Potential energy stored in chemical bonds is known as **chemical energy.** The energy in food is a familiar example. By breaking down the high-energy bonds in the pizza you had for lunch, your body obtains energy to power its activities and functions. Likewise, an automo-bile engine combusts gasoline and releases its chemical energy to propel the car.

**TEMPERATURE** All matter, even the frozen water in the world's ice caps, contains some energy. When we say

## CRUNCH THE NUMBERS

### Calculating Energy Use

Your electricity bill shows that you use 600 kWh of electricity each month. Your refrigerator, which is 15 years old, could be responsible for up to 25 percent of this electricity consumption. Newer refrigerators are more efficient, meaning that they use less energy to do the same amount of work. If you wish to conserve electrical energy and save money, should you replace your refrigerator? How can you compare the energy efficiency of your old refrigerator with that of the more efficient new models?

Your refrigerator uses 500 watts when the motor is running. The motor runs for about 30 minutes per hour (or a total of 12 hours per day).

**1.** How much energy does your current refrigerator use?

0.5 kW × 12 hours/day = 6 kWh/day

6 kWh/day × 365 days/year = 2,190 kWh/year

**2.** How much more efficient is the best new refrigerator compared with your current model?

The best new model uses 400 kWh per year. Your refrigerator uses 2,190 kWh per year.

2,190 kWh/year – 400 kWh/year = 1,790 kWh/year

**3.** Assume that you are paying, on average, $0.10 per kilowatt-hour for electricity. A new refrigerator would cost you $550. You will receive a rebate of $50 from your electric company for purchasing an energy-efficient refrigerator. If you replace your refrigerator, how long will it be before your energy savings compensate you for the cost of the new appliance?

You will save

1,790 kWh/year × $0.10/kWh = $179/year

Dividing $500 by $179 indicates that in less than 3 years, you will recover the cost of the new appliance.

**Your Turn:** Environmental scientists must often convert energy units in order to compare various types of energy. For instance, you might want to compare the energy you would save by purchasing an energy-efficient refrigerator with the energy you would save by driving a more fuel-efficient car. Assume that for the amount you would spend on the new refrigerator ($500), you can make repairs to your car engine that would save you 20 gallons (76 liters) of gasoline per month. (1 liter of gasoline contains the energy equivalent of about 10 kWh.) Use this information and Table 2.1 to convert the quantities of both gasoline and electricity into joules. Compare the energy savings. Which decision would save the most energy?

that energy moves matter, we mean that it is moving the molecules within a substance. The measure of the average kinetic energy of a substance is its **temperature.**

Changes in temperature—and, therefore, in energy—can convert matter from one state to another. At a certain temperature, the molecules in a solid substance start moving so fast that they begin to flow, and the substance melts into a liquid. At an even higher temperature, the molecules in the liquid move even faster, with increasing amounts of energy. Finally the molecules move with such speed and energy that they overcome the forces holding them together and become gases.

## First Law of Thermodynamics: Energy Is Conserved

*Just as matter can neither be created nor destroyed, energy is neither created nor destroyed.* This principle is the **first law of thermodynamics.** Like matter, energy also changes form. So, when water is released from behind a dam, the potential energy of the impounded water becomes the kinetic energy of the water rushing through the gates of the dam.

The first law of thermodynamics dictates that you can't get something from nothing. When an organism needs biologically usable energy, it must convert it from an energy source such as the Sun or food. The potential energy contained in firewood never goes away, but is transformed into heat energy permeating a room when the wood is burned in a fireplace. Sometimes it may be difficult to identify where the energy is going, but it is always conserved.

Look at **FIGURE 2.13**, which uses a car to show the first law in action through a series of energy conversions. Think of the car, including its fuel tank, as a system. The potential energy of the fuel (gasoline) is converted into kinetic energy when the battery supplies a spark in the presence of gasoline and air. The gasoline combusts, and the resulting gases expand, pushing the pistons in the engine—converting the chemical energy

in the gasoline into the kinetic energy of the moving pistons. Energy is transferred from the pistons to the drive train, and from there to the wheels, which propel the car. The combustion of gasoline also produces heat, which dissipates into the environment outside the system. The kinetic energy of the moving car is converted into heat and sound energy as the tires create friction with the road and the body of the automobile moves through the air. When the brakes are applied to stop the car, friction between brake parts releases heat energy. No energy is ever destroyed in this example, but chemical energy is converted into motion, heat, and sound. Notice that some of the energy stays within the system and some (such as the heat from burning gasoline) leaves the system.

## Second Law of Thermodynamics

We have seen how the potential energy of gasoline is transformed into the kinetic energy of moving pistons in a car engine. But as Figure 2.13 shows, some of that energy is converted into a less usable form—in this case, heat. The heat that is created is called waste heat, meaning that it is not used to do any useful work. The **second law of thermodynamics** tells us that *when energy is transformed, the quantity of energy remains the same, but its ability to do work diminishes.*

**ENERGY EFFICIENCY** To quantify this observation, we use the concept of *energy efficiency.* **Energy efficiency** is the ratio of the amount of work that is done to the total amount of energy that is introduced into the system in the first place. Two machines or engines that perform the same amount of work, but use different amounts of energy to do that work, have different energy efficiencies. Consider the difference between modern woodstoves and traditional open fireplaces. A woodstove that is 70 percent efficient might use 2 kg of wood to heat a room to a comfortable 20°C (68°F), whereas a fireplace that is 10 percent efficient would require 14 kg

**Energy Input**

Potential (chemical) energy in gasoline

**Energy Outputs**

**Useful energy:**
Kinetic energy, which moves car

**Waste energy:**
Heat from friction in engine, tires on road, brakes, etc.

Sound energy from tires on road surface

**FIGURE 2.13** **Conservation of energy within a system.** In a car, the potential energy of gasoline is converted into other forms of energy. Some of that energy leaves the system, but all of it is conserved.

**(a)** Traditional fireplace

**(b)** Modern woodstove

**FIGURE 2.14** **Energy efficiency.** (a) The energy efficiency of a traditional fireplace is low because so much heated air can escape through the chimney. (b) A modern woodstove, which can heat a room using much less wood, is much more energy efficient.

to achieve the same temperature—a sevenfold greater energy input (**FIGURE 2.14**).

We can also calculate the energy efficiency of transforming one form of energy into other forms of energy. Let's consider what happens when we convert the chemical energy of coal into the electricity that operates a reading lamp and the heat that the lamp releases. **FIGURE 2.15** shows the process.

A modern coal-burning power plant can convert 1 metric ton of coal, containing 24,000 megajoules (MJ; 1 MJ = 1 million joules) of chemical energy into about 8,400 MJ of electricity. Since 8,400 is 35 percent of 24,000, this means that the process of turning coal into electricity is about 35 percent efficient. The rest of the energy from the coal—65 percent—is lost as waste heat.

In the electrical transmission lines between the power plant and the house, 10 percent of the electrical energy from the plant is lost as heat and sound, so the transport of energy away from the plant is about 90 percent efficient. We know that the conversion of electrical energy into light in an incandescent bulb is 5 percent efficient; again, the rest of the energy is lost as heat. From beginning to end, we can calculate the energy efficiency of converting coal into incandescent lighting by multiplying all the individual efficiencies:

$$0.35 \times 0.90 \times 0.05 = 0.016$$
(1.6% efficiency)

| coal to electricity | × | transport of electricity | × | light bulb efficiency | = | overall efficiency |

**Calculation: (35%) × (90%) × (5%) = 1.6% efficiency**

**FIGURE 2.15** **The second law of thermodynamics.** Whenever one form of energy is transformed into another, some of that energy is converted into a less usable form of energy, such as heat. In this example, we see that the conversion of coal into the light of an incandescent bulb is only 1.6 percent efficient.

(a)

(b)

**FIGURE 2.16** **Energy and entropy.** Entropy increases in a system unless an input of energy from outside the system creates order. (a) In order to reduce the entropy of this messy room, a human must expend energy, which comes from food. (b) A tornado has increased the entropy of this forest system in Wisconsin.

**ENERGY QUALITY** Related to energy efficiency is **energy quality,** the ease with which an energy source can be used for work. A high-quality energy source has a convenient, concentrated form so that it does not take too much energy to move it from one place to another. Gasoline, for example, is a high-quality energy source because its chemical energy is concentrated (about 44 MJ/kg), and because we have technology that can conveniently transport it from one location to another. In addition, it is relatively easy to convert gasoline energy into work and heat. Wood, on the other hand, is a lower-quality energy source. It has less than half the energy concentration of gasoline (about 20 MJ/kg) and is more difficult to use to do work. Imagine using wood to power an automobile. Clearly, gasoline is a higher-quality energy source than wood. Energy quality is one important factor humans must consider when they make energy choices.

**ENTROPY** The second law of thermodynamics also tells us that all systems move toward randomness rather than toward order. This randomness, called **entropy,** is always increasing in a system, unless new energy from outside the system is added to create order.

Think of your bedroom as a system. At the start of the week, your books may be in the bookcase, your clothes may be in the dresser, and your shoes may be lined up in a row in the closet. But what happens if, as the week goes on, you don't expend energy to put your things away (**FIGURE 2.16**)? Unfortunately, your books will not spontaneously line up in the bookcase, your clothes will not fall folded into the dresser, and your shoes will not pair up and arrange themselves in the closet. Unless you bring energy into the system to put things in order, your room will slowly become more and more disorganized.

The energy you use to pick up your room comes from the energy stored in food. Food is a relatively high-quality energy source because the human body easily converts it into usable energy. The molecules of food are ordered rather than random. In other words, food is a low-entropy energy source. Only a small portion of the energy in your digested food is converted into work, however; the rest becomes body heat, which may or may not be needed. This waste heat has a high degree of entropy because heat is the random movement of molecules. Thus, in using food energy to power your body to organize your room, you are decreasing the entropy of the room, but increasing the entropy in the universe by producing waste body heat.

Another example of the second law can be found in the observation that energy always flows from hot to cold. A pot of water will never boil without an input of energy, but hot water left alone will gradually cool as its energy dissipates into the surrounding air. This application of the second law is important in many of the global circulation patterns that are powered by the energy of the Sun.

**GAUGE YOUR PROGRESS**

✓ What is the difference between power and energy? Why is it important to know the difference?

✓ How do potential energy and kinetic energy differ? What is chemical energy?

✓ What are the first and second laws of thermodynamics?

# Energy conversions underlie all ecological processes

Life requires order. If organisms were not made up of molecules organized into structures such as proteins and cells, they could not grow—in fact, they could never develop in the first place. All living things work against entropy by using energy to maintain order.

Individual organisms rely on a continuous input of energy in order to survive, grow, and reproduce. But interactions at levels beyond the organism can also be seen as a process of converting energy into organization. Consider a forest ecosystem. Trees absorb water through their roots and carbon dioxide through their leaves. By combining these compounds in the presence of sunlight, they convert water and carbon dioxide into sugars that will provide them with the energy they need. Trees fight entropy by keeping their atoms and molecules together in tree form, rather than having them dispersed randomly throughout the universe. But then a deer grazes on tree leaves, and later a mountain lion eats the deer. At each step, energy is converted by organisms into work.

The form and amount of energy available in an environment determines what kinds of organisms can live there. Plants thrive in tropical rainforests where there is plenty of sunlight as well as water. Many food crops, not surprisingly, can be planted and grown in temperate climates that have a moderate amount of sunlight. Life is much more sparse at high latitudes, toward the North and South Poles, where less solar energy is available to organisms. The landscape is populated mainly by small plants and shrubs, insects, and migrating animals. Plants cannot live at all on the deep ocean floor, where no solar energy penetrates. The animals that live there, such as eels, anglerfish, and squid, get their energy by feeding on dead organisms that sink from above. Chemical

(a)

(b)

(c)

(d)

**FIGURE 2.17** **The amount of available energy determines which organisms can live in a natural system.** (a) A tropical rainforest has abundant energy available from the Sun and enough moisture for plants to make use of that energy. (b) The Arctic tundra has much less energy available, so plants grow more slowly there and do not reach large sizes. (c) Organisms, such as this squid, living at the bottom of the ocean must rely on dead biological matter falling from above. (d) The energy supporting this deep-ocean vent community comes from chemicals emitted from the vent. Bacteria convert the chemicals into forms of energy that other organisms, such as tube worms, can use.

energy, in the form of sulfides emitted from deep-ocean vents (underwater geysers), supports a plantless ecosystem that includes sea spiders, 2.4 meter (8-foot) tube worms, and bacteria (FIGURE 2.17).

# Systems analysis shows how matter and energy flow in the environment

Why is it important for environmental scientists to study whole systems rather than focusing on the individual plants, animals, or substances within a system? Imagine taking apart your cell phone and trying to understand how it works simply by focusing on the microphone. You wouldn't get very far. Similarly, it is important for environmental scientists to look at the whole picture, not just the individual parts of a system, in order to understand how that system works.

Studying systems allows scientists to think about how matter and energy flow in the environment. In this way, researchers can learn about the complex relationships between organisms and the environment, but more importantly, they can predict how changes to any part of the system—for example, changes in the water level at Mono Lake—will change the entire system.

Systems can be either *open* or *closed*. In an **open system,** exchanges of matter or energy occur across system boundaries. Most systems are open. Even at remote Mono Lake, water flows in, and birds fly to and from the lake. The ocean is also an open system. Energy from the Sun enters the ocean, warming the waters and providing energy to plants and algae. Energy and matter are transferred from the ocean to the atmosphere as energy from the Sun evaporates water, giving rise to meteorological events such as tropical storms, in which clouds form and send rain back to the ocean surface. Matter, such as sediment and nutrients, enters the ocean from rivers and streams and leaves it through geologic cycles and other processes.

In a **closed system,** matter and energy exchanges across system boundaries do not occur. Closed systems are less common than open systems. Some underground cave systems are nearly completely closed systems.

As FIGURE 2.18 shows, Earth is an open system with respect to energy. Solar radiation enters Earth's atmosphere, and heat and reflected light leave it. But because of its gravitational field, Earth is essentially a closed system with respect to matter. Only an insignificant amount of material enters or leaves the Earth system. All important material exchanges occur within the system.

## Inputs and Outputs

By now you have seen numerous examples of both **inputs,** or additions to a given system, and **outputs,** or losses from the system. People who study systems often conduct a **systems analysis,** in which they determine inputs, outputs, and changes in the system under various conditions. For instance, researchers studying Mono Lake might quantify the inputs to that system—such as water and salts—and the outputs—such as water that evaporates from the lake and brine shrimp removed by migratory birds. Because no water flows out of the lake, salts are not removed, and even without the aqueduct, Mono Lake, like other terminal lakes, would slowly

**(a)** Open system

**(b)** Closed system

FIGURE 2.18 **Open and closed systems.** (a) Earth is an open system with respect to energy. Solar radiation enters the Earth system, and energy leaves it in the form of heat and reflected light. (b) However, Earth is essentially a closed system with respect to matter because very little matter enters or leaves the Earth system. The white arrows indicate the cycling of energy and matter.

### The Mystery of the Missing Salt

Before the Los Angeles Aqueduct was built, about 120 billion liters of stream water (31 billion gallons) flowed into Mono Lake in an average year. As a terminal lake, it had no outflow streams. The water level did not rise or fall in an average year. Therefore, the water in the lake had to go somewhere to balance the water coming in; if the system size was not changing, inputs must equal outputs. In this case, roughly the same amount of water that entered the lake must have evaporated. The salt content of the stream water flowing into Mono Lake varied, but a typical liter of stream water averaged 50 mg of salt.

**1.** How much salt entered Mono Lake annually?

To calculate the total amount of salt that entered Mono Lake each year, we can multiply the amount of salt (50 mg) per liter of water by the number of liters of water flowing into the lake (120 billion per year):

$$50 \text{ mg/L} \times 120 \text{ billion L/year} =$$
$$6 \text{ trillion mg/year} = 6 \text{ million kg/year}$$

**2.** The lake today contains about 285 billion kilograms of dissolved salt. At the rate of salt input we have just calculated, how long would it take to accumulate that much salt, starting with zero salt in the lake?

We have just determined that the salt concentration of Mono Lake increases by 6 million kilograms per year. Mono Lake contains approximately 285 billion kilograms of dissolved salts today, so at the rate of stream flow before the diversion, it would have taken about 47,500 years to accumulate that much salt:

$$285 \text{ billion kg} \div 6 \text{ million kg/year} = 47,500 \text{ years}$$

**3.** No water has flowed out of the Mono Lake basin since it was formed about 120,000 years ago. Assume that Earth's climate hasn't changed much over that time. At today's input rate, how much salt should be in the water of Mono Lake today?

$$6 \text{ million kg/year} \times 120,000 \text{ years} = 720 \text{ billion kg}$$

**4.** The salt loads in questions 2 and 3 do not agree. How can we explain the discrepancy?

The lake's towering tufa formations hold the answer: many of the salts (including calcium and sodium) have precipitated—that is, separated—out of the water to form the tufa rock. In this way, the salts have been removed from the water, but not from the Mono Lake system as a whole. Our analysis is complete when we account for the salts removed from the lake as tufa. FIGURE 2.19 summarizes these inputs to and outputs from the Mono Lake system.

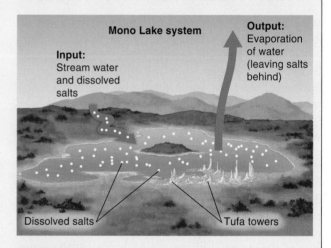

**FIGURE 2.19** Inputs to and outputs from the Mono Lake ecosystem.

become saltier. Crunch the Numbers "The Mystery of the Missing Salt" provides an example of what calculating inputs and outputs can tell us about a system.

## Steady States

At Mono Lake, in any given period, the same amount of water that enters the lake eventually evaporates. In many cases, the most important aspect of conducting a systems analysis is determining whether your system is in **steady state**—that is, whether inputs equal outputs, so that the system is not changing over time. This information is particularly useful in the study of environmental science. For example, it allows us to know whether the amount of a valuable resource or harmful pollutant is increasing, decreasing, or staying the same.

The first step in determining whether a system is in steady state is to measure the amount of matter and energy within it. If the scale of the system allows, we can perform these measurements directly. Consider the leaky bucket shown in FIGURE 2.20. We can measure the amount of water going into the bucket and the amount of water flowing out through the holes. However, some properties of systems, such as the volume of a lake or the size of an insect population, are difficult to measure directly, so we must calculate or estimate the amount of energy or matter stored in the system. We can then use this information to determine the inputs to and outputs from the system to determine whether it is in steady state.

Many aspects of natural systems, such as the water vapor in the global atmosphere, have been in steady state

**Input:** 1 L/second

10 L

**Output:** 1 L/second

**FIGURE 2.20** **A system in steady state.** In this leaky bucket, inputs equal outputs. As a result, there is no change in the total amount of water in the bucket: the system is in steady state.

for at least as long as we have been studying them. The amount of water that enters the atmosphere by evaporation from oceans, rivers, and lakes is roughly equal to the amount that falls from the atmosphere as precipitation. Until recently, the oceans have also been in steady state: the amount of water that enters from rivers and streams has been roughly equal to the amount that evaporates

into the air. One concern about the effects of global climate change is that some global systems, such as the system that includes water balance in the oceans and atmosphere, may no longer be in steady state.

It's interesting to note that one part of a system can be in steady state while another part is not. Before the Los Angeles Aqueduct was built, the Mono Lake system was in steady state with respect to water (the inflow of water equaled the rate of water evaporation), but *not* with respect to salt: salt was slowly accumulating, as it does in all terminal lakes.

## Feedbacks

Most natural systems are in steady state. Why? A natural system can respond to changes in its inputs and outputs. For example, during a period of drought, evaporation from a lake will be greater than precipitation and stream water flowing into the lake. Therefore, the lake will begin to dry up. Soon there will be less surface water available for evaporation, and the evaporation rate will continue to fall until it matches the new, lower precipitation rate. When this happens, the system returns to steady state, and the lake stops shrinking.

Of course, the opposite is also true. In very wet periods, the size of the lake will grow, and evaporation from the expanded surface area will continue to increase until the system returns to a steady state at which inputs and outputs are equal.

Adjustments in input or output rates caused by changes to a system are called *feedbacks*. The term **feedback** means that the results of a process *feed back* into the system to change the rate of that process. Feedbacks, which can be diagrammed as loops or cycles, are found throughout the environment.

There are two kinds of feedback, *negative* and *positive*. In natural systems, scientists most often observe **negative feedback loops,** in which a system responds to a change by returning to its original state, or at least by decreasing the rate at which the change is occurring. FIGURE 2.21a shows a negative feedback loop for Mono Lake: when

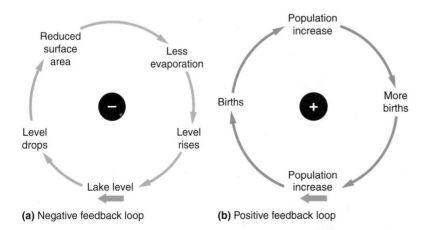

**(a)** Negative feedback loop

Reduced surface area → Less evaporation → Level rises → Lake level → Level drops → Reduced surface area

**(b)** Positive feedback loop

Population increase → More births → Population increase → Births → Population increase

**FIGURE 2.21** **Negative and positive feedback loops.** (a) A negative feedback loop occurs at Mono Lake: when the water level drops, the lake surface area is reduced, and evaporation decreases. As a result of the decrease in evaporation, the lake level rises again. (b) Population growth is an example of positive feedback. As members of a species reproduce, they create more offspring that will be able to reproduce in turn, creating a cycle that increases the population size. The green arrow indicates the starting point of each cycle.

water levels drop, there is less lake surface area, so evaporation decreases as well. With less evaporation, the water in the lake slowly returns to its original volume.

Positive feedbacks also occur in the natural world. **FIGURE 2.21b** shows an example of how births in a population can give rise to a **positive feedback loop.** The more members of a species that can reproduce, the more births there will be, creating even more of the species to give birth, and so on.

It's important to note that *positive* and *negative* here do not mean *good* and *bad;* instead, positive feedback *amplifies* changes, whereas negative feedback *resists* changes. People often talk about the balance of nature. That balance is the logical result of systems reaching a state at which negative feedbacks predominate—although positive feedback loops play important roles in environmental systems as well.

One of the most important questions in environmental science is to what extent Earth's temperature is regulated by feedback loops, and if so, what types, and at what scale. In general, warmer temperatures at Earth's surface increase the evaporation of water. The additional water vapor that enters the atmosphere by evaporation causes two kinds of clouds to form. Low-altitude clouds reflect sunlight back into space. The result is less heating of Earth's surface, less evaporation, and less warming—a negative feedback loop. High-altitude clouds, on the other hand, absorb terrestrial energy that might have otherwise escaped the atmosphere, leading to higher temperatures near Earth's surface, more evaporation of water, and more warming—a positive feedback loop. In the absence of other factors that compensate for or balance the warming, this positive feedback loop will continue making temperatures warmer, driving the system further away from its starting point. This and other potential positive feedback loops may play critical roles in climate change.

The health of many environmental systems depends on the proper operation of feedback loops. Sometimes, natural or anthropogenic factors lead to a breakdown in a negative feedback loop and drive an environmental system away from its steady state. This is particularly true when a new component is added to a system, as with the introduction of an invasive species, or when humans use too much of a natural resource. As you study the exploitation of natural resources, try to determine what factors may be disrupting the negative feedback loops of the systems that provide those resources.

## GAUGE YOUR PROGRESS

✓ What is an open system? What is a closed system?

✓ Why is it important to look at a whole system rather than only at its parts?

✓ What is steady state? What are feedback loops? Why are they important?

# Natural systems change across space and over time

The decline in the water level of Mono Lake was caused by people: humans diverted water from the lake for their own use. Anthropogenic change in an environmental system is often very visible. We see anthropogenic change in rivers that have been dammed, air that has been polluted by automobile emissions, and cities that have encroached on once wild areas.

Differences in environmental conditions affect what grows or lives in an area, creating geographic variation among natural systems. Variations in temperature, precipitation, or soil composition across a landscape can lead to vastly different numbers and types of organisms. In Texas, for example, sycamore trees grow in river valleys where there is plenty of water available, whereas pine trees dominate mountain slopes because they can tolerate the cold, dry conditions there. Paying close attention to these natural variations may help us predict the effect of any change in an environment. So we know that if the rivers that support the sycamores in Texas dry up, the trees will probably die.

Natural systems are also affected by the passage of time. Thousands of years ago, when the climate of the Sahara was much wetter than it is today, it supported large populations of Nubian farmers and herders. Small changes in Earth's orbit relative to the Sun, along with a series of other factors, led to the disappearance of monsoon rains in northern Africa. As a result, the Sahara—now a desert nearly the size of the continental United States—became one of Earth's driest regions. Other, more dramatic changes have occurred on the planet. In the last few million years, Earth has moved in and out of several ice ages; 70 million years ago, central North America was covered by a sea; 240 million years ago, Antarctica was warm enough for 6-foot-long salamander-like amphibians to roam its swamps. Natural systems respond to such changes in the global environment with migrations and extinctions of species as well as the evolution of new species.

Throughout Earth's history, small natural changes have had large effects on complex systems, but human activities have increased both the pace and the intensity of these natural environmental changes, as they did at Mono Lake. Studying variations in natural systems over space and time can help scientists learn more about what to expect from the alterations humans are making to the world today.

## GAUGE YOUR PROGRESS

✓ Give some examples of environmental conditions that might vary among natural systems.

✓ Why is it important to study variation in natural systems over space and time?

South Florida's vast Everglades ecosystem extends over 50,000 km² (12,500,000 acres) (FIGURE 2.22). The region, which includes the Everglades and Biscayne Bay national parks, is home to many threatened and endangered bird, mammal, reptile, and plant species, including the Florida panther (*Puma concolor coryi*) and the Florida manatee (*Trichechus manatus latirostris*). The 4,000 km² (988,000-acre) subtropical wetland area for which the region is best known has been called a "river of grass" because a thin sheet of water flows constantly through it, allowing tall water-tolerant grasses to grow (FIGURE 2.23).

A hundred years of rapid human population growth, and the resulting need for water and farmland, have had a dramatic impact on the region. Flood control, dams,

## Managing Environmental Systems in the Florida Everglades

irrigation, and the need to provide fresh water to Floridians have led to a 30 percent decline in water flow through the Everglades. Much of the water that does flow through the region is polluted by phosphorus-rich fertilizer and waste from farms and other sources upstream. Cattails thrive on the input of phosphorus, choking out other native plants. The reduction in water flow and water quality is, by most accounts, destroying the Everglades. Can we save this natural system while still providing water to the people who need it?

The response of scientists and policy makers has been to treat the Everglades as a set of interacting systems and manage the inputs and outputs of water and pollutants to those systems. The Comprehensive Everglades Restoration Plan of 2000 is a systems-based approach to the region's problems. It covers 16 counties and 46,600 km² (11,500,000 acres) of South Florida. The plan is based on three key steps: increasing water flow into the Everglades, reducing pollutants coming in, and developing strategies for dealing with future problems.

The first step—increasing water flow—will counteract some of the effects of decades of drainage by local communities. Its goal is to provide enough water to support the Everglades' aquatic and marsh organisms. The plan calls for restoring natural water flow as well as natural hydroperiods (seasonal increases and decreases in water flow). Its strategies include removal of over 390 km (240 miles) of inland levees, canals, and water control structures that have blocked this natural water movement.

FIGURE 2.22 **The Florida Everglades Ecosystem.** This map shows the locations of the Florida Everglades, Lake Okeechobee, and the broader Everglades ecosystem, which includes Everglades and Biscayne Bay National Parks and Big Cypress National Preserve.

FIGURE 2.23 **River of grass.** The subtropical wetland portion of the Florida Everglades has been described as a river of grass because of the tall water-tolerant grasses that cover its surface.

Water conservation will also be a crucial part of reaching this goal. New water storage facilities and restored wetlands will capture and store water during rainy seasons for use during dry seasons, redirecting much of the 6.4 billion liters (1.7 billion gallons) of fresh water that currently flow to the ocean every day. About 80 percent of this fresh water will be redistributed back into the ecosystem via wetlands and aquifers. The remaining water will be used by cities and farms. The federal and state governments also hope to purchase nearby irrigated cropland and return it to a more natural state. In 2009, for example, the state of Florida purchased 29,000 ha (71,700 acres) of land from the United States Sugar Corporation, the first of a number of actions that will allow engineers to restore the natural flow of water from Lake Okeechobee into the Everglades. Florida is currently negotiating to purchase even more land from United States Sugar.

To achieve the second goal—reducing water pollution—local authorities will improve waste treatment facilities and place restrictions on the use of agricultural chemicals. Marshlands are particularly effective at absorbing nutrients and breaking down toxins. Landscape engineers have designed and built more than 21,000 ha (52,000 acres) of artificial marshes upstream of the Everglades to help clean water before it reaches Everglades National Park. Although not all of the region has seen water quality improvements, phosphorus concentrations in runoff from farms south of Lake Okeechobee are lower, meaning that fewer pollutants are reaching the Everglades.

The third goal—to plan for the possibility of future problems—requires an **adaptive management plan:** a strategy that provides flexibility so that managers can modify it as future changes occur. Adaptive management is an answer to scientific uncertainty. In a highly complex system such as the Everglades, any changes, however well intentioned, may have unexpected consequences. Management strategies must adapt to the actual results of the restoration plan as they occur. In addition, an adaptive management plan can be changed to meet new challenges as they come. One such challenge is global warming. As the climate warms, glaciers melt, and sea levels rise, much of the Everglades could be inundated by seawater, which would destroy freshwater habitat. Adaptive management essentially means paying attention to what works and adjusting your methods accordingly. The Everglades restoration plan will be adjusted along the way to take the results of ongoing observations into account, and it has put formal mechanisms in place to ensure that this will occur.

The Everglades plan has its critics. Some people are concerned that control of water flow and pollution will restrict the use of private property and affect economic development, possibly even harming the local economy. Yet other critics fear that the restoration project is underfunded or moving too slowly, and that current farming practices in the region are inconsistent with the goal of restoration.

In spite of its critics, the Everglades restoration plan is, historically speaking, a milestone project, not least because it is based on the concept that the environment is made up of interacting systems.

### Reference

Kiker, C., W. Milon, and A. Hodges. 2001. South Florida: The reality of change and the prospects for sustainability. Adaptive learning for science-based policy: The Everglades restoration. *Ecological Economics* 37: 403–416.

## REVISIT THE KEY IDEAS

■ **Define** *systems* **within the context of environmental science.**

Environmental systems are sets of interacting components connected in such a way that changes in one part of the system affect the other parts. Systems exist at multiple scales, and a large system may contain smaller systems within it. Earth itself is a single interconnected system.

■ **Explain the components and states of matter.**

Matter is composed of atoms, which are made up of protons, neutrons, and electrons. Atoms and molecules can interact in chemical reactions in which the bonds between particular atoms may change. Matter cannot be created or destroyed, but its form can be changed.

■ **Distinguish between various forms of energy and discuss the first and second laws of thermodynamics.**

Energy can take various forms, including energy that is stored (potential energy) and the energy of motion (kinetic energy). According to the first law of thermodynamics, energy cannot be created or destroyed, but it can be converted from one form into another. According to the second law of thermodynamics, in any conversion of energy, some energy is converted into unusable waste energy, and the entropy of the universe is increased.

■ **Describe the ways in which ecological systems depend on energy inputs.**

Individual organisms rely on a continuous input of energy in order to survive, grow, and reproduce. More organisms can live where more energy is available.

- Explain how scientists keep track of inputs, outputs, and changes to complex systems.

Systems can be open or closed to exchanges of matter, energy, or both. A systems analysis determines what goes into, what comes out of, and what has changed within a given system. Environmental scientists use systems analysis to calculate inputs to and outputs from a system and its rate of change. If there is no overall change, the system is in steady state. Changes in one input or output can affect the entire system.

- Describe how natural systems change over time and space.

Variation in environmental conditions, such as temperature or precipitation, can affect the types and numbers of organisms present. Short-term and long-term changes in Earth's climate also affect species distributions.

# CHECK YOUR UNDERSTANDING

1. Which of the following statements about atoms and molecules is *correct?*
   (a) The mass number of an element is always less than its atomic number.
   (b) Isotopes are the result of varying numbers of neutrons in atoms of the same element.
   (c) Ionic bonds involve electrons while covalent bonds involve protons.
   (d) Inorganic compounds never contain the element carbon.
   (e) Protons and electrons have roughly the same mass.

2. Which of the following does *not* demonstrate the law of conservation of matter?
   (a) $CH_4 + 2 O_2 \rightarrow CO_2 + 2 H_2O$
   (b) $NaOH + HCl \rightarrow NaCl + H_2O$
   (c) $2 NO_2 + H_2O \rightarrow HNO_3 + HNO_2$
   (d) $PbO + C \rightarrow 2 Pb + CO_2$
   (e) $C_6H_{12}O_6 + 6 O_2 \rightarrow 6 CO_2 + 6 H_2O$

3. Pure water has a pH of 7 because
   (a) its surface tension equally attracts acids and bases.
   (b) its polarity results in a molecule with a positive and a negative end.
   (c) its ability to dissolve carbon dioxide adjusts its natural pH.
   (d) its capillary action attracts it to the surfaces of solid substances.
   (e) its $H^+$ concentration is equal to its $OH^-$ concentration.

4. Which of the following is *not* a type of organic biological molecule?
   (a) Lipids
   (b) Carbohydrates
   (c) Salts
   (d) Nucleic acids
   (e) Proteins

5. A wooden log that weighs 1.00 kg is placed in a fireplace. Once lit, it is allowed to burn until there are only traces of ash, weighing 0.04 kg, left. Which of the following *best* describes the flow of energy?
   (a) The potential energy of the wooden log was converted into the kinetic energy of heat and light.
   (b) The kinetic energy of the wooden log was converted into 0.04 kg of ash.
   (c) The potential energy of the wooden log was converted into 1.00 J of heat.
   (d) Since the ash weighs less than the wooden log, matter was converted directly into energy.
   (e) The burning of the 1.00 kg wooden log produced 0.96 kg of gases and 0.04 kg of ash.

6. Consider a power plant that uses natural gas as a fuel to generate electricity. If there are 10,000 J of chemical energy contained in a specified amount of natural gas, then the amount of electricity that could be produced would be
   (a) greater than 10,000 J because electricity has a higher energy quality than natural gas.
   (b) something less than 10,000 J, depending on the efficiency of the generator.
   (c) greater than 10,000 J when energy demands are highest; less than 10,000 J when energy demands are lowest.
   (d) greater than 10,000 J because of the positive feedback loop of waste heat.
   (e) equal to 10,000 J because energy cannot be created or destroyed.

7. A lake that has been affected by acid rain has a pH of 4. How many more times acidic is the lake water than seawater? (See Figure 2.8 on page 34.)
   (a) 4
   (b) 10
   (c) 100
   (d) 1,000
   (e) 10,000

**8.** An automobile with an internal combustion engine converts the potential energy of gasoline (44 MJ/kg) into the kinetic energy of the moving pistons. If the average internal combustion engine is 10 percent efficient and 1 kg of gasoline is combusted, how much potential energy is converted into energy to run the pistons?
(a) 39.6 MJ
(b) 20.0 MJ
(c) 4.4 MJ
(d) Depends on the capacity of the gas tank
(e) Depends on the size of the engine

**9.** If the average adult woman consumes approximately 2,000 kcal per day, how long would she need to run in order to utilize 25 percent of her caloric intake, given that the energy requirement for running is 42,000 J per minute?
(a) 200 minutes      (d) 0.05 minutes
(b) 50 minutes      (e) 0.012 minutes
(c) 5 minutes

**10.** The National Hurricane Center studies the origins and intensities of hurricanes over the Atlantic and Pacific oceans and attempts to forecast their tracks, predict where they will make landfall, and assess what damage will result. Its systems analysis involves
(a) changes within a closed system.
(b) inputs and outputs within a closed system.
(c) outputs only within an open system.
(d) inputs from a closed system and outputs in an open system.
(e) inputs, outputs, and changes within an open system.

**11.** Based on the graph below, which of the following is the *best* interpretation of the data?

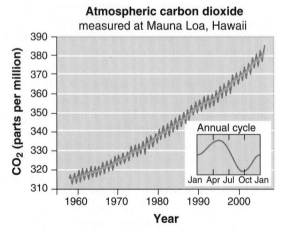

**Atmospheric carbon dioxide**
measured at Mauna Loa, Hawaii

(a) The atmospheric carbon dioxide concentration is in steady state.
(b) The output of carbon dioxide from the atmosphere is greater than the input into the atmosphere.
(c) The atmospheric carbon dioxide concentration appears to be decreasing.
(d) The input of carbon dioxide into the atmosphere is greater than the output from the atmosphere.
(e) The atmospheric carbon dioxide concentration will level off due to the annual cycle.

**12.** The diagram below represents which of the following concepts?

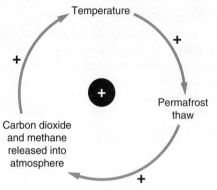

(a) A negative feedback loop, because melting of permafrost has a negative effect on the environment by increasing the amounts of carbon dioxide and methane in the atmosphere.
(b) A closed system, because only the concentrations of carbon dioxide and methane in the atmosphere contribute to the permafrost thaw.
(c) A positive feedback loop, because more carbon dioxide and methane in the atmosphere result in greater permafrost thaw, which releases more carbon dioxide and methane into the atmosphere.
(d) An open system that resists change and regulates global temperatures.
(e) Steady state, because inputs and outputs are equal.

**13.** Which of the following statements about the Comprehensive Everglades Restoration Plan is *not* correct?
(a) Human and natural systems interact because feedback loops lead to adaptations and changes in both systems.
(b) Water conservation will alter land uses and restore populations of aquatic and marsh organisms.
(c) Improvements in waste treatment facilities and restrictions on agricultural chemicals will reduce the nutrients and toxins in the water that reaches the Everglades.
(d) Adaptive management will allow for the modification of strategies as changes occur in this complex system.
(e) The Florida Everglades is a closed system that includes positive and negative feedback loops and is regulated as such.

**14.** Which of the following would represent a system in steady state?
  I   The birth rate of chameleons on the island of Madagascar equals their death rate.
  II   Evaporation from a lake is greater than precipitation and runoff flowing into the lake.
  III   The steady flow of the Colorado River results in more erosion than deposition of rock particles.
(a) I only      (d) I and II
(b) II only      (e) I and III
(c) III only

## APPLY THE CONCEPTS

1. The atomic number of uranium-235 is 92, its half-life is 704 million years, and the radioactive decay of 1 kg of $^{235}U$ releases $6.7 \times 10^{13}$ J. Radioactive material must be stored in a safe container or buried deep underground until its radiation output drops to a safe level. Generally, it is considered "safe" after 10 half-lives.

   (a) Assume that a nuclear power plant can convert energy from $^{235}U$ into electricity with an efficiency of 35 percent, the electrical transmission lines operate at 90 percent efficiency, and fluorescent lights operate at 22 percent efficiency.

       (i) What is the overall efficiency of converting the energy of $^{235}U$ into fluorescent light?

       (ii) How much energy from 1 kg of $^{235}U$ is converted into fluorescent light?

       (iii) Name one way in which you could improve the overall efficiency of this system. Explain how your suggestion would improve efficiency.

   (b) What are the first and second laws of thermodynamics?

   (c) How long would it take for the radiation from a sample of $^{235}U$ to reach a safe level?

2. U.S. wheat farmers produce, on average, 3,000 kg of wheat per hectare. Farmers who plant wheat year after year on the same fields must add fertilizers to replace the nutrients removed by the harvested wheat. Consider a wheat farm as an open system.

   (a) Identify two inputs and two outputs of this system.

   (b) Using one input to and one output from (a), diagram and explain one positive feedback loop.

   (c) Identify two adaptive management strategies that could be employed if a drought occurred.

   (d) Wheat contains about 2.5 kcal per gram, and the average U.S. male consumes 2,500 kcal per day. How many hectares of wheat are needed to support one average U.S. male for a year, assuming that 30 percent of his caloric intake is from wheat?

## MEASURE YOUR IMPACT

**Bottled Water versus Tap Water** A 2007 study traced the energy input required to produce bottled water in the United States. In addition to the energy required to make plastic bottles from PET (polyethylene terephthalate), energy from 58 million barrels of oil was required to clean, fill, seal, and label the water bottles. This is 2,000 times more than the amount of energy required to produce tap water.

In 2007, the population of the United States was 300 million people, and on average, each of those people consumed 114 L (30 gallons) of bottled water. The average 0.6 L (20-ounce) bottle of water cost $1.00. The average charge for municipal tap water was about $0.0004 per liter.

   (a) Complete the following table for the year 2007. Show all calculations.

| Liters of bottled water consumed in 2007 | Liters of bottled water produced per barrel of oil |
| --- | --- |
|  |  |

   (b) How much energy (in barrels of oil) would be required to produce the amount of tap water equivalent to the amount of bottled water consumed in 2007? How many liters of tap water could be produced per barrel of oil?

   (c) Compare the cost of bottled water versus tap water per capita per year.

   (d) Identify and explain one output of the bottled water production and consumption system that could have a negative effect on the environment.

   (e) List two reasons for using tap water rather than bottled water.

## ENGAGE YOUR ENVIRONMENT

How efficient are best-selling automobile models? Compare the energy efficiency and operating costs of several popular car models. See *Engage Your Environment Activity 2 at the end of this book.*

# science applied

# Were We Successful in Halting the Growth of the Ozone Hole?

We rely on refrigeration to keep our foods safe and edible, and on air conditioning to keep us comfortable in hot weather. For many years, the same chemicals that made refrigeration and air conditioning possible were also used in a host of other consumer items, including aerosol spray cans and products such as Styrofoam. These chemicals, called chlorofluorocarbons, or CFCs, were considered essential to modern life, and producing them was a multibillion-dollar industry. CFCs were considered "safe" because they are both nontoxic and nonflammable.

## Why do we need an ozone layer?

In the 1970s, scientists learned that CFCs might be responsible for destroying ozone in the upper atmosphere. This discovery led to great concern because a layer of ozone in the upper atmosphere protects us from high-energy ultraviolet (UV) radiation, which causes sunburns, skin cancer, and cataracts as well as environmental damage. In the 1980s, scientists reported an ozone "hole," or depletion of ozone, over Antarctica and documented dangerous thinning of the ozone layer elsewhere.

The nations of the world faced a critical choice: should they continue to produce and use CFCs, and risk further damage to the ozone layer and the resulting effects on people and natural systems, or should they reduce ozone depletion by discontinuing use of this important class of chemicals? In 1987, the majority of nations chose the latter course. As of this writing, most of the world has stopped using CFCs. But the choice at the time was a difficult one. What were the scientific findings that convinced nations to phase out CFCs, the economic consequences of this important decision, and finally, the impact of the CFC ban on the environment? Have we, indeed, protected the ozone layer?

## How do chlorofluorocarbons damage the ozone layer?

As we saw in Chapter 2, the Sun radiates energy at many different wavelengths, including the ultraviolet range. The ultraviolet wavelengths are further classified into three groups: UV-A, or low-energy ultraviolet radiation, and the shorter, higher-energy UV-B and UV-C wavelengths. UV radiation of all types can damage the tissues and DNA of living organisms. Exposure to UV-B radiation increases the risks of skin cancer and cataracts and suppresses the immune system. Exposure to UV-B is also harmful to the cells of plants and reduces their ability to convert sunlight into usable energy. UV-B exposure can therefore lead to crop losses and effects on entire biological communities. For example, losses of phytoplankton—the microscopic algae that form the base of many marine food chains—can harm fisheries.

Next we examine the chemistry of ozone production and how the introduction of chlorine atoms disturbs ozone's steady state in the stratosphere. Oxygen molecules ($O_2$) are common throughout Earth's atmosphere. When solar radiation hits $O_2$ in the stratosphere, 16 to 50 km (10–31 miles) above Earth's surface, a series of chemical reactions begins that produces a new molecule: ozone ($O_3$).

In the first step, UV-C radiation breaks the molecular bond holding an oxygen molecule together:

$$O_2 + UV\text{-}C \rightarrow O + O \qquad (1)$$

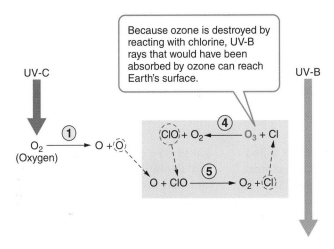

In the absence of stratospheric chlorine (Cl), oxygen and ozone interconvert in a cycle that continuously absorbs UV-B and UV-C radiation.

**(a)** Ozone production and cycling

Because ozone is destroyed by reacting with chlorine, UV-B rays that would have been absorbed by ozone can reach Earth's surface.

**(b)** Effect of chlorine on ozone

**FIGURE SA1.1** **Oxygen-ozone cycles in the stratosphere.** Circled numbers refer to the numbered chemical reactions in the text.

This happens to only a few oxygen molecules at any given time. The vast majority of the oxygen in the atmosphere remains in the form $O_2$.

In the second step, a free oxygen atom (O) produced in reaction 1 encounters an oxygen molecule, and they form ozone. The simplified form of this reaction is written as follows:

$$O + O_2 \rightarrow O_3 \quad (2)$$

Both UV-B and UV-C radiation can break a bond in this new ozone molecule, forming molecular oxygen and a free oxygen atom once again:

$$O_3 + UV\text{-}B \text{ or } UV\text{-}C \rightarrow O_2 + O \quad (3)$$

Thus the formation of ozone in the presence of sunlight and its subsequent breakdown is a cycle (FIGURE SA1.1) that can occur indefinitely as long as there is UV energy entering the atmosphere. Under normal conditions, the amount of ozone in the stratosphere remains at steady state.

However, certain chemicals can promote the breakdown of ozone, disrupting this steady state. Free chlorine (Cl) is one such chemical. The concern over CFCs began when atmospheric scientists realized that CFCs were introducing chlorine into the stratosphere. When chlorine is present, it can attach to an oxygen atom in an ozone molecule, thereby breaking the bond between that atom and the molecule and forming chlorine monoxide (ClO) and $O_2$:

$$O_3 + Cl \rightarrow ClO + O_2 \quad (4)$$

Subsequently, the chlorine monoxide molecule reacts with a free oxygen atom, which pulls the oxygen from the ClO to produce free chlorine again:

$$ClO + O \rightarrow Cl + O_2 \quad (5)$$

Looking at reactions 4 and 5 together, we see that chlorine starts out and ends up as a free Cl atom. In contrast, an ozone molecule and a free oxygen atom are converted into two oxygen molecules. A substance that

**1979**    **1988**    **1998**    **2008**

**Ozone concentration**

Lesser ⟵──────⟶ Greater

**FIGURE SA1.2** **The ozone hole over time.** An area of decreased atmospheric ozone concentration has been forming during the Antarctic spring (September–December) every year since 1979. There has been a decrease in ozone to about one-third of its 1979 concentration.

aids a reaction but does not get used up itself is called a **catalyst.** A single chlorine atom can catalyze the breakdown of as many as 100,000 ozone molecules, until finally one chlorine atom finds another and the process is stopped. The ozone molecules are no longer available to absorb incoming UV-B radiation. As a result, the UV-B radiation can reach Earth's surface and cause biological harm.

## How did nations address the ozone crisis?

In response to the findings described above, the U.S. Environmental Protection Agency banned the use of CFCs in most aerosol sprays in 1978. Policy makers deemed further actions to reduce CFC use too expensive.

By 1986, however, the political climate had changed dramatically. British scientists announced the discovery of a vast ozone "hole" forming seasonally over Antarctica (**FIGURE SA1.2**). This region of unusually low ozone concentrations had not been predicted by scientific models, and the idea of an unexpected hole in the ozone layer captured public attention. Moreover, two important reports appeared in 1985 and 1986, from the World Meteorological Organization and the EPA, that demonstrated an emerging scientific consensus on the magnitude of the ozone depletion problem. Finally, DuPont, the world's leading producer of CFCs, stated that CFC alternatives could be available within 5 years, given the right market conditions.

The issue remained contentious, however. In order to convert to CFC alternatives, many industries would need to be retrofitted with new equipment, and those industries were strongly opposed to the change. In 1987, a trade group called the CFC Alliance estimated that just stopping the *growth* of new CFC production

would cost more than $1 billion and affect 700,000 jobs in the United States. In addition, because chlorine remains in the stratosphere for tens to hundreds of years, some argued that a reduction in CFCs would have minimal short-term benefits for the environment and would result in an improvement only after several decades, not justifying expensive changes now.

In spite of these objections, in 1987, 24 nations signed an agreement called the Montreal Protocol on Substances That Deplete the Ozone Layer. Those nations committed to taking concrete steps to cut the production of CFCs in half by the year 2000. As the scientific case against CFCs strengthened and the economic costs turned out to be less than had been projected, more nations joined the Montreal Protocol, and amendments added in 1990 and 1992 strengthened the treaty by calling for a complete phaseout of CFCs in developed countries by 1996.

Small amounts of CFCs continue to be used in developing countries, and certain agricultural chemicals and CFC replacements can also destroy ozone, although to a lesser degree than CFCs. However, because of the Montreal Protocol, CFC production worldwide had fallen to 2 percent of its peak value by 2004, and chlorine concentrations in the stratosphere are slowly decreasing. Scientists believe that stratospheric ozone depletion will decrease in subsequent decades as chlorine concentrations stabilize. New cases of skin cancer should eventually decrease as well, again after a significant time lapse due to the fact that some cancers take many years to appear.

The Montreal Protocol demonstrated that the manufacturers of products and the nations that used them were willing to make changes in manufacturing processes, and incur economic hardship, in order to protect the environment. Even more importantly, the agreement protects both human health and nonhuman organisms. A 1997 study by the Canadian government estimated that the Montreal Protocol would cost the global economy $235 billion (Canadian dollars) between 1987 and 2060, but would result in benefits worth twice that amount, even before considering the benefits to human health. For example, the study's economists estimated a global savings of almost $200 billion in agriculture because without the Montreal Protocol, the increased UV-B radiation would have damaged crop productivity. They also found that protection of the ozone layer avoided $238 billion in losses to global fisheries that depend on UV-B–sensitive phytoplankton as a food source. Because of its success, policy makers and environmental scientists view the Montreal Protocol as a model for future action on other international environmental problems such as climate change.

## References

Kaniaru, D., ed. 2007. *The Montreal Protocol: Celebrating 20 Years of Environmental Progress—Ozone Layer and Climate Protection*. Cameron May.

Seinfeld, J. H., and S. N. Pandis. 2006. *Atmospheric Chemistry and Physics: From Air Pollution to Climate Change*. 2nd ed. Wiley.

# Ecosystem Ecology: Interactions Between the Living and Nonliving World

## Reversing the Deforestation of Haiti

Even before the devastating earthquake of 2010, life in Haiti was hard. On the streets of the capital city, Port-au-Prince, people would line up to buy charcoal to cook their meals. According to the United Nations, 76 percent of Haitians lived on less than $2.00 a day. Because other forms of cooking fuel, including oil and propane, were too expensive, people turned to the forests, cutting trees to make charcoal from firewood.

Relying on charcoal for fuel has had a serious impact on the forests of Haiti. In 1923, 60 percent of this mountainous funded the planting of 60 million trees there. Unfortunately, the local people can't afford to let them grow while they are in desperate need of firewood and charcoal. A more successful effort has been the planting of mango trees (*Mangifera indica*). A mature mango tree can provide $70 to $150 worth of mangoes annually. Their value provides an economic incentive for allowing trees to reach maturity. The deforestation problem is also being addressed through efforts to develop alternative fuel sources, such as discarded paper processed into dried cakes that can be burned.

> By 2006, more than 9 million people lived in this small nation, and less than 2 percent of its land remained forested.

country was covered in forest. However, as the population grew and demand for charcoal increased, the amount of forest shrunk. By 2006, more than 9 million people lived in this small nation, and less than 2 percent of its land remained forested. Today, most trees in Haiti are cut before they grow to more than a few centimeters in diameter. This rate of deforestation is not sustainable for the people or for the forest.

Deforestation disrupts the ecosystem services that living trees provide. In Chapter 1 we saw some of the consequences of subjecting land to such massive deforestation. When Haitian forests are clear-cut, the land becomes much more susceptible to erosion. When trees are cut, their roots die, and dead tree roots can no longer stabilize the soil. Without roots to anchor it, the soil is eroded away by the heavy rains of tropical storms and hurricanes. Unimpeded by vegetation, the rainwater runs quickly down the mountainsides, dislodging the topsoil that is so important for forest growth. In addition, oversaturation of the soil causes massive mudslides that destroy entire villages.

But the news from Haiti is not all bad. For more than two decades, the U.S. Agency for International Development has

Extensive forest removal is a problem in many developing nations, not just in Haiti. In many places, widespread removal of trees on mountains has led to rapid soil erosion and substantial disruptions of the natural cycles of water and soil nutrients, which in turn have led to long-term degradation of the environment. The results not only illustrate the connectedness of ecological systems, but also show how forest ecosystems, like all ecosystems, can be influenced by human decisions. ■

Deforestation allows water to run rapidly down the mountains, leading to more extreme flooding.

*Sources:* Deforestation exacerbates Haiti floods, *USA Today,* September 23, 2004, http://www.usatoday.com/weather/hurricane/2004-09-23-haiti-deforest_x.htm; Haitians seek remedies for environmental ruin, National Public Radio, July 15, 2009, http://www.npr.org/templates/story/story.php?storyId=104684950.

◀ Deforestation in the mountains of Haiti has disrupted natural cycles.

The collections of living and nonliving components on Earth can be thought of as ecological systems, commonly called *ecosystems*. Ecosystems control the movement of the energy, water, and nutrients that organisms must have to grow and reproduce. Understanding the processes that determine these movements is the goal of ecosystem ecology.

After reading this chapter you should be able to

- list the basic components of an ecosystem.

- describe how energy flows through ecosystems.

- describe how carbon, nitrogen, and phosphorus cycle within ecosystems.

- explain how ecosystems respond to natural and anthropogenic disturbances.

- discuss the values of ecosystems and how humans depend on them.

# Ecosystem ecology examines interactions between the living and nonliving world

The story of deforestation in Haiti reminds us that all the components of an ecosystem are interrelated. An **ecosystem** is a particular location on Earth distinguished by its particular mix of interacting biotic and abiotic components. A forest, for example, contains many interacting biotic components, such as trees, wildflowers, birds, mammals, insects, fungi, and bacteria, that are quite distinct from those in a grassland. Collectively, all the living organisms in an ecosystem represent that ecosystem's biodiversity. Ecosystems also have abiotic components such as sunlight, temperature, soil, water, pH, and nutrients. The abiotic components of the ecosystem determine which organisms can live there.

The components of a particular ecosystem are highly dependent on climate. For example, ecosystems in the dry desert of Death Valley, California, where temperatures may reach 50°C (120°F), are very different from those on the continent of Antarctica, where temperatures may drop as low as −85°C (−120°F). Similarly,

water can range from being immeasurable in deserts to being a defining part of the ecosystem in lakes and oceans. On less extreme scales, small differences in precipitation and the ability of the soil to retain water can favor different terrestrial ecosystem types. Regions with greater quantities of water in the soil can support trees, whereas regions with less water in the soil can support only grasses.

## Ecosystem Boundaries

The biotic and abiotic components of an ecosystem provide the boundaries that distinguish one ecosystem from another. Some ecosystems have well-defined boundaries, whereas others do not. A cave, for example, is a well-defined ecosystem (**FIGURE 3.1**). It contains identifiable biotic components, such as animals and microorganisms that are specifically adapted to live in a cave environment, as well as distinctive abiotic components, including temperature, salinity, and water that flows through the cave as an underground stream. Roosting bats fly out of the cave each night and consume insects. When the bats return to the cave and defecate, their feces provide energy that passes through the relatively few animal species that live in the cave. In many caves, for example, small invertebrate animals

**FIGURE 3.1** **A cave ecosystem.** Cave ecosystems typically have distinct boundaries and are home to highly adapted species.

consume the feces and are in turn consumed by cave salamanders.

The cave ecosystem is relatively easy to study because its boundaries are clear. With the exception of the bats feeding outside the cave, the cave ecosystem is easily defined as everything from the point where the stream enters the cave to the point where it exits. Likewise, many aquatic ecosystems, such as lakes, ponds, and streams, are relatively easy to define because the ecosystem's boundaries correspond to the boundaries between land and water. Knowing the boundaries of an ecosystem makes it easier to identify the system's biotic and abiotic components and to trace the cycling of energy and matter through the system.

In most cases, however, determining where one ecosystem ends and another begins is difficult. For this reason, ecosystem boundaries are often subjective. Environmental scientists might define a terrestrial ecosystem as the range of a particular species of interest, such as the area where wolves roam, or they might define it using topographic features, such as two mountain ranges enclosing a valley. The boundaries of some managed ecosystems, such as national parks, are set according to administrative rather than scientific criteria. Yellowstone National Park, for example, was once managed as its own ecosystem, until scientists began to realize that many species of conservation interest, such as grizzly bears (*Ursus arctos horribilis*), spent time both inside and outside the 1-million-hectare (2.5-million-acre) park. To manage these species effectively, scientists had to think much more broadly: they had to include nearly 20 million hectares (50 million acres) of public and private land outside the park. This larger region was named the Greater Yellowstone Ecosystem. As the name suggests, the actual ecosystem extends well beyond the administrative boundaries of the park (FIGURE 3.2a).

As we saw in Chapter 2, not all ecosystems are as vast as the Greater Yellowstone Ecosystem. Some can be quite small, such as a water-filled hole in a fallen tree trunk or an abandoned car tire that fills with rainwater

**(a)** The Greater Yellowstone Ecosystem

**(b)** A small ecosystem  10 cm

**FIGURE 3.2** **Large and small ecosystems.** (a) The Greater Yellowstone Ecosystem includes the land within Yellowstone National Park and many adjacent properties. (b) Some ecosystems are very small, such as a rain-filled tree hole that houses a diversity of microbes and aquatic insects.

(FIGURE 3.2b). Such tiny ecosystems include all the physical and chemical components necessary to support a diverse set of species, such as microbes, mosquito larvae, and other insects. Therefore, ecosystems can occur in a wide range of sizes.

## Ecosystem Processes

Although it is helpful to divide locations on Earth into distinct ecosystems, it is important to remember that each ecosystem interacts with surrounding ecosystems through the exchange of energy and matter. Organisms, such as bats flying to and from their cave, and chemical elements, such as carbon or nitrogen dissolved in water, move across ecosystem boundaries. As a result, changes in any one ecosystem can ultimately have far-reaching effects on the global environment.

---

**GAUGE YOUR PROGRESS**

✓ What is an ecosystem and what are its components?

✓ How would you know when you left one ecosystem and entered another?

✓ How are ecosystem boundaries imposed by humans sometimes different from natural boundaries?

---

# Energy flows through ecosystems

To understand how ecosystems function and how to best protect and manage them, ecosystem ecologists study not only the biotic and abiotic components that

---

define an ecosystem, but also the processes that move energy and matter within it.

Plants absorb energy directly from the Sun. That energy is then spread throughout an ecosystem as *herbivores* (animals that eat plants) feed on plants and *carnivores* (animals that eat other animals) feed on herbivores. Consider the Serengeti Plain in East Africa, shown in FIGURE 3.3. There are millions of herbivores, such as zebras and wildebeests, in the Serengeti ecosystem, but far fewer carnivores, such as lions (*Panthera leo*) and cheetahs (*Acinonyx jubatus*), that feed on those herbivores. In accordance with the second law of thermodynamics, when one organism consumes another, not all of the energy in the consumed organism is transferred to the consumer. Some of that energy is lost as heat. Therefore, all the carnivores in an area contain less energy than all the herbivores in the same area because all the energy going to the carnivores must come from the animals they eat. To better understand these energy relationships, let's trace this energy flow in more detail.

## Photosynthesis and Respiration

Nearly all of the energy that powers ecosystems comes from the Sun as solar energy, which is a form of kinetic energy. Plants, algae, and other organisms that use the Sun's energy to produce usable forms of energy are called **producers,** or **autotrophs.** Through the process of **photosynthesis,** producers use solar energy to convert carbon dioxide ($CO_2$) and water ($H_2O$) into glucose ($C_6H_{12}O_6$), a form of potential energy that can be used by a wide range of organisms. As we can see in FIGURE 3.4, the process also produces oxygen ($O_2$) as a waste product. That is why plants and other producers are beneficial to our atmosphere: they produce the oxygen we need to breathe.

Producers use the glucose they produce by photosynthesis to store energy and to build structures such

**FIGURE 3.3 Serengeti Plain of Africa.** The Serengeti ecosystem has more plants than herbivores, and more herbivores than carnivores.

**Photosynthesis**
(performed by plants, algae, and some bacteria)

Sun

6 $CO_2$

6 $H_2O$

6 $O_2$

$C_6H_{12}O_6$
(glucose)

Solar energy + 6 $H_2O$ + 6 $CO_2$ ⟶ $C_6H_{12}O_6$ + 6 $O_2$

**Respiration**
(performed by all organisms)

Energy

6 $CO_2$

6 $H_2O$

6 $O_2$

$C_6H_{12}O_6$

Energy + 6 $H_2O$ + 6 $CO_2$ ⟵ $C_6H_{12}O_6$ + 6 $O_2$

**FIGURE 3.4 Photosynthesis and respiration.**
Photosynthesis is a process by which producers use solar energy to convert carbon dioxide and water into glucose and oxygen. Respiration is a process by which organisms convert glucose and oxygen into water and carbon dioxide, releasing the energy needed to live, grow, and reproduce. All organisms, including producers, perform respiration.

as leaves, stems, and roots. Other organisms, such as the herbivores on the Serengeti Plain, eat the tissues of producers and gain energy from the chemical energy contained in those tissues. They do this through **cellular respiration,** a process that unlocks the chemical energy stored in the cells of organisms. Respiration is the opposite of photosynthesis: cells convert glucose and oxygen into energy, carbon dioxide, and water. In essence, they run photosynthesis backward to recover the solar energy stored in glucose.

All organisms—including producers—carry out respiration to fuel their own metabolism and growth. Thus producers both produce and consume oxygen. When the Sun is shining and photosynthesis occurs, producers generate more oxygen via photosynthesis than they consume via respiration, and there is a net production of oxygen. At night, producers only respire, consuming oxygen without generating it. Overall, producers photosynthesize more than they respire. The net effect is an excess of oxygen that is released into the air and an excess of carbon that is stored in the tissues of producers.

## Trophic Levels, Food Chains, and Food Webs

Unlike producers, which make their own food, **consumers,** or **heterotrophs,** are incapable of photosynthesis and must obtain their energy by consuming other organisms. In **FIGURE 3.5,** we can see that heterotrophs fall into several different categories. Heterotrophs that consume producers are called herbivores or **primary consumers.** Primary consumers include a variety of familiar plant- and algae-eating animals, such as zebras,

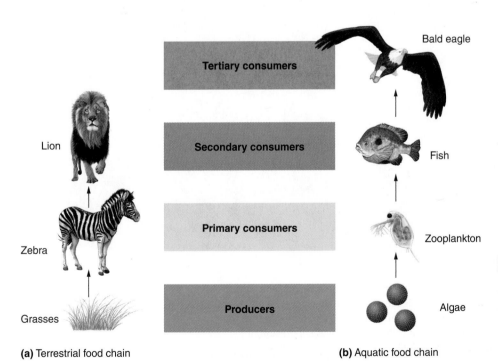

Tertiary consumers

Secondary consumers

Primary consumers

Producers

Bald eagle

Fish

Zooplankton

Algae

Lion

Zebra

Grasses

**(a)** Terrestrial food chain

**(b)** Aquatic food chain

**FIGURE 3.5 Simple food chains.** A simple food chain that links producers and consumers in a linear fashion illustrates how energy and matter move through the trophic levels of an ecosystem.

grasshoppers, and tadpoles. Heterotrophs that obtain their energy by eating other consumers are called carnivores. Carnivores that eat primary consumers are called **secondary consumers.** Secondary consumers include creatures such as lions, hawks, and rattlesnakes. Rarer are **tertiary consumers:** carnivores that eat secondary consumers. Animals such as bald eagles (*Haliaeetus leucocephalus*) can be tertiary consumers: algae (producers) living in lakes convert sunlight into glucose, zooplankton (primary consumers) eat the algae, fish (secondary consumers) eat the zooplankton, and eagles (tertiary consumers) eat the fish. We call these successive levels of organisms consuming one another **trophic levels** (from the Greek word *trophē,* which means "nourishment"). The sequence of consumption from producers through tertiary consumers is known as a **food chain.** In a food chain, energy moves from one trophic level to the next.

A food chain helps us visualize how energy and matter move between trophic levels. However, species in natural ecosystems are rarely connected in such a simple, linear fashion. A more realistic type of model, shown in **FIGURE 3.6**, is known as a **food web.** Food webs take into account the complexity of nature, and they illustrate one of the most important concepts of ecology: that all species in an ecosystem are connected to one another.

Not all organisms fit neatly into a single trophic level. Some organisms, called *omnivores,* operate at several trophic levels. Omnivores include grizzly bears, which eat berries and fish, and the Venus flytrap (*Dionaea muscipula*), which can photosynthesize and also digests insects that become trapped in its leaves. In addition, each trophic level eventually produces dead individuals and waste products that feed other organisms. **Scavengers** are

**FIGURE 3.6** **A simplified food web.** Food webs are more realistic representations of trophic relationships than simple food chains. They include scavengers, detritivores, and decomposers, and they recognize that some species feed at multiple trophic levels. Arrows indicate the direction of energy movement. Actual food webs are even more complex than this one. For instance, in an actual ecosystem, many more organisms are present. In addition, for simplicity, not all possible arrows are shown.

carnivores, such as vultures, that consume dead animals. **Detritivores** are organisms, such as dung beetles, that specialize in breaking down dead tissues and waste products (referred to as *detritus*) into smaller particles. These particles can then be further processed by **decomposers:** the fungi and bacteria that complete the breakdown process by recycling the nutrients from dead tissues and wastes back into the ecosystem. Without scavengers, detritivores, and decomposers, there would be no way of recycling organic matter and energy, and the world would rapidly fill up with dead plants and animals.

## Ecosystem Productivity

The amount of energy available in an ecosystem determines how much life the ecosystem can support. For example, the amount of sunlight that reaches a lake surface determines how much algae can live in the lake. In turn, the amount of algae determines the number of zooplankton the lake can support, and the size of the zooplankton population determines the number of fish the lake can support.

If we wish to understand how ecosystems function, or how to manage and protect them, it is important to understand where the energy in an ecosystem comes from and how it is transferred through food webs. To do this, environmental scientists look at the total amount of solar energy that the producers in an ecosystem capture via photosynthesis over a given amount of time. This measure is known as the **gross primary productivity (GPP)** of the ecosystem.

Note that the term *gross,* as used here, indicates the *total* amount of energy captured by producers. In other words, GPP does not subtract the energy lost when the producers respire. The energy captured minus the energy respired by producers is the ecosystem's **net primary productivity (NPP):**

$$NPP = GPP - \text{respiration by producers}$$

You can think of GPP and NPP in terms of a paycheck. GPP is like the total amount your employer pays you. NPP is the actual amount you take home after taxes are deducted.

GPP is essentially a measure of how much photosynthesis is occurring over some amount of time. Determining GPP is a challenge for scientists because a plant rarely photosynthesizes without simultaneously respiring. However, if we can determine the rate of photosynthesis and the rate of respiration, we can use this information to calculate GPP.

We can determine the rate of photosynthesis by measuring the compounds that participate in the reaction. So, for example, we can measure the rate at which $CO_2$ is taken up during photosynthesis and the rate at which $CO_2$ is produced during respiration. A common approach to measuring GPP is to first measure the production of $CO_2$ in the dark. Because no photosynthesis

occurs in the dark, this measure eliminates $CO_2$ uptake by photosynthesis. Next, we measure the uptake of $CO_2$ in sunlight. This measure gives us the net movement of $CO_2$ when respiration and photosynthesis are both occurring. By adding the amount of $CO_2$ produced in the dark to the amount of $CO_2$ taken up in the sunlight, we can determine the gross amount of $CO_2$ that is taken up during photosynthesis:

$$CO_2 \text{ taken up during photosynthesis} =$$
$$CO_2 \text{ taken up in sunlight} + CO_2 \text{ produced in the dark}$$

In this way, we can derive the GPP of an ecosystem per day within a given area. We can give our answer in units of kilograms of carbon taken up per square meter per day (kg $C/m^2/day$).

Converting sunlight into chemical energy is not an efficient process. As **FIGURE 3.7** shows, of the total amount of solar energy that reaches the producers in an ecosystem—the sunlight on a pond surface, for example—only about 1 percent, on average, is converted into chemical energy via photosynthesis. Most of that solar energy is lost from the ecosystem as heat that returns to the atmosphere. Some of the lost energy consists of wavelengths of light that producers cannot absorb. Those wavelengths are either reflected from the surfaces of producers or pass through their tissues.

The NPP of ecosystems ranges from 25 to 50 percent of GPP, or as little as 0.25 percent of the solar energy

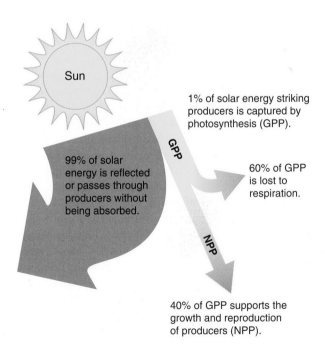

**FIGURE 3.7** **Gross and net primary productivity.** Producers typically capture only about 1 percent of available solar energy via photosynthesis. The energy they capture (gross primary productivity, or GPP) can be divided into energy used for the producers' respiration and energy available for the producers' growth and reproduction (net primary productivity, or NPP).

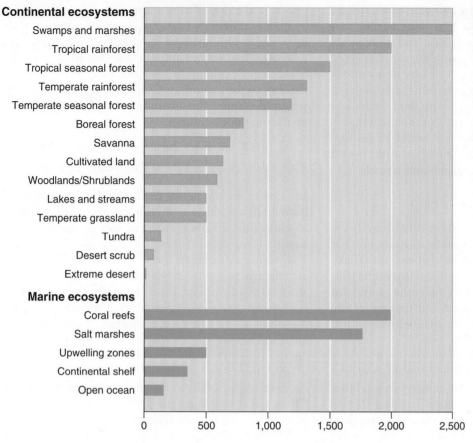

**Continental ecosystems**

Swamps and marshes
Tropical rainforest
Tropical seasonal forest
Temperate rainforest
Temperate seasonal forest
Boreal forest
Savanna
Cultivated land
Woodlands/Shrublands
Lakes and streams
Temperate grassland
Tundra
Desert scrub
Extreme desert

**Marine ecosystems**

Coral reefs
Salt marshes
Upwelling zones
Continental shelf
Open ocean

0    500    1,000    1,500    2,000    2,500

**Net primary productivity (g C/m²/year)**

**FIGURE 3.8** **Net primary productivity varies among ecosystems.** Productivity is highest where temperatures are warm and water and solar energy are abundant. As a result, NPP varies tremendously among different areas of the world. [After R. H. Whittaker and G. E. Likens, Primary production: The biosphere and man, *Human Ecology* 1 (1973): 357–369.]

striking the plant. Clearly, it takes a lot of energy to conduct photosynthesis. Let's look at the math. On average, of the 1 percent of the Sun's energy that is captured by a producer (its individual GPP), about 60 percent is used to fuel the producer's respiration. The remaining energy (its individual NPP) is about 40 percent of the original 1 percent (see Figure 3.7). This 40 percent can be used to support the producer's growth and reproduction. A forest in North America, for example, might have a GPP of 2.5 kg C/m²/year and lose 1.5 kg C/m²/year to respiration by the plants in the forest. Because NPP = GPP − respiration, the NPP of the forest is 1 kg C/m²/year (1.8 pounds C/yard²/year). This means that the plants living in 1 m² of forest will add 1 kg of carbon to their tissues every year by means of growth and reproduction. So, in this example, NPP is 40 percent of GPP.

Measurement of NPP allows us to compare the productivity of different ecosystems, as shown in **FIGURE 3.8**. It is perhaps not surprising that producers grow best in ecosystems where they have plenty of sunlight, lots of available water and nutrients, and warm temperatures, such as tropical rainforests and salt marshes, which are the most productive ecosystems on Earth. Conversely, producers grow poorly in the cold regions of the Arctic, dry deserts, and the dark regions

of the deep sea. In general, the greater the productivity of an ecosystem, the more primary consumers can be supported.

Measuring NPP is also a useful way to measure change in an ecosystem. For example, after a drastic change alters an ecosystem, the amount of stored energy (NPP) tells us whether the new system is more or less productive than the previous system.

## Energy Transfer Efficiency and Trophic Pyramids

The energy in an ecosystem can be measured in terms of **biomass**, which is the total mass of all living matter in a specific area. The net primary productivity of an ecosystem—its NPP—establishes the rate at which biomass is produced over a given amount of time. To analyze the productivity of an ecosystem, scientists calculate the biomass of all individuals accumulated over a given amount of time.

The amount of biomass present in an ecosystem at a particular time is its **standing crop.** It is important to differentiate standing crop, which measures the *amount* of energy in a system at a given time, from productivity, which measures the *rate* of energy production over a span of time. For example, slow-growing forests have low productivity; the trees add only a small amount of biomass through growth and reproduction each year. However, the standing crop of long-lived trees—the biomass of trees that has accumulated over hundreds of years—is quite high. In contrast, the high growth rates of algae living in the ocean make them extremely productive. But because primary consumers eat these algae so rapidly, the standing crop of algae at any particular time is relatively low.

Not all of the energy contained in a particular trophic level is in a usable form. Some parts of plants are not digestible and are excreted by primary consumers. Secondary consumers such as owls consume the muscles and organs of their prey, but they cannot digest bones and hair. Of the food that is digestible, some fraction of the energy it contains is used to power the consumer's day-to-day activities, including moving, eating, and (for birds and mammals) maintaining a constant body temperature. That energy is ultimately lost as heat. Any energy left over may be converted into consumer biomass by growth and reproduction and thus becomes available for consumption by organisms at the next higher trophic level. The proportion of consumed energy that can be passed from one trophic level to another is referred to as **ecological efficiency.**

Ecological efficiencies are fairly low: they range from 5 to 20 percent and average about 10 percent across all ecosystems. In other words, of the total biomass available at a given trophic level, only about 10 percent can be converted into energy at the next higher trophic level. We can represent the distribution of biomass among trophic levels using a **trophic pyramid,** like the one for the Serengeti ecosystem shown in **FIGURE 3.9**. Trophic pyramids tend to look similar across ecosystems. Most energy (and biomass) is found at the producer level, and energy (and biomass) decrease as we move up the pyramid.

The Serengeti ecosystem offers a good example of a trophic pyramid. The biomass of producers (grasses and shrubs) is much greater than the biomass of primary consumers (such as gazelles, wildebeests, and zebras) for which the producers serve as food. Likewise, the biomass of primary consumers is much greater than the biomass of secondary consumers (such as lions and cheetahs). The flow of energy between trophic levels helps to determine the population sizes of the various species within each trophic level. As we saw earlier in this chapter, the number of primary consumers in an area is generally higher than that of the carnivores they sustain.

The principle of ecological efficiency also has implications for the human diet. For example, if all humans were to act only as primary consumers—that is, become vegetarians—we would harvest much more energy from any given area. How would this work?

Suppose an acre of cropland could produce 1,000 kg of soybeans. This food could feed humans directly. Or, if we assume 10 percent ecological efficiency, it could be fed to cattle to produce approximately 100 kg of meat. In terms of biomass, there would be 10 times more food available for humans acting as primary consumers by eating soybeans than for humans acting as secondary consumers by eating beef. However, 1 kg of soybeans actually contains about 2.5 times as many calories as 1 kg of beef. Therefore, 1 acre of land would produce 25 times more calories when used for soybeans than when used for beef. In general, when we act as secondary consumers, the animals we eat require land to support the producers they consume. When we act as primary consumers, we require only the land necessary to support the producers we eat.

**FIGURE 3.9** **Trophic pyramid for the Serengeti ecosystem.** This trophic pyramid represents the amount of energy that is present at each trophic level, measured in joules (J). While this pyramid assumes 10 percent ecological efficiency, actual ecological efficiencies range from 5 to 20 percent across different ecosystems. For most ecosystems, graphing the numbers of individuals or biomass within each trophic level would produce a similar pyramid.

Secondary consumers — 100 J
Primary consumers — 1,000 J
Producers — 10,000 J

**GAUGE YOUR PROGRESS**

✓ Why is photosynthesis an important process?

✓ What determines the productivity of an ecosystem?

✓ How efficiently is energy transferred between trophic levels in an ecosystem?

# Matter cycles through the biosphere

The combination of all ecosystems on Earth forms the *biosphere*. The **biosphere** is the region of our planet where life resides. It forms a 20 km (12-mile) thick shell around Earth between the deepest ocean bottom and the highest mountain peak. Energy *flows through* the biosphere: it enters as energy from the Sun, moves among the living and nonliving components of ecosystems, and is ultimately emitted into space by Earth and its atmosphere. As a result, energy must be constantly replenished by the Sun. Matter, in contrast, does not enter or leave the biosphere, but *cycles within* the biosphere in a variety of forms. As we saw in Chapter 2, Earth is an open system with respect to energy, but a closed system with respect to matter.

The movements of matter within and between ecosystems involve biological, geological, and chemical processes. For this reason, these cycles are known as **biogeochemical cycles.** To keep track of the movement of matter in biogeochemical cycles, we refer to the components that contain the matter, including air, water, and organisms, as *pools.* Processes that move matter between pools are known as *flows.*

All of Earth's living organisms are composed of chemical elements—mostly carbon, hydrogen, nitrogen, oxygen, and phosphorus. Organisms survive by constantly acquiring these various elements, either directly from their environment or by consuming other organisms, breaking down the digestible material, and rearranging the elements into usable compounds. The elements eventually leave the biotic components of the ecosystem when they are excreted as wastes or released by decomposition. Understanding the sources of these elements and how they flow between the biotic and abiotic components of ecosystems helps us to understand how ecosystems function and the ways in which human activities can alter these processes.

The specific chemical forms elements take determine how they cycle within the biosphere. In this section we will look at the elements that are the most important to the productivity of photosynthetic organisms. We will begin with the hydrologic cycle—the movement of water. Then we will explore the cycles of carbon, nitrogen, and phosphorus. Finally, we will take a brief look at the cycles of calcium, magnesium, potassium, and sulfur.

## The Hydrologic Cycle

Water is essential to life. It makes up over one-half of a typical mammal's body weight, and no organism can survive without it. Water allows essential molecules to move within and between cells, draws nutrients into the leaves of trees, dissolves and removes toxic materials, and performs many other critical biological functions. On a larger scale, water is the primary agent responsible for dissolving and transporting the chemical elements necessary for living organisms. The movement of water through the biosphere is known as the **hydrologic cycle.**

FIGURE 3.10 shows how the hydrologic cycle works. Heat from the Sun causes water to evaporate from oceans, lakes, and soils. Solar energy also provides the energy for photosynthesis, during which plants release water from their leaves into the atmosphere—a process known as **transpiration.** The water vapor that enters the atmosphere eventually cools and forms clouds, which, in turn, produce precipitation in the form of rain, snow, and hail. Some precipitation falls back into the ocean and some falls on land.

When water falls on land, it may take one of three distinct routes. First, it may return to the atmosphere by evaporation or, after being taken up by plant roots, by transpiration. The combined amount of evaporation and transpiration, called **evapotranspiration,** is often used by scientists as a measure of the water moving through an ecosystem. Alternatively, water can be absorbed by the soil and percolate down into the groundwater. Finally, water can move as **runoff** across the land surface and into streams and rivers, eventually reaching the ocean—the ultimate pool of water on Earth. As water in the ocean evaporates, the cycle begins again.

The hydrologic cycle is instrumental in the cycling of elements. Many elements are carried to the ocean or taken up by organisms in dissolved form. As you read about biogeochemical cycles, notice the role that water plays in these processes.

### HUMAN ACTIVITIES AND THE HYDROLOGIC CYCLE

Because Earth is a closed system with respect to matter, water never leaves it. Nevertheless, human activities can alter the hydrologic cycle in a number of ways. For example, harvesting trees from a forest can reduce evapotranspiration by reducing plant biomass. If evapotranspiration decreases, then runoff or percolation will increase. On a moderate or steep slope, most water will leave the land surface as runoff. That is why, as we saw at the opening of this chapter, clear-cutting a mountain slope can lead to erosion and flooding. Similarly, paving over land surfaces to build roads, businesses, and homes reduces the amount of percolation that can take place in a given area, increasing runoff and evaporation. Humans can also alter the hydrologic cycle by diverting water from one area to another to provide water for drinking, irrigation, and industrial uses.

## The Carbon Cycle

The elements carbon (C), nitrogen (N), phosphorus (P), potassium (K), magnesium (Mg), calcium (Ca), and sulfur (S) cycle through trophic levels in similar ways. Producers obtain these elements from the atmosphere or as ions dissolved in water. Consumers then obtain these elements by eating producers. Finally, decomposers absorb these elements from dead producers and consumers and their waste products. Through the process of decomposition, they convert the elements into forms that are once again available to producers.

Evaporated water condenses into clouds.

Evaporation from ocean

② Evaporation from soil

Evapotranspiration from plants

Rain clouds

③ Water returns to Earth as precipitation (rain, snow, hail).

Precipitation

③

Surface runoff

④

Infiltration

① Solar energy heats Earth, and causes evaporation.

Ocean

Groundwater

④ Precipitation falling on land is taken up by plants, runs off along the land surface, or percolates into the soil and enters the groundwater.

**FIGURE 3.10** **The hydrologic cycle.** Water moves from the atmosphere to Earth's surface and back to the atmosphere.

Carbon is the most important element in living organisms; it makes up about 20 percent of their total body weight. Carbon is the basis of the long chains of organic molecules that form the membranes and walls of cells, constitute the backbones of proteins, and store energy for later use. Other than water, there are few molecules in the bodies of organisms that do not contain carbon.

FIGURE 3.11 illustrates the six processes that drive the carbon cycle: *photosynthesis, respiration, exchange, sedimentation and burial, extraction,* and *combustion.* These processes can be categorized as either fast or slow. The fast part of the cycle involves processes that are associated with living organisms. The slow part of the cycle involves carbon that is held in rocks, in soils, or as petroleum hydrocarbons (the materials we use as fossil fuels). Carbon may be stored in these forms for millions of years.

Let's take a closer look at the carbon cycle, beginning with photosynthesis. When producers photosynthesize, they take in $CO_2$ and incorporate the carbon into their tissues. Some of this carbon is returned to the atmosphere when organisms respire. It is also returned to the atmosphere after organisms die. In the latter case, the

carbon that was part of the live biomass pool becomes part of the dead biomass pool. Decomposers break down the dead material, returning $CO_2$ to the atmosphere via respiration and continuing the cycle.

A large amount of carbon is exchanged between the atmosphere and the ocean. The amount of $CO_2$ released from the ocean into the atmosphere roughly equals the amount of atmospheric $CO_2$ that diffuses into ocean water. Some of the $CO_2$ dissolved in the ocean enters the food web via photosynthesis by algae. Some combines with calcium ions in the water to form calcium carbonate ($CaCO_3$), a compound that can precipitate out of the water and form limestone and dolomite rock via sedimentation and burial. Although sedimentation is a very slow process, the small amounts of calcium carbonate sediment formed each year have accumulated over millions of years to produce the largest carbon pool in the slow part of the carbon cycle.

A small fraction of the organic carbon in the dead biomass pool is buried and incorporated into ocean sediments before it can decompose into its constituent elements. This organic matter becomes fossilized and,

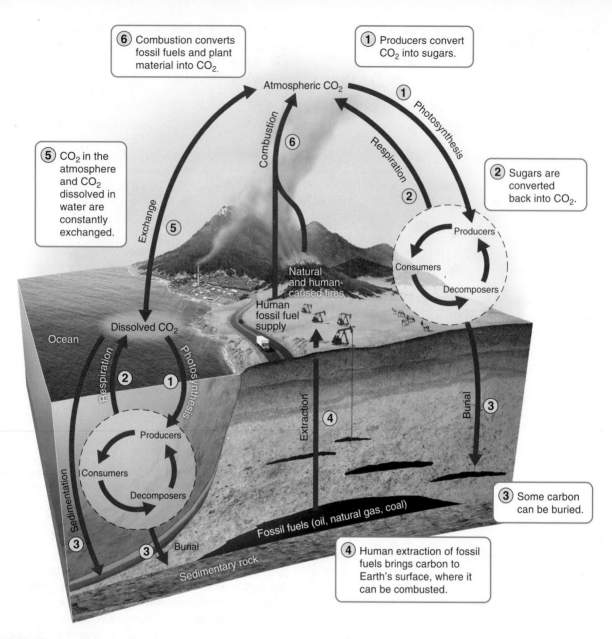

**FIGURE 3.11** **The carbon cycle.** Producers take up carbon from the atmosphere and water via photosynthesis and pass it on to consumers and decomposers. Some inorganic carbon sediments out of the water to form sedimentary rock while some organic carbon may be buried and become fossil fuels. Respiration by organisms returns carbon back to the atmosphere and water. Combustion of fossil fuels and other organic matter returns carbon back to the atmosphere.

over millions of years, some of it may be transformed into fossil fuels. The amount of carbon removed from the food web by this slow process is roughly equivalent to the amount of carbon returned to the atmosphere by weathering of carbon-containing rocks (such as limestone) and by volcanic eruptions, so the slow part of the carbon cycle is in steady state.

The fifth process in the carbon cycle is the extraction of fossil fuels by humans. This process is a relatively recent phenomenon that began when human society started to rely on coal, oil, and natural gas as energy

sources. Extraction by itself does not alter the carbon cycle, however. It is the subsequent step of combustion that makes the difference. Combustion, whether of fossil fuels or of timber in a forest fire, releases carbon into the atmosphere as $CO_2$ or into the soil as ash.

Respiration, decomposition, and combustion operate in very similar ways: all three processes cause organic molecules to be broken down to produce $CO_2$, water, and energy. The difference is that respiration and decomposition are biotic processes, whereas in combustion, the breakdown process occurs abiotically.

In the absence of human disturbance, the exchange of carbon between Earth's surface and atmosphere is in steady state. Carbon taken up by photosynthesis eventually ends up in the soil. Decomposers in the soil gradually release that carbon at roughly the same rate it is added. Similarly, the gradual movement of carbon into the buried or fossil fuel pools is offset by the slow processes that release it. Before the Industrial Revolution, atmospheric carbon concentrations had changed very little for 10,000 years (see Figure 1.8). So, until recently, carbon entering any of these pools was balanced by carbon leaving these pools.

Since the Industrial Revolution, however, human activities have had a major influence on carbon cycling. The best-known and most significant human alteration of the carbon cycle is the combustion of fossil fuels. This process releases fossilized carbon into the atmosphere, which increases atmospheric carbon concentrations and upsets the balance between Earth's carbon pools and the atmosphere. The excess $CO_2$ in the atmosphere acts to increase the retention of heat energy in the biosphere. The result, global warming, is a major concern among environmental scientists and policy makers.

Tree harvesting is another human activity that can affect the carbon cycle. Trees store a large amount of carbon in their wood, both above and below ground. The destruction of forests by cutting and burning increases the amount of $CO_2$ in the atmosphere. Unless enough new trees are planted to recapture the carbon, the destruction of forests will upset the balance of $CO_2$. To date, large areas of forest, including tropical forests as well as North American and European temperate forests, have been converted into pastures, grasslands, and croplands. In addition to destroying a great deal of biodiversity, this destruction of forests has added large amounts of carbon to the atmosphere. The increases in atmospheric carbon due to human activities have been partly offset by an increase in carbon absorption by the ocean. Still, the harvesting of trees remains a concern.

## The Nitrogen Cycle

We have seen that water and carbon, both essential to life, cycle through the biosphere in complex ways. Now we turn to some of the other elements that play an important part in the life of ecosystems. There are six key elements, known as **macronutrients,** that organisms need in relatively large amounts: nitrogen, phosphorus, potassium, calcium, magnesium, and sulfur.

Organisms need nitrogen—the most abundant element in the atmosphere—in relatively high amounts. Because so much of it is required, nitrogen is often a **limiting nutrient** for producers. In other words, a lack of nitrogen constrains the growth of the organism. Adding other nutrients, such as water or phosphorus, will not improve plant growth in nitrogen-poor soil.

Nitrogen is used to form *amino acids,* the building blocks of proteins, and *nucleic acids,* the building blocks of DNA and RNA. In humans, nitrogen makes up about 3 percent of total body weight. The movement of nitrogen from the atmosphere through many transformations within the soil, then into plants, and then back into the atmosphere makes the nitrogen cycle one of the more interesting and complex biogeochemical cycles.

FIGURE 3.12 shows the complex processes of the nitrogen cycle. Although Earth's atmosphere is 78 percent nitrogen by volume, the vast majority of that nitrogen is in a form that most producers cannot use. Nitrogen gas ($N_2$) molecules consist of two N atoms tightly bound together. Only a few organisms can convert $N_2$ gas directly into ammonia ($NH_3$) by a process known as **nitrogen fixation.** This process is the first step in the nitrogen cycle. Nitrogen-fixing organisms include cyanobacteria (also known as blue-green algae) and certain bacteria that live within the roots of legumes (plants such as peas, beans, and a few species of trees). These nitrogen-fixing bacteria possess specialized enzymes that can break the strong $N_2$ bond and add hydrogen ions to form $NH_3$, which is readily converted into its ionic form, ammonium ($NH_4^+$), in the soil.

Nitrogen-fixing organisms use the fixed nitrogen to synthesize their own tissues, then excrete any excess. Cyanobacteria, which are primarily aquatic organisms, excrete excess ammonium ions into the water. Bacteria living within plant roots excrete excess ammonium ions into the plant's root system; the plant, in turn, supplies the bacteria with sugars it produces via photosynthesis.

Nitrogen can also be fixed through two abiotic pathways. First, $N_2$ can be fixed in the atmosphere by lightning or during combustion processes such as fires and the burning of fossil fuels. These processes convert $N_2$ into nitrate ($NO_3^-$), which is usable by plants. The nitrate is carried to Earth's surface in precipitation. Humans have also developed techniques for converting $N_2$ gas into ammonia or nitrate to be used in plant fertilizers. Although these processes require a great deal of energy, humans now fix more nitrogen than is fixed in nature. The development of synthetic nitrogen fertilizers has led to large increases in crop yields, particularly for crops such as corn that require large amounts of nitrogen. Regardless of how nitrogen fixation occurs, the end product—$NH_4^+$ or $NO_3^-$—is a form of nitrogen that can be used by producers.

Once producers obtain fixed nitrogen, they assimilate it into their tissues (step 2 of the cycle). When primary consumers feed on the producers, some of that nitrogen is assimilated into the consumers' tissues, and some is eliminated as waste products. Eventually, both producers and consumers die and decompose. In step 3 of the cycle, a process called *ammonification,* fungal and bacterial decomposers use nitrogen-containing wastes and dead bodies as a food source and excrete ammonium.

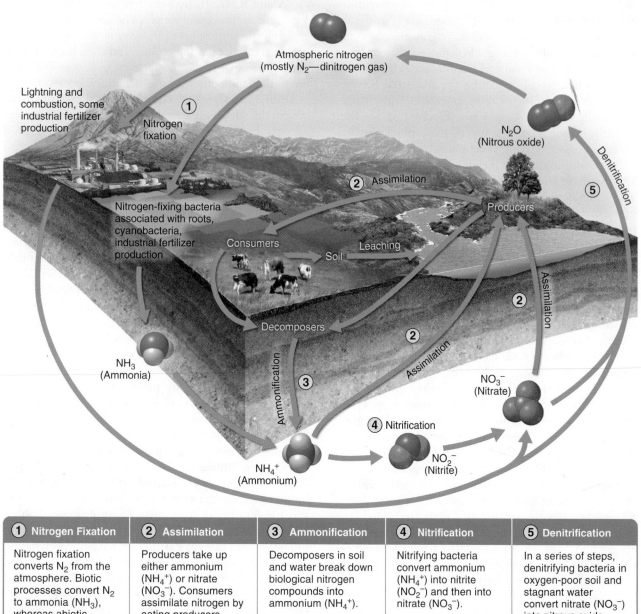

Atmospheric nitrogen
(mostly N₂—dinitrogen gas)

Lightning and combustion, some industrial fertilizer production

① Nitrogen fixation

N₂O (Nitrous oxide)

Denitrification

Nitrogen-fixing bacteria associated with roots, cyanobacteria, industrial fertilizer production

② Assimilation

Producers

⑤

Consumers

Soil   Leaching

Assimilation

NH₃ (Ammonia)

Decomposers

② Assimilation

②

Ammonification

③

Assimilation

②

NO₃⁻ (Nitrate)

NH₄⁺ (Ammonium)

④ Nitrification

NO₂⁻ (Nitrite)

| ① **Nitrogen Fixation** | ② **Assimilation** | ③ **Ammonification** | ④ **Nitrification** | ⑤ **Denitrification** |
|---|---|---|---|---|
| Nitrogen fixation converts N₂ from the atmosphere. Biotic processes convert N₂ to ammonia (NH₃), whereas abiotic processes convert N₂ to nitrate (NO₃⁻). | Producers take up either ammonium (NH₄⁺) or nitrate (NO₃⁻). Consumers assimilate nitrogen by eating producers. | Decomposers in soil and water break down biological nitrogen compounds into ammonium (NH₄⁺). | Nitrifying bacteria convert ammonium (NH₄⁺) into nitrite (NO₂⁻) and then into nitrate (NO₃⁻). | In a series of steps, denitrifying bacteria in oxygen-poor soil and stagnant water convert nitrate (NO₃⁻) into nitrous oxide (N₂O) and eventually nitrogen gas (N₂). |

**FIGURE 3.12** **The nitrogen cycle.** The nitrogen cycle moves nitrogen from the atmosphere and into soils through several fixation pathways, including the production of fertilizers by humans. In the soil, nitrogen can exist in several forms. Denitrifying bacteria release nitrogen gas back into the atmosphere.

Ammonium, in turn, is converted into nitrite ($NO_2^-$) and then into nitrate ($NO_3^-$) by specialized nitrifying bacteria in a two-step process called *nitrification* (step 4 of the cycle). Nitrite is of minor importance in natural ecosystems, but nitrate can be used by producers.

Because negatively charged particles repel one another, negatively charged nitrate ions do not bind easily to soil particles, most of which are negatively charged. As a result, nitrate is readily transported through the soil with water—a process called **leaching.** Leached nitrates eventually settle in the bottom sediments of oceans, lakes, and swamps. Under these conditions, or in waterlogged soils, denitrifying bacteria convert nitrate in a series of steps into the gases nitrous oxide ($N_2O$) and, eventually, $N_2$, which is emitted into the atmosphere. This conversion back into atmospheric $N_2$, a process called *denitrification*, completes the nitrogen cycle.

**EXCESS NITROGEN** Nitrogen is a limiting nutrient in most terrestrial ecosystems, so excess inputs of nitrogen can have consequences in these ecosystems. Adding nitrogen to soils in fertilizers ultimately increases atmospheric concentrations of nitrogen. This nitrogen can be transported through the atmosphere and deposited by rainfall in natural ecosystems that have adapted over time to a particular level of nitrogen availability. The added nitrogen can alter the distribution or abundance of species in those ecosystems.

In one study of nine different terrestrial ecosystems across the United States, scientists added nitrogen fertilizer to some plots and left other plots unfertilized as controls. They found that adding nitrogen reduced the number of species in a plot by up to 48 percent because some species that could survive under low-nitrogen conditions could no longer compete against larger plants that thrived under high-nitrogen conditions. Other studies have documented cases in which plant communities that have grown on low-nitrogen soils for millennia are now experiencing changes in their species composition. An influx of nitrogen due to human activities has favored colonization by new species that are better adapted to soils with higher fertility.

The observation that nutrients can have unintended effects on ecosystems highlights an important principle of environmental science: in ecosystems containing species that have adapted to their environments over thousands of years (or longer), changes in conditions are likely to cause changes in biodiversity as well as in the movement of energy through, and the cycling of matter within, those ecosystems.

## The Phosphorus Cycle

Organisms need phosphorus for many biological processes. Phosphorus is a major component of DNA and RNA as well as ATP, the molecule used by cells for energy transfer. Required by both plants and animals, phosphorus is a limiting nutrient second only to nitrogen in its importance for successful agricultural yields. Thus phosphorus, like nitrogen, is commonly added to soils in the form of fertilizer.

**FIGURE 3.13** shows the processes of the phosphorus cycle. Because this cycle has no gaseous component, atmospheric inputs of phosphorus—which occur when phosphorus is dissolved in rainwater or sea spray—are very small. Phosphorus is not very soluble in water, so much of it precipitates out of solution, forming phosphate ($PO_4^{3-}$)-laden sediments on the ocean floor. Humans mine some of these ancient phosphate sediments for fertilizer. The small amount of phosphorus dissolved in water also means that phosphorus is the primary limiting nutrient in many freshwater and marine food webs.

On land, the major natural source of phosphorus is the weathering of rocks. Negatively charged phosphate ions bind readily to several positively charged minerals found in soil, so phosphorus is not easily leached out of the soil by water. Producers, however, can extract it from the soil, at which point it can move through the food web in a manner similar to other elements.

**EXCESS PHOSPHORUS** Because phosphorus is so tightly held by soils on land, and because much of what enters water precipitates out of solution, very little dissolved phosphorus is naturally available in rivers and streams. As a result, phosphorus is a limiting nutrient in many aquatic systems. Even small inputs of leached phosphorus into these systems can greatly increase the growth of producers. Phosphorus inputs into phosphorus-limited aquatic systems can cause rapid growth of algae, known as an *algal bloom*. Algal blooms quickly increase the amount of biomass in the ecosystem (**FIGURE 3.14**). The algae eventually die, initiating a massive amount of decomposition, which consumes large amounts of oxygen. Thus algal blooms result in *hypoxic* (low-oxygen) conditions that kill fish and other aquatic animals. Hypoxic dead zones occur around the world, including where the Mississippi River empties into the Gulf of Mexico.

Two major sources of phosphorus in waterways are fertilizer-containing runoff from agricultural or residential areas and household detergents. From the 1940s through the 1990s, laundry detergents contained phosphates to make clothes cleaner. The water discharged from washing machines inadvertently fertilized streams, rivers, and lakes. Because ecological dead zones caused by excess phosphorus represent substantial environmental and economic damage, manufacturers stopped adding phosphates to laundry detergents in 1994 and to dishwashing detergents in 2010.

In addition to causing algal blooms, increases in phosphorus concentrations can alter plant communities. We have already seen one example in Chapter 2, in

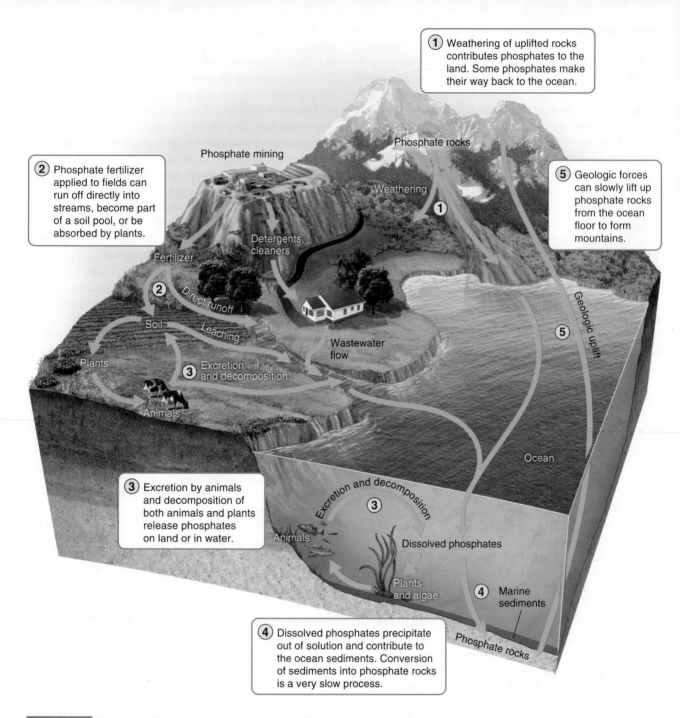

**FIGURE 3.13** **The phosphorus cycle.** The phosphorus cycle begins with the weathering or mining of phosphate rocks and use of phosphate fertilizer, which releases phosphorus into the soil and water. This phosphorus can be used by producers and subsequently moves through the food web. Phosphorus can precipitate out of solution and form sediments, which over time are transformed into new phosphate rocks.

which the "Working Toward Sustainability" feature discussed the deterioration of the environment in the Florida Everglades. Because of agricultural expansion in southern Florida, the water that flows through the Florida Everglades has experienced elevated phosphorus concentrations. This change in nutrient cycling has changed the ecosystem of the Everglades. Over time, cattails have become more common and sawgrass has declined. Animals that depended on sawgrass for food or habitat are no longer favored.

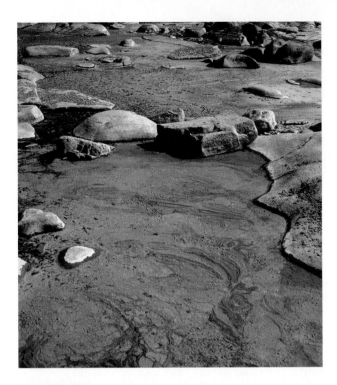

**FIGURE 3.14** | **Algal bloom.** When excess phosphorus enters waterways, it can stimulate a sudden and rapid growth of algae that turns the water bright green. The algae eventually die, and the resulting increase in decomposition can reduce dissolved oxygen to levels that are lethal to fish and shellfish.

## Calcium, Magnesium, Potassium, and Sulfur

Calcium, magnesium, and potassium play important roles in regulating cellular processes and in transmitting signals between cells. Like phosphorus, these macronutrients are derived primarily from rocks and decomposed vegetation. All three can be dissolved in water as positively charged ions: $Ca^{2+}$, $Mg^{2+}$, and $K^+$. None is present in a gaseous phase, but all can be deposited from the air in small amounts as dust.

Because of their positive charges, calcium, magnesium, and potassium ions are attracted to the negative charges present on the surfaces of most soil particles. Calcium and magnesium occur in high concentrations in limestone and marble. Because $Ca^{2+}$ and $Mg^{2+}$ are strongly attracted to soil particles, they are abundant in many soils overlying these rock types. In contrast, $K^+$ is only weakly attracted to soil particles and is therefore more susceptible to being leached away by water moving through the soil. Leaching of potassium can lead to potassium-deficient soils that constrain the growth of plants and animals.

The final macronutrient, sulfur, is a component of proteins and also plays an important role in allowing organisms to use oxygen. Most sulfur exists in rocks and is released into soils and water as these rocks weather over time. Plants absorb sulfur through their roots in the form of sulfate ions ($SO_4^{2-}$), and the sulfur then cycles through the food web. The sulfur cycle also has a gaseous component. Volcanic eruptions are a natural source of atmospheric sulfur in the form of sulfur dioxide ($SO_2$). Human activities also add sulfur dioxide to the atmosphere, especially the burning of fossil fuels and the mining of metals such as copper. In the atmosphere, $SO_2$ is converted into sulfuric acid ($H_2SO_4$) when it mixes with water. The sulfuric acid can then be carried back to the ground when it rains or snows. As humans add more sulfur dioxide to the atmosphere, we cause more acid precipitation, which can negatively affect terrestrial and aquatic ecosystems. Although anthropogenic deposition of sulfur remains an environmental concern, clean air regulations in the United States have significantly lowered these deposits since 1995.

### GAUGE YOUR PROGRESS

✓ What are the dominant elements that make up living organisms?

✓ What role does water play in nutrient cycling?

✓ What are the main similarities and differences among the carbon, nitrogen, and phosphorus cycles?

# Ecosystems respond to disturbance

An event caused by physical, chemical, or biological agents that results in changes in population size or community composition is called a **disturbance.** Natural ecosystem disturbances include hurricanes, ice storms, tsunamis, tornadoes, volcanic eruptions, and forest fires (**FIGURE 3.15**). Anthropogenic ecosystem disturbances include human settlements, agriculture, air pollution, clear-cutting of forests, and the removal of entire mountaintops for coal mining. Disturbances can occur over both short and long time scales. Ecosystem ecologists are often interested in how such disturbances affect the flow of energy and matter through an ecosystem. More specifically, they are interested in whether an ecosystem can resist the impact of a disturbance and whether a disturbed ecosystem can recover its original condition.

In this section we will look at how scientists study disturbance. We will then go on to consider resistance and resilience. Finally, we will apply our knowledge to an important theory about how systems respond to disturbances.

(a)

(b)

**FIGURE 3.15** **Ecosystem disturbance.** The Chandeleur Islands in Louisiana were almost completely submerged by Hurricane Katrina in August 2005. (a) This photo of the islands, taken on July 17, 2001, shows vegetated sand dunes. (b) This photo, taken 2 days after the hurricane made landfall in Louisiana and Mississippi, shows massive erosion and the loss of sand dunes and much of the vegetation.

## Watershed Studies

Understanding the natural rates and patterns of biogeochemical cycling in an ecosystem provides a basis for determining how a disturbance has changed the system. Because it is difficult to study biogeochemical cycles on a global scale, most such research takes place on a smaller scale, at which scientists can measure all of the ecosystem processes. A *watershed* is a common place for scientists to conduct such studies. As shown in **FIGURE 3.16**, a **watershed** is all of the land in a given landscape that drains into a particular stream, river, lake, or wetland.

One of the most thorough studies of disturbance at the watershed scale has been ongoing in the Hubbard Brook ecosystem of New Hampshire since 1962. For almost 50 years, investigators have monitored the hydrological and biogeochemical cycles of six watersheds at Hubbard Brook, ranging in area from 12 to 43 ha

**FIGURE 3.16** **Watershed.** A watershed is the area of land that drains into a particular body of water.

(30 to 106 acres). The soil in each watershed is underlain by impenetrable bedrock, so there is no deep percolation of water; all precipitation that falls on the watershed leaves it either by evapotranspiration or by runoff. Scientists measure precipitation throughout each watershed, and a stream gauge at the bottom of the main stream draining each watershed allows them to measure the amounts of water and nutrients leaving the system.

At Hubbard Brook, researchers investigated the effects of clear-cutting and subsequent suppression of plant regrowth. The researchers cut down the forest in one watershed and used herbicides to suppress the regrowth of vegetation for several years. An adjacent watershed that was not clear-cut served as a control (FIGURE 3.17). The concentrations of nitrate in stream water were similar in the two watersheds before the clear-cutting. Within 6 months after the cutting, the clear-cut watershed showed significant increases in stream nitrate concentrations. With this information, the researchers were able to determine that when trees are no longer present to take up nitrate from the soil, nitrate leaches out of the soil and ends up in the stream that drains the watershed. This study and subsequent research have demonstrated the importance of plants in regulating the cycling of nutrients, as well as the consequences of not allowing new vegetation to grow when a forest is cut.

Studies such as the one done at Hubbard Brook allow investigators to learn a great deal about biogeochemical cycles. We now understand that as forests and grasslands grow, large amounts of nutrients accumulate in the vegetation and in the soil. The growth of forests allows the terrestrial landscape to accumulate nutrients that would otherwise cycle through the system and end up in the ocean. Forests, grasslands, and other terrestrial ecosystems increase the retention of nutrients on land. This is an important way in which ecosystems directly influence their own growing conditions.

## Resistance versus Resilience

Not every ecosystem disturbance is a disaster. For example, a low-intensity fire might kill some plant species, but at the same time it might benefit fire-adapted species that can use the additional nutrients released from the dead plants. So, although the population of a particular producer species might be diminished or even eliminated, the net primary productivity of all the producers in the ecosystem might remain the same. When this is the case, we say that the productivity of the system is *resistant*. The **resistance** of an ecosystem is a measure of how much a disturbance can affect the flows of energy and matter. When a disturbance influences populations and communities, but has no effect on the overall flows of energy and matter, we say that the ecosystem has *high resistance.*

When an ecosystem's flows of energy and matter are affected by a disturbance, environmental scientists often ask how quickly and how completely the ecosystem can recover its original condition. The rate at which an ecosystem returns to its original state after a disturbance is termed **resilience.** A highly resilient ecosystem returns to its original state relatively rapidly; a less resilient ecosystem does so more slowly. For example, imagine that a severe drought has eliminated half the species in an area. In a highly resilient ecosystem, the flows of energy and matter might return to normal in the following year. In a less resilient ecosystem, the flows of energy and matter might not return to their pre-drought conditions for many years.

An ecosystem's resilience often depends on specific interactions of the biogeochemical and hydrologic cycles. For example, in response to anthropogenic increases in global atmospheric $CO_2$ concentrations, there has been an increase in carbon uptake in terrestrial and aquatic ecosystems. The carbon cycle as a whole can thus mitigate some of the changes that we might expect from increases in atmospheric $CO_2$ concentrations, including global climate change. Conversely, when a drought occurs, the soil may dry out and harden so much that when it eventually does rain, the soil cannot absorb as much water as it did before the drought. This change in the soil leads to further drying and an intensification of the drought damage. In this case, the hydrologic cycle does not relieve the effects of the drought; instead, a positive feedback in the system makes the situation worse.

Many anthropogenic disturbances—for example, housing developments, clear-cutting, or draining of wetlands—are so large that they eliminate an entire ecosystem. In some cases, however, scientists can work

**FIGURE 3.17** **Studying disturbance at the watershed scale.** In the Hubbard Brook ecosystem, researchers clear-cut one watershed to determine the importance of trees in retaining soil nutrients. They compared nutrient runoff in the clear-cut watershed with that in a control watershed that was not clear-cut. (The two other watersheds shown in the photo received other experimental treatments.)

Clear-cut watershed        Control watershed

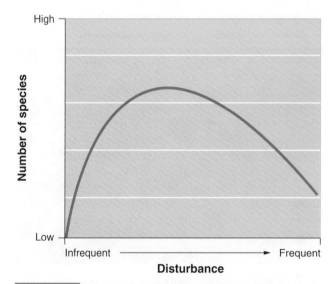

**FIGURE 3.18** **Wetland restoration.** The draining of wetlands can destroy a wetland ecosystem. The damage can be mitigated by using heavy machinery to build new wetlands (inset) that serve the same function.

to reverse these effects and restore much of the original function of the ecosystem (**FIGURE 3.18**). Growing interest in restoring damaged ecosystems has led to the creation of a new scientific discipline called **restoration ecology.** Restoration ecologists are currently working on two high-profile ecosystem restoration projects, in the Florida Everglades and in the Chesapeake Bay, to restore water flows and nutrient inputs that are closer to historic levels so that the functions of these ecosystems can be restored.

## The Intermediate Disturbance Hypothesis

We have seen that not all disturbance is bad. In fact, some level of ecosystem disturbance is natural, and may even be necessary for the maintenance of species diversity. The **intermediate disturbance hypothesis** states that *ecosystems experiencing intermediate levels of disturbance are more diverse than those with high or low disturbance levels.* The graph in **FIGURE 3.19** illustrates this relationship between ecosystem disturbance and species diversity. Ecosystems in which disturbances are rare experience intense competition among species. Because of this, populations of only a few highly competitive species eventually dominate the ecosystem. In places where disturbances are frequent, population growth rates must be high enough to counter the effects of frequent disturbance and prevent species extinction. Research shows that when disturbances occur at some intermediate

frequency, the populations of major competitors never reach a size at which they can dominate an ecosystem, and populations of other species are never driven too close to zero. As a result, we expect to see the highest

**FIGURE 3.19** **Intermediate disturbance hypothesis.** Species diversity is highest at intermediate levels of disturbance. Rare disturbances favor the best competitors, which outcompete other species. Frequent disturbances eliminate most species except those that have evolved to live under such conditions. At intermediate levels of disturbance, species from both extremes can persist.

diversity of species in ecosystems that experience an *intermediate* frequency of disturbance.

**GAUGE YOUR PROGRESS**

✓ Why is Hubbard Brook valuable as a study area? What does it teach us?

✓ What is the difference between resistance and resilience in an ecosystem?

✓ What is the intermediate disturbance hypothesis?

# Ecosystems provide valuable services

Humans rely on only a small number of the millions of species on Earth for our essential needs. Why should we care about the millions of other species in the world? What is the value in protecting biodiversity?

The answer may lie in the type of value a species has for humans. A species may have **instrumental value,** meaning that it has worth as an *instrument* or *tool* that can be used to accomplish a goal. Instrumental values, which include the value of items such as lumber and pharmaceutical drugs, can be thought of in terms of how much economic benefit a species bestows. Alternatively, a species may have **intrinsic value,** meaning that it has worth independent of any benefit it may provide to humans. Intrinsic values include the moral value of an animal's life; they cannot be quantified.

## Instrumental Values of Ecosystems

Ecosystems, as collections of species and as locations for biogeochemical cycling, can have instrumental value, intrinsic value, or both. The instrumental value of ecosystems lies in what economists call *ecosystem services:* the benefits that humans obtain from natural ecosystems. For example, the ability of an agricultural ecosystem to produce food is an important ecosystem service, as is the ability of a wetland ecosystem to filter and clean the water that flows through it. Most economists believe that the instrumental uses of an ecosystem can be assigned monetary values, and they are beginning to incorporate these values into their calculations of the economic costs and benefits of various human activities. However, assigning a dollar value is easier for some categories of ecosystem services than for others.

In 1997, a team of environmental scientists and ecological economists attempted to estimate the total value of ecosystem services to the human economy. They considered replacement value—the cost to replace the services provided by natural ecosystems. They also looked at other factors, such as how property values were affected by location relative to these services—for example, oceanfront housing—and how much time or money people were willing to spend to use these services—for example, whether they were willing to pay a fee to visit a national park. Using this method, researchers estimated that ecosystem services were worth over $30 trillion per year, or more than the entire global monetary economy at that time.

When calculating the instrumental value of ecosystem services, it is helpful to group those services into five categories: *provisions, regulating services, support systems, resilience,* and *cultural services.*

**PROVISIONS** Goods that humans can use directly are called **provisions.** Examples include lumber, food crops, medicinal plants, natural rubber, and furs. Of the top 150 prescription drugs sold in the United States, about 70 percent come from natural sources. For example, Taxol, a potent anticancer drug, was originally discovered in the bark of the Pacific yew (*Taxus brevifolia*), a rare tree that grows in forests of the Pacific Northwest (**FIGURE 3.20**). Within a decade of its approval by the FDA, this single drug accounted for over $1 billion in

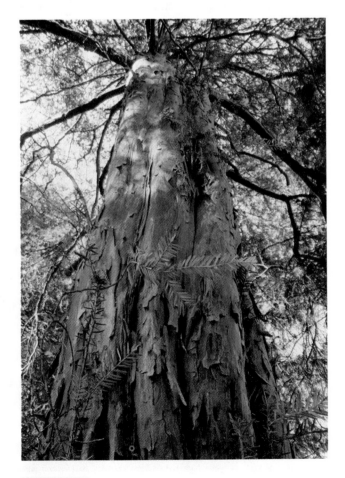

**FIGURE 3.20** **Provisions.** Scientists discovered that the bark of the Pacific yew contains a chemical that has anticancer properties.

### Raising Mangoes

As we saw in this chapter's opening story, farmers in Haiti are being encouraged to plant mango trees because the provisions in the form of fruit are more valuable than the provisions in the form of firewood. A group of Haitian farmers decides to plant mango trees. Mango saplings cost $10 each. Once the trees become mature, each tree will produce $75 worth of fruit per year. A village of 225 people decides to pool its resources and set up a community mango plantation. Their goal is to generate a per capita income of $300 per year for the entire village.

1. How many mature trees will the village need to meet the goal?

Total annual income desired:

$$\$300/person \times 225 \ persons = \$67,500$$

Number of trees needed to produce $67,500 in annual income:

$$\frac{\$67,500}{\$75/tree} = 900 \ trees$$

2. Each tree requires 25 m² of space. How many hectares must the village set aside for the plantation?

$$900 \ trees \times 25 \ m^2 = 22,500 \ m^2 = 2.25 \ ha$$

**Your Turn:** Each tree requires 20 L of water per day during the 6 hot months of the year (180 days). The water must be pumped to the plantation from a nearby stream. How many liters of water are needed each year to water the plantation of 900 trees?

annual sales. However, there is no way to estimate the potential value of natural pharmaceuticals that have yet to be discovered. Our best strategy may be to preserve as much biodiversity as we can to improve our chances of finding the next critical drug.

Because provisions are usually sold in the marketplace, their monetary value is fairly easy to quantify. Crunch the Numbers "Raising Mangoes" provides an opportunity to use provision values to calculate the costs and benefits of human activities.

**REGULATING SERVICES** Natural ecosystems help to regulate environmental conditions. For example, humans currently add about 8 gigatons of carbon to the

**FIGURE 3.21** **Regulating services.** Tropical rainforests play a major role in regulating the amount of carbon in the atmosphere.

atmosphere annually (1 gigaton = 1 trillion kilograms), but only about 4 gigatons of carbon remain there. The rest is removed by natural ecosystems such as tropical rainforests and oceans, providing us with more time to deal with climate change than we would otherwise have (**FIGURE 3.21**). As described earlier in this chapter, ecosystems play important roles in regulating nutrient and hydrologic cycles as well.

**SUPPORT SYSTEMS**  Natural ecosystems provide numerous support services that would be extremely costly for humans to generate. One example is pollination of food crops (**FIGURE 3.22**). The American Institute of Biological Sciences estimates that crop pollination in the United States by native species of bees and other insects, hummingbirds, and bats is worth roughly $3.1 billion in added food production. In addition to providing habitat for animals that pollinate crops, ecosystems provide natural pest control services because they serve as habitat for predators that prey on agricultural pests. Although organic farmers, who rarely use synthetic pesticides, gain the most from these pest controls, conventional agriculture benefits as well.

Healthy ecosystems also filter harmful pathogens and chemicals from water, leaving humans with water that requires relatively little treatment prior to drinking. Without these water-filtering services, humans would have to build many new water treatment facilities using expensive filtration technologies. New York City, for example, draws its water from naturally clean reservoirs in the Catskill Mountains. But residential development and tourism in the area has threatened to increase contamination of the reservoirs with silt and chemicals. Building a filtration plant adequate to address these problems would cost $6 billion to $8 billion. For this reason, New York City and the U.S. Environmental Protection

**FIGURE 3.23**  **Species diversity is an important component of resilience.** This prairie ecosystem contains a high diversity of grasses and wildflowers, including many species of nitrogen-fixing wildflowers. If one nitrogen-fixing species is eliminated, the lost function can be compensated for by other nitrogen-fixing species.

Agency have been working to protect sensitive regions of the Catskills.

**RESILIENCE**  We have already seen that resilience ensures that an ecosystem will continue to exist in its current state, which means it can continue to provide benefits to humans. Resilience depends greatly on species diversity. For example, several different species may perform similar functions in an ecosystem, but differ in their susceptibility to disturbance. If a pollutant kills one plant species that contains nitrogen-fixing bacteria, but not several other plant species that contain nitrogen-fixing bacteria, the ecosystem can continue to fix nitrogen despite the disturbance (**FIGURE 3.23**). Genetic diversity also provides valuable insurance against the loss of ecosystem services.

**CULTURAL SERVICES**  Ecosystems provide cultural or aesthetic benefits to many people. The awe-inspiring beauty of nature has instrumental value because it provides an aesthetic benefit for which people are willing to pay (**FIGURE 3.24**). Similarly, scientific funding agencies may award grants to scientists for research that explores biodiversity with no promise of any economic

**FIGURE 3.22**  **Support systems.** Pollinators such as this honeybee play an essential role in ensuring the pollination of food crops such as cherries.

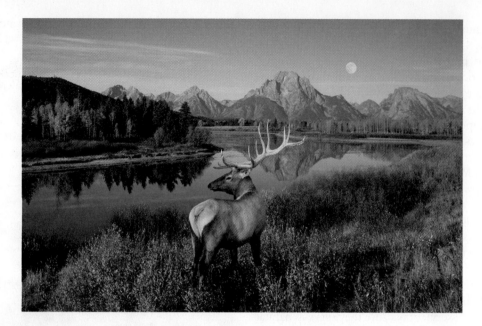

**FIGURE 3.24** **Cultural services.** Many natural areas, such as this scene from the Grand Tetons National Park, provide aesthetic beauty valued by humans.

gain. Nevertheless, the research itself has instrumental value because the scientists and others benefit from the experience by gaining knowledge. While intellectual gain and aesthetic satisfaction may be difficult to quantify, they can be considered cultural services that have instrumental value.

## Intrinsic Values of Ecosystems

Many people believe that ecosystems not only have instrumental value, but also have intrinsic value—that is, that they are valuable independent of *any* benefit to humans. These beliefs may grow out of religious or philosophical convictions. People who believe that ecosystems are inherently valuable may argue that we have a moral obligation to preserve them. They may equate the obligation of protecting ecosystems with our responsibility toward people or animals who might need our help to survive. People who argue that ecosystems have intrinsic value do not necessarily deny that ecosystems also have instrumental value. Rather, they believe that environmental policy and the protection of ecosystems should be driven by this intrinsic value.

> **GAUGE YOUR PROGRESS**
>
> ✓ What factors go into calculating the instrumental value of an ecosystem?
>
> ✓ What are the five categories of ecosystem services?
>
> ✓ How do the instrumental and intrinsic values of ecosystems differ?

 **WORKING TOWARD SUSTAINABILITY**

For a game that is played outdoors on open green courses designed around the contours of the natural landscape, golf has a surprisingly bleak environmental reputation. Golf courses are highly managed ecosystems that cover over 3 million hectares (7.5 million acres) worldwide—an area about the size of Belgium. About two-thirds of this area consists of closely mowed turfgrass. Closely mowed grass has short leaves that cannot gain enough energy from photosynthesis to grow deep roots. This causes a

## Can We Make Golf Greens Greener?

host of problems: The grass dries out easily and has difficulty obtaining soil nutrients. As a result, the grass is susceptible to challenges from weeds, grubs that feed on grass roots, and fungal diseases that can weaken or kill the grass. Collectively, these are rather formidable challenges that are faced by golf course managers worldwide.

To combat these challenges, golf courses use a disproportionate amount of water, fertilizer, and pesticides. Because humans expect to see green, well-manicured golf courses no matter where in the world they are located, golf courses collectively use 9.5 billion liters (2.5 billion gallons) of water annually to keep their grasses green. Much of this water is used in regions where water is already scarce. In addition, providing the grass with sufficient nutrients requires a large amount of fertilizer. Putting greens require as much nitrogen per hectare as corn, the heaviest nitrogen user of all major food crops. If the course requires irrigation soon after the application of fertilizer, or if it rains, up to 60 percent of the fertilizer can be leached into nearby waterways. To maintain a uniform texture on the greens, golf courses use about six times the amount of agricultural pesticides per hectare as do conventional farms. These chemicals include herbicides to remove weeds, insecticides to kill soil-dwelling grubs, and fungicides to control disease.

Since 1991, the Audubon Cooperative Sanctuary Program (ACSP), a partnership between Audubon International and the U.S. Golf Association, has been working to improve the environmental management of golf courses. The ACSP encourages golf course managers to develop courses that perform more like natural ecosystems, with nutrient and water recycling to reduce waste and biodiversity to increase ecosystem resilience. It also educates golf course managers about low-impact pest management, water conservation, and water quality management.

The golf course of the Palisades Country Club in North Carolina was constructed with natural ecosystem services in mind (FIGURE 3.25). To prevent the runoff of nutrients and lawn-care chemicals into nearby waters, the course directs all runoff water through a treatment system. The course was designed to reduce the amount of closely mowed turfgrass. Deep-rooted native grasses surrounding the greens and fairways soak up nutrients and help to direct water underground. As a result, maintenance costs, chemical applications, and time spent using machinery have all declined. Smaller areas of turfgrass also leave space for more native vegetation of various heights, providing better habitats for birds and predatory insects. These consumers keep pest populations low, reducing the need for pesticides. When pesticides are used, they are chosen to protect nontarget wildlife and applied on wind-free days to keep them from spreading beyond where they are needed.

By 2008, more than 2,100 golf courses worldwide had participated in the ACSP. Audubon International found that over 80 percent of the courses in the program reduced the amounts and toxicity of pesticides applied, improved nutrient retention within the course, and used less water for irrigation. The average course in the program saved about 7 million liters (1.9 million gallons) of water per year, and the amount of land area devoted to providing wildlife habitat increased by about 50 percent, from 18 to 27 ha (45 to 67 acres) per 60 ha (150-acre) golf course. Moreover, 99 percent of managers reported that playing quality and golfer satisfaction were maintained or improved.

Even with these changes, golf courses still require large amounts of water, nutrients, fossil fuel energy, and upkeep. Highly managed ecosystems cannot be made input free. However, within these limits, a growing number of courses are attempting to reduce their ecological footprint and make their greens greener.

### References

Audubon Cooperative Sanctuary Program. http://acspgolf.auduboninternational.org/.

Beecham, Tara. 2007. How green is your tee? *Stormwater Features.* http://www.stormh2o.com/september-2007/audubon-program-golf.aspx.

**FIGURE 3.25 Making golf more sustainable.** The Palisades Country Club in North Carolina is making its golf course more environmentally friendly by considering the important roles of natural ecosystem processes.

■ **List the basic components of an ecosystem.**

An ecosystem has both biotic and abiotic components, all of which interact with one another. An ecosystem has characteristic species as well as specific abiotic characteristics such as amount of sunlight, temperature, and salinity. Every ecosystem has boundaries, although they are often subjective.

■ **Describe how energy flows through ecosystems.**

The energy that flows through most ecosystems originates from the Sun. Ecosystems have multiple trophic levels through which energy flows. Producers use solar energy to generate biomass via photosynthesis. That stored energy can be passed on to consumers and decomposers and is ultimately lost as heat. The low efficiency of energy transfer between trophic levels means that only a small fraction of the energy at any trophic level—about 10 percent—is available to be used at the next higher trophic level. Low ecological efficiency results in a large biomass of producers, but a much lower biomass of primary consumers, and an even lower biomass of secondary consumers.

■ **Describe how carbon, nitrogen, and phosphorus cycle within ecosystems.**

In the carbon cycle, producers take up $CO_2$ for photosynthesis and transfer the carbon to consumers and decomposers. Some of this carbon is converted back into $CO_2$ by respiration, while the rest is lost to sedimentation and burial. The extraction and combustion of fossil fuels, as well as the destruction of forests, returns $CO_2$ to the atmosphere. The nitrogen cycle has many steps. Nitrogen is fixed by organisms, lightning, or human activities, then assimilated by organisms. Ammonium is released during decomposition of dead organisms and wastes. Finally, denitrification returns nitrogen to the atmosphere. The phosphorus cycle involves a large pool of phosphorus in rock that can be made available to organisms either by leaching or by mining. Organisms then assimilate it and ultimately transfer it back to the soil via excretion and decomposition.

■ **Explain how ecosystems respond to natural and anthropogenic disturbances.**

Different ecosystems respond to disturbances (both natural and anthropogenic) in different ways. A resistant ecosystem is one that experiences little change in flows of energy and matter after a disturbance. A resilient ecosystem is one that returns rapidly to its original state after a disturbance. Species diversity tends to be highest at intermediate levels of disturbance.

■ **Discuss the values of ecosystems and how humans depend on them.**

Ecosystems have a variety of instrumental values that can directly benefit humans. Ecosystem services include provisions that humans can use directly, such as food and medicine; regulation services that prevent drastic changes in environmental conditions; support systems that provide important services such as pollination and water filtration; resilience that allows ecosystems to continue functioning despite disturbances; and cultural services, including aesthetic value. The intrinsic value of ecosystems derives from the philosophical or religious idea that ecosystems are inherently valuable and that we have a moral obligation to preserve them.

## CHECK YOUR UNDERSTANDING

1. Which of the following is *not* an example of an abiotic component of an ecosystem?
   (a) Water
   (b) Minerals
   (c) Sunlight
   (d) Fungi
   (e) Air

2. Which of the following is *not* characteristic of ecosystems?
   (a) Biotic components
   (b) Abiotic components
   (c) Recycling of matter
   (d) Distinct boundaries
   (e) A wide range of sizes

3. Which biogeochemical cycle(s) does *not* have a gaseous component?
   - I Potassium
   - II Sulfur
   - III Phosphorus
   - (a) II only
   - (b) I and II only
   - (c) III only
   - (d) II and III only
   - (e) I and III only

*For questions 4, 5, and 6, select from the following choices:*
- (a) Producers
- (b) Decomposers
- (c) Primary consumers
- (d) Secondary consumers
- (e) Tertiary consumers

4. At which trophic level are eagles that consume fish that eat algae?

5. At which trophic level do organisms use a process that produces oxygen as a waste product?

6. At which trophic level are dragonflies that consume mosquitoes that feed on herbivorous mammals?

7. Beginning at the lowest trophic level, arrange the following food chain found on the Serengeti Plain of Africa in the correct sequence.
   - (a) Shrubs–gazelles–cheetahs–decomposers
   - (b) Shrubs–decomposers–gazelles–cheetahs
   - (c) Shrubs–decomposers–cheetahs–gazelles
   - (d) Gazelles–decomposers–cheetahs–shrubs
   - (e) Decomposers–cheetahs–shrubs–gazelles

8. Which macronutrient is required by humans in the largest amounts?
   - (a) Calcium
   - (b) Nitrogen
   - (c) Sulfur
   - (d) Potassium
   - (e) Magnesium

9. Roughly what percentage of incoming solar energy is converted into chemical energy by producers?
   - (a) 99
   - (b) 80
   - (c) 50
   - (d) Between 5 and 20
   - (e) 1

10. The net primary productivity of an ecosystem is 1 kg C/m$^2$/year, and the energy needed by the producers for their own respiration is 1.5 kg C/m$^2$/year. The gross primary productivity of such an ecosystem would be
    - (a) 0.5 kg C/m$^2$/year.
    - (b) 1.0 kg C/m$^2$/year.
    - (c) 1.5 kg C/m$^2$/year.
    - (d) 2.0 kg C/m$^2$/year.
    - (e) 2.5 kg C/m$^2$/year.

11. An ecosystem has an ecological efficiency of 10 percent. If the producer level contains 10,000 kilocalories of energy, how much energy does the tertiary consumer level contain?
    - (a) 1 kcal
    - (b) 10 kcal
    - (c) 100 kcal
    - (d) 1000 kcal
    - (e) 10,000 kcal

12. Research at Hubbard Brook showed that stream nitrate concentrations in two watersheds were _____ before clear-cutting, and that after one watershed was clear-cut, its stream nitrate concentration was _____.
    - (a) similar/decreased
    - (b) similar/increased
    - (c) similar/the same
    - (d) different/increased
    - (e) different/decreased

13. Small inputs of this substance, commonly a limiting factor in aquatic ecosystems, can result in algal blooms and dead zones.
    - (a) Dissolved carbon dioxide
    - (b) Sulfur
    - (c) Dissolved oxygen
    - (d) Potassium
    - (e) Phosphorus

14. The anticancer drug Taxol was originally extracted from the bark of the Pacific yew tree. This drug is an example of a type of ecosystem service known as
    - (a) cultural services.
    - (b) support systems.
    - (c) provisions.
    - (d) resilience.
    - (e) regulating services.

15. After a severe drought, the productivity in an ecosystem took many years to return to pre-drought conditions. This observation indicates that the ecosystem has
    - (a) high resilience.
    - (b) low resilience.
    - (c) high resistance.
    - (d) low resistance.
    - (e) equal resilience and resistance.

1. Nitrogen is crucial for sustaining life in both terrestrial and aquatic ecosystems.

   (a) Draw a fully labeled diagram of the nitrogen cycle.

   (b) Describe the following steps in the nitrogen cycle:
       (i) Nitrogen fixation
       (ii) Ammonification
       (iii) Nitrification
       (iv) Denitrification

   (c) Describe one reason why nitrogen is crucial for sustaining life on Earth.

   (d) Describe one way that the nitrogen cycle can be disrupted by human activities.

2. Read the following article written for a local newspaper and answer the questions below.

### Neighbors Voice Opposition to Proposed Clear-Cut

A heated discussion took place last night at the monthly meeting of the Fremont Zoning Board. Local landowner Julia Taylor has filed a request that her 150-acre woodland area be rezoned from residential to multi-use in order to allow her to remove all of the timber from the site. "This is my land, and I should be able to use it as I see fit," explained Ms. Taylor. "In due course, all of the trees will return and everything will go back to the same as it is now. The birds and the squirrels will still be there in the future. I have to sell the timber because I need the extra revenue to supplement my retirement as I am on a fixed income. I don't see what all the fuss is about," she commented.

A group of owners of adjacent properties see things very differently. Their spokesperson, Ethan Jared, argued against granting a change in the current zoning. "Ms. Taylor has allowed the community to use these woods for many years, and we thank her for that. But I hope that the local children will be able to hike and explore the woods with their children as I have done with mine. Removing the trees in a clear-cut will damage our community in many ways, and it could lead to contamination of the groundwater and streams and affect many animal and plant species. Like the rest of us property owners, Ms. Taylor gets her drinking water from a well, and I do not think she has really looked at all the ramifications should her plan go through. We strongly oppose the rezoning of this land—it has a right to be left untouched."

After more than two hours of debate between Ms. Taylor and many of the local residents, the chair of the Zoning Board decided to research the points raised by the neighbors and report on his findings at next month's meeting.

   (a) Name and describe the ecosystem value(s) that are being expressed by Ms. Taylor in her proposal to clear-cut the wooded area.

   (b) Name and describe the ecosystem value(s) that Mr. Jared is placing on the wooded area.

   (c) Provide three realistic suggestions for Ms. Taylor that could provide her with revenue from the property but leave the woods intact.

   (d) Identify and then discuss the validity of the environmental concerns that were raised by Mr. Jared.

## MEASURE YOUR IMPACT

### Atmospheric Carbon Dioxide

   (a) Describe two anthropogenic influences on the carbon cycle that have resulted in the elevation of atmospheric $CO_2$ concentrations.

   (b) Use one of the following carbon calculator Web sites to determine your household carbon emissions. (You may wish to investigate additional Web sites for comparison purposes.)

www.safeclimate.net/calculator/

www.myfootprint.org

   (c) Comment on your calculated carbon footprint estimate. How does your carbon footprint compare with the United States average?

How does your campus rate as wildlife habitat? Explore your campus while noting the food, water, shelter, and space necessary for maintaining insects, birds, and other small animals. In this Engage Your Environment activity you will determine ways in which your campus could better support animal life. *See Engage Your Environment Activity 3 at the end of this book.*

# Global Climates and Biomes: Geographic Variations in Temperature and Precipitation

## Floods, Droughts, and Famines

In April 2003, heavy rains began to fall along the equator in western Kenya, flooding villages, roads, and bridges. More than 70 people died, and hundreds of thousands more had to abandon their homes. In Nairobi, the capital of Kenya, mudslides damaged water pipes that led from a reservoir to a water treatment plant, and more than a million people were affected by the resulting disruption of water supplies. In many parts of the country, displaced flood victims had no clothing, food, or drinkable water. The danger of waterborne diseases such as cholera

What caused western Kenya to suffer unusually heavy rains and floods while northeastern Kenya was hit by a devastating drought? In other words, why didn't rainfall in Kenya occur in predictable patterns during this period?

Global processes drive rainfall patterns. Rainfall in the tropics is closely tied to the seasonal position of the Sun: wherever the Sun strikes Earth most directly, water evaporates and forms clouds that drop heavy rains. The Sun passes directly over the equator around March 21 on its way north and again around September 22 on its way south. Because

> **The Kenyan government confirmed that over 2.5 million people were at risk of starvation and declared the situation a national disaster.**

increased throughout the region. Normally, rainfall in this region of East Africa ends by June and is followed by 3 months of relative dryness. In 2003, however, the rains continued throughout June, July, and August. Widespread flooding gave the exhausted inhabitants neither relief nor the opportunity to begin rebuilding their lives.

That same year, hundreds of kilometers to the northeast, nomadic tribes living in northern Kenya anxiously awaited the summer rains that normally end their dry season. But the rains failed to arrive. Indeed, during the next 3 years, their region experienced a prolonged drought. Crops withered and died, and many people walked 2 hours a day to fetch water—until even those wells were dry. Local newspapers attributed 50 human deaths to the drought, but officials thought the actual number was much higher. The Kenyan government confirmed that over 2.5 million people were at risk of starvation and declared the situation a national disaster. Children were most vulnerable. Poor nutrition and the lack of safe drinking water made them susceptible to chronic diarrhea and dehydration. By the time the rains finally arrived in 2006, decomposing carcasses of cattle, sheep, and goats littered the landscape and threatened the quality of the drinking water.

Kenya is located at the equator, there is typically heavy rainfall during April and May. As the rain clouds continue to move north with the Sun, they drop additional rain on northeastern Kenya. But in April 2003, for reasons that are not fully understood, the rain clouds did not move northward from the equator. Instead, precipitation persisted near the equator, causing severe flooding in western Kenya and a drought in northeastern Kenya. In places such as Kenya that do not have sophisticated water management systems, unanticipated climatic events can have especially devastating consequences. ■

*Sources:* H. Kadomura, Climate anomalies and extreme events in Africa in 2003, including heavy rains and floods that occurred during Northern Hemisphere summer, *African Study Monographs Supplement* 30 (2005): 165–181; A. McFerran, Famine in Kenya: The rains have finally come, but for many it's already too late, *Independent* (London), May 20, 2006.

A Kenyan man walks through a cornfield that has been killed by drought.

◄ Storm clouds gather over an acacia tree on the savanna at Kenya's Masai Mara Game Reserve.

Earth is characterized by patterns of temperature and precipitation. These patterns arise from the circulation of air and ocean water, which is ultimately driven by unequal heating of Earth by the Sun, the rotation of Earth, and Earth's geographic features. Geographic variations in temperature and precipitation have led to the development of distinct terrestrial biomes, which are defined by their unique plant communities, and distinct aquatic biomes, which are defined by their particular physical conditions.

After reading this chapter you should be able to

- explain the forces that drive global circulation patterns and how those patterns determine weather and climate.

- describe the major terrestrial biomes.

- describe the major aquatic biomes.

# Global processes determine weather and climate

An afternoon thunderstorm or a few dry, sunny days—when we talk about weather, we are referring to the short-term conditions of the atmosphere in a local area. These conditions include temperature, humidity, clouds, precipitation, wind speed, and atmospheric pressure.

Weather happens on time scales from seconds to days. **Climate,** on the other hand, is the average weather that occurs in a given region over a long period—typically over several decades. It is not possible to predict weather more than a few days into the future, but we can make general observations about global, regional, and even local climate. For example, areas close to the North and South Poles are considerably colder than those closer to the equator. Similarly, areas at high elevations are colder than nearby sites at lower elevations. Areas near the equator typically receive tremendous amounts of precipitation, whereas regions at latitudes around 30° N and 30° S typically receive very little precipitation.

Regional differences in temperature and precipitation collectively help determine which organisms can survive in each region. To understand these differences, we need to look at the processes that affect the distribution of heat and precipitation across the globe. These processes include *unequal heating of Earth by the Sun, atmospheric convection currents, the rotation of Earth, Earth's orbit around the Sun on a tilted axis,* and *ocean currents.* Before we look at them in detail, however, we need to know something about Earth's atmosphere, where several of these processes take place.

## Earth's Atmosphere

As **FIGURE 4.1** shows, Earth's atmosphere consists of five layers of gases. The pull of gravity on the gas molecules keeps these layers of gases in place. Because gravitational pull weakens as we move farther away from Earth, molecules are more densely packed closer to Earth and less densely packed farther from Earth.

The layer closest to Earth's surface is the **troposphere.** The troposphere extends roughly 16 km (10 miles) above Earth. It is the densest layer of the atmosphere: most of the atmosphere's nitrogen, oxygen, and water vapor occur in this layer. The troposphere is characterized by a great deal of circulation and mixing of liquids and gases, and it is the layer where Earth's weather occurs. Air temperature in the troposphere decreases with distance from Earth's surface and varies with latitude. Temperatures can fall as low as −52°C (−62°F) near the top of the troposphere.

Above the troposphere is the **stratosphere,** which extends roughly 16 to 50 km (10 to 31 miles) above Earth's surface. Because of its greater distance from Earth's gravitational pull, the stratosphere is less dense than the troposphere. In the stratosphere, because UV (ultraviolet) radiation reaches the higher altitudes first and warms them, the higher altitudes are warmer than the lower altitudes. Ozone, a pale blue gas composed of molecules made up of three oxygen atoms ($O_3$), forms a layer within the stratosphere. This ozone layer absorbs most of the Sun's ultraviolet-B (UV-B) radiation and all of its ultraviolet-C (UV-C) radiation. UV radiation can cause DNA damage and cancer in organisms, so the stratospheric ozone layer provides critical protection for our planet. Science Applied 1, "Were We Successful in Halting the Growth of the Ozone Hole?" discusses this topic in more depth.

Beyond the stratosphere are the *mesosphere,* the *thermosphere,* and farthest from Earth, the *exosphere.* Given the weaker gravitational pull on molecules at these greater distances from Earth, the atmospheric pressure and density in each of these layers decreases as it extends out into space. The thermosphere is particularly important to organisms on Earth's surface because of its ability to block harmful X-ray and UV radiation. The thermosphere is also interesting because it contains charged gas molecules that, when hit by solar energy, begin to glow and produce light, in the same way that a light bulb glows when electricity is applied. Because this interaction between solar energy and gas molecules is driven most intensely by magnetic forces at the North Pole and

Exosphere

Thermosphere

Temperature

Mesosphere

Stratosphere

Peak ozone layer

Troposphere

Altitude (km)

600

500

400

300

200

120
100
80
60
40
20
0

Temperature (°C)

−100    −60    0 20 200    1,750

**FIGURE 4.1** The layers of Earth's atmosphere. The troposphere is the atmospheric layer closest to Earth. Because the density of air decreases with altitude, the troposphere's temperature also decreases with altitude. Temperature increases with altitude in the stratosphere because the Sun's UV-B and UV-C rays warm the upper part of this layer. Temperatures in the thermosphere can reach 1,750°C (3,182°F). [After http://www.nasa.gov/audience/forstudents/9-12/features/912_liftoff_atm.html.]

South Pole, the best places to view the phenomenon are at high latitudes. In the northern United States, Canada, and northern Europe, these glowing gases are known as the *northern lights,* or *aurora borealis.* In Australia and southern South America, they are called the *southern lights,* or *aurora australis* (FIGURE 4.2).

## Unequal Heating of Earth

Now that we know something about Earth's atmosphere, we can take a closer look at the processes that affect heat and precipitation distribution. As the Sun's energy passes through the atmosphere and strikes land and water, it warms the surface of Earth. But this warming does not occur evenly across the planet. This uneven warming pattern has three primary causes.

The first cause is variation in the angle at which the Sun's rays strike Earth. As we can see in FIGURE 4.3, in the region nearest to the equator—the tropics—the Sun strikes at a perpendicular, or right, angle. In the mid-latitude and polar regions, the Sun's rays strike at a more oblique angle. As a result, the Sun's rays travel a shorter distance through the atmosphere to reach Earth's surface in the tropics. Because solar energy is lost as it passes through the atmosphere, more solar energy reaches the equator than the mid-latitude and polar regions.

**FIGURE 4.2** **Northern lights.** The glowing, moving lights that are visible at high latitudes in both hemispheres are the product of solar radiation energizing the gases of the thermosphere.

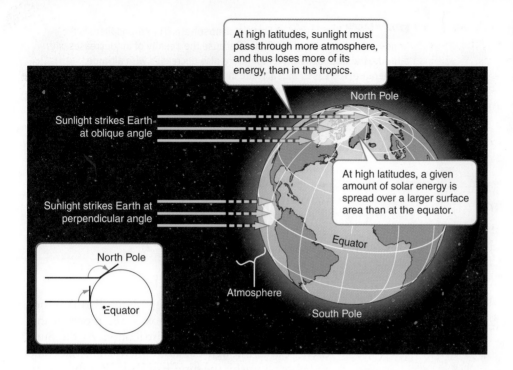

The second cause of the uneven warming of Earth is variation in the amount of surface area over which the Sun's rays are distributed. The perpendicular angle of the Sun's rays in the tropics causes solar energy to be distributed over a smaller surface area there than at higher latitudes. Thus tropical regions receive more solar energy per square meter than mid-latitude and polar regions. You can replicate this phenomenon by shining a flashlight onto a round object, such as a basketball, in a dark room. If you shine the light perpendicular to the surface of the ball, you will create a small circle of bright light. If you shine the flashlight at an oblique angle, you will create an oval pool of dimmer light because the light is distributed over a larger area.

Finally, some areas of Earth reflect more solar energy than others. The percentage of incoming sunlight that

FIGURE 4.4 **Albedo.** The albedo of a surface is the percentage of the incoming solar energy that it reflects.

is reflected from a surface is called its **albedo.** The higher the albedo of a surface, the more solar energy it reflects, and the less it absorbs. A white surface has a higher albedo than a black surface, so it tends to stay cooler. FIGURE 4.4 shows albedo values for various surfaces on Earth. Although Earth has an average albedo of 30 percent, tropical regions with dense green foliage have albedo values of 10 to 20 percent, whereas the snow-covered polar regions have values of 80 to 95 percent.

## Atmospheric Convection Currents

Now that we have explored the reasons for the uneven heating of Earth, we will see how that uneven heating drives the circulation of air in the atmosphere. First, however, we will examine the properties of air that influence its movement.

**PROPERTIES OF AIR** Air has four properties that determine how it circulates in the atmosphere: *density, water vapor capacity, adiabatic heating or cooling,* and *latent heat release.*

The density of air determines its movement: less dense air rises, whereas denser air sinks. At a constant atmospheric pressure, warm air has a lower density than cold air. Because of this density difference, warm air rises, whether in a room in your house or in the atmosphere.

Warm air also has a higher capacity for water vapor than cold air. That is why, in many parts of the world, including North America, hot summer days are associated with high humidity: the warm air contains a lot of water vapor. The maximum amount of water vapor that can be in the air at a given temperature is called its **saturation point.** FIGURE 4.5 shows the relationship between the temperature of air and its saturation point. When the temperature of air falls, its saturation point decreases, water vapor condenses into liquid water, clouds form, and precipitation occurs.

A third important property of air is its response to changes in pressure. As air rises higher in the atmosphere, the pressure on it decreases. The lower pressure allows the rising air to expand in volume, and this expansion lowers the temperature of the air. This process is called **adiabatic cooling.** Conversely, when air sinks toward Earth's surface, the pressure on it increases. The higher pressure forces the air to decrease in volume, and this decrease raises the temperature of the air. This process is called **adiabatic heating.**

The final important property of air is the production of heat when water vapor condenses from a gas to a liquid. As you may know, the Sun provides the energy necessary to evaporate water on Earth's surface and convert it into water vapor, which enters the atmosphere. In the reverse process, when water vapor in the atmosphere condenses into liquid water, energy is released. This phenomenon is known as **latent heat release.** It is important because it means that whenever water

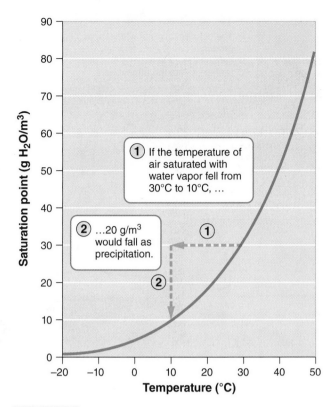

**FIGURE 4.5** **The saturation point of air increases with temperature.** When air cools and its saturation point drops, water vapor condenses into liquid water that forms clouds. These clouds are ultimately the source of precipitation.

vapor in the atmosphere condenses, the air will become warmer, and this warm air will rise.

**FORMATION OF CONVECTION CURRENTS** Now that we understand the four properties of air, we can look at the role of these properties in atmospheric circulation. We can follow the process in FIGURE 4.6, beginning in the tropics.

Atmospheric convection currents are global patterns of air movement that are initiated by the unequal heating of Earth. In the tropics, the warming of humid air at Earth's surface decreases its density. As a result, the air begins to rise. When the air rises, it begins to experience lower atmospheric pressures and adiabatic cooling. The cooling causes the air to reach its saturation point. This leads to condensation, which causes cloud formation and precipitation. Condensation also causes latent heat release, which offsets some of the adiabatic cooling and becomes a strong driving force to make the air expand further and rise more rapidly through the troposphere. Collectively, these processes cause air to rise continuously from Earth's surface near the equator, forming a river of air flowing upward into the troposphere.

Air near the top of the troposphere is chilled by adiabatic cooling. This air contains relatively little water vapor. As warmer air rises from below, this cold, dry air is displaced horizontally both north and south of the equator. This displaced air eventually begins to sink

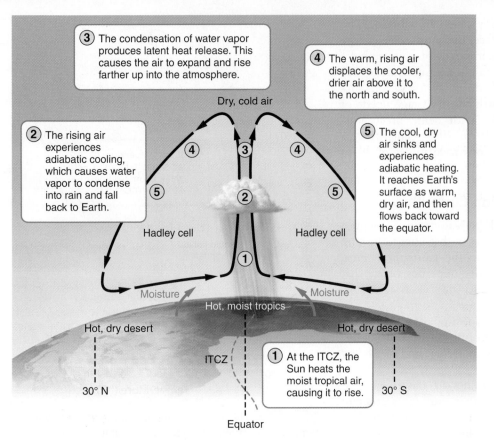

**FIGURE 4.6** **The formation of Hadley cells.** Solar energy warms humid air in the tropics. The warm air rises and eventually cools below its saturation point. The water vapor it contains condenses into clouds and precipitation. The now dry air sinks to Earth's surface at approximately 30° N and S. As the air descends, it is warmed by adiabatic heating. This descent of hot, dry air causes desert environments to develop at those latitudes.

③ The condensation of water vapor produces latent heat release. This causes the air to expand and rise farther up into the atmosphere.

④ The warm, rising air displaces the cooler, drier air above it to the north and south.

② The rising air experiences adiabatic cooling, which causes water vapor to condense into rain and fall back to Earth.

⑤ The cool, dry air sinks and experiences adiabatic heating. It reaches Earth's surface as warm, dry air, and then flows back toward the equator.

Dry, cold air

Hadley cell

Hadley cell

Moisture

Moisture

Hot, moist tropics

Hot, dry desert

Hot, dry desert

① At the ITCZ, the Sun heats the moist tropical air, causing it to rise.

ITCZ

30° N

30° S

Equator

back to Earth's surface at approximately 30° N and S. As the air sinks, it experiences higher atmospheric pressures, and its reduction in volume causes adiabatic heating. By the time the air reaches Earth's surface, it is hot and dry. As a result, regions at 30° N and S are typically hot, dry deserts.

Much of this desert air then moves along Earth's surface toward the equator to replace the air that is rising there, completing the cycle. The convection currents that cycle between the equator and 30° N and S in this way are called **Hadley cells.**

The area of Earth that receives the most intense sunlight, where the ascending branches of the two Hadley cells converge, is called the **intertropical convergence zone** (abbreviated **ITCZ**). It is typified by dense clouds and intense thunderstorm activity. The latitude along which the ITCZ is located is not fixed. Instead, over the course of a year, it moves north and south of the equator, following the path of the Sun's most direct rays. Because Earth's axis of rotation is tilted, the area receiving the most intense sunlight shifts between approximately 23.5° N and 23.5° S as Earth orbits the Sun. As a result, the tropics experience seasonal patterns of precipitation.

Similar to the Hadley cells are the **polar cells.** These convection currents are formed by air that rises at 60° N and S and sinks at the poles (90° N and S). At 60° N and S, the rising air cools, and the water vapor condenses into precipitation. The air dries as it moves toward the poles, where it sinks back to Earth's surface.

At the poles, the air moves back toward 60° N and S, completing the cycle.

Between the Hadley cells and the polar cells lies a third area of air circulation. Air circulation at these latitudes does not form distinct convection cells, but is instead driven by the circulation of the neighboring Hadley cells and polar cells. At Earth's surface, some of the warmer air from the Hadley cells moves toward the poles from 30° N and S, and some of the cooler air from the polar cells moves toward the equator from 60° N and S. This movement not only helps to distribute warm air away from the tropics and cold air away from the poles, but also allows a wide range of warm and cold air currents to circulate between 30° and 60°. In this latitudinal range, which includes most of the United States, wind direction can be quite variable, both at Earth's surface and at the top of the troposphere.

Collectively, these convection currents slowly move the warm air of the tropics toward the mid-latitude and polar regions. This pattern of air circulation is largely responsible for the locations of rainforests, deserts, and grasslands on Earth.

## Earth's Rotation and the Coriolis Effect

So far, we have considered the convection currents that move air around Earth without taking into account the rotation of Earth on its axis. Earth's rotation has an important influence on climate, particularly on the directions of prevailing winds. As Earth rotates, its surface moves much faster at the equator than in mid-latitude and polar regions. This disparity occurs because the planet's circumference is 40,000 km (25,000 miles) at the equator, but decreases to zero at the poles. Imagine traveling around Earth at the equator versus at the poles. FIGURE 4.7 will help you visualize this journey. At the equator, you would have to travel much farther to go all the way around Earth than you would if you were near one of the poles. Now imagine yourself standing still as Earth rotates. Given that a single rotation is completed in 24 hours, you would be traveling much faster standing at the equator than standing near one of the poles. Indeed, Earth's surface moves at 1,670 km (1,038 miles) per hour at the equator, 1,445 km (898 miles) per hour at 30° N or S, and 291 km (181 miles) per hour at 80° N or S.

The faster rotation speeds closer to the equator cause a deflection of objects that are moving directly north or south. Look at the left side of FIGURE 4.8 and imagine that you can stand at the North Pole and throw a ball directly south, all the way down to the equator (0° latitude). If Earth did not rotate, the ball would travel south to the equator in a straight line. But because Earth rotates to the east while the ball is traveling, the ball will land in a location that is west of its intended target. This happens because Earth's surface beneath the ball is moving faster and faster as the ball moves toward the equator. In other words, the path of the ball is deflected with respect to a given location on the globe because Earth is rotating. This deflection of an object's path due to Earth's rotation is called the **Coriolis** (core-ee-OH-lis) **effect.**

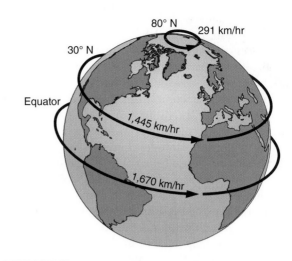

**FIGURE 4.7** **The speed of Earth's rotation varies with latitude.** Because all locations on Earth complete one revolution every 24 hours, and because Earth has a greater circumference near the equator than near the poles, its speed of rotation is much faster at the equator than near the poles.

The prevailing wind systems of the world are produced by a combination of atmospheric convection currents and the Coriolis effect. If Earth did not rotate, the air within each convection cell would simply move directly north or south and cycle back again. Consider, for example, a Hadley cell. Where the air sinks to Earth's surface at 30° latitude, Earth is rotating at 1,445 km per hour. However, as the air travels along Earth's surface toward the equator, Earth's speed of rotation increases to 1,670 km per hour. As a result, the air movement

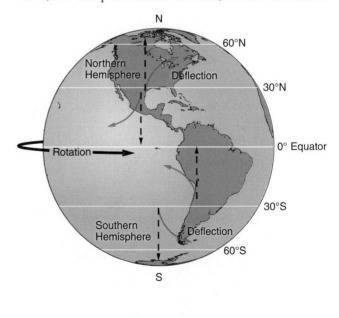

**FIGURE 4.8** **The Coriolis effect.** (a) A ball thrown from the North Pole toward the equator would be deflected to the west by the Coriolis effect. (b) The different rotation speeds of Earth at different latitudes cause a deflection in the paths of traveling objects.

toward the equator is deflected to the west. Because of this deflection, the Hadley cell north of the equator produces prevailing winds along Earth's surface that come from the northeast (called northeast trade winds), whereas the cell south of the equator produces prevailing winds that come from the southeast (called southeast trade winds) (**FIGURE 4.9**).

The Coriolis effect also explains the prevailing wind directions in the mid-latitudes (between 30° and 60°). These winds can be quite variable as a result of the mixing of air currents from the Hadley cells and the polar cells. Closer to 30°, air tends to move along Earth's surface *away* from the equator. If Earth were not rotating, this air would move straight north in the Northern Hemisphere and straight south in the Southern Hemisphere. Given that Earth is rotating faster at 30° than at 60°, this air movement is deflected to the east. The combined effect of the air currents and the Coriolis effect causes regions just north of 30° to experience prevailing winds from the southwest. In the Southern Hemisphere, the prevailing winds are from the northwest. In both cases, these winds are called *westerlies.*

Finally, the Coriolis effect helps us understand the prevailing wind directions in the polar regions. At Earth's surface, the polar cells move air away from the poles and toward 60° latitude. Given that Earth is rotating faster at 60° than at 90°, the air movement is deflected to the west. Because of this, polar winds come out of the northeast in the Northern Hemisphere and out of the southeast in the Southern Hemisphere. These winds are called *easterlies.*

Simply stated, the atmospheric convection currents of tropical and polar latitudes, the mixing of air currents in the mid-latitudes, and the Coriolis effect cause the prevailing wind patterns that occur worldwide, although local features, such as mountain ranges, can alter wind directions significantly.

## Earth's Tilt and the Seasons

As we saw in our discussion of the ITCZ, the latitude receiving the most direct sunlight shifts over the course of the year. Because Earth's axis of rotation is tilted 23.5°, Earth's orbit around the Sun causes most regions of the world to experience seasonal changes in temperature and precipitation. Specifically, when the Northern Hemisphere is tilted toward the Sun, the Southern Hemisphere is tilted away from the Sun, and vice versa.

**FIGURE 4.10** will help us visualize how this works. The Sun's rays strike the equator directly twice a year: once during the March equinox, on March 20 or 21, and again during the September equinox, on September 22 or 23. On those days, virtually all regions of Earth (except those nearest the poles) receive 12 hours of daylight and 12 hours of darkness. For the 6 months between the March and September equinoxes, the Northern Hemisphere tilts toward the Sun, experiencing more hours of daylight than darkness. The opposite is true in the Southern Hemisphere. On June 20 or 21, the Sun is directly above the Tropic of Cancer at 23.5° N latitude. On this day—the June solstice—the Northern Hemisphere experiences more daylight hours than on any other day of the year. For the 6 months between the September and March equinoxes, the Northern Hemisphere tilts away from the Sun, experiencing fewer hours of daylight than darkness. On December 21 or 22—the December solstice—the Sun is directly over the Tropic of Capricorn at 23.5° S latitude. On this day, the Northern Hemisphere experiences its shortest daylight period of the year, and the Southern Hemisphere experiences its longest daylight period of the year.

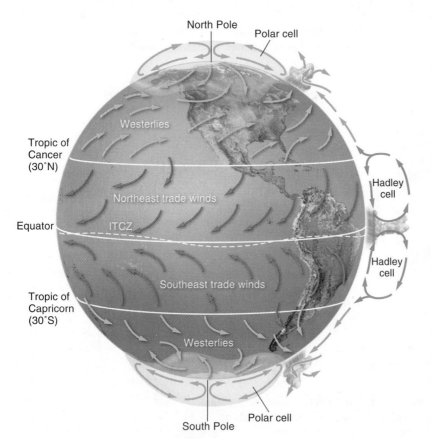

**FIGURE 4.9 Prevailing wind patterns.** Prevailing wind patterns around the world are produced by a combination of atmospheric convection currents and the Coriolis effect.

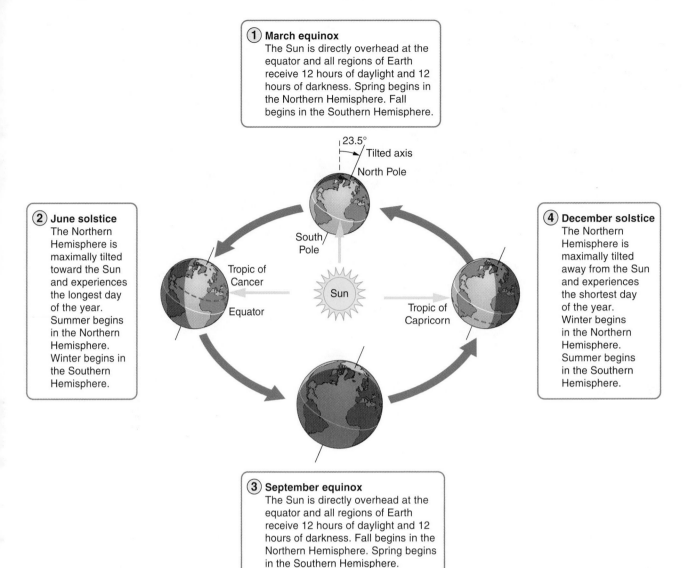

**(1) March equinox**
The Sun is directly overhead at the equator and all regions of Earth receive 12 hours of daylight and 12 hours of darkness. Spring begins in the Northern Hemisphere. Fall begins in the Southern Hemisphere.

23.5°
Tilted axis
North Pole

**(2) June solstice**
The Northern Hemisphere is maximally tilted toward the Sun and experiences the longest day of the year. Summer begins in the Northern Hemisphere. Winter begins in the Southern Hemisphere.

South Pole

Tropic of Cancer

Equator

Sun

Tropic of Capricorn

**(4) December solstice**
The Northern Hemisphere is maximally tilted away from the Sun and experiences the shortest day of the year. Winter begins in the Northern Hemisphere. Summer begins in the Southern Hemisphere.

**(3) September equinox**
The Sun is directly overhead at the equator and all regions of Earth receive 12 hours of daylight and 12 hours of darkness. Fall begins in the Northern Hemisphere. Spring begins in the Southern Hemisphere.

**FIGURE 4.10** **Earth's seasons.** Because Earth's axis of rotation is tilted 23.5°, the latitude that receives the most direct rays of the Sun and the most hours of daylight changes throughout the year as Earth orbits the Sun. Thus Earth's tilt produces predictable seasons. This diagram illustrates the pattern of seasons in the Northern Hemisphere.

## Ocean Currents

So far we have examined four processes that influence Earth's weather and climate: unequal heating of Earth, atmospheric convection currents, the rotation of Earth and the Coriolis effect, and Earth's orbit around the Sun on a tilted axis. The fifth global process that influences weather and climate is the circulation of ocean waters, both at the surface and in the deep ocean.

Ocean currents are driven by a combination of temperature, gravity, prevailing winds, the Coriolis effect, and the locations of continents. As we have already observed, the tropics receive the most direct sunlight throughout the year, and as a result, tropical waters are generally warm. Warm water, like warm air, expands and rises. This process raises the tropical water surface

about 8 cm (3 inches) higher in elevation than mid-latitude waters. While this difference might seem trivial, the slight slope is sufficient for the force of gravity to make water flow away from the equator.

**GYRES** Global prevailing wind patterns play a major role in determining the direction in which ocean surface water moves away from the equator. In the Northern Hemisphere, for example, the trade winds near the equator push water from the northeast to the southwest, and the Coriolis effect deflects this wind-driven current so that water actually moves from east to west. Similarly, when winds in northern mid-latitude regions push water from the southwest to the northeast, the Coriolis effect deflects this current so that water actually moves

**FIGURE 4.11** **Oceanic circulation patterns.** Oceanic circulation patterns are the result of differential heating, gravity, prevailing winds, the Coriolis effect, and the locations of continents. Each of the five major ocean basins contains a gyre driven by the trade winds in the tropics and the westerlies at mid-latitudes. The result is a clockwise circulation pattern in the Northern Hemisphere and a counterclockwise circulation pattern in the Southern Hemisphere. Along the west coasts of many continents, currents diverge and cause the upwelling of deeper and more fertile water.

from west to east. FIGURE 4.11 shows the overall effect: ocean surface currents rotate in a clockwise direction in the Northern Hemisphere and in a counterclockwise direction in the Southern Hemisphere. These large-scale patterns of water circulation are called **gyres.**

Gyres redistribute heat in the ocean, just as atmospheric convection currents redistribute heat in the atmosphere. Cold water from the polar regions moves along the west coasts of continents, and the transport of cool air from immediately above these waters causes cooler temperatures on land. For example, the California Current, which flows south from the North Pacific along the coast of California, causes coastal areas of California to have cooler temperatures than areas at similar latitudes on the east coast of the United States. Similarly, warm water from the tropics moves along the east coasts of continents, and the transport of warm air from immediately above these waters causes warmer temperatures on land.

**UPWELLING** Ocean currents also help explain why some regions of the ocean support highly productive ecosystems. Along the west coasts of most continents, for example, the surface currents diverge, or separate from one another, causing deeper waters to rise and replace the water that has moved away (see Figure 4.11). This upward movement of water toward the surface is called

**upwelling.** The deep waters bring with them nutrients from the ocean bottom that support large populations of producers. The producers, in turn, support large populations of fish that have long been important to commercial fisheries.

**THERMOHALINE CIRCULATION** Another oceanic circulation pattern, **thermohaline circulation,** drives the mixing of surface water and deep water. Scientists believe this process is crucial for moving heat and nutrients around the globe.

Thermohaline circulation appears to be driven by surface waters that contain unusually large amounts of salt. As Figure 4.11 shows, warm currents flow from the Gulf of Mexico to the very cold North Atlantic. Some of this water freezes or evaporates, and the salt that remains behind increases the salt concentration (salinity) of the water. This cold, salty water is relatively dense, so it sinks to the bottom of the ocean, mixing with deeper ocean waters. Two processes—the sinking of cold, salty water at high latitudes and the rising of warm water near the equator—create the movement necessary to drive a deep, cold current that slowly moves past Antarctica and northward to the northern Pacific Ocean, where it returns to the surface and then makes its way back to the Gulf of Mexico. This global round trip, traced in FIGURE 4.12, can take hundreds of

| ① Warm water flows from the Gulf of Mexico to the North Atlantic, where some of it freezes and evaporates. | ② The remaining water, now saltier and denser, sinks to the ocean bottom. | ③ The cold water travels along the ocean floor, connecting the world's oceans. | ④ The cold, deep water eventually rises to the surface and circulates back to the North Atlantic. |

**FIGURE 4.12** **Thermohaline circulation.** The sinking of dense, salty water in the North Atlantic drives a deep, cold current that moves slowly around the world.

years to complete. Thermohaline circulation helps to mix the water of all the oceans.

**HEAT TRANSPORT** Ocean currents can affect the temperature of nearby landmasses. For example, the ocean current known as the Gulf Stream originates in the tropics near the Gulf of Mexico and flows northeast across the Atlantic Ocean toward western Europe (see Figure 4.11). As it transports warm tropical waters northward, the Gulf Stream brings vast amounts of heat energy to cooler regions, moderating temperatures in latitudes that would otherwise be much colder. For instance, England's average winter temperature is approximately 20°C (36°F) warmer than that of Newfoundland, Canada, which is located at a similar latitude but receives cold ocean currents from the North Atlantic.

One of the present concerns about global warming is that increased air temperatures could accelerate the melting of glaciers in the Northern Hemisphere, which could make the waters of the North Atlantic less salty and thus less likely to sink. Such a change could potentially shut down thermohaline circulation and stop the transport of warm water to western Europe, making it a much colder place.

**EL NIÑO–SOUTHERN OSCILLATION** Earth's atmosphere and ocean interact in complex ways. Periodically, approximately every 3 to 7 years, these interactions cause surface currents in the tropical Pacific Ocean to reverse direction. **FIGURE 4.13** shows this process in action. First, the trade winds near South America weaken. This weakening allows warm equatorial water from the western Pacific to move eastward toward the west coast of South America. The movement of warm water and air toward South America suppresses upwelling off the coast of Peru and decreases productivity there, reducing fish populations near the coast. This phenomenon is called *El Niño* ("the baby boy") because it often begins around the December 25 Christmas holiday. El Niño can last from a few weeks to a few years. These periodic changes in winds and ocean currents are collectively called the **El Niño–Southern Oscillation,** or **ENSO.** Globally, the impact of ENSO includes cooler and wetter conditions in the southeastern United States

1. During most years, trade winds push surface water from east to west.

2. Deep water moves upward (upwelling) to replace surface water that has moved westward.

3. During El Niño years, trade winds weaken or reverse direction; warm surface water moves from west to east.

4. The warm surface water builds up along the coast of South America and prevents upwelling of the deep cold water.

**FIGURE 4.13** **The El Niño–Southern Oscillation.** (a) In a normal year, trade winds push warm surface waters away from the coast of South America and promote the upwelling of water from the ocean bottom. (b) In an El Niño year, the trade winds weaken or even reverse direction, allowing warm waters to build up along the west coast of Peru.

and unusually dry weather in southern Africa and Southeast Asia.

## Rain Shadows

Although many processes that affect weather and climate operate on a global scale, local features, such as mountain ranges, can also play a role. Air moving inland from the ocean often contains a large amount of water vapor. As shown in **FIGURE 4.14**, when this air meets the *windward* side of a mountain range—the side facing the wind—it rises and begins to experience adiabatic cooling. Because water vapor condenses as air cools, clouds form and precipitation falls. As is the case in Hadley cells, this condensation causes latent heat release, which helps to accelerate the upward movement of the air. Thus the presence of the mountain range causes large amounts of precipitation to fall on its windward side. The cold, dry air then travels to the other side of the mountain range—called the *leeward* side—where it descends and experiences higher pressures, which cause adiabatic heating. This now warm, dry air produces arid

**FIGURE 4.14** **Rain shadow.** Rain shadows occur where humid winds blowing inland from the ocean meet a mountain range. On the windward (wind-facing) side of the mountains, air rises and cools, and large amounts of water vapor condense to form clouds and precipitation. On the leeward side of the mountains, cold, dry air descends, warms via adiabatic heating, and causes much drier conditions.

conditions on the leeward side of the range, forming a region called a **rain shadow.** It is common to see lush vegetation on the windward side of a mountain range and very dry conditions on the leeward side.

In tropical areas, rain shadows tend to be on the western sides of mountain ranges because of the prevailing trade winds moving from east to west. In mid-latitude zones, such as North America, rain shadows are commonly on the eastern sides of mountain ranges because the prevailing westerlies move from west to east. We can readily see this effect in the Sierra Nevada range in the western United States: the western side of the range receives large amounts of precipitation and supports lush vegetation, but the eastern side—an area called the Great Basin—is quite dry.

---

**GAUGE YOUR PROGRESS**

✓ What is the difference between weather and climate?

✓ What effect does Earth's rotation have on atmospheric circulation and ocean currents?

✓ In what ways are atmospheric and oceanic circulation patterns similar? How are they different?

---

# Variations in climate determine the dominant plant growth forms of terrestrial biomes

Climate affects the distribution of species around the globe. The deserts of the American Southwest and the Kalahari Desert in Africa, for example, tend to have high temperatures and little precipitation, and only species that are well adapted to hot and dry conditions can survive there. A very different set of organisms survives in cold, snowy places.

Scientists have long recognized that organisms possess distinct growth forms, many of which represent adaptations to local temperature and precipitation patterns. For example, if we were to examine the plants from all the deserts of the world, we would find many cactuslike species. Plants that look like cacti in North American deserts are indeed members of the cactus family, but those in the Kalahari Desert are members of the euphorb family. These two distantly related groups of species look similar because they have evolved similar adaptations to hot, dry environments—including the ability to store large amounts of water in their tissues and a waxy coating to reduce water loss (**FIGURE 4.15**).

(a)

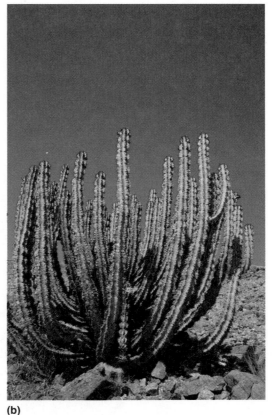
(b)

**FIGURE 4.15** **Cacti and euphorbs exhibit similar growth forms.** These two plant families are not closely related, but they have evolved many similar adaptations that allow them to live in hot, dry environments. (a) Organ pipe cactus (*Stenocereus thurberi*) in Arizona. (b) Euphorbia (*E. damarana*) in Namibia.

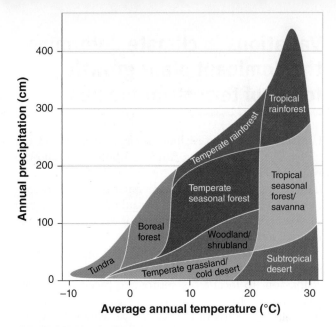

**FIGURE 4.16** **Biomes.** Biomes are categorized by particular combinations of average annual temperature and annual precipitation. [After R. H. Whitaker, *Communities and Ecosystems,* 1975. Modified from R. E. Ricklefs, *The Economy of Nature* (New York: W. H. Freeman, 2000.]

Despite these common adaptations to desert conditions, the two plant families have distinctive flowers and spines, and only the euphorb family produces a milky sap. These differences help to confirm that, while the two groups may look superficially similar, they are not closely related genetically. Likewise, mature tropical rainforests anywhere in the world have tall trees with buttressed roots for support, even though most of the tree species in the Asian tropics are only distantly related to those in the Americas.

The presence of similar plant growth forms in areas possessing similar temperature and precipitation patterns allows scientists to categorize terrestrial geographic regions known as *biomes.* **Biomes** have a particular combination of average annual temperature and annual precipitation and contain distinctive plant growth forms that are adapted to that climate. **FIGURE 4.16** shows the range of biomes on Earth in the context of precipitation and temperature. For example, boreal and tundra biomes have average annual temperatures below 5°C (41°F), whereas temperate biomes have average annual temperatures between 5°C and 20°C (68°F), and tropical biomes have average annual temperatures above 20°C. Within each of these temperature ranges, we can observe a wide range of precipitation. **FIGURE 4.17** is a map showing the distribution of biomes around the world.

Note that although terrestrial biomes are categorized by plant growth forms, the animal species living in different biomes are often quite distinctive as well. For example, rodents inhabiting deserts around the world have a number of adaptations for hot, dry climates, including highly efficient kidneys that allow very little water loss via urination.

Climate diagrams such as those shown in **FIGURE 4.18** can be a helpful way to visualize regional patterns of temperature and precipitation. By graphing the average monthly temperature and precipitation, these diagrams illustrate how the conditions in a biome vary during a typical year. They also indicate when the temperature is warm enough for plants to grow—that is, the months when it is above 0°C (32°F), known as the *growing season.* In Figure 4.18a, we can see that the growing season is mid-March through mid-October.

In addition to the growing season, climate diagrams can show the relationship between precipitation,

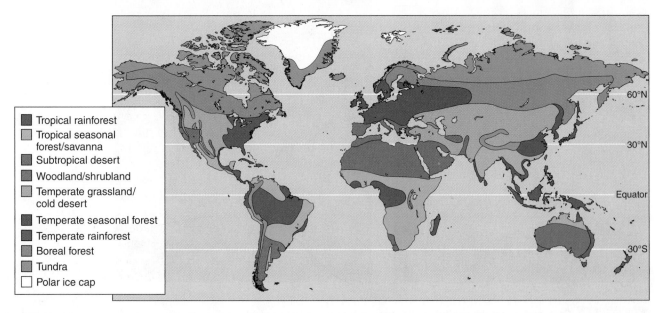

Tropical rainforest

Tropical seasonal forest/savanna

Subtropical desert

Woodland/shrubland

Temperate grassland/cold desert

Temperate seasonal forest

Temperate rainforest

Boreal forest

Tundra

Polar ice cap

**FIGURE 4.17** **Locations of the world's biomes.**

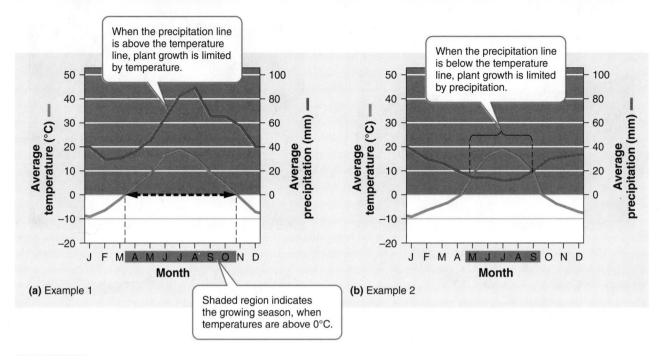

**FIGURE 4.18** **Climate diagrams.** Climate diagrams display monthly temperature and precipitation values, which help determine the productivity of a biome.

temperature, and plant growth. For every 10°C (18°F) temperature increase, plants need 20 mm (0.8 inches) of additional precipitation each month to supply the extra water demand that warmer temperatures cause. As a result, plant growth can be limited either by temperature or by precipitation. In Figure 4.18a, the precipitation line is above the temperature line during all months. This means that water supply exceeds demand, so plant growth is more constrained by temperature than by precipitation. In Figure 4.18b, we see a different scenario. When the precipitation line intersects the temperature line, the amount of precipitation available to plants equals the amount of water lost by plants via evapotranspiration. At any point where the precipitation line is below the temperature line, water demand exceeds supply. In this situation, plant growth will be constrained more by precipitation than by temperature.

Climate diagrams also help us understand how humans use different biomes. For example, areas of the world that have warm temperatures, long growing seasons, and abundant rainfall are generally highly productive and so are well suited to growing many crops. Warm regions that have less abundant precipitation are suitable for growing grains such as wheat and for grazing domesticated animals, including cattle and sheep. Colder regions are often best used to grow forests for harvesting lumber.

We can divide terrestrial biomes into three categories: tundra and boreal forest, temperate, and tropical. Within these three categories are a total of nine biomes. We will examine each of these biomes in turn, looking

at its temperature and precipitation patterns, geographic distribution, and typical plant growth forms.

## Tundra

The **tundra** is cold and treeless, with low-growing vegetation (**FIGURE 4.19**★). In winter, the soil is completely frozen. *Arctic tundra* is found in the northernmost regions of the Northern Hemisphere in Russia, Canada, Scandinavia, and Alaska. *Antarctic tundra* is found along the edges of Antarctica and on nearby islands. At lower latitudes, *alpine tundra* can be found on high mountains, where high winds and low temperatures prevent trees from growing.

The tundra's growing season is very short, usually only about 4 months during summer, when the polar region is tilted toward the Sun and the days are very long. During this time the upper layer of soil thaws, creating pools of standing water that are ideal habitat for mosquitoes and other insects. The underlying subsoil, known as **permafrost,** is an impermeable, permanently frozen layer that prevents water from draining and roots from penetrating. Permafrost, combined with the cold temperatures and short growing season, prevents deep-rooted plants such as trees from living in the tundra.

While the tundra receives little precipitation, there is enough to support some plant growth. The characteristic plants of this biome, such as small woody shrubs, mosses, heaths, and lichens, can grow in shallow,

★ The diagrams in Figures 4.19 through 4.27 are after http://climatediagrams.com.

**FIGURE 4.19** Tundra biome.

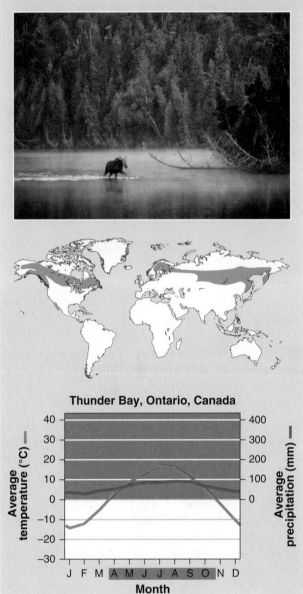

**FIGURE 4.20** Boreal forest biome.

waterlogged soil and can survive short growing seasons and bitterly cold winters. At these cold temperatures, chemical reactions occur slowly, and as a result, dead plants and animals decompose slowly. This slow rate of decomposition results in the accumulation of organic matter in the soil over time and relatively low levels of soil nutrients.

## Boreal Forest

**Boreal forests** (sometimes called *taiga*) are forests made up primarily of *coniferous* (cone-bearing) evergreen trees that can tolerate cold winters and short growing seasons (**FIGURE 4.20**). "Evergreen" trees appear green year-round because they drop only a fraction of their needles each year. Boreal forests are found between about 50° and 60° N in Europe, Russia, and North

America. This subarctic biome has a very cold climate, and plant growth is more constrained by temperature than by precipitation.

As in the tundra, cold temperatures and relatively low precipitation make decomposition in boreal forests a slow process. In addition, the waxy needles of evergreen trees contain compounds that are resistant to decomposition. As a result of the slow rate of decomposition and the low nutrient content of the needles, boreal forest soils are covered in a thick layer of organic material, but are poor in nutrients.

These factors—cold temperatures, low precipitation, and nutrient-poor soil—determine the species of plants that can survive in boreal forests. In addition to coniferous trees such as pine, spruce, and fir, some deciduous trees, such as birch, maple, and aspen, can also be found in this biome. The needles of coniferous trees can tolerate below-freezing conditions, but the deciduous trees drop all their leaves in autumn before the subfreezing temperatures of winter have a chance to damage them. When the weather warms, the deciduous trees produce new leaves and grow rapidly.

Because their soils are poor and their growing season is short, boreal forests are poorly suited for agriculture. However, these forests serve as an important source of trees for pulp, paper, and building materials. As a result, many have been extensively logged.

## Temperate Rainforest

Moving to the mid-latitudes, where the climate is more temperate (with average annual temperatures between 5°C and 20°C), we find a range of temperate biomes, including temperate rainforest, temperate seasonal forest, woodland/shrubland, and temperate grassland/cold desert.

Moderate temperatures and high precipitation typify **temperate rainforests** (FIGURE 4.21). The temperate rainforest is a coastal biome. It can be found along the west coast of North America from northern California to Alaska, in southern Chile, on the west coast of New Zealand, and on the island of Tasmania, which is off the coast of Australia. Ocean currents along these coasts help to moderate temperature fluctuations, and ocean water provides a source of water vapor. The result is relatively mild summers and winters, compared with other biomes at similar latitudes, and a nearly 12-month growing season. In the temperate rainforest, winters are rainy and summers are foggy.

The combination of mild temperatures and high precipitation supports the growth of very large trees. In North America, the most common temperate rainforest trees are coniferous species, including fir, spruce, cedar, and hemlock as well as some of the world's tallest trees: the coastal redwoods (*Sequoia sempervirens*). These immense trees can live hundreds to thousands of years and achieve heights of 90 m (295 feet) and diameters of 8 m (26 feet). Because many of these large tree species are attractive sources of lumber, much of this biome has been logged and subsequently converted into single-species tree plantations.

As we have already seen, coniferous trees produce needles that are slow to decompose. The relatively cool temperatures in the temperate rainforest also favor slow decomposition, although it is not nearly as slow as in boreal forest and tundra. The nutrients released are rapidly taken up by the trees or leached down through the soil by the abundant rainfall, which leaves the soil low in nutrients. Ferns and mosses, which can survive in nutrient-poor soil, are commonly found living under the enormous trees.

## Temperate Seasonal Forest

**Temperate seasonal forests** are more abundant than temperate rainforests (FIGURE 4.22). They are found in the eastern United States, Japan, China, Europe, Chile, and eastern Australia. Temperate seasonal forests receive over 1 m (39 inches) of precipitation annually. Away from the moderating influence of the ocean, these forests experience much warmer summers and colder winters than temperate rainforests. They are dominated by broadleaf deciduous trees such as beech, maple, oak, and hickory, although some coniferous tree species may also be present. Because of the predominance of deciduous trees, these forests are also called *temperate deciduous forests*.

The warm summer temperatures in temperate seasonal forests favor rapid decomposition. In addition, the leaves shed by broadleaf trees are more readily decomposed than the needles of coniferous trees. As a result, the soils of temperate seasonal forests generally contain more nutrients than those of boreal forests. Their higher soil fertility, combined with their longer growing season, means that temperate seasonal forests have greater productivity than boreal forests.

Because the temperate seasonal forest is so productive, it has

**Nanaimo Departure Bay, British Columbia, Canada**

FIGURE 4.21 Temperate rainforest biome.

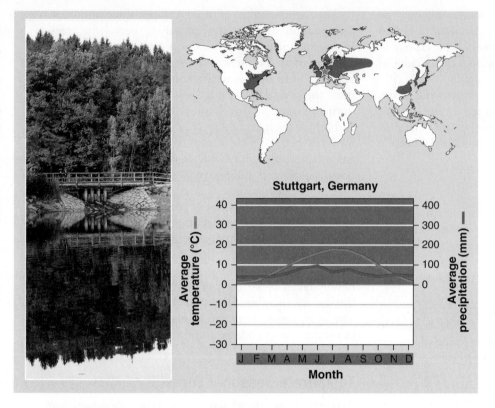

## Temperate Grassland/Cold Desert

The **temperate grassland/cold desert** biome has the lowest average annual precipitation of any temperate biome (**FIGURE 4.24**). Temperate grasslands are found in the Great Plains of North America (where they are called *prairies*), in South America (*pampas*), and in central Asia and eastern Europe (*steppes*). Cold, harsh winters and hot, dry, summers characterize this biome. Thus, as in the woodland/ shrubland biome, plant growth is constrained by insufficient precipitation in summer and cold temperatures in winter. Fires are common, as the dry and frequently windy conditions fan flames ignited by lightning. Although estimates vary, it is thought that, historically, large wildfires occurred in this biome every few years, sometimes burning as much as 10,000 ha (nearly 25,000 acres) in a single fire.

**FIGURE 4.22** Temperate seasonal forest biome.

historically been one of the first biomes to be converted to agriculture on a large scale. When European settlers arrived in North America, they cleared large areas of the eastern forests for agriculture and dwellings.

## Woodland/Shrubland

The **woodland/shrubland** biome is found on the coast of southern California (where it is called *chaparral*), in southern South America (*matorral*), in southwestern Australia (*mallee*), in southern Africa (*fynbos*), and in a large region surrounding the Mediterranean Sea (*maquis*) (**FIGURE 4.23**). This biome is characterized by hot, dry summers and mild, rainy winters. There is a 12-month growing season, but plant growth is constrained by low precipitation in summer and by relatively low temperatures in winter.

The hot, dry summers of the woodland/shrubland biome favor the natural occurrence of wildfires. Plants of this biome are well adapted to both fire and drought. Many plants quickly resprout after a fire. Others produce seeds that open only upon exposure to the intense heat of a fire. Typical plants of this biome include drought-resistant shrubs such as yucca, scrub oak, and sagebrush. Soils in this biome are low in nutrients because of leaching by the winter rains. As a result, the major agricultural uses of this biome are grazing animals and growing drought-tolerant deep-rooted crops, such as grapes to make wine.

Typical plants of temperate grasslands include grasses and nonwoody flowering plants. These plants are generally well adapted to wildfires and frequent grazing by animals. Their deep roots store energy to enable quick regrowth. Within this biome, the amount of rainfall determines which plants can survive in a region. In the North American prairies, for example, nearly 1 m (39 inches) of rain falls per year on the eastern edge of the biome, supporting grasses that can grow up to 2.5 m (8 feet) high. Although these *tallgrass prairies* receive sufficient rainfall for trees to grow, frequent wildfires keep trees from encroaching. In fact, the Native American people are thought to have intentionally kept the eastern prairies free of trees by using controlled burning. To the west, annual precipitation drops to 0.5 m (20 inches), favoring the growth of grasses less than 0.5 m (20 inches) tall. These *shortgrass prairies* are simply too dry to support trees or tall grasses. Farther west, in the rain shadow of the Rocky Mountains, annual precipitation continues to decline to 0.25 m (10 inches). In this region, the shortgrass prairie gives way to *cold desert*.

Cold deserts, also known as *temperate deserts,* have even sparser vegetation than shortgrass prairies. Cold deserts are distinct from subtropical deserts in that they have much colder winters and do not support the characteristic plant growth forms of hot deserts, such as cacti and euphorbs.

The combination of a relatively long growing season and rapid decomposition that adds large amounts of

**FIGURE 4.23** Woodland/shrubland biome.

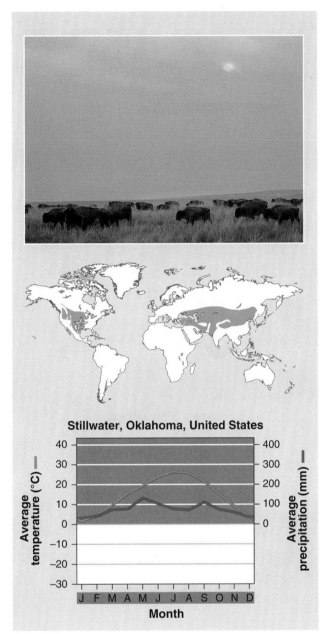

**FIGURE 4.24** Temperate grassland/cold desert biome.

nutrients to the soil makes temperate grasslands very productive. More than 98 percent of the tallgrass prairie in the United States has been converted to agriculture. The less productive shortgrass prairie is predominantly used for growing wheat and grazing cattle.

## Tropical Rainforest

In the tropics, average annual temperatures exceed 20°C. Here we find the tropical biomes: tropical rainforests, tropical seasonal forests/savannas, and subtropical deserts.

**Tropical rainforests** lie within approximately 20° N and S of the equator (**FIGURE 4.25**). They are found in Central and South America, Africa, Southeast Asia, and northeastern Australia. They are also found on large tropical islands, where the oceans provide a constant source of atmospheric water vapor.

The tropical rainforest biome is warm and wet, with little seasonal temperature variation. Precipitation occurs frequently, although there are seasonal patterns in precipitation that depend on when the ITCZ passes overhead. Because of the warm temperatures and abundant rainfall, productivity is high, and decomposition is extremely rapid. The lush vegetation takes up nutrients quickly, leaving few nutrients to accumulate in the soil. Because of its high productivity, approximately 24,000 ha (59,500 acres) of tropical rainforest are cleared each year for agriculture. But the high rate of

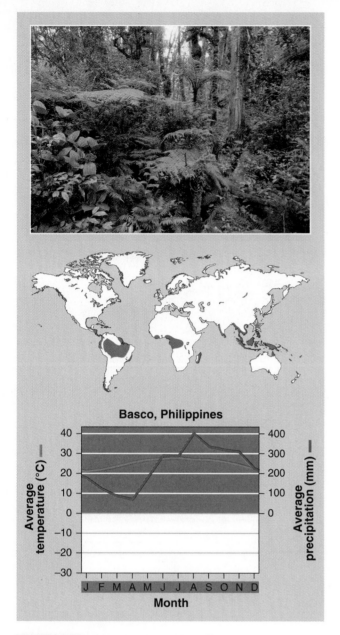

**FIGURE 4.25** Tropical rainforest biome.

aquatic ecosystems far above the forest floor. Numerous species of woody vines (also called *lianas*) are rooted in the soil, but climb up the trunks of trees and often into the canopy.

## Tropical Seasonal Forest/Savanna

**Tropical seasonal forests** and **savannas** (FIGURE 4.26) are marked by warm temperatures and distinct wet and dry seasons. This seasonal pattern is caused by the seasonal movement of the ITCZ, which, because it tracks the seasonal movement of the most intense sunlight, passes overhead and drops precipitation only during summer. The trees drop their leaves during the dry season as an adaptation to survive the drought conditions, then produce new leaves during the wet season. Thus these forests are also called *tropical deciduous forests.*

Tropical seasonal forests are common in much of Central America, on the Atlantic coast of South America, in southern Asia, in northwestern Australia, and in sub-Saharan Africa. Areas with moderately long dry seasons support dense stands of shrubs and trees. In areas with the longest dry seasons, the tropical seasonal climate leads to the formation of *savannas,* relatively open landscapes dominated by grasses and scattered deciduous trees. Common plants in this biome include acacia and baobab trees. Grazing and fire discourage the growth of many smaller woody plants and keep the savanna landscape open. The presence of trees and a warmer average annual temperature distinguish savannas from grasslands.

The warm temperatures of the tropical seasonal forest/savanna biome promote decomposition, but the low amounts of precipitation constrain plants from using the soil nutrients that are released. As a result, the soils of this biome are fairly fertile and can be farmed. Their fertility has resulted in the conversion of large areas of tropical seasonal forest and savanna into agricultural fields and grazing lands. For example, over 99 percent of the tropical seasonal forest of Pacific Central America and the Atlantic coast of South America has been converted to human uses, including agriculture and grazing.

## Subtropical Desert

At roughly 30° N and S, hot temperatures, extremely dry conditions, and sparse vegetation prevail (FIGURE 4.27). This latitudinal band of **subtropical deserts,** also known as *hot deserts,* includes the Mojave Desert in the southwestern United States, the Sahara Desert in Africa, the Arabian Desert of the Middle East, and the Great Victoria Desert of Australia. Cacti, euphorbs, and succulent plants are well adapted to this biome. To prevent water loss, the leaves of desert plants may be small, nonexistent, or modified into spines, and the outer layer of the plant is thick, with few pores for water and

decomposition causes the soils to lose their fertility quickly. As a result, farmers growing crops on tropical soils often have to keep moving to newly deforested areas.

Tropical rainforests contain more biodiversity per hectare than any other terrestrial biome and contain up to two-thirds of Earth's terrestrial species. These forests have several distinctive layers of vegetation. Large trees form a forest *canopy* that shades the underlying vegetation. Several layers of successively shorter trees make up the *subcanopy,* also known as the *understory.* Attached to the trunks and branches of the trees are *epiphytes,* plants that hold small pools of water that support small

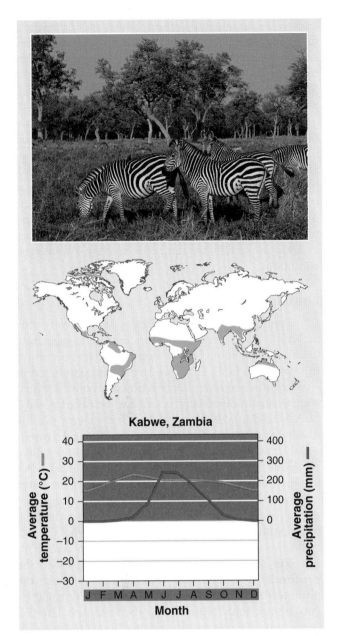

FIGURE 4.26 Tropical seasonal forest/savanna biome.

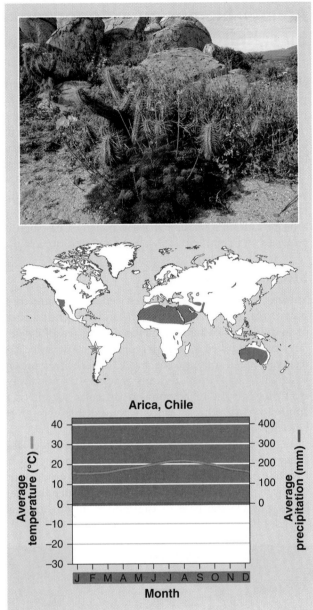

FIGURE 4.27 Subtropical desert biome.

air exchange. Most photosynthesis occurs along the plant stem, which stores water so that photosynthesis can continue even during very dry periods. To protect themselves from herbivores, desert plants have developed defense mechanisms such as spines to discourage grazing.

When rain does fall, the desert landscape is transformed. Annual plants—those that live for only a few months, reproduce, and die—grow rapidly during periods of rain. In contrast, perennial plants—those that live for many years—experience spurts of growth when it rains, but then exhibit little growth during the rest of the year. The slow overall growth of perennial plants in

subtropical deserts makes them particularly vulnerable to disturbance, and they have long recovery times.

### GAUGE YOUR PROGRESS

✓ What characteristics are used to identify terrestrial biomes?

✓ What are some of the ways that humans use different terrestrial biomes?

✓ What characteristics of a terrestrial biome determine its productivity?

# Aquatic biomes are categorized by salinity, depth, and water flow

Whereas terrestrial biomes are categorized by dominant plant growth forms, aquatic biomes are categorized by physical characteristics such as salinity, depth, and water flow. Temperature is an important factor in determining which species can survive in a particular aquatic habitat, but it is not a factor used to categorize aquatic biomes.

Aquatic biomes fall into two broad categories: freshwater and marine. Freshwater biomes include streams, rivers, lakes, and wetlands. Saltwater biomes, also known as *marine* biomes, include shallow marine areas such as estuaries and coral reefs as well as the open ocean.

## Streams and Rivers

Streams and rivers are characterized by flowing fresh water that may originate from underground springs or as runoff from rain or melting snow (FIGURE 4.28). Streams (also called *creeks*) are typically narrow and carry relatively small amounts of water. Rivers are typically wider and carry larger amounts of water. It is not always clear, however, at what point a particular stream, as it combines with other streams, becomes large enough to be called a river.

As water flow changes, biological communities also change. Most streams and many rapidly flowing rivers have few plants or algae to act as producers. Instead, inputs of organic matter from terrestrial biomes, such as fallen leaves, provide the base of the food web. This organic

**FIGURE 4.28** **Streams and rivers.** Streams and rivers are freshwater aquatic biomes that are characterized by flowing water. This photo shows Berea Falls on the Rocky River in Cleveland, Ohio.

matter is consumed by insect larvae and crustaceans such as crayfish, which then provide food for secondary consumers such as fish. As fast-moving streams combine to form larger rivers, the water flow typically slows, sediments and organic material settle to the bottom, and rooted plants and algae are better able to grow.

Fast-moving streams and rivers typically have stretches of turbulent water called *rapids,* where water and air are mixed together. This mixing allows large amounts of atmospheric oxygen to dissolve into the water. Such high-oxygen environments support fish species such as trout and salmon that need large amounts of oxygen. Slower-moving rivers experience less mixing of air and water. These lower-oxygen environments favor species such as catfish that can better tolerate low-oxygen conditions.

## Lakes and Ponds

Lakes and ponds contain standing water, at least some of which is too deep to support emergent vegetation (plants that are rooted to the bottom and emerge above the water's surface). Lakes are larger than ponds, but as with streams and rivers, there is no clear point at which a pond is considered large enough to be called a lake.

As FIGURE 4.29 shows, lakes and ponds can be divided into several distinct zones. The **littoral zone** is the shallow area of soil and water near the shore where algae and emergent plants such as cattails grow. Most photosynthesis occurs in this zone. In the open water, or **limnetic zone,** rooted plants can no longer survive; floating algae called **phytoplankton** are the only photosynthetic organisms. The limnetic zone extends as deep as sunlight can penetrate. Very deep lakes have a region of water below the limnetic zone, called the **profundal zone.** Because sunlight does not reach the profundal zone, producers cannot survive there, so nutrients are not easily recycled into the food web. Bacteria decompose the detritus that reaches the profundal zone, but they consume oxygen in the process. As a result, dissolved oxygen concentrations are not sufficient to support many large organisms. The muddy bottom of a lake or pond beneath the limnetic and profundal zones is called the **benthic zone.**

## Freshwater Wetlands

**Freshwater wetlands** are aquatic biomes that are submerged or saturated by water for at least part of each year, but shallow enough to support emergent vegetation throughout. They support species of plants that are specialized to live in submerged or saturated soils.

Freshwater wetlands include swamps, marshes, and bogs. *Swamps* are wetlands that contain emergent trees, such as the Great Dismal Swamp in Virginia and North Carolina and the Okefenokee Swamp in Georgia and Florida (FIGURE 4.30a). *Marshes* are wetlands that contain primarily nonwoody vegetation, including cattails

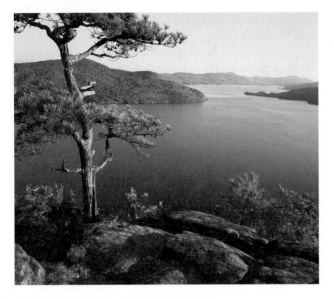

**(a)** Lake George, Adirondack Park, New York

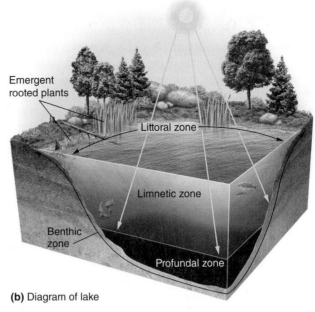

**(b)** Diagram of lake

**FIGURE 4.29** **Lakes and ponds.** (a) Lake George, in the Adirondack Park and Preserve, New York State. (b) In lakes and ponds, at least some of the standing water is too deep for emergent vegetation to grow.

and sedges (**FIGURE 4.30b**). *Bogs,* in contrast, are very acidic wetlands that typically contain sphagnum moss and spruce trees (**FIGURE 4.30c**).

Freshwater wetlands are among the most productive biomes on the planet, and they provide several critical ecosystem services. For example, wetlands can take in

**(a)**

**(b)**

**(c)**

**FIGURE 4.30** **Freshwater wetlands.** Freshwater wetlands have soil that is saturated or covered by fresh water for at least part of the year and are characterized by particular plant communities. (a) In this swamp in southern Illinois, bald cypress trees emerge from the water. (b) This marsh in south-central Wisconsin is characterized by cattails, sedges, and grasses growing in water that is not acidic. (c) This bog in northern Wisconsin is dominated by sphagnum moss as well as shrubs and trees that are adapted to acidic conditions.

large amounts of rainwater and release it slowly into the groundwater or into nearby streams, thus reducing the severity of floods and droughts. Wetlands also filter pollutants from water, recharging the groundwater with clean water. Many bird species depend on wetlands during migration or breeding. As many as one-third of all endangered bird species in the United States spend some part of their lives in wetlands, even though this biome makes up only 5 percent of the nation's land area. Unfortunately, more than half of the freshwater wetland area in the United States has been drained for agriculture or development or to eliminate breeding grounds for mosquitoes and various disease-causing organisms.

## Salt Marshes

**Salt marshes** are found along the coast in temperate climates (FIGURE 4.31). Like freshwater marshes, they contain nonwoody emergent vegetation. The salt marsh is one of the most productive biomes in the world. Many salt marshes are found in *estuaries,* which are areas along the coast where the fresh water of rivers mixes with salt water from the ocean. Because rivers carry large amounts of nutrient-rich organic material, estuaries are extremely productive places for plants and algae, and the abundant plant life helps filter contaminants out of the water. Salt marshes provide important habitat for spawning fish and shellfish; two-thirds of marine fish and shellfish species spend their larval stages in estuaries.

## Mangrove Swamps

**Mangrove swamps** occur along tropical and subtropical coasts. Like freshwater swamps, they contain trees whose roots are submerged in water (FIGURE 4.32). Unlike most trees, however, mangrove trees are salt tolerant. They often grow in estuaries, but they can also be found along shallow coastlines that lack inputs of fresh water. The trees help to protect those coastlines from erosion and storm damage. Falling leaves and trapped organic material produce a nutrient-rich environment. Like salt marshes, mangrove swamps provide sheltered habitat for fish and shellfish.

## Intertidal Zone

The **intertidal zone** is the narrow band of coastline that exists between the levels of high tide and low tide (FIGURE 4.33). Intertidal zones range from steep, rocky areas to broad, sloping mudflats. Environmental conditions in this biome are relatively stable when it is submerged during high tide. But conditions can become quite harsh during low tide, when organisms are exposed to direct sunlight, high temperatures, and desiccation. Moreover, waves crashing onto shore can make it a challenge for organisms to hold on and not get washed away. Intertidal zones are home to a wide variety of organisms that have adapted to these conditions, including barnacles, sponges, algae, mussels, crabs, and sea stars.

## Coral Reefs

**Coral reefs,** which are found in warm, shallow waters beyond the shoreline, represent Earth's most diverse marine biome (FIGURE 4.34). Corals are tiny animals that secrete a layer of limestone (calcium carbonate) to form an external skeleton. The animal living inside this

**FIGURE 4.31** **Salt marsh.** The salt marsh is a highly productive biome typically found in temperate regions where fresh water from rivers mixes with salt water from the ocean. This salt marsh is in Plum Island Sound in Massachusetts.

**FIGURE 4.32** **Mangrove swamp.** Salt-tolerant mangrove trees, such as these in Everglades National Park, are important in stabilizing tropical and subtropical coastlines and in providing habitat for marine organisms.

FIGURE 4.33 **Intertidal zone.** Organisms that live in the area between high and low tide, such as these giant green sea anemones (*Anthopleura xanthogrammica*), goose barnacles (*Lepas anserifera*), and ochre sea stars (*Pisaster ochraceus*), must be highly tolerant of the harsh, desiccating conditions that occur during low tide. This photo was taken at Olympic National Park, Washington.

*high CO₂ content*

tiny skeleton is essentially a hollow tube with tentacles that draw in plankton and detritus. Corals live in water that is relatively poor in nutrients and food. This is possible because of their relationship with single-celled algae that live within the tissues of the corals. When a coral digests the food it captures, it releases $CO_2$ and nutrients. The algae use the $CO_2$ during photosynthesis to produce sugars. The nutrients stimulate the algae to release their sugars to the coral. The coral gains energy in the form of sugars, and the algae obtain $CO_2$, nutrients, and a safe place to live within its tiny limestone skeleton. But this association with photosynthetic algae means that corals can live only in shallow waters where light can penetrate.

Although each individual coral is tiny, most corals live in vast colonies. As individual corals die and decompose, their limestone skeletons remain. Thus, over time, these skeletons accumulate and develop into coral reefs, which can become quite massive. The Great Barrier Reef of Australia, for example, covers an area of 2,600 km² (1,600 mile²). A tremendous diversity of other organisms, including fish and invertebrates, use the structure of the reef as both a refuge in which to live and a place to find food. At the Great Barrier Reef there are more than 400 species of coral, 1,500 species of tropical fish, and 200 species of birds.

Coral reefs are currently facing a wide range of challenges, including pollutants and sediments that make it difficult for the corals to survive. Coral reefs also face the growing problem of **coral bleaching,** a phenomenon in which the algae inside the corals die. Without the algae, the corals soon die as well, and the reef turns white. Scientists believe that the algae are dying from a combination of disease and environmental changes, including lower ocean pH and abnormally high water temperatures. Coral bleaching is a serious

FIGURE 4.34 **Coral reef.** The skeletons of millions of corals build reefs that serve as home to a great variety of other marine species. Diana's hogfish (*Bodianus diana*) and other animals inhabit this reef of soft coral (*Dendronephthya* sp.) in the Red Sea, Egypt.

Intertidal zone

High tide

Low tide

200 m

Photic zone

Benthic zone

Aphotic zone

problem: without the corals, the entire coral reef biome is endangered.

## The Open Ocean

Away from the shoreline in deeper water, sunlight can no longer reach the ocean bottom. The exact depth of penetration by sunlight depends on a number of factors, including the amounts of sediment and algae suspended in the water, but it generally does not exceed 200 m (approximately 650 feet).

Like a pond or lake, the ocean can be divided into zones. These zones are shown in FIGURE 4.35. The upper layer of water that receives enough sunlight to allow photosynthesis is the **photic zone,** and the deeper layer of water that lacks sufficient sunlight for photosynthesis is the **aphotic zone.** The ocean floor is called the *benthic zone.*

In the photic zone, algae are the major producers. They form the base of a food web that includes tiny zooplankton, fish, and whales. Given the lack of light in the aphotic zone, there are no photosynthetic producers there. However, there are some species of bacteria that can use the energy contained in the bonds of methane and hydrogen sulfide, which are both found in the deep ocean, to generate energy via **chemosynthesis** rather than photosynthesis. These bacteria form the base of a deep-ocean food web that includes animals such as tube worms (see Figure 2.17d). The aphotic zone also contains a variety of organisms that can generate their own light (a phenomenon called bioluminescence) to help them feed in the dark waters. These organisms include several species of crustaceans, jellyfish, squid, and fish.

The patterns that emerge as we study the terrestrial and aquatic biomes highlight the fact that regional variations in global climate have a major effect on the types of organisms that can live in different parts of the world. Among the terrestrial biomes, temperature and precipitation affect the rate of decomposition of dead organisms and the productivity of the soil. Understanding these patterns helps us understand how humans have come to use the land in different ways: growing crops in regions with enough water and a sufficient growing season, grazing domesticated animals in drier areas, and harvesting lumber from forests. Among the aquatic biomes, differences in water flow, depth, and salinity help us understand why different species of producers and consumers, including commercially important species of fish and shellfish, live in different regions of the world.

### GAUGE YOUR PROGRESS

✓ How are aquatic biomes categorized? Why are they categorized differently than terrestrial biomes?

✓ What are the different zones of lakes and the open ocean, and what defines them?

✓ How does water depth or flow influence the organisms that live in an aquatic biome?

## Is Your Coffee Made in the Shade?

Around the world, people enjoy drinking coffee. In the United States alone, 54 percent of adults drink coffee every day, at an average of 3.1 cups per day. Worldwide, people buy 7.7 billion kilograms (16.9 billion pounds) of coffee beans each year. In short, coffee is an important part of many people's lives. But have you ever thought about where your coffee comes from?

Coffee beans come from several species of shrubs that historically grew in the tropical rainforests of Ethiopia. The coffee plant naturally grows under the shade of the tropical rainforest canopy. In the fifteenth century, coffee was brought to the Middle East and eventually spread throughout the world. Because of its popularity, coffee is now farmed in many places around the world, including South America, Africa, and Southeast Asia.

As farmers began cultivating coffee, they grew it like many other crops, by clearing large areas of rainforest and planting coffee bushes close together in large open fields. Because the coffee plants' native habitat was a shady forest, coffee farmers found that they had to construct shade over the plants to prevent them from becoming sunburned in the intense tropical sunlight. Over the past several decades, however, breeders have developed more sunlight-tolerant plants. These new varieties can not only handle intense sunlight, but can also produce many more coffee beans per plant.

An interesting thing happened as coffee was transformed from a plant that was naturally scattered throughout a diverse rainforest to one that was grown as a single species in large numbers in open fields: the coffee fields became attractive targets for insect pests and diseases. Farmers have applied a variety of pesticides to combat these pests. This use of pesticides has increased the cost of farming coffee, poisoned workers, and polluted the environment. Given the world's demand for coffee, what other options do coffee farmers have?

Some coffee farmers thought back to the natural environment in which coffee grows and wondered if they could farm coffee under more natural conditions. Such coffee, called shade-grown coffee, is grown in one of three ways: by planting coffee bushes in an intact rainforest, by planting the bushes in a rainforest that has had some of the trees removed, or by planting the bushes in a field alongside trees that produce other marketable products, including fruit (FIGURE 4.36). Coffee bushes grown in this way attract fewer pests, so less money is needed to buy and apply pesticides, and there is less risk to workers and the nearby soil and water. Using these methods, coffee can be grown while still preserving some of the plant diversity of the rainforest. And the coffee often tastes better. The density of coffee plants is lower in these more diverse landscapes, however, which means that only about one-third as much coffee is produced per hectare. So, while there are cost savings, the yield is lower. Economically, this means that owners of shade-grown coffee farms need to charge higher prices to match the profits of other farms.

How can farmers producing shade-grown coffee stay in business? A number of environmental groups that want to preserve biodiversity in tropical rainforests have stepped in to help. Researchers found that shade-grown coffee farms provided habitat for approximately 150 species of rainforest birds, whereas open-field coffee farms provided habitat for only 20 to 50 bird species. Not surprisingly, researchers also found that other groups of animals were more diverse on shade-grown coffee farms. In response to these findings, the Smithsonian Migratory Bird Center in Washington, D.C., developed a program

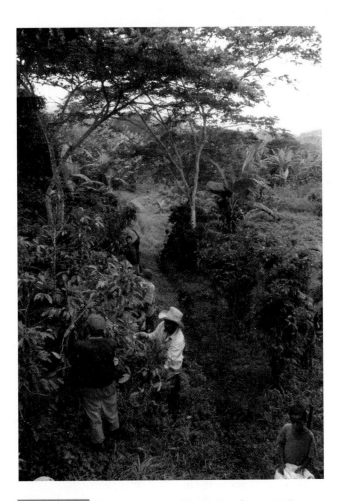

**FIGURE 4.36** **Shade-grown coffee in Honduras.** Coffee grown in the shade requires less pesticide, helps to preserve the plant diversity of the rainforest, and even tastes better.

to offer a "Bird Friendly" seal of approval to coffee farmers who were producing shade-grown coffee. Combined with an advertising campaign that explained the positive effect of shade-grown coffee on biodiversity, this seal of approval alerted consumers to make a conscious choice

A bird-friendly certification label.

about the impact that their favorite beverage was having on rainforests. The Arbor Day Foundation, an environmental organization that promotes the planting of trees, also joined the effort by selling its own brand of shade-grown coffee.

Perhaps the greatest impact occurred when the Starbucks Coffee Company began selling shade-grown coffee that received a seal of approval from Conservation International, another conservation organization. Although it was originally planned as a short-term offering, customer responses were so positive that the coffee has been added as a permanent Starbucks product. Over the past 10 years, it has become clear that when consumers are informed about how coffee is grown, many people are willing to choose the shade-grown varieties, even if it means spending a bit more money to reduce their impact on the tropical rainforest biome.

### References

Philpott, S. M., et al. 2008. Biodiversity loss in Latin American coffee landscapes: Review of the evidence on ants, birds, and trees. *Conservation Biology* 22:1093–1105.

Smithsonian Migratory Bird Center. Coffee Drinkers and Bird Lovers. http://nationalzoo.si.edu/ConservationAndScience/MigratoryBirds/Coffee/lover.cfm.

## REVISIT THE KEY IDEAS

■ **Explain the forces that drive global circulation patterns and how those patterns determine weather and climate.**

Global climate patterns are driven by a combination of unequal heating of Earth by the Sun, atmospheric convection currents, the rotation of Earth and the Coriolis effect, Earth's orbit around the Sun on a tilted axis, and ocean currents. The unequal heating of Earth is the driver of atmospheric convection currents. These air circulation patterns are further modified by the deflecting action of the Coriolis effect. The tilt of Earth's axis of rotation causes seasonal changes in temperature and precipitation as Earth orbits the Sun. Ocean currents are driven by a combination of temperature, gravity, prevailing winds, the Coriolis effect, and the locations of continents. Together, prevailing winds and ocean currents distribute heat and precipitation around the globe.

■ **Describe the major terrestrial biomes.**

Terrestrial biomes are distinguished by a particular combination of average annual temperature and annual precipitation and by plant growth forms that are adapted to these conditions. Terrestrial biomes can be broken down into three groups: those in cold, polar regions with average annual temperatures of less than 5°C (tundra and boreal forest), those in temperate regions at mid-latitudes that have average annual temperatures between 5°C and 20°C (temperate rainforest, temperate seasonal forest, woodland/shrubland, and temperate grassland/cold desert), and those in tropical regions that have average annual temperatures of more than 20°C (tropical rainforest, tropical seasonal forest/savanna, and subtropical forest).

■ **Describe the major aquatic biomes.**

Aquatic biomes are categorized by their physical characteristics, including salinity, depth, and water flow. Freshwater aquatic biomes include streams and rivers, lakes and ponds, and freshwater wetlands. Marine biomes include salt marshes, mangrove swamps, shallow ocean biomes (intertidal zones and coral reefs), and the open ocean.

1. In which layer of Earth's atmosphere does most weather occur?
   (a) Troposphere
   (b) Stratosphere
   (c) Mesosphere
   (d) Thermosphere
   (e) Lithosphere

2. Which of the following *best* explains why polar regions are colder than tropical regions?
   (a) Polar regions have lower albedo values.
   (b) Polar regions receive less solar energy per unit of surface area.
   (c) Tropical regions receive less direct sunlight throughout the year.
   (d) Sunlight travels through more atmosphere and loses more energy in tropical regions.
   (e) Tropical regions rotate at a faster speed than polar regions.

3. Which of the following statements about patterns of temperature and precipitation is *not* correct?
   (a) The air in a Hadley cell rises where sunlight strikes Earth most directly.
   (b) The greatest amount of precipitation occurs at the intertropical convergence zone.
   (c) The air in a Hadley cell descends near 30° N and S, causing the formation of deserts.
   (d) The air of a polar cell rises near 60° latitude.
   (e) Along Earth's surface, the air of a Hadley cell moves toward the equator.

4. Which of the following processes is *not* characteristic of oceanic circulation?
   (a) Counterclockwise gyres in the Northern Hemisphere
   (b) Slow thermohaline circulation of surface and deep ocean waters
   (c) Unequal heating of tropical versus polar ocean waters
   (d) El Niño–Southern Oscillation
   (e) Coriolis effect

5. Which of the following statements about rain shadows is *correct?*
   (a) They occur on the western sides of mountain ranges in the Northern Hemisphere.
   (b) Air gains water vapor as it rises.
   (c) As air rises over a mountain range, water vapor condenses into precipitation.

   (d) They occur on the eastern sides of mountain ranges in the Southern Hemisphere.
   (e) The rain shadow side of a mountain range receives the most rain.

6. Why do scientists use dominant plant growth forms to categorize terrestrial biomes?
   (a) Plants with similar growth forms are always closely related genetically.
   (b) Different plant growth forms indicate climate differences, whereas different animal forms do not.
   (c) Plants from similar climates evolve different adaptations.
   (d) Similar plant growth forms are found in climates with similar temperatures and amounts of precipitation.
   (e) Similar plant growth forms exist in terrestrial and aquatic biomes.

7. Which information is *not* found in climate diagrams?
   (a) Average annual temperature
   (b) Seasonal changes in temperature
   (c) Average annual humidity
   (d) The months when plant growth is limited by precipitation
   (e) The length of the growing season

8. Which of the following statements about tundras and boreal forests is *correct?*
   (a) Both are characterized by slow plant growth, so there is little accumulation of organic matter.
   (b) Tundras are warmer than boreal forests.
   (c) Boreal forests have shorter growing seasons than tundras.
   (d) Plant growth in both biomes is limited by precipitation.
   (e) Boreal forests have larger dominant plant growth forms than tundras.

9. Which of the following statements about temperate biomes is *not* correct?
   (a) Temperate biomes have average annual temperatures above 20°C.
   (b) Temperate rainforests receive the most precipitation, whereas cold deserts receive the least precipitation.
   (c) Temperate rainforests can be found in the northwestern United States.
   (d) Temperate seasonal forests are characterized by trees that lose their leaves.
   (e) Temperate shrublands are adapted to frequent fires.

10. Which of the following statements about tropical biomes is *correct?*
    (a) Tropical seasonal forests are characterized by evergreen trees.
    (b) Tropical rainforests have the highest precipitation due to the proximity of the ITCZ.
    (c) Savannas are characterized by the densest forests.
    (d) Tropical rainforests have the slowest rates of decomposition due to high rainfall.
    (e) Subtropical deserts have the highest species diversity.

11. Which of the following statements about aquatic biomes is *correct?*
    (a) They are characterized by dominant plant growth forms.
    (b) They can be categorized by temperature and precipitation.
    (c) Lakes contain littoral zones and intertidal zones.
    (d) Freshwater wetlands have emergent plants in their deepest areas, whereas ponds and lakes do not.
    (e) Coral reefs have the lowest diversity of species.

## APPLY THE CONCEPTS

1. As the greenhouse effect continues to warm the planet slowly, the glaciers of Greenland are melting at a rapid rate. Scientists are concerned that this melting may dilute the salt water in that region of the ocean enough to shut down thermohaline circulation. Use what you know about climate to answer the following questions.
   (a) Explain how shutting down thermohaline circulation would affect the temperature of western Europe.
   (b) Explain the consequences such a temperature change might have for agriculture in western Europe.
   (c) Explain why there would be large populations of fish along the west coasts of most continents.
   (d) Explain how shutting down thermohaline circulation would affect the transport of nutrients among the oceans of the world.

2. A number of Earth's features determine the locations of biomes around the world.
   (a) Explain why the regions of the world that receive the most direct sunlight contain tropical rainforests.
   (b) Describe the role that the movement of the ITCZ over the year plays in creating seasonal forests in tropical regions.
   (c) Identify the mechanisms by which albedo and the angle of the Sun's rays cause colder temperatures to occur on Earth near the North and South Poles.

## MEASURE YOUR IMPACT

**How Much Paper Do You Use?** The forested biomes of the world contain dozens of species of trees that are used for making paper products such as cardboard, newspaper, tissues, and paper for writing and printing. The average person in the United States consumes about 300 kg (661 pounds) of paper products per year, and 30 percent of those products consist of paper for writing and printing.
   (a) How much paper does the average person in the United States use annually for writing and printing?
   (b) Given that the U.S. population was 308 million people in 2009, how many kilograms of total paper products are used annually by the U.S. population?
   (c) If the average tree produces 30 kg (66 pounds) of writing and printing paper, how many trees would be used annually for this paper if all of the paper came from trees?

   (d) If 50 percent of all writing and printing paper could be made from 100 percent recycled paper, how many fewer trees would be needed for making paper?
   (e) About 18 million students attend college in the United States annually, and each of these students purchases new textbooks containing an average of 18 kg (40 pounds) of paper each year. If recycled paper is not used to make these textbooks, how many trees would be consumed annually to make textbooks?

How might water shortages in your community affect you personally? In this Engage Your Environment exercise you will examine your local climate, assess its vulnerability to water shortages, and examine your personal use of water resources. *See Engage Your Environment Activity 4 at the end of this book.*

# Evolution and Biodiversity: Origin and Diversification of Organisms

## The Dung of the Devil

From 1918 to 1920, the world experienced a flu outbreak of unprecedented scale. Known as the Spanish flu, the disease had a devastating effect. Mortality estimates from that time vary, but somewhere between 20 million and 100 million people died worldwide, including more than 600,000 people in the United States. During the height of the outbreak, reports stated that some people in China had found the roots of a particular plant beneficial in fighting the flu. The plant (*Ferula assafoetida*) had a pleasant smell when cooked, but

the source of two drugs used to treat childhood leukemia and Hodgkin's disease. The mayapple (*Podophyllum peltatum*), a common herb of the eastern United States, is the source of two other anticancer drugs. Many new medicines, including anti-inflammatory, antiviral, and antitumor drugs, have come from a variety of invertebrate animals that inhabit coral reefs, including sponges, corals, and sea squirts. Of the most promising current candidates for new drugs, 70 percent were first discovered in plants, animals, and microbes. Unfortunately, many species that are either known or suspected sources of

> ### The Dung of the Devil has the potential to produce a new pharmaceutical drug to fight H1N1 flu epidemics.

the raw sap from the roots had a foul smell that inspired the plant's common name, the Dung of the Devil.

The Dung of the Devil story does not end in 1920. It turns out that Spanish flu was caused by an H1N1 virus that is closely related to the H1N1 virus that caused the "swine flu" outbreak of 2009–2010. Scientists in China recalled that people had used the plant to fight the Spanish flu 80 years ago, so they decided to explore its potential to combat the modern H1N1 flu virus. They found that extracts from the plant had strong antiviral properties, stronger even than those of contemporary antiviral drugs. Thus the Dung of the Devil has the potential to produce a new pharmaceutical drug to fight future H1N1 flu epidemics.

The Dung of the Devil is just one of the organisms from which humans have extracted life-saving drugs. Willow trees from temperate forests were the original source of salicylic acid, from which aspirin is derived. More recently, wild plants have provided several important medicines for treating a variety of cancers. For example, the rosy periwinkle (*Catharanthus roseus*), found only in the tropical forests of Madagascar, is

drugs are being lost to deforestation, agriculture, and other human activities. At the same time, indigenous people with knowledge about medicinal uses of the natural drugs in their environment are being forced to relocate, and their knowledge may soon be lost.

There are millions of species on Earth, only a small fraction of which has been screened for useful drugs. It is likely that many more medicines could be found in living organisms. The continual discovery of new drugs in organisms around the world, including the Dung of the Devil, makes yet another convincing argument for conserving Earth's biodiversity. ■

*Sources:* C. L. Lee et al., Influenza A (H1N1) antiviral and cytotoxic agents from *Ferula assafoetida, Journal of Natural Products* 72 (2009): 1568–1572; D. Newman and G. M. Cragg, Natural products as sources of new drugs over the last 25 years, *Journal of Natural Products* 70 (2007): 461–477.

The rosy periwinkle is a source of new drugs that fight childhood leukemia and Hodgkin's disease.

◀ The plant known as the Dung of the Devil, discovered as a treatment for the Spanish flu of 1918, may also be a remedy for the H1N1 virus.

Biodiversity is an important indicator of environmental health. A rapid decline of biodiversity in an ecosystem indicates that it is under stress. The biodiversity on Earth today is the result of evolution and extinction. Understanding these processes helps us to understand past and present environmental changes and their effects.

After reading this chapter you should be able to

- explain the concept of biodiversity and how it is measured.
- describe the ways in which evolution can occur.
- explain how environmental change affects speciation and extinction.
- explain the concept of an ecological niche.

# Earth is home to a tremendous diversity of species

A short walk through the woods, a corner lot, or the city park makes one thing clear: life comes in many forms. A small plot of untended land or a tiny pond contains dozens, perhaps hundreds, of different kinds of plants and animals visible to the naked eye as well as thousands of different kinds of microscopic organisms. In contrast, a carefully tended lawn or a commercial timber plantation usually supports only a few types of grasses or trees (FIGURE 5.1). The total number of

*organisms* in the plantation or lawn may be the same as the number in the pond or the untended plot, but the number of *species* will be far smaller.

As you will recall from Chapter 1, we can think about biodiversity at three different scales (FIGURE 5.2). Within a given region, for example, the variety of ecosystems is a measure of **ecosystem diversity.** Within a given ecosystem, the variety of species constitutes **species diversity.** Within a given species, we can think about the variety of genes as a measure of **genetic diversity.** Every individual organism is distinguished from every other organism, at the most basic level, by how different their genes are. Because genes

(a)

(b)

**FIGURE 5.1** **Species diversity varies among ecosystems.** (a) Natural forests contain a high diversity of tree species. (b) In forest plantations, in which a single tree species has been planted for lumber and paper products, species diversity is low.

**(a)** Ecosystem diversity

**(b)** Species diversity

**(c)** Genetic diversity

**FIGURE 5.2** **Levels of biodiversity.** Biodiversity exists at three scales. (a) Ecosystem diversity is the variety of ecosystems within a region. (b) Species diversity is the variety of species within an ecosystem. (c) Genetic diversity is the variety of genes among individuals of a species.

form the blueprint for an organism's traits, the diversity of genes on Earth ultimately helps determine the species diversity and ecosystem diversity on Earth. In other words, all three scales of biodiversity contribute to the overall biodiversity of the planet. As we will see in later chapters, the diversity of genes, species,

and ecosystems plays a major role in the normal functioning of natural systems, including the ecosystem services humans depend on.

## How Many Species Live on Earth?

Recall from Chapter 1 that a *species* is defined as a group of organisms that is distinct from other such groups in terms of size, shape, behavior, or biochemical properties, and that can interbreed with other individuals in the group to produce viable offspring. This last requirement is important because sometimes individuals from different species can mate, but they do not produce offspring that survive.

The number of species in any given place is the most common measure of biodiversity. But estimating the total number of species on Earth is a challenge. Many species are easy to find, such as the birds or small mammals you might see in your neighborhood. Others are not so easy to find. Some species are active only at night, live in inaccessible locations such as the deep ocean, or require a microscope to be seen. To date, scientists have named approximately 2 million species, which means the total must be larger than that.

The insect group contains more species than most other groups, so scientists reason that if we could get a good estimate for the number of insect species, we would have a much better sense of how many species there are in total. In one study, researchers fumigated the canopies of a single tree species in the tropical rainforest and then collected all the dead insects that fell from the trees onto a tarp placed on the ground. From this collection of dead insects, they counted the number of beetle species that fed on only this one tree species (**FIGURE 5.3**). By multiplying this number of beetle species by the total number of tropical tree species, they estimated that there were perhaps 8 million species of beetles in the tropics that feed on a single species of tree. Because beetles make up about 40 percent of all insect species, and because insect species in the forest canopy tend to be about twice as numerous as insect species on the forest floor, the researchers suggested that a reasonable estimate for the total number of tropical insect species might be 30 million. More recent work has indicated that this number is probably too high. Current estimates for the total number of species on Earth range between 5 million and 100 million, but most scientists estimate that there are about 10 million species. Crunch the Numbers "Estimating Biodiversity" shows you a simple way to do a biodiversity calculation.

## Species Richness and Species Evenness

Given that species are not uniformly distributed, the number of species on Earth is not a useful indicator

**(a)**

**(b)**

FIGURE 5.3 **Cataloging tropical insect species.** (a) By fumigating a tree with an insecticide, scientists can collect all insects that live on the tree. (b) The various species of insects can be identified to determine the species richness of the tree.

of local or regional species diversity. To measure species diversity at these scales, environmental scientists have developed two measures: *species richness* and *species evenness.*

The number of species in a given area, such as a pond, the canopy of a tree, or a plot of grassland, is known as **species richness.** Species richness is used to give an approximate sense of the biodiversity of a particular

## CRUNCH THE NUMBERS

### Estimating Biodiversity

Imagine that you are a scientist and have been given the task of estimating the number of insect species on a tropical island. You have neither the time nor the money to conduct a search for every such species on the island, so you decide to collect all of the beetles that live on a single tree by fumigating the tree with an insecticide. You use this information to estimate the species richness of all insects on the island.

Fumigation produces 120 species of beetles. Of this total, 100 beetle species live on all tree species and 20 beetle species are specialized to live only on the species of tree used in your sample. If we assume that every tree species has 20 beetle species that are specialized to live only on it and there are 10 species of trees on the island, how many total beetle species live on the trees of the island?

Total number of beetle species on the trees =
(20 species that specialize on a single tree species × 10 tree species) + 100 species that live on all trees

Total number of beetle species on the trees =
(20 × 10) + 100 = 300

If beetles represent 40 percent of all insect species on the trees, how many total insect species live on the trees?

Total number of insect species on the trees =

$$\frac{\text{Total number of beetle species on the trees}}{40\%}$$

Total number of insect species on the trees = $\frac{300}{40\%}$ = 750

**Your Turn:** If the number of insect species living on the trees is twice the number of insect species living on the ground, how many total insect species are living on the island?

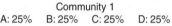

Community 1
A: 25%   B: 25%   C: 25%   D: 25%

Community 2
A: 70%   B: 10%   C: 10%   D: 10%

**FIGURE 5.4** **Measures of species diversity.** Species richness and species evenness are two different measures of species diversity. Although both communities contain the same number of species, community 1 has a more even distribution of species and is therefore more diverse than community 2.

place. However, we may also want to know the relative proportions of individuals within the different species. **Species evenness** tells us whether a particular ecosystem is numerically dominated by one species or whether all of its species have similar abundances. An ecosystem has high species evenness if its species are all represented by similar numbers of individuals. An ecosystem has low species evenness if one species is represented by many individuals while others are represented by few individuals. In the latter case, there is effectively less diversity.

Scientists evaluating the biodiversity of an area must often evaluate both species richness and species evenness. Consider the two forest communities, community 1 and community 2, shown in **FIGURE 5.4**. Both forests contain 20 trees that are distributed among four species. In community 1, each species is represented by 5 individuals. In community 2, one species is represented by 14 individuals and each of the other three species is represented by 2 individuals. Although the species richness of the two forests is identical, the four species are more evenly represented in community 1. That forest therefore has greater species evenness and is considered to be more diverse.

Because species richness or evenness often declines after a human disturbance, knowing the species richness and species evenness of an ecosystem gives environmental scientists a baseline they can use to determine how much that ecosystem has changed.

## Determining Evolutionary Relationships Among Species

Scientists organize species into categories that indicate how closely related they are to one another. The branching patterns of evolutionary relationships are called **phylogenies.** Phylogenies can be described with a diagram like the one shown in **FIGURE 5.5**, called a *phylogenetic tree.*

The relatedness of the species in a phylogeny is determined by the similarity of their traits: the more similar the traits of two species, the more closely related the two species are assumed to be. Historically, scientists used mostly morphological traits, including a large number of bone measurements, to measure similarity. Today, scientists base phylogenies on a variety of characteristics, including morphology, behavior, and genetic similarity.

> **GAUGE YOUR PROGRESS**
>
> ✓ Why is it challenging to determine the number of species on Earth?
>
> ✓ Why are estimates of species diversity valuable to environmental scientists?
>
> ✓ What is the difference between species richness and species evenness? Why are they both important measures?

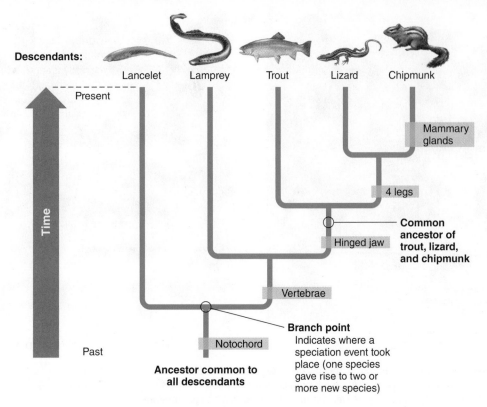

**Descendants:**

Lancelet    Lamprey    Trout    Lizard    Chipmunk

Present

Time

Mammary glands

4 legs

Common ancestor of trout, lizard, and chipmunk

Hinged jaw

Vertebrae

Branch point
Indicates where a speciation event took place (one species gave rise to two or more new species)

Notochord

Past

**Ancestor common to all descendants**

**FIGURE 5.5** **A phylogenetic tree.** Phylogenies are based on the similarity of traits among species. Scientists can assemble phylogenetic trees that indicate how different groups of organisms are related and show where speciation events have occurred. The brown boxes indicate when major morphological changes evolved over evolutionary time.

# Evolution is the mechanism underlying biodiversity

Earth's biodiversity is the product of **evolution,** which can be defined as a change in the genetic composition of a population over time.

Evolution can occur at multiple levels. Evolution below the species level, such as the evolution of different varieties of apples or potatoes, is called **microevolution.** In contrast, when genetic changes give rise to new species, or to new *genera, families, classes,* or *phyla*—larger categories of organisms into which species are organized—we call the process **macroevolution**. Among these many levels of macroevolution, the term *speciation* is restricted to the evolution of new species.

Evolution depends on genetic diversity, so we begin our discussion of evolution with a look at how genetic diversity is created.

## Creating Genetic Diversity

**Genes** are physical locations on chromosomes within each cell of an organism. An organism's genes determine the range of possible traits (physical or behavioral characteristics) that it can pass down to its offspring.

The complete set of genes in an individual is called its **genotype.**

Two processes that create genetic diversity are *mutation* and *recombination*. DNA is copied millions of times during an organism's lifetime as cells grow and divide. An occasional mistake in the copying process produces a random change, or **mutation,** in the genetic code. Environmental factors, such as ultraviolet radiation from the Sun, can also cause mutations. When mutations occur in cells responsible for reproduction, such as the eggs and sperm of animals, those mutations can be passed on to the next generation.

Most mutations are detrimental. Many mutations cause the offspring that carry them to die before they are born. The effects of some mutations are less severe, but can still be detrimental. For example, some dusky-headed conures (*Aratinga weddellii*) have a mutation that makes these normally green-feathered parrots produce feathers that appear to be blue (**FIGURE 5.6**). In the wild, individuals with this mutation have a poor chance of survival because their blue feathers stand out against the green vegetation and make them conspicuous to predators.

Sometimes a mutation improves an organism's chances of survival or reproduction. If such a mutation is passed along to the next generation, it adds new genetic diversity to the population. Some mosquitoes,

FIGURE 5.6 **Most mutations are detrimental.** A mutation in the genetic code of the dusky-headed conure causes these normally green-feathered parrots to develop feathers that appear blue. In nature the mutation makes individuals more conspicuous and prone to predation.

for example, possess a mutation that makes them less vulnerable to insecticides. In areas that are sprayed with insecticides, such mutations improve an individual mosquito's chance of surviving and reproducing.

The second way in which genetic diversity is created is through *recombination*. In plants and animals, genetic **recombination** occurs as chromosomes are duplicated during reproductive cell division and a piece of one chromosome breaks off and attaches to another chromosome. This process does not create new genes, but it does bring together new combinations of genes on a chromosome and can therefore produce novel traits. For example, the human immune system must battle a large variety of viruses and bacteria that regularly attempt to invade the body. Recombination allows new gene combinations to come together, providing new immune defenses that may prove to be effective against the invading organisms.

**GENOTYPES VERSUS PHENOTYPES** An individual's genotype serves as the blueprint for the complete set of traits that organism may potentially possess. An individual's **phenotype** is the actual set of traits expressed in that individual. Among these traits are the individual's anatomy, physiology, and behavior. The color of your eyes, for example, is your phenotype, whereas the genes you possess that code for eye color are a part of your genotype. Changes in the genotype due to mutation or recombination can produce important changes in an individual's phenotype.

In some cases, an individual's phenotype is determined almost entirely by its genes. For instance, a person who inherits the genes for brown eyes will have brown eyes, regardless of where that person lives. Most phenotypes, however, are the product of an individual's environment as well as its genotype. For example, in many turtle and crocodile species, the temperature of eggs during incubation determines whether the offspring will hatch as males or females. The water flea, a tiny animal that lives in ponds and lakes, offers another interesting example. The body shape of the water flea depends on whether or not a young individual smells predators in its environment (FIGURE 5.7). If predators are absent, the water flea develops a relatively small head and tail spine. If predators are present, however, the water flea develops a much larger head and a long tail spine. Although the larger head and longer tail spine help prevent the water flea from being eaten, they come at the cost of slower reproduction. As a result, it is beneficial to the water flea not to produce these defenses in the absence of predators, when they

FIGURE 5.7 **Environmental effects on phenotype.** Water fleas raised in the absence of predators produce relatively small heads and short tail spines (left), whereas individuals raised in the presence of predators produce relatively large heads and long tail spines (right).

are not needed. By being able to respond to changing environmental conditions, organisms such as the water flea can improve their ability to survive and reproduce in any environment.

## Evolution by Artificial Selection

Evolution occurs in three primary ways: by *artificial selection,* by *natural selection,* and by *random processes.* Let's look at each of these mechanisms of evolution in turn.

Humans have long influenced evolution by breeding plants and animals for traits we desire. For example, all breeds of domesticated dogs belong to the same species as the gray wolf, *Canis lupus,* yet dogs exist in an amazing variety of sizes and shapes, ranging from toy poodles to Siberian huskies. **FIGURE 5.8** shows the phylogenetic relationships among the wolf and different breeds of domestic dogs that were bred from the wolf by humans. Beginning with the domestication of wolves, dog breeders have selectively bred individuals that had particular qualities they desired, including body size, body shape, and coat color. After many generations of breeding, the selected traits became more and more exaggerated until breeders felt satisfied that the desired characteristics of a new dog breed had been achieved. As a result of this carefully controlled breeding, we have a tremendous variety of dog sizes, shapes, and colors today. Yet dogs remain a single species: all dog breeds can still mate with one another and produce viable offspring.

When humans determine which individuals breed, typically with a preconceived set of traits in mind, we call the process **evolution by artificial selection.** Artificial selection has produced numerous breeds of horses, cattle, sheep, pigs, and chickens with traits that humans find useful or aesthetically pleasing. Most of our modern agricultural crops are also the result of many years of careful breeding. For example, starting with a single species of wild mustard, *Brassica oleracea,* plant breeders have produced a variety of food crops, including cabbage, cauliflower, broccoli, Brussels sprouts, kale, and kohlrabi (**FIGURE 5.9**).

As useful as artificial selection has been to humans, it can also produce a number of unintended results. For example, farmers often use herbicides to kill weeds. However, as we cover larger and larger areas with herbicides, there is an increasing chance that at least one weed will possess a mutation that allows it to survive the herbicide application. If that one mutant plant passes on its herbicide resistance to its offspring, we will have artificially selected for herbicide resistance in that weed. This process is occurring in many parts of the world, where increased use of the popular herbicide Roundup (chemical name: glyphosate) has led to the evolution of several species of Roundup-resistant weeds. A similar process has occurred in hospitals, where the use of antibiotics and antibacterial cleaners has caused artificial selection of harmful drug-resistant bacteria. These examples underscore the importance of understanding the mechanisms of evolution and the ways in which humans can purposefully or inadvertently direct the evolution of organisms.

## Evolution by Natural Selection

The process of *evolution by natural selection* is similar to the process of evolution by artificial selection, with one key difference. In **evolution by natural selection,** the environment determines which individuals survive and

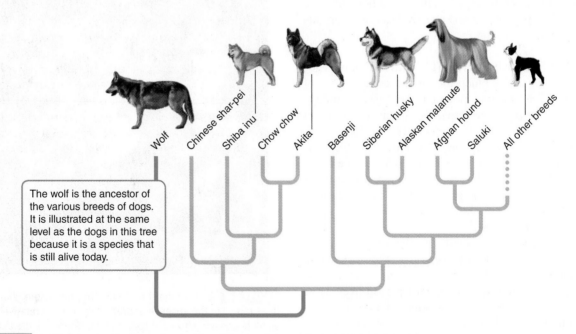

The wolf is the ancestor of the various breeds of dogs. It is illustrated at the same level as the dogs in this tree because it is a species that is still alive today.

Wolf | Chinese shar-pei | Shiba inu | Chow chow | Akita | Basenji | Siberian husky | Alaskan malamute | Afghan hound | Saluki | All other breeds

**FIGURE 5.8** **Artificial selection on animals.** The diversity of domesticated dog breeds is the result of artificial selection on wolves. [After H. G. Parker et al., *Science* 304 (2004): 1160–1164.]

**FIGURE 5.9** **Artificial selection on plants.** Plant breeders have produced a wide range of edible plants from a single species of wild mustard.

reproduce. Simply put, members of a population naturally vary in their traits. Certain combinations of traits make individuals better able to survive and reproduce. As a result, the genes that produce those traits are more common in the next generation.

Prior to the mid-nineteenth century, the idea that species could evolve over time had been suggested by a number of scientists and philosophers. However, the concept of evolution by natural selection did not become synthesized into a unifying theory until two scientists, Alfred Wallace (1823–1913) and Charles Darwin (1809–1882), independently put the various pieces together.

Of the two scientists, Charles Darwin is perhaps the better known. At age 22, he became the naturalist on board HMS *Beagle,* a British survey ship that sailed around the world from 1831 to 1836. During his journey, Darwin made many observations of trait variation across a tremendous variety of species. In addition to observing living organisms, he found fossil evidence of a large number of extinct species. He also recognized that organisms produce many more offspring than are needed to replace the parents, and that most of these offspring do not survive. Darwin questioned why, out of all the species that had once existed on Earth, only a small fraction had survived. Similarly, he wondered why, among all the offspring produced in a population in a given year, only a small fraction survived to the next year. During the decades following his voyage, he developed his ideas into a robust theory. His *On the Origin of Species by Means of Natural Selection,* published in 1859, changed the way people thought about the natural world.

The key ideas of *Darwin's theory of evolution by natural selection* are the following:

- Individuals produce an excess of offspring.

- Not all offspring can survive.

- Individuals differ in their traits.

- Differences in traits can be passed on from parents to offspring.

- Differences in traits are associated with differences in the ability to survive and reproduce.

**FIGURE 5.10** shows how this process works using the example of body size in crustaceans.

Both artificial and natural selection begin with the requirement that individuals vary in their traits and that these variations are capable of being passed on to the next generation. In both cases, parents produce more offspring than necessary to replace themselves, and some of those offspring either do not survive or do not reproduce. But in the case of artificial selection, humans decide which individuals get to breed, based on which individuals possess the traits that tend toward some predetermined goal, such as a curly coat or large size. Natural selection does not select for specific traits that tend toward some predetermined goal. Rather, natural selection favors any combination of traits that improves an individual's **fitness:** its ability to survive and reproduce. Traits that improve an individual's fitness are called **adaptations.**

Natural selection can favor multiple solutions to a particular environmental challenge, as long as each solution improves an individual's ability to survive and reproduce. For example, all plants living in the desert face the challenge of low water availability in the soil, but different species have evolved different solutions to this common challenge. Some species have evolved

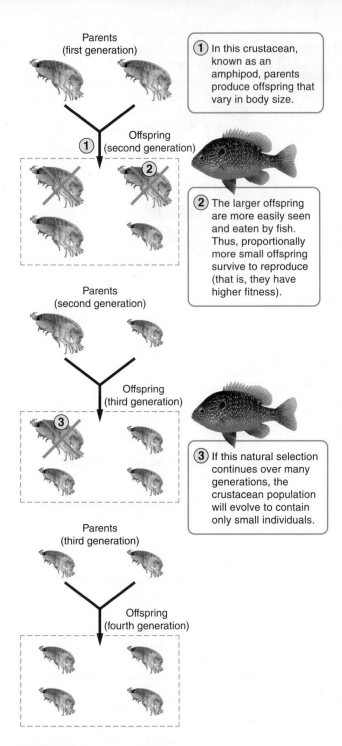

**Parents (first generation)**

① In this crustacean, known as an amphipod, parents produce offspring that vary in body size.

**Offspring (second generation)**

② The larger offspring are more easily seen and eaten by fish. Thus, proportionally more small offspring survive to reproduce (that is, they have higher fitness).

**Parents (second generation)**

**Offspring (third generation)**

③ If this natural selection continues over many generations, the crustacean population will evolve to contain only small individuals.

**Parents (third generation)**

**Offspring (fourth generation)**

**FIGURE 5.10** **Natural selection.** All species produce an excess number of offspring. Only those offspring having the fittest genotypes will pass on their genes to the next generation.

large taproots to draw water from deep in the soil. Other species have evolved the ability to store excess water during infrequent rains. Still other species have evolved waxy or hairy leaf surfaces that reduce water loss. Each of these very different adaptations allows the plants to survive and reproduce in a desert environment (FIGURE 5.11).

## Evolution by Random Processes

Whereas natural selection is an important mechanism of evolution, evolution can also occur by random, or nonadaptive, processes. In these cases, the genetic composition of a population still changes over time, but the changes are not related to differences in fitness among individuals. These random processes, illustrated in FIGURE 5.12, are *mutation, genetic drift, bottleneck effects,* and *founder effects.*

**MUTATION** As mentioned previously, mutations occur randomly. If they are not lethal, they can add to the genetic variation of a population, as shown in Figure 5.12a. The larger the population, the more opportunities there will be for mutations to appear within it. Over time, as the number of mutations accumulates in the population, evolution occurs.

**(a)**

**(b)**

**(c)**

**FIGURE 5.11** **Adaptations.** Desert plants have evolved several different adaptations to their desert environment. (a) The old man of the desert *(Cephalophorus senilis)* has hairs that reduce water loss. (b) The mammillaria cactus *(Mammillaria albiflora)* has a large taproot to draw water from deep in the soil. (c) The waxy outer layers of *Aloe vera* reduce water loss. (Inset) Close-up of waxy layer on an aloe.

**GENETIC DRIFT** **Genetic drift** is a change in the genetic composition of a population over time as a result of random mating. Like mutation, genetic drift is a non-adaptive, random process. It can be particularly important in small populations, in which the random events that affect which individuals mate can most easily alter the genetic composition of the next generation (Figure 5.12b). Imagine a small population of five animals, in which two individuals carry genes that produce black hair and three individuals carry genes that produce white hair. If, by chance, the individuals carrying the genes for black hair fail to find a mate, those genes will not be passed on. The next generation will be entirely white-haired, and the black-haired phenotype will be lost. In

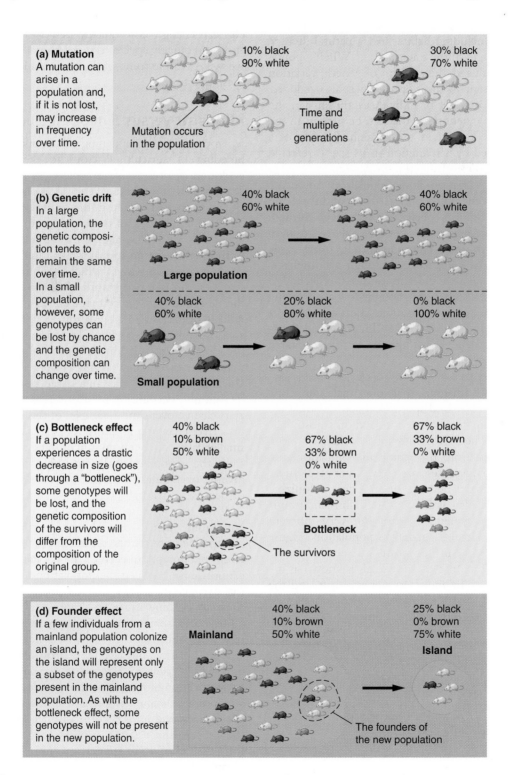

**(a) Mutation**
A mutation can arise in a population and, if it is not lost, may increase in frequency over time.

Mutation occurs in the population

Time and multiple generations

10% black
90% white

30% black
70% white

**(b) Genetic drift**
In a large population, the genetic composition tends to remain the same over time. In a small population, however, some genotypes can be lost by chance and the genetic composition can change over time.

40% black
60% white

40% black
60% white

**Large population**

40% black
60% white

20% black
80% white

0% black
100% white

**Small population**

**(c) Bottleneck effect**
If a population experiences a drastic decrease in size (goes through a "bottleneck"), some genotypes will be lost, and the genetic composition of the survivors will differ from the composition of the original group.

40% black
10% brown
50% white

67% black
33% brown
0% white

67% black
33% brown
0% white

**Bottleneck**

The survivors

**(d) Founder effect**
If a few individuals from a mainland population colonize an island, the genotypes on the island will represent only a subset of the genotypes present in the mainland population. As with the bottleneck effect, some genotypes will not be present in the new population.

**Mainland**

40% black
10% brown
50% white

25% black
0% brown
75% white

**Island**

The founders of the new population

**FIGURE 5.12** **Evolution by random processes.** Populations can evolve by four random processes: (a) mutation, (b) genetic drift, (c) bottleneck effects, and (d) founder effects.

this case, the genetic composition of the population has changed, and the population has therefore evolved. The cause underlying this evolution is random: the failure to find a mate has nothing to do with hair color.

**BOTTLENECK EFFECT** A drastic reduction in the size of a population—known as a *population bottleneck*—may change its genetic composition. There are many reasons why a population might experience a drastic reduction in its numbers, including habitat loss, a natural disaster, hunting, or changes in the environment. When a population's size is reduced, the amount of genetic variation that can be present in the population is also reduced. The smaller the number of individuals, the smaller the number of unique genotypes that can be present in the population. A reduction in the genetic diversity of a population caused by a reduction in its size is referred to as the **bottleneck effect** (Figure 5.12c).

Low genetic variation in a population can cause several problems, including increased risk of disease and low fertility. In addition, species that have been through a population bottleneck are often less able to adapt to future changes in their environment. In some cases, once a species has been forced through a bottleneck, the resulting low genetic diversity causes it to decline to extinction. Such declines are thought to be occurring in a number of species today. The cheetah (*Acinonyx jubatus*), for example, has relatively little genetic variation due to a bottleneck that appears to have occurred 10,000 years ago.

**FOUNDER EFFECT** Imagine that one male and one female of a particular bird species happen to be blown off their usual migration route and land on a hospitable oceanic island. These two individuals will have been drawn at random from the mainland population, and the genotypes they possess are only a subset of those in the original mainland population. These colonizing individuals, or *founders,* will give rise to an island population that has a genetic composition very different from that of the original mainland population (Figure 5.12d). Such a change in a population descended from a small number of colonizing individuals is known as the **founder effect.** Like mutation, genetic drift, and the bottleneck effect, the founder effect is a random process that is not based on differences in fitness.

---

### GAUGE YOUR PROGRESS

✓ What is evolution, and what are the three main ways in which it occurs?

✓ How are artificial and natural selection similar? How are they different?

✓ How does evolution lead to biodiversity?

---

# Speciation and extinction determine biodiversity

Over time, speciation has given rise to the millions of species present on Earth today. Beyond knowing how many species exist, environmental scientists are also interested in understanding how quickly existing species can change, how quickly new species can evolve, and how quickly species can go extinct. In this section we will examine some of the tools available to explore these questions.

## Allopatric versus Sympatric Speciation

Microevolution is happening all around us, from the breeding of agricultural crops to the unintentional evolution of drug-resistant bacteria in hospitals to the bottleneck that reduced genetic variation in the cheetah. But how do we move from the evolution of genetically distinct populations of a species to the evolution of genetically distinct species? That is, how do we move from microevolution to macroevolution?

One common way in which evolution creates new species is through **geographic isolation.** As we saw in the description of the founder effect, a subset of individuals from a larger population may colonize a new area of habitat that is physically separated from that of the rest of the population. For example, a group of birds might colonize a new island created by an erupting volcano. Alternatively, a single large population might be split into two smaller populations as geographic barriers change over time. Examples of this scenario would be a river changing its course, a large lake forming two smaller lakes, or a new mountain range rising. In such cases, the genotypes of the isolated populations might diverge over time, either because of random processes or because of natural selection.

If the two separated habitats differ in environmental conditions, such as temperature, precipitation, or the occurrence of predators, natural selection will favor different phenotypes in each of the habitats. If individuals cannot move between the populations, then over time, the two geographically isolated populations will continue to become more and more genetically distinct. FIGURE 5.13 shows how this happens. Eventually, the two populations will be separated not only by geographic isolation, but also by **reproductive isolation:** they will become so different that even if the physical barrier were removed, they could no longer interbreed and produce viable offspring. At this point, the two populations will have become distinct species. Because this process of speciation requires geographic isolation, it is called **allopatric speciation** (from the Greek *allos,* meaning "other," and *patris,* meaning "fatherland").

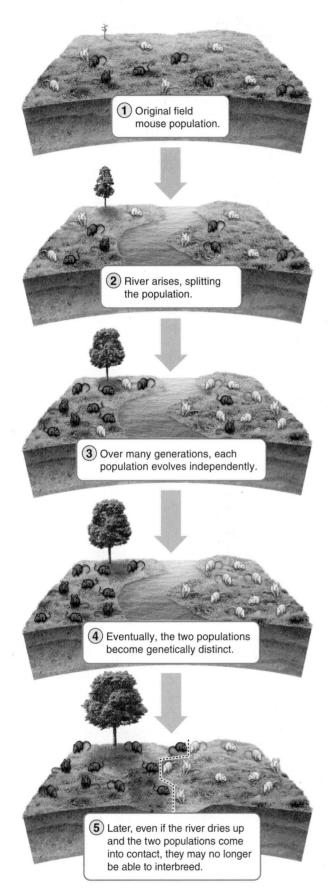

**1** Original field mouse population.

**2** River arises, splitting the population.

**3** Over many generations, each population evolves independently.

**4** Eventually, the two populations become genetically distinct.

**5** Later, even if the river dries up and the two populations come into contact, they may no longer be able to interbreed.

**FIGURE 5.13** **Allopatric speciation.** Geographic barriers can split populations. Natural selection may favor different traits in the environment of each isolated population, resulting in different adaptations. Over time, the two populations may become so genetically distinct that they are no longer capable of interbreeding.

Allopatric speciation is thought to be responsible for the diversity of the group of birds known as Darwin's finches. When Charles Darwin visited the Galápagos Islands, located just west of Ecuador, he noted a large variety of finch species, each of which seemed to live in different habitats or eat different foods. Research on these birds has demonstrated that they all share a common ancestor from the mainland that colonized the islands long ago. **FIGURE 5.14** is a phylogenetic tree for these finches. As Darwin discovered, over a few million years, the finches that were geographically isolated on different islands became genetically distinct and eventually became reproductively isolated.

Allopatric speciation is thought to be the most common way in which evolution generates new species. However, it is not the only way. **Sympatric speciation** is the evolution of one species into two species in the absence of geographic isolation. It usually happens through a process known as *polyploidy*. Most organisms are *diploid:* they have two sets of chromosomes. In polyploidy, the number of chromosomes increases to three, four, or even six sets. Such increases can occur during the division of reproductive cells, either accidentally in nature or as a result of deliberate human actions. Plant breeders, for example, have found several ways to interrupt the normal cell division process. Polyploid organisms include some species of snails and salamanders, 15 percent of all flowering plant species, and a wide variety of agricultural crops such as bananas, strawberries, and wheat. As **FIGURE 5.15** shows for wheat, polyploidy often results in larger plants and larger fruits.

The key feature of polyploid organisms is that once they become polyploid, they generally cannot interbreed with their diploid ancestors. Therefore, at the instant polyploidy occurs, the polyploid and diploid organisms are reproductively isolated from each other and are therefore distinct species, even though they continue to live in the same place.

## The Pace of Evolution

How long does evolution take? A significant change in a species' genotype and phenotype, such as an adaptation to a completely different food source, can take anywhere from hundreds to millions of years. The average global rate of evolution is about one new species every 3 million years. Sometimes evolution can occur rapidly, as in the case of the cichlid fishes of Lake Tanganyika, one of

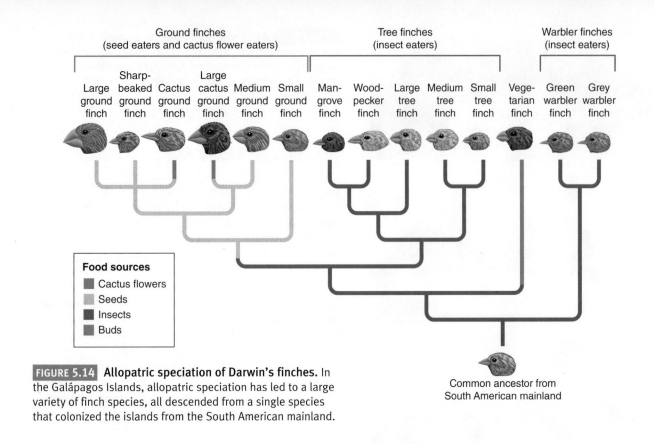

Ground finches
(seed eaters and cactus flower eaters)

Tree finches
(insect eaters)

Warbler finches
(insect eaters)

Large ground finch
Sharp-beaked ground finch
Cactus ground finch
Large cactus ground finch
Medium ground finch
Small ground finch
Man-grove finch
Wood-pecker finch
Large tree finch
Medium tree finch
Small tree finch
Vege-tarian finch
Green warbler finch
Grey warbler finch

**Food sources**
- ■ Cactus flowers
- ■ Seeds
- ■ Insects
- ■ Buds

Common ancestor from
South American mainland

**FIGURE 5.14** **Allopatric speciation of Darwin's finches.** In the Galápagos Islands, allopatric speciation has led to a large variety of finch species, all descended from a single species that colonized the islands from the South American mainland.

**(a)** Einkorn wheat    **(b)** Durum wheat    **(c)** Common wheat

**FIGURE 5.15** **Sympatric speciation.** Flowering plants such as wheat commonly form new species through the process of polyploidy, an increase in the number of sets of chromosomes beyond the normal two sets. (a) The ancestral einkorn wheat (*Triticum boeoticum*) has two sets of chromosomes and produces small seeds. (b) Durum wheat (*Triticum durum*), which is used to make pasta, was bred to have four sets of chromosomes and produces medium-sized seeds. (c) Common wheat (*Triticum aestivum*), which is used mostly for bread, was bred to have six sets of chromosomes and produces the largest seeds. [After http://zr.molbiol.ru/poaceae_znachenije.html.]

the African Great Lakes (**FIGURE 5.16**). Evidence indicates that the roughly 200 different species of cichlids in the lake evolved from a single ancestral species over a period of several million years. During this period, some cichlid species specialized to become insect eaters, others became fish eaters, and still others evolved to eat invertebrates such as snails and clams.

As quickly as the cichlids of Lake Tanganyika evolved, the pupfishes of the Death Valley region of California and Nevada evolved even more rapidly. In the 20,000 to 30,000 years since the large lakes of the region were reduced to isolated springs, several species of pupfish have evolved.

The ability of a species to survive an environmental change depends greatly on how quickly it evolves the adaptations needed to thrive and reproduce under the new conditions. If a species cannot adapt quickly enough, it will go extinct. **FIGURE 5.17** shows the influence of four factors on successful adaptation: *rate of environmental change, genetic variation, population size,* and *generation time.*

**RATE OF ENVIRONMENTAL CHANGE** To survive a rapid environmental change, a population must evolve quickly. An abandoned coal mine, for example, can produce highly acidic water that enters streams and causes a large drop in stream pH in only a few decades. This drop in pH is too rapid for most stream species to adapt to, and

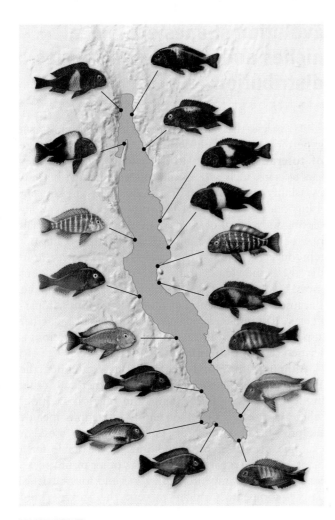

**FIGURE 5.16** **Rapid evolution.** The cichlid fishes of Lake Tanganyika have evolved approximately 200 distinct and colorful species in the relatively short period since the lake formed in eastern Africa. The location where each species can be found in the lake is indicated by a corresponding black dot on the map. [After http://www.uni-graz.at/~sefck/.]

many stream species have gone locally extinct under these circumstances. If the rate of change in acidity had been slower, more species might have adapted to the change by evolving greater tolerance for the acidic conditions.

**GENETIC VARIATION** Species with high genetic variation contain individuals with a wide variety of phenotypes. Their variation makes it more likely that at least some individuals will be well suited to the new environmental conditions. Species that have very large populations across wide geographic areas—such as the white-tailed deer (*Odocoileus virginianus*), which lives across much of North America—typically have the greatest amount of genetic variation. Such high genetic variation allows more rapid evolution by natural selection.

**POPULATION SIZE** Although small populations generally have less genetic variation than large ones, if a beneficial

mutation does happen to occur in a small population, it can spread more quickly. In a small population, an individual with a mutation will breed with a number of mates that represents a relatively large proportion of the entire population. In a larger population, the mutation would spread more slowly because the individual with the mutation would breed with the same number of mates, but those mates would represent a very small fraction of the population. Small populations are also more likely than large ones to undergo rapid evolution by random processes such as genetic drift.

**GENERATION TIME** Species that become reproductively mature in a short time tend to evolve more quickly

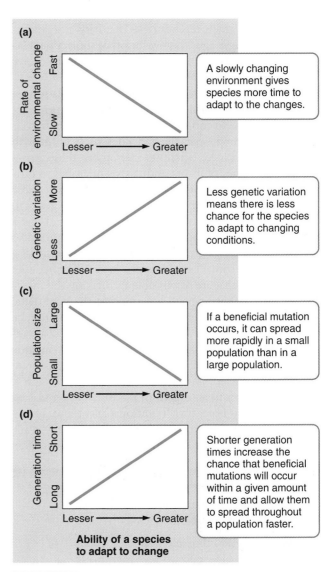

**FIGURE 5.17** **Factors determining the pace of evolution.** A species can better adapt to an environmental change if (a) the rate of the environmental change is relatively slow, (b) the population has high genetic variation for selection to act on, (c) the population is relatively small, and (d) the generation time is short.

than species that require a long time to mature. If it takes 100 generations to produce a change in the genetic composition of a species, then that change will occur more quickly if the time between generations is a few weeks than if it is a few years.

The pace of evolution by artificial selection can be incredibly fast. Such rapid evolution is occurring in many species of commercially harvested fish, including the Atlantic cod (*Gadus morhua*). Intensive fishing over several decades has targeted the largest adults, selectively removing most of those individuals from the population, and therefore also removing the genes that produce large adults. Because larger fish tend to reach sexual maturity later, the genes that code for a later onset of sexual maturity have also been removed. As a result, after just a few decades of intensive fishing, the Atlantic cod population has evolved to reach reproductive maturity at a smaller size and a younger age. This evolution of shorter generation times also means that the cod may be able to evolve even faster in the future.

Evolution occurs even more rapidly in populations of *genetically modified organisms*. Using **genetic engineering** techniques, scientists can now copy genes from a species with some desirable trait, such as rapid growth or disease resistance. Scientists can insert these genes into other species of plants, animals, or microbes to produce **genetically modified organisms.** When those organisms reproduce, they pass on the inserted genes to their offspring. For example, scientists have found that a soil bacterium (*Bacillus thuringiensis*) naturally produces an insecticide as a defense against being consumed by insects in the soil. Plant breeders have identified the bacterial genes that are responsible for making the insecticide, copied those genes, and inserted them into the genomes of crop plants. Such crops naturally produce their own insecticide, which makes them less attractive to insect herbivores. Common examples include Bt-corn and Bt-cotton, so named because they contain genes from *Bacillus thuringiensis*. As you might guess, inserting genes into an organism is a much faster way to produce desired traits than traditional plant and animal breeding, which can only select from the naturally available variation in a population.

---

**GAUGE YOUR PROGRESS**

✓ How does geographic isolation lead to reproductive isolation?

✓ What factors influence a species' chances of adapting successfully to a change in its environment?

✓ Why is the pace of human-driven evolution faster than that of natural evolutionary processes?

---

# Evolution shapes ecological niches and determines species distributions

Every species has an optimal environment in which it performs particularly well. All species have a **range of tolerance,** or limits to the abiotic conditions they can tolerate, such as extremes of temperature, humidity, salinity, and pH. **FIGURE 5.18** illustrates this concept using one environmental factor: temperature. As conditions move further away from the ideal, individuals may be able to survive, and perhaps even grow, but not reproduce. As conditions continue to move away from the ideal, individuals can only survive. If conditions move beyond the range of tolerance, individuals will die. Because the combination of abiotic conditions in a particular environment fundamentally determines whether a species can persist there, the suite of ideal conditions is termed the **fundamental niche** of the species.

Although the fundamental niche establishes the abiotic limits on a species' persistence, there are often biotic factors that further limit the physical locations where it can live. Common biotic limitations include the presence of competitors, predators, and diseases. For example, even if abiotic conditions are favorable for a plant species in a particular location, other plant species may be better competitors for water and soil nutrients. Those competitors might prevent the species from growing in that environment. Similarly, even if a small rodent can tolerate the temperature and humidity of a tropical forest, if a deadly disease is present, the disease might kill much of the rodent population and effectively prevent the species from persisting in the forest. Thus, although the match between a species' traits and the range of possible abiotic conditions defines the species' fundamental niche, the addition of biotic factors more narrowly defines the parts of the fundamental niche that a species actually uses. The range of abiotic and biotic conditions under which a species actually lives is called its **realized niche.** Once we understand what contributes to the realized niche of a species, we have a better understanding of the species' **distribution,** or the areas of the world in which the species lives.

When we examine the realized niches of species in nature, we see that some species can live under a very wide range of abiotic or biotic conditions, whereas others can live only under a very narrow range of conditions. For example, some insects, such as the meadow spittlebug *Philaenus spumarius,* feed on a wide variety of plant species. Because they can live in a variety of habitats or feed on a variety of species, organisms such as the meadow spittlebug are considered **niche generalists.** However, other insects, such as the skeletonizing leaf beetle *Trirhabda virgata,* feed on only a single species

FIGURE 5.18 **Fundamental niche.** All species have an ideal range of abiotic conditions, such as temperature, under which their members can survive, grow, and reproduce. Under more extreme conditions, their ability to perform these essential functions declines.

or genus of plant. Because organisms such as the skeletonizing leaf beetle are specialized to live in a specific habitat or feed on a small group of species, they are called **niche specialists.** Niche specialists can persist quite well when environmental conditions remain relatively constant, but they are vulnerable to extinction if conditions change because the loss of a favored habitat or food source leaves them with few alternatives for survival (FIGURE 5.19). In contrast, niche generalists should fare better under changing conditions because they have a number of alternative habitats and food sources available.

## Environmental Change and Species Distributions

Given what we know about the adaptations of species to environmental conditions, we should not be surprised that changes in environmental conditions have the potential to affect species' distributions. How do we determine whether or not this can actually happen?

Some evidence is found in the layers of sediments that have accumulated over time at the bottoms of modern lakes. Each sediment layer contains pollen from

**(a)**                                                                 **(b)**

FIGURE 5.19 **Generalists and specialists.** (a) Some organisms, such as this meadow spittlebug, are niche generalists that have broad diets and wide habitat preferences. (b) Other organisms, such as this skeletonizing leaf beetle, are niche specialists with narrow diets and highly specific habitat preferences.

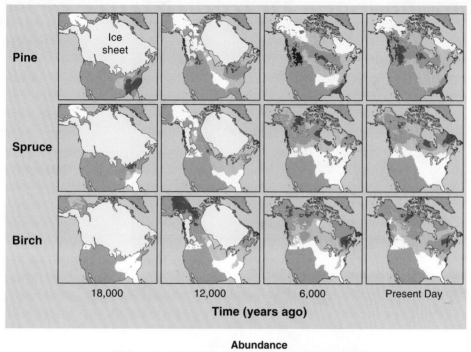

Pollen recovered from lake sediments indicates that plant species moved north as temperatures warmed following the retreat of the glaciers, beginning about 12,000 years ago. Areas shown in color or white were sampled for pollen, whereas areas shown in gray were not sampled. [After http://veimages.gsfc.nasa.gov//3453/boreal_model.gif.]

plants that lived in the region when the sediments were deposited. In some cases, this pollen record goes very far back. For example, lakes in much of northern North America formed 12,000 years ago at the ending of the last ice age, when temperatures warmed and the glaciers slowly retreated to the north. The retreating glaciers left behind a great deal of barren land, which was quickly colonized by plants, including trees. Some of the pollen produced by these trees fell into lakes and was buried in the lake sediments. Scientists can determine the ages of these sediment layers with carbon dating (as described in Chapter 2). Furthermore, because each tree species has uniquely shaped pollen, they can determine when particular tree species arrived near a particular lake and how the entire community of plant species changed over time. FIGURE 5.20 shows pollen records for three tree species in North America. These pollen records make it clear that changes in climatic conditions after the ice age produced substantial changes in the distributions of plants over time. As plants moved north with the retreat of the glaciers, animals followed.

Because past changes in environmental conditions have led to changes in the distributions of species, it is reasonable to ask whether current and future changes in environmental conditions might cause changes in species distributions. For example, as the global climate warms, some areas of the world are expected to receive less precipitation, while other areas are predicted to receive more. If our predictions of future environmental changes are correct, and if we have a good understanding of the niche requirements of many species, we should be able to predict how the distribution of species will change in the future. North American trees, for example, are expected to have more northerly distributions with future increases in global temperatures. As FIGURE 5.21 shows, the red pine (*Pinus resinosa*) is expected to move north to Canada and be nearly absent from the United States by 2100, whereas the loblolly pine (*Pinus taeda*) is expected to move from its far southern distribution and become common throughout the eastern half of the United States.

Species vary in their ability to move physically across the landscape as the environment changes. Some species are highly mobile at particular life stages: adult birds and wind-dispersed seeds, for example, move across the landscape easily. Other organisms, such as the desert tortoise (*Gopherus agassizii*), are slow movers. Furthermore, the movements of many species may be impeded by anthropogenic obstacles, including roads and dams. It remains unclear how species that face these challenges will shift their distributions as global climate change occurs.

## Environmental Change and Species Extinctions

If environmental conditions change, species that cannot adapt to the changes or move to more favorable environments will eventually go extinct. The average life span of a species appears to be only about 1 million to 10 million years. In fact, 99 percent of the species that have ever lived on Earth are now extinct.

There are several reasons why species might go extinct. First, there may be no favorable environment

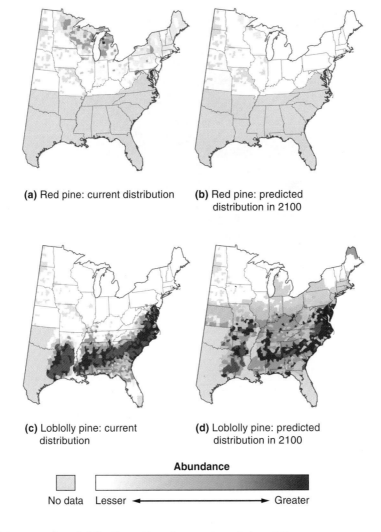

**(a)** Red pine: current distribution

**(b)** Red pine: predicted distribution in 2100

**(c)** Loblolly pine: current distribution

**(d)** Loblolly pine: predicted distribution in 2100

**Abundance**

No data    Lesser ⟵——————⟶ Greater

**FIGURE 5.21** **Predicting future species distributions.** Based on our knowledge of the niche requirements of different tree species, we can predict how their distributions might change as a result of future changes in environmental conditions. [After http://www.fs.fed.us/ne/delaware/atlas/web_atlas.html#.]

that is geographically close enough to move to. This is the current situation of the polar bear (*Ursus maritimus*), which depends on sea ice as a vital habitat for hunting seals, one of its main prey. Because of rising global temperatures, the Arctic sea ice now melts 3 weeks earlier than it did 20 years ago, leaving less time for the bears to hunt. As a result, polar bears observed near Hudson Bay in Canada are in poorer condition than they were 30 years ago, with males weighing an average of 67 kg (150 pounds) less.

Even if there is an alternative favorable environment to which a species can move, it may already be occupied by other species against which the moving populations cannot successfully compete. For example, the predicted northern movement of the loblolly pine, shown in Figure 5.21, might not happen if another pine tree species in the northern United States is a better competitor and prevents the loblolly pine from surviving in that area.

Finally, an environmental change may occur so rapidly that the species does not have time to evolve new adaptations.

**THE FOSSIL RECORD** Much of what we know about the evolution of life is based on **fossils,** the remains of organisms that have been preserved in rock. Most dead organisms decompose rapidly, and the elements they contain are recycled; in this case, nothing of the organism is preserved. Occasionally, however, organic material is buried and protected from decomposition by mud or other sediments. That material may eventually become *fossilized,* or hardened into rocklike material, as it is buried under successive layers of sediment (**FIGURE 5.22**). When these layers are uncovered, they reveal a record of at least some of the organisms that existed at the time the sediments were deposited. Because of the way layers of sediment are deposited on top of one another over time, the oldest fossilized

**FIGURE 5.22** **Fossils.** Fossils, such as this fish discovered in Fossil Butte National Monument in Wyoming, are a record of evolution.

organisms are found in the deepest layers of the fossil record. Thus we can use the fossil record to determine when different species existed on Earth.

**THE FIVE GLOBAL MASS EXTINCTIONS** Throughout Earth's history, individual species have evolved and gone extinct at random intervals. But the fossil record has revealed five periods of global **mass extinction,** in which large numbers of species went extinct over relatively short periods of time. The times of these mass extinctions are shown in **FIGURE 5.23.** Note that because species are not always easy to discriminate in the fossil record, scientists count the number of genera, rather than species, that once roamed Earth but are now extinct.

The greatest mass extinction on record took place 251 million years ago. Roughly 90 percent of marine species and 70 percent of land vertebrates went extinct

during this time. The cause of this mass extinction is not known.

A better-known mass extinction occurred at the end of the Cretaceous period (65 million years ago), when roughly one-half of Earth's species, including the dinosaurs, went extinct. The cause of this mass extinction has been the subject of great debate, but there is now a near consensus that a large meteorite struck Earth and produced a dust cloud that circled the planet and blocked incoming solar radiation. The result was an almost complete halt to photosynthesis, and thus an almost total lack of food at the bottom of the food chain. Among the few species that survived was a small squirrel-sized primate that was the ancestor of humans.

Many scientists view extinctions as the ultimate result of change in the environment. Environmental scientists can learn about the potential effects of both large

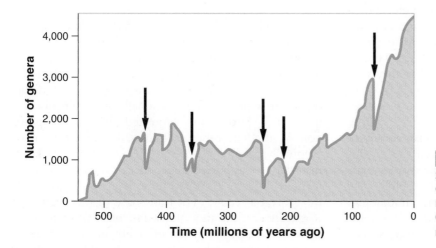

**FIGURE 5.23** **Mass extinctions.** Five global mass extinction events have occurred since the evolution of complex life roughly 500 million years ago. [After GreenSpirit, http://www.greenspirit.org.uk/resources/TimeLines, jpg.]

and small environmental changes by studying historic environmental changes and applying the lessons learned to help predict the effects of the environmental changes that are taking place on Earth today.

**THE SIXTH MASS EXTINCTION** During the last two decades, scientists have reached a consensus that we are currently experiencing a sixth global mass extinction of a magnitude within the range of the previous five mass extinctions. Estimates of extinction rates vary widely, ranging from 2 percent to as many as 25 percent of species going extinct by 2020. However, in contrast to some previous mass extinctions, there is agreement among scientists that the current mass extinction has human causes. These wide-ranging causes include habitat destruction, overharvesting, introductions of invasive species, climate change, and emerging diseases. We will examine all of these factors in detail in Chapters 18 and 19. Because much of the current environmental change caused by human activities is both dramatic and sudden, environmental scientists contend that many species may not be able to move or adapt in time to avoid extinction.

The recovery of biodiversity from earlier mass extinctions took about 10 million years, an unthinkably long time from a human perspective. Recovery from the present mass extinction could take just as long: 500,000 human generations. Much of the current debate among environmental scientists and government officials centers on the true magnitude of this crisis and on the costs of reducing the human impact on extinction rates.

### GAUGE YOUR PROGRESS

✓ How do fundamental niches and realized niches differ?

✓ How does environmental change determine species distribution? When does it lead to extinction?

✓ How are human activities affecting extinction rates, and why is their impact a particular concern?

# WORKING TOWARD SUSTAINABILITY

For over 50 years, The Nature Conservancy (TNC) has protected biodiversity using a simple strategy: buy it. The Conservancy uses grants and donations to purchase privately owned natural areas or to buy development rights to those areas. TNC owns over 0.8 million hectares (2 million acres) of land and has protected over 46 million hectares (115 million acres) of land by buying development rights. As a nonprofit, nongovernmental organization, TNC has great flexibility to use innovative conservation and restoration techniques on natural areas in its possession.

TNC focuses its efforts on areas containing rare species or high biodiversity, including the Florida Keys in southern Florida and Santa Cruz Island in California. Recently, it has set its sights on the oceans, including coastal marine ecosystems. Coastal ecosystems have experienced steep declines in the populations of many fish and shellfish, including oysters, clams, and mussels, due to a combination of overharvesting and pollution (FIGURE 5.24). By preserving these coastal ecosystems, TNC hopes to create reserves that will serve as breeding grounds for declining populations of overharvested species. In this way, protecting a relatively small area of ocean will benefit the much larger unprotected areas, and even the very industries that have led to the population declines.

## Buying the Oceans?

Shellfish are particularly valuable in many coastal ecosystems because they are filter feeders: they remove tiny organisms, including algae, from large quantities of water, cleaning the water in the process. However, shellfish worldwide have been harvested

**FIGURE 5.24 Buying the oceans.** Because the ocean floor cannot be privately owned, The Nature Conservancy has implemented a plan to lease the harvesting rights to imperiled areas and then either not harvest shellfish in the area or harvest in a sustainable way.

unsustainably, leading to a cascade of effects throughout many coastal regions. For example, oyster populations in the Chesapeake Bay were once sufficient to filter the water of the entire bay in 3 to 6 days. Now there are so few individuals that it would take a year for them to filter the same amount of water. As a result, the bay has become much murkier, and excessive algae have led to lowered oxygen levels that make the bay less hospitable to fish.

Conserving marine ecosystems is particularly challenging because private ownership is rare. State and federal governments generally do not sell areas of the ocean. Instead, they have allowed industries to lease the harvesting or exploitation rights to marine resources such as oil, shellfish, and physical space for marinas and aquaculture. So how can TNC protect coastal ecosystems if they cannot buy an area of the ocean? The Nature Conservancy's strategy is to purchase harvesting and exploitation rights and use them as a conservation tool. In some cases, TNC will not harvest any shellfish in order to allow the populations to rebound. In many cases, the leases require at least some harvesting, and TNC hopes to demonstrate sustainable management practices that will serve as an example of how shellfish harvests can be conducted while restoring the shellfish beds.

In 2002, TNC acquired the rights to 4,650 ha (11,500 acres) of oyster beds in New York's Great South Bay, along the southern edge of Long Island. These rights, which were donated to TNC by the Blue Fields Oyster Company, were valued at $2 million. TNC plans to begin by developing restoration strategies. After the oyster populations have rebounded, TNC hopes to engage in sustainable harvesting over part of this area and conduct research in the rest of it. TNC has similar projects under way off the coasts of Virginia, North Carolina, and Washington State. In California, TNC has purchased trawling permits, and by allowing them to go unused, has secured a no-trawl area the size of Connecticut. By 2009, TNC had accumulated the rights to 10,000 ha (25,000 acres) of marine fisheries along the coasts of the United States. In the future, they hope to lease these permits to other harvesters of fish and shellfish that will use sustainable practices.

**References**

The Nature Conservancy. 2002. Leasing and Restoration of Submerged Lands: Strategies for Community-Based, Watershed-Scale Conservation.

The Nature Conservancy. 2006. Annual Report. http://www .nature.org/aboutus/annualreport/file/annualreport2008.pdf.

## REVISIT THE KEY IDEAS

■ **Explain the concept of biodiversity and how it is measured.**

Biodiversity exists at three scales: ecosystem diversity, species diversity, and genetic diversity. Environmental scientists measure species diversity both by the number of species in a particular location (species richness) and by how evenly individuals are distributed among those species (species evenness). The greater the number of species, and the more even the distribution, the higher the diversity.

■ **Describe the ways in which evolution can occur.**

Evolution can occur through artificial selection, natural selection, or random processes. Artificial selection occurs when humans determine which individuals will mate and pass on their genes to the next generation to achieve a predetermined suite of traits. Natural selection does not favor a predetermined suite of traits, but simply favors those individuals that are best able to survive and reproduce. Random processes (mutation, genetic drift, bottleneck effects, and founder effects) do not favor a

predetermined suite of traits, nor do they favor individuals with the highest fitness.

■ **Explain how environmental change affects speciation and extinction.**

Allopatric and sympatric speciation are two ways in which new species can evolve. Four factors that affect a species' ability to adapt to environmental change are the rate of environmental change, the amount of genetic variation within the species, population size, and generation time.

■ **Explain the concept of an ecological niche.**

Evolution by natural selection favors combinations of traits that perform well under particular environmental conditions. As a result, each species has a range of preferred abiotic conditions that constitute its fundamental niche. This fundamental niche is further restricted by biotic factors, including competition, predation, and disease, to form the species' realized niche. Changes in environmental conditions therefore have the potential to change species' distributions.

1. Which of the following is *not* a measure of biodiversity?
   (a) Economic diversity
   (b) Ecosystem diversity
   (c) Genetic diversity
   (d) Species diversity
   (e) Species richness

2. The table below represents the number of individuals of different species that were counted in three forest communities. Which of the following statements *best* interprets these data?

| Species | Community A | Community B | Community C |
|---|---|---|---|
| Deer | 95 | 20 | 10 |
| Rabbit | 1 | 20 | 10 |
| Squirrel | 1 | 20 | 10 |
| Mouse | 1 | 20 | 10 |
| Chipmunk | 1 | 20 | 10 |
| Skunk | | | 10 |
| Opossum | | | 10 |
| Elk | | | 10 |
| Raccoon | | | 10 |
| Porcupine | | | 10 |

   (a) Community A has greater species evenness than Community B.
   (b) Community A has greater species richness than Community B.
   (c) Community B has greater species evenness than Community C.
   (d) Community C has greater species richness than Community A.
   (e) Community A has greater species evenness than Community C.

3. Which of the following is an example of artificial selection?
   (a) Cichlids have diversified into nearly 200 species in Lake Tanganyika.
   (b) Thoroughbred racehorses have been bred for speed.
   (c) Whales have evolved tails that help propel them through water.
   (d) Darwin's finches have beaks adapted to eating different foods.
   (e) Ostriches have lost the ability to fly.

4. The yellow perch (*Perca flavescens*) is a fish that breeds in spring. A single female can produce up to 40,000 eggs at one time. This species is an example of which of the key ideas of Darwin's theory of evolution by natural selection?

   (a) Individuals produce an excess of offspring.
   (b) Humans select for predetermined traits.
   (c) Individuals vary in their phenotypes.
   (d) Phenotypic differences in individuals can be inherited.
   (e) Different phenotypes have different abilities to survive and reproduce.

5. In 2002, Peter and B. Rosemary Grant studied a population of Darwin's finches on one of the Galápagos Islands that feed on seeds of various sizes. After a drought that caused only large seeds to be available to the birds, they found that natural selection favored those birds that had larger beaks and bodies. Once the rains returned and smaller seeds became much more abundant, however, natural selection favored those birds that had smaller beaks and bodies. Which of the following processes is the *best* interpretation of this scenario?
   (a) Genetic drift            (d) Macroevolution
   (b) Founder effect           (e) Bottleneck effect
   (c) Microevolution

6. The northern elephant seal (*Mirounga angustirostris*) was once hunted to near extinction. Only 20 animals remained alive in 1890. After the species was protected from hunting, its population grew to nearly 30,000 animals, but the large population possesses very low genetic variation. Which of the following processes is the *best* interpretation of this scenario?
   (a) Evolution by natural selection
   (b) Evolution by artificial selection
   (c) Evolution by the founder effect
   (d) Evolution by the bottleneck effect
   (e) Evolution by genetic drift

7. Which of the following statements is *not* correct?
   (a) Most speciation is thought to occur via allopatric speciation.
   (b) Polyploidy is an example of sympatric speciation.
   (c) Speciation can be caused by either natural selection or random processes.
   (d) Geographic isolation can eventually lead to reproductive isolation.
   (e) Speciation cannot occur without geographic isolation.

8. Which of the following allows more rapid evolution?
   (a) Long generation times
   (b) Rapid environmental change
   (c) Large population sizes
   (d) Low genetic variation
   (e) High genetic variation

9. Which of the following conditions does *not* define the fundamental niche of a species?
   (a) Humidity
   (b) Predators
   (c) Temperature
   (d) Salinity
   (e) pH

10. Some scientists estimate that the current global extinction rate is about 30,000 species per year. If there are currently 10,000,000 species on Earth, how long will it take to destroy all of Earth's biodiversity?
    (a) Less than 100 years
    (b) Between 100 and 300 years
    (c) Between 300 and 500 years
    (d) Between 500 and 700 years
    (e) Between 700 and 1,000 years

## APPLY THE CONCEPTS

1. Look at the photograph below and answer the following questions.

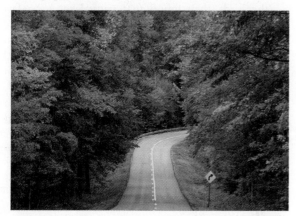

**Forest A**                                    **Forest B**

   (a) Explain how this human impact on a forest ecosystem might affect the ability of some species to move to more suitable habitats as Earth's climate changes.
   (b) Propose and explain *one* alternative plan that could have preserved this forest ecosystem.
   (c) Distinguish between the terms *microevolution* and *macroevolution*. Explain how the organisms in forest A could evolve into species different from those in forest B.

2. Read the following article, which appears courtesy of The University of Texas Health Science Center at San Antonio, and answer the questions that follow.

---

### Drug-Resistant *E. coli* and *Klebsiella* Bacteria Found in Hospital Samples and Elsewhere in U.S.

A research team from The University of Texas Health Science Center at San Antonio, examining bacterial isolates obtained in hospital and non-hospital clinical settings between 2000 and 2006, has identified drug-resistant strains of *E. coli* and *Klebsiella* bacteria in more than 50 blood, urine and respiratory samples. These resistant strains, which resemble bacteria reported in Latin America, Asia and Europe, were thought to be rare in the U.S.

"This antibiotic resistance problem is likely to become widespread," said paper co-author Jan Evans Patterson, M.D., professor of medicine, infectious diseases and pathology at the UT Health Science Center. "It affects the way we will treat infections in the future. In the past, we were concerned with antibiotic resistance in the hospital primarily, but in this review many of the strains we detected were from the community. This tells us antibiotic resistance is spreading in the community, as well, and will affect how we choose antibiotics for outpatient infections."

If the trend continues, it may become difficult to select appropriate antibiotic therapy for urinary tract infections, for example. "The trend over the last decade has been to treat urinary infections empirically, to pick the drug that has worked," said James Jorgensen, Ph.D., professor of pathology, medicine, microbiology and clinical laboratory sciences at the Health Science Center. "Now it is important for physicians to culture the patient's urine to be sure they have selected the right antibiotic. The top three drugs that are often prescribed may not be effective with these resistant bacteria."

   (a) Explain how drug-resistant strains of bacteria could evolve in a hospital.
   (b) According to the article, what is it that the scientists are now concerned about that they were not concerned about in the past?
   (c) Explain how new drugs could be viewed as restricting the fundamental niche of a particular bacterial species.
   (d) Propose two possible solutions to the current problem of drug–resistant bacteria.

## MEASURE YOUR IMPACT

**The True Cost of a Green Lawn** One area of potential biodiversity improvement that many people overlook is their own lawn. Mowing a lawn and applying herbicides and fertilizer typically reduces plant diversity to only a few species of grasses. In contrast, unmowed fields can contain dozens of plant species, including many species of wildflowers that are not only aesthetically pleasing but also promote a high diversity of animal species.

(a) Given that approximately 85 million Americans have a lawn, and that the average lawn size is 0.08 ha (0.2 acres), how much total land area is composed of lawns in the United States?

(b) If every lawn owner set aside 10 percent of his or her lawn and let it grow into an area of natural wildflowers, how many hectares of this higher-biodiversity land would be added?

(c) Given that lawn owners spend a total of $40 billion on professional lawn care services each year, what would be the annual savings on lawn care services if 10 percent of all lawns were set aside to grow natural wildflowers?

(d) Approximately 2.2 billion liters (0.6 billion gallons) of gasoline are used annually for lawn mowers. If gas costs $0.80 per liter ($3.00 per gallon), how many total dollars could be saved in the United States each year if lawn owners stopped mowing 10 percent of their lawns?

(e) What do these data suggest about the economics of preserving biodiversity?

## ENGAGE YOUR ENVIRONMENT

Does species richness differ in areas of human disturbance? In this activity you'll survey two habitats, one disturbed and one undisturbed, and you will compare the species richness of both. *See Engage Your Environment Activity 5 at the end of this book.*

# science applied

# How Should We Prioritize the Protection of Species Diversity?

As a result of human activities, we have seen a widespread decline in biodiversity across the globe. Many people agree that we should try to slow or even stop this loss. But how do we proceed? Ideally, we might want to preserve all biodiversity. In reality, preserving biodiversity requires compromises. For example, in order to preserve the biodiversity of an area, we might have to set aside land that would otherwise be used for housing developments, shopping malls, or strip mines. If we cannot preserve all biodiversity, how do we decide which species receive our attention?

In 1988, Oxford University professor Norman Myers noted that much of the world's biodiversity is concentrated in areas that make up a relatively small fraction of the globe. Part of the reason for this uneven pattern of biodiversity is that so many species are *endemic species*. **Endemic species** are species that live in a very small area of the world and nowhere else, often in isolated locations such as the Hawaiian Islands. Because they are home to so many endemic species, these isolated areas end up containing a high proportion of all the species found on Earth. Myers called these areas **biodiversity hotspots.**

Scientists originally identified 10 biodiversity hotspots, including Madagascar, western Ecuador, and the Philippines. Myers argued that these 10 areas were in need of immediate conservation attention because human activities there could have disproportionately large negative effects on the world's biodiversity. A year later, the group Conservation International adopted Myers's concept of biodiversity hotspots to guide its conservation priorities. As of 2010, Conservation International had identified the 34 biodiversity hotspots shown in FIGURE SA2.1. Although these hotspots collectively represent only 2.3 percent of the world's land area, 50 percent of all plant species and 42 percent of all vertebrate species are confined to these areas. As a result of this categorization, major conservation organizations have adjusted their funding priorities and are spending hundreds of millions of dollars to conserve these areas. What does environmental science tell us about the hotspot approach to conserving biodiversity?

## What makes a hotspot hot?

Since Norman Myers initiated the idea of biodiversity hotspots, scientists have debated which factors should be considered most important when deciding where to focus conservation efforts. For example, most scientists agree that species richness is an important factor. There are more than 1,300 bird species in the small nation of Ecuador—more than twice the number of bird species living in the United States and Canada. For this reason, protecting a habitat in Ecuador has the potential to save many more bird species than protecting the same amount of habitat in the United States. From this point of view, the choice to protect areas with a lot of species makes sense.

Identifying biodiversity hotspots is challenging, however, because scientists have not yet discovered and identified all the species on Earth. Because the distribution of plant diversity is typically much better known than that of animal diversity, the most practical way to

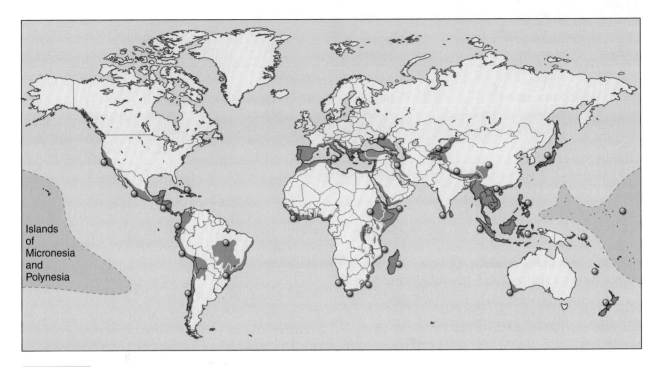

**FIGURE SA2.1 Biodiversity hotspots.** Conservation International has identified 34 biodiversity hotspots that have at least 1,500 endemic plant species and a loss of at least 70 percent of all vegetation.

identify hotspots has been to locate areas containing high numbers of endemic plant species. It is reasonable to expect that areas with high plant diversity will contain high animal diversity as well. Conservation International requires that areas it identifies as hotspots contain at least 1,500 endemic plant species. By conserving plant diversity, the hope is that we will simultaneously conserve animal diversity, especially for those groups, such as insects, that are poorly cataloged.

In addition to requiring high numbers of endemic plant species, Conservation International requires its hotspots to have lost more than 70 percent of the vegetation that contains those endemic plant species. In this way, high-diversity areas with a high level of habitat loss receive the highest conservation priority. High-diversity areas that are not being degraded receive lower conservation priority.

## What else can make a hotspot hot?

The number of endemic species in an area is undoubtedly important in identifying biodiversity hotspots, but other scientists have argued that this criterion alone is not enough. They suggest that we also consider the total number of species in an area or the number of species currently threatened with extinction in an area.

Would all three approaches identify similar regions of conservation priority? A recent analysis of birds suggests they would not. When scientists identified bird diversity hotspots using each of the three criteria—endemic species, total species richness, and threatened species—their results, shown in Table SA2.1, identified very different areas. Scientists using the three criteria to identify hotspots of mammal diversity reached the same conclusion.

As we can see in Table SA2.1, some areas of the world that have high species richness do not contain high numbers of endemic species. The Amazon, for example, has a high number of bird species, but not a particularly high number of endemic bird species compared with other more isolated regions of the world, such as the islands of the Caribbean. Similarly, areas with high numbers of threatened or endangered species do not always have high numbers of endemic species. These findings highlight the critical problem of deciding whether conservation efforts should be focused on areas containing the greatest number of species, areas containing the greatest number of threatened species, or areas containing the greatest number of endemic species. All three approaches are reasonable.

In addition to considering species diversity, some scientists have argued that we must also consider the size

| Table SA2.1 | Biodiversity hotspots for birds, identified by three criteria. | | |
|---|---|---|---|
| Rank | Total number of species | Number of endemic species | Number of threatened species |
| 1 | Andes | Andes | Andes |
| 2 | Amazon Basin | New Guinea and Bismarck Archipelago | Amazon Basin |
| 3 | Western Great Rift Valley | Panama and Costa Rica highlands | Guyana highlands |
| 4 | Eastern Great Rift Valley | Caribbean | Himalayas |
| 5 | Himalayas | Lesser Sunda Islands | Atlantic coastal forest, Brazil |

*Source:* Data from C. D. Orme et al., Global hotspots of species richness are not congruent with endemism or threat, *Nature* 436 (2005): 1016–1019.

of the human population in diverse areas. For example, we might expect that natural areas containing more people face a greater probability of being affected by human activities. Furthermore, if we wish to project into the future, we must consider not only the size of the human population today, but also the expected size of the human population several decades from now. Scientists have found that many hotspots for endemic species have human population densities that are well above the world's average. Whereas the world has an average human population density of 42 people per square kilometer, the average hotspot has a human population of 73 people per square kilometer. Such places may be at a higher risk of degradation from human activities. This risk should be incorporated into our choices of priority areas for conservation, and it should motivate us to promote smarter development that does not come at the cost of species diversity.

## What are the costs and benefits of conserving biodiversity hotspots?

A focus on regions containing large numbers of species places a clear priority on preserving the largest number of species within a given region. However, it does not explicitly consider the likelihood of succeeding in this goal, nor does it consider the costs associated with the effort. For example, there may be many ways of helping a species persist in an area, including buying habitat, entering into agreements with landowners not to develop their land, or removing threats such as invasive species. Each option will have a different impact on the number of species that will be helped, and each option will have different costs of implementation. Given the limited funds that are available for protecting species, it is certainly worth comparing the expected costs and benefits of different options, both within and among biodiversity hotspots. In this way, we have the potential to maximize the return on our conservation investment.

## What about biodiversity coldspots?

The concept of biodiversity hotspots assumes that our primary goal is to protect the maximum number of species. That goal is an admirable one, but it could come at the cost of many important ecosystems that do not fall within hotspots. Yellowstone National Park, for example, has a relatively low diversity of species, yet it is one of the few places in the United States that contains remnant populations of large mammals, including wolves, grizzly bears, and bison. Does this mean that places such as Yellowstone National Park should receive decreased conservation attention?

Biodiversity coldspots also provide ecosystem services that humans value at least as much as species diversity. For example, wetlands in the United States are incredibly important for flood control, water purification, wildlife habitat, and recreation. Many wetlands, however, have relatively low plant diversity and, as a result, would not be identified as biodiversity hotspots. It is true that increased species richness leads to improved ecosystem services, but only as we move from very low species richness to moderate species richness. Moving from moderate species richness to high species richness generally does not further improve the functioning of an ecosystem. Since very high species diversity is not expected to provide any substantial improvement in ecosystem function, protecting more and more species produces diminishing returns in terms of protecting ecosystem services. Hence, if our primary goal is to preserve the functioning of the ecosystems that improve our lives, we do not necessarily need to preserve every species in those ecosystems.

## How can we reach a resolution?

During the past two decades, it has become clear that scientists and policy makers need to set priorities for the conservation of biodiversity. No single criterion may be agreed upon by everyone. However, it is important to appreciate the bias of each approach and to consider the possible unintended consequences of favoring some geographic regions over others. Our investment in conservation cannot be viewed as an all-or-nothing choice of some areas over others. Instead, our decisions must take into account the costs and benefits of alternative conservation strategies and incorporate current and future threats to both species diversity and ecosystem function. In this way, we can strike a balance between our desire to preserve Earth's species diversity and our desire to protect the functioning of Earth's ecosystems.

### References

Kareiva, P., and M. Marvier. 2003. Conserving biodiversity coldspots. *American Scientist* 91:344–351.

Myers, N., et al. 2000. Biodiversity hotspots for conservation priorities. *Nature* 403:853–858.

Orme, C. D., et al. 2005. Global hotspots of species richness are not congruent with endemism or threat. *Nature* 436:1016–1019.

# Population and Community Ecology: Distribution and Abundance of Species

## New England Forests Come Full Circle

When the Pilgrims arrived in Massachusetts in 1620, they found immense areas of undisturbed temperate seasonal forest containing a variety of tree species, including sugar maple (*Acer saccharum*), American beech (*Fagus grandifolia*), white pine (*Pinus strobus*), and eastern hemlock (*Tsuga canadensis*). Over the next 200 years, settlers cut down most of the trees to clear land for farming and to build houses. This deforestation peaked in the 1800s, at which point up to 80 percent of all New

beetle (*Microrhopala vittata*) that specializes in eating goldenrods. Periodic outbreaks of leaf beetles in the abandoned fields of New England dramatically reduced the populations of goldenrods. With fewer goldenrods, other plant species could compete and prosper.

The complex interactions among populations of goldenrods, insects, and other species created an ever-changing ecosystem. For example, as the leaf beetle population increased in the community, so did the populations of predators and parasites that fed on them. As predators and parasites reduced

> ## The old stone walls are the only evidence that this forest was once farmland.

England forests had been cleared. Between 1850 and 1950, however, many people abandoned their New England farms to take jobs in the growing textile industry. Others moved to the Midwest, where farmland was considerably less expensive.

What happened to the former farmland is a testament to the resilience of the forest ecosystem. The transformation began shortly after the farmers left. Seeds of grasses and wildflowers were carried to the abandoned fields by birds or blown there by the wind. Within a year, the fields were carpeted with a large variety of plant species. Eventually, a single group of plants—the goldenrods—came to dominate the fields by growing taller and outcompeting other species of plants for sunlight. The other species remained in the fields, but they were not very abundant. Nevertheless, the dominance of the goldenrods was short-lived.

Goldenrods and other wildflowers play an important part in old field communities by supporting a diverse group of plant-eating insects. Some of these herbivorous insects are generalists that feed on a wide range of plant species, while others specialize on only a small number of plant species. The number of individuals of each insect species varies from year to year, and occasionally some species experience very large population increases, or *outbreaks*. One such species is a leaf

the leaf beetle population, the goldenrod population began to rebound. As goldenrod populations surged, they once again caused the other plant species to decline in numbers.

Over time, tree seeds arrived, and tree seedlings began to grow. Their presence changed the species composition of the old fields once again. One species in particular, the fast-growing white pine, eventually came to dominate. The pine trees cast so much shade that the goldenrods and other sunlight-loving plant species could not survive.

White pines dominated the old field communities until humans began harvesting them for lumber in the 1900s. Just as the reduction of goldenrod populations made room for other plant species, logging of the white pines made room for broadleaf tree species. Two of these broadleaf species, American beech and sugar maple, are dominant in New England forests today. Thus the abandoned New England fields were slowly transformed into communities that resemble the original forests of centuries ago, with ▶

Goldenrod.

a mix of pines, hemlocks, and broadleaf trees. The old stone walls are the only evidence that this forest was once farmland.

The story of the New England forests shows us that populations can increase or decrease dramatically over time. It also illustrates how species interactions within a community can alter species abundance. Finally, it demonstrates how human activity can alter the distribution and diversity of species within an ecosystem. ■

*Sources:* W. P. Carson and R. B. Root, Herbivory and plant species coexistence: Community regulation by an outbreaking phytophagous insect, *Ecological Monographs* 70 (2000): 73–99; T. Wessels, *Reading the Forested Landscape* (Countryman Press, 1997).

## UNDERSTAND THE KEY IDEAS

There are clear patterns in the distribution and abundance of species across the globe. Understanding the factors that generate these patterns can help us find ways to preserve global biodiversity. These factors include the ways in which populations increase and decrease in size and the ways in which species interact with one another in their communities.

After reading this chapter you should be able to

■ list the levels of complexity found in the natural world.

■ contrast the ways in which density-dependent and density-independent factors affect population size.

■ explain growth models, reproductive strategies, survivorship curves, and metapopulations.

■ describe species interactions and the roles of keystone species.

■ discuss the process of ecological succession.

■ explain how latitude, time, area, and distance affect the species richness of a community.

# Nature exists at several levels of complexity

A New England forest is a wonderful reminder of the intricate complexity of the natural world. As FIGURE 6.1 shows, the environment around us exists at a series of increasingly complex levels: individuals, populations, communities, ecosystems, and the biosphere. The simplest level is the individual—a single organism. Natural selection operates at the level of the individual because it is the individual that must survive and reproduce.

The second level of complexity is the *population*. A **population** is composed of all individuals that belong to the same species and live in a given area at

**Individual**
Survival and reproduction— the unit of natural selection

**Population**
Population dynamics— the unit of evolution

**Community**
Interactions among species

**Ecosystem**
Flow of energy and matter

**Biosphere**
Global processes

**FIGURE 6.1** **Levels of complexity.** Environmental scientists study nature at several different levels of complexity, ranging from the individual organism to the biosphere. At each level, scientists focus on different processes.

a particular time. Evolution occurs at the level of the population. Scientists who study populations are also interested in the factors that cause the number of individuals to increase or decrease.

As we saw in Chapter 1, some populations, such as mosquito larvae living in a puddle of water, inhabit small areas, whereas others, such as the white-tailed deer (*Odocoileus virginianus*) that range across much of North America, inhabit large areas. The boundaries of a population are rarely clear and may be set arbitrarily by scientists. For example, depending on what we want to learn, we might study the entire population of white-tailed deer in North America, or we might focus on the deer that live within a single state, or even within a single forest.

The third level of complexity is the *community*. A **community** incorporates all of the populations of organisms within a given area. Like those of a population, the boundaries of a community may be defined by the state or federal agency responsible for managing it. Scientists who study communities are generally interested in how species interact with one another. Many communities are named for the species that are visually dominant. In the New England forest, for instance, we can talk about the maple-beech-hemlock community that made up the original forest or the goldenrod community that occupied the abandoned fields.

As we saw in Chapter 4, when terrestrial communities in different parts of the world experience similar patterns of temperature and precipitation, those communities can be grouped into *biomes* that contain plants with similar growth forms. Temperate seasonal forests around the world, for example, all contain deciduous trees. However, their actual tree species composition varies from community to community. Thus the temperate seasonal forests of the eastern United States and Europe may at first appear to be very similar, but they contain different tree species.

Communities exist within an *ecosystem,* which consists of all of the biotic and abiotic components in a particular location. Ecosystem ecologists study flows of energy and matter, such as the cycling of nutrients through the system.

The largest and most complex system environmental scientists study is the *biosphere,* which incorporates all of Earth's ecosystems. Scientists who study the biosphere are interested in the movement of air, water, and heat around the globe.

**GAUGE YOUR PROGRESS**

✓ What levels of complexity make up the biosphere?

✓ What do scientists study at each level of complexity?

✓ How do populations and communities differ?

# Population ecologists study the factors that regulate population abundance and distribution

Populations are *dynamic*—that is, they are constantly changing. As **FIGURE 6.2** shows, the exact size of a population is the difference between the number of inputs to the population (births and immigration) and outputs from the population (deaths and emigration) within a given time period. If births and immigration exceed deaths and emigration, the population will grow. If deaths and emigration exceed births and immigration, the population will decline and, over time, will eventually go extinct. The study of factors that cause populations to increase or decrease is the science of **population ecology.**

There are many circumstances in which scientists find it useful to identify the factors that influence population size over time. For example, in the case of endangered species such as the California condor (*Gymnogyps californianus*), knowing the factors that affect a species' population size has helped us implement measures to improve its survival and reproduction. Similarly, knowing the factors that influence the population size of a pest species can help us control it. For instance, population ecologists are currently studying the emerald ash borer (*Agrilus planipennis*), an invasive species in the American Midwest that is causing widespread deaths of ash trees. Once we understand the population ecology of this destructive insect, we can begin to discover and develop strategies to control or eradicate it.

**FIGURE 6.2** **Population inputs and outputs.** Populations increase in size due to births and immigration and decrease in size due to deaths and emigration.

## Population Characteristics

To understand how populations change over time, we must first examine some basic population characteristics. These characteristics are population *size, density, distribution, sex ratio,* and *age structure.*

**POPULATION SIZE Population size** (*N*) is the total number of individuals within a defined area at a given time. For example, the California condor once ranged throughout California and the southwestern United States. Over the past two centuries, however, a combination of poaching, poisoning, and accidents (such as flying into electric power lines) greatly reduced the population's size. By 1987, there were only 22 birds remaining in the wild. Scientists who realized that the species was nearing extinction decided to capture all the wild birds and start a captive breeding program in zoos. As a result of captive breeding and other conservation efforts, the condor population size had increased to more than 300 by 2009.

**POPULATION DENSITY Population density** is the number of individuals per unit area (or volume, in the case of aquatic organisms) at a given time. Knowing a population's density, in addition to its size, can help scientists estimate whether a species is rare or abundant. For example, the density of coyotes (*Canis latrans*) in some parts of Texas might be only 1 per square kilometer, but in other parts of the state it might be as high as 12 per square kilometer. Scientists also study population density to determine whether a population in a particular location is so dense that it might outstrip its food supply.

Population density can be a particularly useful measure for wildlife managers who must set hunting or fishing limits on a species. For example, managers may divide the entire population of an animal species that is hunted or fished into *management zones*. These management zones may be human-defined areas, such as counties, or areas with natural boundaries, such as the major water bodies in a state. Wildlife managers might offer more hunting or fishing permits for high-density zones and fewer permits for low-density zones.

**POPULATION DISTRIBUTION** In addition to knowing a population's size and density, population ecologists are interested in knowing how a population occupies space. **Population distribution** is a description of how individuals are distributed with respect to one another. FIGURE 6.3 shows three types of population distributions. In some populations, such as a population of trees in a natural forest, the distribution of individuals is *random* (Figure 6.3a). In other words, there is no pattern to the locations where individual trees grow.

In other populations, such as a population of trees in a plantation, the distribution of individuals is *uniform*. In

**(a)** Random distribution

**(b)** Uniform distribution

**(c)** Clumped distribution

**FIGURE 6.3 Population distributions.** Populations in nature distribute themselves in three ways. (a) Many of the tree species in this New England forest are randomly distributed, with no apparent pattern in the locations of individuals. (b) Territorial nesting birds, such as these Australasian gannets (*Morus serrator*), exhibit a uniform distribution, in which all individuals maintain a similar distance from one another. (c) Many pairs of eyes are better than one at detecting approaching predators. The clumped distribution of these meerkats (*Suricata suricatta*) provides them with extra protection.

other words, individuals are evenly spaced (Figure 6.3b). Uniform distributions are common among territorial animals, such as nesting birds that defend similar-sized areas around their nests. Uniform distributions are also observed among plants that produce toxic chemicals to prevent other plants of the same species from growing close to them.

In still other populations, the distribution of individuals is *clumped* (Figure 6.3c). Clumped distributions, which are common among schooling fish, flocking birds, and herding mammals, are often observed when living in large groups enhances feeding opportunities or protection from predators.

**POPULATION SEX RATIO** The **sex ratio** of a population is the ratio of males to females. In most sexually reproducing species, the sex ratio is usually close to 50:50. Sex ratios can be far from equal in some species, however. In fig wasps, for example, there may be as many as 20 females for every male. Because the number of offspring produced is primarily a function of how many females there are in the population, knowing a population's sex ratio helps scientists estimate the number of offspring a population will produce in the next generation.

**POPULATION AGE STRUCTURE** Many populations are composed of individuals of varying ages. A population's **age structure** is a description of how many individuals fit into particular age categories. Knowing a population's age structure helps ecologists predict how rapidly a population can grow. For instance, a population with a large proportion of old individuals that are no longer capable of reproducing, or with a large proportion of individuals too young to reproduce, will produce far fewer offspring than a population that has a large proportion of individuals of reproductive age.

## Factors That Influence Population Size

Factors that influence population size can be classified as *density dependent* or *density independent*. We will look at each type in turn.

**DENSITY-DEPENDENT FACTORS** **Density-dependent factors** influence an individual's probability of survival and reproduction in a manner that depends on the size of the population. The amount of available food, for example, is a density-dependent factor. Because a smaller population requires less total food, food scarcity will have a greater negative effect on the survival and reproduction of individuals in a large population than in a small population.

In 1932, Russian biologist Georgii Gause conducted a set of experiments that demonstrated how food supply controls population growth. Gause monitored population growth in two species of *Paramecium* (a type of single-celled aquatic organism) living under ideal conditions in test tubes. Each day he added a constant amount of food. As the graph in FIGURE 6.4a shows, both species of *Paramecium* initially experienced rapid population growth. Over time, however, the rate of growth began to slow. Eventually, the population sizes reached a plateau and remained there for the rest of the experiment.

Gause suspected that *Paramecium* population growth was limited by food supply. To test this hypothesis, he conducted a second experiment in which he doubled the amount of food he added to the test tubes. Again,

both species of *Paramecium* experienced rapid population growth early in the experiment, and the rate of growth slowed over time. However, as FIGURE 6.4b shows, their maximum population sizes were approximately double those observed in the first experiment.

Gause's results confirmed that food is a *limiting resource* for *Paramecium*. A **limiting resource** is a resource that a population cannot live without and which occurs in quantities lower than the population would require to increase in size. If a limiting resource decreases, so does the size of a population that depends on it. For terrestrial plant populations, water and nutrients such as nitrogen and phosphorus are common limiting resources. For animal populations, food, water, and nest sites are common limiting resources.

To better understand density dependence, let's consider a situation in which there is a moderate amount of food available for animals to eat. At low population densities, only a few individuals share this limiting resource, and each individual has access to sufficient quantities. As a result, each individual in the population survives and reproduces well, and the population grows rapidly. At high population densities, however, many more individuals must share the food, so each individual receives a smaller share. With a smaller share, each individual has a lower probability of surviving, and those that do survive produce fewer offspring. As a result, the population grows slowly. In this example, the ability to survive and reproduce depends on the density of the population, so population growth is rapid at low population densities but slow at high population densities.

Similarly, in Gause's *Paramecium* experiments, food was the limiting resource. Population growth slowed as population size increased because there was a limit to how many individuals the food supply could sustain. This limit is called the **carrying capacity** of the environment and denoted as **K.** Knowing the carrying capacity for a species, and what its limiting resource is, helps us predict how many individuals an environment can sustain. This is true whether those individuals are paramecia, cows, or humans.

**DENSITY-INDEPENDENT FACTORS** Density-independent **factors** have the same effect on an individual's probability of survival and amount of reproduction at any population size. A tornado, for example, can uproot and kill a large number of trees in an area. However, a given tree's probability of being killed does not depend on whether it resides in a forest with a high or low density of other trees. Other density-independent factors include hurricanes, floods, fires, volcanic eruptions, and other climatic events. An individual's likelihood of mortality increases during such an event regardless of whether the population happens to be at a low or high density.

Bird populations are often regulated by density-independent factors. For example, in the United Kingdom, a particularly cold winter can freeze the surfaces

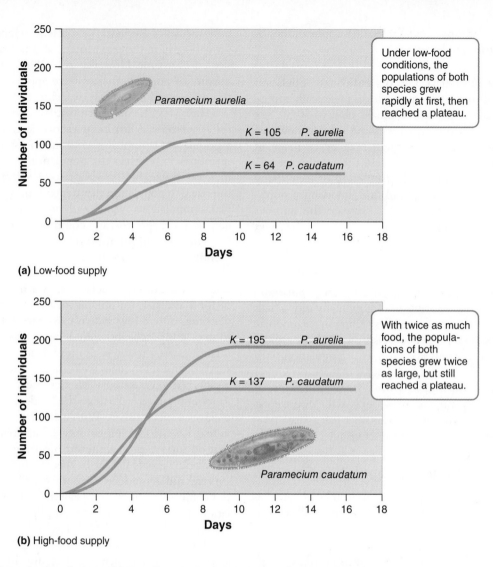

**(a)** Low-food supply

**(b)** High-food supply

**FIGURE 6.4** **Gause's experiments.** (a) Over time, the population sizes of two species of *Paramecium* initially increased, but then leveled off as their food supply became limiting. (b) When Gause doubled their food supply, both species attained populations sizes that were nearly twice as large. [After Gause 1932.]

of ponds, making amphibians and fish inaccessible to wading birds such as herons. With their food supply cut off, herons have an increased risk of starving to death, regardless of whether the heron population is at a low or a high density.

> **GAUGE YOUR PROGRESS**
>
> ✓ What factors regulate the size of a population?
>
> ✓ What did Gause discover in his classic experiments?
>
> ✓ What is the difference between density-dependent and density-independent factors that influence population size? Give an example of each.

# Growth models help ecologists understand population changes

Scientists often use models to help them explain how things work and to predict how things might change in the future. Population ecologists use population growth models that incorporate density-dependent and density-independent factors to explain and predict changes in population size. Population growth models are important tools for population ecologists, whether they are protecting an endangered condor population, managing a commercially harvested fish species, or controlling an insect pest. In this section we will look at several growth models and other tools for understanding changes in population size.

## The Exponential Growth Model

*Population growth models* are mathematical equations that can be used to predict population size at any moment in time. The **growth rate** of a population is the number of offspring an individual can produce in a given time period, minus the deaths of the individual or its offspring during the same period. Under ideal conditions, with unlimited resources available, every population has a particular maximum potential for growth, which is called the **intrinsic growth rate** and denoted as $r$. When there is plenty of food available, for example, white-tailed deer can give birth to twin fawns, domesticated hogs (*Sus domestica*) can have litters of 10 piglets, and American bullfrogs (*Rana catesbeiana*) can lay up to 20,000 eggs. Under these ideal conditions, the number of deaths also decreases. Together, a high number of births and a low number of deaths produce a high population growth rate. Under less than ideal conditions, when resources are limited, the population's growth rate will be lower than its intrinsic growth rate because individuals will produce fewer offspring (or forego breeding entirely) and the number of deaths will increase.

If we know the intrinsic growth rate of a population ($r$) and the number of reproducing individuals that are currently in the population ($N_0$), we can estimate the population's future size ($N_t$) after some period of time ($t$) has passed. To do this, we can use the **exponential growth model,**

$$N_t = N_0 e^{rt}$$

where e is the base of the natural logarithms (the $e^x$ key on your calculator, or 2.72) and $t$ is time. This equation tells us that, under ideal conditions, the future size of the population ($N_t$) depends on the current size of the population ($N_0$), the intrinsic growth rate of the population ($r$), and the amount of time ($t$) over which the population grows.

When populations are not limited by resources, their growth can be very rapid, as more births occur with each step in time. When graphed, the exponential growth model produces a **J-shaped curve,** as shown in FIGURE 6.5.

One way to think about exponential growth in a population is to compare it to annual interest payments in a bank account. Let's say you put $1,000 in a bank account at an annual interest rate of 5 percent. After a year, assuming you did not withdraw any of your money, you would earn 5 percent of $1,000, which is $50. The account would then show a balance of $1,050. In the second year, again assuming no withdrawals, the 5 percent interest rate would be applied to the new amount of $1,050, and would generate $52.50 in interest. In the tenth year, the account would

**FIGURE 6.5** **The exponential growth model.** When populations are not limited by resources, their growth can be very rapid. More births occur with each step in time, creating a J-shaped growth curve.

generate $77.57 in interest. Moving forward to the twentieth year, the same 5 percent interest rate would produce a much larger increase of $126.35.

Applying an annual rate of growth to an increasing amount, whether money in a bank account or a population of organisms, produces rapid growth over time. Exponential growth is density independent because no matter how much money you have in the account, the value will grow by the same percentage every year. Crunch the Numbers "Calculating Exponential Growth" gives a step-by-step example to show how this principle works.

The exponential growth model is an excellent starting point for understanding population growth. Indeed, there is solid evidence that real populations—even small ones—can grow exponentially, at least initially. However, no population can experience exponential growth indefinitely. In Gause's experiments with *Paramecium,* the two populations initially grew exponentially until they approached the carrying capacity of their test-tube environment, at which point their growth slowed and eventually leveled off to reflect the amount of food that was added daily. We turn next to another model that gives a more complete view of population growth.

## The Logistic Growth Model

The exponential growth model describes a continuously increasing population that grows at a fixed rate. But populations do not experience exponential growth indefinitely. For this reason, ecologists have modified the exponential growth model to incorporate environmental limits on population growth, including limiting resources. The **logistic growth model** describes a population whose growth is initially exponential, but

## Calculating Exponential Growth

Consider a population of rabbits that has an initial population size of 10 individuals ($N_0 = 10$). Let's assume that the intrinsic rate of growth for a rabbit is $r = 0.5$ (or 50 percent), which means that each rabbit produces a net increase of 0.5 rabbits each year. With this information, we can predict the size of the rabbit population 1 year from now:

$$N_t = N_0\, e^{rt}$$
$$N_t = 10 \times e^{0.5 \times 1}$$
$$N_t = 10 \times e^{0.5}$$
$$N_t = 10 \times 1.6$$
$$N_t = 16 \text{ rabbits}$$

We can then ask how large the rabbit population will be after 5 years:

$$N_t = 10 \times e^{0.5 \times 5}$$
$$N_t = 10 \times e^{2.5}$$
$$N_t = 10 \times 12.2$$
$$N_t = 122 \text{ rabbits}$$

We can also project the size of the rabbit population 10 years from now:

$$N_t = 10 \times e^{0.5 \times 10}$$
$$N_t = 10 \times e^{5}$$
$$N_t = 10 \times 148.4$$
$$N_t = 1{,}484 \text{ rabbits}$$

These data are graphed in FIGURE 6.6.

**FIGURE 6.6** **Rabbit population growth.** Graphing the data from Crunch the Numbers "Calculating Exponential Growth" gives us a clearer sense of how rapid exponential growth can be. When the rabbit population is small, small numbers of rabbits are added in a single year. The larger the population grows, however, the more rabbits are added each year.

**Your Turn:** Now assume that the intrinsic rate of growth is 1.0 for rabbits. Calculate the predicted size of the rabbit population after 1, 5, and 10 years.

---

slows as the population approaches the carrying capacity of the environment ($K$). As we can see in FIGURE 6.7, if a population starts out small, its growth can be very rapid.

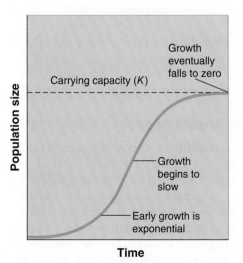

**FIGURE 6.7** **The logistic growth model.** A small population initially experiences exponential growth. As the population becomes larger, however, resources become scarcer, and the growth rate slows. When the population size reaches the carrying capacity of the environment, growth stops. As a result, the pattern of population growth follows an S-shaped curve.

As the population size nears about one-half of the carrying capacity, however, the population's growth begins to slow. As the population size approaches the carrying capacity, the population stops growing. When graphed, the logistic growth model produces an **S-shaped curve.** We observed this pattern in Gause's *Paramecium* experiments: at the carrying capacity, the populations stopped growing and remained at a constant size.

The logistic growth model is used to predict the growth of populations that are subject to density-dependent constraints, such as increased competition for food, water, or nest sites, as the population grows. Because density-independent factors such as hurricanes and floods are inherently unpredictable, the logistic growth model does not account for them.

## Variations on the Logistic Growth Model

One of the assumptions of the logistic growth model is that the number of offspring individuals produce depends on the current population size and the carrying capacity of the environment. However, many species of mammals mate during the fall or winter, and the number of offspring that develop depends on the food supply at the time of mating. Because these offspring are not actually born until the following spring, there is a risk that food availability will not match the new population size. If there is less food available in the spring than needed to feed the offspring, the population will experience an **overshoot** by becoming larger than the spring carrying capacity. As a result, there will not be enough food for all the individuals in the population, and the population will experience a **die-off,** or population crash.

The reindeer (*Rangifer tarandus*) population on St. Paul Island in Alaska is a good example of this pattern. After a small population of 25 reindeer was introduced to the island in 1910, it grew exponentially until it reached more than 2,000 reindeer in 1938. After 1938, the population crashed to only 8 animals, most likely because the reindeer ran out of food. FIGURE 6.8 shows these changes graphically.

Such die-offs can take a population well below the carrying capacity of the environment. In subsequent cycles of reproduction, the population may grow large again. FIGURE 6.9 illustrates the recurring cycle of overshoots and die-offs that causes populations to oscillate

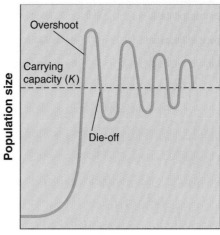

**FIGURE 6.9** **Population oscillations.** Some populations experience recurring cycles of overshoots and die-offs that lead to a pattern of oscillations around the carrying capacity of their environment.

around the carrying capacity. In many cases, these oscillations decline over time and approach the carrying capacity.

So far we have considered only how populations are limited by resources such as food, water, and the availability of nest sites. Predation may play an important additional role in limiting population growth. A classic example is the relationship between snowshoe hares (*Lepus americanus*) and the lynx (*Lynx canadensis*) that prey on them in North America. Trapping records from the Hudson's Bay Company, which purchased hare

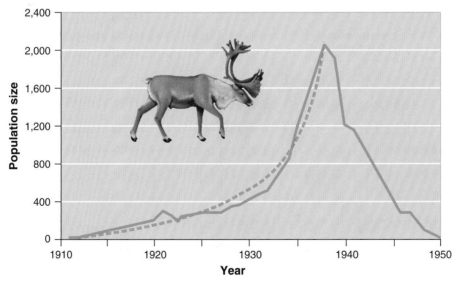

**FIGURE 6.8** **Growth and decline of a reindeer population.** Humans introduced 25 reindeer to St. Paul Island, Alaska, in 1910. The population initially experienced rapid growth (blue line) that approximated a J-shaped exponential growth curve (orange line). In 1938, the population crashed, probably because the animals exhausted the food supply. [Data from V. B. Scheffer, The rise and fall of a reindeer herd, *Scientific Monthly* (1951): 356–362.]

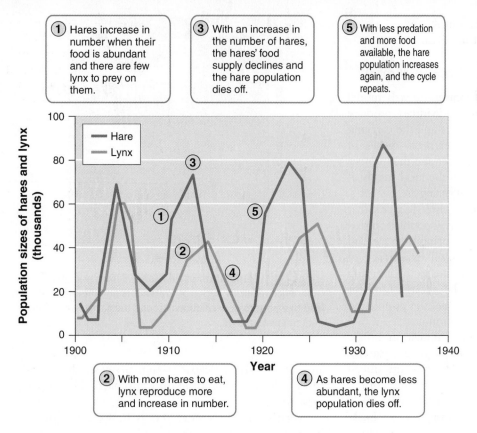

① Hares increase in number when their food is abundant and there are few lynx to prey on them.

③ With an increase in the number of hares, the hares' food supply declines and the hare population dies off.

⑤ With less predation and more food available, the hare population increases again, and the cycle repeats.

**FIGURE 6.10** **Population oscillations in lynx and hares.** Population sizes of predatory lynx and their hare prey have been estimated from records of the number of hare and lynx pelts purchased by the Hudson's Bay Company. Both species exhibit repeated oscillations of abundance, with the lynx population peaking 1 to 2 years after the hare population. [Data from Hudson's Bay Company.]

② With more hares to eat, lynx reproduce more and increase in number.

④ As hares become less abundant, the lynx population dies off.

and lynx pelts for nearly 90 years in Canada, indicate that the populations of both species cycle over time. **FIGURE 6.10** shows how this interaction works. The lynx population peaks 1 or 2 years after the hare population peaks. As the hare population increases, it provides more prey for the lynx, and thus the lynx population begins to grow. As the hare population reaches a peak, food for hares becomes scarce, and the hare population dies off. The decline in hares leads to a subsequent decline in the lynx population. The low lynx numbers reduce predation, and the hare population increases again.

A similar case of predator control of prey populations has been observed on Isle Royale, Michigan, an island in Lake Superior. Wolves (*Canis lupus*) and moose (*Alces alces*) have coexisted on Isle Royale for several decades. Starting in 1981, the wolf population declined sharply, probably as a result of a deadly canine virus. As wolf predation on moose declined, the moose population grew rapidly until it ran out of food. It then experienced a large die-off beginning in 1995. **FIGURE 6.11** shows the changes in both populations over a 50-year period.

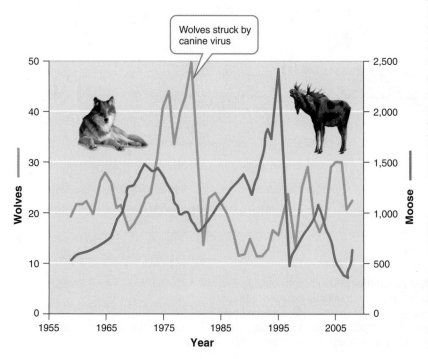

Wolves struck by canine virus

**FIGURE 6.11** **Predator control of prey populations.** As the population of wolves on Isle Royale succumbed to a canine virus, their moose prey experienced a dramatic population increase. [Data from J. A. Vucetich and R. O. Peterson, *Ecological Studies of Wolves on Isle Royale: Annual Report 2007–2008*, School of Forest Resources and Environmental Science, Michigan Technological University.]

## Reproductive Strategies and Survivorship Curves

Population size most commonly increases through reproduction. Population ecologists have identified a range of reproductive strategies in nature.

**K-SELECTED SPECIES** Some species have a low intrinsic growth rate, which causes their populations to increase slowly until they reach the carrying capacity of the environment. As a result, the abundance of such species is determined by the carrying capacity, and their population fluctuations are small. Because carrying capacity is denoted as $K$ in population models, such species are referred to as **K-selected species.**

K-selected species have certain traits in common. For instance, K-selected animals are typically large organisms that reach reproductive maturity relatively late, produce a few, large offspring, and provide substantial parental care. Elephants, for example, do not become reproductively mature until they are 13 years old, breed only once every 2 to 4 years, and produce only one calf at a time. Large mammals and most birds are K-selected species. For environmental scientists interested in biodiversity management or protection, K-selected species pose a challenge because their populations grow slowly. In practical terms, this means that an endangered K-selected species cannot respond quickly to efforts to save it from extinction.

**r-SELECTED SPECIES** At the opposite end of the spectrum are those species that have a high intrinsic growth rate because they reproduce often and produce large numbers of offspring. Because intrinsic growth rate is denoted as $r$ in population models, such species are referred to as **r-selected species.** In contrast to K-selected species, populations of r-selected species do not typically remain near their carrying capacity, but instead exhibit rapid population growth that is often followed by overshoots and die-offs. Among animals, r-selected species tend to be small organisms that reach reproductive maturity relatively early, reproduce frequently, produce many small offspring, and provide little or no parental care. House mice (*Mus musculus*), for example, become reproductively mature at 6 weeks of age, can breed every 5 weeks, and produce up to a dozen offspring at a time. Other r-selected organisms include small fishes, many insect species, and weedy plant species. Many organisms that humans consider to be pests, including cockroaches, dandelions, and rats, are r-selected species.

Table 6.1 summarizes the traits of K-selected and r-selected species. These two concepts represent opposite ends of a wide spectrum of reproductive strategies. Most species fall somewhere in between these two extremes, and many exhibit combinations of traits from the two extremes. For example, tuna and redwood trees are both long-lived species that take a long time to reach reproductive maturity. Once they do, however, they produce millions of small offspring that receive no parental care.

In addition to different reproductive strategies, species have distinct patterns of survival over time. These patterns can be plotted on a graph as **survivorship curves,** as shown in FIGURE 6.12. There are three basic types of survivorship curves. K-selected species such as elephants, whales, and humans have high survival rates throughout most of their life span. As they approach old age, however, they start to die in large numbers. This pattern is called a *type I survivorship curve.* In contrast, r-selected species such as mosquitoes and dandelions experience low survivorship early in life, and few individuals reach adulthood. This pattern is called a *type III survivorship curve.* Many other species, such as corals and squirrels, experience a relatively constant decline in survivorship throughout their life span. This pattern is called a *type II survivorship curve.*

| TABLE 6.1 | Traits of K-selected and r-selected species | |
|---|---|---|
| Trait | K-selected species | r-selected species |
| Life span | Long | Short |
| Time to reproductive maturity | Long | Short |
| Number of reproductive events | Few | Many |
| Number of offspring | Few | Many |
| Size of offspring | Large | Small |
| Parental care | Present | Absent |
| Population growth rate | Slow | Fast |
| Population regulation | Density dependent | Density independent |
| Population dynamics | Stable, near carrying capacity | Highly variable |

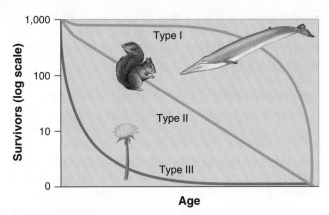

**FIGURE 6.12** **Survivorship curves.** Different species have distinct patterns of survivorship over the life span. Species range from exhibiting excellent survivorship until old age (type I curve) to exhibiting a relatively constant decline in survivorship over time (type II curve) to having very low rates of survivorship early in life (type III curve). *K*-selected species tend to exhibit type I curves, whereas *r*-selected species tend to exhibit type III curves.

## Metapopulations

Cougars (*Puma concolor*)—also called mountain lions or pumas—once lived throughout North America, but because of habitat destruction and overhunting, they are now found primarily in the remote mountain ranges of the western United States. In New Mexico, cougar populations are distributed in patches of mountainous habitat scattered across the desert landscape. These mountain habitats allow the cats to avoid human activities and provide them with reliable sources of water and prey such as mule deer (*Odocoileus hemionus*).

Because areas of desert separate the cougar's mountain habitats, we can consider the cougars of each mountain range to be a distinct population. Each population has its own dynamics based on local abiotic conditions and prey availability: large mountain ranges support large cougar populations and smaller mountain ranges support smaller cougar populations. However, cougars sometimes move between mountain ranges, often using strips of natural habitat that connect the separated populations. Such strips of habitat—called **corridors**—provide some connectedness among the populations, as FIGURE 6.13 shows. A group of spatially distinct populations that are connected by occasional movements of individuals between them is called a **metapopulation.**

The connectedness among the populations within a metapopulation is an important part of each population's overall persistence. Small populations are more likely than large ones to go extinct. First, as we have seen, small populations contain relatively little genetic variation and therefore may not be able to adapt to changing environmental conditions. Second, they are more vulnerable than large populations to density-independent catastrophes such as particularly harsh winters. In a metapopulation, occasional immigrants from larger nearby populations can add to the size of a small population and introduce new genetic diversity. Both of these factors help to reduce the risk of extinction.

Metapopulations can also provide a species with some protection against threats such as diseases. A disease could cause a population living in a single large habitat patch to go extinct. But if a population living in an isolated habitat patch is part of a much larger metapopulation, then a disease could wipe out that isolated population, but immigrants from other populations could later recolonize the patch and help the species to persist.

Because many habitats are naturally patchy across the landscape, many species are part of metapopulations.

**FIGURE 6.13** **A cougar metapopulation.** Populations of cougars live in separate mountain ranges in New Mexico. Occasionally, however, individuals move between mountain ranges. These movements can recolonize mountain ranges with extinct populations and add individuals and genetic diversity to existing populations.

For instance, numerous species of butterflies specialize on plants with patchy distributions. Some amphibians live in isolated wetlands, but occasionally disperse to other wetlands. The number of species that exist as metapopulations is growing because human activities have fragmented habitats, dividing single large populations into several smaller populations. Identifying and managing metapopulations is thus an increasingly important part of protecting biodiversity.

**(a)** *P. aurelia* grown separately

---

**(b)** *P. caudatum* grown separately

---

> ### GAUGE YOUR PROGRESS
>
> ✓ What happens if you alter the *r* or *K* terms in the logistic growth model?
>
> ✓ What did scientists learn from the records of the Hudson's Bay Company?
>
> ✓ What is a metapopulation? How do metapopulations contribute to the preservation of biodiversity?

# Community ecologists study species interactions

So far we have been exploring the factors that determine population size, as well as ways to model or predict how populations will grow or decline. However, the size of a population tells us nothing about what determines the distribution of populations across the planet.

The distribution of species around the world is determined by three factors. As we learned in Chapter 5, the presence of a species in a given area is influenced by the species' fundamental niche, or the *range of abiotic conditions that it can tolerate*. In addition, for a species to be present in an area, it must be *able to disperse* to that area. Kangaroos, for example, are unable to disperse from Australia to North America without human intervention. Even though North America contains habitats that might be favorable for kangaroos, the ocean serves as an effective barrier to kangaroo dispersal. The third factor in determining a species' distribution is *interactions with other species*. These interactions fall into four categories: *competition, predation, mutualism,* and *commensalism*. The study of these interactions, which determine the survival of a species in a habitat, is the science of **community ecology.**

## Competition

In 1934, two years after his experiments on logistic growth in populations of *Paramecium*, Georgii Gause studied how different *Paramecium* species affected each other's population growth. **FIGURE 6.14** shows the results of his experiments. When the two species—*P. caudatum* and *P. aurelia*—were grown in separate laboratory cultures,

**(c)** *P. aurelia* and *P. caudatum* grown together

**FIGURE 6.14** **Competition for a limiting resource.** When Gause grew two species of *Paramecium* separately, both achieved large population sizes. However, when the two species were grown together, *P. aurelia* continued to grow well, while *P. caudatum* declined to extinction. These experiments demonstrated that two species competing for the same limiting resource cannot coexist. [Data from Gause 1934.]

each species thrived and reached a relatively high population size within 10 days. When the two species were grown together, however, *P. aurelia* continued to thrive, but *P. caudatum* declined to extinction. **Competition,** the struggle of individuals to obtain a limiting resource, caused *P. aurelia* to do better. Gause's observations, combined with additional experiments with other organisms, led researchers to formulate the **competitive exclusion principle,** which states that *two species competing for the same limiting resource cannot coexist.* Under a given set of environmental conditions, when two species have the same realized niche, one species will perform better and will drive the other species to extinction.

We see competition at work throughout nature. For example, many plant species influence the distribution and abundance of other plant species. Goldenrods are dominant in the old fields of New England because of their superior competitive ability: they can grow taller than other wildflowers and obtain more of the available sunlight. Similarly, the wild oat plant (*Avena fatua*) can outcompete crop plants on the Great Plains of North America because its seeds ripen earlier, permitting oat seedlings to start growing before the other species.

Competition for a limiting resource can lead to **resource partitioning,** in which two species divide a resource based on differences in the species' behavior or morphology. In evolutionary terms, when competition reduces the ability of individuals to survive and reproduce, natural selection will favor individuals that overlap less with other species in the resources they use. **FIGURE 6.15** shows how this process works. Let's imagine

two species of birds that eat seeds of different sizes. In species 1, represented by blue, some individuals eat small seeds and others eat medium seeds. In species 2, represented by yellow, some individuals eat medium seeds and others eat large seeds. As a result, some individuals of both species compete for medium seeds. This overlap is represented by green. If species 1 is the better competitor for medium seeds, then individuals of species 2 that compete for medium seeds will have poor survival and reproduction. After several generations, species 2 will evolve to contain fewer individuals that feed on medium seeds. This process of resource partitioning reduces the amount of competition between the two species.

Species in nature reduce resource overlap in several ways. **FIGURE 6.16** shows examples of each. If two species reduce competition by utilizing the same resource but at different times, they are exhibiting *temporal resource partitioning.* For example, wolves and coyotes that live in the same territory are active at different times of day to reduce the overlap in the times that they are hunting (Figure 6.16a). Similarly, some plants reduce competition for pollinators by flowering at different times of the year.

If two species reduce competition by using different habitats, they are exhibiting *spatial resource partitioning.* Desert plant species have evolved a variety of different

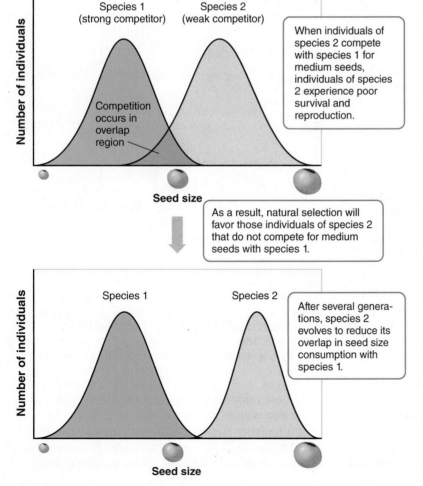

When individuals of species 2 compete with species 1 for medium seeds, individuals of species 2 experience poor survival and reproduction.

As a result, natural selection will favor those individuals of species 2 that do not compete for medium seeds with species 1.

After several generations, species 2 evolves to reduce its overlap in seed size consumption with species 1.

**FIGURE 6.15** **The evolution of resource partitioning.** When two species overlap in their use of a limiting resource, selection favors those individuals of each species whose use of the resource overlaps the least with that of the other species. Over many generations, the two species can evolve to reduce their overlap and thereby partition their use of the limiting resource.

**Wolf**

**Coyote**

**(a)** Temporal resource partitioning

Tarbush

Black grama grass

1 m

2 m

3 m

**(b)** Spatial resource partitioning

**Black-footed ferret**

**Ermine**

**(c)** Morphological resource partitioning

**FIGURE 6.16** **Three types of resource partitioning.** Species that face competition for a shared limiting resource may evolve to partition the resource and reduce competition. (a) Wolves and coyotes that live together exhibit both spatial and temporal resource partitioning. These two species try to stay out of each other's territories. In addition, wolves tend to be more active in the early evening, whereas coyotes tend to be more active from midnight until dawn. (b) Plants in many desert habitats exhibit spatial resource partitioning. Tarbush grows very deep roots to reach deep sources of water, whereas grama grass has an extensive shallow root system that allows it to take up water rapidly during rare rain events. (c) Species of predatory weasels exhibit morphological resource partitioning. Differences in skull and tooth size allow each species to specialize on different sizes of prey.

root systems that reduce competition for water and soil nutrients; some have very deep roots and others have shallow roots. Comparing the two desert plants shown in Figure 6.16b, we see that black grama grass (*Bouteloua eriopoda*) has shallow roots that extend over a large area to capture rainwater, whereas tarbush (*Flourensia cernua*)

sends roots deep into the ground to tap deep sources of water.

A third method of reducing resource overlap is by *morphological resource partitioning:* the evolution of differences in body size or shape. As we saw in Chapter 5, Charles Darwin observed morphological resource partitioning among the finches of the Galápagos Islands (see Figure 5.14). The 14 species of finches on the islands probably descended from a single finch species that colonized the islands from mainland South America. This ancestral finch species would have consumed seeds on the ground. Over time, however, the finches have evolved into 14 species with uniquely shaped beaks that allow each species to eat different foods. The beaks of some species are well suited for crushing seeds, whereas the beaks of others are better suited for catching insects. This morphological resource partitioning reduces competition among the finch species. We can see the same process at work with the two weasel skulls shown in Figure 6.16c. The black-footed ferret (*Mustela nigripes*) and the ermine (*Mustela erminea*) have evolved differences in skull size and tooth size to specialize on different types of prey.

## Predation

The word *predation* might conjure up images of large fearsome carnivores tearing at the flesh of their prey. But in the broadest sense, **predation** refers to the use of one species as a resource by another species. Organisms of all sizes may be predators, and their effects on their prey vary widely. Predators can be grouped into four categories:

- **True predators** typically kill their prey and consume most of what they kill. True predators include African lions (*Panthera leo*) that eat gazelles and great horned owls (*Bubo virginianus*) that eat small rodents.

- **Herbivores** consume plants as prey. They typically eat only a small fraction of an individual plant without killing it. The gazelles of the African plains are well-known herbivores, as are the various species of deer in North America.

- **Parasites** live on or in the organism they consume, referred to as their *host.* Because parasites typically consume only a small fraction of their host, a single parasite rarely causes the death of its host. Parasites include tapeworms that live in the intestines of animals as well as the protists that live in the bloodstream of animals and cause malaria. Parasites that cause disease in their host are called **pathogens.** Pathogens include viruses, bacteria, fungi, protists, and wormlike organisms called *helminths.*

(a)

(b)

(c)

(d)

**FIGURE 6.17** **Prey defenses.** Predation has favored the evolution of fascinating antipredator defenses. (a) The camouflage of this stone flounder (*Kareius bicoloratus*) makes it difficult for predators to see it. (b) Sharp spines protect this porcupine (*Erethizon dorsatum*) from predators. (c) The poison dart frog (*Epipedobates bilinguis*) has a toxic skin. (d) This nontoxic frog (*Allobates zaparo*) mimics the appearance of the poison dart frog.

- **Parasitoids** are organisms that lay eggs inside other organisms. When the eggs hatch, the parasitoid larvae slowly consume the host from the inside out, eventually leading to the host's death. Parasitoids include certain species of wasps and flies.

As we learned from the lynx-hare interactions in Canada and the wolf-moose interactions on Isle Royale, predators can play an important role in controlling the abundance of prey in nature. In Sweden, red foxes (*Vulpes vulpes*) prey on several species of hares and grouse. In the 1970s and 1980s, the fox population was reduced by a disease called mange. With fewer foxes around, the foxes' prey increased in abundance: populations of grouse doubled in size, and populations of hares increased sixfold. The predators in all of these examples play a critical role in their ecosystems: that of regulating prey populations.

To avoid being eaten or harmed by predators, many prey species have evolved defenses. These defenses may be behavioral, morphological, or chemical, or may simply mimic another species' defense. FIGURE 6.17 shows several examples of antipredator defenses. Animal prey commonly use behavioral defenses, such as hiding and reduced movement, so as to attract less attention from predators. Other prey species have evolved impressive morphological defenses, including camouflage to help them hide from predators and spines to help deter predators. Many plants, for example, have evolved spines that deter herbivores from grazing on their leaves and fruits. Similarly, many animals, such as porcupines, stingrays, and puffer fish, have spines that deter predator attack.

Chemical defenses are another common mechanism of protection from predators. Several species of insects, frogs, and plants emit chemicals that are toxic or distasteful to their predators. Many of these prey are also brightly colored, and predators learn to recognize and avoid consuming them. In some cases, prey have not evolved toxic defenses of their own, but instead have evolved to mimic other prey species that do possess chemical defenses. By mimicking toxic species, these nontoxic species can fool predators into not attacking them.

## Mutualism

The third type of interspecific interaction, **mutualism,** benefits two interacting species by increasing both species' chances of survival or reproduction. Each species in a mutualistic interaction is ultimately assisting the other species in order to benefit itself. If the benefit is too small, the interaction will no longer be worth the cost of helping the other species. Under such conditions, natural selection will favor individuals that no longer engage in the mutualistic interaction.

There are many types of mutualistic interactions. One of the most ecologically important is the relationship between plants and their pollinators, which include birds, bats, and insects. The plants depend on the pollinators for their reproduction, and the pollinators depend on the plants for food. In some cases, one pollinator species might visit many species of plants. In other cases, many plant species are visited by a range of pollinator species. In still other cases, one animal species pollinates only one plant species, and that plant is pollinated only by that animal species. For example, there are about 900 species of fig trees, and almost every one is pollinated by a particular species of fig wasp.

One well-studied example of plant-animal mutualism is the interaction between acacia trees and several species of *Pseudomyrmex* ants in Central America. The acacia tree supplies the ants with food and shelter, and the ants protect the tree from herbivores and competitors. The ants eat nectar produced by the tree in special structures called nectaries, and they live in the tree's large thorns, which have soft cores that the ants can

**FIGURE 6.18** **Mutualism.** Acacia trees and *Pseudomyrmex* ants have each evolved adaptations that enhance their mutualistic interaction. The trees' thorns serve as nest sites for the ants (inset) and provide them with food in the form of nectar produced in specialized nectaries (arrow). In return, the ants protect the trees from herbivores and competing plants.

easily hollow out (FIGURE 6.18). The ants protect the tree by attacking anything that falls on it—for example, by stinging intruding insects that might eat the acacia's leaves and destroying vines that might shade the tree's branches. Through natural selection over generations of close association, the ants and the trees have both evolved traits that make this mutualism work. The ants

have evolved several behavioral adaptations: living in the hollowed-out thorns of the tree, extensive patrolling of branches and leaves, and attacking all foreign animals and plants regardless of whether they are useful to the ants as food. For its part, the acacia tree has evolved specialized thorns, year-round leaf production, and modified leaves and nectaries that provide food for the ants.

Two other well-known mutualistic relationships are those found in coral reefs and in lichens. As we saw in Chapter 4, algae live within the tiny coral animals that build coral reefs. Their relationship is critical: if the algae die, the coral will die as well. Many of the lichens that grow on the surfaces of rocks and trees are made up of an alga and a fungus (and often bacteria) living in close association. The fungus provides many of the nutrients the alga needs, and the alga provides carbohydrates for the fungus via photosynthesis (FIGURE 6.19).

## Commensalism

**Commensalism** is a type of relationship in which one species benefits but the other is neither harmed nor helped. Commensalisms include birds using trees as perches and fish using coral reefs as places to hide from predators.

Commensalism, mutualism, and parasitism are all *symbiotic* relationships. A **symbiotic** relationship is the relationship of two species that live in close association with each other.

Interactions among species are important in determining which species can live in a community. Table 6.2 summarizes these interactions and the effects they have on each of the interacting species, whether positive (+), negative (−), or neutral (0). Competition for a limiting resource has a negative effect on both of the competing species. In contrast, predation has a positive effect on the predator, but a negative effect on the prey. Mutualism has positive effects on both interacting species.

**FIGURE 6.19** **Lichens.** Lichens, such as this species named "British soldiers" (*Cladonia cristatella*), are composed of a fungus and an alga that are tightly linked together in a mutualistic interaction. The fungus provides nutrients to the alga, and the alga provides carbohydrates for the fungus via photosynthesis.

| TABLE 6.2 | Interactions between species and their effects | | |
|---|---|---|---|
| Type of interaction | | Species 1 | Species 2 |
| Competition | | − | − |
| Predation | | + | − |
| Mutualism | | + | + |
| Commensalism | | + | 0 |

Commensalism has a positive effect on one species and no effect on the other species.

## Keystone Species

As we have discussed, interspecific interactions can affect the abundance and distribution of species in communities. In most cases, a given species has an effect on a small number of other species, but not on the entire community. As a result, the extinction of a single species usually does not affect the long-term stability of a community or ecosystem. Other species at the same trophic level, or species from adjacent areas, can usually provide the links necessary for energy and matter to flow.

Sometimes, however, the loss of one species from a community can have a disproportionately large effect on the entire community. A **keystone species** is a species that plays a role in its community that is far more important than its relative abundance might suggest. The name *keystone species* is a metaphor that comes from architecture. As **FIGURE 6.20** shows, in a stone arch, the keystone is the single center stone that supports all the other stones. Without the keystone, the arch would collapse. Typically, the most abundant species or the major energy producers, while vital to the health of a community, are not keystone species. Keystone species typically exist in low numbers. They may be predators, sources of food, mutualistic species, or providers of some other essential service.

The role of keystone predators was well demonstrated in a classic experiment conducted in intertidal communities off the coast of Washington State, whose results are shown in **FIGURE 6.21**. These intertidal communities include mobile animal species such as sea stars and snails as well as dozens of other species that make their living attached to the rocky substratum, including mussels, barnacles, and algae. When sea stars (*Pisaster ochraceus*) are present, they prey on mussels (*Mytilus californianus*). This predation continually clears spaces where other species can attach to the rocks. When researchers removed the sea stars from the community, the mussels were no longer subject to predation by sea stars, and they outcompeted the other species in the community. The mussels became numerically dominant, while 25 other species declined in abundance. Thus the predatory sea star, while not particularly numerous, played a key role in reducing the abundance of a superior competitor—the mussel—and allowing inferior competitors to persist.

This phenomenon, called **predator-mediated competition,** is common in nature. We observed it in this chapter's opening story, where we described the effects of outbreaks of the leaf beetle that specializes on goldenrods in the old fields of New England. An outbreak of these herbivorous beetles reduces populations of goldenrods, which are superior competitors, and increases the abundance of many wildflower species that are inferior competitors.

A species that provides food for a community at times when food is scarce may also be a keystone species. For example, plants that produce nectar and certain fruits, including figs, make up less then 1 percent of the plant diversity in the tropical forests of Central and South America. In most years, there is a 3-month period in which the most abundant plant species do not produce enough food to support the community's herbivores. During this period of food scarcity, the herbivores rely on the less abundant fruits and nectar for food, making the plants that produce them keystone species in this community.

Some species are considered keystone species because of the importance of their mutualistic interactions with other species. For example, most animal pollinators are abundant or provide a service that can be duplicated by other species. Some communities, however, rely on relatively rare pollinator species, which are therefore keystone species. On many South Pacific islands, a species of bat known as the flying fox (*Pteropus vampyrus*) is the only pollinator and seed disperser for hundreds of tropical plant species. Flying foxes have been hunted for food to near extinction. Without their pollinating and seed-dispersing functions, many plant species may become extinct, dramatically changing these island communities.

Mycorrhizal fungi are also keystone mutualists. These fungi are found on and in the roots of many plant species, where they increase the plants' ability to extract nutrients from the soil.

**FIGURE 6.20** **Keystone.** Keystone species get their name from the keystone of an arch. Without the keystone in place, the arch would fall apart.

Keystone

**FIGURE 6.21** **Keystone predators.** Sea stars are keystone predators in their rocky intertidal communities in Washington State. When sea stars are present, they consume mussels, which are strong competitors for space. This predation creates open spaces that inferior competitors can colonize. As a result, the diversity of species is high. In the absence of sea stars, the mussels dominate the surfaces of the intertidal rocks, and the diversity of species declines dramatically. [Data from T. T. Paine, Intertidal community structure: Experimental studies on the relationship between a dominant competitor and its principal predator, *Oecologia* 15 (1974): 93–120.]

They play a critical role in the growth of plant species, which in turn provide habitat and resources for other members of a forest or field community.

Finally, a keystone species may create or maintain habitat for other species. Such species are referred to as **ecosystem engineers.** The North American beaver (*Castor canadensis*) is a prime example. Although they make up only a small percentage of the total biomass of the North American forest, beavers have a critical role in the forest community. They build dams that convert narrow streams into large ponds, thereby creating new habitat for pond-adapted plants and animals **FIGURE 6.22**.

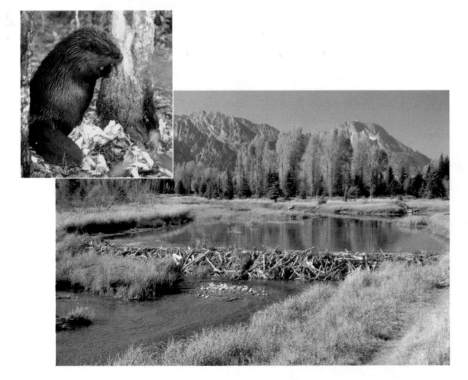

**FIGURE 6.22** **Ecosystem engineers.** Beavers are considered a keystone species because of the role they play in creating new pond and wetland habitat. For a species that is not particularly abundant, the beaver has a strong influence on the presence of other organisms in the community.

These ponds also flood many hectares of forest, causing the trees to die and creating habitat for animals that rely on dead trees. Several species of woodpeckers and some species of ducks make their nests in cavities that are carved into the dead trees. Alligators play a similar role in their communities by digging deep "gator holes" in summer. These holes serve as critical sources of water for many other animals during dry months.

# The composition of a community changes over time

Even without human activity, natural communities do not stay the same forever. Change in the species composition of communities over time is a perpetual process in nature. One type of change that occurs in virtually every community is **ecological succession,** the predictable replacement of one group of species by another group of species over time. Depending on the community type, ecological succession can occur over time spans from decades to centuries. In terrestrial communities, ecological succession can be *primary* or *secondary,* depending on the starting point of the community.

## Primary Succession

**Primary succession** occurs on surfaces that are initially devoid of soil, such as an abandoned parking lot, newly exposed rock left behind after a glacial retreat, or newly cooled lava.

FIGURE 6.23 shows the process of primary succession. It begins when bare rock is colonized by organisms such as algae, lichens, and mosses—organisms that can survive with little or no soil. As these early-successional species grow, they excrete acids that allow them to take up nutrients directly from the rock. The resulting chemical alteration of the rock also makes it more susceptible to erosion. When the algae, lichens, and mosses die, they become the organic matter that mixes with minerals eroded from the rock to create new soil.

Over time, soil develops on the bare surface, and it becomes a hospitable environment for plants with deep

Exposed rocks

Lichens and mosses

Annual weeds

Perennial weeds and grasses

Shrubs

Aspen, cherry, and young pine forest

Beech and maple broadleaf forest

**Time**

FIGURE 6.23 **Primary succession.** Primary succession occurs in areas devoid of soil. Early-arriving plants and algae can colonize bare rock and begin to form soil, making the site more hospitable for other species to colonize later. Over time, a series of distinct communities develops. In this illustration, representing an area in New England, bare rock is initially colonized by lichens and mosses and later by grasses, shrubs, and trees.

root systems. Mid-successional plants such as grasses and wildflowers are easily dispersed to such areas. These species are typically well adapted to exploiting open, sunny areas and are able to survive in the young, nutrient-poor soil. The lives and deaths of these mid-successional species gradually improve the quality of the soil by increasing its ability to retain nutrients and water. As a result, new species colonize the area and outcompete the mid-successional species.

The type of community that eventually develops is determined by the temperature and rainfall of the region. In the United States, succession produces forest communities in the East, grassland communities in the Midwest, and shrubland communities in the Southwest. In some areas, the number of species increases as succession proceeds. In others, late-successional communities have fewer species than early-successional communities.

A dramatic example of primary succession occurred in Indonesia in 1883, when the island volcano Krakatau erupted and covered the island with lava and hot ash. The ash killed all living organisms and buried the soil on the island. By late 1884, algae and grasses had become established on the rock and ash. They were followed just a few years later by trees and shrubs. The island forest continued to undergo succession as additional species of trees and shrubs established themselves over time. Because the number of species on Krakatau has not reached the number living on undisturbed neighboring islands of similar size, it is likely that primary succession continues there today.

## Secondary Succession

**Secondary succession** occurs in areas that have been disturbed but have not lost their soil. Secondary succession follows an event, such as a forest fire or hurricane, that removes vegetation but leaves the soil mostly intact. Secondary succession also occurs on abandoned agricultural fields, such as the New England farms we discussed in this chapter's opening story.

**FIGURE 6.24** shows the process of secondary succession. It usually begins with rapid colonization by plants that can easily disperse to the disturbed area. The first to arrive are typically plants such as grasses and wildflowers that have light, wind-borne seeds. As in primary succession, these species are eventually replaced by species that are better competitors for sunlight, water, and soil nutrients. In regions that receive sufficient rainfall, trees replace grasses and flowers. The first tree species to colonize are those that can disperse easily and grow rapidly. For example, the seeds of aspen trees are carried on the wind, and cherry tree seeds are carried by birds that consume the fruit and excrete the seeds on the ground. Trees such as aspen and cherry are often called

Annual weeds

Perennial weeds and grasses

Shrubs

Aspen, cherry, and young pine forest

Beech and maple broadleaf forest

**Time**

**FIGURE 6.24** **Secondary succession.** Secondary succession occurs where soil is present, but all plants have been removed. Early-arriving plants set these areas on a path of secondary succession. Secondary succession in a New England forest begins with grasses and wildflowers, which are later replaced by trees.

**pioneer species** because of their ability to colonize new areas rapidly and grow well in full sunshine. As the pioneer trees increase in number and grow larger, however, they start to shade one another, reducing one another's growth and fitness. Other species that are more shade tolerant, including many species of beech and maple, survive and grow well in the shade of the pioneer tree species. These shade-tolerant species grow up through the pioneer canopy, eventually outcompeting the pioneer trees and dominating the forest community.

The process of secondary succession happens on many spatial scales, from small-scale disturbances such as a single treefall to huge areas cleared by natural disasters such as hurricanes or wildfires. The process is similar in both cases. For example, if a large tree falls and creates an opening for light to reach the forest floor, the seedlings of trees that grow rapidly in full sunlight outcompete other seedlings, and secondary succession begins again.

Historically, ecologists have described succession as ending with a *climax stage*. In forests, for example, ecologists considered the oldest forests to be *climax forests*. It is now recognized that, because natural disturbances such as fire, wind, and outbreaks of insect herbivores are a regular part of most communities, a late-successional stage is typically not final because at any moment it can be reset to an earlier stage.

## Aquatic Succession

Disturbances also create opportunities for succession in aquatic environments. For example, the rocky intertidal zone along the Pacific coast of North America is exposed to the air during low tide and lies under water during high tide. From time to time, major storms turn over rocks or otherwise clear them of living things. These bare rocks can then be colonized through the process of primary succession. Diatoms and short-lived species of red and green algae are usually the first to arrive. Not long afterward, the rocks are colonized by barnacles and several species of long-lived red algae. This mid-successional stage develops rather quickly, in about 15 months. Finally, if the rock is not disturbed again for 3 years or more, the area may become

dominated by several species of attached barnacles and mussels.

**FIGURE 6.25** shows the pattern of succession in freshwater lakes. A glacier may carve out a lake basin, scouring it of sediments and vegetation. Over time, algae and aquatic plants colonize the lake. The growth of plants and algae and the erosion of surrounding rock and soil slowly fill the basin with sediment and organic matter, making it increasingly shallow. This process may take hundreds or thousands of years. Eventually, the lake may be filled in completely and become a terrestrial habitat.

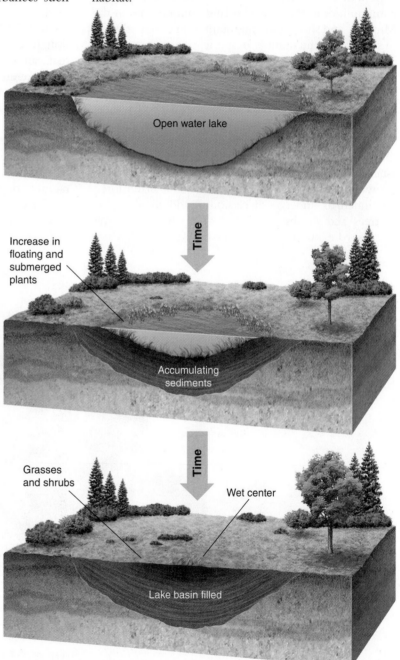

**FIGURE 6.25** **Succession in lakes.** Over hundreds to thousands of years, lakes are filled with sediments and slowly become terrestrial habitats.

# The species richness of a community is influenced by many factors

As we have seen, species are not distributed evenly on Earth. They are organized into biomes by global climate patterns and into communities whose composition changes regularly as species interact. In a given region within a biome, the number and types of species present are determined by three basic processes: colonization of the area by new species, speciation within the area, and losses from the area by extinction. The relative importance of these processes varies from region to region and is influenced by four factors: *latitude, time, habitat size,* and *distance from other communities.*

## Latitude

As we move from the equator toward the North or South Pole, the number of species declines. For example, the southern latitudes of the United States support more than 12,000 species of plants, whereas a similar-sized area in northern Canada supports only about 1,700 species. This latitudinal pattern is also observed among birds, reptiles, amphibians, and insects. For more than a century, scientists have sought to understand the reasons for this pattern, yet those reasons remain unclear.

## Time

Patterns of species richness are also regulated by time. The longer a habitat exists, the more colonization, speciation, and extinction can occur there. For example, at more than 25 million years old, Lake Baikal in Siberia is one of the oldest lakes in the world. Its benthic zone is home to over 580 species of invertebrates. By contrast, only 4 species of invertebrates inhabit the benthic zone of the Great Slave Lake in northern Canada—a lake that is similar in size and latitude to Lake Baikal, but is only a few tens of thousands of years old. This difference suggests that older communities have had more opportunities for speciation.

## Habitat Size and Distance: The Theory of Island Biogeography

The final two factors that influence the number and types of species are the size of the habitat and the distance of that habitat from a source of colonizing species. These factors are the basis for the **theory of island biogeography,** which demonstrates the dual importance of habitat size and distance in determining species richness.

Larger habitats typically contain more species. In FIGURE 6.26, for example, we can see that larger islands of reed habitat in Hungary contain a greater number of bird species than smaller islands. There are three reasons for this pattern. First, dispersing species are more likely to find larger habitats than smaller habitats, particularly when those habitats are islands. Second, at any given latitude, larger habitats can support more species than smaller habitats. Larger habitats are capable of supporting larger populations of any given species, and larger populations are less prone to extinction. Third, larger habitats often contain a wider range of environmental conditions, which in turn provide more niches that support a larger number of species. A wider range of environmental conditions also provides greater opportunities for speciation over time.

The distance between a habitat and a source of colonizing species is the final factor that affects the species richness of communities. For example, oceanic islands that are more distant from continents generally have fewer species than islands that are closer to continents. Distance matters because many species can disperse short distances, but only a few can disperse long

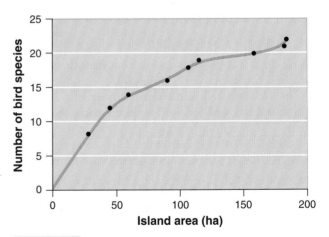

**FIGURE 6.26** **Habitat size and species richness.** Species richness increases as the size of the habitat increases. In this example, researchers counted the number of bird species that inhabited reed islands in Lake Velence, Hungary. As island area increased, the number of bird species initially rose quickly and then began to slow. [Data from A. Baldi and T. Kisbenedek, Bird species numbers in an archipelago of reeds at Lake Velence, Hungary, *Global Ecology and Biogeography* 9 (2000): 451–461.]

distances. In other words, if two islands are the same size, the nearer island should accumulate more species than the farther island because it has a higher rate of immigration by new species.

## Conservation and Island Biogeography

The effects of colonization, speciation, and extinction on the species richness of communities have important implications for conservation. The theory of island biogeography was originally applied to oceanic islands, but it has since been applied to "habitat islands" within a continent, such as the "islands" of protected habitat within national parks. These habitat islands are often surrounded by less hospitable habitats that have been dramatically altered by human activities. If we wish to

set aside natural habitat for a given species or group of species, we need to consider both the size of the protected area and the distance between the protected area and other areas that could provide colonists.

### GAUGE YOUR PROGRESS

✓ How are species distributed globally, and what processes are responsible for these patterns?

✓ What are the four factors that determine the number of species found in a community?

✓ What does the theory of island biogeography describe?

# WORKING TOWARD SUSTAINABILITY

## Bringing Back the Black-Footed Ferret

Throughout the western United States, the Great Plains were once covered with large congregations of prairie dogs. Spanning several different species, these prairie dog "towns" consisted of networks of underground tunnels. They were great attractions for out-of-town tourists, but the local ranchers who had to live with the prairie dogs were not nearly as fond of the little rodents. Prairie dogs are effective herbivores that consume a great deal of plant biomass, including agricultural crops. The ranchers viewed prairie dogs as competitors for their crops and sought a variety of ways, such as poisoning, to eradicate them. An unintended consequence of their winning the battle against the prairie dogs was the near extinction of another species, the black-footed ferret (FIGURE 6.27).

The black-footed ferret is a member of the weasel family and the only species of ferret native to North America. It lives in burrows and preys on prairie dogs. In fact, a single ferret can consume 125 to 150 prairie dogs a year. Although prairie dogs had been the ferrets' main source of food for millions of years, things started to change when settlers brought their plows to the Great Plains. Plowing the ground for agriculture destroyed many prairie dog towns. Other prairie dog towns were poisoned. Together, the plowing and poisoning efforts have reduced the prairie dog population by 98 percent.

As we would expect in a density-dependent scenario, the reduction in the prairie dog population reduced the carrying capacity of the environment for the black-footed ferret. In addition, the poisoning campaign poisoned ferrets as well as prairie dogs. Because both

small and large populations of ferrets were poisoned, and a large proportion of ferrets died in both cases, the poisoning had a density-independent effect on the ferret population.

In 1967, the black-footed ferret was officially listed as an endangered species, and the last known population died out in 1974. People feared that the black-footed ferret was extinct. In 1981, however, a small population of 130 ferrets was discovered in Wyoming. Conservation efforts began, but a highly lethal disease known as canine distemper passed through the population, reducing the entire species to a mere 18 animals.

FIGURE 6.27 **Black-footed ferret.** Once critically endangered, populations of the black-footed ferret are rebounding as a result of collaborative conservation efforts.

As part of a collaboration between federal biologists, private landowners, and several zoos, all 18 ferrets were immediately brought into captivity in the hope that a captive breeding program might be able to bring their numbers back. The black-footed ferret is a K-selected species: the ferrets breed once a year and have 3 to 4 offspring, for which they provide a great deal of parental care. As a result, population increases could not occur rapidly. Nevertheless, the captive population grew to 120 animals by 1989.

At this point, the captive breeding program was considered successful enough to begin reintroducing the species into the wild. However, the biologists first had to consider exactly where they wanted to reintroduce the ferrets. They understood the risks of introducing all the animals into a single site: another round of disease could kill the entire population. To manage the reintroduction process, the biologists decided to make use of the metapopulation concept. They reintroduced the ferrets in several places across the Great Plains, choosing locations with healthy prairie dog populations that had a low risk of extermination.

From 1991 to 2008, black-footed ferrets were reintroduced at 18 sites, from west Texas all the way up to Montana. These efforts are clearly paying off. Today, approximately 1,000 ferrets live in the wild, and hundreds more are part of the ongoing captive breeding program. But the rescue effort is not over. During years of high rainfall, outbreaks of plague, a bacterial disease carried by fleas, kill many ferrets. To combat plague, biologists have been dusting prairie dog towns with an insecticide powder that kills fleas. They are also working on a new vaccine against plague. Their goal is to achieve a wild population of 1,500 ferrets, with more than 30 animals at each of 10 reintroduction sites. If the biologists meet this goal, the ferret can be moved from the endangered species list to the less perilous classification of "threatened."

### References

Black-footed Ferret Recovery Implementation Team. The Black-footed Ferret Recovery Program. http://www.blackfootedferret.org/index.htm.

Robbins, J. 2008. Efforts on 2 fronts to save a population of ferrets. *New York Times,* July 15.

## REVISIT THE KEY IDEAS

■ **List the levels of complexity found in the natural world.**

Nature exists at several levels of complexity: individuals, populations, communities, ecosystems, and the biosphere.

■ **Contrast the ways in which density-dependent and density-independent factors affect population size.**

Density-dependent factors influence an individual's probability of survival and reproduction in a manner that is related to the size of the population. Density-independent factors have the same effect on an individual's probability of survival and reproduction in populations of any size.

■ **Explain growth models, reproductive strategies, survivorship curves, and metapopulations.**

The exponential growth model describes rapid growth under ideal conditions when resources are not limited. The logistic growth model incorporates density-dependent factors that cause population growth to slow down as populations approach their carrying capacity. Organisms have a range of reproductive patterns. At the extremes are r-selected species, which experience rapid population growth rates, and K-selected species, which experience high survivorship and slow population growth rates. Patterns of survivorship over the life span can be graphically represented as type I, II, and III survivorship curves. Metapopulations are groups of spatially distinct populations that are connected by occasional movements of individuals. These movements reduce the probability of any of the populations going extinct.

■ **Describe species interactions and the roles of keystone species.**

Competition is an interaction between two species that share a limiting resource. Predation is an interaction in which one species consumes part or all of another species. Mutualism is an interaction in which two species provide fitness benefits to each other. Keystone species are species that have an effect on their community that is disproportionate to their abundance.

■ **Discuss the process of ecological succession.**

Ecological succession is the predictable replacement of one group of species by another group of species over time. In terrestrial communities, there are two kinds of ecological succession: primary succession and secondary succession.

■ **Explain how latitude, time, area, and distance affect the species richness of a community.**

Species richness typically increases as we move from the poles toward the equator, with the length of time a habitat has existed, with increased habitat size, and with shorter distances from sources of colonizing species.

1. Which of the following is *not* an example of a density-independent factor?
   (a) Drought
   (b) Competition
   (c) Forest fire
   (d) Hurricane
   (e) Flood

2. As the size of a white-tailed deer population increases,
   (a) the carrying capacity of the environment for white-tailed deer will be reduced.
   (b) a volcanic eruption will have a greater proportional effect than it would on a smaller population.
   (c) the effect of limiting resources will decrease.
   (d) the number of gray wolves, a natural predator of white-tailed deer, will increase.
   (e) white-tailed deer are more likely to become extinct.

3. The following graph of the population growth of Canada geese in Ohio between 1955 and 2002 can *best* be described as

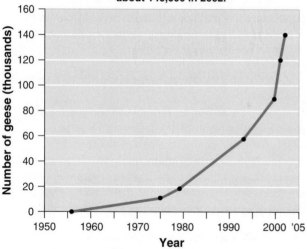

**Resident Canada goose population in Ohio increased from 20 birds introduced in 1956 to about 140,000 in 2002.**

   (a) an exponential growth curve.
   (b) a logistic growth curve.
   (c) a stochastic growth curve.
   (d) oscillation between overshoot and die-off.
   (e) approaching the carrying capacity.

4. Which of the following is *not* a statement of the logistic growth model?
   (a) Population growth is limited by density-dependent factors.
   (b) A population will initially increase exponentially and then level off as it approaches the carrying capacity of the environment.
   (c) Future population growth cannot be predicted mathematically.
   (d) Population growth slows as the number of individuals approaches the carrying capacity.
   (e) A graph of population growth produces an S-shaped growth curve over time.

5. Which of the following characteristics are typical of *r*-selected species?
   I They produce many offspring in a short period of time.
   II They have very low survivorship early in life.
   III They take a long time to reach reproductive maturity.

   (a) I only          (d) I and II
   (b) II only         (e) II and III
   (c) III only

6. A high intrinsic growth rate would most likely be characteristic of
   (a) a *K*-selected species such as elephants.
   (b) an *r*-selected species such as the American bullfrog.
   (c) a *K*-selected species that lives near its carrying capacity.
   (d) a species that is near extinction.
   (e) a species with a low reproductive rate that takes a long time to reach reproductive maturity.

7. Which of the following descriptions *best* matches the survivorship curves presented below?

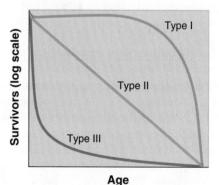

   (a) Type I could represent the house mouse, which is a typical *r*-selected species.
   (b) Type II could be the average of *r*-selected and *K*-selected species in a specific area.

(c) Type III could represent elephants, which are typical *K*-selected species.

(d) Type I could represent an oak tree species that experiences very low survivorship early and late in life.

(e) Type II could represent a coral species that has a constant decline in survivorship throughout its life.

8. In the coniferous forests of Oregon, eight species of woodpeckers coexist. Four species select their nesting sites based on tree diameter. The fifth species nests only in fir trees that have been dead for at least 10 years. The sixth species also nests in fir trees, but only in live or recently dead trees. The two remaining species nest in pine trees, but each selects trees of different sizes. This pattern is an example of

(a) resource partitioning.

(b) commensalism.

(c) true predation.

(d) predator-mediated competition.

(e) a keystone species.

9. In the Antarctic food web below, the arrows point from a food source to a consumer. Which of the following statements is *not* a conclusion that can be drawn from this food web?

(a) Penguins, other birds, leopard seals, and elephant seals compete for fish.

(b) Carnivorous zooplankton and herbivorous zooplankton interact mutualistically.

(c) Leopard seals are predators and penguins are their prey.

(d) Krill are omnivores.

(e) Smaller toothed whales prey on baleen whales, several species of seals, and penguins.

10. Which of the following statements about ecological succession is *correct*?

(a) Secondary succession is followed by primary succession.

(b) Primary succession occurs over a shorter time span than secondary succession.

(c) Succession is influenced by competition for limiting resources such as available soil, moisture, and nutrients.

(d) In forest succession, less shade-tolerant trees replace more shade-tolerant trees.

(e) Forest fires and hurricanes lead to primary succession because a soil base still exists.

11. Which of the following sequences of secondary succession would be likely to occur in abandoned farmland in the eastern United States?

(a) Bare soil, lichens, mosses, grasses, deciduous trees

(b) Bare rock, lichens, mosses, grasses, shrubs, mixed shade- and sunlight-tolerant trees

(c) Bare soil, grasses and wildflowers, shrubs, shade-tolerant trees

(d) Bare soil, grasses and wildflowers, shrubs, sunlight-tolerant trees, shade-tolerant trees

(e) Bare rock, grasses and wildflowers, lichens, mosses, shrubs, coniferous trees

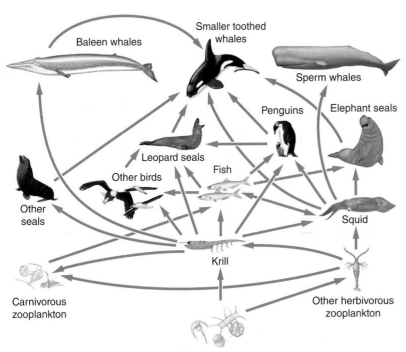

12. The theory of island biogeography suggests that species richness is affected by which of the following factors?

    I   Island distance from mainland
    II  How the island is formed
   III  Island size

(a) I only
(b) II only
(c) III only
(d) I and III
(e) II and III

## APPLY THE CONCEPTS

1. The California Department of Fish and Game is developing a plan to connect mountain "habitat islands" that are separated by open areas of flat, arid land in the deserts of southeastern California. These mountain areas are habitats for desert bighorn sheep (*Ovis canadensis*), which move extensively among the islands through habitat corridors. The habitat corridors provide opportunities for recolonization, seasonal migration, and maintenance of genetic variation among the metapopulation of desert bighorn sheep.

   (a) Explain what is meant by a metapopulation and how it relates to the desert bighorn sheep.
   (b) Identify *two* density-dependent factors and *one* density-independent factor that could affect the populations of desert bighorn sheep.
   (c) Explain the consequences to the desert bighorn sheep population if the plan to connect the mountain habitat islands is not implemented.
   (d) Explain how the theory of island biogeography applies to the mountainous areas of southeastern California.

2. Read the following information, which was posted in the great apes exhibit of the Fremont Zoo, and answer the questions that follow.

The Western lowland gorilla (*Gorilla gorilla*) lives in the moist tropical rainforests of western Africa. The gorillas' primary diet consists of fruits, leaves, foliage, and sometimes ants and termites. Occasionally they venture onto farms and feed on crops. Their only natural enemy is the leopard, the only animal other than humans that can successfully kill an adult gorilla. Disease, particularly the Ebola virus, is threatening to decimate large populations of these gorillas in Congo.

The Western lowland gorilla has a life span of about 40 years and produces one offspring every 4 years. Males usually mate when they are 15 years old. Females reach sexual maturity at about 8 years of age, but rarely mate before they are 10 years old.

   (a) Based on the information above, is the Western lowland gorilla an *r*-selected or a *K*-selected species? Provide evidence to support your answer.
   (b) Identify and explain *four* community interactions that involve the Western lowland gorilla.
   (c) Explain what is meant by secondary succession and describe how it may be initiated in a tropical forest.

## MEASURE YOUR IMPACT

**The Living Planet Index** In 2006, the World Wildlife Fund and its partners published an analysis of species survival rates between 1970 and 2005. Using published data on 1,477 animal species that had been monitored in five regions of the world, they developed a "Living Planet Index," which estimated whether animal diversity in these regions was increasing or decreasing. The year 1970 was used as a baseline (that is, the index is set to 1 for 1970).

| Region of the world | Index (1970) | Index (1985) | Index (2000) | Index (2005) |
|---|---|---|---|---|
| Terrestrial temperate areas | 1.00 | 0.98 | 1.05 | 1.04 |
| Terrestrial tropical areas | 1.00 | 0.91 | 0.55 | 0.54 |
| Freshwater temperate areas | 1.00 | 1.16 | 0.71 | — |
| Freshwater tropical areas | 1.00 | 0.95 | 0.65 | — |
| Marine areas | 1.00 | 1.00 | 0.85 | 0.72 |

*Source:* World Wildlife Fund, *2010 and Beyond: Rising to the Biodiversity Challenge*, http://www.wwf.org.uk/filelibrary/pdf/2010_and_beyond.pdf.

(a) Graph the changes in the Living Planet Index over time for each of the five regions.

(b) Globally, which region of the world has experienced the greatest decline in biodiversity? Overall, has the decline been greater in temperate or tropical areas?

(c) The World Wildlife Fund attributes the decline of biodiversity to habitat destruction, overharvesting, and pollution. Based on the above data, identify *three* actions that individuals could take to slow the decline of species richness.

## ENGAGE YOUR ENVIRONMENT

How would you describe species abundance and distribution in a local ecosystem? In this Engage Your Environment Activity, you will sample three plots of land in a small vegetated area and determine the abundance and distribution of species in each. *See Engage Your Environment Activity 6 at the end of this book.*

## The Environmental Implications of China's Growing Population

Human population size, affluence, and resource consumption all have interrelated impacts on the environment. The example of China is striking. With 1.3 billion people—20 percent of the world's population—China is the world's most populous nation. Because of its rapid economic development, it is expected to soon become the world's largest economy. Once-scarce consumer goods such as automobiles and refrigerators are becoming increasingly commonplace in China. Although the United States, with 307 million of 1.6 births per woman. If its current population dynamics continue, China's population may begin to decline by 2040.

Population control is only one part of the picture, however. Even if China's population were to stop growing today, the country's resource consumption would continue to increase as standards of living improve. Greater numbers of Chinese people are purchasing cars, home appliances, and other material goods that are common in Western nations. All of these products require resources to produce and use. Manufacturing a refrigerator requires mining and processing raw

### The Chinese are facing considerable environmental challenges as their affluence increases.

people, is currently the world's largest consumer of resources and the greatest producer of many pollutants, China may soon surpass the United States in both consumption and pollution. It is already the largest emitter of carbon dioxide and sulfur dioxide, and it consumes one-third of commercial fish and seafood. The Chinese are facing considerable environmental challenges as their affluence increases.

Managing the presence of humans on Earth sustainably requires addressing both population growth and resource consumption. China has already taken dramatic steps to limit its population growth. Since the 1970s, China has had a "one-child" policy. Couples that restrict themselves to a single child are rewarded financially, while those with three or more children face sanctions, such as a 10 percent salary reduction. Chinese officials use numerous tools—many controversial—to meet population targets, including abortions, sterilizations, and the designation of certain pregnancies as "illegal."

China is one of only a few countries where government-mandated population control measures have significantly reduced population growth. After decades of having one of the world's highest fertility rates, China now has a fertility rate materials such as steel and copper, producing plastic from oil, and using large quantities of electricity. Having a refrigerator in the home increases daily electricity demand. All of these processes generate carbon dioxide, air and water pollution, and other waste products.

A look at a typical Chinese city street is evidence of the country's growing affluence. Between 1985 and 2002, China's population increased 30 percent, but the number of vehicles used in China grew by over 500 percent, from 3 million to 20 million. Today, China has about 25 million private cars on the road. By 2020, China will have 140 million vehicles, according to Chinese government estimates. China is already the second largest consumer of petroleum (after the United States), and concentrations of urban air pollutants, such as carbon monoxide and photochemical smog, are on the rise. ▶

A traffic jam in Beijing, China.

◀ There are more than 250 million children under the age of 15 in China today.

There is some good news, however. China already has higher fuel efficiency standards for cars than the United States, and it is quickly becoming a leader in the manufacturing of renewable energy technologies. Nevertheless, in 2006, 16 of the 20 most polluted cities in the world were in China, according to the World Bank.

China's influence on the environment is dramatic because of its size and its increasing industrial activity. However, increasing industrial activity is occurring in many other parts of the world as human populations and consumption increase. What will be the environmental impact of humans in 2020, and what can we do to reduce it? ■

*Sources:* H. Kan, B. Chen, and C. Hong, Health impact of outdoor air pollution in China: Current knowledge and future research needs, *Environmental Health Perspectives* 117 (2009): A1870; J. Liu and J. Diamond, China's environment in a globalizing world, *Nature* 435 (2005): 1179–1186.

## UNDERSTAND THE KEY IDEAS

Human population growth and associated resource consumption have large impacts on the environment. In 2010, Earth's human population was 6.8 billion and growing by 212,000 people per day. At the same time, people in many parts of the world are using more resources than ever before.

After reading this chapter you should be able to

■ describe the potential limits to human population growth.

■ describe important aspects of global and national population growth using demographic terminology and tools.

■ evaluate the social, economic, and environmental factors that have contributed to decreasing growth rates in many countries.

■ analyze relationships among changes in population size, economic development, and resource consumption at global and local scales.

■ explain how people have attempted to harmonize economic development with sustainable development.

# Scientists disagree on Earth's carrying capacity

Every 5 days, the global human population increases by roughly a million lives: 1.8 million infants are born and 800,000 people die. The human population has not always grown at this rate, however. As **FIGURE 7.1** shows, until a few hundred years ago, the human population was stable: deaths and births occurred in roughly equal numbers. This situation changed about 400 years ago, when agricultural output and sanitation began to improve. Better living conditions caused death rates to fall, but birth rates remained relatively high. This was the beginning of a period of rapid population growth resulting in the current human population of 6.8 billion people.

As we saw in Chapter 6, under ideal conditions, all populations grow exponentially. In most cases, exponential growth halts when an environmental limit is reached. The limiting factor can be a scarcity of resources such as food or water or an increase in predators, parasites, or diseases. Limiting factors determine the carrying capacity of a habitat. Are human populations constrained by limiting factors, as other populations are?

Environmental scientists have differing opinions on Earth's carrying capacity

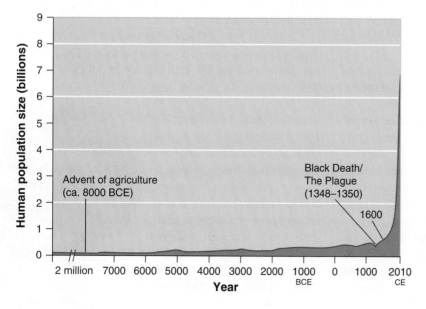

**FIGURE 7.1** **Human population growth.** The global human population has grown more rapidly in the last 400 years than at any other time in history.

for humans. Some scientists believe we have outgrown, or eventually will outgrow, the available supply of food, water, timber, fuel, and other resources. One of the first proponents of the notion that the human population could exceed Earth's carrying capacity was English clergyman and professor Thomas Malthus. In 1798, Malthus observed that the human population was growing exponentially, while the food supply we rely on was growing linearly. In other words, the food supply increases by a fixed amount each year, while the human population increases in proportion to its own increasing size. Malthus concluded that eventually the human population size would exceed the food supply. FIGURE 7.2a shows this projection graphically. A number of environmental scientists today subscribe to Malthus's view that humans will eventually reach their carrying capacity, after which the rate of population growth will decline.

Other scientists do not believe that Earth has a fixed carrying capacity for humans. They argue that the growing population of humans provides an increasing supply of intellect that leads to increasing amounts of innovation. By employing creativity, humans can alter Earth's carrying capacity. This is one of the fundamental ways in which humans differ from most other species on Earth.

For example, in the past, whenever the food supply seemed small enough to limit the human population, major technological advances increased food production. This progression began thousands of years ago. The development of arrows made hunting more efficient, which allowed hunters to feed a larger number of people. Early farmers increased crop yields by plowing by hand and later with oxen- or horse-driven plows. More recently, mechanical harvesters made farming even more efficient. Each of these inventions increased the planet's carrying capacity for humans, as FIGURE 7.2b shows.

The ability of humans to innovate in the face of challenges has led some scientists to expect that we will continue to make technological advances indefinitely into the future. This expectation is reasonable, but questions remain. Based on our history, should we assume that humans will continue to find ways to feed a growing population? Are there other limits to human population growth? How do we know if we have exceeded Earth's carrying capacity?

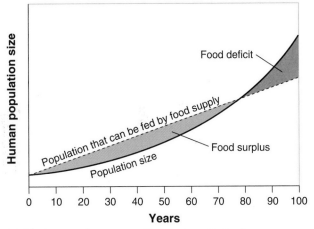

**(a)** No significant improvement in agricultural technology

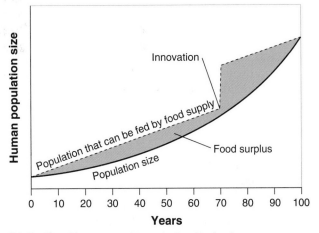

**(b)** Significant improvement in agricultural technology

**FIGURE 7.2** **A theoretical model of food supply and population size.** (a) In a theoretical 100-year period *without* significant improvements in agricultural technology, the human population grows exponentially, while the food supply grows linearly. Consequently, a food surplus is followed by a food deficit. (b) In a theoretical 100-year period *with* significant improvements in agricultural technology, the food supply increases suddenly. Consequently, there is a continuing food surplus.

**GAUGE YOUR PROGRESS**

✓ What determines the carrying capacity of a habitat?

✓ How might humans differ from other organisms in terms of carrying capacity?

✓ Will humans exceed Earth's carrying capacity? What evidence can you use to justify your argument?

# Many factors drive human population growth

In order to understand the impact of the human population on the environment, we must first understand what drives human population growth. The study of human populations and population trends is called **demography,** and scientists in this field are called **demographers.** By analyzing specific data such as birth rates, death rates, and migration rates, demographers can offer insights—some of them surprising—into how and why human populations change and what can be done to influence rates of change.

## Changes in Population Size

We can view the human population as a system with inputs and outputs, like all biological systems. If there are more births than deaths, the inputs are greater than the outputs, and the system expands. For most of human history, total births have slightly outnumbered total deaths, resulting in very slow population growth. If the reverse had been true, the human population would have decreased and would have eventually become extinct.

When demographers look at population trends in individual countries, they take into account the **immigration** of people into a country and **emigration** out of the country. As FIGURE 7.3 shows, this means that inputs to the population include births and immigration, and outputs, or decreases, include deaths and emigration. When inputs to the population are greater than outputs, the growth rate is positive. Conversely, if outputs are greater than inputs, the growth rate is negative.

Demographers use specific measurements to determine yearly birth and death rates. The **crude birth rate (CBR)** is the number of births per 1,000 individuals per year. The **crude death rate (CDR)** is the number of deaths per 1,000 individuals per year. Worldwide, there were 20 births and 8 deaths per 1,000 people in 2009. We do not factor in migration for the global population because, even though people move from place to place, they do not leave Earth. Thus, in 2009, the global population increased by 12 people per 1,000 people. This rate can be expressed mathematically as

$$\frac{\text{Global population}}{\text{growth rate}} = \frac{[\text{CBR} - \text{CDR}]}{10}$$

$$= \frac{[20 - 8]}{10}$$

$$= 1.2 \text{ percent}$$

Note that because the birth and death rates are expressed per 1,000 people, we divide by 10 in order to represent the value as a percentage, a unit of expression that is familiar to most people.

To calculate the population growth rate for a single nation, we take immigration and emigration into account:

$$\frac{\text{Nat. pop.}}{\text{growth rate}} = \frac{[(\text{CBR} + \text{immigr.}) - (\text{CDR} + \text{emigr.})]}{10}$$

If we know the growth rate of a population and assume that growth rate is constant, we can calculate the number of years it takes for a population to double, known as its **doubling time.** Because growth rates may change in future years, we can never determine a country's doubling time with certainty. We can say that a population will double in a certain number of years *if* the growth rate remains constant.

The doubling time can be approximated mathematically using a formula called the *rule of 70:*

$$\text{Doubling time (in years)} = \frac{70}{\text{growth rate}}$$

Therefore, a population growing at 2 percent per year will double every 35 years:

$$\frac{70}{2} = 35 \text{ years}$$

Note that this is true of any population growing at 2 percent per year, regardless of the size of that population. At a 2 percent growth rate, a population of 50,000 people will increase by 50,000 in 35 years, and a population of 50 million people will increase by 50 million in 35 years.

As we saw in Figure 7.1, Earth's population has doubled several times since 1600. It is almost certain, however, that Earth's population will not double again. FIGURE 7.4 shows the current projections through the year 2100. Most demographers believe that the human population will be somewhere between 8.1 billion and 9.6 billion in 2050 and will stabilize between 6.8 billion and 10.5 billion by roughly 2100.

**Population size**

FIGURE 7.3 **The human population as a system.** We can think of the human population as a system, with births and immigration as inputs and deaths and emigration as outputs.

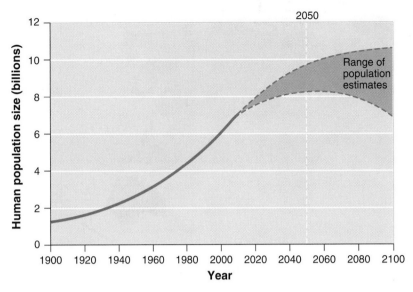

**FIGURE 7.4** **Projected world population growth.** Demographers project that the global human population will be between 8.1 billion and 9.6 billion by 2050. By 2100, it is projected to be between 6.8 billion and 10.5 billion. The dashed lines represent estimated values. [After Millennium Ecosystem Assessment, 2005.]

## Fertility

To understand more about the role births play in population growth, demographers look at the **total fertility rate (TFR),** an *estimate* of the average number of children that each woman in a population will bear throughout her childbearing years, between the onset of puberty and menopause. In the United States in 2008, for example, the TFR was 2.1, meaning that, on average, each woman of childbearing age would have just over 2 children. Note that, unlike CBR and CDR, TFR is not calculated per 1,000 people. Instead, it is a measure of births per woman.

To gauge changes in population size, demographers also calculate **replacement-level fertility,** the TFR required to offset the average number of deaths in a population so that the current population size remains stable. Typically, replacement-level fertility is just over 2 children, with 2 being the number of offspring that will replace the parents who conceived them when those parents die. However, replacement-level fertility also depends on rates of *prereproductive mortality,* or death before a person has children, which depend on a country's economic status.

In **developed countries**—countries with relatively high levels of industrialization and income—we typically see a replacement-level fertility of 2.1. In **developing countries,** however, relatively low levels of industrialization and incomes of less than $3 per person per day are the norm. In these countries, mortality among young people tends to be higher, and a TFR of greater than 2.1 is needed to achieve replacement-level fertility.

When TFR is equal to replacement-level fertility and immigration and emigration are equal, a country's population is stable. A country with a TFR of less than 2.1 and no net increase from immigration is likely to experience a population decrease because that TFR is below replacement-level fertility. In contrast, a devel-

oped country with a TFR of more than 2.1 and no net decrease from emigration is likely to experience population growth because that TFR is above replacement-level fertility.

## Life Expectancy

To understand more about the outputs in a human population system, demographers study the human life span. **Life expectancy** is the average number of years that an infant born in a particular year in a particular country can be expected to live, given the current average life span and death rate in that country. Life expectancy is generally higher in countries with better health care. A high life expectancy also tends to be a good predictor of high resource consumption rates and environmental impacts. **FIGURE 7.5** shows life expectancies around the world.

Life expectancy is often reported in three different ways: for the overall population of a country, for males only, and for females only. For example, in 2008, global life expectancy was 69 years overall, 67 years for men, and 70 years for women. In the United States, life expectancy was 78 years overall, 75 years for men, and 81 years for women. In general, human males have higher death rates than human females, leading to a shorter life expectancy for men. In addition to biological factors, men have historically tended to face greater dangers in the workplace, made more hazardous lifestyle choices, and been more likely to die in wars. Cultures have changed over time, however, and as more and more women enter the workforce and the armed forces, the life expectancy gap between men and women will probably decrease.

**INFANT AND CHILD MORTALITY** The availability of health care, access to good nutrition, and exposure to pollutants are all factors in life expectancy, *infant mortality,* and *child*

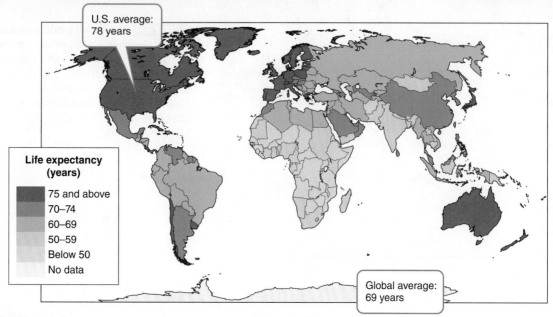

**FIGURE 7.5** **Average life expectancies around the world.** Life expectancy varies significantly by continent and in some cases by country. [After CountryWatch, http://www.countrywatch.com/facts/facts_default.aspx?type=image&img=LEAG.]

*mortality.* The **infant mortality** rate is defined as the number of deaths of children under 1 year of age per 1,000 live births. The **child mortality** rate is defined as the number of deaths of children under age 5 per 1,000 live births. **FIGURE 7.6** shows infant mortality rates around the world.

If a country's life expectancy is relatively high and its infant mortality rate is relatively low, we might predict that the country has a high level of available health care, an adequate food supply, potable drinking water, sanitation, and a moderate level of pollution. Conversely, if its life expectancy is relatively low and its infant mortality rate is relatively high, it is likely that the country's population does not have sufficient health care or sanitation and that potable drinking water and food are in limited supply. Pollution and exposure to other environmental

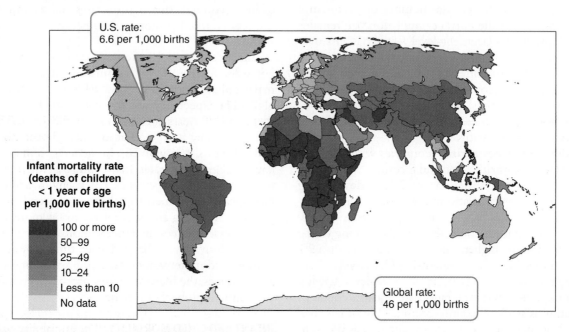

**FIGURE 7.6** **Infant mortality around the world.** Infant mortality rates are lower in developed countries and higher in developing countries. [After http://www.mapsofworld.com/infant-mortality-rate-map.htm#bot.]

hazards may also be high. In 2009, the global infant mortality rate was 46. In the United States, the infant mortality rate was 6.6. In other developed countries, such as Sweden and France, the infant mortality rate was even lower (2.5 and 3.6, respectively). Availability of prenatal care is an important predictor of infant mortality rate. For example, the infant mortality rate is 99 in Liberia and 50 in Bolivia, both countries where many women do not have good access to prenatal care.

Sometimes, life expectancy and infant mortality in a given sector of a country's population differ widely from life expectancy and infant mortality in the country as a whole. In this case, even when the overall numbers seem to indicate a high level of health care throughout the country, the reality may be starkly different for a portion of its population. For example, whereas the infant mortality rate for the U.S. population as a whole is 6.6, it is 13.6 for African Americans, 8.1 for Native Americans, and 5.8 for Caucasians. This variation in infant mortality rates is probably related to socioeconomic status and varying access to adequate nutrition and health care. These differences are often issues of *environmental justice,* a topic we discuss in more detail in Chapter 20.

**AGING AND DISEASE** Even with a high life expectancy and a low infant mortality rate, a country may have a high crude death rate, in part because it has a large number of older individuals. The United States, for example, has a higher standard of living than Mexico, which is consistent with its higher life expectancy and lower infant mortality rate. At the same time, the United States has a much higher CDR, at 8 deaths per 1,000 people on average, than Mexico, which has 5 deaths per 1,000 people on average. This is because the United States has a much larger elderly population, with 13 percent of its population aged 65 years or older, compared with the 6 percent of the population aged 65 or older in Mexico.

Disease is an important regulator of human populations. According to the World Health Organization, infectious diseases—those caused by microbes that are transmissible from one person to another—are the second biggest killer worldwide after heart disease. In the past, tuberculosis and malaria were two of the infectious diseases responsible for the greatest number of human deaths. Today, the human immunodeficiency virus (HIV), which causes acquired immune deficiency syndrome (AIDS), is responsible for more deaths annually than either tuberculosis or malaria. Between 1990 and 2007, AIDS-related illnesses killed more than 22 million adults and children. Because HIV disproportionately infects people aged 15 to 49—the most productive years in a person's life span—HIV has had a more disruptive effect on society than other illnesses that affect the very young and the very old.

HIV has a significant effect on infant mortality, child mortality, population growth, and life expectancy. In

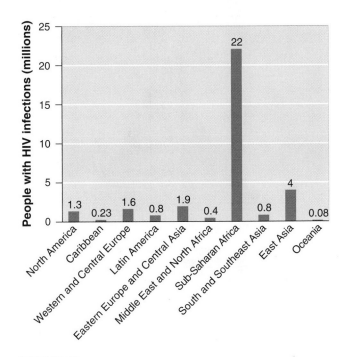

**FIGURE 7.7** **HIV infection worldwide.** Worldwide, about 33 million people are living with HIV, two-thirds of them in sub-Saharan Africa. [Data from World Health Organization, UNAIDS.]

Lesotho, in southern Africa, where 23 percent of the adult population is infected with HIV, life expectancy fell from 63 years in 1995 to 40 years in 2009.

As **FIGURE 7.7** shows, approximately 33 million people were living with HIV in 2009, 22 million of them in sub-Saharan Africa. The annual number of deaths due to AIDS reached a peak of 2.1 million in 2005 and has decreased since then.

## Age Structure

Demographers use data on age to predict how rapidly a population will increase and what its size will be in the future. A population's *age structure* describes how its members are distributed across age ranges, usually in 5-year increments. **Age structure diagrams,** examples of which are shown in **FIGURE 7.8**, are visual representations of age structure within a country for males and females. Each horizontal bar of the diagram represents a 5-year age group. The total area of all the bars in the diagram represents the size of the whole population.

Every nation has a unique age structure, but we can group countries very broadly into three categories. A country with many more younger people than older people has an age structure diagram that is widest at the bottom and smallest at the top, as shown in Figure 7.8a. This type of age structure diagram, called a **population pyramid,** is typical of developing countries, such as Venezuela and India. The wide base of the graph compared with the levels above it indicates that the population will grow because a large number of

**(a)** India

**(b)** United States

**(c)** Germany

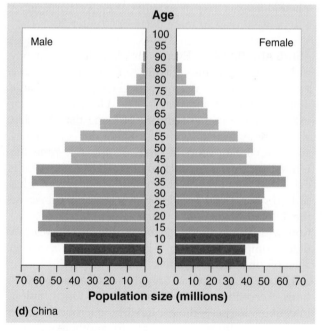

**(d)** China

**FIGURE 7.8** **Age structure diagrams.** The horizontal axis of the age structure diagram shows the population size in millions for males and females in each 5-year age group shown on the vertical axis. (a) A population pyramid illustrates a rapidly growing population. (b) A column-shaped age structure diagram indicates population stability. (c) In some developed countries, the population is declining. (d) China's population control measures will eventually lead to a population decline. [After http://www.census.gov/ipc/www/idb/pyramids.html.]

females aged 0 to 15 have yet to bear children. Even if each one of these future potential mothers has only two children, the population will grow simply because there are increasing numbers of women able to give birth.

The population pyramid also illustrates that it takes time for actions that attempt to reduce births to catch up with a growing population. This phenomenon, called **population momentum,** is the reason a popu-

lation keeps on growing after birth control policies or voluntary birth reductions have begun to lower the crude birth rate of a country. Eventually, over several generations, those actions will bring the population to a more stable growth rate.

A country with little difference between the number of individuals in younger age groups and in older age groups has an age structure diagram that looks more like a column than a pyramid from age 0 through age 50, as shown in Figure 7.8b. If a country has few individuals in the younger age classes, we can deduce that it has slow population growth or is approaching no growth at all. The United States, Canada, Australia, Sweden, and many other developed countries have this type of age structure diagram. A number of developing countries that have recently lowered their growth rates should begin to show this pattern within the next 10 to 15 years.

A country with a greater number of older people than younger people has an age structure diagram that resembles an inverted pyramid. Such a country has a total fertility rate below 2.1 and a decreasing number of females within each younger age range. Such a population will continue to shrink. Italy, Germany, Russia, and a few other developed countries display this pattern, seen in Figures 7.8c and 7.8d. China is in the very early stages of showing this pattern.

## Migration

Regardless of its birth and death rates, a country may experience population growth, stability, or decline as a result of migration. A country with a relatively low CBR but a high immigration rate may still experience population growth. For example, the United States has a TFR of 2.1 and an age structure diagram that is approaching a column shape, but it has a high rate of immigration. As a result, the U.S. population will probably increase by 44 percent by 2050.

**Net migration rate** is the difference between immigration and emigration in a given year per 1,000 people in a country. A positive net migration rate means there is more immigration than emigration, and a negative net migration rate means the opposite. For example, approximately 1 million people immigrate to the United States each year, and only a small number emigrate. With a U.S.

population of 300 million, these rates are equal to 3.3 immigrants per 1,000 people.

Canada has a net migration rate of 7 per 1,000 and a TFR of 1.6 (*below* replacement-level fertility). Therefore, both the United States and Canada will experience net population growth over the next few decades, but the growth will come from immigration rather than from births originating within the existing population. (Crunch the Numbers "Calculating Population Growth" shows how this can happen.) In countries with a negative net migration rate and a low TFR, the population actually decreases over time. Very few countries fit this model. One is the country of Georgia, in western Asia. It has a growth rate of 0.2 percent, a TFR of 1.4, and a net migration rate of −5 per 1,000. Georgia is projected to have a 24 percent population decrease by 2050.

Some countries experience both high levels of emigration and a high CBR. Fiji, for example, has a growth rate of 1.7 percent, a TFR of 2.6, and a net migration rate of −8 per 1,000. With a population of 800,000, Fiji loses 6,400 people to emigration every year. However, the increase in population due to more births than deaths will exceed the losses from emigration, assuming the net migration rate remains the same. Thus, even though the net migration rate reduces the impact of its growth rate and TFR, Fiji's population is still projected to increase by 12 percent over the next 40 years.

The movement of people around the world does not affect the total number of people on the planet. But for many reasons, migration is still an important issue in environmental science. The movement of people displaced because of disease, natural disasters, environmental problems, or conflict can create crowded, unsanitary conditions, food and water shortages, and refugee camps, all of which can easily become humanitarian and environmental health issues. The movement of people from developing countries to developed countries tends to

---

### CRUNCH THE NUMBERS

#### Calculating Population Growth

New Zealand has a population of 4.3 million people, a total fertility rate (TFR) of 2.1, and a net migration rate of 2 per 1,000. How many people will New Zealand gain next year as a result of immigration? If the TFR stays the same for the next century, and the net migration rate stays the same as well, what will be the annual rate of increase in the population?

$$\text{Net migration rate} = \frac{\text{(number of immigrants)}}{\text{(number of people in the population)}}$$

A TFR of 2.1 for a developed country suggests that the country is at replacement-level fertility and thus that the population is stable. The migration rate suggests that

$$\frac{2}{1,000} = \frac{X}{4,300,000}$$

and therefore that

$$X = 8,600 \text{ people/year}$$

So, 8,600 people are added to New Zealand each year by migration. If there is no growth due to biological replacement, then the rate of increase is

$$\frac{8,600 \text{ people/year}}{4,300,000 \text{ people}} = 0.2\%/\text{year}$$

**Your Turn:** To calculate doubling time, we use the rule of 70

$$\left( \frac{70}{\%\ \text{growth rate}} = \text{doubling time} \right)$$

How many years will it take for the New Zealand population to double if the population increases due to migration only?

increase the ecological footprint of those people. Over time, immigrants typically adopt the lifestyle and consumption habits of their new country. A person who migrates from Mexico to the United States is likely to use more resources as a U.S. resident than as a resident of Mexico, simply because the United States has a more affluent lifestyle.

**GAUGE YOUR PROGRESS**

✓ What are the main factors that influence human population growth?

✓ How does age structure influence the population growth rate?

✓ How do a country's total fertility rate and net migration rate determine population growth?

# Many nations go through a demographic transition

Before 1600, the global average population growth rate was typically below 0.1 percent per year, meaning that the time it would have taken for the population to double was more than 700 years. Then, as technology, health care, and sanitation improved, the growth rate increased dramatically. By 1965, the doubling time for the global population was 42 years. Since then, the global population growth rate has slowed, and most demographers believe that Earth's population is likely to level off by 2100 without ever doubling again.

Demographers need to understand why the rate of population growth fluctuated so much in the past and whether there are lessons they can learn from those fluctuations that will help us understand the future. Developing and testing theories is one way to gain this understanding.

## The Theory of Demographic Transition

Historically, nations that have gone through similar processes of economic development have experienced similar patterns of population growth. Scientists who studied the population growth patterns of European countries in the early 1900s described a four-phase process they referred to as a *demographic transition*. The **theory of demographic transition** says that *as a country moves from a subsistence economy to industrialization and increased affluence, it undergoes a predictable shift in population growth*. The four phases of a demographic transition are shown in **FIGURE 7.9**. At the beginning of the transition, called *phase 1,* the country experiences slow growth or no growth. This phase is followed by rapid growth in *phase 2.* Many countries in Africa and some Asian countries, such as China and India, are classified as being in

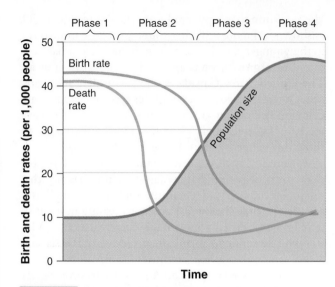

**FIGURE 7.9** **Demographic transition.** The theory of demographic transition models the way that birth, death, and growth rates for a nation change with economic development. Phase 1 is a preindustrial period characterized by high birth rates and high death rates. In phase 2, as the society begins to industrialize, death rates drop rapidly, but birth rates do not change. Population growth is the greatest at this point. In phase 3, birth rates decline for a variety of reasons. In phase 4, the population stops growing and sometimes begins to decline as birth rates drop below death rates.

phase 2. Countries such as the United States, Canada, and Australia are in *phase 3,* which is characterized by population stabilization. In *phase 4,* the population declines. Some western European nations are currently in this phase.

The theory of demographic transition, while helpful as a learning tool, does not adequately describe the population growth patterns of some developing countries today or in the last quarter-century. Birth and death rates have declined rapidly in a number of developing countries as a result of a variety of factors that are not yet entirely understood. In some developing countries the government has taken measures to improve health care and sanitation and promote birth control, in spite of the country's poverty. For example, in Nicaragua, the second poorest country in the Western Hemisphere, birth and death rates declined rapidly in the 1980s and 1990s, and there has been no increase in the 0 to 5-year age range since 1990. Nicaragua displays the birth rate and death rate patterns of a phase 4 country although its industrial activity is that of a developing country. In contrast, China is economically and industrially ahead of Nicaragua, and arguably behind the United States, yet its population growth rate is slightly lower than that of the United States, in large part because of the Chinese government's one-child policy.

Despite the limitations of the theory of demographic transition, it is worth examining in more detail because it allows us to understand the way some countries will influence the environment as they undergo growth and development.

## PHASE 1: SLOW POPULATION GROWTH

Phase 1 represents a population that is nearly at steady state. The size of the population will not change very quickly because high birth rates and high death rates offset one another. In other words, crude birth rate roughly equals crude death rate. This pattern is typical of countries before they begin to modernize. In these countries, life expectancy for adults is relatively short due to difficult and often dangerous working conditions. The infant mortality rate is also high because of disease, lack of health care, and poor sanitation, and as a result, many families have more children than they expect will live to adulthood. In a subsistence economy, where most people are farmers, having numerous children is an asset. Children can do jobs such as collecting firewood, tending crops, watching livestock, and caring for younger siblings. With no social security system, parents also count on having many children to care for them when they become old.

Western Europe and the United States were in phase 1 before the Industrial Revolution, from the late eighteenth to the mid-nineteenth century. Today, crude birth rates exceed crude death rates in almost every country, so even the poorest nations have moved beyond phase 1. However, an increase in crude death rates due to war, famine, and diseases such as AIDS has pushed some countries back in the direction of phase 1. Lesotho (CBR = 25, CDR = 23) is an example of a country that may have recently moved backward into phase 1.

## PHASE 2: RAPID POPULATION GROWTH

In phase 2, death rates decline while birth rates remain high, and as a result, the population grows rapidly. As a country modernizes, better sanitation, clean drinking water, increased access to food and goods, and access to health care, including childhood vaccinations, all reduce the infant mortality rate and CDR. However, the CBR does not markedly decline. Couples continue to have large families because it takes at least one generation, if not more, for people to notice the decline in infant mortality and adjust to it. This is another example of population momentum. It also takes time to implement educational systems and birth control measures.

A phase 2 country is in a state of imbalance: births outnumber deaths. India is in phase 2 today. The United States population exhibited a phase 2 population pyramid in the early twentieth century when there were high birth rates, high death rates, and a large total fertility rate.

## PHASE 3: STABLE POPULATION GROWTH

A country enters phase 3 as its economy and educational system improve. In general, as family income increases, people have fewer children, as FIGURE 7.10 shows. As a result, the CBR begins to fall. Phase 3 is typical of many developed countries, including the United States and Canada.

Why do people produce fewer children as their income increases? As societies transition from subsistence farming to more complex economic specializations, having large numbers of children may become a financial burden rather than an economic benefit. Relative affluence, more time spent pursuing education, and the availability of birth control increase the likelihood that people will choose to have smaller families. However, it is important to note that cultural, societal, and religious norms may also play a role in birth rates.

As birth rates and death rates decrease in phase 3, the system returns to a steady state. Population growth levels off during this phase, and population size does not change very quickly, because low birth rates and low death rates cancel each other out.

## PHASE 4: DECLINING POPULATION GROWTH

Phase 4 is characterized by declining population size and often by a relatively high level of affluence and economic

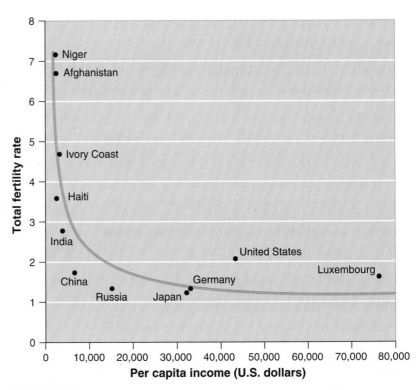

FIGURE 7.10 Total fertility rate versus per capita income. Wealthier nations tend to have lower total fertility rates. [Data from http://www.gapminder.org, 2007 data.]

FIGURE 7.11 **Some countries have very large elderly populations.** Here, men and women gather at a senior residence home in Hamburg, Germany.

development. Japan, the United Kingdom, Germany, Russia, and Italy are phase 4 countries, with the CBR well below the CDR.

The declining population in phase 4 means fewer young people and a higher proportion of elderly people (**FIGURE 7.11**). This demographic shift can have important social and economic effects. With fewer people in the labor force and more people retired or working part-time, the ratio of dependent elderly to wage earners increases, and pension programs and social security services put a greater tax burden on each wage earner. There may be a shortage of health care workers to care for an aging country. Governments may encourage immigration as a source of additional workers. In some countries, such as Japan, the government provides economic incentives to encourage families to have more children in order to offset the demographic shift.

Recent studies on demographic shifts in highly developed countries suggest that the TFR actually increases after reaching a low point between 1.2 and 1.5. The reasons for the increase are unclear, but it appears that when a population becomes affluent and well educated, it becomes somewhat easier for women to raise children, and they choose to do so in slightly greater numbers. Such a pattern is occurring in Norway, Italy, the United States, and other developed nations.

## Family Planning

We have already observed that as family income increases, people tend to have fewer children. In fact, there is a link between higher levels of education and affluence among females, in particular, and lower birth rates. As the education levels of women increase and women earn incomes of their own, fertility generally decreases. Even in developed countries where the TFR

has increased slightly since hitting its apparent low point, women have fewer children than those in developing countries. Educated and working women tend to have fewer children than other women, and many delay having children because of the demands of school and work. A higher age at first reproduction means that a woman is likely to have fewer children in her lifetime.

In addition to these factors, women with more education and income tend to have more access to information about methods of birth control, they are more likely to interact with their partners as equals, and they may choose to practice *family planning* with or without the consent of their partners. **Family planning** is regulation of the number or spacing of offspring through the use of birth control. When women have the option to use family planning, crude birth rates tend to drop. **FIGURE 7.12** shows how female education levels correlate with crude birth rates. In Ethiopia, for instance, women with a secondary school education or higher have a TFR of 2.0, whereas the TFR among uneducated women is 6.1.

There have been many examples of effective family planning campaigns in the last few decades. In the 1980s, Kenya had one of the highest growth rates in the world, and its TFR was almost 8. By 1990, its TFR was about 4—one-half of the previous rate. Kenya's government achieved these dramatic results by implementing an active family planning campaign. The campaign, which began in the 1970s, encouraged smaller families. Advertising directed toward both men and women emphasized that overpopulation led to both unemployment and harm to the natural environment. The ads also promoted condom use.

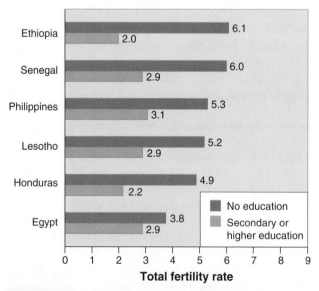

FIGURE 7.12 **Total fertility rate for educated and uneducated women in six countries.** Fertility is strongly related to female education in many developing countries. [After Population Reference Bureau, 2007 World Population Data Sheet, http://www.prb.org/pdf07/07WPDS_Eng.pdf.]

Thailand also successfully used family planning campaigns to lower its growth rate and TFR. Beginning in 1971, national population policy encouraged married couples to use birth control. Contraceptive use increased from 15 to 70 percent, and within 15 years the population growth rate fell from 3.2 to 1.6 percent. Today, Thailand's growth rate is 0.6 percent, among the lowest in Southeast Asia. Some of the credit for this hugely successful growth rate reduction is given to a creative and charismatic government official who gained a great deal of attention, in part by handing out condoms in public places.

# Population size and consumption interact to influence the environment

Population size is a critical factor in the impact of humans on Earth. The amount of resources each person uses is another critical factor. Every human exacts a toll on the environment by eating, drinking, generating waste, and consuming products. Even relatively simple foods such as beans and rice require energy, water, and mineral resources to produce and prepare. Raising and preparing meat requires even more resources; animals require crops to feed them as well as water and energy resources. Building homes, manufacturing cars, and making clothing and consumer products all require energy, water, wood, steel, and other resources. These and many other human activities contribute to environmental degradation.

As we have seen, population and economic development are not equally distributed around the world. Thus the human impact on natural resources is also unequally distributed.

## Economic Development

Of Earth's 6.8 billion human inhabitants, 5.6 billion live in developing countries, and only 1.3 billion live in developed countries. As **FIGURE 7.13** shows, 9 of the 12 most populous nations on Earth are developing countries. The population disparity between rich and poor countries has accelerated in recent decades. **FIGURE 7.14** charts the relationship between economic development and population growth rate for developing nations. Populations in developing parts of the world have continued to grow rapidly, at an average rate of 1.5 percent per year. At the same time, populations in the developed world have almost leveled off, with an average growth rate of 0.2 percent per year. Impoverished countries are increasing their populations more rapidly than affluent countries.

In addition to the disparity in population growth between developed countries and developing countries, their difference in resource use is striking. As we saw in Chapter 1, although only one-fifth of the human population lives in developed countries, those people consume much more than one-half of the world's energy and resources. In fact, one person in a developed country may have 2 to 10 times the environmental impact

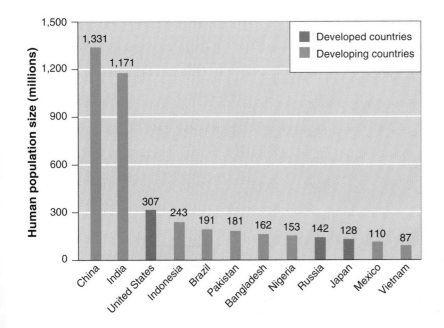

**FIGURE 7.13** **The 12 most populous countries in the world.** China and India are by far the largest nations in the world. Only 3 of the 12 most populous countries are developed nations. [Data from Population Reference Bureau, 2008 data.]

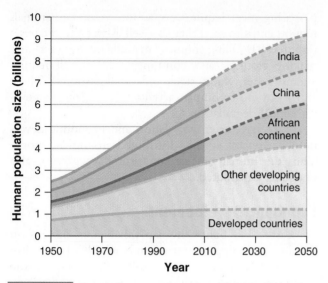

**FIGURE 7.14** **Population growth past and future.** Population growth in developed countries has mostly ceased, while that in developing countries is slowing, but is expected to continue beyond 2050. [After United Nations Population Division.]

of a person in a developing country. For example, the average ecological footprint for the world's 30 wealthiest countries is 6.4 ha (15.8 acres) per capita. Contrast that figure with the average ecological footprint for the 50 poorest countries: 1.0 ha (2.5 acres) per capita.

Calculating the per capita ecological footprint for a country provides a way to measure the effect of **affluence**—money, goods, or property—on the planet. **FIGURE 7.15** shows some examples of ecological footprints for selected countries. The world average ecological footprint is 2.7 ha (6.7 acres) per capita. The United States has the largest ecological footprint of any nation, at 9.0 ha (22 acres) per capita. China's footprint is 1.8 ha (4.5 acres) per capita. Haiti, the poorest country in the Western Hemisphere, has a footprint of 0.5 ha (1.2 acres) per capita. In other words, a person living in

the United States has more than 5 times the environmental impact of a person living in China and 18 times that of a person living in Haiti.

In addition to looking at per capita data, it is useful to examine the footprints of entire countries. We can do this by multiplying a country's per capita ecological footprint by the number of people in the country. We find that the United States has a footprint of 2,810 million hectares (6,944 million acres). China's footprint is 2,790 million hectares (6,894 million acres), and Ivory Coast (in western Africa) has a footprint of 18.6 million hectares (46.0 million acres). Looked at another way, the United States, with only one-fourth the population of China, has an ecological footprint comparable to China's because of its high levels of consumption. However, China's rapid development, as described in this chapter's opening story, means that its ecological footprint is likely to exceed that of the United States within the next 10 years.

## The IPAT Equation

The total environmental impact of 6.8 billion people is hard to appreciate and even more difficult to quantify. Some people consume large amounts of resources and have a negative impact on environmental systems, while others live much more lightly on the land (**FIGURE 7.16**). Living lightly can be intentional, as when people in the developed world make an effort to live "green," or sustainably. It can also be unintentional, as when poverty prevents people from acquiring material possessions or building homes.

To estimate the impact of human lifestyles on Earth, environmental scientists Barry Commoner, Paul Ehrlich, and John Holdren developed the **IPAT equation**:

$$Impact = Population \times Affluence \times Technology$$

Although it is written mathematically, the IPAT equation is a conceptual representation of the three major

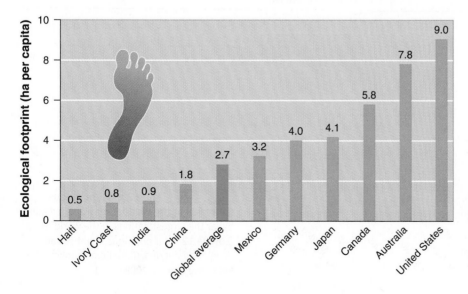

**FIGURE 7.15** **Per capita ecological footprints.** Many countries exceed the global average footprint of 2.7 ha per capita. [Data from Global Footprint Network, 2009 Data Sheet.]

(a)

(b)

**FIGURE 7.16** **Material possessions.** Most families in developing countries have few possessions compared with their counterparts in developed countries. In each of these photos, the members of a family are shown outside their home with all of their possessions. (a) A typical Thai family. (b) A typical Japanese family.

factors that influence environmental impact. *Impact* in this context is the overall environmental effect of a human population multiplied by affluence, multiplied by technology. It is useful to look at each of these factors individually.

*Population* has a straightforward effect on impact. All else being equal, two people consume twice as much as one. Therefore, when we compare two countries with similar economic circumstances, the one with more people is likely to have a larger impact on the environment.

*Affluence* is created by economic opportunity and does not have as simple a relationship to impact as population does. One person in a developed country can have a greater impact than *two or more* people in a developing country. A family of four in the United States that owns two large sport utility vehicles and lives in a spacious home with a lawn and swimming pool uses a much larger share of Earth's resources than a Bangladeshi family of four living in a two-room apartment and traveling on bicycles and buses. The more affluent a society or individual is, the higher the environmental impact.

The effect of *technology* is even more complicated. Technology can either degrade the environment or create solutions to minimize our impact on the environment. For example, the manufacturing of the chlorofluorocarbons (CFCs) that resulted in safe and effective refrigeration and air conditioning (benefiting human health) inadvertently led to ozone destruction in the stratosphere (see Science Applied 1 "Were We Successful in Halting the Growth of the Ozone Hole?"). By contrast, the hybrid electric car helps to *reduce* the impact of the automobile on the environment because it has greater fuel efficiency than a conventional internal combustion vehicle. The IPAT equation originally used the

term *technology,* but some scientists now use the term *destructive technology* to differentiate it from beneficial technologies such as the hybrid electric car.

## Local, Global, and Urban Impacts

Impacts on the environment may occur locally—within the borders of a region, city, or country—or they may be global in scale. The scale of impact depends on the economy and degree of development of a society. For example, a person may create a local impact by using products created within the country's borders. That same person may create a global impact by using materials that are imported. Both local and global impacts can be harmful to the environment in different ways.

In general, highly localized impacts are typical of rural, agriculturally based societies, while global impacts are typical of affluent or urban societies that tend to be more focused on industrial production and high technology. For example, more than half the ecological footprint of the United States comes from its use of fossil fuels, of which approximately 60 percent are imported. China's ecological footprint has doubled since 1970, and its ratio of local to global impact has shifted. Most of its ecological footprint used to be driven by demand for food, fiber crops such as cotton, hemp, and flax, and woody biomass. However, in recent years, China's demand for fossil fuels has increased dramatically. In contrast, Ivory Coast is a developing country in which a significant proportion of the economy is devoted to small-scale and subsistence agriculture. Ivory Coast has an ecological footprint of 0.8 ha (2.0 acres) per capita, almost all of which comes from demand for food, fiber, and woody biomass. Very little comes from the demand for fossil fuels because most people in Ivory Coast use little or no fossil fuel.

**LOCAL IMPACTS** Most of the materials that are consumed in developing countries are produced locally. While this may benefit the local economy, it can lead to regional overuse of resources and environmental degradation. We saw an example of such overuse in Chapter 3's opening story, which described deforestation in Haiti.

Two commonly overused resources are land itself and woody biomass from trees and other plants. A growing population requires increasing amounts of food. In developing countries that do not import their food, local demand for agricultural land increases with population size. To put more land into cultivation, farmers may convert forests or natural grasslands into cropland. The United Nations Food and Agriculture Organization estimates that in Brazil, approximately 3 million hectares (7.4 million acres) were cleared per year between 2000 and 2005, some for small-scale agriculture and some for industrial production of soybeans, sugarcane, and corn. The local environmental impacts of converting land to agriculture include erosion, soil degradation, and habitat loss (**FIGURE 7.17**).

Farmers may produce more food not only by cultivating more land, but also by increasing the *yield,* or productivity, of cultivated land through fertilization, irrigation, extending the time each year that the land is under cultivation, or rotating among different crops. Increasing crop yields can also have adverse local environmental effects, including soil degradation and water pollution.

**GLOBAL IMPACTS** Agriculture has global as well as local impacts, whether it occurs in developed or developing countries. Conversion of land to agriculture reduces the uptake of atmospheric carbon dioxide by plants, which affects the global carbon cycle. In addition, an increase in the use of fertilizers made from fossil fuels increases the release of greenhouse gases into the atmosphere.

Families in suburban areas of developed countries such as the United States consume far fewer local resources than rural families in developing countries, but they have a much greater impact on the global environment. In general, populations with large global impacts tend to deplete more environmental resources. Much of their impact comes from their consumption of imported energy sources such as oil and other imported resources such as food. When people are affluent, they can afford to import bananas, fish, and coffee from other countries, drive long distances in automobiles built in factories hundreds or thousands of miles away, and live in homes surrounded by lawns that require large quantities of water, fertilizer, and pesticides. Simply transporting the food, water, fertilizer, and pesticides requires an expenditure of energy, usually in the form of fossil fuels. Building suburbs often means replacing natural areas or agricultural lands with lawns and asphalt.

At the same time, developed countries benefit from technologies that actually reduce their local environ-

**FIGURE 7.17** **Deforested land in Brazil** Clearing forests and grasslands for agriculture can lead to erosion, soil degradation, and habitat loss. This photo shows erosion caused by clear-cutting in Brazil.

mental impact. For example, sewage treatment and household solid waste collection somewhat offset the environmental impact of consumption by diminishing the degradation of the environment with waste. Despite these technological advantages, affluent suburban living has the greatest impact of all lifestyles on the environment.

**URBAN IMPACTS** Urban populations represent one-half of the human population, but consume three-fourths of Earth's resources. While definitions vary by country, an **urban area,** according to the U.S. Census Bureau, is one that contains more than 386 people per square kilometer (1,000 people per square mile). New York City is the most densely populated city in the United States, with 10,400 people per square kilometer (27,000 people per square mile). Mumbai, India, is the most densely populated city in the world, with 23,000 people per square kilometer (60,000 people per square mile).

More than 75 percent of people in developed countries live in urban areas, as **FIGURE 7.18** shows, and that number is expected to increase slightly over the next 20 years. In developing countries, 44 percent of people live in urban areas, but that number is increasing more

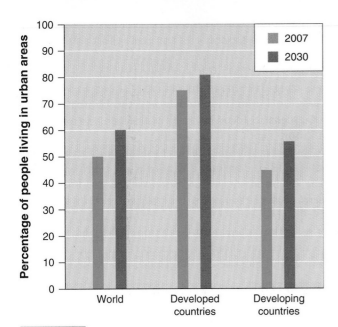

**FIGURE 7.18** **Urban growth.** More than one-half of the world's population will live in urban settings by 2030. [Data from United Nations Population Fund.]

Urban living in both developed and developing countries presents environmental challenges. Most developed countries employ city planning to some degree. As urban areas expand, experts design and install public transportation, water and sewer lines, and other municipal services. In addition, while urban areas produce greater amounts of solid waste, pollution, and carbon dioxide emissions than suburban or rural areas, they tend to have smaller per capita ecological footprints. There are many reasons for this difference, including greater access to public transportation and nearby services such as shopping.

In developing countries, the relatively affluent portions of urban areas have safe drinking water, sewage treatment systems, and systems for disposal of household solid waste to minimize their impact on the surrounding environment. However, many less affluent urban residents have no access to these services. Rapid urbanization in the developing world often results in an influx of the very poor. These poor often cannot afford permanent housing and instead construct temporary shelters with whatever materials are available to them, including mud, cardboard, or plastic. Whether they are squatter settlements, shantytowns, or slums, these overcrowded and underserved living situations are a common fact of life in most cities in the developing world. The United Nations organization UN-HABITAT estimates that 1 billion people live in squatter settlements and other similar areas throughout the world. Most residents live in housing structures without flooring, safe walls and

rapidly than in developed countries and will probably reach 56 percent by 2030. Table 7.1 shows that, of the 20 largest cities in the world, 16 are in developing countries. Worldwide, almost 5 billion people are expected to live in urban areas by 2030.

| TABLE 7.1 | The 20 largest urban areas in the world | |
| --- | --- | --- |
| **Rank** | **City, country** | **Population (millions)** |
| 1 | Tokyo, Japan | 35.7 |
| 2 | New York–Newark, United States | 19.0 |
| 3 | Mexico City, Mexico | 19.0 |
| 4 | Mumbai, India | 19.0 |
| 5 | São Paulo, Brazil | 18.9 |
| 6 | Delhi, India | 16.0 |
| 7 | Shanghai, China | 15.0 |
| 8 | Kolkata, India | 14.8 |
| 9 | Dacca, Bangladesh | 13.5 |
| 10 | Buenos Aires, Argentina | 12.8 |
| 11 | Los Angeles–Long Beach–Santa Ana, United States | 12.5 |
| 12 | Karachi, Pakistan | 12.1 |
| 13 | Cairo, Egypt | 11.9 |
| 14 | Rio de Janeiro, Brazil | 11.8 |
| 15 | Osaka–Kobe, Japan | 11.3 |
| 16 | Beijing, China | 11.1 |
| 17 | Manila, Philippines | 11.1 |
| 18 | Moscow, Russia | 10.4 |
| 19 | Istanbul, Turkey | 10.0 |
| 20 | Paris, France | 9.90 |

*Source:* United Nations Population Division.
*Note:* Data are from 2007 and contain the areas defined by the United Nations as "urban agglomerations."

**FIGURE 7.19 Shantytowns.** A squatter settlement on the outskirts of Lima, Peru. Such shantytowns expand outward from the urban center.

ceilings, or such basic amenities as water, sanitation, or health care (**FIGURE 7.19**).

## The Impact of Affluence

The most commonly used measure of a nation's wealth is its **gross domestic product (GDP),** the value of all products and services produced in a year in that country. GDP is made up of four types of economic activity: *consumer spending, investments, government spending,* and *exports minus imports.*

A country's per capita GDP often correlates with its pollution levels. At very low levels of per capita GDP, industrial activity is too low to produce much pollution; the country uses very little fossil fuel and generates relatively little waste. Many developing countries fit this pattern.

As GDP increases, a nation begins to be able to afford to burn fossil fuels, especially coal, which is a relatively inexpensive fuel but a large emitter of pollution. The country may also rely on rudimentary, inefficient equipment that emits large amounts of pollutants. It is at this point in its development that a country emits pollution at the highest levels. The United States fit this pattern during the twentieth century. China, which is going through a similar rapid industrialization, currently relies on coal as its primary energy source. Many people view this shift as a trade-off that occurs as a country's GDP increases: breathing dirty air poses a different risk to human health than poverty. It is not always a more desirable risk, but many people would choose it over poverty.

As a nation's GDP increases further, it may reach a turning point. It can afford to purchase equipment that burns fossil fuels more efficiently and cleanly, which helps to reduce the amounts and types of pollution

generated. People may also be willing to expend resources and support government efforts to regulate polluting industries. Wealthier societies are also able to afford better policing and enforcement mechanisms, ensuring that environmental regulations are followed. Western European countries and the United States fit this scenario today. We will explore this turning point in more depth in Chapter 20.

Some environmental scientists argue that increasing the GDP of developing nations is the best way to save the environment, for at least two reasons. First, as we have seen, rising income generally correlates with falling birth rates, and reducing population size should lead to a reduction in environmental impact. Second, wealthier countries can afford to make environmental improvements and increase their efficiency of resource use.

### GAUGE YOUR PROGRESS

✓ What is the IPAT equation? What does it describe?

✓ How do local and global environmental impacts differ? Where do we tend to see one versus the other?

✓ How does a country's degree of development influence its environmental impact?

# Sustainable development is a common, if elusive, goal

Economic development, as we saw in Chapter 1, is improvement in human well-being through economic advancement. As this chapter shows, economic development has a strong influence on a society's environmental impact. Many people believe that we cannot have both economic development and environmental protection. However, a growing number of social and natural scientists maintain that, in fact, sustainable economic development is possible.

As we saw in Chapter 1, sustainable development goes beyond economic development to meet the essential needs of people in the present without compromising the ability of future generations to meet their needs. In other words, sustainable development strives to improve standards of living—which involves greater expenditures of energy and resources—without causing additional environmental harm. Many cities in Scandinavia have achieved something close to sustainable development. For example, Övertorneå, a Swedish town of 5,600 people, recently revived the local economy by focusing its economic activities on renewable energy, free public transportation, organic agriculture, and land preservation. The city still has an ecological footprint, of course, but it is much smaller, and thus much more sustainable, than it was previously.

How can sustainable development be achieved? There are no simple answers to that question, nor is there a single path that all people must follow. The Millennium Ecosystem Assessment project, completed in 2005, offers some insights. This project's reports constitute a global analysis of the effects of the human population on ecosystem services such as clean water, forest products, and natural resources. They are also a blueprint for sustainable development. The reports, prepared at the request of the United Nations, concluded that human demand for food, water, lumber, fiber, and fuel has led to a large and irreversible loss of biodiversity. The Millennium Ecosystem Assessment drew several other conclusions:

- Ecosystem sustainability will be threatened if the human population continues along its current path of resource consumption around the globe.

- The continued alterations to ecosystems that have improved human well-being (greater access to food, clean water, suitable housing) will also exacerbate poverty for some populations.

- If we establish sustainable practices, we may be able to improve the standard of living for a large number of people.

The project's reports state that "human actions are depleting Earth's natural capital, putting such strain on the environment that the ability of the planet's ecosystems to sustain future generations can no longer be taken for granted." They further suggest that sustainability, as well as sustainable development, will be achieved only with a broader and accelerated understanding of the connections between human systems and natural systems. This means that governments, nongovernmental organizations, and communities of people will have to work together to raise standards of living while understanding the impacts of those improvements on the local, regional, and global environment.

**GAUGE YOUR PROGRESS**

✓ What is the difference between economic development and sustainable development?

✓ How is the Millennium Ecosystem Assessment project a blueprint for sustainable development?

✓ What factors have we discussed that make sustainable development difficult to achieve?

#  WORKING TOWARD SUSTAINABILITY

India is the world's second most populous country and will probably overtake China as the most populous country by 2030. Since the 1950s, India has tried various methods of population control, steadily reducing its growth rate to the current 1.5 percent. Although this growth rate would suggest below-replacement-level fertility, India's population will continue to increase for some time because of its large number of young people. The Population Reference Bureau estimates that India's population will not stabilize until after 2050 and may reach the 2 billion mark in the next century.

In the 1960s, India attempted to enforce nationwide population control through sterilization. Massive protests against this coercive approach led to changes in policy. Since the 1970s, India has emphasized family planning and reproductive health, although each Indian state chooses its own approach to population stabilization. Some states have had tremendous

## Gender Equity and Population Control in Kerala

success in lowering growth rates. Kerala, in southwestern India, is one of those states (FIGURE 7.20).

The state of Kerala is about twice the size of Connecticut. Connecticut has a population of roughly 3.5 million people, however, while Kerala has a population of almost 30 million people. Kerala's population density is about 820 people per square

**FIGURE 7.20** College students in Kerala, India.

kilometer (2,100 people per square mile), almost 3 times that of India as a whole, and 25 times that of the United States. Until 1971, Kerala's population was growing even faster than the population in the rest of India, in large part because the state government had implemented an effective health care system that decreased infant and adult mortality rates.

Since 1971, Kerala's birth rate, like its mortality rate, has fallen to levels similar to those in North America and other industrialized countries. The current total fertility rate in Kerala is about 2.0—lower than the U.S. TFR of 2.1 and much lower than the TFR of 2.7 for India as a whole. As a result, Kerala's population has stabilized. How did the state's population stabilize despite its poverty?

A special combination of social and cultural factors seems to be responsible for Kerala's sustainable population growth. Kerala has emphasized "the three Es" in its approach: education, employment, and equality. In addition to its good health care system, Kerala has good schools that support a literacy rate of over 90 percent, the highest in India. Furthermore, unlike those in the rest of India, literacy rates in Kerala are almost identical for males and females. In other parts of India and in most other developing countries, women attend school, on average, only half as long as men do. As we have learned, higher education levels for women lead to greater female empowerment and increased use of family planning. In Kerala, 63 percent of women use contraceptives,

compared with 48 percent in the rest of India, and many women delay childbirth and join the workforce before deciding to start families. Kerala has a strong matriarchal tradition in which women are highly valued and their education is encouraged. Education and empowerment allow these women to be in a better position to make decisions regarding family size.

The World Bank estimates that if the rest of the developing world had followed Kerala's lead 30 years ago and equalized education for men and women, current TFR throughout the developing world would be close to replacement-level fertility. Evidence for this view can be found in the reduced fertility rates of other relatively poor countries where women and men have equal status, such as Sri Lanka and Cuba. In fact, *gender equity* was made a cornerstone of an international program to control population growth at the 1994 Third International Conference on Population and Development in Cairo, Egypt.

### References

Franke, R. W., and B. H. Chasin. 2005. Kerala: Radical reform as development in an Indian state. In *The Anthropology of Development and Globalization: From Classical Political Economy to Contemporary Neoliberalism,* ed. M. Edelman and A. Haugerud. Blackwell.

Pulsipher, L. M., and A. Pulsipher. 2008. *World Regional Geography.* 4th ed. W. H. Freeman.

## REVISIT THE KEY IDEAS

■ **Describe the potential limits to human population growth.**

Scientists disagree about the size of Earth's carrying capacity for humans. Some scientists believe we have already exceeded that carrying capacity. Others believe that innovative approaches and new technologies will allow the human population to continue to grow beyond the environmental limits currently imposed by factors such as the supply of food, water, and natural resources.

■ **Describe important aspects of global and national population growth using demographic terminology and tools.**

The human population is currently 6.8 billion people, and it is growing at a rate of about 212,000 people per day. If we think of the human population—as a whole or in individual countries—as a system, there are more inputs—births and immigration—than outputs—deaths and emigration. To understand changes in population size, demographers measure crude birth rate, crude death rate, total fertility rate, replacement-level fertility, life

expectancy, infant and child mortality, age structure, and net migration rate.

■ **Evaluate the social, economic, and environmental factors that have contributed to decreasing growth rates in many countries.**

A number of countries have undergone a demographic transition as their economies modernized. Economic development generally leads to increased affluence, increased education, less need for children to help their families generate subsistence income, and increased family planning. These factors have reduced the average size of families in developed countries, which leads to slower population growth. Eventually, population size may even decline.

■ **Analyze relationships among changes in population size, economic development, and resource consumption at global and local scales.**

Most population growth today is occurring in developing countries. Only one-fifth of the global population lives in developed countries, but those countries consume more than half the world's energy and resources. One person in a developed country may have 2 to 10 times the environmental impact of a person from a developing country. The IPAT equation states that the environmental impact of a population is a result of population size, affluence, and technology. A relatively small population can have a high environmental impact if its affluence leads to high consumption and extensive use of destructive technology. However, an affluent nation can more easily take measures to reduce its environmental impact through the use of technology that counters pollution and increases the efficiency of resource use. Rural populations tend to have a high local environmental impact but a low global environmental impact.

■ **Explain how people have attempted to harmonize economic development with sustainable development.**

Sustainable development attempts to raise standards of living without increasing environmental impact. The Millennium Ecosystem Assessment is a blueprint for sustainable development.

## CHECK YOUR UNDERSTANDING

1. A metropolitan region of 100,000 people has 2,000 births, 500 deaths, 200 emigrants, and 100 immigrants over a 1-year period. Its population growth rate is
   (a) 1.2 percent.
   (b) 1.4 percent.
   (c) 1.6 percent.
   (d) 1.8 percent.
   (e) 2.0 percent.

2. Which of the following pairs of indicators *best* reflects the availability of health care in a country?
   (a) Crude death rate and growth rate
   (b) Crude death rate and crude birth rate
   (c) Growth rate and life expectancy
   (d) Infant mortality rate and crude death rate
   (e) Infant mortality rate and life expectancy

3. Which of the following characteristics are typical of developed countries?
   I    High technology use
   II   Low GDP
   III  Small-scale sustainable agriculture

   (a) I only
   (b) II only
   (c) I and III only
   (d) II and III only
   (e) I, II, and III

4. In 2010, the population of Earth was about _____ billion, with about _____ billion living in China.
   (a) 6.1, 1.1
   (b) 6.1, 1.2
   (c) 6.3, 1.3
   (d) 6.8, 1.2
   (e) 6.8, 1.3

5. A country with an age structure diagram like the one below is most likely experiencing
   (a) a high life expectancy.
   (b) slow population growth.
   (c) a short doubling time.
   (d) a low infant mortality rate.
   (e) replacement-level fertility.

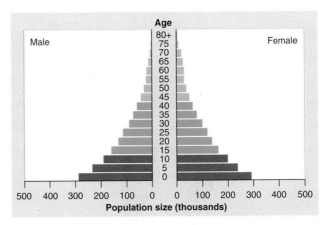

6. Which of the following statements about total fertility rate is *correct?*
   (a) TFR is equal to the crude birth rate minus the crude death rate.
   (b) TFR is the average number of children each woman must have to replace the current population.
   (c) TFR is generally higher in developed countries than in developing countries.
   (d) TFR is equal to the growth rate of a country.
   (e) TFR is the average number of children each woman will give birth to during her childbearing years.

7. Even if a country reduces its birth rate and maintains replacement-level fertility, its population will still continue to grow for several decades because of
   (a) lower death rates.
   (b) increased income.
   (c) population momentum.
   (d) better health care.
   (e) increased life expectancy.

8. At current growth rates, which country will probably be the most populous in the world after 2050?
   (a) China
   (b) Brazil
   (c) India
   (d) Indonesia
   (e) United States

9. What percentage of the world population lives in developing countries?
   (a) 34
   (b) 50
   (c) 66
   (d) 82
   (e) 98

10. Which of the following countries *best* exemplifies phase 4 of a demographic transition?
    (a) Argentina
    (b) China
    (c) India
    (d) Japan
    (e) Mexico

## APPLY THE CONCEPTS

1. Answer the following questions about the theory of demographic transition.
   (a) Draw a fully labeled diagram that shows how birth and death rates change as a country undergoes the four phases of a demographic transition.
   (b) For each of the phases labeled in your diagram, explain the changes occurring in each phase and describe what is causing them.

   (c) Describe a strategy that a government might implement to slow its population growth that could be utilized by a country undergoing a demographic transition. Explain how your proposed strategy would work, and describe one potential drawback to its implementation.

2. Look at the age structure diagrams for country A and country B below and answer the following questions.

Country A

Country B

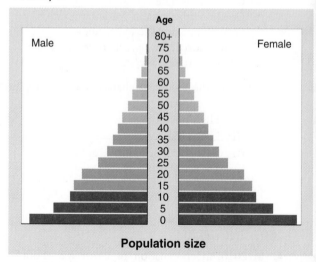

(a) What observations and educated predictions can you make about the following characteristics of country A?
   (i) The age structure of its population
   (ii) The total fertility rate of the country
   (iii) The life expectancy of the population
   (iv) The growth rate and doubling time of the population

(b) Describe one socioeconomic feature of country A.
(c) Explain how country B differs from country A in terms of
   (i) the age structure of its population.
   (ii) its infant mortality rate.
   (iii) its rate of population growth.

## MEASURE YOUR IMPACT

**National Footprints**  The following table shows the average per capita ecological footprint for the world as a whole and for several countries. It also gives population sizes and the area of usable (biologically productive) land available in each case.

| Country | Average footprint (hectares per capita) | Population (millions) | Usable land (millions of hectares) | Total footprint (millions of hectares) | Deficit/ surplus (millions of hectares) |
|---------|------|-------|--------|--------|--------|
| World | 2.7 | 6,800 | 11,400 | _____ | _____ |
| United States | 9.0 | 307 | 1,350 | _____ | _____ |
| Japan | 4.1 | 128 | 77 | _____ | _____ |
| China | 1.8 | 1,330 | 1,100 | _____ | _____ |
| India | 0.8 | 1,170 | 460 | _____ | _____ |

*Source:* Data from Global Footprint Network for the year 2009.

(a) Complete the table by calculating the land area that would be required for each population, given its footprint, and determine how large a deficit or surplus of usable land each region has per person.

(b) If the entire global population lived at the level of consumption found in the United States, it would require 9.0 hectares per person × 6.8 billion people, or 61.2 billion hectares. Given that there are only 11.4 billion usable hectares on Earth, calculate how many "Earths" would be required for the entire human population to live sustainably at the level of consumption of the United States. Repeat the calculation for the other countries.

(c) Which do you think is a more reasonable measure of a country's environmental impact, whether the people in that country are living sustainably within their own borders, or whether all of the people in the world could consume at that country's level?

## ENGAGE YOUR ENVIRONMENT

How can census data be used to estimate future needs in your state or community? In this Engage Your Environment exercise you will project changes in numbers of schoolchildren over the next 10 years by analyzing U.S. Census Bureau data from selected states. You will also project the number of schools and teachers that will be needed in the coming decade. *See Engage Your Environment Activity 7 at the end of this book.*

# How Can We Manage Overabundant Animal Populations?

The growing human population has caused population declines in many animal species, as we saw in Chapter 5. At the same time, other species have flourished as the result of human alterations of the landscape. Increases in the populations of these species have had many consequences. How have these increases happened, and what, if anything, should we do about them?

## What causes animal population explosions?

The white-tailed deer is a high-profile example of dramatic animal population growth in the United States. Before European colonists arrived, its densities were approximately 4 to 6 deer per square kilometer (10 to 15 deer per square mile). In many areas today, however, deer densities are up to 20 times higher. In New York State, for example, unregulated hunting and the clearing of preferred deer habitats for farming had nearly eliminated the deer population by 1900, but the population has since grown exponentially to 1 million deer.

This population growth can be explained by processes we have discussed in Chapters 6 and 7. For example, white-tailed deer prefer to live in early-successional fields containing wildflowers and shrubs. Abandoned farms that have undergone secondary succession during the twentieth century, such as the old fields of New England that we described in Chapter 6, are prime deer habitat. Such increases in habitat and food supply have increased the carrying capacity of the environment for deer. With a higher carrying capacity, density-dependent controls on the population are weaker, and reproduction increases.

In addition to providing them with habitat and food, humans have eliminated the deer's major predators: wolves and cougars. During much of the twentieth century, regulated hunting took the place of natural predators and continued to hold deer populations in check. Over the last few decades, however, the number of deer hunters has steadily declined, allowing the deer population to increase further. In short, a higher carrying capacity and a lower mortality caused by predators and hunters have allowed deer populations to grow exponentially and exceed their historic carrying capacity (FIGURE SA3.1).

The white-tailed deer is not the only species whose populations are exploding. Animal overpopulation is a global issue. Many large herbivorous mammals have benefited from anthropogenic alterations of the environment, including predator extinctions, reduced hunting, increased habitat, and increased food supplies provided by agriculture and by people who feed them. The sika deer (*Cervus nippon*) of Nara, Japan, for example, lives in an area that has been protected as a natural monument since World War II. Many factors have contributed to a population explosion of this species. Human visitors to the area feed the deer, hunting is restricted, and Japan's native wolves—natural predators of the deer—are extinct. In Australia, eastern grey kangaroos (*Macropus giganteus*), which normally consume wild grasses, are also consuming agricultural crops (FIGURE SA3.2). Their natural predator, the Tasmanian wolf (*Thylacinus cynocephalus*), is now extinct. In South Africa, the population of African elephants (*Loxodonta africana*) doubled within 10 years of their being granted legal protection from hunting.

Large herbivorous mammals are not the only animals that experience overpopulation. Florida's 500,000 monk parakeets (*Myiopsitta monachus*) are the offspring of individuals that have escaped from the pet trade since the 1960s. Other former pets, including dogs and cats,

**FIGURE SA3.1** **Deer overpopulation.** As humans have provided increased food and habitat for white-tailed deer and simultaneously removed their major predators, deer populations have increased to the point that they exceed the historic carrying capacity.

quickly establish feral populations when released into the wild. Populations of Canada geese (*Branta canadensis*) have grown along with the number of agricultural fields and golf courses on which they can feed.

## What effects do overabundant species have on communities?

Populations that exceed their historic carrying capacity can cause widespread problems. Large populations of herbivore species such as deer, elephants, and kangaroos can overgraze the plants in their community, leaving little food for other herbivorous species. In Australia, grazing by the overabundant kangaroos removes many of the plants needed by other herbivores, including several species of endangered animals that depend on these plants for food and habitat.

These large animal populations also have direct effects on humans. Herbivorous mammals consume agricultural crops, landscape plants, and valuable wild plant species. In the United States, for example, the deer

**FIGURE SA3.2** **Overabundant kangaroos.** Abnormally high populations of kangaroos consume all of their natural food supply and consequently move into human-dominated landscapes, where they come into conflict with humans.

population so greatly exceeds the land's natural carrying capacity that it is now difficult to regenerate several valuable forest tree species, including northern red oak (*Quercus rubra*) and sugar maple (*Acer saccharum*). Deer also consume suburban landscape plants and damage $60 million worth of agricultural crops each year.

Wildlife overabundance also increases wildlife-human encounters, such as deadly collisions of vehicles with deer or kangaroos. More frequent contact with wildlife also increases the risk of humans contracting diseases that the animals carry. White-tailed deer, for example, carry ticks that can be infected by the bacterium that causes Lyme disease. Nearly 10,000 people in the United States contracted Lyme disease in 1992. This number climbed to more than 27,000 in 2007.

### How can we control overabundant species?

Controlling the population of any species involves issues that are ecological, political, moral, and ethical. From an ecological perspective, it is clear how to control a wildlife population: reduce the available food and habitat to lower the carrying capacity, compensate for the missing predators by killing individuals in the population, or slow the population's ability to reproduce. On a local scale, it can be helpful to discourage people from feeding wildlife. On a larger scale, however, reducing food and habitat is often the least viable option because it would typically require restricting the access of large herbivores to millions of hectares of land.

In 2009, the Australian government ordered the army to shoot 6,000 kangaroos as a way to control the population. This controversial decision was opposed by animal rights activists and by many Australians who consider the kangaroo a national symbol that should be protected. The activists obtained a court injunction to stop the culling of the species, but after a delay, the kangaroo cull proceeded. Throughout the world, even when all parties agree that the overabundance of a species poses major problems for humans and for the environment, culling a herd through shooting is an increasingly unpopular strategy.

A strategy that has more public support is birth control. This approach is common in cities, where many feral cats are trapped, neutered, and returned to the wild. Researchers are also investigating ways to give animals contraceptive drugs in order to reduce their populations over time. The drugs being developed differ in their modes of action, ranging from chemicals that prevent animals from making critical reproductive hormones to chemicals that prevent sperm from fertilizing an egg. Contraceptive strategies are currently being tested on a variety of species, including deer, kangaroos, elephants, and monk parakeets (FIGURE SA3.3).

Reducing animal overpopulation with contraceptives has widespread public appeal because it eliminates the need to kill animals. Getting the contraceptive drugs into the animals, however, presents substantial challenges. Drugs can be administered to some animals via dart guns. For others, such as monk parakeets, researchers are working on ways to administer medications in food. Some contraceptive drugs last for a single breeding season and require a new dose each season, which can

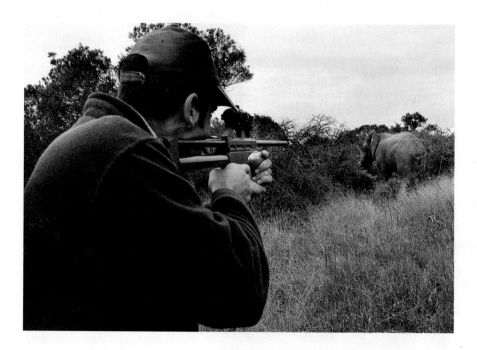

**FIGURE SA3.3** **Administering contraceptives to wild animals.** In South Africa, a wildlife biologist shoots a dart containing contraceptives at a female elephant. The goal is to reduce the overabundant elephant population and reduce the destructive effects of these elephants on the habitat.

be difficult and expensive. The cost of giving an animal a single dose of a contraceptive drug ranges from $100 to $500 per animal.

The issue of overabundant wildlife highlights a number of other environmental issues. The underlying causes of animal overpopulation are typically associated with human population increases. These human-caused wildlife population explosions can, in turn, have substantial negative effects on humans and their environment. Although there are several potential solutions to the problem of animal overpopulation, successful solutions must take into account the public's affection for wildlife, its opposition to killing animals, and its frequent lack of knowledge about the negative effects of overabundant species. As we saw in the case of China, enforced infertility can be an effective, though controversial, strategy for controlling human populations. Many people view enforced infertility as an acceptable strategy for controlling wildlife, although it is not yet effective.

References

Broache, A. 2005. Oh deer! *Smithsonian,* October.

*New York Times,* 2004. Putting nature on the pill. August 31.

# Earth's Resources: Geologic Processes, Soil, and Minerals

## Are Hybrid Electric Vehicles as Environmentally Friendly as We Think?

Many people in the environmental science community believe that hybrid electric vehicles (HEV) and all-electric vehicles are some of the most exciting innovations of the last decade. Cars that use electric power or a combination of electricity and gasoline are much more efficient in their use of fuel than similarly sized internal

(22 pounds) of lanthanum. Mining these elements involves pumping acids into deep boreholes to dissolve the surrounding rock and then removing the resulting acid and mineral slurry. Lithium is extracted from certain rocks, and lithium carbonate is extracted from brine pools and mineral springs adjacent to or under salt flats. Both extraction procedures are types of surface mining, which can have severe environmental

> Even though they reduce our consumption of liquid fossil fuels, hybrid electric vehicles do come with environmental trade-offs.

combustion (IC) automobiles. Depending on whether the car is a hybrid, a plug-in HEV, or an all-electric vehicle, it may travel up to twice as far on a tank of gasoline as an IC car or even use no gasoline at all.

Even though they reduce our consumption of liquid fossil fuels, hybrid electric vehicles do come with environmental trade-offs. The construction of these vehicles uses scarce metals, including neodymium, lithium, and lanthanum. Neodymium is needed to form the magnets used in the electric motors, and lithium and lanthanum are used in the cars' compact high-performance batteries. At present, there appears to be just enough lanthanum available in the world to meet the demand of the Toyota Prius HEV, which has a projected production of almost 1 million vehicles in 2011. Toyota obtains its lanthanum from China. There are also supplies of lanthanum in various geologic deposits in California, Australia, Bolivia, Canada, and elsewhere, but most of these deposits have not yet been developed for mining. Until this happens, many scientists believe that the production of HEVs and all-electric vehicles will be limited by the availability of lanthanum.

In addition to the scarcity of these metals, we have to consider how we acquire them. A typical Toyota Prius HEV uses approximately 1 kg (2.2 pounds) of neodymium and 10 kg

impacts. The holes, open pits, and ground disturbance created by the mining of these minerals provide the opportunity for air and water to react with other minerals in the rock, such as sulfur, to form an acidic slurry. As this acid mine drainage flows over the land or underground toward rivers and streams, it dissolves metals and other elements. As a result, water near surface mining operations is highly acidic (sometimes with a pH of 2.5 or lower) and may contain harmful levels of dissolved metals and minerals.

Wherever it occurs, mining has a number of environmental consequences, including the creation of holes in the ground and road construction, both of which lead to fragmentation and alteration of habitat, erosion, and contamination of water supplies. So, while current HEV technology may reduce our dependence on certain fossil fuels, it increases our dependence on other limited resources that must be extracted from the ground.

Why are some of Earth's mineral resources so limited? Why do certain elements occur in some locations and not in others? What processes create minerals and other Earth ▶

A Toyota Prius hybrid electric vehicle.

◀ A lithium mine in Bolivia.

materials? Understanding the answers to these questions requires knowing more about Earth systems—the topic of this chapter. ■

*Sources:* M. Armand and J.-M. Tarascon, Building better batteries, *Nature* 451 (2008): 652–657; K. Senior, Hybrid cars threatened by rare-metal shortages, *Frontiers in Ecology*, October 7, 2009, p. 402.

## UNDERSTAND THE KEY IDEAS

Almost all of the mineral resources on Earth accumulated when the planet formed 4.6 billion years ago. But Earth is a dynamic planet. Earth's geologic processes form and break down rocks and minerals, drive volcanic eruptions and earthquakes, determine the distribution of scarce mineral resources, and create the soil in which plants grow. Human extraction and use of these resources has environmental consequences.

After reading this chapter you should be able to

■ describe the formation of Earth and the distribution of critical elements on Earth.

■ define the theory of plate tectonics and discuss its importance in environmental science.

■ describe the rock cycle and discuss its importance in environmental science.

■ explain how soil forms and describe its characteristics.

■ explain how elements and minerals are extracted for human use.

# The availability of Earth's resources was determined when the planet formed

Nearly all the elements found on Earth today are as old as the planet itself. FIGURE 8.1 illustrates how Earth formed roughly 4.6 billion years ago from cosmic dust in the solar system. The early Earth was a hot, molten sphere. For a period of time, additional debris from the formation of the Sun bombarded Earth. As this molten material slowly cooled, the elements within it separated into layers based on their mass. Heavier elements such as iron sank toward Earth's center, and lighter elements such as silica floated toward its surface. Some gaseous elements left the solid planet and became part of Earth's atmosphere. Although asteroids occasionally strike Earth today, the bombardment phase of planet formation has largely ceased, and the elemental composition of Earth has stabilized. In other words, the elements and minerals that were present when the planet formed are all that we have.

These elements and minerals are distributed unevenly around the globe.

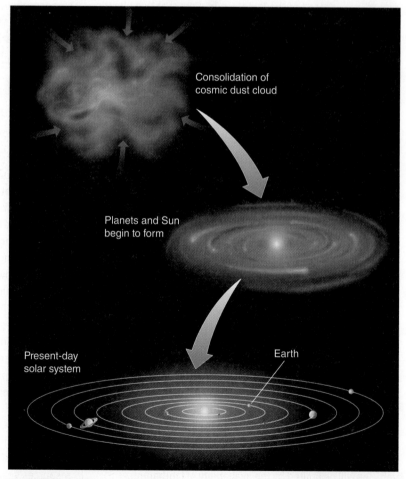

Consolidation of cosmic dust cloud

Planets and Sun begin to form

Present-day solar system

Earth

**FIGURE 8.1** **Formation of Earth and the solar system.** The processes that formed Earth 4.6 billion years ago determined the distribution and abundance of elements and minerals today.

**(a)** Earth's vertical zonation

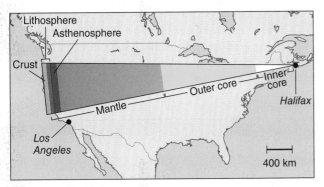

**(b)** Scale of Earth's layers

**FIGURE 8.2** **Earth's layers.** (a) Earth is composed of concentric layers. (b) If we were to slice a wedge from Earth, it would cover the width of the United States.

Some minerals, such as silicon dioxide—the primary component of sand and glass—are readily available worldwide on beaches and in shallow marine and glacial deposits. Others, such as diamonds, which are formed from carbon that has been subjected to intense pressure, are found in relatively few isolated locations. Over the course of human history, this uneven geographic distribution has driven many economic and political conflicts.

Because Earth's elements settled into place based on their mass, the planet is characterized by distinct vertical zonation. If we could slice into Earth, as shown in **FIGURE 8.2**, we would see concentric layers composed of various materials. The innermost zone is the planet's **core,** over 3,000 km (1,860 miles) below Earth's sur-

face. The core is a dense mass largely made up of nickel and some iron. The inner core is solid, and the outer core is liquid. Above the core is the **mantle,** containing molten rock, or **magma,** that slowly circulates in convection cells, much as the atmosphere does. The **asthenosphere,** located in the outer part of the mantle, is composed of semi-molten, *ductile* (flexible) rock. The brittle outermost layer of the planet, called the **lithosphere** (from the Greek word *lithos,* which means "rock"), is approximately 100 km (60 miles) thick. It includes the solid upper mantle as well as the **crust,** the chemically distinct outermost layer of the lithosphere. It is important to recognize that these regions overlap: the lowest part of the lithosphere is also the uppermost portion of the mantle.

The lithosphere is made up of several large and numerous smaller *plates,* which overlie the convection cells within the asthenosphere. Over the crust lies the thin layer of soil that allows life to exist on the planet. The crust and overlying soil provide most of the chemical elements that comprise life.

Because Earth contains a finite supply of mineral resources, we cannot extract resources from the planet indefinitely. In addition, once we have mined the deposits of resources that are most easily obtained, we must use more energy to extract the remaining resources. Both of these realities provide an incentive for us to minimize our use of mineral resources and to reuse and recycle them when possible.

**GAUGE YOUR PROGRESS**

✓ How did Earth form?

✓ What is the composition of each of Earth's layers?

✓ What is the practical significance today of the way Earth's resources were distributed when the planet formed and cooled?

# Earth is dynamic and constantly changing

Earth's geologic cycle consists of three major processes: the *tectonic cycle,* the *rock cycle,* and *soil formation.* In the following sections, we'll look at each of these processes in turn.

## Convection and Hot Spots

One of the critical consequences of Earth's formation and elemental composition is that the planet remains very hot at its center. The high temperature of Earth's outer core and mantle is thought to be a result of the radioactive decay of various isotopes of elements such as potassium, uranium, and thorium, which releases heat. The heat causes plumes of hot magma to well upward from the mantle. These plumes produce **hot spots:** places where molten material from the mantle reaches the lithosphere. As we shall see in the following pages, hot spots are an important component of Earth dynamics. The heat from Earth's core also creates convection cells in the mantle similar to those described in Chapter 4 in our discussion of atmospheric circulation (Figure 4.6 provides a good review). Mantle convection drives the continuous change, creation, and renewal of Earth materials in the lithosphere.

## Theory of Plate Tectonics

Prior to the 1900s, scientists believed that Earth's major features—such as continents and oceans—were fixed in place. In 1912, a German meteorologist named Alfred Wegener published a revolutionary hypothesis proposing that the world's continents had once been joined in a single landmass, which he called "Pangaea" (FIGURE 8.3). His evidence included observations of identical rock formations on both sides of the Atlantic Ocean, as shown in Figure 8.3a. The positions of these formations suggested that a single supercontinent may have broken up into separate landmasses. Fossil evidence also

suggested that a single large continent existed in the past. Today, we can find fossils of the same species on different continents that are separated by oceans; Figure 8.3b shows one example.

For a long time, scientists resisted the idea that Earth's lithosphere could move laterally. Geologists in the early twentieth century went to great lengths to find alternative explanations for Wegener's observations. However, following the publication of Wegener's hypothesis, scientists found additional evidence that Earth's landmasses had existed in several different configurations over time. Eventually, a new idea was proposed: the theory of **plate tectonics,** which states that *Earth's lithosphere is divided into plates, most of which are in constant motion.* The **tectonic cycle** is the sum of the processes that build up and break down the lithosphere.

We now know that the lithosphere is broken into a number of plates. *Oceanic plates* lie primarily beneath the oceans, whereas *continental plates* lie beneath landmasses. The crust of oceanic plates is dense and rich in iron. The crust of continental plates generally contains more silicon dioxide, which is much less dense than iron. Thus the continental plates are lighter and typically rise above the oceanic plates. FIGURE 8.4 identifies Earth's major plates.

Oceanic and continental plates "float" on top of the denser material beneath them. Their slow movements

**(a)** Rock formations

Ancient rock assemblages

Continental shelf

**(b)** Fossil evidence

Fossils of the freshwater reptile *Mesosaurus* have been found in South America and Africa.

*Mesosaurus*

FIGURE 8.3 **Evidence of drifting continents.** Several lines of evidence show that the current landmasses were once joined together in a single supercontinent. (a) Identical rock formations are found on both sides of the Atlantic Ocean. (b) Fossils of the same species have been collected from different continents.

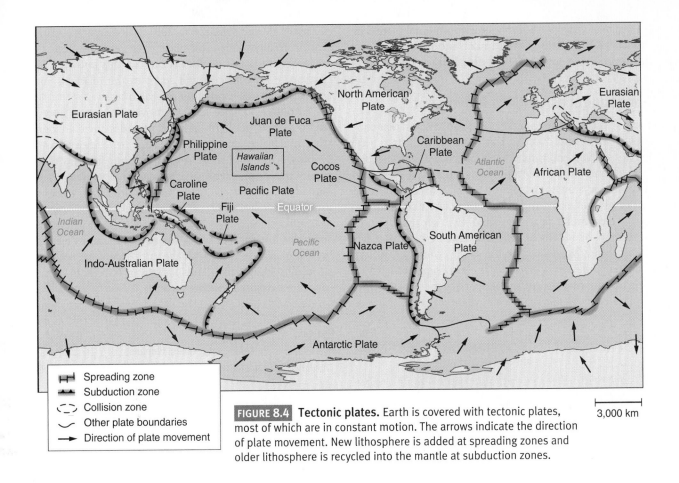

**FIGURE 8.4** **Tectonic plates.** Earth is covered with tectonic plates, most of which are in constant motion. The arrows indicate the direction of plate movement. New lithosphere is added at spreading zones and older lithosphere is recycled into the mantle at subduction zones.

Legend:
- ⊢⊣ Spreading zone
- ▲▲ Subduction zone
- ⊂⊃ Collision zone
- ⌣ Other plate boundaries
- → Direction of plate movement

3,000 km

are driven by convection cells in Earth's mantle. As the plates move, the continents slowly drift. As oceanic plates move apart, rising magma forms new oceanic crust on the seafloor at the boundaries between those plates. This process, called *seafloor spreading,* is shown in **FIGURE 8.5**. Where oceanic plates meet continental plates, old oceanic crust is pulled downward, beneath the continental lithosphere, and the heavier oceanic

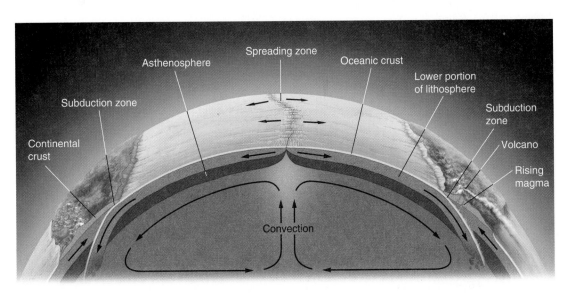

**FIGURE 8.5** **Convection and plate movement.** Convection in the mantle causes oceanic plates to spread apart as new rock rises to the surface at spreading zones. Where oceanic and continental plate margins come together, older oceanic crust is subducted.

plate slides underneath the lighter continental plate. This process of one plate passing under another, also shown in Figure 8.5, is known as **subduction.**

## Consequences of Plate Movement

Earth's history is measured using the *geologic time scale,* shown in FIGURE 8.6. As the continents drifted over Earth's surface, their climates changed, geographic barriers formed or were removed, and species evolved, adapted, or slowly or rapidly went extinct. In some places, as the plates moved, a continent that straddled two plates broke apart, creating two separate smaller continents or islands in different climatic regions. Over time, species that existed on the single continent evolved into two different species on the two separated islands (recall our discussion of allopatric speciation from Chapter 4). The fossil record tells us how species adapted to the changes that took place over geologic time. Climate scientists and ecologists can use this knowledge to anticipate how species will adapt to the relatively rapid climate changes happening on Earth today.

Although the rate of plate movement is too slow for us to notice, geologic activity provides vivid evidence that the plates are in motion. As a plate moves over a geologic hot spot, heat from the rising mantle plume melts the crust, forming a **volcano:** a vent in Earth's surface that emits ash, gases, and molten lava. Volcanoes are a natural source of atmospheric carbon dioxide, particulates, and metals. Over time, as the plate moves past the hot spot, it can leave behind a trail of extinct volcanic islands, each with the same chemical composition. The Hawaiian Islands, shown in FIGURE 8.7, are an excellent example of this pattern.

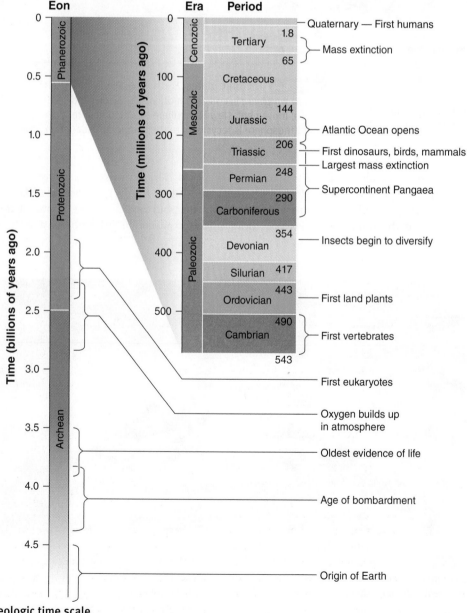

**FIGURE 8.6** The geologic time scale.

FIGURE 8.7 **Plate movement over a hot spot.** The Hawaiian Islands were formed by volcanic eruptions as the Pacific Plate traveled over a geologic hot spot. The chain of inactive volcanoes to the northwest of Hawaii shows that those locations used to be over the hot spot. Numbers indicate how long ago each area was located over the hot spot (in millions of years).

The island of Niihau was formed 4.89 million years ago when that region of the Pacific Plate was over the hot spot.

Movement of Pacific Plate

○ Position of hot spot

A new island (Loihi) is forming over the current location of the hot spot.

100 km

## Types of Plate Contact

Many other geologic events occur at the zones of contact resulting from the movements of plates relative to one another. These zones of plate contact can be classified into three types: *divergent plate boundaries, convergent plate boundaries,* and *transform fault boundaries* (FIGURE 8.8).

Beneath the oceans, plates move away from each other at **divergent plate boundaries,** as illustrated in Figure 8.8a. At these boundaries, oceanic plates move apart as if on a giant conveyer belt. As magma from the mantle reaches Earth's surface and pushes upward and outward, new rock is formed. This phenomenon, called **seafloor spreading,** creates new lithosphere and brings important elements such as copper, lead, and silver to the surface of Earth. However, this new rock is typically under the deep ocean. Over tens to hundreds of millions of years, as the tectonic cycle continues, some of that material forms new land containing these valuable resources.

Clearly, if tectonic plates are diverging, and if the surface of Earth has a finite area, the plates must be moving together somewhere. **Convergent plate boundaries** form where plates move toward one another and collide, as shown in Figure 8.8b. The plates generate a great deal of pressure as they push against one another. If the margin of an oceanic plate composed of dense, iron-rich rock collides with that of a lighter continental plate, the oceanic plate margin, and the plate, will slide under the continental crust, pushing up the lighter plate. This subduction of the heavier plate is responsible for the formation of long, narrow coastal mountain

**(a)** Divergent plate boundary

**(b)** Convergent plate boundary

**(c)** Transform fault boundary

FIGURE 8.8 **Types of plate boundaries.** (a) At divergent plate boundaries, plates move apart. (b) At convergent plate boundaries, plates collide. (c) At transform fault boundaries, plates slide past each other.

**(a)** Formation of the Himalayas

**(b)** The Himalayas from space

**FIGURE 8.9** **Collision of two continental plates.** (a) The Himalayan mountain range was formed when the collision of two continental plates forced the margins of both plates upward. (b) A satellite image of this mountain range, which includes the highest mountains on Earth.

ranges, such as the Andes in South America. Because the subducted plate will melt, rising magma may be the source of new volcano formation.

If two continental plates meet, both plate margins may be lifted, forming a mid-continental mountain range such as the Himalayas in Asia. **FIGURE 8.9** shows how this process works.

When plates move sideways past each other, the result is a **transform fault boundary,** as shown in Figure 8.8c.

Most plates and continents move at about the rate your fingernails grow: roughly 36 mm, or 1.4 inches, per year. While this movement is far too slow to notice on a daily basis, the two plates underlying the Atlantic Ocean have spread apart and come together twice over the past 500 million years, causing Europe and Africa to collide with North America and South America and separate from them again. Crunch the Numbers "Plate Movement" (on page 216) shows how we can calculate the time it takes for plates to move.

## Faults and Earthquakes

Although the plates are always in (slow) motion, their movement is not necessarily smooth. A **fault** is a fracture in rock across which there is movement. **Fault zones,** large expanses of rock where movement has occurred, form in the brittle upper lithosphere where two plates meet. The rock near the plate margins becomes fractured and deformed from the immense pressures exerted by plate movement. Imagine rubbing two rough and jagged rocks past each other. The rocks would resist that movement and get stuck together. The rock along a fault is jagged and thus resists movement as the plates

attempt to move. Eventually, however, the mounting pressure overcomes the resistance, and the plates give way, slipping quickly. The result is an *earthquake.*

**Earthquakes** occur when the rocks of the lithosphere rupture unexpectedly along a fault. The plates can move up to several meters in just a few seconds. Earthquakes are common in fault zones, which are also called areas of **seismic activity. FIGURE 8.10** shows one such area, the San Andreas Fault in California, which

**FIGURE 8.10** **A fault zone.** Many areas of seismic activity, including the San Andreas Fault in California, are characterized by a transform fault. The epicenter of an earthquake is the point on Earth's surface directly above the location where the rock ruptures.

**FIGURE 8.11** **Locations of earthquakes and volcanoes.** A "Ring of Fire" circles the Pacific Ocean along plate boundaries. Other zones of seismic and volcanic activity, including hot spots, are also shown on this map.

is a transform fault. The **epicenter** of an earthquake is the exact point on the surface of Earth directly above the location where the rock ruptures, as also shown in Figure 8.10.

Earthquakes are a direct result of the movement of plates and their contact with each other. Volcanic eruptions happen when molten magma beneath the crust is released to the atmosphere. Sometimes the two events are observed together, most often along plate boundaries where tectonic activity is high. **FIGURE 8.11** shows one example, in which earthquake locations and volcanoes form a circle of tectonic activity, called the "Ring of Fire," around the Pacific Ocean.

## The Environmental and Human Toll of Earthquakes and Volcanoes

Plate movements, volcanic eruptions, seafloor spreading, and other tectonic processes bring molten rock from deep beneath Earth's crust to the surface, and subduction sends surface crust deep into the mantle. This tectonic cycle of surfacing and sinking is a continuous Earth process. When humans live in close proximity to areas of seismic or volcanic activity, however, the results can be dramatic and devastating.

Earthquakes occur many times a day throughout the world, but most are so small that humans do not feel them. The magnitude of an earthquake is reported on the **Richter scale,** a measure of the largest ground movement that occurs during an earthquake. The Richter scale, like the pH scale described in Chapter 2, is logarithmic. On a logarithmic scale, a value increases by a factor of 10 for each unit increase. Thus a magnitude 7.0 earthquake, which causes serious damage, is 10 times greater than a magnitude 6.0 earthquake and 1,000 ($10^3$) times greater than a magnitude 4.0 earthquake, which only some people can feel or notice. Worldwide, there may be as many as 800,000 small earthquakes of magnitude 2.0 or less per year, but an earthquake of magnitude 8.0 occurs only approximately once every 10 years.

Even a moderate amount of Earth movement can be disastrous. Moderate earthquakes (defined as magnitudes 5.0 to 5.9) lead to collapsed structures and buildings, fires, contaminated water supplies, ruptured dams, and deaths. Loss of life is more often a result of the proximity of large population centers to the epicenter than of the magnitude of the earthquake. The quality of building construction in the affected area is also an important factor in the amount of damage that occurs. In 2008, a magnitude 7.9 earthquake in the southwestern region of Sichuan Province, China, killed more than 69,000 people. The epicenter was near a populated area where many buildings were probably not built to withstand a large earthquake. In 2010, a magnitude 7.0 earthquake

### Plate Movement

Los Angeles is 630 km (380 miles) southeast of San Francisco. The plate under Los Angeles is moving northward at about 36 mm per year relative to the plate under San Francisco. Given this average rate of plate movement, how long will it take for Los Angeles to be located next to San Francisco?

The distance traveled is 630 km, and the net distance moved is 36 mm per year. We can use this formula to determine the answer:

$$\text{time} = \frac{\text{distance}}{\text{rate}}$$

$$630 \text{ km} = 630,000 \text{ m} = 630,000,000 \text{ mm}$$

$$\frac{630,000,000 \text{ mm}}{36 \text{ mm/year}} = 17,500,000 \text{ years}$$

We are left with an answer in numbers of years:

$$630 \text{ km} \left(\frac{1,000 \text{ m}}{1 \text{ km}}\right) \left(\frac{1,000 \text{ mm}}{1 \text{ m}}\right) \left(\frac{1 \text{ yr}}{36 \text{ mm}}\right) = 17,500,000 \text{ yr}$$

It will take about 18 million years for Los Angeles to be located alongside San Francisco.

**Your Turn:** How long will it take for a plate that moves at 20 mm per year to travel the distance of one football field? Note that a football field is 91.44 m (100 yards) long.

---

in Haiti killed more than 200,000 people. Many of the victims were trapped under collapsed buildings (FIGURE 8.12).

Extra safety precautions are needed when dangerous materials are used in areas of seismic activity. Nuclear power plants are designed to withstand significant ground movement and are programmed to shut down if movement above a certain threshold occurs. The World Nuclear Association estimates that 20 percent of nuclear power plants operate in areas of significant seismic activity. Between 2004 and 2009, in four separate incidents, nuclear power plants in Japan shut down operation because of ground movement that exceeded the threshold.

Volcanoes, when active, can be equally disruptive and harmful to human life. Active volcanoes are not distributed randomly over Earth's surface; 85 percent of them occur along plate boundaries. As we have seen, volcanoes

**FIGURE 8.12 Earthquake damage in Haiti.** The 2010 earthquake in Haiti killed more than 200,000 people and destroyed most of the structures in the capital, Port-au-Prince.

**FIGURE 8.13** **A volcanic eruption.** This 2001 explosive eruption from the Etna volcano in Italy threatened several nearby towns.

can also occur over hot spots. Depending on the type of volcano, an eruption may eject cinders, ash, dust, rock, or lava into the air (**FIGURE 8.13**). Volcanoes can result in loss of life, habitat destruction and alteration, reduction in air quality, and many other environmental consequences.

The world gained a new awareness of the impact of volcanoes when eruptions from a volcano in Iceland disrupted air travel to and from Europe in April 2010. Ash from the eruption entered the atmosphere in a large cloud and was spread over a wide area by the prevailing winds. The ash contained small particles of silicon dioxide, which have the potential to damage airplane engines. Air travel was suspended in many parts of Europe, and millions of travelers were stranded. This may have been the greatest travel disruption ever to have been caused by a volcano.

**GAUGE YOUR PROGRESS**

✓ What is the theory of plate tectonics?

✓ What are some environmental consequences of the tectonic cycle?

✓ How do earthquakes and volcanoes occur?

# The rock cycle recycles scarce minerals and elements

The second part of the geologic cycle is the **rock cycle:** the constant formation and destruction of rock. The rock cycle is the slowest of all of Earth's cycles. Environmental scientists are most often concerned with the part of the rock cycle that occurs at or near Earth's surface.

*Rock,* the substance of the lithosphere, is composed of one or more *minerals.* **Minerals** are solid chemical substances with uniform (often crystalline) structures that form under specific temperatures and pressures. They are usually compounds, but may be composed of a single element such as silver or gold. Some examples of common minerals are shown in **FIGURE 8.14**.

(a)

(b)

(c)

**FIGURE 8.14** **Some common minerals.** (a) Pyrite ($FeS_2$), also called "fool's gold." (b) Graphite, a form of carbon (C). (c) Halite, or table salt (NaCl).

# Formation of Rocks and Minerals

**FIGURE 8.15** shows the processes of the rock cycle. Rock forms when magma from Earth's interior reaches the surface, cools, and hardens. Once at Earth's surface, rock masses are broken up, moved, and deposited in new locations by processes such as weathering and erosion. New rock may be formed from the deposited material. Eventually, the rock is subducted into the mantle, where it melts and becomes magma again. The rock cycle slowly but continuously breaks down rock and forms new rock.

While magma is the original source of all rock, there are three major ways in which the rocks we see at Earth's surface can form: directly from molten magma, by compression of sediments—materials deposited by wind, water, or glaciers—and by exposure of rocks and other Earth materials to high temperatures and pressures. These three modes of formation lead to three distinct rock types: *igneous, sedimentary,* and *metamorphic rock.*

**IGNEOUS ROCKS Igneous rocks** are those that form directly from magma. They are classified by their chemical composition as *basaltic* or *granitic,* and by their mode of formation as *intrusive* or *extrusive.*

*Basaltic rock* is dark-colored rock that contains minerals with high concentrations of iron, magnesium, and calcium. It is the dominant rock type in the crust of oceanic plates. Granitic rock is lighter-colored rock made up of the minerals feldspar, mica, and quartz, which contain elements such as silicon, aluminum, potassium, and calcium. It is the dominant rock type in the crust of continental plates. When granitic rock breaks down due to weathering, it forms sand. Soils that develop from granitic rock tend to be more permeable than those that develop from basaltic rock, but both types of rock can form fertile soil.

**Intrusive igneous rocks** form within Earth as magma rises up and cools in place underground. **Extrusive igneous rocks** form when magma cools above Earth's surface, as when it is ejected from a volcano or released by seafloor spreading. Extrusive rocks cool rapidly, so their minerals have little time to expand into large individual crystals. The result is fine-grained, smooth rock types such as obsidian. Both extrusive and intrusive rocks can be either granitic or basaltic in composition.

The formation of igneous rock often brings to the surface rare elements and metals that humans find economically valuable, such as the lanthanum described in this chapter's opening story. When rock cools, it is subject to stresses that cause it to break. Cracks that occur in this way, known as **fractures,** can occur in any kind of rock. Water from Earth's surface running through fractures may dissolve valuable metals, which may precipitate out in the fractures to form concentrated deposits called

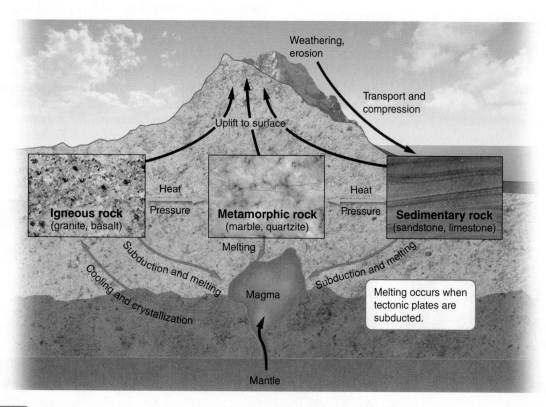

**FIGURE 8.15** **The rock cycle.** The rock cycle slowly but continuously forms new rock and breaks down old rock. Three types of rock are created in the rock cycle: igneous rock is formed from magma; sedimentary rock is formed by the compression of sedimentary materials; and metamorphic rock is created when rocks are subjected to high temperatures and pressures.

*veins.* These deposits are important sources of gold- and silver-bearing ores as well as rare metals such as tantalum, which is used to manufacture electronic components of cell phones.

**SEDIMENTARY ROCKS Sedimentary rocks** form when sediments such as muds, sands, or gravels are compressed by overlying sediments. Sedimentary rock formation occurs over long periods when environments such as sand dunes, mudflats, lake beds, or landslide-prone areas are buried and the overlying materials create pressure on the materials below. The resulting rocks may be uniform in composition, such as sandstones and mudstones that formed from ancient oceanic or lake environments. Alternatively, they may be highly heterogeneous, such as conglomerate rocks formed from mixed cobbles, gravels, and sands.

Sedimentary rocks hold the fossil record that provides a window into our past. When layers of sediment containing plant or animal remains are compressed over eons, those organic materials may be preserved, as described in Chapter 5.

**METAMORPHIC ROCKS Metamorphic rocks** form when sedimentary rocks, igneous rocks, or other metamorphic rocks are subjected to high temperatures and pressures. The pressures that form metamorphic rock cause profound physical and chemical changes in the rock. These pressures can be exerted by overlying rock layers or by tectonic processes such as continental collisions, which cause extreme horizontal pressure and distortion. Metamorphic rocks include stones such as slate and marble as well as anthracite, a type of coal. Metamorphic rocks have long been important building materials in human civilizations because they are structurally strong and visually attractive.

## Weathering and Erosion

Most rock forms beneath Earth's surface under intense heat, pressure, or both. When rock is exposed at Earth's surface, it begins to break down through the processes of *weathering* and *erosion*. These processes are components of the rock cycle, returning chemical elements and rock fragments to the crust by depositing them as sediments via the hydrologic cycle. This physical breakdown and chemical alteration of rock begins the cycle all over again, as shown in Figure 8.15. Without this part of the rock cycle, elements would never be recycled.

**WEATHERING** Weathering occurs when rock is exposed to air, water, certain chemical compounds, or biological agents such as plant roots, lichens, and burrowing animals. There are two major categories of weathering—*physical* and *chemical*—that work in combination to degrade rocks.

**Physical weathering** is the mechanical breakdown of rocks and minerals (**FIGURE 8.16**). Physical weathering can be caused by water, wind, or variations in temperature such as seasonal freeze-thaw cycles. When water works its way into cracks or fissures in rock, it can remove loose material and widen the cracks (Figure 8.16a). When water freezes in the cracks, the water expands, and the pressure of its expansion can force rock to break. Different responses to temperature can cause two minerals within a rock to expand and contract differently, which also results in the splitting or cracking of rocks. Coarse-grained rock formed by slow cooling or metamorphism tends to weather more quickly than fine-grained rock formed by rapid cooling or metamorphism.

Biological agents can also cause physical weathering. Plant roots can work their way into small cracks in rocks and pry them apart (Figure 8.16b). Burrowing animals may also contribute to the breakdown of rock material, although their contributions are usually minor. However it occurs, physical weathering exposes more surface area and makes rock more vulnerable to further degradation. By producing more surface area for chemical weathering processes to act on, physical weathering increases the rate of chemical weathering.

(a)

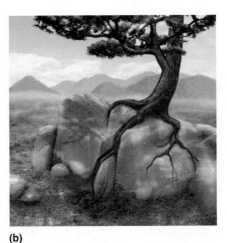
(b)

**FIGURE 8.16** **Physical weathering.** (a) Water can work its way into cracks in rock, where it can wash away loose material. When the water freezes and expands, it can widen the cracks. (b) Growing plant roots can force rock sections apart.

**Chemical weathering** is the breakdown of rocks and minerals by chemical reactions, the dissolving of chemical elements from rocks, or both. It releases essential nutrients from rocks, making them available for use by plants and other organisms.

Chemical weathering is most important on newly exposed minerals, known as *primary minerals*. It alters primary minerals to form *secondary minerals* and the ionic forms of their constituent chemical elements. For example, when feldspar—a mineral found in granitic rock—is exposed to natural acids in rain, it forms clay particles and releases ions such as potassium, an essential nutrient for plants. Lichens can break down rock in a similar way by producing weak acids. Their effects can commonly be seen on soft gravestones and masonry. Rocks that contain compounds that dissolve easily, such as calcium carbonate, tend to weather quickly. Rocks that contain compounds that do not dissolve readily are often the most resistant to chemical weathering.

Recall from Chapter 2 that solutions can be basic or acidic. Depending on the starting chemical composition of rock and the pH of the water that comes in contact with it, hundreds of different chemical reactions can take place. For example, as we saw in Chapter 2, carbon dioxide in the atmosphere dissolves in water vapor to create a weak solution of carbonic acid. When waters containing carbonic acid flow into limestone-rich geologic regions, they dissolve the limestone (which is composed of calcium carbonate) and create spectacular cave systems (**FIGURE 8.17**).

**FIGURE 8.18** **Erosion.** Some erosion, such as the erosion that created these formations in the Badlands of South Dakota, occurs naturally as a result of the effects of water, glaciers, or wind. The Badlands are the result of the erosion of softer sedimentary rock types, such as shales and clays. Harder rocks, including many types of metamorphic and igneous rocks, are more resistant to erosion.

Some chemical weathering is the result of human activities. For example, sulfur emitted into the atmosphere as a result of fossil fuel combustion combines with oxygen to form sulfur dioxide. That sulfur dioxide reacts with water vapor in the atmosphere to form sulfuric acid, which then causes *acid precipitation*. **Acid precipitation,** also called **acid rain,** is responsible for rapid degradation of old statues, gravestones, limestone, and marble. When acid precipitation falls on soil, it can promote chemical weathering of certain minerals in the soil, releasing elements that may then be taken up by plants or leached from the soil into groundwater and streams.

Chemical weathering, either by natural processes or due to acid precipitation, can contribute additional elements to an ecosystem. Understanding whether the rate of weathering is fast or slow helps researchers assess how rapidly soil fertility can be renewed in an ecosystem. In addition, because the chemical reactions involved in the weathering of certain granitic rocks consume carbon dioxide from the atmosphere, weathering can actually reduce atmospheric carbon dioxide concentrations.

**EROSION** Whereas physical and chemical weathering result in the breakdown and chemical alteration of rock, **erosion** is the physical removal of rock fragments (sediment, soil, rock, and other particles) from a landscape or ecosystem. Erosion is usually the result of two mechanisms. Wind, water, and ice transport soil and other Earth materials by downslope creep under the force of gravity. Living organisms, such as animals that burrow under the soil, also cause erosion. After

**FIGURE 8.17** **Chemical weathering.** Water that contains carbonic acid wears away limestone, sometimes forming spectacular caves.

eroded material has traveled a certain distance from its source, it accumulates. **Deposition** is the accumulation or depositing of eroded material such as sediment, rock fragments, or soil.

Erosion is a natural process: streams, glaciers, and wind-borne sediments continually carve, grind, and scour rock surfaces (**FIGURE 8.18**). In many places, however, human land uses contribute substantially to the rate of erosion. Poor land use practices such as deforestation, overgrazing, unmanaged construction activity, and road building can create and accelerate erosion problems. Furthermore, erosion usually leads to deposition of the eroded material somewhere else, which may cause additional environmental problems. We discuss human-caused erosion further in Chapters 10 and 11.

---

**GAUGE YOUR PROGRESS**

✓ What is the rock cycle?

✓ What are the three basic rock types, and how is each formed?

✓ What is the difference between weathering and erosion? Why are both processes important?

---

# Soil links the rock cycle and the biosphere

So far, we have discussed two parts of the geologic cycle: the tectonic cycle and the rock cycle. The third part of the geologic cycle—*soil formation*—takes place at Earth's surface. **Soil** is a mix of geologic and organic components. Soil forms a dynamic membrane that covers much of Earth's land surface, connecting the overlying biology to the underlying geology.

As we can see in **FIGURE 8.19**, soil has a number of functions that benefit organisms and ecosystems. Soil is a medium for plant growth. It also serves as the primary filter of water as water moves from the atmosphere into rivers, streams, and groundwater. Soil provides habitat for a wide variety of living organisms—from bacteria, algae, and fungi to insects and other animals—and thus contributes greatly to biodiversity. Collectively, these soil organisms act as recyclers of organic matter. In the process of using dead organisms and wastes as an energy source, they break down organic detritus and release mineral nutrients and other materials that benefit plants. Finally, soil and the organisms within it filter chemical

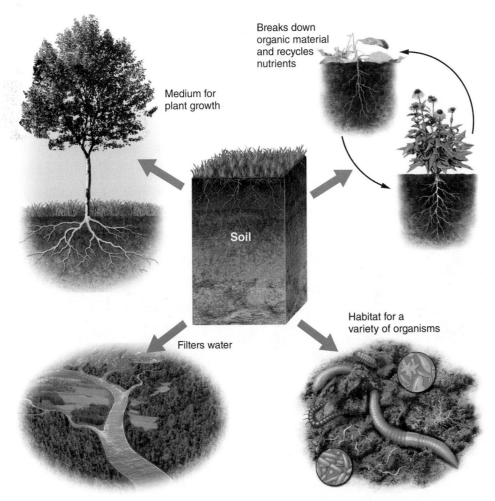

Breaks down organic material and recycles nutrients

Medium for plant growth

Soil

Filters water

Habitat for a variety of organisms

**FIGURE 8.19** **Ecosystem services provided by soil.** Soil serves as a medium for plant growth, as a habitat for other organisms, and as a recycling system for organic wastes. Soil also helps to filter and purify water.

compounds deposited by air pollution and by household sewage systems, retaining some of these materials and releasing some to the atmosphere above or to the groundwater below.

In order to appreciate the role of soil in ecosystems, we need to understand how and why soil forms and what happens to soil when humans alter it.

## The Formation of Soil

It takes hundreds to thousands of years for soil to form. Soil is the result of physical and chemical weathering of rocks and the gradual accumulation of detritus from the biosphere. We can determine the specific properties of a soil if we know its parent rock type, the amount of time during which it has been forming, and its associated biotic and abiotic components.

The processes that form soil work in two directions simultaneously. The breakdown of rocks and primary minerals by weathering provides the raw material for soil from below. The deposition of organic matter from organisms and their wastes contributes to soil formation from above. What we normally think of as "soil" is a mix of these mineral and organic components. A poorly developed (or "young") soil has substantially less organic matter and fewer nutrients than a more developed ("mature") soil. Very old soils may also be nutrient poor because, over time, plants remove many of the essential nutrients, while others are leached away by water. FIGURE 8.20 shows the stages of soil development from rock to mature soil.

Five factors determine the properties of soils: *parent material, climate, topography, organisms,* and *time*. None of these factors alone can determine soil properties. They work simultaneously, and thus they must be considered together, rather than individually.

**PARENT MATERIAL** A soil's **parent material** is the rock material underlying it from which its inorganic components are derived. Different soil types arise from different parent materials. For example, a quartz sand (made up of silicon dioxide) parent material will give rise to a soil that is nutrient poor, such as those along the Atlantic coast of the United States. By contrast, a soil that has calcium carbonate as its parent material will contain an abundant supply of calcium, have a high pH, and may also support high agricultural productivity. Such soils are found in the area surrounding Lake Champlain in northern New York and Vermont, and in many other locations.

**CLIMATE** The second factor influencing soil formation is climate. Soils do not develop well when temperatures are below freezing because decomposition of organic matter and water movement are both extremely slow in frozen or nearly frozen soils. Therefore, soils at high latitudes of the Northern Hemisphere are composed largely of organic material in an undecomposed state, as we saw in Chapter 4. In contrast, soil development in the humid tropics is accelerated by the rapid weathering of rock and soil minerals, the leaching of nutrients, and the decomposition of organic detritus. Climate also has an indirect effect on soil formation through its influence on the type of vegetation that develops, and thus on the type of detritus left after the vegetation dies.

**TOPOGRAPHY** *Topography,* the surface slope and arrangement of a landscape, is the third factor that influences

Parent rock is weathered and fragments move upward.

Organic material accumulates as plants and other organisms die.

Greater amounts of organic material are present in a mature soil.

Immature soil

Young soil

**Time**

Mature soil

FIGURE 8.20 **Soil formation.** Soil is a mixture of organic and inorganic matter. The breakdown of rock and primary minerals from the parent material provides the inorganic matter. The organic matter comes from organisms and their wastes.

soil formation. Soils that form on steep slopes are constantly subjected to erosion and, on occasion, more drastic mass movements of material such as landslides. In contrast, soils that form at the bottoms of steep slopes may continually accumulate material from higher elevations and become quite deep.

**ORGANISMS** Many organisms influence soil formation. Plants remove nutrients from soil and excrete organic acids that speed chemical weathering. Animals that tunnel or burrow, such as earthworms, gophers, and voles, mix the soil, distributing organic and mineral matter uniformly throughout. Humans have dramatic effects on soils, as we will see below and in Chapters 11 and 14.

**TIME** Finally, the amount of time a soil has spent developing is important in determining its properties. As soils age, they develop a variety of characteristics. The grassland soils that support much of the food crop and livestock feed production in the United States are relatively old soils. Because they have had continual inputs of organic matter for hundreds of thousands of years from the grassland and prairie vegetation growing above them, they have become deep and fertile. Other soils that are equally old, but with less productive communities above them and perhaps greater quantities of water moving through them, can become relatively infertile.

## Soil Horizons and Properties

The five factors that influence soil formation vary across the landscape, so they lead to the formation of different soil types in different locations. Soils with different properties serve different functions for humans. For example, some soil types are good for growing crops, others for building a housing development. Therefore, to understand and classify soil types, we need to understand both soil horizons and soil properties.

**SOIL HORIZONS** As soils form, they develop characteristic **horizons,** or layers, like those visible in **FIGURE 8.21**. The specific composition of those horizons depends largely on climate, vegetation, and parent material.

At the surface of many soils is a layer of organic detritus such as leaves, needles, twigs, and even animal bodies, all in various stages of decomposition. This horizon, called the **O horizon,** or organic horizon, is most pronounced in forest soils and is also found in some grasslands.

In a soil that is mixed, either naturally or by human agricultural practices, the top layer of soil is the **A horizon,** also known as **topsoil.** The A horizon is a zone of organic material and minerals that have been mixed together.

In some acidic soils, an **E horizon**—a zone of leaching, or *eluviation*—forms under the O horizon or, less often, the A horizon. When present, it always occurs above the B horizon. When an E horizon is present, iron, aluminum, and dissolved organic acids from the overlying horizons are transported through and removed from the E horizon and then deposited in the B horizon, where they accumulate.

All soils have a **B horizon.** The B horizon, commonly known as *subsoil,* is composed primarily of mineral material with very little organic matter. If there are nutrients in the soil, they will be present in the B horizon.

The **C horizon,** the least weathered soil horizon, always occurs beneath the B horizon. The C horizon is similar to the parent material.

Each of these horizons lends certain physical, chemical, and biological properties to the soil within which it exists. These properties are extremely important in understanding the role of soils in environmental science.

**PHYSICAL PROPERTIES OF SOIL** *Sand, silt,* and *clay* are mineral particles of different sizes. The **texture** of a soil is determined by the percentages of sand, silt, and clay

**O horizon:** Organic matter in various stages of decomposition

**A horizon (topsoil):** Zone of overlying organic material mixed with underlying mineral material

**B horizon (subsoil):** Zone of accumulation of metals and nutrients

**C horizon (subsoil):** Least-weathered portion of the soil profile, similar to the parent material

Horizons

O
.1 m
A
.3 m

B

In some soils, an E horizon occurs beneath either the O or the A horizon.

1 m

C

1.3 m

**FIGURE 8.21** **Soil horizons.** All soils have horizons, or layers, which vary depending on soil-forming factors such as climate, organisms, and parent material.

40% sand
40% silt
20% clay

Percentage of clay

Percentage of silt

Clay

Sandy clay

Silty clay

Clay loam

Silty clay loam

Sandy clay loam

Sandy loam

Loam

Silt loam

Loamy sand

Sand

Silt

Percentage of sand

**(a)** Soil texture chart

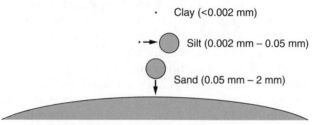

Clay (<0.002 mm)

Silt (0.002 mm – 0.05 mm)

Sand (0.05 mm – 2 mm)

**(b)** Relative soil particle sizes
(magnified approximately 100 times)

**FIGURE 8.22** **Soil properties.** (a) Soils consist of a mixture of clay, silt, and sand. The relative proportions of these particles determine the texture of the soil. (b) The relative sizes of sand, silt, and clay.

it contains. We can plot those percentages on a triangle-shaped diagram like the one in **FIGURE 8.22** to identify and compare soil types. For each location on the diagram, there are three determinants: the percentages of sand, silt, and clay. A point in the middle of the "loam" category (approximately at the "a" in "loam" in Figure

8.22) represents a soil that contains 40 percent sand, 40 percent silt, and 20 percent clay. We can determine this by following the lines from that point to the scales on each of the three sides of the triangle in turn. If you are not certain which line to follow from a given point, always follow the line that leads to the lower value. The sum of sand plus silt plus clay will always be 100 percent. Conversely, in the laboratory, a soil scientist can determine the percentage of each component in a soil sample and then plot the results. This allows her to give a name to the soil, such as "silty clay loam."

The *porosity* of soil—how quickly soil drains—depends on its texture, as **FIGURE 8.23** shows. Sand particles—the largest of the three components—pack together loosely. Water can move easily between the particles, making sand quick to drain and quick to dry out. Soils with a high proportion of sand are also easy for roots to penetrate, making sandy soil somewhat advantageous for growing plants such as carrots and potatoes. Clay particles—the smallest of the three components—pack together much more tightly than sand particles. As a result, there is less pore space in a clay-dominated soil, and water and roots cannot easily move through it. Silt particles are intermediate in size and in their ability to drain or retain water.

The best agricultural soil is a mixture of sand, silt, and clay. This mixture promotes balanced water drainage and retention. In natural ecosystems, however, various herbaceous and woody plants have adapted to growing in wet, intermediate, and dry environments, so there are plants that thrive in soils of virtually all textures.

Soil texture can have a strong influence on how the physical environment responds to environmental pollution. For example, the groundwater of western Long Island in New York State has been contaminated by toxic chemicals discharged from local industries. One reason for the contamination is that Long Island is dominated by sandy soils that readily allow surface water to drain into the groundwater. Soil usually serves as a filter that removes pollutants from the water moving through it, but because sandy soils are so porous, pollutants move

Time = zero     1 hour     100 days     100 years

Water

Sand, silt, or clay

Sand

Silt

Clay

**FIGURE 8.23** **Soil porosity.** The porosity of soil depends on its texture. Sand, with its large, loosely packed particles, drains quickly. Clay drains much more slowly.

through them quickly and therefore are not filtered effectively.

Clay is particularly useful where a potential contaminant needs to be contained. Many modern landfills are lined with clay, which helps keep the contaminants in solid waste from leaching into the soil and groundwater beneath the landfill.

**CHEMICAL PROPERTIES OF SOIL** Chemical properties are also important in determining how a soil functions. Clay particles contribute the most to the chemical properties of a soil because of their ability to attract positively charged mineral ions, referred to as *cations*. Because clay particles have a negative electrical charge, cations are *adsorbed*—held on the surface—by the particles. The cations can be subsequently released from the particles and used as nutrients by plants.

The ability of a particular soil to adsorb and release cations is called its **cation exchange capacity (CEC).** CEC is sometimes referred to as the nutrient holding capacity. The overall CEC of a soil is a function of the amount and types of clay particles present. Soils with high CECs have the potential to provide essential cations to plants and therefore are desirable for agriculture. If a soil is more than 20 percent clay, however, its water retention becomes too great for most crops and many other types of plants. In such waterlogged soils, plant roots are deprived of oxygen. Thus there is a trade-off between CEC and porosity.

Another important chemical property of soil involves the relationship between soil bases and soil acids. Calcium, magnesium, potassium, and sodium are collectively called soil bases because they can neutralize or counteract soil acids such as aluminum and hydrogen. Soil acids are generally detrimental to plant nutrition, while soil bases tend to promote plant growth. With the exception of sodium, all the soil bases are essential for plant nutrition. **Base saturation** is a measure of the proportion of soil bases to soil acids, expressed as a percentage.

Because of the way they affect nutrient availability to plants, CEC and base saturation are important determinants of overall ecosystem productivity. If a soil has a high CEC, it can retain and release plant nutrients. If it has a relatively high base saturation, its clay particles will hold important plant nutrients such as calcium, magnesium, and potassium. A soil with both high CEC and high base saturation is likely to support high productivity.

**BIOLOGICAL PROPERTIES OF SOIL** As we have seen, a diverse group of organisms populates the soil. **FIGURE 8.24** shows a representative sample. Three groups of organisms account for 80 to 90 percent of the biological activity in soils: fungi, bacteria, and protozoans (certain single-celled organisms). Rodents and earthworms

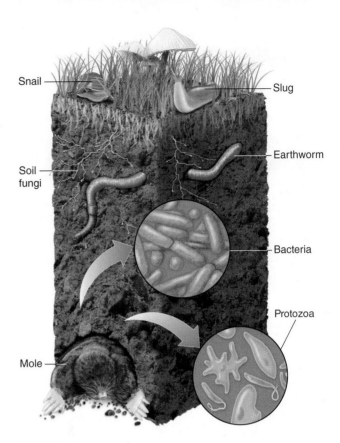

**FIGURE 8.24** **Soil organisms.** Bacteria, fungi, and protozoans account for 80 to 90 percent of soil organisms. Also present are snails, slugs, insects, earthworms, and rodents.

contribute to soil mixing and the breakdown of large organic materials into smaller pieces. Some soil organisms, such as snails and slugs, are herbivores that eat plant roots as well as aboveground parts of plants. However, the majority of soil organisms are detritivores, which consume dead plant and animal tissues and recycle the nutrients they contain. Some soil bacteria also fix nitrogen, which, as we saw in Chapter 3, is essential for plant growth.

## Soil Degradation and Erosion

For centuries, the use and overuse of land for agriculture, forestry, and other human activities has led to significant **soil degradation:** the loss of some or all of the ability of soils to support plant growth. One of the major causes of soil degradation is soil erosion, which occurs when topsoil is disturbed—for example, by plowing—or vegetation is removed, allowing the soil to be eroded by water or wind (**FIGURE 8.25**). Topsoil loss can happen rapidly, in as little as a single growing season, but it takes centuries for the lost topsoil to be replaced. Compaction of soil by machines, humans, and livestock can alter its properties and reduce its ability to retain moisture. Compaction and drying of soil can, in turn, reduce the amount of vegetation that grows in the soil and thereby increase erosion. Intensive agricultural use and irrigation can deplete soil nutrients, and the application of agricultural pesticides can pollute the soil.

---

**GAUGE YOUR PROGRESS**

✓ How do soils form?

✓ What are the roles of soils in ecosystems?

✓ How do a soil's physical and chemical properties influence its role as a medium for plant growth?

---

# The uneven distribution of mineral resources has social and environmental consequences

The tectonic cycle, the rock cycle, and soil formation and erosion all influence the distribution of rocks and minerals on Earth. These resources, along with fossil fuels, exist in finite quantities, but are vital to modern human life. In this section we will discuss some important nonfuel mineral resources and how humans obtain them. (We will discuss fuel resources in Chapters 12 and 13.) Some of these resources are abundant, whereas others are rare and extremely valuable.

As we saw at the beginning of this chapter, early Earth cooled and differentiated into distinct vertical zones. Heavy elements sank toward the core, and lighter elements rose toward the crust. **Crustal abundance** is the average concentration of an element in the crust.

Looking at **FIGURE 8.26**, we can see that four elements—oxygen, silicon, aluminum, and iron—constitute over 88 percent of the crust. However, the chemical composition of the crust in any one location is highly variable.

Environmental scientists and geologists study the distribution and types of mineral resources around the planet in order to locate them and manage their extraction or conservation. **Ores** are concentrated accumulations of minerals from which economically valuable materials can be extracted. Ores are typically characterized by the presence of valuable metals, but accumulations of other valuable materials, such as salt or sand, can also be considered ores. **Metals** are elements with properties that allow them to conduct electricity and heat energy and perform other important functions. Copper, nickel, and aluminum are common examples of metals. In rock, they exist in varying concentrations, usually in association with elements such as sulfur, oxygen, and silicon. Some metals, such as gold, exist naturally in a pure form.

Ores are formed by a variety of geologic processes. Some ores form when magma comes into contact with water, heating the water and creating a solution from which metals precipitate. Other ores form after the deposition of igneous rock. Some ores occur in relatively small areas of high concentration, such as veins. Others, called *disseminated deposits,* occur in much larger areas of rock, often in lower concentrations. Still others, such as copper, can be deposited throughout a large area as well as in veins. Nonmetallic mineral resources, such as clay, sand, salt, limestone, and phosphate, typically occur in concentrated deposits. These deposits occur as a result of their chemical or physical separation from other materials by water, in conjunction with the tectonic and rock cycles. Some ores, such as *bauxite,*

FIGURE 8.26 **Elemental composition of Earth's crust.** Oxygen is the most abundant element in the crust. Silicon, aluminum, and iron are the next three most abundant elements.

| TABLE 8.1 | Approximate supplies of metal reserves remaining | |
| --- | --- | --- |
| Metal | Global reserves remaining (years) | U.S. reserves remaining (years) |
| Iron (Fe) | 120 | 40 |
| Aluminum (Al) | 330 | 2 |
| Copper (Cu) | 65 | 40 |
| Lead (Pb) | 20 | 40 |
| Zinc (Zn) | 30 | 25 |
| Gold (Au) | 30 | 20 |
| Nickel (Ni) | 75 | 0 |
| Cobalt (Co) | 50 | 0 |
| Manganese (Mn) | 70 | 0 |
| Chromium (Cr) | 75 | 0 |

*Sources:* S. Marshak, *Earth: Portrait of a Planet,* 3rd ed. (W. W. Norton, 2007); U.S. Geological Survey Mineral Commodity Summaries, http://minerals.er.usgs.gov/minerals/pubs/mcs/.

the ore in which aluminum is most commonly found, are formed by intense chemical weathering in tropical regions.

The global supply of mineral resources is difficult to quantify. Because private companies hold the rights to extract certain mineral resources, information about the exact quantities of resources is not always made public. The publicly known estimate of how much of a particular resource is available is based on its **reserve:** the known quantity of the resource that can be economically recovered. Table 8.1 lists the estimated number of years of remaining supply of some of the most important metal resources commonly used in the United

States, assuming that rates of use do not change. Some important metals, such as tantalum, have never existed in the United States. The United States has used up all of its reserves of some other metals, such as nickel, and must now import those metals from other countries.

## Types of Mining

Mineral resources are extracted from Earth by mining the ore and separating any other minerals, elements, or residual rock away from the sought-after element or mineral. Two kinds of mining take place on land: *surface mining* and *subsurface mining* (FIGURE 8.27). Each method has environmental, human, and social benefits and costs.

**SURFACE MINING** A variety of surface mining techniques can be used to remove a mineral or ore deposit that is close to Earth's surface. **Strip mining,** the removal of "strips" of soil and rock to expose ore, is used when the desired ore is relatively close to Earth's surface and runs parallel to it, as is often the case for deposits of sedimentary materials such as coal and sand. In these situations, miners remove a large volume of material, extract the resource, and return the unwanted waste material, called **mining spoils** or **tailings,** to the hole created during the mining. A variety of strategies can be used to restore the affected area to something close to its condition before mining began.

FIGURE 8.27 **Surface and subsurface mining.** Surface mining methods include strip, open pit, mountaintop removal, and placer mining.

**Open-pit mining,** the creation of a large pit or hole in the ground that is visible from Earth's surface, is used when the resource is close to the surface but extends beneath the surface both horizontally and vertically. Copper mines are usually open-pit mines. One of the largest open-pit mines in the world is the Kennecott Bingham Canyon mine near Salt Lake City, Utah. This copper mine is 4.4 km (2.7 miles) across and 1.1 km (0.7 miles) deep.

In **mountaintop removal,** miners remove the entire top of a mountain with explosives. Large earth-moving equipment removes the resource and deposits the tailings in lower-elevation regions nearby, often in or near rivers and streams.

**Placer mining** is the process of looking for metals and precious stones in river sediments. Miners use the river water to separate heavier items, such as diamonds, tantalum, and gold, from lighter items, such as sand and mud. The prospectors in the California gold rush were placer miners, and the technique is still used today.

**SUBSURFACE MINING** Surface mining techniques are ineffective when the desired resource is more than 100 m (328 feet) below Earth's surface. In such cases, miners must turn to **subsurface mining.** Typically, a subsurface mine begins with a horizontal tunnel dug into the side of a mountain or other feature containing the resource. From this horizontal tunnel, vertical shafts are drilled, and elevators are used to bring miners down to the resource and back to the surface. The deepest mines on Earth are up to 3.5 km (2.2 miles) deep. Coal, diamonds, and gold are some of the resources removed by subsurface mining.

## Mining, Safety, and the Environment

The extraction of mineral resources from Earth's crust has a variety of environmental impacts and human health consequences. Virtually all mining requires the construction of roads, which can result in soil erosion, damage to waterways, and habitat fragmentation. In addition, all types of mining produce tailings. Some tailings contaminate land and water with acids and metals (Table 8.2).

In mountaintop removal, the mining spoils are typically deposited in the adjacent valleys, sometimes blocking or changing the flow of rivers. Mountaintop removal is used primarily in coal mining and is safer for workers than subsurface coal mining. In terms of environmental damage, mining companies do sometimes make efforts to restore the mountain to its original shape. However, there is considerable disagreement about whether these reclamation efforts are effective. Damage to streams and nearby groundwater during mountaintop removal cannot be completely rectified by the reclamation process.

Placer mining can also contaminate large portions of rivers, and the areas adjacent to the rivers, with sediment and chemicals. In certain parts of the world, the toxic metal mercury is used in placer mining of gold and silver. Mercury is a highly *volatile* metal; that is, it moves easily among air, soil, and water. Mercury is harmful to plants and animals and can damage the central nervous system in humans; children are especially sensitive to its effects.

The environmental impacts of subsurface mining may be less apparent than the visible scars left behind by surface mining. One of these impacts is acid mine drainage. To keep underground mines from flooding, water must be continually pumped out. Water from mining operations can have an extremely low pH. This acid mine drainage lowers the pH of nearby soils and streams, causing severe damage to the ecosystem.

Subsurface mining is a dangerous occupation. Hazards to miners include accidental burial, explosions, and fires. In addition, the inhalation of gases and particles

| TABLE 8.2 | Types of mining operations and their effects | | | | |
|---|---|---|---|---|---|
| Type of mining operation | Effects on air | Effects on water | Effects on soil | Effects on biodiversity | Effects on humans |
| Surface mining | Significant dust from earth-moving equipment | Contamination of water that percolates through tailings | Most soil removed from site; may be replaced if reclamation occurs | Habitat alteration and destruction over the surface areas that are mined | Minimal in the mining process, but air quality and water quality can be adversely affected near the mining operation |
| Subsurface mining | Minimal dust at the site, but emissions from fossil fuels used to power mining equipment can be significant | Acid mine drainage as well as contamination of water that percolates through tailings | | Road construction to mines fragments habitat | Occupational hazards in mine; possibility of death or chronic respiratory diseases such as black lung disease |

over long periods can lead to a number of occupational respiratory diseases, including black lung disease and asbestosis, a form of lung cancer. In the United States, between 1900 and 2006, more than 11,000 coal miners died in underground coal mine explosions and fires. A much larger number died from respiratory diseases. Today, there are relatively few deaths per year in coal mines in the United States, in part because of improved work safety standards and in part because there is much less subsurface mining in the country today. In other countries, especially China, mining accidents remain fairly common.

As human populations grow and developing nations continue to industrialize, the demand for mineral resources continues to increase. But as the most easily mined mineral resources are depleted, extraction efforts become more expensive and environmentally destructive. The ores that are easiest to reach and least expensive to remove are always recovered first. When those sources are exhausted, mining companies turn to deposits that are more difficult to reach. These extraction efforts result in greater amounts of mining spoils and more of the environmental problems we have already noted. Therefore, an important part of living sustainably will be learning to use and reuse limited mineral resources more efficiently.

## Mining Legislation

Governments have sought to regulate the mining process for many years. Early mining legislation was primarily focused on promoting economic development, but later legislation became concerned with worker safety as well as environmental protection. The effectiveness of these mining laws has varied.

The Mining Law of 1872 was passed by the U.S. Congress to regulate the mining of silver, copper, and gold ore as well as fuels, including natural gas and oil, on federal lands. This law, also known as the General Mining Act, allowed individuals and companies to recover ores or fuels from federal lands. The law was written primarily to encourage development and settlement in the western United States and, as a result, contains very few environmental protection provisions.

The Surface Mining Control and Reclamation Act of 1977 regulates surface mining of coal and the surface effects of subsurface coal mining. The act mandates that land be minimally disturbed during the mining process and reclaimed after mining is completed. Mining legislation does not regulate all of the mining practices that can have harmful effects on air, water, and land, however. In later chapters we will learn about other U.S. legislation that does, to some extent, address these issues, including the Clean Air Act, Clean Water Act, and Superfund Act.

---

**GAUGE YOUR PROGRESS**

✓ Why are economically valuable mineral resources distributed unevenly on the planet?

✓ Describe the various types of surface mining operations.

✓ What are the consequences of surface mining versus subsurface mining, and how has mining legislation tried to reduce those impacts?

---

# WORKING TOWARD SUSTAINABILITY

One of the environmental impacts of surface mining is the amount of land surface it disturbs. Once a mining operation is finished, the mining company can either leave the land disturbed or try to restore it to its original condition. In the United States, the Surface Mining Control and Reclamation Act of 1977 requires coal mining companies to restore the lands they have mined. Regulations also require other types of mining operations to do some level of restoration.

## Mine Reclamation and Biodiversity

A disturbed ecosystem can return to a state similar to its original condition only if the original physical, chemical, and biological properties of the land are re-created. Therefore, reclamation after mining involves several steps. First, the mining company must fill in the hole or depression it has created in the landscape. Second, the fill material must be shaped to follow the contours of the land that existed before mining began.

Usually, the mining company has scraped off the topsoil that was on the land and put it aside. This topsoil must be returned and spread over the landscape. Finally, the land must be replanted. In order to re-create the communities of organisms that inhabited the area before mining, the vegetation planted on the site must be native to the area and foster the process of natural succession. Properly completed reclamation makes the soil physically stable so that erosion does not occur and water infiltration and retention can proceed as they did before mining. The reclaimed material must be relatively free of metals, acids, and other compounds that could potentially leach into nearby bodies of water.

Many former mining areas, for one reason or another, have not been reclaimed properly. However, there are an increasing number of stories of reclamation efforts that have achieved conditions equal to those that existed prior to the mining operation. The Trapper

**FIGURE 8.28** **Mining and reclamation.** At a strip mine operation in Colorado, not far from the Trapper Mine: (a) active mining underway; (b) restored native habitat after reclamation.

(b)

(a)

Mine in Craig, Colorado, and other mines like it, illustrate reclamation success stories (**FIGURE 8.28**).

The Trapper Mine produces about 2 million tons of coal per year, which is sold to a nearby electricity generation plant. Although all coal mining operations are required to save excavated rock and topsoil, managers at the Trapper Mine have stated that they are meticulous about saving all the topsoil they remove. The rock they save is used to fill the lower portions of excavated holes. Workers then install drainage pipes and other devices to ensure proper drainage of water. The topsoil that has been set aside is spread over the top of the restored ground and contoured as it was before mining. Trapper Mine reclamation staff then replant the site with a variety of native species of grasses and shrubs, including the native sagebrush commonly found in the high plateaus of northwestern Colorado.

Whereas government officials, and even the Colorado branch of the Sierra Club, have expressed approval of the Trapper Mine reclamation process, perhaps the strongest evidence of its success is the wildlife that now inhabits the formerly mined areas. The Columbian sharp-tailed grouse (*Tympanuchus phasianellus columbianus*), a threatened bird species, has had higher annual survival and fertility rates on the reclaimed mine land than on native habitat in other parts of Colorado. Populations of elk, mule deer, and antelope have all increased on reclaimed mine property.

Reclamation issues must be addressed for each particular area where mining occurs. As we have seen, subsurface and surface water runoff in certain locations can contain high concentrations of toxic metals and acids. In other situations, although certain native species may increase in abundance after reclamation, other native species may decrease. Nevertheless, with supervision, skill, and enough money to pay for the proper reclamation techniques, a reclaimed mining area can become a satisfactory or even an improved habitat for many species.

**References**

Department of the Interior, Office of Surface Mining Reclamation and Enforcement. http://www.osmre.gov/.

Raabe, S. 2002. Nature's comeback: Trapper Mine reclamation attracts wildlife, wins praise. *Denver Post,* November 12, p. C01.

## REVISIT THE KEY IDEAS

- **Describe the formation of Earth and the distribution of critical elements on Earth.**

Earth formed from cosmic dust in the solar system. As it cooled, heavier elements, such as iron, sank toward the core, while lighter elements, such as silica, floated toward the surface. These processes have led to an uneven distribution of elements and minerals throughout the planet.

- **Define the theory of plate tectonics and discuss its importance in environmental science.**

Earth is overlain by a series of plates that move at rates of a few millimeters per year. Plates can move away from each other, move toward each other, or slide past each other. One plate can be subducted under another. These tectonic processes create mountains, earthquakes, and volcanoes.

- **Describe the rock cycle and discuss its importance in environmental science.**

Rocks are made up of minerals, which are formed from the various chemical elements in Earth's crust. The processes of the rock cycle lead to the formation, breakdown, and recycling of rocks. Weathering and erosion break down and move rock material, releasing chemical elements and contributing to soil formation.

- **Explain how soil forms and describe its characteristics.**

Soil forms as the result of physical and chemical weathering of rocks and the gradual accumulation of organic detritus from the biosphere. The factors that determine soil properties are parent material, climate, topography, soil organisms, and time. The relative abundances of sand, silt, and clay in a soil determine its texture.

- **Explain how elements and minerals are extracted for human use.**

Concentrated accumulations of minerals from which economically valuable materials can be extracted are called ores. Ores are removed by surface or subsurface mining operations. Surface mining generally results in more environmental impacts, whereas subsurface mining is more dangerous to miners. With the exception of coal mining, legislation directly related to mining does not address most environmental considerations.

## CHECK YOUR UNDERSTANDING

1. As Earth slowly cooled,
   (a) lighter elements sank to the core and heavier elements moved to the surface.
   (b) lighter elements mixed with heavier elements and sank to the core.
   (c) lighter elements mixed with heavier elements and moved to the surface.
   (d) lighter elements moved to the surface and heavier elements sank to the core.
   (e) lighter and heavier elements dispersed evenly from the core to the surface.

2. The correct vertical zonation of Earth above the core is
   (a) asthenosphere-mantle-soil-lithosphere.
   (b) asthenosphere-lithosphere-mantle-soil.
   (c) mantle-asthenosphere-lithosphere-soil.
   (d) mantle-lithosphere-soil-asthenosphere.
   (e) soil-mantle-lithosphere-asthenosphere.

3. Evidence for the theory of plate tectonics includes
   I  large deposits of copper ore.
   II  identical rock formations on both sides of the Atlantic.
   III  fossils of the same species on distant continents.

   (a) I only          (d) I and II only
   (b) I and III only  (e) I, II, and III
   (c) II and III only

4. Which type of mining is usually most directly harmful to miners?
   (a) Mountaintop removal
   (b) Open-pit mining

   (c) Placer mining
   (d) Strip mining
   (e) Subsurface mining

5. Measured on the Richter scale, an earthquake with a magnitude of 7.0 is _____ times greater than an earthquake with a magnitude of 2.0.
   (a) 10
   (b) 100
   (c) 1,000
   (d) 10,000
   (e) 100,000

*For questions 6 to 9, select from the following choices:*
   (a) Seismic activity center
   (b) Divergent plate boundary
   (c) Convergent plate boundary
   (d) Transform fault boundary
   (e) Epicenter

6. At which type of boundary do tectonic plates move sideways past each other?

7. At which type of boundary does subduction occur?

8. At which type of boundary does seafloor spreading occur?

9. Which term refers to the point on Earth's surface directly above an earthquake?

10. Fossil records are found in
   (a) intrusive rock.
   (b) extrusive rock.
   (c) igneous rock.
   (d) metamorphic rock.
   (e) sedimentary rock.

11. Which of the following statements about soil is *not* correct?
    (a) Soil always filters out pollutants that pass through it.
    (b) Soil is the medium for plant growth.
    (c) Soil is a primary filter of water.
    (d) A wide variety of organisms live in soil.
    (e) Soil plays an important role in biogeochemical cycles.

12. The soil horizon commonly known as *subsoil* is the
    (a) A horizon.
    (b) O horizon.
    (c) B horizon.
    (d) C horizon.
    (e) E horizon.

## APPLY THE CONCEPTS

1. The rock cycle plays an important role in the recycling of Earth's limited amounts of mineral resources.
    (a) The mineral composition of rock depends on how it is formed. Name the *three* distinct rock types, explain how each rock type is formed, and give *one* specific example of a rock of each type.
    (b) Explain *either* the physical *or* the chemical weathering process that leads to the breakdown of rocks.
    (c) Describe the natural processes that can lead to the formation of soil, and discuss how human activities can accelerate the loss of soil.

2. Read the following article that appeared on the Web site of the organization "Eatwild," at http://www.eatwild.com/environment.html, and answer the questions that follow.

---

### Grassfarming Benefits the Environment

#### Grazing better for the soil than growing grain

Six Minnesota pasture-based ranchers asked researchers to compare the health of their soil with soil from neighboring farms that produced corn, soybean, oats, or hay. At the end of four years of monitoring, researchers concluded that the carefully managed grazed land had

- 53% greater soil stability
- 131% more earthworms
- Substantially more organic matter
- Less nitrate pollution of groundwater
- Improved stream quality
- Better habitat for grassland birds and other wildlife

Depending on the way that cattle are managed, they can either devastate a landscape or greatly improve the health of the soil. ["Managed Grazing as an Alternative Manure Management Strategy," Jay Dorsey, Jodi Dansingburg, Richard Ness, USDA-ARS, Land Stewardship Project.]

### Pasture reduces topsoil erosion by 93 percent

Currently, the United States is losing three billion tons of nutrient-rich topsoil each year. Growing corn and soy for animal feed using conventional methods causes a significant amount of this soil loss. Compared with row crops, pasture reduces soil loss by as much as 93 percent. [Ontario Ministry of Agriculture and Food, Robert P. Stone and Neil Moore, Fact Sheet 95-089.]

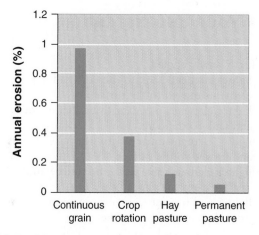

(a) Explain the consequences of the observations in the studies described and comment on the long-term sustainability of each agricultural practice.
    (i) What roles do earthworms play in maintaining soil stability?
    (ii) How does the presence of organic matter benefit the soil?
    (iii) Suggest why there is less nitrate contamination of groundwater from the permanent pastures.
    (iv) In what ways could the stream quality have improved?

(b) Discuss *one* negative consequence of grazing cattle on pastureland instead of feeding them on grain.

(c) Describe *one* potential negative effect on rivers and streams of grazing cattle on pastureland.

## MEASURE YOUR IMPACT

**What is the impact of your diet on soil dynamics?** In the landmark 1997 report "Livestock Production: Energy Inputs and the Environment," Cornell University ecologist David Pimentel wrote that feeding grain to cattle consumes more resources than it yields, accelerates soil erosion, and reduces the supply of food for the world's people. Some highlights of the report include the following:

- Each year, an estimated 41 million tons of plant protein is fed to U.S. livestock to produce an estimated 7 million tons of animal protein for human consumption. About 26 million tons of the livestock feed comes from grains and 15 million tons from forage crops. For every kilogram of high-quality animal protein produced, livestock are fed nearly 6 kg of plant protein. The 7 billion animals consume five times as much grain as the entire U.S. human population.

- Every kilogram of beef produced takes 100,000 liters of water. Some 900 liters of water go into producing a kilogram of wheat. Potatoes are even less "thirsty," at 500 liters per kilogram.

- About 90 percent of U.S. cropland is losing soil to erosion at 13 times the rate of soil formation. Soil loss is most severe in some of the richest farming areas: Iowa, for example, loses topsoil at 30 times the rate of soil formation. Iowa has lost one-half of its topsoil in 150 years of farming. That soil took thousands of years to form.

Over the course of 1 week, make a daily record of what you eat and drink. At the end of the week, answer the following questions:

(a) Evaluate the components of your diet for the week. How many portions of animal protein did you eat each day?

(b) Most agricultural fields receive inputs of phosphorus, calcium, and magnesium, which are usually obtained by mining rocks containing those elements, grinding them up, and adding them to fertilizers. Assess the likely impact of this practice on the demand for certain rocks and on soil dynamics.

(c) Describe changes you could make to your diet to minimize the impacts you cited above.

(d) How do you think your diet would compare to that of a person in a developing country? How would their ecological footprint compare to yours? *Hint:* You may have to draw upon previous chapters you have read as well as this chapter to answer this question.

## ENGAGE YOUR ENVIRONMENT

How does remote sensing help us view environmental change? Explore remote sensing data from areas around the globe that have undergone dramatic change. *See Engage Your Environment Activity 8 at the end of this book.*

# Water Resources: Supply, Distribution, and Use

## Dams and Salmon on the Klamath River

The Klamath River is a beautiful stretch of water that runs for 400 km (250 miles) from southern Oregon through northern California, where it empties into the ocean. The Klamath River was once a spectacular habitat for salmon—the third largest salmon fishery on the West Coast. During the past 100 years, however, the Klamath River has seen many changes. It has been dammed for electricity, and farmers have settled

dams generate less pollution than other sources of electricity, these dams provided a substantial benefit to the region. But their presence imposed a barrier to the natural migration of salmon. In addition, the formation of large pools behind the dams caused the water temperature to rise into a range that is not favorable to salmon.

As the decades passed, the multiple uses of the Klamath River began to come into conflict. In addition, a gradual

### The Klamath River was once a spectacular habitat for salmon.

the land and diverted water for irrigation. The salmon population has been greatly reduced, and local Native American tribes and the commercial fishing industry have suffered as a result.

The story of the Klamath River began thousands of years ago, when six tribes of Native Americans were the sole inhabitants of the region. For more than 300 generations, these tribes relied on salmon as an important part of their diet. Historically, salmon eggs hatched in the upstream portion of the river. Young fish migrated down the river into the ocean, where they spent several years growing into mature fish. Once mature, they migrated up the river, returning to the same place where they had hatched to lay their own eggs. Several species of salmon were abundant in the river. An estimated 800,000 mature chinook salmon (*Oncorhynchus tshawytscha*)—which can grow as large as 45 kg (100 pounds)—migrated up the river each year.

This migratory behavior worked well for thousands of years, until the region began to change during the twentieth century. In the early 1900s, the U.S. Bureau of Reclamation decided to install canals to drain two large lakes that were the source of the river. The exposed lake bottoms were turned into farmland, and the water that drained into the canals provided a constant source of irrigation for this new farmland. In addition, four hydroelectric dams were built on the river between 1908 and 1962. These dams generated enough electricity to power 70,000 homes. Because hydroelectric

warming of temperatures in the region caused less snow to fall in the mountains. Less winter snow meant less water for the river. Although there was now less water available, the same amount continued to be withdrawn for agricultural use. These withdrawals reduced the river's flow to low levels that created an additional threat to the salmon. In 1997, the coho salmon (*Oncorhynchus kisutch*) was given protection under the U.S. Endangered Species Act. This protection meant that water withdrawals for irrigation could not reduce the river's flow below a minimal level. That restriction, in turn, prevented the farmers from receiving enough water to irrigate their crops, which threatened their livelihood.

The debate among the various interest groups intensified. Native Americans, the hydroelectric company, farmers, the commercial fishing industry, ▶

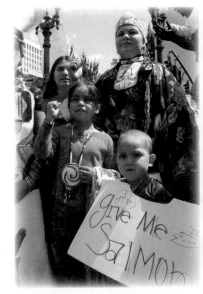

The multiple uses of the water in the Klamath River have led to conflicts over water rights, including Native Americans who have depended on the river for salmon for many centuries.

◀ The beautiful Klamath River flows from southern Oregon and down through northern California.

conservationists, and multiple government agencies all had different interests and conflicting points of view about the best policy for water management in the region.

In 2002, the migrating salmon experienced a massive die-off in the river. This catastrophe was caused by unusually warm water, the growth of toxic algae in the dammed waters, and diseases. The sight of 30,000 large, dead chinook salmon brought worldwide attention to the Klamath River. Nearly 30 interest groups began to meet to find a solution to the river's complex problems.

In 2009, the parties reached a decision. The hydroelectric company, which realized it would soon have to spend $300 million on dam renovations, agreed to remove the dams. The farmers agreed to conserve water by planting less water-thirsty crops and updating their irrigation technology. These changes are expected to improve river flow, thereby improving

conditions for the salmon and those who depend on them. The agreement was signed in 2010, and the dams are scheduled to be removed in 2020. Once the project is completed, it will represent the largest dam removal in history.

Conflicts over the use of water resources are found throughout the world as human populations grow and place ever-increasing demands on limited water resources. The compromise reached on the Klamath has the potential to serve as a model for resolving such conflicts in the future. ■

Sources: R. Rymer, Reuniting a river, *National Geographic,* December 2008, http://ngm.nationalgeographic.com/2008/12/klamath-river/rymer-text/1; C. Sullivan, Landmark agreement to remove four Klamath River dams, *New York Times,* September 30, 2009, http://www.nytimes.com/gwire/2009/09/30/30greenwire-landmark-agreement-to-remove-4-klamath-river-d-72992. html?scp=2&sq=dam%20removal&st=cse.

## UNDERSTAND THE KEY IDEAS

Climate variation causes some regions of the world to possess abundant supplies of water, whereas other regions have very little water. Around the world, growing human populations are facing issues related to water availability.

After reading this chapter you should be able to

- identify Earth's natural sources of water.

- discuss the ways in which humans manage water distribution.

- describe the major human uses of water.

- identify the factors that will affect the future availability of water.

# Water is abundant, but usable water is rare

Nearly 70 percent of Earth's surface is covered by water. As **FIGURE 9.1** shows, this water is found in five main repositories. The vast majority of Earth's water, more than 97 percent, is found in the oceans as salt water. The remainder—less than 3 percent—is fresh water, the type of water that can be consumed by humans. Of this small percentage of Earth's water that is fresh water, nearly one-fourth resides underground. The remaining three-fourths is above ground, but most of it is in the form of ice and glaciers and so is generally not available for human consumption. Very small fractions of Earth's

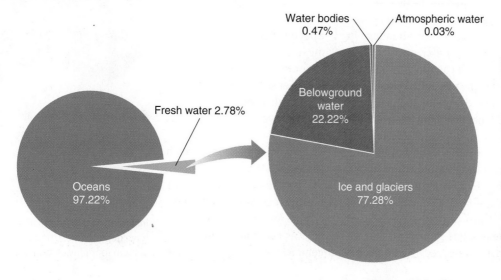

**FIGURE 9.1** **Distribution of water on Earth.** Fresh water represents less than 3 percent of all water on Earth, and only about three-fourths of that fresh water is surface water. Most of that surface water is frozen as ice and in glaciers. Therefore, less than 1 percent of all water on the planet is accessible for use by humans. [After R. W. Christopherson, *Geosystems,* 7 ed. Pearson/Prentice Hall, 2009.]

Recharge area

Flowing artesian well

Confining layer (impermeable)

Unconfined aquifer

Confined aquifer

Water table

Well

**FIGURE 9.2** **Aquifers.** Aquifers are sources of usable groundwater. Unconfined aquifers are rapidly recharged by water that percolates downward from the land surface. Confined aquifers are capped by an impermeable layer of rock or clay, which can cause water pressure to build up underground. Artesian wells are formed when a well is drilled into a confined aquifer and the natural pressure causes water to rise toward the ground surface.

aboveground water are found in the atmosphere and in the form of water bodies such as streams, rivers, wetlands, and lakes.

## Groundwater

Groundwater exists in the multitude of small spaces found within permeable layers of rock and sediment called **aquifers.** FIGURE 9.2 shows how different types of aquifers are situated. Many aquifers are porous rock covered by soil. Because water can easily flow in and out of such aquifers, they are called **unconfined aquifers.** In contrast, some aquifers are surrounded by a layer of impermeable rock or clay. These aquifers are called **confined aquifers** because the impermeable, or confining, layer impedes water flow to or from the aquifer. The uppermost level at which the water in a given area

fully saturates the rock or soil is called the **water table.** The water table is considered to be the surface of the groundwater in an area.

Water from precipitation can percolate through the soil and work its way into an aquifer. This input process is known as **groundwater recharge.** If water falls on land that contains a confined aquifer, however, it cannot penetrate the impermeable layer of rock. Therefore, a confined aquifer cannot be recharged unless the impermeable layer has an opening at the land's surface that can serve as a *recharge area.*

Aquifers serve as important sources of fresh water for many organisms. Plant roots can access the groundwater in an aquifer and draw down the water table. Water from some aquifers naturally percolates up to the ground surface as **springs** (FIGURE 9.3). Springs serve as a natural source of water for freshwater aquatic biomes,

**FIGURE 9.3** **Natural spring.** When an aquifer has an opening at the land surface, the water can flow out to form a spring. Springs, such as this one in New Mexico, can be important sources of water for organisms and serve as the initial water source for many streams and rivers.

and they can be used directly by humans as sources of drinking water. Humans discovered centuries ago that water can also be obtained from aquifers by digging a well—essentially a hole in the ground. Most modern wells contain pumps that move the water up to the ground surface against the force of gravity. The water in some confined aquifers, however, is under tremendous pressure due to the impermeable layer of rock that surrounds it. Drilling a hole into a confined aquifer releases the pressure on the water, allowing it to burst out of the aquifer and rise up in the well, as shown in Figure 9.2. Such wells are called **artesian wells.** If the pressure is sufficiently great, the water can rise all the way up to the ground surface, in which case no pump is required to extract the water from the ground.

The age of water in aquifers varies, as does the rate at which aquifers are recharged. The water in an unconfined aquifer may originate from water that fell to the ground last year or even last week. This direct and rapid connection with the surface is one of the reasons why water from unconfined aquifers is much more likely

to be contaminated with chemicals released by human activities. Confined aquifers, on the other hand, are generally recharged very slowly, perhaps over 10,000 to 20,000 years. For this reason, water from a confined aquifer is usually much older, and less likely to be contaminated by anthropogenic chemicals, than water from an unconfined aquifer.

Large-scale use of water from a confined aquifer is unsustainable because the withdrawal of water is not balanced by recharge. The largest aquifer in the United States is the massive Ogallala aquifer in the Great Plains. Large amounts of water have been withdrawn from this aquifer for household, agricultural, and industrial uses. Unfortunately, the aquifer is recharged slowly, and the rate of recharge is not keeping pace with the rate of water withdrawal, as **FIGURE 9.4** shows. For this reason, the Great Plains region could run out of water during this century.

**FIGURE 9.5** shows what happens when more water is withdrawn from an aquifer than enters the aquifer. As the water table drops farther from the ground surface,

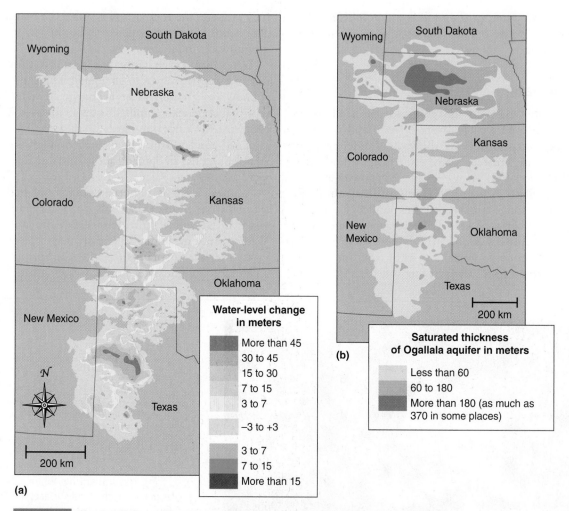

**FIGURE 9.4** **The Ogallala aquifer.** The Ogallala aquifer, also called the High Plains aquifer, is the largest in the United States, with a surface area of about 450,000 km² (174,000 miles²). (a) The change in water level from 1950 to 2005, mostly due to withdrawals for irrigation that have exceeded the aquifer's rate of recharge. (b) The current thickness of the aquifer.

Well  Well  Well

Water table

**(a)** Before heavy pumping

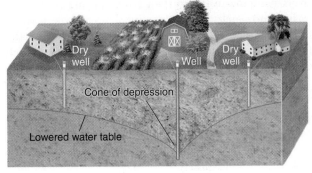

Dry well  Well  Dry well

Cone of depression

Lowered water table

**(b)** After heavy pumping

**FIGURE 9.5** **Cone of depression.** (a) When a deep well is not heavily pumped, the recharge of the water table keeps up with the pumping. (b) In contrast, when a deep well pumps water from an aquifer more rapidly than it can be recharged, it can form a cone of depression in the water table and cause nearby shallow wells to go dry.

springs that once bubbled up to the surface no longer emerge, and spring-fed streams dry up. In addition, some shallow wells no longer reach the water table. The water table adjacent to a well where water is being rapidly withdrawn is lowered the most, creating a **cone of depression:** an area where there is no longer any groundwater. In other words, rapid pumping of a deep well can cause adjacent, shallower wells to go dry.

In some situations, pumping fresh water out of wells faster than the aquifer can be recharged compromises water quality. **FIGURE 9.6** shows how this can happen when wells are drilled near coastlines. Some coastal regions have an abundance of fresh water underground. This groundwater exerts downward pressure that prevents the adjacent ocean water from flowing into aquifers and mixing with the groundwater. As humans drill thousands of wells along these coastlines, however, rapid pumping draws down the water table, reduces the depth of the groundwater, and thereby lessens that pressure.

The adjacent salt water is then able to infiltrate the area of rapid pumping, making the water in the wells salty. This process, called **saltwater intrusion,** is a common problem in coastal areas.

## Surface Water

Surface water is the fresh water that exists above the ground, including streams, rivers, ponds, lakes, and wetlands. As mentioned in Chapter 3, these waters represent ecosystems that perform a number of important functions.

The world's three largest rivers, as measured by the volume of water they carry, are the Amazon in South America, the Congo in Africa, and the Yangtze in China. Early human civilizations typically settled along major rivers not only because these waterways served as important means of transportation, but also because the land surrounding rivers is often highly fertile. Most

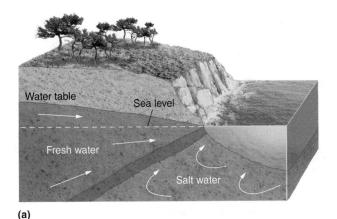

Water table  Sea level

Fresh water  Salt water

**(a)**

Well

Water table  Sea level

Fresh water  Saltwater intrusion

**(b)**

**FIGURE 9.6** **Saltwater intrusion.** (a) When there are few wells along a coastline, the water table remains high and the resulting pressure prevents salt water from intruding. (b) In contrast, rapid pumping of wells drilled in aquifers along a coastline can lower the water table. Lowering the water table reduces water pressure in the aquifer, allowing the nearby salt water to move into the aquifer and contaminate the well water with salt.

rivers naturally overflow their banks during periods of spring snowmelt or heavy rainfall. The excess water spreads onto the land adjacent to the river, called the **floodplain.** These floodwaters deposit nutrient-rich sediment on the floodplain, improving the fertility of the soil. The ancient Egyptians built a large and prosperous civilization by using the Nile River for commerce and its fertile floodplains for agriculture. The use of rivers as a means of transport continues today. Many large cities in North America, including Pittsburgh, New Orleans, and St. Louis, developed on the banks of major rivers.

Ponds and lakes typically form in depressions in the landscape that are filled by precipitation, runoff that is not absorbed by the surrounding landscape, and groundwater that flows into the depression. Determining Earth's largest lakes depends on whether we consider surface area, depth, or volume. FIGURE 9.7 shows comparisons based on each of these factors. Some lakes, such as Lake Victoria in Africa, have large surface areas but are not especially deep, whereas other lakes, such as Lake Baikal in Asia, do not have especially large surface areas, but are incredibly deep and therefore have a large volume.

Lakes can be created by a variety of processes, including tectonic activity and glaciation. When tectonic activity causes areas of land to rise up, it can isolate a region of the ocean. Because lakes formed in this way are ocean-derived, they have higher salinities than other lakes. The Caspian Sea in western Asia is an excellent example of a salty lake that was formed by tectonic uplift. Tectonic activity can also cause land to split open into long, deep fissures that subsequently fill with water. Some of the world's largest lakes have formed this way, including Lake Victoria and Lake Tanganyika in Africa and Lake Baikal in Asia. Another way that lakes are formed is through the movement of glaciers. Over thousands of years, glacial movement scrapes large depressions in the land that subsequently fill with water. This process is thought to have played a major role in the creation of the Great Lakes in North America.

Lakes are classified by their level of primary productivity. Lakes that have low productivity due to low amounts of nutrients such as phosphorus and nitrogen in the water are called **oligotrophic** lakes. In contrast, lakes with a moderate level of productivity are called **mesotrophic** lakes, and lakes with a high level of productivity are called **eutrophic** lakes.

Freshwater wetlands play important, if sometimes unrecognized, roles in water distribution and regulation, as we saw in Chapter 4. During periods of heavy rainfall that might otherwise lead to flooding, freshwater wetlands—as well as salt marshes and mangrove swamps—can absorb and store the excess water and release it slowly, thereby reducing the likelihood of a flood.

Wetlands are measured by surface area. There is some debate about which freshwater wetland is the world's largest, but most researchers consider the Pantanal in South America to be the largest wetland and the Florida Everglades to be the second largest.

## Atmospheric Water

Although the atmosphere contains only a very small percentage of the water on Earth, that atmospheric water is essential to global water distribution. People living in arid regions rely heavily on precipitation, in the form of rain and snow, for their water needs. In certain regions, such as East Africa, annual rainfall follows a fairly regular pattern: April, May, October, and November are normally rainy months. However, even areas with predictable rainfall patterns can experience unexpected droughts. In the fall of 2000, the United Nations Food and Agriculture Organization reported that a succession of droughts had put 16 million people at risk of starvation in eastern Africa. In 2004, the worst drought in more than a decade adversely affected many parts of southern Africa, destroying crops, killing cattle, and causing millions of people to go hungry.

In addition to the direct losses of human lives, livestock, and crops that they cause, droughts have long-term

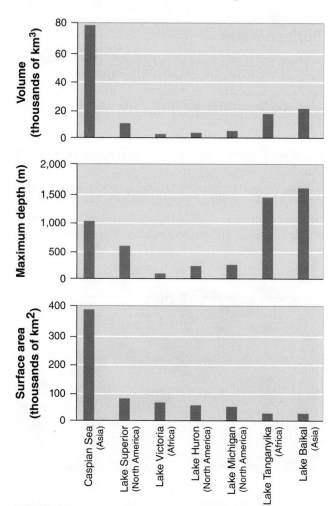

**FIGURE 9.7** **The world's largest lakes.** Measurements of surface area, depth, and volume are shown for each lake. [Data from http://wldb.ilec.or.jp/LakeDB2/.]

effects on soil. Because the cycling of many nutrients important to ecosystem productivity, such as nitrogen and phosphorus, depends on the movement of water, droughts affect nutrient cycling and thus soil fertility. Furthermore, prolonged droughts can dry out the soil to such an extent that the fertile upper layer—the topsoil—blows away in the wind. As a result, the land may become useless for agriculture for decades or longer. Finally, soil that has become severely parched may harden and become impermeable. In this case, when the rains finally do arrive, the water runs off over the land surface rather than soaking in, eroding the topsoil in the process.

Human activities can contribute to the negative effects of droughts. In the midwestern United States in the 1920s and 1930s, for example, the amount of conversion of native grasslands into wheat fields increased dramatically. Because wheat fields are more susceptible to soil erosion by wind than native grasslands, this conversion produced many more hectares of erodible land. In the early 1930s, a severe and prolonged drought caused a decade of crop failures. With few crop plants alive to hold the soil in place, the drought led to massive dust storms like the one pictured in FIGURE 9.8, as winds picked up the parched topsoil and blew it away. Indeed, one dust storm, on April 14, 1935, was so severe that the Sun was entirely blocked out in the southern Great Plains, leading to the name "Black Sunday." Topsoil from these dust storms traveled as far as Washington, D.C., and earned the southern Great Plains the nickname of "the Dust Bowl." The decade of dust storms and the resulting losses of topsoil led to large-scale human migration out of the region and impacted the economy of the entire nation. Today, improved farming practices have reduced the susceptibility of soil to erosion by wind, making major dust storms in the Great Plains and elsewhere much less likely.

Many drought-prone areas of the world rarely experience high amounts of rainfall, but when it does happen, severe flooding can occur. California, for example, is a relatively dry region of the United States. Occasionally, however, California experiences heavy rainstorms. There is no system in place to capture this water, so the state can have serious flooding problems.

Just as human activities such as farming influence the impact of droughts, human activities can worsen the risk of flooding. Flooding occurs when water input exceeds the ability of an area to absorb that water. In healthy natural areas, porous soil and wetlands soak up excess rainwater. In areas where the soil has been baked hard by drought, however, or in urban and suburban areas with large areas of **impermeable surfaces**—pavement or buildings that do not allow water penetration—the water from heavy rainfall events cannot soak into the ground. Instead, storm waters run off over impermeable soil or paved surfaces and into storm sewers or nearby streams. Excess water that the ground does not absorb fills streams and rivers, which overflow their banks and may flood lowland areas. Like droughts, floods can lead to crop and property damage as well as losses of animal and human lives.

### GAUGE YOUR PROGRESS

✓ What are the primary repositories of fresh water on Earth? Which of these repositories is the largest?

✓ What is the difference between a confined and an unconfined aquifer? How do their recharge rates differ?

✓ How do human activities worsen the effects of droughts and floods?

**FIGURE 9.8** **The Dust Bowl.** A farmhouse in Stratford, Texas, just prior to being hit by a massive dust storm in 1935. Poor agricultural practices combined with prolonged droughts can produce severe dust storms that carry away topsoil and rob the land of its fertility.

# Humans can alter the availability of water

As we have just seen, the availability of water around the world can be unpredictable, yet water is critical for the survival of living things. Humans have learned to live with variations in water availability in several ways. We can channel the flow of rivers with levees, block the flow of rivers with dams to store water, divert water from rivers and lakes and transport it to distant locations, and even obtain fresh water by removing the salt from salt water. In this section we will look at each of these water distribution methods and examine their costs and benefits.

## Levees

Before the intervention of humans, most rivers periodically overflowed their banks. These occasional overflows of nutrient-rich water made their floodplains particularly fertile. We have seen that throughout history, humans have taken advantage of these fertile floodplains for agriculture. In more recent times, humans have sought ways to prevent flooding so they could develop floodplain land for residential and commercial use. One way to prevent flooding is by constructing a **levee,** an enlarged bank built up on each side of the river. The Mississippi River has the largest system of levees in the world. This river is enclosed by more than 2,400 km (1,500 miles) of levees that offer flood protection to more than 6 million hectares (15 million acres) of floodplains.

The use of levees has produced several major challenges. First, natural floodwaters can no longer add fertility to floodplains by depositing sediments there, so the fertility of these lands is reduced. Second, because the sediments do not leave the river, they are carried farther downstream and settle out where the river enters the ocean. Third, levees may prevent flooding at one location, but by doing so, they force floodwater farther downstream, where it can cause even worse flooding in the next city or town. Finally, the building of levees encourages development in floodplains, but these areas still occasionally flood. This practice raises the question of whether development should be allowed in these areas where flooding remains a high risk.

When floodwaters become too high, levees can either collapse due to the tremendous pressure of the water, or water can come over the top and quickly erode a large hole in the levee. Both events result in massive flooding (**FIGURE 9.9**). One of the most famous levee failures occurred in New Orleans in 2005, as the result of Hurricane Katrina. The levees could not contain the storm surge and heavy rainfall associated with the hurricane. The water overtopped nearly 50 levees, quickly washing out the banks of dirt and flooding many of the neighborhoods that the levees were built to protect. The hurricane and the devastating floodwaters

**FIGURE 9.9** **Levees.** Levees are built to prevent rivers from flowing over their banks and onto the floodplain. In 2008, this levee, just north of St. Louis, Missouri, collapsed and allowed the floodwaters to spread over the surrounding fields.

caused more than 1,800 deaths and more than $80 billion in damage to homes and businesses.

## Dikes

**Dikes** are similar to levees, but they are typically built to prevent ocean waters from flooding adjacent land. Dikes are common in northern Europe, where large areas of farmland lie below sea level. Perhaps the best-known dikes are those of the Netherlands, where dikes have been used for nearly 2,000 years. Approximately 27 percent of the land in the Netherlands is below sea level. The dikes, combined with pumps that move any water that does intrude back out to the ocean, have allowed the country to develop areas that would otherwise be under water. The original pumps were powered by windmills, which is why we often associate the Netherlands with wind energy. Today, however, the water is pumped with electric and diesel pumps.

## Dams

A **dam** is a barrier that runs across a river or stream to control the flow of water. Water is stored behind the dam in a large body of water called a **reservoir.** Dams hold water for a wide variety of purposes, including human consumption, generation of electricity, flood control, and recreation. In 2009, there were 845,000 dams in the world and 82,642 dams in the United States. The largest reservoir system in the United States is along the Missouri River. With six dams and reservoirs in Montana, North Dakota, South Dakota, and Nebraska, this system stores nearly 90 trillion liters (24 trillion gallons) of water.

For centuries, humans have used dams to do work, from turning waterwheels that operated grain mills to powering modern turbines that generate electricity in hydroelectric plants. Although hydroelectric dams (discussed further in Chapter 13) are some of the world's

**FIGURE 9.10** **Dams.** Dams serve a wide variety of purposes, including flood control and electricity generation. The world's largest dam, the Three Gorges Dam on the Yangtze River in China, serves both purposes.

largest, they represent only 3 percent of all dams in the United States. In contrast, 18 percent of U.S. dams were built with the primary purpose of flood control—to reduce or prevent flooding farther down the river. Another 38 percent of U.S. dams were constructed primarily for recreation. Dams have also been built to create scenic lakes for housing developments.

Dams provide substantial benefits to humans, but they come with financial, societal, and environmental costs. The world's largest dam, the Three Gorges Dam built across the Yangtze River in China, illustrates these costs and benefits well. This massive structure, pictured in FIGURE 9.10, is 2 km (1.3 miles) wide, 185 m (610 feet) high, and has created a reservoir 660 km (410 miles) long behind it. The Three Gorges Dam took 13 years to construct, from 1993 to 2006, and it took another 2 years for the reservoir to fill with water. Dam building, like other large-scale construction, uses large amounts of energy and materials and displaces people. The Three Gorges Dam and the reservoir it created flooded 13 cities, 140 towns, and 1,350 villages, forcing more than 1.3 million people to be relocated. The benefits of the dam include the large amount of electricity generated through hydroelectric power rather than through the burning of fossil fuels and the reduction of the seasonal flooding that has plagued cities and villages downstream of the dam, killing more than 1 million people during the past 100 years.

One of the most common environmental problems associated with dams is the interruption of the natural flow of water to which many organisms are adapted. For migrating fish such as salmon, dams represent an insurmountable obstacle to breeding. The loss of these fish can have a cascading effect on other organisms that depend on them, such as bears that feed on migrating salmon. To alleviate this problem, **fish ladders,** which are built like a set of stairs with water flowing over them, have been added to some dams. Migrating fish can swim up the fish ladders and reach their traditional breeding grounds (FIGURE 9.11).

Using dams to prevent seasonal flooding has other consequences for the ecology of an area. Seasonal flooding scours out pools and shorelines, and this disturbance favors their colonization by certain plants and animals.

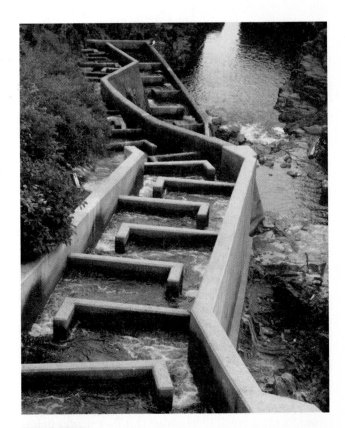

**FIGURE 9.11** **Fish ladder.** Because dams are an impediment to fish such as salmon that migrate upstream to breed, fish ladders like this one in New Brunswick, Canada, have been designed to allow the fish to get around the dam and continue on their traditional path of migration.

Some of the species that are highly dependent on such disturbances are very rare. In recent years, managers have begun to experiment with releasing large amounts of water from reservoirs to simulate seasonal flooding. The results have been very promising. In some areas where dams are no longer needed, they have been removed to restore the natural flow of water, as described in this chapter's opening story. In these cases, much of the natural ecology has been quickly restored.

## Aqueducts

Levees, dams, and dikes are designed to hold water back from its normal course. For centuries, however, people have needed not only to store water, but also to move it. **Aqueducts** are canals or ditches used to carry water from one location to another. Typically, aqueducts remove water from a lake or river and transport this water to where it is needed. Although aqueducts are famously associated with the Roman Empire, their use dates back to at least the seventh century BCE in Greece. The earliest aqueducts were made of limestone, but modern aqueducts include concrete canals and pressurized steel pipes laid above or under the ground. These structures are more efficient water carriers than the ditches and stone-lined canals of historic times. Older aqueducts, many of which are still in use, can lose as much as 55 percent of the water they carry through leakage or evaporation. These losses can be particularly troublesome in water-poor areas such as the arid regions of Israel and Jordan, where water is already scarce.

In the United States, two of the country's largest cities, New York and Los Angeles, depend on aqueducts for their daily water needs. The Catskill Aqueduct brings clean, fresh water over 200 km (120 miles) from the streams and lakes of the Catskill Mountains to the city of New York. The Colorado River Aqueduct is a canal that carries water 400 km (250 miles) from the Colorado River to Los Angeles (**FIGURE 9.12**).

As with many of our answers to natural resource supply challenges, there are trade-offs and environmental costs associated with aqueducts. Although bringing water to cities from pristine areas may ensure its cleanliness, the construction of aqueducts for this purpose not only requires a great deal of money, but also disturbs and fragments natural habitats. Even if an aqueduct is buried as an underground pipeline, a great deal of disruption occurs during the construction process. Moreover, diverting water from a natural river means that there is less water flowing where water has flowed for millennia. Some major rivers in the United States, including the Colorado River and the Rio Grande, now frequently have periods during which so much water is removed from them, at multiple locations, that they go dry before they reach the ocean.

Some water diversion projects have international impacts. India and Bangladesh, for example, share nearly 250 rivers that originate in the Himalayas of India and

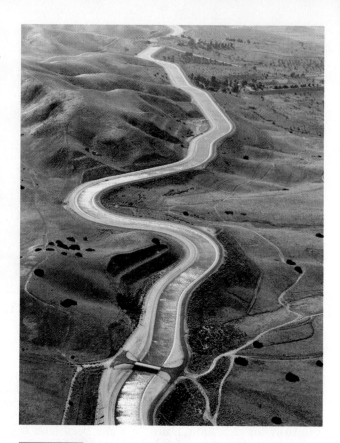

**FIGURE 9.12** **Aqueducts.** Aqueducts deliver water from places where it is abundant to places where it is needed. This section of the Colorado River Aqueduct, which diverts water from the Colorado River to Los Angeles, passes through the Mojave Desert.

flow into Bangladesh. In 2009, India proposed a large-scale water diversion project, involving more than 50 rivers, that has caused Bangladesh some major concerns. By diverting water from these rivers, India stands to gain 170 billion liters (55 billion gallons) of water each year for agriculture and household use. However, Bangladesh is situated downstream of these river diversions. Bangladeshi officials worry that the project will end up dramatically reducing the flow of river water into their country, affecting local fish populations and river navigability for commerce. Even farther downstream, the project may reduce the flow of fresh water into estuaries. When less fresh water enters an estuary, ocean water moves farther up into the river. This influx of seawater raises the salinity of the estuary and harms aquatic life that is not adapted to such salty water.

The most infamous river diversion project happened in the 1950s, when the Soviet Union diverted two rivers that fed the Aral Sea in Central Asia. Diversion of the rivers dramatically decreased freshwater input into the Aral Sea. The salinity of the remaining lake water increased, destroying the local fish populations. Although the Aral Sea was originally the fourth largest lake in the world in terms of surface area, the diversion project reduced its surface area by more than 60 percent,

**FIGURE 9.13 Consequences of river diversion.** Diverting river water can have devastating impacts downstream. The Aral Sea, on the border of Kazakhstan and Uzbekistan, was once the world's fourth largest lake. Since the two rivers that fed the lake were diverted, its surface area has declined by 60 percent, and the lake has become split into two parts: the North and South Aral seas.

as shown in **FIGURE 9.13.** The withdrawals of water caused the Aral Sea to split into two parts: the North Aral Sea and the South Aral Sea. Dust storms have eroded soil, salt, and pesticide residues from the dry lake bed and carried particles of these substances into the atmosphere. The lake's size reduction has also affected the local climate. Without the climate-moderating effect of a large water body, summers in the region are now hotter, and winters are colder. While efforts are being made to save the North Aral Sea, saving the South Aral Sea has been deemed too expensive, and all efforts have been abandoned. The drying of the South Aral Sea continues today: the shallower eastern half of the South Aral Sea is predicted to be completely dry in 10 years.

## Desalination

One of the ways in which some water-poor countries are obtaining fresh water is by removing the salt from salt water, a process called **desalination,** or *desalinization* (**FIGURE 9.14**). The salt water usually comes from the ocean, but it can also come from salty inland lakes. Currently, the countries of the Middle East produce 50 percent of the world's desalinated water.

The two most common desalination technologies are distillation and reverse osmosis. In *distillation* (Figure 9.14a), heat is used to boil water, which leaves its salts behind as it evaporates. The steam that is produced is captured and condensed, yielding pure water. A great deal of energy is required to boil the water and then condense it, so

**(a)** Distillation

① Seawater flows into chamber.

② Heating element boils water, creating steam.

③ Cool seawater in condensing coil causes steam to condense.

④ Salt-free water flows out of chamber.

⑤ Brine (very salty water) flows out of chamber.

**(b)** Reverse osmosis

① Seawater flows into chamber.

② Pressure is applied to the water.

③ Under pressure, water is pushed through a semipermeable membrane but salt is not.

④ Salt-free water flows out of chamber.

⑤ Brine (very salty water) flows out of chamber.

**FIGURE 9.14 Desalination technologies.** Salt water can be converted into fresh water in one of two ways. (a) Distillation uses heat to convert pure water into steam that is later condensed, leaving the salt behind. (b) Reverse osmosis uses pressure to force pure water through a semipermeable membrane, leaving the salt behind.

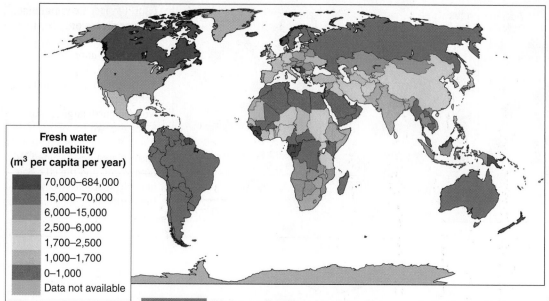

**Fresh water availability (m³ per capita per year)**

- 70,000–684,000
- 15,000–70,000
- 6,000–15,000
- 2,500–6,000
- 1,700–2,500
- 1,000–1,700
- 0–1,000
- Data not available

**FIGURE 9.15** **Water availability per capita** The amount of water available per person varies tremendously around the world. North Africa and the Middle East are the regions with the lowest amounts of available fresh water.

distillation can be a monetarily and environmentally expensive process.

In *reverse osmosis* (Figure 9.14b), water is forced through a thin semipermeable membrane at high pressure. Water can pass through the membrane, but salt cannot. Reverse osmosis is a newer technology that is more efficient and often less costly than distillation. However, the liquid that remains, called brine, has a very high salt concentration. It cannot be deposited on land because its high salt content would contaminate the soil, potentially harming plant and animal life. If dumped into a bay or coastal area, it can harm fish and other aquatic life. Typically, brine is returned to the open ocean, though its high salt concentrations can still cause localized harm to ocean life.

Ultimately, all water management systems require a large investment to build, maintain, and repair. This means that many water-poor countries are not able to implement large-scale methods of water management to provide for their needs. Water availability varies greatly around the world, as **FIGURE 9.15** shows. The Middle Eastern and North African countries represent about 5 percent of the world's population, yet they have less than 1 percent of the fresh water available for drinking. The United Nations estimates that nearly 1.2 billion people live in regions that have a scarcity of water.

### GAUGE YOUR PROGRESS

✓ How do levees, dikes, dams, and aqueducts differ from one another? What is the primary purpose of each?

✓ Why is it necessary to desalinize water?

# Water is used by humans for agriculture, industry, and household needs

A human can survive without food for 3 weeks or more, but cannot survive without water for more than a few days. In addition, water is essential for producing food. In fact, 70 percent of the world's freshwater use is for agriculture. The remaining 30 percent is split between industrial and household uses, the proportion of which varies from country to country. On average, experts estimate that about 20 percent of the world's freshwater use is for industry and about 10 percent is for household uses.

As we have seen with so many other resources, the daily use of water per capita varies dramatically among the nations of the world. **FIGURE 9.16** shows total daily per capita use of fresh water for a number of countries.

## Agriculture

As we have just noted, the largest use of water worldwide is for agriculture. During the last 50 years, as agricultural output has kept up with human population growth, the amount of water used for irrigation throughout the world has more than doubled. Indeed, producing a metric ton of grain (1,000 kg, or 2,200 pounds) requires more than 1 million liters of water (264,000 gallons). Together, India, China, the United States, and Pakistan account for more than half the irrigated land in the world. In the United States, approximately one-third of all freshwater use is for irrigation. As we saw in

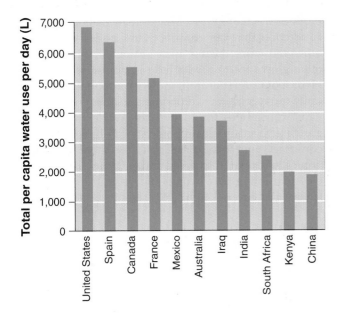

FIGURE 9.16 **Total per capita water use per day.** The total water use per person for agriculture, industry, and households varies tremendously by country.

Chapter 8 (see "Measure Your Impact" on page 233), raising livestock for meat also requires vast quantities of water. For example, producing 1 kg (2.2 pounds) of beef in the United States requires about 11 times more water than producing 1 kg of wheat.

**IRRIGATION** Since agriculture is the greatest consumer of fresh water throughout the world, agriculture presents the greatest potential for conserving water by changing irrigation practices. There are four major techniques for irrigating crops: furrow irrigation, flood irrigation, spray irrigation, and drip irrigation (**FIGURE 9.17**).

The oldest technique is *furrow irrigation* (Figure 9.17a), which is easy and inexpensive. The farmer digs trenches, or furrows, along the crop rows and fills them with water, which seeps into the ground and provides moisture to plant roots. Furrow irrigation is about 65 percent efficient, meaning that 65 percent of the water is accessible to the plants; the other 35 percent either runs off the field or evaporates.

A second technique is *flood irrigation* (Figure 9.17b), which involves flooding an entire field with water

**(a)** Furrow irrigation

**(b)** Flood irrigation

**(c)** Spray irrigation

**(d)** Drip irrigation

FIGURE 9.17 **Irrigation techniques.** Several techniques are used for irrigating agricultural crops, each with its own set of costs and benefits.

and letting the water soak in evenly. This technique is generally more disruptive to plant growth than furrow irrigation, but is also slightly more efficient, ranging from 70 to 80 percent efficiency.

A third technique is *spray irrigation* (Figure 9.17c), which is more expensive than furrow or flood irrigation and also uses a fair amount of energy. Water is pumped from a well into an apparatus that contains a series of spray nozzles that spray water across the field, like giant lawn sprinklers. The advantage of spray irrigation is that it is 75 to 95 percent efficient.

The fourth technique is *drip irrigation* (Figure 9.17d), which uses a slowly dripping hose that is either laid on the ground or buried beneath the soil. Drip irrigation using buried hoses is over 95 percent efficient. It has the added benefit of reducing weed growth because the surface soil remains dry, which discourages weed germination. Drip irrigation systems are particularly useful in fields containing perennial crops such as orchard trees, where the hoses do not have to be moved for annual plowing of the field.

Efficient irrigation technology benefits the environment by reducing water consumption and by reducing the amount of energy needed to deliver the water. However, as with all human activities, the trade-offs involved in each irrigation technique (water costs, energy costs, and equipment costs) need to be weighed to determine the best solution for each situation.

**HYDROPONIC AGRICULTURE** For some crops, *hydroponic agriculture* is an alternative to traditional irrigation. **Hydroponic agriculture** is the cultivation of crop plants under greenhouse conditions with their roots immersed in a nutrient-rich solution, but no soil (**FIGURE 9.18**). Water not taken up by the plants can be reused, so this method uses up to 95 percent less water than traditional irrigation techniques. Hydroponic operations can

produce more crops per hectare than traditional farms. In addition, they can grow crops under ideal conditions, grow the crops during every season of the year, and often grow the crops with little or no use of pesticides. Hydroponic agriculture is growing in popularity, especially for vegetables; hydroponic tomatoes, for example, have won awards for their flavor. While the costs of growing vegetables hydroponically can be a bit higher, consumers appear to be willing to pay more for hydroponically grown vegetables. Indeed, one company in Georgia now grows hydroponic vegetables in a huge facility that covers 129 ha (318 acres).

## Industry

Water is required for many industrial processes, such as generating electricity, cooling machinery, and refining metals and paper.

In the United States, approximately one-half of all water used goes toward generating electricity. It is important to distinguish between water that is withdrawn from and returned to its source and water that is withdrawn and consumed. For example, water that passes through a turbine at a hydroelectric dam is withdrawn from a reservoir to generate electricity, but that water is not consumed. It passes from the reservoir through the turbine and back to the river flowing away from the dam. That is not to say that this process has no effect on the environment. We discussed the ecological impacts of damming rivers earlier in this chapter. In addition, passing millions of gallons of water through a spinning turbine kills millions of fish each year.

Some processes that generate electricity do consume water. This means that some of the water used is not returned to the source from which it was removed, but instead enters the atmosphere as water vapor. Thermoelectric power plants, including the many plants that generate heat using coal or nuclear reactors, are large consumers of water. These plants use heat to convert water into steam, which is then used to turn turbines. The steam needs to be cooled and condensed before it can be returned to the water's source. In many plants, this cooling is accomplished using massive cooling towers. If you have ever seen a nuclear power plant, even from a distance, you may have seen the large plumes of water vapor rising up for thousands of meters from the cooling towers. The towers allow much of the steam from the plant to condense and cool into liquid water, but a large fraction of it is lost to the atmosphere. This water vapor represents the water that is consumed by nuclear reactors (**FIGURE 9.19**).

The refining of metals and paper also requires large amounts of water. Copper, used extensively for electrical wiring, requires 440 L (116 gallons) of water per kilogram to refine. Aluminum, used in products as diverse as automobiles and aluminum foil for cooking, requires 410 L (108 gallons) per kilogram. Steel, used to manufacture home appliances, cars, buildings, and other prod-

**FIGURE 9.18** **Hydroponic agriculture.** These tomatoes are growing in a greenhouse with their roots immersed in a solution of water and nutrients. By recycling the water, hydroponic operations can use up to 95 percent less water than traditional farms.

**FIGURE 9.19** Water consumption by a nuclear power plant. When nuclear reactors, such as this one in Germany, heat water to make the steam that turns electrical turbines, the steam must be cooled back down. During the cooling of the steam, a great deal of water vapor is released to the atmosphere. This water vapor represents consumption of water by the plant.

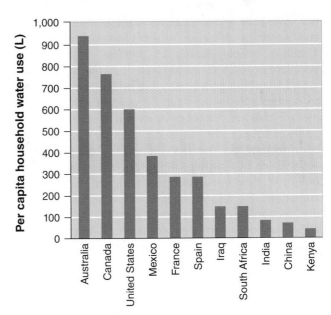

**FIGURE 9.20** Household per capita water use per day. When only household water use, rather than total water use (as in Figure 9.16), is considered, the ranking of many countries changes. [Data from A. K. Chapagain and A. Y. Hoekstra, *Water Footprints of Nations,* Vol. 1, Main Report, UNESCO-IHE. Research Report Series #16, 2004.]

ucts, requires 260 L (68 gallons) per kilogram. When we use paper, we indirectly use water. A kilogram of paper requires 125 L (33 gallons) of water to manufacture. All of this means that a great deal of water is used in industry. As we will see later in this chapter, there are opportunities to reduce the use of water during these processes.

## Households

According to the U.S. Geological Survey, approximately 10 percent of all water used in the United States is used in homes. The quantity of water used in households depends greatly on the types of infrastructure available in a country. Households in less developed countries generally do not have the appliances and bathroom fixtures that are common in more developed countries. As a result, per capita household water use varies dramatically among nations. For example, on average, an individual in the United States uses 595 L (157 gallons) per day, whereas an average individual in Kenya uses only 41 L (11 gallons) per day. FIGURE 9.20 shows per capita daily household water use for the United States and 10 other countries.

Indoor use of water is quite similar across the United States, as households across the country are typically equipped with bathrooms, washing machines, and cooking appliances. FIGURE 9.21 shows the fraction of indoor water use that goes to each of these functions. Of all household water used indoors, 41 percent is used for flushing toilets, 33 percent for bathing, 21 percent for laundry, and 5 percent for cooking and drinking.

Outdoor water use—for watering lawns, washing cars, and filling swimming pools—varies tremendously across the United States by region. In California, for example, the typical household uses about six times more water outdoors than the typical Pennsylvania family.

Although drinking water represents a relatively small percentage of household water use, it is a particularly important one. If you live in a developed country, you

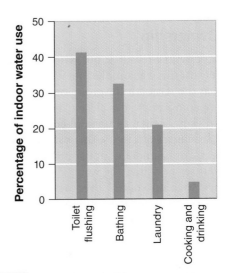

**FIGURE 9.21** Indoor household water use. Most water used indoors is used in the bathroom. [Data from U.S. Environmental Protection Agency, 2003. http://esa21.kennesaw.edu/activities /water-use-overview-epa.pdf.]

may not have given much thought to the availability and safety of drinking water. However, more than 1 billion people—nearly 15 percent of the world's population—lack access to clean drinking water. Every year, 1.8 million people die of diarrheal diseases related to contaminated water, and 90 percent of those people are children under 5 years old. This means that 5,000 people in the world die each day in large part because they do not have access to clean water. As we will see in Chapter 14, in which we discuss water pollution, developed countries typically have modern sanitation systems, stronger environmental laws, and the technologies to remove harmful contaminants and waterborne pathogens from public water supplies. Poor conditions in many developing countries do not provide safeguards that ensure the availability of clean drinking water.

---

**GAUGE YOUR PROGRESS**

✓ What are the dominant uses of water by humans?

✓ How do different irrigation methods influence water use?

✓ Why does it require so much more water to produce 1 kg of beef than 1 kg of grain?

---

# The future of water availability depends on many factors

The future of water availability will depend on many things, including how we resolve issues of water ownership and how we improve water conservation and develop new water-saving technologies as the world's population grows.

## Water Ownership

Water is an essential resource, but who actually owns it? It turns out that this is a complex question. In a particular area, such as the Klamath River region described in this chapter's opening story, it is clear that multiple interest groups can claim a right to use and consume the water. However, having a right to use the water is not the same as owning it. For example, regional and national governments often set priorities for water distribution, but of course they have no control over whether a particular year will bring an abundance of rain and snow. In California, the state government promises specific amounts of water for cities, suburbs, farmers, and fish, but collectively, these promises far exceed the actual amount of water that is available in many years.

Throughout the world, the issues of water rights and ownership have created many conflicts. Earlier in this chapter we discussed India's plan to divert water from rivers that flow from the Himalayas into Bangladesh

(FIGURE 9.22). Since the water originates in India, does India own the water? Does Bangladesh have any legitimate claim to some of this water that its people have relied upon for millennia? In the water-poor Middle East, water rights have long contributed to political tensions. In both the 1967 Arab-Israeli War and the 1980s Iran-Iraq War, disputes over water use added to the conflict. Water experts predict that as populations in these arid regions continue to grow, conflicts over water will increase.

One solution that has been proposed by economists is to allow all interested parties to openly compete for water and let market forces determine its price. In this way, they argue, water could be owned, but the true value of water would be realized and paid for. In 1981, a free-market system of water distribution was initiated in Chile. While water is public property, Chile allows individuals and corporations to buy and sell water. The idea behind this approach is that having to buy water will encourage more efficient use. While market forces can be useful in determining the appropriate distribution of water among competing needs, government oversight helps to ensure that the needs of the people and the environment are balanced with the needs of private corporations. As we will see in the Science Applied section that follows this chapter ("Is There a Way to Resolve the California Water Wars?"), this more balanced approach is now being attempted.

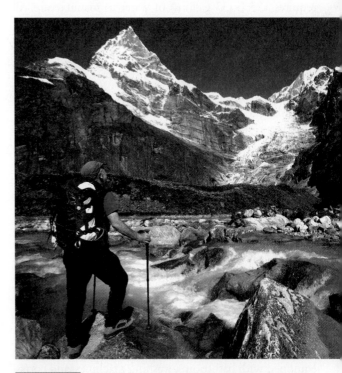

**FIGURE 9.22 A Himalayan river.** The government of India is proposing to divert water from some of the rivers flowing out of the Himalayas to provide more water for its citizens. However, rivers, such as this one flowing down from Mera Peak in Nepal, subsequently flow into Bangladesh, raising interesting questions regarding who owns the water.

### Selecting the Most Cost-Efficient Washing Machine

Suppose you move into a new house and need to purchase a washing machine. You've heard that the new front-loading machines use less water, but that they are more expensive. A traditional washing machine costs $400, whereas the water-efficient washing machine costs $800. How do you choose?

1. If the average family washes 5 loads of laundry per week, how many loads does the family wash per year?

   5 loads/week × 52 weeks/year = 260 loads/year

2. The traditional machine uses 150 L of water per load of laundry, and the efficient machine uses 75 L. How many liters would the efficient washing machine save per year?

   (150 L − 75 L)/load × 260 loads/year = 19,500 L/year

3. The average cost of water in the United States is $0.40 for every 1,000 L. How much money would you save each year by using the more efficient washing machine?

   $0.40/1,000 L × 19,500 L/year = $7.80/year

**Your Turn:** How many years would you need to operate the more efficient washing machine before you began to save money compared to operating a cheaper, traditional machine? (Note that this problem considers only the cost of purchasing the water. However, you must also consider the cost of heating the water. If half of the loads are washed with hot or warm water, for example, the payback time is reduced.)

## Water Conservation

Ultimately, we all have to share a finite amount of water. In some regions, such as much of the northeastern United States, water is abundant. In other regions, where water is scarcer, water conservation is a necessity.

In recent years, many developed countries have begun to find ways to use water more efficiently through technological improvements in water fixtures, faucets, and washing machines. In 1994, new federal standards were issued for toilets and showerheads in the United States. For example, a toilet manufactured before 1994 typically uses 27 L (7 gallons) per flush, but toilets sold after January 1994 must use 6 L (1.6 gallons) or less per flush, representing a 78 percent reduction in water use. Australia and some countries in Europe and Asia have moved to dual-flush toilets. First invented by an Australian company in 1980, this type of toilet allows the user to push one button for a normal 6 L flush to remove solid waste, but another button produces a much more efficient 3 L (0.8-gallon) flush to remove liquid waste. Although dual-flush toilets have not yet been accepted by consumers in the United States, consumers have embraced improved efficiencies in showerheads and washing machines. For example, a 10-minute shower with an older showerhead might use 150 L (40 gallons) of water, but revised federal standards for new reduced-flow showerheads call for a 10-minute shower to use no more than 95 L (25 gallons)—a 37 percent reduction in water use. Washing machines are not subject to the new federal standards, but newer, more efficient front-loading washing machines are now available to consumers. However, these more efficient machines also cost nearly twice as much to purchase. Crunch the Numbers "Selecting the Most Cost-Efficient Washing Machine" looks at the costs and benefits of these newer, more energy-efficient washing machines.

In countries where a percentage of the population is particularly wealthy, household water uses include watering lawns and filling swimming pools, both of which require large amounts of water. In some regions of the United States, homeowners have been encouraged, or even required, to plant vegetation that is appropriate to the local habitat. For example, the city of Las Vegas, Nevada, paid homeowners to remove water-intensive turf grass from their lawns and replace it with more water-efficient native landscaping. Vegetation substitution can result in a savings of 2,000 L of water per square meter (520 gallons per 10 square feet) of lawn per year (FIGURE 9.23).

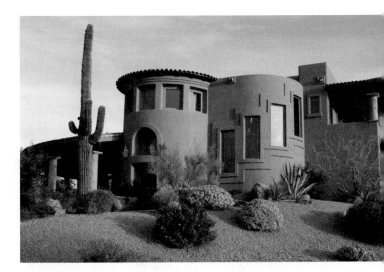

**FIGURE 9.23** **Landscaping in the desert.** By using plants that are adapted to a desert environment, homeowners in Arizona can greatly reduce the need for irrigation, resulting in a considerable reduction in water use compared with growing grass.

One of the best ways to reduce water use and consumption is to produce more efficient manufacturing equipment. In the United States, businesses and factories have achieved more sustainable water use in the last 15 years, mainly through the introduction of equipment that either uses less water or reuses water. Industries that need cooling water for machinery, for example, have converted from once-through systems (systems that bring in water for cooling and pump the heated water back into the environment) to recirculating water systems.

There are some simple ways to conserve water that can be used throughout the world. For example, the impervious surfaces of buildings represent a substantial water collecting surface. A simple gutter system can be used to collect rainwater and channel it into rain barrels or, for greater capacity, a large underground water tank. As we will see in Chapter 14, some countries are now using wastewater for irrigation after sending it through a sewage treatment process.

The world's growing population and the associated expansion of irrigated agriculture have increased global water withdrawals more than fivefold in the last hundred years. Global water use is expected to continue to grow with the human population through the early part of this century. But, as **FIGURE 9.24** shows, despite the United States' growing population, water withdrawals in the United States have leveled off since they peaked in 1980. This is largely a result of greater water use efficiency in agricultural irrigation, electricity generation, and household use. Reductions in water use are projected to continue until at least 2020, when all the estimated gains from water-saving devices and technologies will begin to level off unless newer, even more efficient technologies are developed.

**FIGURE 9.24** Water withdrawals in the United States from 1950 to 2005. Although the U.S. population continues to increase, the country's water use has leveled off due to the use of water-efficient technologies. [After U.S. Geological Survey, 2009.]

### GAUGE YOUR PROGRESS

✓ Why is water ownership a complex issue?

✓ What are some of the ways that humans can conserve water?

✓ How does economic development influence water use?

##  WORKING TOWARD SUSTAINABILITY

### Is the Water in Your Toilet Too Clean?

In certain parts of the world, such as the United States, sanitation regulations impose such high standards on household wastewater that we classify relatively clean water from bathtubs and washing machines as contaminated. This water must then be treated as sewage. We also use clean, drinkable water to flush our toilets and water our lawns. Can we combine these two observations to come up with a way to save water? One idea that is gaining popularity throughout the developed world is to reuse some of the water we normally discard as waste.

This idea has led creative homeowners and plumbers to identify two categories of wastewater in the home: gray water and contaminated water. **Gray water** is defined as the wastewater from baths, showers, bathroom sinks, and washing machines. Although no one would want to drink it, gray water is perfectly suitable for watering lawns and plants, washing cars, and flushing toilets. In contrast, water from toilets, kitchen sinks, and dishwashers contains a good deal of waste and contaminants and should therefore be disposed of in the usual fashion.

Around the world, there are a growing number of commercial and homemade systems in use for storing gray water to flush toilets and water lawns or gardens. For example, a Turkish inventor has designed a household system allowing the homeowner to pipe wastewater from the washing machine to a storage tank that dispenses this gray water into the toilet bowl with each flush (FIGURE 9.25).

**FIGURE 9.25** **Reusing gray water.** A Turkish inventor has designed a washing machine that pipes the relatively clean water left over from a washing machine, termed gray water, to a toilet, where it can be reused for flushing. Such technologies can reduce the amount of drinkable water used and the volume of water going into sewage treatment plants.

Many cities in Australia have considered the use of gray water as a way to reduce withdrawals of fresh water and reduce the volume of contaminated water that requires treatment. The city of Sydney estimates that 70 percent of the water withdrawn in the greater metropolitan area is used in households, and that perhaps 60 percent of that water becomes gray water. The Sydney Water utility company estimates that the use of gray water for outdoor purposes could save up to 50,000 L (13,000 gallons) per household per year.

Unfortunately, many local and state regulations in the United States and around the world do not allow use of gray water. Some localities allow the use of gray water only if it is treated, filtered, or delivered to lawns and gardens through underground drip irrigation systems to avoid potential bacterial contamination. Arizona, a state in the arid Southwest, has some of the least restrictive regulations. As long as a number of guidelines are followed, homeowners are permitted to reuse gray water. In 2009, in the face of a severe water shortage, California reversed earlier restrictions on gray water use and agreed to allow gray water to be used for irrigating lawns and trees. Given that the typical household produces 227,000 L (60,000 gallons) of gray water per year, using gray water for irrigation presents a major opportunity for water conservation throughout the world.

### References

Gelt, J. Home use of graywater, rainwater conserves water—and may save money. University of Arizona Water Research Resources Center. http://ag.arizona.edu/AZWATER/arroyo/071rain.html.

Oasis Design. Grey Water Policy Center. http://www.oasisdesign.net/greywater/law.

---

## REVISIT THE KEY IDEAS

- **Identify Earth's natural sources of water.**

Most water on Earth resides in the oceans. Of the relatively small proportion that is fresh water, nearly three-fourths is tied up as ice and glaciers, leaving a small amount remaining in groundwater, streams, rivers, lakes, and wetlands. All of these sources of fresh water can be used by humans. Atmospheric water is an additional source of water, but its availability may vary seasonally as well as from year to year. Human activities can contribute to the negative effects of drought and flooding.

- **Discuss the ways in which humans manage water distribution.**

Humans have created a variety of ways to store and divert water, including levees, dikes, dams, and aqueducts. Each of these water distribution technologies has important benefits, but can also have negative environmental impacts. Humans have also developed technologies for the desalination of salt water.

- **Describe the major human uses of water.**

Water is used in agriculture, industry, and households. Agricultural uses of water include several different methods of irrigation as well as the developing field of hydroponic agriculture. Industrial uses of water include the generation of electricity, the refining of metals and paper, and the cooling of machinery. In households, water is used primarily in bathrooms and for washing clothes. Per capita water use varies tremendously by country. Developed countries tend to use more water than developing countries, where many people have access to only a few liters of water per day.

▪ **Identify the factors that will affect the future availability of water.**

The future of water availability depends on water ownership, water conservation, economic development, and global change. Water ownership is a highly complex issue that involves the market value of water and our need to ensure that adequate supplies are available. Water conservation efforts include improvements in agricultural irrigation techniques, the increased use of recycled water in industrial processes, more efficient household appliances, planting less water-demanding landscapes, and simple water collection devices that collect rainwater and allow recovery and reuse of gray water.

## CHECK YOUR UNDERSTANDING

1. What percentage of Earth's water is fresh water?
   (a) 3 percent
   (b) 10 percent
   (c) 50 percent
   (d) 90 percent
   (e) 97 percent

2. Which of the following contrasts between confined and unconfined aquifers is *correct*?
   (a) Confined aquifers are more rapidly recharged.
   (b) Only confined aquifers can produce artesian wells.
   (c) Only unconfined aquifers are overlain by a layer of impermeable rock.
   (d) Only unconfined aquifers have a water table above them.
   (e) Only unconfined aquifers can be drilled for wells to extract water.

3. Which of the following statements about surface waters is *not* correct?
   (a) Historically, most rivers regularly spilled over their banks.
   (b) The largest river in the world is the Amazon River.
   (c) Levees are used to make reservoirs.
   (d) Dikes are human-made structures that keep ocean water from moving inland.
   (e) Wetlands play an important role in reducing the likelihood of flooding.

4. Which of the following statements about dams is *not* correct?
   (a) Dams are used to reduce the risk of flooding.
   (b) Dams can cause increased water temperatures.
   (c) The water held back by a dam is called a reservoir.
   (d) Fish ladders allow migrating fish to move past dams.
   (e) Most dams are built to generate electricity.

5. Which of the following statements about aqueducts is *correct*?
   (a) Aqueducts designed as open canals can lose a lot of water via evaporation.
   (b) Aqueducts are a modern invention.
   (c) Aqueducts do not divert water from lakes.
   (d) Aqueducts do not affect the amount of water remaining in rivers.
   (e) Aqueducts move water from locations where the demand for water is high.

6. Which of the following statements about desalination is *correct*?
   (a) Distillation requires more energy than reverse osmosis.
   (b) The brine left over from desalination is not harmful when returned to the ocean.
   (c) Large-scale desalination of water is affordable to all nations.
   (d) Desalination of ocean water is not yet a feasible endeavor.
   (e) Most desalination occurs in the Middle East.

7. Which of the following lists of household water uses is in the correct order, from highest to lowest?
   (a) Toilet > laundry > cooking and drinking > bathing
   (b) Toilet > bathing > laundry > cooking and drinking
   (c) Laundry > bathing > toilet > cooking and drinking
   (d) Bathing > toilet > laundry > cooking and drinking
   (e) Bathing > toilet > cooking and drinking > laundry

8. Which of the following lists of agricultural irrigation techniques is in the correct order, from least efficient to most efficient?
   (a) Drip irrigation, furrow irrigation, flood irrigation, spray irrigation
   (b) Spray irrigation, furrow irrigation, flood irrigation, drip irrigation
   (c) Furrow irrigation, flood irrigation, spray irrigation, drip irrigation
   (d) Furrow irrigation, spray irrigation, drip irrigation, flood irrigation
   (e) Furrow irrigation, flood irrigation, drip irrigation, spray irrigation

9. Which of the following statements about the industrial use of water is *not* correct?
   (a) It is used to refine metals.
   (b) It is used to create steam.
   (c) It is important in generating electricity.
   (d) It plays a role in making paper products.
   (e) Its use is becoming less efficient.

10. Which of the following is *not* a water conservation technique?
    (a) Flood irrigation
    (b) Reduced-flow showerheads
    (c) Recycling of industrial water
    (d) Dual-flush toilets
    (e) Front-loading washing machines

## APPLY THE CONCEPTS

1. An important source of fresh water is groundwater. Answer the following questions about human use of groundwater.
   (a) Using a cross section of the ground, draw the water table, an unconfined aquifer, and a confined aquifer.
   (b) Explain the factors that create an artesian well.
   (c) Describe the concerns that scientists have regarding the pumping of water from confined aquifers.
   (d) Explain how cones of depression near the coasts of continents can lead to saltwater intrusion of water wells.

2. Answer the following questions about the world's approximately 845,000 dams.
   (a) What are the benefits that dams provide to humans?
   (b) What are the costs of dams to human society?
   (c) What are the negative impacts of dams on the environment?
   (d) If dams have both costs and benefits, how should we decide when a dam should be built?

## MEASURE YOUR IMPACT

**Saving Water** The need to conserve water is widespread throughout the world. People who live in developed countries such as the United States can make some simple changes to conserve water.
   (a) An older showerhead uses 150 L for a 10-minute shower, whereas a reduced-flow showerhead uses only 95 L. If you take one shower per day, how many liters of water would you save in 1 year if you replaced your older showerhead?
   (b) An older toilet uses 27 L per flush, whereas a replacement toilet uses only 6 L per flush. If you flush the toilet four times per day, how many liters of water would you save in 1 year if you replaced your older toilet?
   (c) Now let's assume that you have a newer toilet that uses 6 L per flush, but you are considering upgrading to a dual-flush toilet. If you flush the toilet four times per day, but three of those times you could use the 3 L flush, how many liters of water would you save in 1 year?
   (d) If you made all three of the changes described in questions a–c, how much water, in total, would you save from going down the drain?
   (e) If you lived in a family of four people, and you paid $1 for every 1,000 L of water your family used, how much money, in total, could your family save in 1 year with these water conservation improvements?

## ENGAGE YOUR ENVIRONMENT

How can you enjoy the convenience of safe, fresh water while minimizing the cost and plastic waste that comes with relying solely on bottled water? Examine your personal use of bottled water and propose ways to reduce the impact of bottled water consumption on your college campus. *See Engage Your Environment Activity 9 at the end of this book.*

# science applied

## Is There a Way to Resolve the California Water Wars?

Pick up any tomato, broccoli spear, or carrot in the grocery store, and chances are it came from California. California's $37 billion agricultural industry is more than twice the size of the agricultural industry of any other state, and it accounts for nearly one-half of all the fruits, nuts, and vegetables grown in the United States. California's temperatures are ideal for many crops, and its mild climate permits a long growing season. But water is a scarce resource there. Most of California's growing regions lack the necessary water for farming. To remedy this situation, the state has developed an extensive irrigation system that moves water from areas where it is abundant to other areas where it is scarce.

Although agriculture in California has been an economic success, it has increased competition among the state's residents for water. Nearly two-thirds of those residents live in southern California, where residential and business demand for water has long exceeded the locally available supply. Los Angeles, for example, has been importing water from other areas of the state for more than 100 years. As the state's population continues to grow at about 1 percent per year, demand for water will also grow.

Wildlife also depends on a readily available supply of fresh water. In many regions of California, diverting water away from streams and rivers threatens the habitats of fish and other aquatic organisms, including several endangered species. Salmon, for instance, depend on the presence of cold, clean water for laying their eggs. In the delta region where the San Joaquin and Sacramento rivers come together just east of San Francisco, so much water has been pumped out to be sent to southern California that the fish have experienced dramatic declines.

### What historical factors have led to California's water wars?

At 144 trillion liters (38 trillion gallons) of fresh water per day, California's water use is more than twice that of any other state. The vast majority of that water is used in the Central Valley region, which produces more than 80 percent of California's farm income. In 1935, the U.S. Department of Reclamation began an immense endeavor called the Central Valley Project, which diverts 8.6 trillion liters (2.3 trillion gallons) of water per year from rivers and lakes throughout the state for both agricultural and municipal uses (FIGURE SA4.1).

While the project made agriculture possible in many of southern California's desert regions, it also locked into place long-term contracts that have subsidized several decades of cheap water for farmers, who pay rates that are only 5 to 10 percent of those paid by southern California towns and cities. Interestingly, northern California also has farms, but much of its abundant water is sent to southern California farms and municipalities.

In 2005, the federal government renewed nearly 200 large water supply contracts with farmers. These contracts were for 25 years of water, with an option available to extend them another 25 years, at prices so low that the payments did not even cover the cost of the electricity used to transport the water. Organizations ranging from the Cato Institute to the Natural Resources Defense Council protested this decision. The controversy raised the question of whether farmers should be given cheap access to water to protect

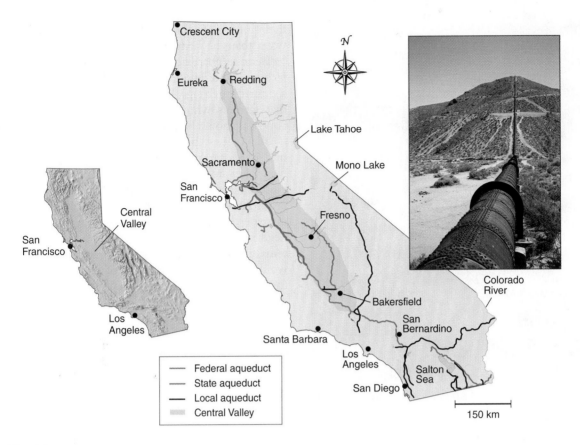

**FIGURE SA4.1** **Water distribution in California.** To supply water to the agricultural areas in the Central Valley and to supply drinking water to municipalities, the state has developed an elaborate system of aqueducts. These aqueducts can be open canals or pipes laid above or below the ground.

In the map legend:
— Federal aqueduct
— State aqueduct
— Local aqueduct
Central Valley

the existing agricultural industry, given that agriculture consumes the majority of water used in the state, or be required to pay the same price for water as everyone else.

## How might farmers respond to increased water prices?

What would happen if the federal government raised the price of water for California farmers? If water were more expensive, farmers would face four choices: raise the prices of their crops, switch to less water-intensive crops, use more efficient irrigation methods, or sell their farms. Each of these options would have different effects on the state's environment and economy.

One obvious way to offset higher water prices would be for farmers to charge more for the produce they grow. For example, if California broccoli growers had to pay the same price for water as municipalities, that cost would add about 13 cents to the price of broccoli per pound—an increase of less than 10 percent. This action would not change the environmental threat of water depletion, but would have a small negative effect on the consumer's wallet.

Other farmers faced with higher water prices might decide to switch to crops that require less water. For example, alfalfa, which uses 25 percent of California's irrigation water and has a low resale value compared with other crops, would no longer be a profitable crop. A farm using 296 million liters (78 million gallons) of water can produce about $60,000 worth of alfalfa per year, according to the Natural Resources Defense Council. If farmers had to purchase water at the unsubsidized rate, that water would cost them $168,000; if the alfalfa continued to sell for only $60,000, the farmers would lose more than $100,000 a year growing alfalfa. Tomatoes, broccoli, and potatoes use about half as much water per acre as alfalfa and have a higher resale value (FIGURE SA4.2). Just by replacing alfalfa with these crops, California would reduce its statewide water use by 20 percent.

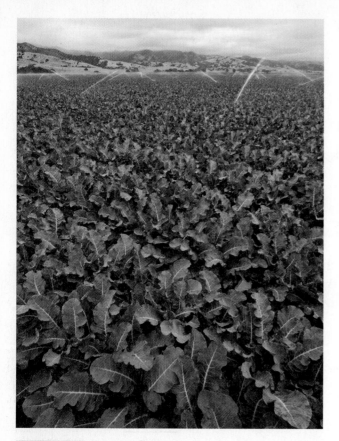

**FIGURE SA4.2 Irrigating broccoli.** By growing crops such as potatoes, tomatoes, and broccoli that use less water and have a higher resale value than crops such as alfalfa, less water would be needed for California agriculture.

This action would have a positive environmental impact, since less overall water would be needed for agriculture in the state.

Investment in new irrigation technology that uses less water would also reduce a farmer's costs. As we saw in Chapter 8, the four major irrigation techniques differ in cost and efficiency. If farmers had to purchase water at the unsubsidized rate, converting to more efficient irrigation techniques could rapidly pay for itself, as the initial investment in equipment would be offset by savings on water. As long as the cost of water continues to be subsidized by the government, farmers have little incentive to upgrade their irrigation technology.

If water prices rise, not all farmers will be able to manage the new costs. This recognition brings us to the final option: some farmers could be driven out of business and sell their land. This outcome would probably reduce the amount of food grown in the region. Whether conversion from farming to other uses would reduce water demand would depend on how the land was used. For example, if it became a housing development in which residents practiced water conservation, then the residents would use much less water per acre

than the farmers had. However, if the residents chose to plant and water large lawns and were not careful about their indoor use of water, they could use even more water than the farmers.

Because charging more for water could force some farmers out of business, many politicians are reluctant to support this change. In response, some economists argue that farmers who possess long-term subsidized water contracts should be allowed to sell their water allotment to the highest bidder, something that is not allowed under the current law. In this scenario, a farmer could buy water from the government at the contracted low cost and then sell that water to a city at a substantial profit rather than using it on the farm. Farmers would increase their income, cities would get the water they need without taking additional water from natural systems, and the government agencies that sell the water would get the same amount of money that they do today. Many critics argue, however, that the government should not be giving preferential water rates to a few users who can then make money on the water.

## How have the water wars played out in the political arena?

The California water wars have developed because of the large number of competing interests involved, including northern California farmers and municipalities that are losing water, southern California farmers and municipalities that need more water, endangered species that are declining due to the diversion of water, and beautiful natural areas that have experienced reduced water capacity, increased salinity, or gone completely dry.

Competition for California's water has been marked by political maneuvering and numerous lawsuits. In the 1960s, for example, there was a large push to build dams that would create reservoirs to collect water. Today there are about 1,200 dams in the state. Because of the money and time needed to construct new dams, many Californians do not consider more dams to be a viable solution to today's water problems.

The past few years have witnessed a number of lawsuits that have produced very interesting court decisions. For example, the Natural Resources Defense Council won a lawsuit regarding the 2005 water contracts for farmers in the eastern San Joaquin Valley, who use water diverted from the San Joaquin River. The judge ruled that the contracts were illegal because they did not consider the potential impact of diverting water on the endangered species that rely on the river.

In 2007, a federal court was asked to consider the impact that the large, powerful pumps used to divert water from the San Joaquin and Sacramento delta region were having on local fish populations—particularly on the delta smelt (*Hypomesus transpacificus*), a federally protected fish whose population in the delta region was declining. The court ruled that pumping from the delta region—the source of much of the state's fresh water—must be reduced. This decision came during a multi-year drought, which made its impact on southern California even larger than it would have been otherwise. A number of California congressmen accused the state and federal governments of caring more about fish than people.

The problem of water scarcity in California rose to a crisis level in June 2008, when Governor Arnold Schwarzenegger declared that the state was officially in a drought. As a result of a lack of rain and a reduced amount of snow in the Sierra Nevada, the spring of 2008 was the driest in 88 years. The California legislature passed a bill to cut water use by 20 percent. But the legislature put most of the responsibility for reducing water use on residential users, despite the fact that agriculture uses a much larger portion of the state's water. Some municipalities started rationing water, and a number of municipalities began enforcing a 2001 state law that requires new building developments to show plans that ensure a 20-year water supply.

As the drought continued into 2009, additional ideas to promote water conservation were proposed. In June 2009, for example, the Los Angeles Department of Water and Power announced that it would credit customers with $11 per square meter of turfgrass replaced with drought-resistant plants or mulch.

In November 2009, the California legislature passed a series of bills to address the water crisis. The $40 billion package called for the restoration of the San Joaquin and Sacramento delta ecosystem as well as the building of new dams and a series of water conservation efforts to reduce water demand. A few days later, the governor signed the bill into law.

## What can be done?

Given the scarcity of water in California, it stands to reason that its sustainable use is an important topic with many environmental, economic, and even political implications. Economists generally argue that all users should be charged the same amount for a resource so that only those who value it most will use it. Doing so would provide increased motivation for more efficient water use by farms, residences, and businesses. Increased water costs could be helpful in reducing the demand for water, but we cannot forget the simultaneous need for all users to better conserve water.

Scenarios similar to the California water wars are being played out in many other parts of the world. Competition for water exists not only among farmers and municipalities, but also among neighboring countries. Indeed, disagreements over water have turned into real wars between nations. If the competing interests in California can devise creative solutions to the state's water problems, those solutions can serve as a model for the rest of the world so that we can continue to have economic prosperity without severely harming the environment.

## References

Martin, G. 2005. The California water wars: Water flowing to farms, not fish. *San Francisco Chronicle,* October 23.

Steinhauer, J. 2008. Governor declares drought in California. *New York Times,* June 5.

# Land: Public and Private

## Who Owns a Tree? Julia Butterfly Hill versus Maxxam

For most of its history, northern California's Pacific Lumber Company was a leader in environmental stewardship. It pioneered the practice of selectively cutting some trees on its land while leaving nearby trees intact. Unlike the usual practice of clear-cutting—clearing all the trees from an area of land—selective cutting allows the roots of the remaining trees to

The ultimate goal of a tree sit is to remain in a tree long enough to save it from logging. However, many such actions merely delay the inevitable. Still, activists support tree sits because they gain time for further legal proceedings and increase media awareness. But Julia Butterfly Hill proved indomitable. In her most famous tree sit, she occupied Luna, a 55-meter (180-foot) redwood tree that was nearly a thousand

> Julia Butterfly Hill proved indomitable. In her most famous tree sit, she occupied Luna, a 55-meter (180-foot) redwood tree that was nearly a thousand years old.

retain water and hold soil in place. In addition, the surviving trees allow forest regrowth that more closely mimics natural succession after a disturbance. Pacific Lumber developed a 100-year sustainable logging plan that ensured relatively healthy forests as well as job security for the loggers and their families.

In 1986, however, Pacific Lumber was purchased by new owners and renamed Maxxam. To increase revenue from the company's land holdings, the new management clear-cut hundreds of thousands of acres of redwood forests. Because the new harvesting method removed vegetation and tree roots, Maxxam's practices left the forests susceptible to soil erosion and landslides. On December 31, 1996, an immense landslide began on a steeply sloped Maxxam site above the town of Stafford, California. The landslide destroyed a number of homes and drastically altered the natural environment.

The adjacent uncut land, owned by Maxxam, was home to the largest remaining intact forest of ancient redwood trees and to a number of endangered species. Environmental activist Julia Butterfly Hill was appalled by Maxxam's forest management practices. In response, she joined a "tree sit" on Maxxam's land organized by the direct action group Earth First!

years old, and resolved to remain there until Maxxam agreed to spare the tree. She spent 2 years in the tree without coming down once. She stayed through gale-force winds, two winters, intimidation tactics, the death of another activist, and the cutting of most of the trees in the stand that she occupied. Julia's efforts were rewarded when Maxxam agreed to protect Luna, and a 61-meter (200-foot) buffer around it, in perpetuity. In a separate, complex deal that included many stipulations about logging practices, Maxxam and another company also agreed to sell the U.S. government 3,035 hectares (7,500 acres) of ancient forest for $480 million. That property has now become the Headwaters Forest Reserve.

Hill's actions cost Maxxam millions of dollars in delays and lost revenue and brought unsustainable logging practices to the attention of many people. Because of its large debt load, Maxxam filed for bankruptcy protection in January 2007, and its depleted assets were reorganized and transferred to a new company in 2008. Although ▶

Julia Butterfly Hill sitting in Luna.

◀ Headwaters Forest Reserve.

Maxxam's logging practices appeared to have placed profit before sustainability, the new company has stated that it will return to more sustainable practices with the trees that remain on its holdings.

The conflict over logging practices raises many questions. Do citizens of the United States have the right to influence what activities occur on private lands? What if the land is public? Was Julia Butterfly Hill a hero or a villain? Some see her as having made a personal sacrifice to save an ancient tree and to bring the issue of unsustainable logging practices to the public's attention. Others see her as a lawbreaker who trespassed on private property in an effort to prevent a legal activity. ■

*Sources:* J. B. Hill, *The Legacy of Luna: The Story of a Tree, a Woman, and the Struggle to Save the Redwoods* (HarperCollins, 2000); H. Sims, The scion, *North Coast Journal,* July 20, 2006.

## UNDERSTAND THE KEY IDEAS

The issues involved in land use are not simple, and there are no easy solutions to many of the conflicts that arise. In this chapter we will begin to explore how our use of land affects the environment and what we can do to minimize its negative impacts. We will look at land use and management in both the public and private sectors, and we will explore the topic of sustainable land use practices. We will also look at how land use has changed with shifting and growing populations.

After reading this chapter you should be able to

- describe the concepts of the tragedy of the commons and maximum sustainable yield, and explain how they pertain to land use.
- describe the function, operation, and efficacy of the four major public land management agencies in the United States.
- understand the causes and consequences of urban sprawl.
- describe approaches and policies that promote sustainable land use.

# Human land use affects the environment in many ways

Agriculture, housing, recreation, industry, mining, and waste disposal are all uses of land that have benefits to humans. But, as we have seen in previous chapters, some of these activities also have negative consequences. Extensive logging may lead to mudslides. Deforestation of large areas contributes to climate change and many other environmental problems. Changes to the landscape are the single largest cause of species extinctions today. Paving over land surfaces reroutes water runoff. The paved surfaces also absorb heat from the sun and reradiate it, creating urban "heat islands." Overuse of farmland can lead to soil degradation and water pollution (FIGURE 10.1).

Humans value land for what it can provide: food, shelter, and natural resources. In addition, many people recognize that land has intrinsic value apart from its instrumental, or monetary, value. Every human use of land alters it in some way. Furthermore, individual activities on any parcel of land can have wide-ranging effects on other lands. For this reason, communities around the world use laws, regulations, and other methods to influence or regulate private and public land use.

As we saw in this chapter's opening story, people do not always agree on land use and management priorities. Do we save a beautiful, ancient stand of trees, or do we harvest the trees in order to gain benefits in the form of jobs, profit, structures made of wood, and economic development? Such conflicts can arise from both public and private land uses.

Let's begin with three concepts that are essential for understanding land use: the *tragedy of the commons, externalities,* and *maximum sustainable yield.*

## The Tragedy of the Commons

In certain societies, land was viewed as a common resource: anyone could use land for foraging, growing crops, felling trees, hunting, or mining. As populations increased, such common lands tended to become degraded—overgrazed, overharvested, and deforested. In 1968, ecologist Garrett Hardin brought the issue of overuse of common resources to the attention of the broader scientific community when he described the **tragedy of the commons:** the tendency of a shared, limited resource to become depleted because people act from self-interest for short-term gain. Hardin observed that when many individuals share a common resource without agreement on or regulation of its use, it is likely to become overused very quickly.

For example, imagine a communal pasture on which many farmers graze their sheep. At first, no single farmer appears to have too many sheep. But because an individual farmer gains from raising as many sheep as possible, each farmer may be tempted to continue adding sheep to the pasture. However, if the total number of sheep

**FIGURE 10.1** **Some negative consequences of land use by humans.** (a) A mudslide caused by logging and poor land management. (b) Habitat conversion and land degradation due to shifting agriculture. (c) Logging and other habitat alteration can adversely affect many species, including this spotted owl. (d) Extensive paving of land due to the growth of cities reduces the amount of land available for vegetation and water infiltration.

owned by all the farmers continues to grow, it will soon exceed the carrying capacity of the land. The sheep will overgraze the common pasture to the point at which plants will not have a chance to recover. The common land will be degraded, and the sheep will no longer have an adequate source of nourishment, as **FIGURE 10.2**

Use of the commons is below the carrying capacity of the land. All users benefit.

If one or more users increase the use of the commons beyond its carrying capacity, the commons becomes degraded. The cost of the degradation is incurred by all users.

Unless environmental costs are accounted for and addressed in land use practices, eventually the land will be unable to support the activity.

**FIGURE 10.2** **The tragedy of the commons.** If the use of common land is not regulated in some way—by the users or by a government agency—that land can easily be degraded to the point at which it can no longer support that use.

shows. Over a longer period, the entire community will suffer. When the farmers make decisions that benefit only their own short-term gain and do not consider the common good, everybody loses.

The tragedy of the commons applies not only to agriculture, but to any publicly available resource that is not regulated, including land, air, and water. For example, the use of many global fisheries as commons has led to the overexploitation and rapid decline of many commercially harvested fish species, and has upset the balance of entire marine ecosystems.

## Externalities

The tragedy of the commons is the result of an economic phenomenon called a *negative externality*. More generally, an **externality** is a cost or benefit of a good or service that is not included in the purchase price of that good or service. For example, if a bakery moves into the building next to you and you wake up every morning to the delicious smell of freshly baked bread, you are benefiting from a positive externality. On the other hand, if the bakers arrive at three in the morning and make so much noise that they interrupt your sleep, and you are not as productive at your job the next day, you are suffering from a negative externality.

In environmental science, we are especially concerned with negative externalities because they so often lead to serious environmental damage for which no one is held legally or financially responsible. For example, if one farmer grazes too many sheep in a common pasture, his action will ultimately result in more total harm than total benefit. But, as long as the land continues to support grazing, the individual farmer will not have to pay for the harm he is causing—ultimately, this cost is *externalized* to the other farmers. If the farmer responsible for the extra sheep had to bear the cost of his overuse of the land, he would not graze the extra sheep on the commons; the cost of doing so would exceed the benefit. From this example, we can see that in order to calculate the true cost of using a resource, we must always include the externalized cost. In other words, we must account for any potential harm that comes from the use of that resource.

Some economists maintain that private ownership can prevent the tragedy of the commons. After all, a landowner is much less likely to overgraze his own land than common land. Regulation is another approach. For example, a local government could prevent overuse of a common pasture by passing an ordinance that permits only a certain number of sheep to graze there.

Challenging the idea that government regulation is necessary, Professor Elinor Ostrom of Indiana University showed that many commonly held resources can be managed effectively at the community level or by user institutions. Ostrom's work, for which she was awarded the 2009 Nobel Prize in Economics, has shown that self-regulation by resource users can prevent the tragedy of the commons.

## Maximum Sustainable Yield

When we want to obtain the maximum amount of a resource, we need to know how much of a given plant or animal population can be harvested without harming the resource as a whole.

Imagine a situation in which deer hunting in a public forest is unregulated. Each hunter is free to harvest as many deer as possible. As a result of unlimited hunting, the deer population could be depleted to the point of endangerment. This, in turn, would disrupt the functioning of the forest ecosystem. On the other hand, if hunting were prohibited entirely, the deer herd might grow so large that there would not be enough food in the forests and fields to support it. In extreme cases, such as that of the reindeer of St. Paul Island (summarized graphically in Figure 6.8), the population could grow unchecked until it crashed due to starvation.

Some intermediate amount of hunting will leave enough adult deer to reproduce at a rate that will maintain the population, but not so many that there is too much competition for food. This intermediate harvest is called the *maximum sustainable yield*. Specifically, the **maximum sustainable yield (MSY)** of a renewable resource is the maximum amount that can be harvested without compromising the future availability of that resource. In other words, it is the maximum harvest that will be adequately replaced by population growth.

MSY varies case by case, but a reasonable starting point is to assume that population growth is the fastest at about one-half the carrying capacity of the environment, as shown on the S-shaped curve in FIGURE 10.3. (You may wish to review the logistic growth model

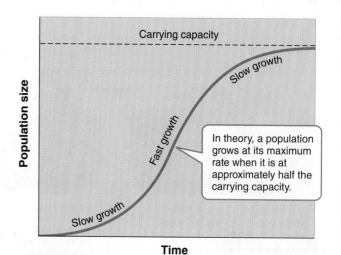

**FIGURE 10.3** **Maximum sustainable yield.** Every population has a point at which a maximum number of individuals can be harvested sustainably. That point is often reached when the population size is about one-half the carrying capacity.

described in Chapter 6.) Looking at the graph, we can see that at a small population size, the growth curve is shallow and growth is relatively slow. As the population increases in size, the slope of the curve is steeper, indicating a faster growth rate. As the population size approaches the carrying capacity, the growth rate slows. The MSY is the amount of harvest that keeps the resource population at about one-half the carrying capacity of the environment.

Forest trees, like animal populations, have a maximum sustainable yield. Loggers may remove a particular fraction of the trees at a site in order to allow a certain amount of light to penetrate to the forest floor and reach younger trees. If they cut too many trees, an excess of sunlight will penetrate and dry the forest soil. This drying may create conditions inhospitable to tree germination and growth, thus inhibiting adequate regeneration of the forest.

In theory, harvesting the maximum sustainable yield will permit an indefinite use without depletion of the resource. In reality, it is very difficult to calculate MSY with certainty because in a natural ecosystem, it is difficult to obtain necessary information such as birth rates, death rates, and the carrying capacity of the system. Once an MSY calculation is made, we still cannot know if a yield is truly sustainable until months or years later, when we can evaluate the effect of the harvest on reproduction. By that time, if the harvest rate has been too great, it is too late to prevent harm to the population.

**GAUGE YOUR PROGRESS**

✓ Why do humans value land?

✓ What is the tragedy of the commons? What is an externality?

✓ What is maximum sustainable yield?

# Public lands are classified according to their use

All countries have public lands, which they manage for a variety of purposes, including environmental protection. The 2003 United Nations List of Protected Areas—the most recent global study of protected areas—includes almost 1.7 billion hectares (4.2 billion acres) of land in a variety of categories. Given that Earth's total land area is about 15 billion hectares (37 billion acres), this means that approximately 11 percent, or one-ninth, of Earth's land area is protected in one way or another. These areas, as well as protected marine areas, are shown on the map in **FIGURE 10.4**. Let's look at both international and national categories of land protection.

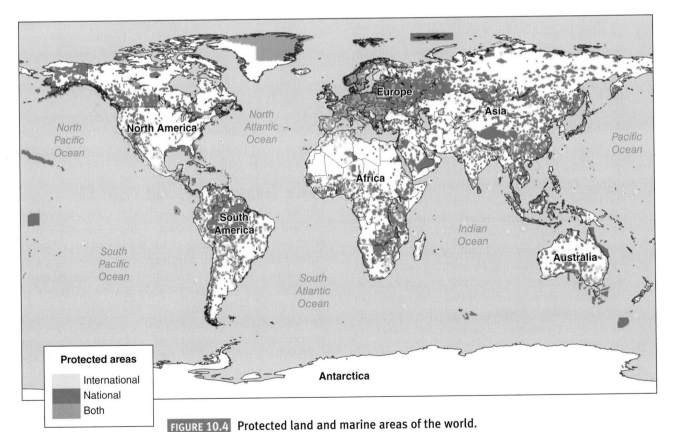

**Protected areas**
- International
- National
- Both

**FIGURE 10.4** Protected land and marine areas of the world.
[From http://www.wdpa.org/wdpamapflex.aspx.]

## International Categories of Public Lands

The 2003 United Nations List of Protected Areas classifies protected public lands into six categories according to how they are used:

- *National Parks.* There are roughly 3,400 national parks in the world, covering more than 400 million hectares (1 billion acres). This means that national parks make up about 2.7 percent of Earth's land area. National parks are managed for scientific, educational, and recreational use, and sometimes for their beauty or unique landforms. In most cases, they are not used for the extraction of resources such as timber or ore. Some of the most famous national parks in the world are found in Africa. These parks include Amboseli National Park in Kenya and Kruger National Park in South Africa. Parks like these generally exist to protect animal species such as elephants, rhinoceroses, and lions, as well as areas of great natural beauty. They also generate tourism, which can be a large source of income. On the negative side, in order to create and maintain national parks, governments have sometimes evicted and excluded indigenous human populations from the land. For example, in the winter of 2009, a new round of evictions from the Mau Forest in the Rift Valley of Kenya led to the displacement of 20,000 families. Such programs continue to generate controversy in Kenya and other countries.

- *Managed Resource Protected Areas.* This classification allows for the sustained use of biological, mineral, and recreational resources. In most countries, these areas are managed for multiple uses. There are approximately 4,100 such sites in the world, encompassing more than 440 million hectares (1.1 billion acres). In the United States, national forests are one example of this kind of area.

- *Habitat/Species Management Areas.* These areas are actively managed to maintain biological communities, for example with fire prevention or predator control. There are approximately 27,600 of these sites, covering more than 300 million hectares (740 million acres). Karelia, a part of Russia bordering Finland, has one of the highest proportions of protected areas in Europe: 5 percent of its total area. Of this total, more than one-half consists of habitat or species management areas that are actively managed for hunting and conservation.

- *Strict Nature Reserves and Wilderness Areas.* These areas are established to protect species and ecosystems. There are approximately 6,000 such sites worldwide, covering more than 200 million hectares (490 million acres). The Chang Tang Reserve, on the Tibetan Plateau in China, was set aside to protect a number of species—including the declining population of wild yak—from hunting, habitat destruction, and hybridization with domesticated animals.

- *Protected Landscapes and Seascapes.* These areas combine the nondestructive use of natural resources with opportunities for tourism and recreation. Orchards, villages, beaches, and other such areas make up the 6,500 such sites worldwide, which cover more than 100 million hectares (250 million acres). Among these protected areas is the Batanes Protected Landscape and Seascape in the northernmost islands of the Philippines, home to several endemic plant and animal species as well as important marine habitats.

- *National Monuments.* These areas are set aside to protect unique sites of special natural or cultural interest. There are almost 20,000 national monuments and landmarks in the world, covering nearly 28 million hectares (69 million acres). Most of these are established to protect historical landmarks, such as the Arc de Triomphe in Paris, France.

## Public Lands in the United States

In the United States, publicly held land may be owned by federal, state, or local governments. Of the nation's land area, 42 percent is publicly held—a larger percentage than in any other nation. The federal government is by far the largest single landowner in the United States: it owns 240 million hectares (600 million acres), or roughly 25 percent of the country (FIGURE 10.5). Most of this land—55 percent—is in the 11 western continental states, and an additional 37 percent is in Alaska. Less than 10 percent of federal land is located in the Midwest and on the East Coast.

PUBLIC LAND CLASSIFICATIONS Public lands in the United States include rangelands, national forests, national parks, national wildlife refuges, and wilderness areas. Since the founding of the nation, many different individuals and groups have expressed interest in using these public lands. However, most environmental policies, laws, and management plans have been based, at least partially, on the *resource conservation ethic,* which calls for policy makers to consider the instrumental value of nature (recall our discussion of ecosystem services and instrumental value in Chapter 3). The **resource conservation ethic** states that people should maximize resource use based on the greatest good for everyone. In conservation and land use terms, it has meant that areas are preserved and managed for economic, scientific, recreational, and aesthetic purposes.

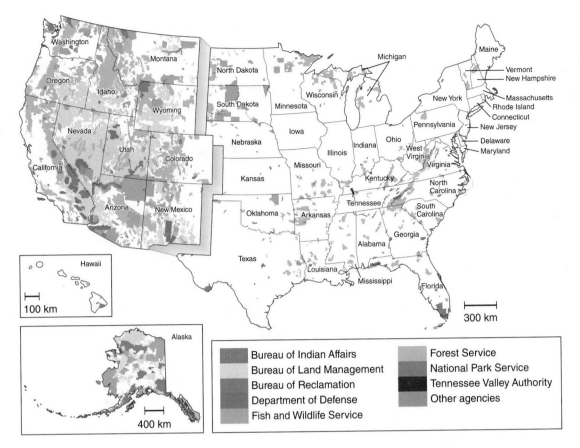

FIGURE 10.5 Federal lands in the United States. Approximately 42 percent of the land in the United States is publicly owned, and 25 percent of the nation's land is owned by the federal government. [After http://nationalatlas.gov.]

Of course, many of these purposes conflict. In order to manage competing interests, the U.S. government has, for decades, adopted the principle of *multiple use* in managing its public resources. Some public lands are in fact classified as **multiple-use lands,** and may be used for recreation, grazing, timber harvesting, and mineral extraction. Others are designated as protected lands in order to maintain a watershed, preserve wildlife and fish populations, or maintain sites of scenic, scientific, and historical value.

**LAND USE AND FEDERAL AGENCIES** As shown in FIGURE 10.6, land in the United States, both public and private, is used for many purposes. These uses can be divided into a number of categories. The probable use of public land determines how it is classified and which federal agency will manage it. More than 95 percent of all federal lands are managed by four federal agencies (see Figure 10.5): the Bureau of Land Management (BLM), the United States Forest Service (USFS), the National Park Service

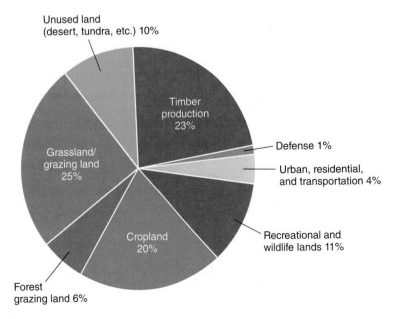

FIGURE 10.6 Land use in the United States. Public and private land in the United States is used for many purposes. [After R. N. Lubowski et al., Major land uses in the United States, Economic Research Service, USDA, 2002.]

(NPS), and the Fish and Wildlife Service (FWS). BLM, USFS, and NPS lands are typically classified as multiple-use lands because most, and sometimes all, public uses are allowed on them.

Although individual tracts may differ, the following are typical divisions of public land uses:

- BLM lands: grazing, mining, timber harvesting, and recreation
- USFS lands: timber harvesting, grazing, and recreation
- NPS lands: recreation and conservation
- FWS lands: wildlife conservation, hunting, and recreation

**GAUGE YOUR PROGRESS**

✓ What are the main uses of public lands in the United States?

✓ How do human land use decisions influence categories of public land classification?

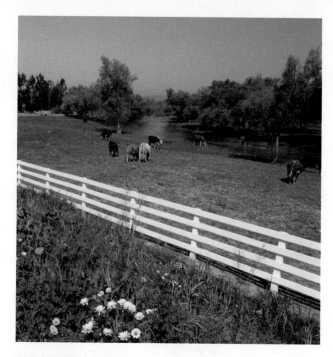

**FIGURE 10.7** **Overgrazed rangeland.** Overgrazing can rapidly strip an area of vegetation.

# Land management practices vary according to land use

Now that we have a basic picture of how public land is classified and of the relationship between public land use and management agencies in the United States, let's turn to some of the specific issues involved in managing different types of public lands. Note that many of the management issues we discuss here apply to private lands as well.

## Rangelands

**Rangelands** are dry, open grasslands. They are used primarily for cattle grazing, which is the most common use of land in the United States. Rangelands are semiarid ecosystems and are therefore particularly susceptible to fires and other environmental disturbances. If humans overuse rangelands, they can easily lose biodiversity.

Like most human activities, livestock grazing has mixed environmental effects. One environmental benefit of grazing is that ungulates—hoofed animals such as cattle and sheep—can be raised on lands that are too dry to farm. In addition, grazing these animals uses less fossil fuel energy than raising them in feedlots. However, improperly managed livestock can damage stream banks and pollute surface waters. Grazing too many animals can quickly denude a region of vegetation (**FIGURE 10.7**). Loss of vegetation leaves the land exposed to wind erosion and makes it difficult for soils to absorb and retain water when it rains.

Many environmental scientists argue that rangeland ecosystems are too fragile for multiple uses. Certain environmental organizations have suggested that as much as 55 percent of U.S. rangeland soils are in poor or very poor condition, due in large part to overgrazing. However, the BLM, which manages most public rangelands in the United States, has maintained that the percentage is not nearly that high. Reconciling this difference is a challenge because of the many factors that influence how soil condition is determined.

The Taylor Grazing Act of 1934 was passed to halt overgrazing. It converted federal rangelands from a commons to a permit-based grazing system. The goal of a permit-based system is to limit the number of animals grazing in a particular area and thereby avoid a tragedy of the commons situation. However, critics maintain that the low cost of the permits continues to encourage overgrazing. In 2006, the federal government spent seven times more money managing its rangelands than it received in permit fees. Thus, in effect, grazing is subsidized with federal funds.

The BLM focuses on mitigating the damage caused by grazing. The agency considers "rangeland health" when it sets guidelines for grazing. For example, state and regional rangeland managers are required to ensure healthy watersheds, maintain ecological processes such as nutrient cycles and energy flow, preserve water quality, maintain or restore habitats, and protect endangered species. However, these managers are not given detailed guidance, and the BLM regulations do not require the involvement of environmental scientists. This omission gives the managers wide latitude to set their own guidelines and standards. As a result, BLM regulations are not consistently successful in preserving vulnerable rangeland ecosystems.

## Forests

**Forests** are dominated by trees and other woody vegetation. Approximately 73 percent of the forests used for commercial timber operations in the United States are privately owned. Commercial logging companies are allowed to use U.S. national forests, usually in exchange for a *royalty*—a percentage of their revenues. Many national forests were originally established to ensure a steady and reliable source of timber. As with grazing, the federal government typically spends more money managing the timber program and building and maintaining logging roads than it receives from these royalties.

**TIMBER HARVEST PRACTICES** The two most common ways in which trees are harvested for timber production, both of which are illustrated in **FIGURE 10.8**, are *clear-cutting* and *selective cutting*. **Clear-cutting** involves removing all, or almost all, the trees within an area (Figure 10.8a). It is the easiest harvesting method and, in most cases, the most economical. When a *stand,* or cluster, of trees has been clear-cut, foresters often replant or reseed the area. Often the entire area will be replanted at the same time, so all the resulting trees will be the same age. Because they are exposed to full sunlight, clear-cut tracts of land are ideal for fast-growing tree species that achieve their maximum growth rates with large amounts of direct sunlight. Other species may not be so successful, which can lead to a reduction in overall biodiversity. However, if a commercially valuable tree species constitutes only a small fraction of a stand of trees, it may not be economically efficient to clear-cut the entire stand. This is particularly true in many tropical forests, where valuable species constitute only a small percentage of the trees and are mixed in with many other species.

Clear-cutting, especially on slopes, increases wind and water erosion, which causes the loss of soil and nutrients (**FIGURE 10.9**). Erosion also adds silt and sediment to nearby streams, harming aquatic populations. In addition, the denuded slopes are prone to dangerous mudslides like the Stafford, California, mudslide described in this chapter's opening story. Clear-cutting also increases the amount of sunlight that reaches rivers and streams. The increased sunlight raises water temperatures, which can adversely affect certain aquatic species. Even the replanting process can have negative environmental consequences. Timber companies often use fire or herbicides to remove bushy vegetation before a clear-cut is replanted. These practices reduce soil quality, and herbicides may contaminate water that runs off into streams and rivers. Many environmental scientists

**(a)** Clear-cutting

**(b)** Selective cutting

**FIGURE 10.8** **Timber harvest practices.** (a) Clear-cutting removes most, if not all, trees from an area and is often coupled with replanting. The resulting trees are all the same age. (b) In selective cutting, single trees or small numbers of trees are harvested. The resulting forest consists of trees of varying ages.

identify clear-cutting as a cause of habitat alteration and destruction and forest fragmentation. These effects, in turn, lead to decreased biodiversity and sometimes to a lower aesthetic value for the affected forest.

**Selective cutting** (Figure 10.8b) removes single trees or relatively small numbers of trees from among many in a forest. This method creates many small

**FIGURE 10.9** Clear-cut forest near Eureka, California. A clear-cut on a steep slope increases the likelihood of erosion and delayed regeneration of vegetation.

openings in a stand where trees can reseed or young trees can be planted, so the regenerated stand contains trees of different ages. Because seedlings and young trees must grow next to larger, older trees, selective cutting produces optimum growth only among shade-tolerant tree species.

The environmental impact of selective cutting is less extensive than that of clear-cutting. However, many of the negative environmental impacts associated with logging remain the same. For example, whether a company uses clear-cutting or selective cutting, it will need to construct logging roads to carry equipment and workers into the area to be harvested. These roads fragment the forest habitat, leading to species diversity changes, and compact the soil, leading to nutrient loss and reductions in water infiltration.

A third approach to logging removes trees from the forest in ways that do not unduly affect the viability of other, noncommercial tree species. Known as **ecologically sustainable forestry,** this approach has a goal of maintaining all species—both plants and animals—in as close to a natural state as possible (**FIGURE 10.10**). Some loggers have even returned to using animals such as horses to pull logs in order to reduce soil compaction, although the costs of such methods make it difficult to compete economically with mechanized logging practices.

## LOGGING, DEFORESTATION, AND REFORESTATION

Approximately 30 percent of all commercial timber in the world is produced in the United States and Canada. Compared with South America and Africa, forest losses in these two major timber-producing countries have been relatively small over the last several decades. Still, timber production presents important ecological challenges in these countries.

Throughout this chapter we have seen examples of the conflicts over land use created by competing interests and values. Perhaps nowhere is this conflict so clear as in the case of logging. For example, timber

**FIGURE 10.10** Sustainable forestry. Logging without the use of fossil fuels, as is done in this forest in Oregon, further enhances the sustainability of a forestry project.

production has always been a part of the mission of the USFS. Yet maintaining biodiversity is an equally important goal. It would seem that we can't have both.

All logging disrupts habitat and usually has an effect, either negative or positive, on plant and animal species. One such species is the marbled murrelet (*Brachyramphus marmoratus*). This bird spends most of its life along the coastal waters of the Pacific Northwest, but it nests in coastal redwood forests. With the intensive logging of these forests, the marbled murrelet has become endangered. The tree sit described in this chapter's opening story brought this species, among others, to the public's attention.

Logging often replaces complex forest ecosystems with **tree plantations:** large areas typically planted with a single rapidly growing tree species. These same-aged stands can be easily clear-cut for commercial purposes, such as pulp and wood, and then replanted. Because of this cycle of planting and harvesting, tree plantations never develop into mature, ecologically diverse forests. If too many planting and harvesting cycles occur, the soil may become depleted of important nutrients such as calcium.

Since 1982, federal regulations have required the USFS to provide appropriate habitat for plant and animal communities while also meeting multiple-use goals. However, these regulations fail to specify how biodiversity protection should be achieved or how the results should be quantified, leaving the USFS to choose its own approach to biodiversity management. Critics charge that the USFS is not adequately protecting biodiversity and forest ecosystems. The USFS maintains that it is doing the best it can at meeting many different objectives.

**FIRE MANAGEMENT** In many ecosystems, fire is a natural process that is important for nutrient cycling and regeneration. As discussed briefly in Chapter 3, when fires periodically move through an ecosystem, they liberate nutrients tied up in dead biomass. In addition, areas where vegetation is killed by the fire provide openings for early-successional species.

Humans have followed a variety of management policies with respect to fire. For many years, managers of forest ecosystems, including the USFS, did everything they could to suppress fires. This strategy led to the accumulation of large quantities of dead biomass on the forest floor. Eventually, this fuel built up until a large fire became inevitable. One method for reducing the accumulation of dead biomass is a **prescribed burn,** in which a fire is deliberately set under controlled conditions. Prescribed burns help reduce the risk of uncontrolled natural fires and provide some of the other benefits of fire as well. More recently, forest managers have allowed certain natural fires to burn. This policy appears to have been accepted in many parts of the United States, as long as human life and property are not threatened.

Probably the best-known forest fires in the United States are those that occurred in Yellowstone National Park in the summer of 1988, the driest year on record at the park. A combination of human activity and lightning set off multiple fires (FIGURE 10.11). Over 25,000 people participated in fighting the fires. While this effort saved human lives and property, the firefighters had little impact on the fires. When the fires were over, more than one-third of Yellowstone National Park had burned. Initially, many people were outraged that the NPS and others had "allowed" the park to burn. However, within

FIGURE 10.11 Yellowstone fires of 1988. As can be seen from the map, extensive areas of the park were burned in this exceptionally hot and dry year.

Boundary
Lakes
Fire
- - - - - Roads

40 km

a few years, it became clear that the fires had created new, nutrient-rich habitat for early-successional plant species, which attracted elk and other herbivores. Ultimately, researchers and forest managers concluded that the Yellowstone fires of 1988 provided many benefits to the Yellowstone ecosystem. Today, a typical visitor viewing a portion of the park that burned in 1988 probably wouldn't even know that there had been a major fire so recently.

## National Parks

As we have already noted, many national parks were established to preserve scenic views and unusual landforms. Today, national parks are managed for scientific, educational, aesthetic, and recreational use. Since Yellowstone National Park was founded in 1872, 58 national parks have been established in the United States. The NPS manages a total of 391 national parks and other areas, such as historical parks and national monuments, and the list continues to grow.

**THE GOALS OF NATIONAL PARK MANAGEMENT** Management of national parks, like that of national forests, is based on the multiple-use principle. Unlike the national forests, the U.S. national parks were set aside specifically to protect ecosystems. In establishing Yellowstone National Park, Congress mandated the Interior Department to regulate the park in a manner consistent with the preservation of timber resources, mineral resources, and "natural curiosities." However, it did not require a management process based on ecological principles.

It was not until the 1960s that ecology became a focus of national park management. In 1963, an advisory board on wildlife management presented the Leopold Report to Secretary of the Interior Stewart Udall. This report established the guiding principles of national park management today. It proposed that the primary purpose of NPS should be to maintain the parks in the same biotic condition in which they were first found by European settlers. To this end, the authors believed that NPS should focus its efforts on conservation and on protection of wildlife species and their habitats. The report claimed that human activity had severely affected normal ecological processes in the parks and that active intervention was required to achieve the goal of a return to a more "natural" state.

Today, NPS applies environmental science to maintain biodiversity and ecosystem function in all national parks. As we saw in the example of the Yellowstone fires, its policies continue to be controversial. Each national park adapts U.S. policy to its specific needs. In parks with high levels of endemic biodiversity, for example, management focuses on conserving endemic species. For instance, the Channel Islands National Park off the coast of southern California is an important breeding ground for many seabirds, including a pelican species that is rare elsewhere in the western United States.

**FIGURE 10.12** **ATV-caused damage near Olympic National Park in Washington State.** Although today many national parks limit ATV use, the conflict between those who want to use the parks for this form of recreation and those who wish to protect biodiversity remains.

Because of this special feature, the park is managed primarily to conserve its biodiversity. Other national parks balance biodiversity protection with recreational use.

**NATIONAL PARKS AND HUMAN ACTIVITIES** Reducing the impact of human activities both outside and inside the park borders is the primary challenge of most national parks throughout the world. Air and water pollution from distant sources can reduce biodiversity as well as recreational value and economic opportunities. Ongoing development adjacent to park boundaries can be particularly problematic. In many locations, national parks have become islands of biodiversity amid increasing human development, but even their protected status cannot defend them completely from invasive species and other problems, such as pollution. These external threats require large-scale evaluation, planning, and management that extend beyond park borders.

National parks are also victims of their popularity. Although the park system was established in part to make areas of great beauty accessible to people, human overuse can harm the very environment that people visit to enjoy. For example, all-terrain vehicles (ATVs) are a major cause of air and noise pollution in national parks, as well as a direct cause of habitat destruction (**FIGURE 10.12**). Today, many parks strictly limit or even ban the use of ATVs. Still, park managers grapple with how to determine appropriate limits on human activity. In many cases, the trade-off between short-term recreational uses and long-term protection of biodiversity does not permit an easy answer.

## Wildlife Refuges and Wilderness Areas

**National wildlife refuges** are the only federal public lands managed for the primary purpose of protecting wildlife. The Fish and Wildlife Service manages more than 450 national wildlife refuges and 28 waterfowl production areas on 34.4 million hectares (85 million acres) of publicly owned land.

**National wilderness areas** are set aside with the intent of preserving large tracts of intact ecosystems or landscapes. Sometimes only a portion of an ecosystem is included. Wilderness areas are created from other public lands, usually national forests or rangelands, and are managed by the same federal agency that managed them prior to their designation as wilderness.

National wilderness areas allow only limited human use and are designated as roadless. Although logging, road building, and mining are banned in national wilderness areas, roads that existed before the designation sometimes remain in use, and activities, such as mining, that were previously permitted on the land are allowed to continue. More than 38.5 million hectares (95 million acres) of federal land, 60 percent of which is in Alaska, are classified as wilderness.

## Federal Regulation of Land Use

Government regulation can influence the use of private as well as public lands. The 1969 **National Environmental Policy Act (NEPA)** mandates an environmental assessment of all projects involving federal money or federal permits. Along with other major laws of the 1960s and 1970s, such as the Clean Air Act, the Clean Water Act, and the *Endangered Species Act,* NEPA creates an environmental regulatory process designed to ensure protection of the nation's resources.

Before a project can begin, NEPA rules require the project's developers to file an **environmental impact statement (EIS).** An EIS typically outlines the scope and purpose of the project, describes the environmental context, suggests alternative approaches to the project, and analyzes the environmental impact of each alternative. NEPA does not require that developers proceed in the way that will have the least environmental impact. However, in some situations, NEPA rules may stipulate that building permits or government funds be withheld until the developer submits an **environmental mitigation plan** stating how it will address the project's environmental impact. In addition, preparation of the EIS sometimes uncovers the presence of endangered species in the area under consideration. When this occurs, the protection measures of the **Endangered Species Act,** a 1973 law designed to protect species from extinction, are applied.

Members of the public are entitled to give input into the environmental assessment, and decision makers are required to respond. And, although developers are not obligated to act in accordance with public wishes, in practice, public concern often improves the project's outcome. For this reason, attending information sessions and providing input is a good way for concerned citizens to learn more about local land use decisions and to help reduce the environmental impact of land development.

**GAUGE YOUR PROGRESS**

✓ What are the ways in which timber is harvested in U.S. forests, and how do they compare in terms of their environmental impact?

✓ What is the significance of the National Wilderness Area designation for parts of federally owned lands?

✓ What is NEPA, and what is an environmental impact statement (EIS)?

# Residential land use is expanding

While many public lands are located in relatively rural areas, there is a very different kind of land use pressure in locations close to cities. In the last 50 years, the greatest percentage of population growth in the United States has occurred in two classes of communities: *suburban* and *exurban.* **Suburban** areas surround metropolitan centers and have low population densities compared with those urban areas. **Exurban** areas are similar to suburban areas, but are unconnected to any central city or densely populated area. Since 1950, more than 90 percent of the population growth in metropolitan areas has occurred in suburbs, and two out of three people now live in suburban or exurban communities.

**FIGURE 10.13** shows U.S. population trends since 1950. The population of cities has grown somewhat in absolute terms, but has declined as a percentage of the U.S. population. The rural population has been declining since 1900, and a century later it made up less than

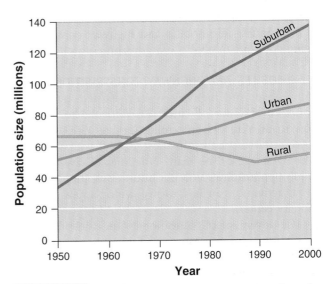

**FIGURE 10.13** Distribution of suburban, urban, and rural populations in the United States between 1950 and 2000. The shift in population from rural to suburban areas can be clearly seen in this graph. [After http://www.sdi.gov/curtis/Pop_Trends.html (Figure 1.4).]

a fifth of the total U.S. population. These population shifts have brought with them a new set of environmental problems, including *urban sprawl* and *urban blight*. Attempts to find creative solutions to these problems have been increasingly successful.

## Urban Sprawl

If you have ever been to a strip mall, you are familiar with the phenomenon known as urban sprawl, which occurs when populations shift away from rural and urban areas and into suburban and exurban areas. **Urban sprawl** is the creation of urbanized areas that spread into rural areas and remove clear boundaries between the two. The landscape in these areas is characterized by clusters of housing, retail shops, and office parks, which are separated by miles of road. Large feeder roads and parking lots that separate "big box" retail stores from the road discourage pedestrian traffic.

Urban sprawl has had a dramatic environmental impact. Dependence on the automobile has led to suburban residents driving more than twice as much as people who live in cities. Between 1950 and 2000, the number of vehicle miles traveled per person in U.S. suburban areas tripled. Because suburban house lots tend to be significantly larger than urban parcels, suburban communities also use more than twice as much land per person as urban communities. Urban sprawl tends to occur at the edge of a city, often replacing farmland and increasing the distance between farms and consumers. In its most recent survey, the U.S. Department of Agriculture estimated that between 1992 and 1997, U.S. farmland was being converted to residential uses at a rate of 500,000 hectares (1.2 million acres) per year.

## Causes and Effects of Sprawl

There are four main causes of urban sprawl in the United States: automobiles and highway construction, living costs, urban blight, and government policies.

**AUTOMOBILES AND HIGHWAY CONSTRUCTION** Before automobiles and highway systems existed, transportation into and out of cities was difficult: horses were slow, roads were bad, and trolley services rarely went far beyond the city limits. In those days, if you wanted to take advantage of the many amenities of city life, such as job opportunities, cultural institutions, shopping, and social activities, you had to live within a few miles of the city center.

The advent of the automobile, and the subsequent development of the interstate highway system in the 1950s and 1960s, changed everything. Today we think nothing of working in the city during the day and commuting home to the suburbs at night. And if you live in the suburbs but want to get into the city on a Saturday night for dinner, a concert, or a movie, that's no problem either. With rapid, comfortable transportation between urban and suburban areas, it became possible to work or play in the city and return to a large home in a quiet neighborhood. For the first time in history, people could enjoy the best of both worlds.

**LIVING COSTS** Many people find suburban living more desirable than city living because they can obtain more land and a larger house for the same amount of money. Because land is readily available in the suburbs, it is relatively inexpensive compared with land in the city. For the cost of a tiny one-bedroom condominium in a desirable section of a city, you may be able to purchase a five-bedroom house with a big yard in the suburbs. In addition, because suburban governments usually provide fewer public services than cities do, tax rates in the suburbs are likely to be lower. For these reasons, moving to the suburbs allows some people to enjoy a higher standard of living than they could afford with the same income in the city.

On the other hand, those with lower incomes may not be able to afford these benefits. Although single-family homes in the suburbs are likely to be less expensive than desirable housing in the city, they are still out of reach for many people. Most suburban communities have little, if any, low-income or "affordable" housing. Because single-family homes are usually the only housing option available in the suburbs, those with lower incomes are excluded. Furthermore, in many suburban locations, it is difficult to commute to the city without a car, which compounds the difficulties for lower-income individuals. Even when public transportation is available, commuting costs can be high.

**URBAN BLIGHT** As population shifts to the suburbs, the city's revenue from sources such as property, sales, and service taxes begins to shrink. At the same time, the cost of maintaining urban services, including public transportation, police and fire protection, and social services, remains stable. Faced with declining tax receipts, cities are forced to reduce services, raise tax rates, or both. As services decline, crime rates may increase, either because police resources are stretched thin or because conditions for lower-income residents decline even further. Infrastructure also deteriorates, leading to a decline in the quality of the built environment. These problems, combined with higher taxes, make cities less attractive places to live, and those who can afford to move away are more likely to do so.

In addition, as the population shifts to the suburbs, jobs and services follow. Suburbanization has spawned suburban office parks, which have led to an increase in suburb-to-suburb commuting. Commuting patterns develop around cities rather than into and out of them. These new traffic patterns make it more difficult to provide public transportation to the spreading region. As wealthy and middle-income people leave cities, urban retail stores lose customers. As stores close, people have

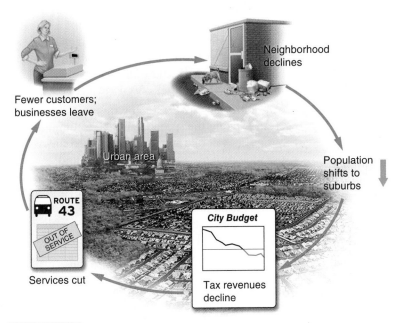

**FIGURE 10.14** **Urban blight.** As people move away from a city to suburbs and exurbs, the city often deteriorates, causing yet more people to leave. This cycle is an example of a positive feedback system. The green arrow indicates the starting point of the cycle.

fewer reasons to go to the city to shop, further decreasing the customer base for the remaining stores. This cascade of effects leads to the positive feedback loop shown in FIGURE 10.14. This loop creates **urban blight:** the degradation of the built and social environments of the city that often accompanies and accelerates migration to the suburbs.

Historically, urban blight has contributed to racial segregation. In the 1950s and 1960s, when migration to the suburbs began in earnest, those leaving the cities for the suburbs were predominantly middle- or upper-income Caucasians. This so-called "white flight" resulted in highly concentrated minority populations in city centers and almost entirely white populations in the suburbs. Over time, the disparity of opportunity increased because higher property tax revenues in suburban communities often allowed for better schools. Very recently, some of this segregation has begun to decline as Caucasians return to the cities and more minorities, after having accumulated wealth, move out to the suburbs. Nevertheless, large racial disparities remain.

**GOVERNMENT POLICIES** Urban sprawl has also been influenced by federal and local laws and policies, including the Highway Trust Fund, zoning laws, and subsidized mortgages.

The **Highway Trust Fund,** begun by the Highway Revenue Act of 1956 and funded by a federal gasoline tax, pays for the construction and maintenance of roads and highways. We have already seen that highways allow people to live farther from where they work. FIGURE 10.15 shows the resulting positive feedback loop. More highways mean more driving and more gasoline purchases, which lead to more gasoline tax receipts, and so on. As people move farther away from their jobs, traffic congestion increases, and roads are expanded. But the new, even larger roads encourage even more people to live farther away from work. This cycle exemplifies a phenomenon known as **induced demand,** in which an increase in the supply of a good causes demand to grow.

Governments may use *zoning* to address issues of traffic congestion, urban sprawl, and urban blight. **Zoning** is a planning tool developed in the 1920s to separate industry and business from residential neighborhoods and create quieter, safer communities. Governments that use zoning can classify land areas into "zones" in which certain land uses are restricted. For instance, zoning ordinances might prohibit developers from building a factory or a strip mall in a residential area or a multi-dwelling apartment building in a single-home neighborhood. Nearly all metropolitan governments across the United States have adopted zoning.

**FIGURE 10.15** **Induced demand as a cause of traffic congestion and urban sprawl.** The use of gasoline tax money to build highways leads to development of suburbs and traffic congestion, at which point yet more money is spent on highways to alleviate the congestion. The green arrow indicates the starting point of the cycle.

Zoning often regulates much more than land use. The number of parking spaces a building must have, how far from the street a building must be placed, or even the size and location of a home's driveway are among the development features that zoning may stipulate. Zoning laws have been helpful in addressing issues of safety and sometimes in minimizing environmental damage caused by new construction. One negative aspect of zoning, however, is that it generally prohibits suburban neighborhoods from developing a traditional "Main Street" with shops, apartments, houses, and businesses clustered together. Many communities are now attempting to incorporate *multi-use zoning* into their municipal plans. **Multi-use zoning** allows retail and high-density residential development to coexist in the same area. However, most zoning in the United States continues to promote automobile-dependent development.

The federal government has also played a large part in encouraging the growth of suburbs through the Federal Housing Administration (FHA). Congress established the FHA during the Great Depression of the 1930s, in part to jump-start the economy by creating more demand for new housing. Through the FHA, people could apply for federally subsidized mortgages that offered low interest rates. This program allowed many people who otherwise could not afford a house to purchase one. However, the FHA financed mortgages only for homes in financially "low risk" areas. Inevitably, these areas were almost always the newly built, low-density suburbs. When, at the end of World War II, the GI Bill extended generous credit terms to war veterans, the suburban housing boom continued. The greater availability of cars as well as mass-produced housing, which allowed developers to sell newly convenient suburban homes at a steep discount, made the suburbs even more attractive.

## Smart Growth

People are beginning to recognize and address the problems of urban sprawl. One way they are doing so is through the principles of *smart growth*. **Smart growth** focuses on strategies that encourage the development of sustainable, healthy communities. The Environmental Protection Agency lists 10 basic principles of smart growth:

1. *Mixed land uses.* Smart growth mixes residential, retail, education, recreation, and business land uses. Mixed-use development allows people to walk or bicycle to various destinations and encourages pedestrians to be in a neighborhood at all times of the day, increasing safety and interpersonal interactions.

2. *Create a range of housing opportunities and choices.* By providing housing for people of all income levels, smart growth counters the concentration of poverty in failing urban neighborhoods. Mixed housing also allows more people to find jobs near where they live, improves schools, and generates strong support for neighborhood transit stops, commercial centers, and other services.

3. *Create walkable neighborhoods.* Walkable neighborhoods are created by mixing land uses, reducing the speed of traffic, encouraging businesses to build stores directly up to the sidewalk, and placing parking lots behind buildings. In neighborhoods that encourage walking, people use their cars less, which both reduces fossil fuel use and traffic congestion and provides health benefits. Communities with more pedestrians tend to see more interaction among neighbors because people stop to talk with each other. This, in turn, creates opportunities for civic engagement. When people interact, the environment usually benefits.

4. *Encourage community and stakeholder collaboration in development decisions.* There is no single "right" way to build a neighborhood; residents and **stakeholders**—people with an interest in a particular place or issue—need to work together to determine how their neighborhoods will appear and be structured.

5. *Take advantage of compact building design.* Smart growth incorporates multistory buildings and parking garages—as opposed to sprawling lots—to reduce a neighborhood's environmental footprint and protect more open space. Ideally, shops, cafés, and small businesses should be easily accessible to pedestrian traffic on the ground floor near

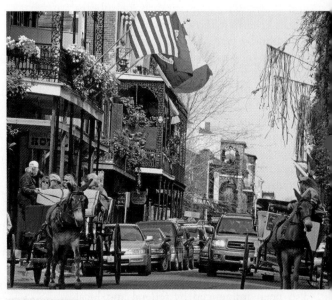

**FIGURE 10.16** **The French Quarter of New Orleans.** One principle of smart growth is to foster communities with a strong sense of place. The French Quarter of New Orleans, Louisiana, is known for its architecture, food, and, especially, music.

sidewalks, with two or three stories of apartments and offices above.

6. *Foster distinctive, attractive communities with a strong sense of place.* A **sense of place** is the feeling that an area has a distinct and meaningful character. Many cities have such unique neighborhoods. For example, the French Quarter of New Orleans (FIGURE 10.16) has a sense of place that adds to the quality of life for people living there. Smart growth attempts to foster this sense of place through development that fits into the neighborhood.

7. *Preserve open space, farmland, natural beauty, and critical environmental areas.* Working farmland is a source of fresh local produce and other goods. Open space provides opportunities for recreation and enjoyment. Land protected from development through restricted growth also provides habitats for a variety of species.

8. *Provide a variety of transportation choices.* **Transit-oriented development (TOD)** attempts to focus dense residential and retail development around stops for public transportation, giving people convenient alternatives to driving. Bicycle racks and safe roads, pleasant sidewalks for walking, frequent bus service, and light rail can all aid in this goal (FIGURE 10.17). Car-sharing networks such as Zipcar can provide easy access to a fleet of rental automobiles. This increasingly popular service reduces the need for private car ownership where public transportation is not available.

9. *Strengthen and direct development toward existing communities.* Development that fills in vacant lots within existing communities, rather than expanding into new land outside the city, is known as **infill.** This type of development can help to reinvigorate urban neighborhoods that are caught in a vicious cycle of depopulation and blight and can protect rural lands from sprawl. Some cities, such as Portland, Oregon, have had success with **urban growth boundaries,** which place restrictions on development outside a designated area.

10. *Make development decisions predictable, fair, and cost-effective.* One reason why suburban developments in all regions of the country often look the same is that standardized designs allow developers to move rapidly through the permitting process. A streamlined approval process that encourages smart growth could increase the number of individualized plans rather than promoting cookie-cutter development.

Of course, no individual new development or neighborhood plan is likely to incorporate all of these ideals, but as a guide to thinking about how to build

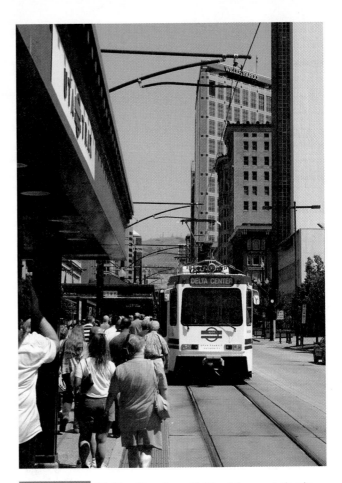

**FIGURE 10.17** **Light rail system.** Light rail is one option in transit-oriented development (TOD), which strives to develop communities with a denser mix of retail and residential components around a convenient public transportation system.

communities, the smart growth concept has been quite successful.

Smart growth can have important environmental benefits. Compact development reduces the amount of impervious surface, reducing runoff and flooding downstream. A 2000 study found that smart growth in New Jersey would reduce water pollution by 40 percent compared with the more common, dispersed growth pattern. By mixing uses and providing transportation options, smart growth also reduces fossil fuel consumption. A 1999 EPA study found that infill development can reduce miles driven by as much as 58 percent. A 2005 study in Seattle found that residents of neighborhoods incorporating just a few of the techniques to make nonauto travel more convenient traveled 26 percent fewer vehicle miles than residents of more dispersed, less-connected neighborhoods.

---

**GAUGE YOUR PROGRESS**

✓ What are urban sprawl and smart growth?

✓ How can zoning help reduce urban sprawl?

---

The Dudley Street area of Roxbury and North Dorchester, in Boston, was once a prime example of urban blight. In the 1980s, years of urban decay and the loss of large numbers of primarily Caucasian families to the suburbs had left 21 percent of the Dudley Street neighborhood vacant—amounting to 1,300 abandoned parcels. Almost 30 percent of the neighborhood's residents had an income below the federal poverty level, making the neighborhood one of the poorest in Boston. Fires were a particular problem; in some cases, arsonists attempted to gain insurance money on homes that could not be sold. The residents felt that they were being ignored by City Hall. Some suggested that this was because 96 percent of the neighborhood's residents were members of ethnic minority groups.

In 1984, residents banded together to turn their neighborhood around. They formed the Dudley Street Neighborhood Initiative (DSNI), designed to allow the residents to move toward a common vision for a sustainable neighborhood. Participants chose a large Board of Directors—34 members—so that they would hear a diversity of perspectives and, in coming to consensus, ensure that decisions had broad support. The DSNI also obtained something no other neighborhood organization has: the power of *eminent domain*. **Eminent domain** allows a government to acquire property at fair market value even if the owner does not wish to sell it. It is frequently used to acquire land for highway projects, but also has been used recently, and controversially, in urban redevelopment.

By 1987, DSNI had worked with community members to develop a comprehensive revitalization plan,

## The Dudley Street Neighborhood

which has been periodically updated since then. The main goals of the DSNI plan are

- to rehabilitate existing housing
- to construct homes that are affordable according to criteria set by residents
- to assemble parcels of vacant land for redevelopment, using the power of eminent domain if necessary
- to plan environmentally sound, affordable development that is physically attractive
- to convert some vacant properties into safe play areas, gardens, and facilities that the entire community can enjoy
- to run a full summer camp program for area children
- to develop strong public and private partnerships to ensure the economic vitality of the neighborhood
- to increase both the economic and political power of residents

DSNI has had many successes. Its first major action was to force the city to remove trash, appliances, and abandoned cars that littered the streets and vacant lots. The city also cleaned up two illegal dumps in the area. Residents planted community gardens to grow produce and flowers. DSNI successfully reduced drug dealing in the neighborhood park, although that remains a constant struggle. And, perhaps most significantly, its work led to the construction of 300 new homes on formerly vacant lots. New residents help to add vitality to the neighborhood, reversing the cycle of depopulation and business closure.

It is evident that the individuals involved in DSNI have taken many of the principles of smart growth to heart. The community mixes residential with retail development, and residents live within walking distance of a grocery store, ethnic markets, and other amenities (**FIGURE 10.18**). Moreover, since the founding of DSNI, two small manufacturing businesses—a furniture maker and an electronics company—have moved into the neighborhood, providing additional jobs within walking distance for residents.

The Dudley Street neighborhood still has a relatively low per capita income, and its development choices have not been without controversy. However, it serves as an example of one of hundreds of neighborhoods that have begun to turn the positive feedback loop of urban decay into one of urban renewal and hope.

**FIGURE 10.18** **The Dudley Street neighborhood.** This neighborhood in Boston, Massachusetts, was once a symbol of urban decay. Today it is a thriving urban community that has adopted many of the principles of smart growth.

**References**

Benfield, F. K., J. Terris, and N. Vorsanger. 2001. *Solving Sprawl: Models of Smart Growth in Communities Across America*. Island Press.

Dudley Street Neighborhood Initiative. http://www.dsni.org.

- **Describe the concepts of the tragedy of the commons and maximum sustainable yield and explain how they pertain to land use issues.**

Individuals have no incentive to conserve common resources when they do not bear the cost of using those resources. A cost or benefit not included in the price of a good is an externality. The lack of incentive to conserve common resources leads to overuse of these resources, which may be degraded if their use is not regulated. The maximum sustainable yield is the largest amount of a renewable resource that can be harvested indefinitely. Harvesting at the MSY keeps the resource population at about one-half the carrying capacity of the environment. However, uncertainty about population dynamics can lead to a miscalculation of the MSY and overharvesting.

- **Describe the function, operation, and efficacy of the four major public land management agencies in the United States.**

In the United States, public land is managed for multiple uses, including grazing, timber harvesting, recreation, and wildlife conservation. The Bureau of Land Management manages rangeland, which is used for grazing. Grazing is subsidized with federal funds, and some lands are overgrazed. The United States Forest Service manages national forests, which are used for timber harvesting, recreation, and other uses. Timber can be harvested by clear-cutting or selective cutting, both of which have environmental impacts, or by ecologically sustainable forestry methods. National parks, managed by the National Park Service, were created primarily for preservation of their scenery and unique landforms, although scientific, educational, and recreational uses have become more important over time. The Fish and Wildlife Service manages national wildlife refuges, which are designed to protect wildlife.

- **Understand the causes and consequences of urban sprawl.**

Causes of urban sprawl include the development of the automobile, construction of highways, less expensive land at the urban fringe, and urban blight. Government institutions and policies, such as the federal Highway Trust Fund, zoning, and subsidized mortgages also contribute to the problem. The result of urban sprawl is automobile dependence, traffic congestion, and social isolation including less involvement in community affairs.

- **Describe approaches and policies that promote sustainable use of land.**

Smart growth is one possible response to urban sprawl. It advocates more compact, mixed-use development that encourages people to walk, bicycle, or use public transportation. Smart growth not only consumes less land than more typical, dispersed development, but has numerous other environmental benefits.

## CHECK YOUR UNDERSTANDING

1. Which of the following is *not* an example of the tragedy of the commons?
   (a) Overgrazing by sheep on community-owned pastures
   (b) Depletion of fish stocks in international waters
   (c) Automobile congestion in Yellowstone National Park
   (d) Depletion of soil minerals by farmers on private land
   (e) Tropical deforestation due to clearing land for agriculture and then moving on to another location

2. In the accompanying graph of the population growth of the common pheasant, one of the world's most hunted birds, *X* represents
   (a) carrying capacity.
   (b) maximum sustainable yield.
   (c) resource depletion.
   (d) endangered species designation.
   (e) population overshoot.

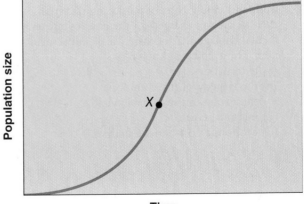

3. Under the provisions of the National Environmental Policy Act (NEPA), which of the following would require the preparation of an environmental impact statement (EIS)?
   (a) The construction of a house on privately owned land
   (b) The paving of a parking lot for a local business
   (c) The expansion of an interstate highway
   (d) The planting of trees in front of City Hall
   (e) The revision of local zoning ordinances

4. Federally owned land in the United States can *best* be described as
   (a) 25 percent of all land, with the majority of it in the west.
   (b) 42 percent of all land, with the majority of it in the east.
   (c) 28 percent of all land, with most of it in Texas.
   (d) 20 percent of all land, with 10 percent of it in the west.
   (e) 35 percent of all land, with the majority of it in the east.

5. The four major public land management agencies in the United States operate under the principle of multiple use. Which of the following uses is common to all four agencies' lands?
   (a) Hunting
   (b) Mining
   (c) Grazing
   (d) Timber harvesting
   (e) Recreation

6. For many years, forest fires were suppressed to protect lives and property. This policy has led to
   (a) a buildup of dead biomass that can fuel larger fires.
   (b) many forest species being able to live without having their habitats destroyed.
   (c) increased solar radiation in most ecosystems.
   (d) soil erosion on steep slopes.
   (e) economic instability.

7. When we purchase an item, we are charged for the labor and supply costs of producing that item. However, we are not charged for the costs of any environmental damage that occurred in manufacturing that item. Those costs are known as
   (a) externalities.
   (b) the tragedy of the commons.
   (c) the maximum sustainable yield.
   (d) marginal costs.
   (e) economic cost-benefit analysis.

8. Which of the following is *not* an environmental consequence of clear-cutting?
   (a) Increased soil erosion and sedimentation in nearby streams
   (b) Decreased biodiversity due to habitat fragmentation
   (c) Increased fish populations due to the influx of nutrients into streams
   (d) Decreased tree species diversity due to the loss of shade-tolerant species
   (e) Stands of same-aged trees

9. Which of the following are environmental impacts of urban sprawl?
   I   Greater reliance on the automobile and increased fossil fuel consumption
   II  Increased consumption of land for housing and highway construction
   III Loss of valuable farmlands
   (a) I only
   (b) II only
   (c) I and II only
   (d) II and III only
   (e) I, II, and III

10. Which of the following was a significant cause of urban sprawl over the past 50 years?
    (a) Migration of people from rural areas to large central cities
    (b) Increased availability of public transportation
    (c) Lower property taxes in urban areas
    (d) Use of the federal gasoline tax to construct and maintain highways
    (e) Improved infrastructure and reduced crime rates in urban areas

11. Which of the following is *not* an environmental benefit of smart growth?
    (a) Reduced flooding
    (b) Increased impervious surfaces
    (c) Reduced fossil fuel consumption
    (d) Increased open space
    (e) Decreased water pollution

1. The property pictured below is the Farm Barn at Shelburne Farms, a National Historic Landmark, non-profit environmental education center, and 1,400-acre working farm on the shores of Lake Champlain. However, for the sake of this exercise, let's assume that the property pictured below belongs to the federal government.

(a) Identify and explain which of the four public land management agencies would be involved in managing this public land.

(b) Applying any three of the basic principles of smart growth, explain how the private land surrounding this federally owned property might be developed to minimize environmental impacts.

(c) Define *environmental impact statement* and describe one condition under which an EIS might be required for the use of either the privately owned or federally owned lands associated with this tract.

2. The town of Fremont met recently to discuss the pros and cons of protecting prairie dogs. Prairie dogs are burrowing rodents the size of rabbits that live in colonies underground in grasslands and prairies. Their numbers have been greatly reduced over the last few decades. Dr. Masser, a local biologist, pointed out that prairie dogs are an important part of the prairie food web, as they are prey for many birds and mammals. Without federal protection from both the Bureau of Land Management and the U.S. Fish and Wildlife Service, they could become extinct in a few years. Dr. Masser also explained that 2 of the 5 species of prairie dogs are already listed as either threatened or endangered. Local ranchers disagreed. Mr. Smith stated that he will continue to poison or shoot the prairie dogs on his land because they destroy the grasses that are needed by his livestock, and he encouraged the BLM to do the same on public lands.

(a) Explain the tragedy of the commons in general terms. Then, using the information you just read about the prairie dog and any other relevant information, incorporate the town of Fremont's discussion into your explanation.

(b) Identify and discuss *one* argument in favor of preserving western grasslands as habitat for prairie dogs and *one* argument in favor of maintaining those grasslands for the grazing of livestock.

(c) Identify *one* action that the Bureau of Land Management and *one* action that the U.S. Fish and Wildlife Service could take to resolve this land use conflict.

## MEASURE YOUR IMPACT

**The Costs of Commuting** Imagine that you are one of 1,000 people who used to live in a downtown area and walked or took public transportation to work. Now you and the others have moved to the suburbs and drive 20 km per day each way.

(a) How many additional kilometers are being driven each day by those 1,000 people?

(b) How many additional kilometers are being driven each week by those 1,000 people? (Assume that each person works 5 days a week.)

(c) How many additional kilometers are being driven each year by those 1,000 people?

(d) If the average car gets 10 km per liter, how many liters of gasoline will be used by those 1,000 people each year?

(e) Assuming that you are not going to move back to an urban area or change your job, what measures could you take to lessen your impact on the environment?

## ENGAGE YOUR ENVIRONMENT

Do you live in a smart growth community? Evaluate the transportation options and open space in your community.

*See Engage Your Environment Activity 10 at the end of this book.*

# Agriculture: Feeding the World

## A Farm Where Animals Do Most of the Work

It is common for a farmer growing juicy tomatoes or sweet apples to become a local hero, but how often does a farmer become known throughout an entire country?

Joel Salatin of Swoope, Virginia, has been featured in a best-selling book and two major motion pictures. He is the author of numerous popular books, including *Holy Cows* nutritious. In nature, Salatin says, birds follow herbivores to take advantage of their waste, so that's what he does on his farm.

Salatin uses other animals to take care of a different manure problem: the buildup of cow manure during winter, when his cattle are kept in a barn. Some farmers push that manure out the doorway and end up with huge manure piles

> ## Salatin understands that farming, done properly, has to replicate many of the processes that exist in nature.

*and Hog Heaven: The Food Buyer's Guide to Farm Friendly Food* and *Everything I Want to Do Is Illegal: War Stories From The Local Food Front.* His farming has drawn the attention of authors, filmmakers, and tens of thousands of other people because he raises vegetables and livestock in a sustainable, organic way. Salatin understands that farming, done properly, has to replicate many of the processes that exist in nature.

Salatin told author Michael Pollan that he is a "grass farmer." Salatin understands that grass is at the foundation of the trophic pyramid. On his Polyface Farm, he grows grass that is harvested, dried by the Sun, and turned into hay, which he feeds to cattle. The cattle are processed into beef and consumed by humans. But that is only one piece of the intricate food chain that Salatin has created on his farm—all without chemical pesticides or synthetic fertilizers.

As cattle graze a field, they leave behind large piles of manure. After a few days, these piles of manure become the energy source for large numbers of grubs and fly larvae. Before the larvae can hatch and produce a fly problem on the farm, however, Salatin brings in his "sanitation crew"—not people spraying pesticides, but chickens, which eat the larvae and, in the process, spread the nitrogen-rich cow manure and deposit some of their own. Throughout this process, the chickens continue to lay eggs, which are tasty and

in the spring. Salatin has a different approach. He layers straw (dried stalks of plants—in essence, another type of grass) and wood chips (from another producer, trees) on top of the manure and allows the floor of his barn to rise higher and higher during winter. Occasionally he throws down some corn. During the winter, the manure-straw-corn layer decomposes and produces heat, which allows the cattle to use less energy and food to stay warm. In spring, when the cattle go out to pasture, he brings pigs into the barn. Pigs have an excellent sense of smell, and they dig up soil—or in this case, layers of manure and straw—with their snouts. One of their favorite foods is corn, and they find fermented corn that has been layered in a manure pile especially delectable. Within a few weeks, the pigs provide Salatin with a barn full of thoroughly mixed compost that required no machinery to make: his animals have done the work for him while they gained nutrients and calories. ▶

The "sanitation crew" cleans up after the cattle.

◀ Joel Salatin at his Polyface Farm.

These are just two examples of the way one farmer uses natural processes and knowledge of plants, animals, and the natural world to grow food for human consumption. Most agricultural methods, especially the modern ones practiced in the developed countries of the world, tend to fight the basic laws of ecology to provide large amounts of food for humans. These modern methods also use large amounts of fossil fuel in every step of the process, from the use of machinery in plowing and harvesting crops to the production of fertilizers and pesticides to the transportation and processing of the food. In contrast, Joel Salatin and other farmers interested in sustainable farming are finding ways to feed the human population by working with, not against, the natural world. ■

*Sources:* M. Pollan, *The Omnivore's Dilemma,* Penguin, 2007; R. Kenner (director/producer) and E. Schlosser (producer), *Food Inc.,* Magnolia Home Entertainment, 2009.

## UNDERSTAND THE KEY IDEAS

A rapidly growing human population needs increasing amounts of food to sustain it. Since agriculture was first practiced more than 10,000 years ago, farmers have been able to increase the world's agricultural output with improved technology and more efficient use of resources. Today, improvements in food production are also being realized through direct manipulation of genetic material in crops. While these improvements have been beneficial to humanity, they also have environmental costs.

After reading this chapter you should be able to

■ describe human nutritional needs and the challenges of overcoming hunger and malnutrition.

■ explain the development of modern industrial agriculture, the role played by inputs such as irrigation water, fertilizers, and pesticides, and the environmental consequences of modern farming methods.

■ identify the benefits and costs of using genetically modified organisms in agriculture.

■ describe alternatives to industrial farming methods.

■ explain the environmental impacts of various approaches to raising and harvesting meat and fish.

# Human nutritional requirements are not always satisfied

Humans survived by hunting and gathering until 10,000 years ago—plus or minus a few thousand years—when agriculture began. Before that time, our ancestors were subject to natural and human-caused variations in the abundance of wild animals and plants. Starvation was common, and, in times of scarcity, very few people received an adequate diet. Advancements in agricultural methods have greatly improved the human diet over the last 10,000 years. In particular, tremendous gains in agricultural productivity and food distribution were made in the 20th century. But despite these advances, many people throughout the world still do not get adequate nutrition. As many as 24,000 people starve to death each day—8.8 million people each year.

As **FIGURE 11.1** shows, about 1 billion people worldwide lack access to adequate amounts of food. That number had been declining for decades, but has been increasing again since 1996. The recent increase is due in part to a decrease in government assistance for agricultural development in much of the developing world over the last 15 years. More recently, it has been made worse by an increase in fuel prices and the global economic downturn.

In this section we will look at basic human nutritional needs, how those needs are met, and the reasons why they are not always met. In the following sections we will discuss various agricultural methods and their environmental implications.

## Nutritional Requirements

Chronic hunger, or **undernutrition,** means not consuming enough calories to be healthy. Food calories are

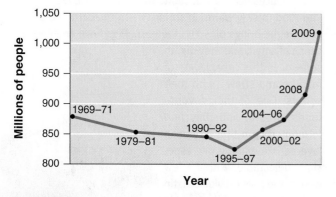

**FIGURE 11.1** **Global undernutrition.** The number of undernourished people declined throughout the 1970s and 1980s, reaching a low of about 800 million in 1996. Since that time, it has risen. [After http://www.imf.org/external/pubs/ft/fandd/2010/03/picture.htm.]

converted into usable energy by the human body. Not receiving enough food calories leads to an energy deficit. An average person needs approximately 2,200 kilocalories per day, though this amount varies with gender, age, and weight. A long-term food deficit of only 100 to 400 kilocalories per day—in other words, an intake of 100 to 400 kilocalories less than one's daily need—deprives a person of the energy needed to perform daily activities and makes the person more susceptible to disease. In children, undernutrition can lead to improper brain development and lower IQ.

In addition, a person lacking sufficient food calories is probably lacking sufficient protein and other nutrients. According to the World Health Organization (WHO), 3 billion people—nearly one-half of the world's population—are **malnourished;** that is, regardless of the number of calories they consume, their diets lack the correct balance of proteins, carbohydrates, vitamins, and minerals, and they experience malnutrition.

The Food and Agriculture Organization of the United Nations (FAO) defines **food security** as the condition in which people have access to sufficient, safe, and nutritious food that meets their dietary needs for an active and healthy life. *Access* refers to the economic, social, and physical availability of food. **Food insecurity** refers to the condition in which people do not have adequate access to food.

**Famine** is a condition in which food insecurity is so extreme that large numbers of deaths occur in a given area over a relatively short period. Actual definitions of famine vary widely depending on the agency using the term. One relief agency defines a famine as an event in which there are more than 5 deaths per day per 10,000 people due to lack of food. According to this definition, there is an annual mortality rate of 18 percent during a famine. Famines are often the result of crop failures, sometimes due to drought, although famines can have social and political causes as well.

Even when people have access to sufficient food, a deficit in just one essential vitamin or mineral can have drastic consequences. The WHO estimates that more than 250,000 children per year become blind due to a vitamin A deficiency. Iron deficiency, known as **anemia,** is the most widespread nutritional deficiency in the world. The WHO estimates that there are 3 billion anemic people in the world—many in developing countries, but also in developed countries. Lack of dietary iron is the major cause of anemia, but there are other causes, including malaria, AIDS, and parasite infestations. An increase in the ingestion of iron-rich foods, such as certain grains, herbs, vegetables, and meats, can reduce anemia.

In the last few decades, one other form of malnutrition has been increasing. **Overnutrition,** the ingestion of too many calories and improper foods, causes a person to become overweight. The WHO estimates that there are over 1 billion people in the world who are overweight, and that roughly 300 million of those people are *obese,* meaning they are more than 20 percent above their ideal weight. Overnutrition is a type of malnutrition that puts people at risk for a variety of diseases, including Type 2 diabetes, hypertension, heart disease, and stroke. While overnutrition is common in developed countries such as the United States, it can also coexist with malnutrition in developing countries. Childhood obesity is a related condition that has started to occur in greater numbers. Overnutrition is often a function of the availability and affordability of certain kinds of foods. For example, overnutrition in the United States has been attributed in part to the easy availability and low cost of processed foods containing ingredients such as high-fructose corn syrup, which are often high in calories and lacking in other nutritional content.

Humans eat a variety of foods, but *grains*—the seed-like fruits of corn (maize), rice, wheat, and rye, among others—make up the largest component of the human diet. Worldwide, there are roughly 50,000 edible plant species, but just three of them—corn, rice, and wheat—constitute 60 percent of human energy intake. Globally, we produce over 300 kg (660 pounds) of grain per person per year, followed by meat at 40 kg (88 pounds) per person per year, and fish at about 21 kg (46 pounds) per person per year. **Meat,** the second largest component of the human diet, is usually defined as livestock (beef, veal, pork, and lamb) and poultry (chicken, turkey, and duck). As income increases with economic growth, people tend to add more meat to their diet. As FIGURE 11.2 shows, meat consumption has been increasing both globally and in the United States.

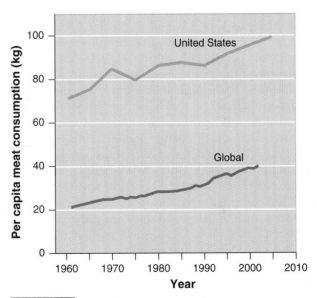

**FIGURE 11.2** **Annual meat consumption.** Per capita meat consumption has increased, both globally and in the United States. [After http://earthtrends.wri.org.]

## Reasons for Undernutrition and Malnutrition

Currently, the world's farmers grow enough grain to feed at least 8 billion people, which would appear to be more than enough for the world's population of 6.8 billion. And grain is only a little more than one-half of the food we produce, in terms of calorie content or biomass. Why, then, are undernutrition and malnutrition common in so many countries of the world? There are many ways to examine this question, and the answers are equally complex.

The primary reason for undernutrition and malnutrition is *poverty:* the lack of resources that allow one access to food. According to many food experts, starvation on a global scale is the result of unequal food distribution rather than absolute food scarcity. In other words, the food exists, but not everyone has access to it. This may mean that people cannot afford to buy the food they need, which is a problem that cannot be solved just by producing more grain.

In addition, political and economic factors play an important role. For example, refugee populations that have fled their homes due to war or natural disasters do not have access to food that they may have grown and stored, but left behind when they became refugees. Lack of an adequate food supply has led to political unrest many times in modern history. People without the means to feed themselves or their families are likely to resort to crime or violence in an attempt to improve their situation. Most recently, a rise in food prices in 2008 led to food riots in Haiti, Egypt, Ivory Coast, Cameroon, Yemen, and elsewhere. Poor governance and political unrest can lead to inadequate food supplies as well, so they can be both causes and effects of undernutrition and malnutrition.

Food researchers have also observed that large amounts of agricultural resources are diverted to feed livestock and poultry rather than people. In fact, roughly 40 percent of the grain grown in the world is used to feed livestock. In the United States, the two largest agricultural crops, corn and soybeans, are grown primarily as animal feed. When these foods are fed to livestock, the low efficiency of energy transfer causes much of the energy they contain to be lost from the system, as we saw in Chapter 3. Ultimately, perhaps only 10 to 15 percent of the calories in grain or soybeans fed to cattle are converted into calories in beef. If people ate producers, such as grains and soybeans, rather than primary consumers, such as cattle, it is possible that more food would be available for people.

FIGURE 11.3 shows global grain production since 1950. A large number of factors influence grain production, including the amount of land under cultivation, global weather and precipitation patterns, world prices for grain, and the productivity of the land on which grain is being grown. As Figure 11.3a shows, grain production has been steadily increasing, though there

(a) Total grain production

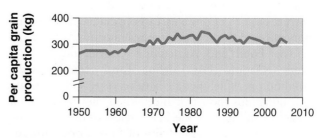

(b) Per capita grain production

**FIGURE 11.3** **Global grain production, 1950–2006.** (a) Global grain production grew rapidly from 1950 through the mid-1980s. Growth has continued since then, but it has slowed. (b) Global per capita grain production has leveled off and may be starting to decline. [After http://earthtrends.wri.org.]

have been some years within the last decade with no increase. Per capita grain production, however, has been level since 1980 and may actually be declining. No one knows exactly why global per capita grain production has not continued to increase. Some environmental scientists suggest that we have finally reached the limit of our ability to supply the human population with food. Others blame some of the political and social factors mentioned above.

In Chapter 7 we learned that the human population is expected to be close to 9 billion, or even larger, by 2050. To feed everyone in the future, we will have to put more land into agricultural production, improve crop yields, reduce consumption of meat, harvest more from the world's fisheries, or use some combination of these strategies.

Experts disagree on whether it is feasible to greatly expand food production. Optimists point to new techniques for crop development as well as unused land in tropical rainforests and grasslands, while pessimists argue that climate change, decreasing biodiversity, dwindling supplies of water for irrigation, diminishing topsoil, and other factors may make it difficult to produce more crops. To understand what we can do to improve food security in all nations of the world and reduce the environmental consequences of growing food, we will need to examine the farming methods we use today and the trade-offs involved in using those methods.

**GAUGE YOUR PROGRESS**

✓ What are undernutrition and malnutrition?

✓ What is food insecurity?

✓ What are some of the possible reasons for food insecurity in the world today?

# The Green Revolution and industrial farming methods have transformed agriculture

Roughly 10,000 years ago, people began to domesticate and raise animals, cultivate the soil, and domesticate, or modify by selective breeding, certain wild plant species to turn them into crops. The advent of agriculture was a major event in human history because it enabled people to move beyond a subsistence level of existence. At the same time, it has had some negative consequences.

The abundance of food supplied by agriculture is one factor that has led to the exponential growth of the human population, as we saw in Chapter 7. Deliberate cultivation of food also initiated a level of environmental degradation never before experienced on Earth. Some people speak of agriculture collectively as "the single most harmful human action that has taken place." While this may be an overstatement, it is true that an increase in food production leads to a positive feedback loop: with more food come more people, who require even more food production.

In the twentieth century, farming became more mechanized, and the use of fossil fuel energy increased. These changes have led to increasing food output as well as a variety of environmental impacts. **Industrial agriculture,** or **agribusiness,** applies the techniques of the Industrial Revolution—mechanization and standardization—to the production of food. Today's modern agribusinesses are quite different from the small family farms that dominated agriculture a few decades ago. We will begin this section by considering the ways in which modern industrial agriculture is dependent on energy, then describe the activities those additional energy inputs allow.

## The Energy Subsidy in Agriculture

A great deal of energy, in addition to solar energy, goes into growing, harvesting, processing, and preparing food. These energy inputs include both fossil fuel energy and human energy. The energy input per calorie of food produced is called the **energy subsidy.** In other words, if we use 5 calories of energy to produce food, and we

receive 1 calorie of energy when we eat that food, then the food has an energy subsidy of 5. We can think of this in another way if we use mass of inputs and outputs as a substitute measure for energy. It may take 20 kg (44 pounds) of grain to feed to cattle to produce 1 kg (2.2 pounds) of beef (an energy subsidy of 20). It takes only 2.8 kg (6 pounds) of grain to feed to chickens to produce 1 kg (2.2 pounds) of chicken meat (an energy subsidy of 2.8). This analogy is not perfect, because the energy content of 1 kg of beef is greater than that of 1 kg of grain. Nevertheless, it should help you understand the relationship between inputs and outputs in producing food. There are other considerations related to producing food, such as how much land is needed. This concept was part of the ecological footprint described in Chapter 1. Crunch the Numbers "Land Needed for Food" shows you how to calculate the different land requirements for obtaining sufficient calories from corn and from beef.

As **FIGURE 11.4** shows, traditional small-scale agriculture requires a relatively small energy subsidy: it uses few energy inputs per calorie of food produced. By contrast, food writer Michael Pollan estimates that if you eat the average modern U.S. diet obtained from a typical supermarket, containing primarily foods produced

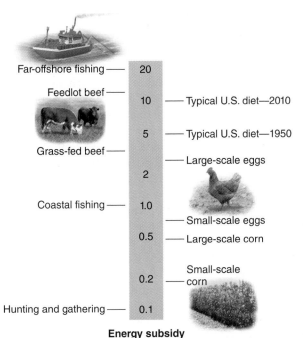

**FIGURE 11.4** **Energy subsidies for various methods of food production and diets.** Energy input per calorie of food obtained is greater for modern agricultural practices than for traditional agriculture. Energy inputs for hunting and gathering and for small-scale food production are mostly in the form of human energy, whereas fossil fuel energy is the primary energy subsidy for large-scale modern food production. All values are approximate, and for any given method there is a large range of values.

### Land Needed for Food

We have seen that raising beef requires more resources than growing corn. Let's look at some of the actual numbers.

On farms in the midwestern United States, a hectare of land yields roughly 370 bushels of corn (equivalent to 150 bushels per acre). A *bushel* consists of 1,250 ears of corn, and each ear typically contains 80 kilocalories. Assume that a person eats only corn and requires 2,000 kilocalories per day. Although this assumption is not very realistic, it allows an approximation of how much land it would take to feed that person.

The person's food requirement is

$$2{,}000 \text{ kilocalories/day} \times 365 \text{ days/year}$$
$$= 730{,}000 \text{ kilocalories/year}$$

A hectare of corn produces

$$370 \text{ bushels/hectare} \times 1{,}250 \text{ ears/bushel}$$
$$\times 80 \text{ kilocalories/ear} = 37{,}000{,}000 \text{ kilocalories/hectare}$$

Thus, one person eating only corn can obtain sufficient calories from 0.02 ha (0.05 acres) of land.

What if that person ate only beef? We have seen that it takes 20 kg of grain to produce 1 kg of beef. So it would take 20 times as much land, or 0.4 ha (1 acre), to feed a person who ate only beef.

What happens if we extend this analysis to a global scale? If Earth has about 1.5 billion hectares (3.7 billion acres) of land suitable for growing food, is there sufficient land on Earth to feed all 6.8 billion inhabitants of the planet if they all ate a diet of only beef?

$$6.8 \text{ billion people} \times 0.4 \text{ ha/person} = 2{,}720{,}000{,}000 \text{ ha}$$

So 2.72 billion hectares (6.72 billion acres) would be needed, and the answer is no.

**Your Turn:** How many people eating a beef-only diet could Earth support?

---

by modern agricultural methods, there is a 10-calorie energy input for every calorie you eat. As this difference shows, food choices are energy choices.

Most of the energy subsidies in modern agriculture are in the form of fossil fuels, which are used to produce fertilizers and pesticides, to operate tractors, to pump water for irrigation, and to harvest food and prepare it for transport. Other energy subsidies take place off the farm. For example, the average food item travels 2,000 km (1,240 miles) from the farm to your plate, so we often spend far more energy on transporting food than we get from the food itself. The Department of Energy reports that in the United States, 17 percent of total commercial energy use goes into growing, processing, transporting, and cooking food. Those of us eating a "supermarket" diet in the developed world are highly dependent on fossil fuel for our food; the modern agricultural system would not work without it.

How did we get to this point? In the twentieth century, the agricultural system was dramatically transformed from a system of small farms relying mainly on human labor and with relatively low fossil fuel inputs to a system of large industrial operations with fewer people and much more machinery. This shift in farming methods, known as the **Green Revolution,** involved new management techniques and mechanization as well as the triad of fertilization, irrigation, and improved crop varieties. These changes increased food production dramatically, and farmers were able to feed many more people.

The Green Revolution began with the work of crop scientists, particularly the American scientist Norman Borlaug (1914–2009), who won the Nobel Peace Prize for his contribution to increasing the world food supply. Through intensive breeding, Borlaug and other agricultural researchers developed new strains of wheat that produced higher yields and were disease resistant. They also used fertilizers and irrigation to improve yields. In the 1940s, researchers brought these techniques to developing nations such as Mexico and the Philippines to help them increase their agricultural output to feed their growing populations. From the 1950s through the 1970s, many countries, particularly those in the developing world, underwent similar shifts in the way they farmed.

The upward trend in world grain production that can be seen in Figure 11.3a is the result of the Green Revolution. From the mid-1960s through the mid-1980s, world grain production doubled. By the 1990s, there were at least 18 organizations around the world promoting and developing Green Revolution techniques, including mechanization, irrigation, use of fertilizer, monocropping, and use of pesticides. However, the Green Revolution has also had negative environmental impacts. Let's examine some Green Revolution practices in more detail.

**MECHANIZATION** Farming involves many kinds of work. Fields must be plowed, planted, irrigated, weeded, protected from pests, harvested, and prepared for the next season. Even after harvesting, crops must be dried, sorted, cleaned, and prepared for market in almost as many different ways as there are different crops. Machines do not necessarily do this work better than humans or animals, but it can be economically advantageous to replace humans or animals with machinery, particularly if fossil fuels are abundant, fuel prices are relatively low, and labor prices are relatively high. In developed countries, where wages are relatively high, less than 5 percent of the workforce works in agriculture. In developing countries, where wages tend to be much lower, 40 to 75 percent of the working population is employed in agriculture.

Since the advent of mechanization, large farms producing staple crops such as beans or corn have generally been more profitable than small farms. The reason for this is **economies of scale,** by which we mean that the average costs of production fall as the output increases. For example, a new combine harvester—a machine that harvests the crop and separates out the grain or seed for transport—costs between $150,000 and $400,000. This large up-front expenditure is a good investment for a large farm, where the cost of the machine is justified by the profits on the increased production that comes from using it. A small farm, however, does not have enough land to be able to recoup the cost of such expensive equipment. Because of economies of scale, profits tend to increase with size, and large agricultural operations generally outcompete small ones. As a result, between 1950 and 2000, the average farm size in Iowa more than doubled, from about 70 ha (173 acres) to over 140 ha (346 acres).

Mechanization also means that single-crop farms are generally more efficient than farms that grow many crops. Mechanized crop planting and picking require specialized, expensive equipment that is specific to each crop, so planting and harvesting only a single type of crop reduces equipment costs.

**IRRIGATION** Irrigation systems like those described in Chapter 9 can increase crop growth rates or even enable crops to grow where they could not otherwise be grown (**FIGURE 11.5**). For example, irrigation has transformed approximately 400,000 ha (1 million acres) of former desert in the Imperial Valley of southeastern California into a major producer of fruits and vegetables. In other situations, irrigation can allow productive land to become extremely productive land. One estimate suggests that, whereas 16 percent of the world's agricultural land is irrigated, that land produces 40 percent of the world's food.

While irrigation has many benefits, including more efficient use of water in some places where it is scarce,

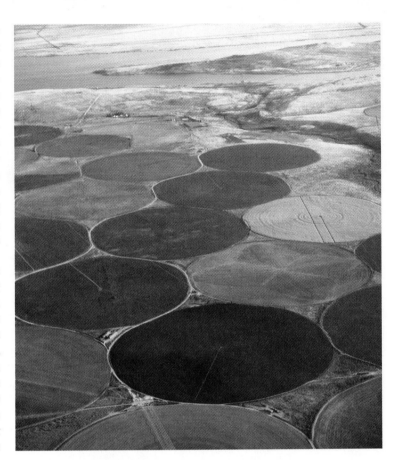

**FIGURE 11.5** **Irrigation circles.** The green circles in this aerial photograph from Oregon are obvious evidence of irrigation.

it can have a number of negative consequences over time. It can deplete groundwater and draw down aquifers, as we saw in the case of the Ogallala aquifer in Chapter 9. In coastal areas, it can promote saltwater intrusion into freshwater wells. It can also contribute to soil degradation through *waterlogging* and *salinization*. **Waterlogging,** which occurs when soil remains under water for prolonged periods, impairs root growth because roots cannot get oxygen. **Salinization** occurs when the small amounts of salts in irrigation water become highly concentrated on the soil surface through evaporation. As **FIGURE 11.6** illustrates, these salts can eventually reach toxic levels and impede plant growth.

**FERTILIZERS** Agriculture removes organic matter and nutrients from soil, and if these materials are not replenished, they can be quickly depleted. Because industrial agriculture keeps soil in constant production, it requires large amounts of fertilizers to replace lost organic matter and nutrients. Fertilizers contain essential nutrients for plants—primarily nitrogen, phosphorus, and potassium—and they foster plant growth where one or more of these nutrients is lacking. There are two types of fertilizers used in agriculture: *organic* and *synthetic*.

FIGURE 11.6 **Irrigation-induced salinization and waterlogging.** Over time, irrigation can degrade soil by leaving a layer of highly concentrated salts at the soil surface and water-logged soil below.

When irrigation water evaporates, the salts are left behind on the soil surface.

Irrigation water

Salts

Transpiration

Salts

Evaporation

Water table

Less permeable soil layer

**Time**

Water table

Waterlogged soil

Salt concentrations eventually increase to levels that are toxic to plants.

Root growth is impaired because roots cannot get oxygen.

As their name suggests, **organic fertilizers** are composed of organic matter from plants and animals. They are typically made up of animal manure that has been allowed to decompose. Traditional farmers often spread animal manure and crop wastes onto fields to return some of the nutrients that were removed from those fields when crops were harvested.

**Synthetic,** or **inorganic,** fertilizers are produced commercially. Nitrogen fertilizers are often produced by combusting natural gas, which allows nitrogen from the atmosphere to be fixed and captured in fertilizer. Fertilizers produced in this way are highly concentrated, and their widespread use has increased crop yields tremendously since the Green Revolution began. Synthetic fertilizers have many advantages over organic fertilizers. They are designed for easy application, their nutrient content can be targeted to the needs of a particular crop or soil, and plants can easily absorb

them, even in poor soils. Worldwide, synthetic fertilizer use increased from 20 million metric tons in 1960 to 160 million metric tons in 2007. It is arguable that without synthetic fertilizers, we could not feed all the people in the world.

Despite these advantages, synthetic fertilizers can have several adverse effects on the environment. The process of manufacturing synthetic fertilizers uses fossil fuel energy. Producing nitrogen fertilizer is an especially energy-intensive process. Furthermore, whereas synthetic fertilizers are more readily available for plant uptake than organic fertilizers, they are also more likely to be carried by runoff into adjacent waterways and aquifers. In surface waters, this nutrient runoff can cause algae and other organisms to proliferate. After these organisms die, they decompose and reduce oxygen levels in the water, a process we will learn more about when we look at water pollution in Chapter 14. Finally,

**FIGURE 11.7** **Fertilizer runoff after fertilization.** The proximity of this drainage ditch to an agricultural field may lead to runoff of fertilizer during a heavy rainstorm.

synthetic fertilizers do not add organic matter to the soil, as organic fertilizers do. As we saw in Chapter 8, organic matter contributes many beneficial properties to soil, including increased cation exchange capacity and water retention abilities.

In the United States, fertilizer use, and consequently nutrient runoff, is somewhat less than in other nations with similar agricultural output. Still, large amounts of nitrogen and other nutrients run into waterways in intensively farmed regions such as California's Central Valley, farming regions along the East Coast, and the Mississippi River watershed (**FIGURE 11.7**).

**MONOCROPPING** Both the mechanization of agriculture and the use of synthetic fertilizers encourage large plantings of a single species or variety, a practice known as **monocropping.** Monocropping is the dominant agricultural practice in the United States, where wheat and cotton are frequently grown in monocrops of 405 hectares (1,000 acres) or more (**FIGURE 11.8**).

Monocropping has greatly improved agricultural productivity. This technique allows large expanses of land to be planted, and then harvested, all at the same time. With the use of large machinery, the harvest can be obtained easily and efficiently. If fertilizer or pesticide treatments are required, those treatments can also be applied uniformly over large fields, which, because they are planted with the same crop, have the same pesticide or nutritional needs.

Despite the benefits of increased efficiency and productivity, monocropping can lead to environmental degradation. First, soil erosion can become a problem.

**FIGURE 11.8** **Monocropping.** This large field in Oregon contains only one crop species: soybeans. There are both advantages and disadvantages to monocropping.

Because fields that are monocropped are readied for planting or harvesting all at once, soil will be exposed over many hectares at the same time. On a 405-hectare (1,000-acre) field that has not yet been planted, wind can blow for over 1.6 km (1 mile) without encountering anything but bare soil. Under these circumstances, the wind can gain enough speed to carry dry soil away from the field. Certain farmland in the United States loses an average of 1 metric ton of topsoil per hectare (2.5 metric tons per acre) per year to wind erosion. As we saw in Chapter 8, this topsoil contains important nutrients, and its loss can reduce productivity.

Monocropping also makes crops more vulnerable to attack by pests. Large expanses of a single plant species represent a vast food supply for any pests that specialize on that particular plant. Such pests will establish themselves in the monocrop and reproduce rapidly. Their populations may experience exponential growth similar to what we saw in Georgii Gause's *Paramecium* populations supplied with unlimited food (see Figure 6.4). Natural predators may not be able to respond rapidly to the exploding pest population. Many predators of crop pests, such as ladybugs and parasitic wasps, are attracted to the pests that feed on monocrops. But these predators also rely on non-crop plants for habitat. Monocropping removes habitat for predators that might otherwise control the pest population.

PESTICIDES Because of the increased pest populations that monocropping encourages, as well as for other reasons, the use of pesticides has become routine and widespread in modern industrial agriculture. **Pesticides** are substances, either natural or synthetic, that kill or control organisms that people consider pests. In the United States, over 227 million kilograms (500 million pounds) of pesticides are applied to food crops, cotton, and fruit trees. The United States accounts for about one-third of worldwide pesticide use.

**Insecticides** target species of insects and other invertebrates that consume crops, and **herbicides** target plant species that compete with crops. Some pesticides are **broad-spectrum** pesticides, meaning that they kill many different types of pests, and some are **selective** pesticides that focus on a narrower range of organisms. The broad-spectrum insecticide dimethoate, for example, kills almost any insect or mite (mites are not insects, but rather relatives of ticks and spiders), while the more selective acequinocyl kills only mites.

The application of pesticides is a rapid, relatively easy response to an infestation of pests on an agricultural crop. In many cases, a single application can significantly reduce a pest population. By preventing crop damage, pesticides allow greater crop yields on less land, thereby reducing the area disturbed by agriculture. Thus the application of pesticides has made agriculture more efficient.

But pesticides, like other industrial agricultural practices, present some environmental problems. One of these problems is that pesticides injure or kill organisms other than their intended targets. Some pesticides, such as dichlorodiphenyltrichloroethane, also known as DDT, are **persistent,** meaning that they remain in the environment for a long time. In 1972, DDT was banned in the United States, in part because it was found to build up over time in the fatty tissues of predators through a process called **bioaccumulation.** DDT is a fat-soluble chemical that is not easily flushed from the body, but instead accumulates in fatty tissues. Whenever an organism containing the pesticide is eaten, the chemical is transferred to the consumer. This process eventually leads to very high pesticide concentrations at high trophic levels. Thus, even if pesticides are not concentrated enough to affect the organisms that initially consume them, they may affect organisms higher up the food chain, such as birds of prey and humans. Bioaccumulation is discussed further in Chapter 17.

Other pesticides, such as the herbicide glyphosate, known by the trade name Roundup, are **nonpersistent,** meaning that they break down relatively rapidly, usually in weeks to months. Nonpersistent pesticides have fewer long-term effects, but they must be applied more often, so their overall environmental impact is not always lower than that of persistent pesticides.

Another disadvantage of pesticide use is that pest populations may evolve resistance to pesticides over time, as described in Chapter 5. Pest populations are usually large and thus contain significant genetic diversity. In those vast gene pools, there are usually a few individuals that are not as susceptible to a pesticide as others, and those individuals may survive an initial application of the pesticide. Those surviving individuals are said to be **resistant** to the pesticide. If the pesticide is successful in reducing the pest population, then in the next generation, the fraction of resistant individuals in that population will increase. As time goes by, resistant individuals will make up a larger and larger fraction of the population. Often the resistance becomes more effective as well, making the pesticide significantly less useful. At this point, crop scientists and farmers must search for a new pesticide. The cycle of pesticide development, followed by pest resistance, followed by new pesticide development, and so on—called the **pesticide treadmill**—is an example of a positive feedback system, as shown in FIGURE 11.9.

Pesticides can cause even wider environmental effects. They may kill organisms that benefit farmers, such as predatory insects that eat crop pests, pollinator insects that pollinate crop plants, and plants that fix nitrogen and improve soil fertility. Furthermore, chemical pesticides, like fertilizers, can run off into surrounding surface waters and potentially enter groundwater, a problem we will look at in detail in Chapter 14. The toxicity of

Develop a new pesticide

Apply pesticide again, with little result.

Crop infested with pests

Apply pesticide.

The proportion of resistant individuals in the pest population grows.

Most pests die, but some resistant individuals survive.

**FIGURE 11.9** **The pesticide treadmill.** Over time, pest populations evolve resistance to pesticides, requiring farmers to use higher doses or to develop new pesticides.

pesticides to farmworkers has been well documented. The risk to humans who ingest food that has been treated with pesticides is a subject of some debate and will be covered in Chapter 17.

**GAUGE YOUR PROGRESS**

✓ Explain the energy subsidy in agriculture.

✓ What are the major features of the Green Revolution?

✓ What are the pros and cons of pesticide use?

# Genetic engineering is revolutionizing agriculture

As we saw in Chapter 5, humans have modified plants and animals by artificial selection for thousands of years. Many crop species, for example, have been modified to increase their output of seeds or fruits. The modern techniques of genetic engineering, however, go far beyond these traditional practices. Scientists today can isolate a specific gene from one organism and transfer it into the genetic material of another, often very different, organism, producing a genetically modified organism, or GMO. By manipulating specific genes, agricultural

scientists can rapidly produce organisms with desirable traits that may be impossible to develop by traditional breeding techniques.

## The Benefits of Genetic Engineering

Genetically modified crops and livestock offer the possibility of greater yield and food quality, reductions in pesticide use, and higher profits for the agribusinesses that use them. They are also seen as a way to help reduce world hunger by increasing food production and reducing losses to pests and varying environmental conditions.

**INCREASED CROP YIELD AND QUALITY** Genetic engineering can increase food production in several ways. It can create strains of organisms that are resistant to pests and harsh environmental conditions such as drought or high salinity. In addition, agricultural scientists have begun to engineer plants to produce essential nutrients for humans. For example, agricultural scientists have inserted a gene for the production of vitamin A into rice plants, creating new seeds known as golden rice (**FIGURE 11.10**). Although golden rice is still an experimental product, some scientists hope that it will help reduce the incidence of blindness resulting from vitamin A deficiency. Crop plants, animals, and bacteria have also been modified to produce pharmaceuticals and other compounds, a process that can make these products

**(a)**  **(b)**

**FIGURE 11.10** **White rice and golden rice.** (a) Crop scientists have inserted a gene that produces vitamin A into white rice. (b) The resulting genetically modified rice is called golden rice.

far less expensive to produce. A number of projects are under way to create genetically modified animals for food production, including a salmon that grows to its full size of 3.6 kg (8 pounds) in 18 months—half the growing time of an unmodified fish.

**POTENTIAL CHANGES IN PESTICIDE USE** Genetic engineering for resistance to pests could reduce the need for pesticides. Corn, for example, is subject to attacks from the bollworm, European corn borer (*Ostrinia nubilalis*), and other lepidopteran (butterfly and moth) larvae. *Bacillus thuringiensis* is a natural soil bacterium that produces a toxin that can kill lepidopterans. The bacterium's insecticidal gene, known as *Bt,* has been inserted into the genetic material of corn plants, resulting in a genetically modified plant that produces a natural insecticide in its leaves. By 2009, 63 percent of the land area planted with corn in the United States was planted with Bt corn. Growers of Bt corn have been able to reduce the amount of synthetic pesticide used on their corn crops.

A similar technique has been used to create crop plants that are resistant to the herbicide Roundup. The "Roundup Ready" gene allows growers to spray the herbicide on their fields to control the growth of weeds without harming the crop plants. It is now widely used in corn, soybean, and cotton plants. The success of no-till agriculture, which we will discuss later in this chapter, rests largely on the use of herbicides and the introduction of herbicide-resistant crops.

**INCREASED PROFITS** Because pesticides can be a significant expense on any farm, genetically modified crops have the potential to reduce expenses. In addition, because GMO crops often produce greater yields, there is also the potential for an increase in revenues. Both of these changes can lead to higher incomes for farmers, lower food prices for consumers, or both.

## Concerns about Genetically Modified Organisms

Industrial agriculture relies more heavily on genetically modified crops each year. In 2009, 63 percent of the corn, 91 percent of the soybeans, and 71 percent of the cotton planted in the United States came from genetically modified seeds. However, many European countries, as well as a number of people in the United States, question the safety of GMOs. Genetically modified crops and livestock are the source of considerable controversy. Concerns that have been raised include their safety for human consumption and their effects on biodiversity. Regulation of GMOs is also an issue, both in the United States and abroad.

**SAFETY FOR HUMAN CONSUMPTION** Some people are concerned that the ingestion of genetically modified foods may be harmful to humans, although so far there is little evidence to support these concerns. Researchers are studying the possibility that GMOs may cause allergic reactions when people eat a food containing genes transferred from another food to which they are allergic.

**EFFECTS ON BIODIVERSITY** There is some concern that if genetically modified crop plants are able to breed with their wild relatives—as many domesticated crop plants are—the newly added genes will spread to the wild plants. The spread of such genes might then alter or eliminate natural plant varieties. Examples of GMOs crossing with wild relatives do exist. Because of these concerns, attempts have been made to introduce buffer zones around genetically modified crops.

The use of genetically modified seeds is contributing to a loss of genetic diversity among food crops. As with any reduction in biodiversity, we cannot know what beneficial genetic traits might be lost. For example, researchers recently discovered that one variety of sorghum, a cereal grass grown for its sweet juice extract, has a natural genetic variation that gives it resistance to a pest called the greenbug. This particular variety has since been used extensively to confer greenbug resistance on the U.S. sorghum crop. If growers had been growing only one or two genetically modified varieties of sorghum, this naturally resistant variety might have been eliminated before its beneficial trait was discovered.

**REGULATION OF GENETICALLY MODIFIED ORGANISMS** Currently, there are no regulations in the United States that mandate the labeling of genetically modified foods. Opponents of labeling argue that labeling of foods containing GMOs might suggest to consumers that there is something wrong with GMOs. They also argue that such labeling would be too difficult because small amounts of GMO materials are found throughout the U.S. agricultural system. Those who want to avoid consuming GMOs can safely purchase organic food; the federal definition of "organic" excludes genetically modified foods (see the Science Applied section that follows this chapter).

The U.S. government has not yet approved any genetically modified animals for market, but applications for a number of genetically modified animals are currently under consideration by the Food and Drug Administration.

**GAUGE YOUR PROGRESS**

✓ What is a genetically modified organism?

✓ What are the major benefits of using GMOs?

✓ What are the disadvantages of using GMOs?

# Alternatives to industrial farming methods are gaining more attention

Industrial agriculture has been so successful in reducing labor inputs, and has therefore become so widespread, that we generally call this type of farming **conventional agriculture.** However, in situations in which the cost of labor is not the most important consideration, traditional farming techniques may be economically successful as well. Small-scale farming is common in the developing world, where labor is less expensive than machinery and fossil fuels. In these countries, there are still many farmers growing crops on small plots of land. Traditional farming methods that differ from those of industrial agriculture include shifting agriculture and nomadic grazing, which are sometimes not sustainable, and more sustainable methods such as intercropping and agricultural forestry.

## Shifting Agriculture and Nomadic Grazing

Locations with a moderately warm climate and relatively nutrient-poor soils, such as the rainforests of Central and South America, lend themselves to *shifting agriculture.* In these environments, a large fraction of the nutrients are contained within the vegetation. **Shifting agriculture** involves clearing land and using it for only a few years until the soil is depleted of nutrients. This traditional method of agriculture uses a technique sometimes called "slash-and-burn," in which existing trees and vegetation are cut down, placed in piles, and burned (**FIGURE 11.11**). The resulting ash is rich in potassium, calcium, and magnesium, which makes the soil more fertile. However, these nutrients are usually depleted after a few years. If the deforestation occurs in an area of heavy rainfall, nutrients may be washed away, along with some of the soil, which further reduces nutrient availability. After a few years, the farmer usually moves on to another plot and repeats the process.

If a plot is used for a few years and then abandoned for a number of decades, over time, the soil may recover its organic content and nutrient supplies, and the vegetation may have a chance to regrow. However, population pressures may cause the land to be used too frequently to allow for its full recovery. In that case, soil productivity can decrease rapidly, leaving the land suitable only for animal grazing. In addition, the burning process *oxidizes* carbon, meaning that it converts it into the oxide compounds carbon monoxide (CO) and carbon dioxide ($CO_2$). In this way, carbon from the vegetation and the soil is released into the atmosphere and ultimately contributes to higher atmospheric $CO_2$ concentrations.

In semiarid environments, dry, nutrient-poor soils can be easily degraded by agriculture to the point at which they are no longer viable for any production at

**FIGURE 11.11** **Shifting agriculture.** This forest has been cleared for agriculture. Clearing land often involves burning it, which may make nutrients and soils vulnerable to erosion.

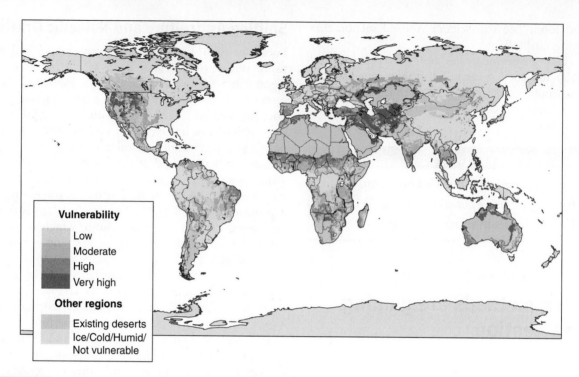

**FIGURE 11.12** **Vulnerability to desertification.** Certain regions of the world are much more vulnerable to desertification than others. [After www.fao.org.]

all. Irrigation can cause salinization, and topsoil is eroded away because the shallow roots of annual crops fail to hold it in place. This process is called **desertification.** The world map in **FIGURE 11.12** shows the parts of the world that are most vulnerable to desertification. Desertification is occurring most rapidly in Africa, where the Sahara is expanding at a rate of up to 50 km (31 miles) per year. Unsustainable farming practices in northern China are also leading to rapid desertification there.

The only sustainable way for people to use soil types with very low productivity is **nomadic grazing,** in which they move herds of animals, often over long distances, to seasonally productive feeding grounds. If grazing animals move from region to region without lingering in any one place too long, the vegetation can usually regenerate.

Shifting agriculture and nomadic grazing worked well under the conditions in which they were first developed—namely, low human population densities and subsistence farming—but as populations increase, both forms of traditional agriculture become less sustainable. Sometimes the relocation of people for political or other reasons can cause traditionally sustainable agricultural techniques to become unsustainable. For example, in the early part of the twentieth century, many subsistence farmers in Central America were relocated away from rich floodplains to mountainous areas. The plowing methods that worked in flat areas could not be sustained in the mountains because they caused severe erosion.

## Sustainable Agriculture

Is it possible to produce enough food to feed the world's population without destroying the land, polluting the environment, or reducing biodiversity? **Sustainable agriculture** fulfills the need for food and fiber while enhancing the quality of the soil, minimizing the use of nonrenewable resources, and allowing economic viability for the farmer. It emphasizes the ability to continue agriculture on a given piece of land indefinitely through conservation and soil improvement. Sustainable agriculture often requires more labor than industrial agriculture, which makes it more expensive in places where labor costs are high. Practitioners of sustainable agriculture consider the improved long-term productivity of the land to be worth this extra cost, however.

Many of the practices used in sustainable agriculture are traditional farming methods (**FIGURE 11.13**). Subsistence farmers in India, Kenya, or Thailand typically use animal and plant wastes as fertilizer because they cannot obtain or afford synthetic fertilizers. Such traditional farmers may also practice **intercropping** (Figure 11.13a), in which two or more crop species are planted in the same field at the same time to promote a synergistic interaction between them. For example, corn, which requires a great deal of nitrogen, can be planted along with peas, a nitrogen-fixing crop. **Crop rotation** achieves the same effect by rotating the crop species in a field from season to season. For example, peas can be

**(a)** Intercropping

**(c)** Contour plowing

**(b)** Agroforestry

**FIGURE 11.13** **Sustainable farming methods.** A variety of farming methods can be used to improve agricultural yield and retain soil and nutrients.

planted in a field for one year, leaving excess nitrogen in the soil to nourish the corn crop that is planted there in the following year.

Intercropping trees with vegetables—a practice that is sometimes called **agroforestry**—allows vegetation of different heights, including trees, to act as windbreaks and catch soil that might otherwise be blown away, greatly reducing erosion (Figure 11.13b). The trees not only protect the vegetable crops and the soil, but also provide fruit and firewood.

Alternative methods of land preparation and use can also help to conserve soil and prevent erosion. For instance, **contour plowing**—plowing and harvesting parallel to the topographic contours of the land—helps prevent erosion by water while still allowing for the practical advantages of plowing (Figure 11.13c). Some farmers plant an autumn crop, such as winter wheat, that

will sprout before frost sets in, so that the land does not remain bare between regular plantings.

## No-Till Agriculture

As we learned in Chapter 8, soils may take hundreds or even thousands of years to develop as organic matter accumulates and soil horizons form. Conventional agriculture relies on *plowing* and *tilling,* processes that physically turn the soil upside down and push crop residues under the topsoil, thereby killing weeds and insect pupae. Critics argue, however, that plowing and tilling have negative effects on soils. Every time soil is plowed or tilled, soil particles that were attached to other soil particles or to plant roots are disturbed and broken apart and become more susceptible to erosion. In addition, repeated plowing increases the exposure

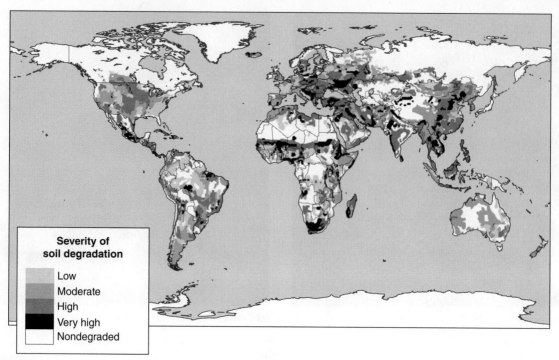

**FIGURE 11.14** **Global distribution of soil degradation.** Soil degradation is a global problem caused by overgrazing and deforestation as well as agricultural mismanagement. [After United Nations Environment Programme.]

Severity of
soil degradation

Low
Moderate
High
Very high
Nondegraded

of organic matter deep in the soil to oxygen. This exposure leads to oxidation of organic matter, a reduction in the organic matter content of the soil, and an increase in atmospheric $CO_2$ concentrations. Tilling, in addition to irrigation and overproduction, has led to severe soil degradation in many parts of the world. The world map in **FIGURE 11.14** indicates areas of severe soil degradation.

**No-till agriculture** is designed to avoid the soil degradation that comes with conventional agricultural techniques. Farmers using this method leave crop residues in the field between seasons (**FIGURE 11.15**). The intact roots hold the soil in place, reducing both wind and water erosion, and the undisturbed soil is able to regenerate natural soil horizons. No-till agriculture also reduces emissions of $CO_2$ because the intact soil undergoes less oxidation. In many cases, however, in order for no-till agriculture to be successful, farmers must apply herbicides to the fields before, and sometimes after, planting so that weeds do not compete with the crops. Therefore, the downside of no-till methods is an increase in the use of herbicides.

**FIGURE 11.15** **No-till agriculture.** Rows of soybeans emerge between the residues of a corn crop left over from the previous season.

## Integrated Pest Management

Another alternative agricultural practice, **integrated pest management (IPM),** uses a variety of techniques designed to minimize pesticide inputs. These techniques include crop rotation and intercropping, the use of pest-resistant crop varieties, creating habitats for predators of pests, and limited use of pesticides.

Crop rotation and the use of pest-resistant crop varieties prevent pest infestations. Crop rotation can foil insect pests that are specific to one crop that may have laid eggs in the soil. It can also hinder crop-specific diseases that may survive on infected plant material from the previous season. Intercropping, as stated earlier, also makes it harder for specialized pests that succeed best with only one crop present to establish themselves. Farmers can also provide habitat for species that prey on crop pests. Agroforestry encourages the presence of insect-eating birds (although birds can also damage some crops), and many herbs and flowers attract beneficial insects (FIGURE 11.16).

Although IPM does use pesticides, it limits applications of pesticides through very careful observation. Farmers regularly inspect their crops for the presence of insect pests or other potential crop hazards. Pest infestations can thus be caught in early stages and treated using natural controls or smaller doses of pesticides than would otherwise be needed. These more targeted methods of pest control can result in significant savings on pesticides as well as improved yields. FIGURE 11.17, from a case study of IPM training in Indonesia, shows the difference IPM can make. Farmers who learned how to determine whether a pesticide application was warranted cut their pesticide applications—and their expenditures on pesticides—in half (Figure 11.17a). Yields also improved after farmers learned IPM methods (Figure 11.17b).

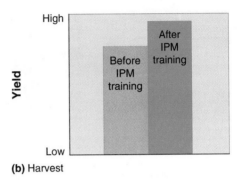

**FIGURE 11.17** **Effects of IPM training.** (a) IPM training of farmers in Indonesia led to a significant reduction in pesticide applications. (b) Yield improvements also occurred after the training because of the additional attention the farmers gave to their crops. [Modified after www.fao.org.]

When farmers take time to inspect their fields carefully, as required by IPM methods, they often notice other crop needs, and this additional attention improves overall crop management. The trade-off for these benefits is that farmers must be trained in IPM methods and must spend more time inspecting their crops. But once farmers are trained, the extra income and reduced costs associated with IPM often outweigh the extra time they must spend in the field. IPM has been especially successful in many parts of the developing world, where the high-input industrial farming model is not feasible because labor costs are low and farmers lack financial resources.

## Organic Agriculture

**Organic agriculture** is the production of crops without the use of synthetic pesticides or fertilizers. Organic agriculture follows several basic principles:

- Use ecological principles and work with natural systems rather than dominating them.
- Keep as much organic matter and as many nutrients in the soil and on the farm as possible.
- Avoid the use of synthetic fertilizers and pesticides.

**FIGURE 11.16** **Beneficial insect habitat.** Practitioners of integrated pest management often provide habitat for insects that prey on crop pests. This wasp is laying eggs in a caterpillar, which it has paralyzed.

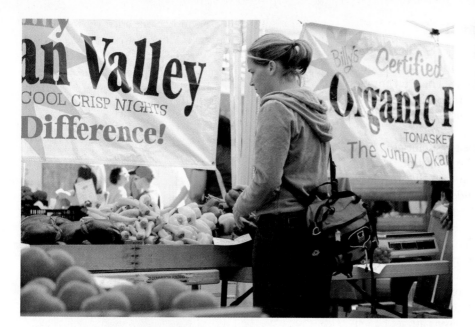

FIGURE 11.18 **Organic farming.** Many of the consumers at this farmers' market are willing to pay higher prices for organically grown food.

- Maintain the soil by increasing soil mass, biological activity, and beneficial chemical properties.
- Reduce the adverse environmental effects of agriculture.

In the developed world, organic farming has increased in popularity over the past three decades. The U.S. Organic Foods Production Act (OFPA) was enacted as part of the 1990 farm bill to establish uniform national standards for the production and handling of foods labeled organic. The Science Applied section that follows this chapter, "How Do We Define Organic Products?" discusses organic food labeling in more detail.

For a long time, organic farms were inevitably small. Today, this is not necessarily the case. However, most organic farmers plant diverse crops and encourage beneficial insects, and to do this, they must keep their farms relatively small. Organic farmers also manage the soil carefully because if they lose soil nutrients and see a decline in the health of their crops, they have fewer options than conventional farmers have. All of these practices usually increase labor costs significantly. However, the farmers can recoup these extra labor costs by selling their harvest at a premium price to consumers who prefer to buy organic food (FIGURE 11.18).

Organic agriculture is not without some adverse environmental consequences. Because they do not use herbicides, organic farmers are less likely than conventional farmers to be able to use no-till methods successfully. And alternative pest control methods are not always environmentally friendly. For example, in order to keep crops such as carrots weed-free, organic farmers may treat the soil with a propane flamer before planting. This technique protects carrots without herbicides, but uses fossil fuel.

### GAUGE YOUR PROGRESS

✓ What are the goals of alternative agricultural methods?

✓ Describe the benefits and disadvantages of alternatives to industrial farming methods.

✓ What are the basic principles of organic agriculture?

## Modern agribusiness includes the farming of meat and fish

In order to remain economically viable and feed large numbers of people, modern agriculture in the United States has become larger and more mechanized. While many of the goals described in the preceding sections remained in place as agricultural activities moved to a larger scale, some have been abandoned. In particular, as agriculture developed to supply meat and poultry to large numbers of people at a low cost, the primary goal became faster growth of animals.

### High-Density Animal Farming

In 2009, according to the U.S. Department of Agriculture, more than 150 million animals were slaughtered for red meat, along with billions of chickens, turkeys, and ducks. Many of these animals were raised in *feedlots,* or **concentrated animal feeding operations (CAFOs),** which are large indoor or outdoor structures designed for maximum output (FIGURE 11.19). This type of high-density animal farming is used for beef cattle, dairy cows, hogs, and poultry, all of which are confined or allowed very little room for movement during all or part of their life cycle. A CAFO may contain as many as 2,500 hogs or 55,000 turkeys in a single building. By keeping animals confined, farmers minimize land costs, improve feeding efficiency, and increase the fraction of food energy that goes into the production of animal body mass. The animals are given antibiotics and nutrient supplements to reduce the risk of adverse health effects and diseases, which would normally be high in such highly concentrated animal populations.

High-density animal farming has many environmental and health consequences. There is evidence that antibiotics given to confined animals are contributing to an increase in antibiotic-resistant strains of microorganisms that can affect humans. Waste disposal is another serious problem. An average CAFO produces over 2,000 tons of manure annually, or about as much as would be produced by a town of 5,000 people. The waste is usually used to fertilize nearby agricultural fields, but if overapplied, it can cause the same nutrient runoff problems as synthetic fertilizer. Sometimes animal wastes are stored in lagoons adjacent to feedlots. During heavy rainstorms, runoff from these lagoons can contaminate nearby waterways. Animal wastes have also been dumped, either inadvertently or intentionally, into natural waters. The U.S. Environmental Protection Agency has concluded that chicken, hog, and cattle waste has caused pollution along 56,000 km (35,000 miles) of rivers in 22 states and has caused some degree of groundwater contamination in 17 states.

**MORE SUSTAINABLE ANIMAL FARMING** Not all meat comes from CAFOs. Free-range chicken and beef are becoming increasingly popular in the United States. Some people find it more ethically acceptable to eat a chicken or cow that has wandered free than one that has spent its entire life confined in a small space. *Free-range* meat, if properly produced, is more likely to be sustainable than meat produced in CAFOs. Because the animals are not as likely to spread disease as when they are kept in close quarters, the use of antibiotics and other medications can be reduced or eliminated. The animals graze or feed on the natural productivity of the land, with little or no supplemental feeding, so less fossil fuel goes into the raising of free-range meat. Finally, manure and urine are dispersed over the range area and are naturally processed by detritivores and decomposers in the soil. As a result, there is no need to treat and dispose of massive quantities of manure. On the negative side, free-range operations use more land than CAFOs do, and the cost of meat produced using these techniques is usually significantly higher.

## Harvesting of Fish and Shellfish

Fish is the third major source of food for humans, after grain and meat. In many coastal areas, particularly in Asia and Africa, fish accounts for nearly all of the animal protein that some people consume. As **FIGURE 11.20** shows, the global production of fish has increased by about 20 percent since 1980. This increase masks two divergent trends: a rapid increase in farmed fish production and a decrease in wild fish caught in the world's oceans.

A **fishery** is a commercially harvestable population of fish within a particular ecological region. The tragedy of the commons, which we learned about in Chapter 10, is a concept that is particularly applicable to ocean fisheries. Since fish do not live their entire lives within national borders, if one individual, or even an entire country, limits its catch, others are likely to make up the difference.

**FIGURE 11.20** Global fish production (caught from the sea) and *aquaculture*. Global fish production has increased by more than 20 percent since 1980, primarily as a result of the large increase in aquaculture. Global fish production, as shown in the figure, includes both wild-caught fish and aquaculture-raised fish. [After K. M. Brander, *PNAS* 104 (2007): 19709–19714.]

No country has an incentive to protect fish stocks or to attempt to replenish them because fish in the ocean do not belong to any one individual or nation. As a result, competition for fish has led to a precipitous decline in fish populations.

In 2003, the journal *Nature* reported a dramatic decline in the number of large predatory ocean fish caught over the past 50 years, even though both the number of fishing vessels and the amount of time spent fishing had increased over that period. Fishers are working harder, but catching fewer fish. A study in 2006 found that 30 percent of fisheries worldwide had experienced a 90 percent decline in fish populations. The decline of a fish population by 90 percent or more is referred to as **fishery collapse.**

Ocean harvests used to be limited by the difficulty of finding fish in a vast ocean as well as by the limited capacity of small boats and nets, but this is no longer the case. Current fishing methods make it easy to catch large numbers of fish. Factory ships can stay at sea for months at a time, processing and freezing their harvest without having to return to port. Most marine fish are now caught either by large nets pulled behind one of these ships or by very long fishing lines bearing hundreds or even thousands of baited hooks. Fishers in pursuit of high-value species such as tuna use spotter planes and sonar to locate schools, which they encircle with "purse-seine" nets. These nets can capture up to 3,000 adult tuna at a time—almost a million pounds of fish. Fish species that live on or close to the ocean bottom, as well as many shellfish, are caught in dragnets, which are weighted so that they can be pulled across the ocean floor.

Large-scale, high-tech fishing can adversely affect both target and nontarget species. Dragnets can damage ocean-bottom habitats by scouring them of coral, sea sponges, and plants. Many commercially important fish are keystone species, so a decline or loss of their populations can have cascading effects on other marine species. Intensive fishing leads to the loss of juvenile fish of the target species as well as to the loss of noncommercial species that are accidentally caught by nets and lines. This unintentional catch of nontarget species, referred to as **bycatch,** has significantly reduced populations of fish species such as sharks and has endangered other organisms such as sea turtles. Some countries require the use of technology that minimizes bycatch or harm to endangered species.

**MORE SUSTAINABLE FISHING** In the interest of creating and supporting sustainable fisheries, many countries around the world have developed fishery management plans, often in cooperation with one another. International cooperation is particularly important because fish migrate across national borders, some marine ecosystems span national borders, and many of the world's most important fisheries lie in international waters.

The northwestern Atlantic fisheries, for example, comprise several continental shelf ecosystems that stretch from the northeastern United States to southeastern Canada. Historically, these fisheries were among the world's most productive. However, overfishing by international fleets of factory ships led to a catastrophic depletion of fish stocks, particularly of cod and pollock, by the early 1990s, as **FIGURE 11.21** shows. The fisheries were forced to close because of the depleted stocks, and the Canadian and U.S. governments imposed a moratorium on bottom fishing in the area. The majority of the Atlantic Canadian fisheries are still closed today. Most fisheries in the United States are now open, at least for part of the year.

In response to the fishery collapse, and in order to restore the depleted stocks and manage the ecosystem as a whole, the U.S. Congress passed the Sustainable Fisheries Act in 1996. This act shifted fisheries management from a focus on economic sustainability to an increasingly conservation-minded, species-sustainability approach. The act calls for the protection of critical marine habitat, which is important for both commercial fish species and nontarget species. For many commercial species considered to be in danger, such as cod, a "sustainable" fishery means no fishing until populations recover.

One successful fishery management plan was developed in Alaska, where the commercial salmon fishery declined rapidly between 1940 and 1970. Managers first tried to increase salmon populations by limiting the fishing season, but by 1970 the season lasted less than a week, and so many fishers participated that populations continued to drop.

In 1973, fishery managers introduced a system of **individual transferable quotas (ITQs).** Before the start of each salmon season, fishery managers establish a total allowable catch and distribute or sell quotas to individual fishers or fishing companies, favoring those that have a long-term history in the fishery. Fishers with ITQs have a secure right to catch their quota, so they have no need to spend money on bigger boats and better equipment in order to outcompete others.

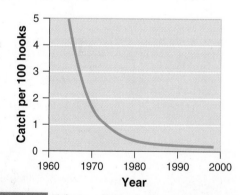

**FIGURE 11.21** Fishery collapse in the northwestern Atlantic Ocean. Most Canadian fisheries have still not opened since this collapse. U.S. fisheries are open for at least part of the year.

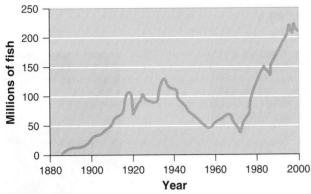

**FIGURE 11.22** **Commercial salmon harvest in the Alaska fishery.** After a peak harvest in 1940, overfishing led to a decline in the number of fish caught. In 1973, fishery managers introduced a system of individual transferable quotas. By 1980, the fishery had rebounded.

If fishers cannot catch enough salmon to remain economically viable, they can sell all or part of their quota to another fisher. **FIGURE 11.22** shows the results: since the beginning of the ITQ program, the salmon population and harvest have increased—at times very rapidly—and costs to fishers have been reduced.

In Alaska, ITQs are sold primarily to small family-run fishing operations. In New Zealand, however, ITQs have been used effectively to control overfishing by large fishing companies. The ITQ system is being used successfully in many other fisheries around the world.

Not all fisheries are declining, but it is often difficult for consumers to know which fish are being overharvested and which are not. To help consumers make more sustainable fish choices, the Environmental Defense Fund and other organizations have compiled a list of popular food fish, dividing them into three categories depending on how sustainable their stocks are. "Best" choices include wild Alaskan salmon and farmed rainbow trout. "Worst" choices include shark and Chilean sea bass.

## Aquaculture

The demand for fish has increased even as wild fish catches have been falling. In response, many scientists, government officials, and entrepreneurs have been developing ways to increase the production of seafood through **aquaculture:** the farming of aquatic organisms such as fish, shellfish, and seaweeds. Aquaculture involves constructing an aquatic ecosystem by stocking the organisms, feeding them, and protecting them from diseases and predators. It usually requires keeping the organisms in enclosures (**FIGURE 11.23**), and it may require providing them with food and antibiotics. Almost all of the catfish and trout eaten in the United States, as well as half of the shrimp and salmon, are produced by aquaculture.

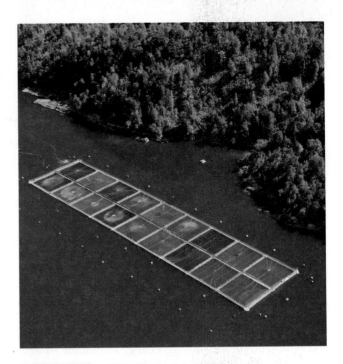

**FIGURE 11.23** **A salmon farming operation in Chile.** Uneaten food and waste released from salmon farms can cause significant nutrient input into natural marine ecosystems.

Proponents of aquaculture believe it can alleviate some of the human-caused pressure on overexploited fisheries while providing much-needed protein for the more than 1 billion undernourished people in the world. Aquaculture also has the potential to boost the economies of many developing countries.

Critics of aquaculture point out that it can create many environmental problems. In a typical aquaculture facility, clean water is pumped in at one end of a pond or marine enclosure, and wastewater containing feces, uneaten food, and antibiotics is pumped back into the river or ocean at the other end. The wastewater may also contain bacteria, viruses, and pests that thrive in the high-density habitat of aquaculture facilities and can infect wild fish and shellfish populations. In addition, fish that escape from aquaculture facilities may harm wild fish populations by competing with them, interbreeding with them, or spreading diseases and parasites. Overall, however, aquaculture has many promising characteristics as a means of sustainable food production.

---

**GAUGE YOUR PROGRESS**

✓ What are some of the ways in which modern agribusiness produces meat and fish?

✓ How does the concept of the tragedy of the commons apply to fisheries?

✓ What is aquaculture?

## Wes Jackson and the Land Institute

Whenever soil is plowed and prepared for another growing season, soil erosion is a problem. One long-term approach to reducing agriculture-related erosion is being explored by plant geneticist Wes Jackson, president of the Land Institute in Salina, Kansas. Jackson maintains that the single most beneficial new development in agriculture would be the development of food crops that do not need to be replanted every year.

**Annual** plants, such as wheat and corn, live only one season and must be replanted each year, which causes enormous disruption to the soil. In contrast, **perennial** plants live for multiple years, so there is no need to plow the field each year for replanting. Perennials have a longer growing season than annuals: they can continue producing roots and storing energy even after harvest, and they often emerge earlier in spring than annuals. In addition, they can rely on root systems that were established in previous years, so they can allocate more resources in the current year to the production of stems, fruits, and seeds. All of these characteristics make them more productive than annuals.

Researchers at the Land Institute are exploring a variety of ways to grow crops in systems that mimic the natural world. Using a combination of conventional selective breeding and technology, they are attempting to convert annual species such as wheat, sorghum, sunflowers, and corn into perennials. At the same time, they are collecting wild perennials such as wheatgrass, domesticating them, and selecting for higher seed yield, size, and quality (FIGURE 11.24). Through these efforts, researchers hope to assemble communities of perennial plants, animals, fungi, and microorganisms that will be stable, productive, and resistant to insect pests and diseases.

The ultimate goal of the Land Institute is to develop sustainable crops that will produce sufficient amounts of food for harvest by humans, reduce soil erosion, and reduce or eliminate the need for synthetic fertilizers, pesticides, and irrigation. The researchers have already managed to double the size of seeds of a species of wheatgrass and increase its seed production by 20 percent without losing other important qualities of the plant. They have also had some success in crossing specially treated domesticated sorghum with perennial sorghum varieties to produce progeny that are both fertile and perennial.

Critics contend that it may take many decades to develop perennial seed crops with usable products.

**FIGURE 11.24** **Annuals and perennials.** The root system of annual wheat (*left*) is much smaller than that of intermediate wheatgrass, a perennial (*right*).

Jackson doesn't seem overly concerned about this. According to Jackson, "if you're working on something you can finish in your lifetime, you're not thinking big enough."

### References

Cox, T. S., et al. 2006. Prospects for developing perennial grain crops. *BioScience* 56: 649–659.

The Land Institute: http://www.landinstitute.org.

■ **Describe human nutritional needs and the challenges of overcoming hunger and malnutrition.**

Humans require both a certain number of calories each day and a balance of proteins and other nutrients. Undernutrition is a condition in which not enough food calories are ingested each day. Malnutrition is a condition in which the diet lacks the correct balance of proteins, carbohydrates, vitamins, and minerals.

■ **Explain the development of modern industrial agriculture, the role played by inputs such as irrigation water, fertilizers, and pesticides, and the environmental consequences of modern farming methods.**

In the twentieth century, the Green Revolution transformed agriculture from a system of small farms relying mainly on human labor into a system of large industrial operations relying mainly on machinery run by fossil fuels. This shift has resulted in larger farms and monocropping. Irrigation can increase crop yields dramatically, but it can also draw down aquifers and lead to soil degradation through waterlogging and salinization. Fertilizers can also increase crop yields dramatically, but can run off into surface waters and cause damage to ecosystems. Pests can become resistant to pesticides over time, resulting in a cycle known as the pesticide treadmill. Pesticides can also have negative environmental consequences, such as losses of beneficial nontarget organisms, human health concerns, and surface water contamination.

■ **Identify the benefits and costs of using genetically modified organisms in agriculture.**

Agricultural scientists are using genetic engineering to produce genetically modified organisms with desirable traits. GMOs can increase yields and reduce the use of pesticides. Concerns have been raised, however, about their safety for human consumption and their effects on biodiversity.

■ **Describe alternatives to industrial farming methods.**

Traditional farming techniques such as intercropping, crop rotation, agroforestry, and contour plowing can sometimes improve agricultural yields and conserve soil and other resources. No-till agriculture is another way to reduce soil erosion and degradation. Integrated pest management reduces the use of pesticides, thus saving money and reducing environmental damage. IPM requires more labor, however, and practitioners must be trained to identify potential hazards to their crops. Organic agriculture focuses on maintaining the soil and avoids the use of synthetic fertilizers and pesticides. This approach often results in more labor-intensive, smaller farms, where many alternative agricultural techniques must be applied.

■ **Explain the environmental impacts of various approaches to raising and harvesting meat and fish.**

Concentrated animal feeding operations and aquaculture facilities allow for efficient animal growth and inexpensive food production, but can have negative environmental impacts in the form of concentrated animal waste and the introduction of antibiotics and other waste products into the environment. Most fisheries in North America have been overharvested and some are now recovering.

1. Which of the following does *not* explain the rise of the modern farming system?
   (a) The cost of labor varies from country to country.
   (b) Small farms are usually more profitable than large farms.
   (c) Irrigation contributes to greater crop yields.
   (d) Fertilizers improve crop yields and are easy to apply.
   (e) Mechanization facilitates monocropping and improves profits.

2. Irrigation can result in which of the following environmental problems?
   - I Reduction of evaporation rates
   - II Accumulation of salts in soil
   - III Waterlogging of soil and plant roots

   (a) I only
   (b) II only
   (c) III only
   (d) I and II
   (e) II and III

3. The use of synthetic fertilizers increases crop yields, but also
   (a) destroys the nitrifying bacteria in the soil.
   (b) increases fish populations in nearby streams.
   (c) decreases phosphorus concentrations in the atmosphere.
   (d) increases nutrient runoff into bordering surface waters.
   (e) slows the release of organic nutrients from compost.

4. Which of the following statements *best* describes the pesticide treadmill?
   (a) Broad-spectrum pesticides degrade into selective pesticides, thereby killing a wide range of insect pests over a long period.
   (b) Pesticides accumulate in the fatty tissues of consumers and increase in concentration as they move up the food chain.
   (c) Some pest populations evolve resistance to pesticides, which become less effective over time, so that new pesticides must be developed.
   (d) Beneficial insects and natural predators are killed at a faster rate than the pest insects.
   (e) Testing of the toxicity of pesticides to humans cannot keep pace with the discovery and production of new pesticides.

5. In which of the following ways did the Green Revolution increase food production?
   - I The development of disease-resistant and high-yielding crop plants
   - II Monocropping and the widespread use of machinery
   - III The application of fertilizers and the use of irrigation techniques

   (a) I only
   (b) II only
   (c) III only
   (d) I and III
   (e) I, II, and III

6. Which of the following is *not* a traditional farming technique that is used in sustainable agriculture?
   (a) Nomadic herding
   (b) Intercropping
   (c) Crop rotation
   (d) Agroforestry
   (e) Contour plowing

7. Which of the following is an environmental advantage of no-till agriculture?
   (a) The use of herbicides improves the stability of the soil.
   (b) Migratory bird populations are reduced.
   (c) The undisturbed soil is less susceptible to erosion.
   (d) The crop residues reduce the soil profile.
   (e) The concentration of $CO_2$ in the fields is increased.

8. Which of the following practices is *not* a part of integrated pest management?
   (a) Crop rotation
   (b) Elimination of pesticides
   (c) Use of pest-resistant crops
   (d) Introduction of predators
   (e) Frequent inspection of crops

9. Farmers who practice organic agriculture have less of an impact on the environment than farmers who practice industrial agriculture because they
   (a) use no-till agriculture exclusively.
   (b) import soil to maintain soil fertility.
   (c) maintain large farms with a single crop.
   (d) avoid pesticides and synthetic fertilizers.
   (e) have lower labor costs.

10. Critics of using genetically modified organisms as food crops warn of which of the following dangers?

    I   Introduction of new allergens into the food supply

    II  Loss of genetic diversity in food crops

    III  Decreases in food production worldwide

(a) I only         (d) I and II

(b) II only        (e) I and III

(c) III only

11. Concentrated animal feeding operations (CAFOs) can *best* be described as

(a) facilities where a large number of animals are housed and fed in a confined space.

(b) a method of producing more meat at a higher cost.

(c) a means of producing great quantities of manure to fertilize fields organically.

(d) an experimental plan to test the effectiveness of antibiotics.

(e) the storing and compacting of grain for use as a nutrient supplement for cattle.

12. Which of the following is *not* an environmental or health problem that has been associated with CAFOs?

(a) The increase of antibiotic-resistant bacteria potentially harmful to humans

(b) The overgrazing of large tracts of land

(c) The runoff of animal wastes into natural waters

(d) The production of huge quantities of manure, creating a waste disposal problem

(e) The use of grain as feed, reducing the food supply available to humans

13. The data presented below, compiled by the FAO, describe world fish production from 1950 to 2003. Based on these data, which of the following statements *best* describes the global trend in fish production?

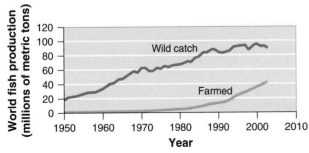

(a) The U.S. Sustainable Fisheries Act of 1996 has successfully banned the catch of threatened fish species.

(b) The use of individual transferrable quotas (ITQs) has led to the overfishing of wild species.

(c) Fish production through aquaculture exceeds fish production in the ocean.

(d) As the human population continues to increase, the per capita wild catch also continues to increase.

(e) As the human population continues to increase, per capita farmed fish production also continues to increase.

14. Which of the following is an environmental effect of bycatch?

(a) Juvenile fish are too small and slip through the nets, thereby ensuring the next generation of commercial fish.

(b) Predators caught in the nets feed on the commercially important fish also caught, thereby reducing the total catch.

(c) Nontarget fish populations have declined.

(d) Cod populations have increased due to the removal of competitor species.

(e) Endangered species, such as sea turtles, are being marketed through the ITQ system.

15. It has been projected that aquaculture could supply about one-half the global demand for seafood by 2025. However, future production through aquaculture could be limited because

(a) concentrated waste from aquaculture facilities contaminates rivers and oceans.

(b) raising fish in a protected environment could lead to a fish population overshoot.

(c) the economies of developing countries would be negatively affected.

(d) ocean fishing operations would go out of business.

(e) the ITQ system would prohibit their expansion.

1. Maintaining dairy cattle in a CAFO requires large quantities of water and produces vast quantities of wastewater and manure. The increasing number of CAFO dairies in eastern New Mexico and West Texas is contributing to significant groundwater contamination and the depletion of the Ogallala aquifer. Consider a CAFO with 1,000 dairy cows and answer the following questions.

   (a) According to the U.S. Department of Agriculture, the average dairy cow can consume up to 200 L of water daily. An additional 120 L per cow per day is required to wash the milking equipment and milking area. How many liters of water are required to operate this dairy daily?

   (b) According to the U.S. Environmental Protection Agency, the average dairy cow produces 55 kg of wet manure daily. How many metric tons would this dairy produce each day (1,000 kg = 1 metric ton)?

   (c) The recommended maximum manure application rate for a farm site with low runoff potential is approximately 275 kg per hectare. What size farm would be required to apply all of the manure produced by this dairy each day?

   (d) Identify one environmental problem that could be caused if the rate of runoff of nitrogen from the manure into natural waters became excessively high. Describe one way to prevent this problem.

   (e) If this same dairy herd was raised in a free-range operation, identify and explain one environmental problem, other than water pollution, that would be minimized or eliminated.

2. The following results of a survey of 113 people conducted in Trinidad, West Indies, appeared in a 2006 edition of the *British Food Journal*:

| Questions | Very likely (%) | Somewhat likely (%) | Not too likely (%) | Not at all likely (%) |
|---|---|---|---|---|
| Would you be willing to purchase these foods (corn oil, potatoes, tomatoes, rice) which are bought regularly if they were genetically modified? | 15.9 | 31.9 | 38.9 | 7.1 |
| Would you purchase GM foods if they were considered healthier than conventionally produced foods? | 30.1 | 34.5 | 25.6 | 9.7 |
| Would you be willing to purchase GM foods if they were safe? | 15.9 | 31.9 | 38.9 | 7.1 |

   (a) Considering the "very" and "somewhat" responses as one group and the "not too" and "not at all" responses as another group, what percentage of the survey respondents would probably purchase GM foods? What percentage would not?

   (b) Referring to the second survey question, identify and explain two possible benefits of the inclusion of GM foods in the diet that might convince people to purchase these products.

   (c) Referring to the third survey question, identify and explain two possible dangers of the use of GM foods that might convince people not to purchase these products.

   (d) Identify one U.S. federal agency that is responsible for the regulation of GMOs.

## MEASURE YOUR IMPACT

**The Ecological Footprint of Food Consumption** The following table is a compilation of data from a study conducted by the Gembloux Agricultural University, which profiled food consumption in the United States in 2004, and from a report that provided the ecological footprint for various food items.

| Food item | Amount consumed (kg/person/year) | Ecological footprint (hectares/1,000 kg/year) | Ecological footprint based on consumption (hectares/year) |
|---|---|---|---|
| Beef | 43 | 15.7 | |
| Poultry | 53 | 1.6 | |
| Pork | 52 | 1.9 | |
| Milk | 83 | 1.4 | |
| Cheese | 15 | 11.1 | |
| Yogurt | 4 | 1.7 | |
| Butter | 2 | 11.5 | |
| Potatoes | 61 | 0.3 | |
| Vegetables | 130 | 0.4 | |
| Fruit | 70 | 0.5 | |

*Source:* Data from B. Duquesne, S. Matendo, and P. Lebailly, Profiling food consumption: Comparison between USA and EU, http://agriculture.wallonie.be/apps/spip_wolwin/IMG/pdf/Gblx.pdf; A. Collins and R. Fairchild, Sustainable food consumption at a sub-national level: An ecological footprint, nutritional and economic analysis, *Journal of Environmental Policy & Planning* 9 (2007): 5–30.

(a) For each food item listed above, calculate the ecological footprint for the amount consumed and complete the last column of the table.

(b) Calculate the total ecological footprint for an individual in the United States consuming all of these food items. How does this footprint compare with the world ecological footprint for food consumption, which is 0.9 ha per person per year?

(c) Identify three ways in which you can reduce your personal ecological footprint for food consumption.

## ENGAGE YOUR ENVIRONMENT

How can you use "food miles" and other energy considerations to choose which fresh fruits and vegetables to buy? In this Engage Your Environment exercise you will consider trade-offs involved in buying produce grown locally versus produce grown in other regions or countries. *See Engage Your Environment Activity 11 at the end of this book.*

# science applied

## How Do We Define Organic Food?

If you've ever spent time roaming through the aisles of a grocery store, you have undoubtedly seen food labeled "organic" (**FIGURE SA5.1**). You may even have purchased organic food, despite the fact that it is generally more expensive than conventionally grown food. Some people prefer organic food because they believe it is healthier and tastes better. Others buy it because they think organic food is safer to eat. Still others believe organic food is produced in a way that is healthier for the environment and safer for farmworkers. Are these beliefs accurate? What exactly does the organic food label mean?

### How did the organic food movement begin?

The notion of organic food emerged in the 1940s, when farmers first started using synthetic pesticides in agriculture. Three decades later, interest in organic food increased as concerns about environmental contaminants became more widespread. Organic food enthusiasts wanted food that was free of chemicals, including pesticides. At the same time, the popularity of food grown locally on small farms was increasing. The push for organic produce in the 1970s was part of a counterculture movement against the farming practices of large-scale commercial agriculture. As author Michael Pollan has documented, organic food was perceived as the food of people rebelling from the mainstream culture.

Today, organic food is much more common and appeals to people from all walks of life. Indeed, organic farming has increased rapidly during the past decade. According to the U.S. Department of Agriculture, the number of acres of organic farmland quadrupled between 1990 and 2005. The states that grow the most organic food are California, Wisconsin, North Dakota,

Minnesota, and Montana. In 2008, there were about 1.9 million ha (4.8 million acres) of organic cropland and pastureland in the United States. But although organic food has been the fastest-growing category of food in the nation, it still represents less than 1 percent of all cropland and pastureland and only about 3 percent of all food sales.

### What does it mean to be organic?

An organic farmer is fundamentally concerned with the health of the soil. He or she will work hard to ensure that the organic matter content, base saturation, and cation exchange capacity of the soil will increase each year. Actions that will degrade the soil or promote erosion are carefully avoided.

The rise in popularity of organic products has brought increased pressure to define what exactly it means to be organic. Without government guidelines, anyone could claim to be selling organic food. In 1990, the U.S. Department of Agriculture (USDA) began to develop guidelines for organic food. Its goal was to set national organic certification standards that would assure consumers that the food was produced and processed with minimal use of chemicals. Food that met these standards could carry the USDA Organic seal (**FIGURE SA5.2**).

In 1995, the USDA determined that "organic agriculture is an ecological production management system that promotes and enhances biodiversity, biological cycles, and soil biological activity. It is based on minimal use of off-farm inputs and on management practices that restore, maintain, and enhance ecological harmony." What does this statement mean in practice? First, it means that to be considered organic, food must be

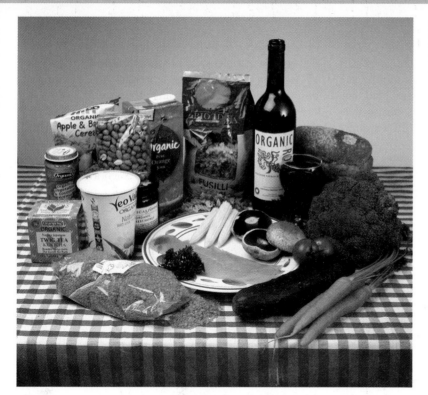

FIGURE SA5.1 **Organic food.**
Supermarkets today carry a wide
variety of organic foods certified by
the U.S. Department of Agriculture.

produced and processed with minimal use of pesticides. It also means that organic meat, eggs, and dairy products come from animals that have not been given antibiotics or growth hormones. Finally, it means that food labeled organic cannot have been fertilized with synthetic fertilizers or sewage sludge. Rather than relying on chemicals that come from outside the farm, organic farmers must use nonchemical methods of pest control and fertilization, as discussed in Chapter 11.

FIGURE SA5.2 **The USDA Organic seal.** Food that meets USDA organic certification standards can carry the USDA Organic seal.

Pest control is perhaps one of the greatest challenges in organic farming. Agricultural pests include weeds that compete with the crops, insects that eat the crops, and fungal pathogens that can kill the crops. Organic farmers are required to try to control pests without synthetic pesticides. Such efforts can include mechanically destroying weeds and using plant-derived insecticides or the release of insect enemies to combat the pests. If these tactics fail, organic farmers are permitted to use a synthetic pesticide from a relatively small list of approved chemicals.

Having standards for organic food is effective only if those standards are enforced. One of the important functions of the USDA's National Organic Program is to conduct inspections of organic food producers and processors to ensure that organic food meets the national standards. Since 1990, the government has allowed inspections to be conducted by independent companies that are hired by producers and processors to inspect organic farms and factories to ensure that the organic standards are being met.

As part of this effort, the certifying companies were required to conduct spot checks. In 2010, the USDA announced that it had failed to enforce the spot check requirement and that, as a result, many of the largest independent certifiers had been conducting spot checks on organic food only if they suspected a problem. When problems were not suspected, producers could operate for years without any spot checks. In fact, some growers had been selling nonorganic food under the USDA Organic seal and obtaining a higher price for their products. In response, the federal government announced in 2010 that the funding and staff of the National Organic Program would nearly double to ensure that the certifying companies were conducting the required spot checks.

## Does organic food mean family farms?

Most organic food production was originally conducted on small family farms. As we saw in Chapter 11, organic farming methods can require more labor, time, and money than conventional methods, so the price of organic food can be significantly higher. As farmers have learned, however, many consumers are willing to pay a premium for what they believe to be a premium product. This consumer behavior has attracted the attention of large factory farms owned by corporations, which have begun to conduct large-scale organic agriculture.

Although organic farms have a history of being small farms, there is no inherent reason why large farms cannot grow organic food (FIGURE SA5.3). An increasing number of corporations have decided to enter the business of organic food production. One of the most significant events to drive this movement occurred in 2006, when Wal-Mart, the largest grocery retailer in the United States, announced that it would begin selling a substantial number of organic foods in its stores at prices only slightly higher than those of nonorganic foods. Wal-Mart's expansion into this market has brought organic food to a much larger number of consumers. As a result, demand for organic food has increased, and more agribusinesses are producing it. Increased production of organic food by larger, more efficient farms has driven down the cost of organic food.

As more factory farms entered the organic food business, there was increasing pressure to permit the use of some nonorganic and synthetic chemicals during the production stage, when the food from the field is turned into a form that can be sold in the grocery store. Proponents of the rule change, including agribusinesses, argued that many processed foods, such as frozen meals, could not be made in an organic version unless the rules about additives were changed. Opponents argued that if processed foods could not be made without additives, then processed foods should not be allowed to be certified as organic. In 2002, the USDA designated several nonorganic and synthetic additives as permissible in organic processed foods. Permitted nonorganic additives include citric acid, pectin, and cornstarch. Synthetic chemicals permitted in the processing of organic foods include alcohol (used for disinfecting machinery) and aspirin (used for medical treatment of animals).

## What doesn't the organic label mean?

The USDA definition of "organic" does not match what many people think or hope "organic" should mean. For example, many people associate organic food with small family farms, but as we have seen, organic food can be grown on farms of all sizes. The USDA Organic seal therefore tells us nothing about the size of the farm that grew the food. Nor does the seal tell us anything about its safety, including whether it contains fewer dangerous food-related pathogens than nonorganic food. Nor does it indicate whether crop plants were genetically modified, or whether eggs, meat, and dairy products came

**FIGURE SA5.3** **Large-scale organic farming.** Although organic farms traditionally have been small-scale family farms, some organic farms, such as this organic dairy in California, operate at much larger scales.

from animals that were treated humanely or allowed to roam freely outdoors, also known as "free range." It tells us nothing about where the food was grown, or about labor conditions, such as compensation for the farmworkers who produced it. Furthermore, there are currently no standards for certifying nonagricultural products, such as seafood or cotton, as organic. Although you may see these products labeled "organic," there is currently no agreed-upon definition for such products. Similarly, companies sometimes label their products "natural," a word that seems to imply "organic," but has no standard definition.

In response to public demand, farmers sometimes use food labels that go beyond the federal requirements for organic certification. They may include information about humane treatment of animals, the conditions and wages of the farmworkers, and whether genetically modified organisms were used. Such detailed labels have seen growing popularity.

Although the USDA organic certification has a somewhat narrow definition, it does offer a number of important benefits. There is no evidence that organic food is healthier for humans to eat, but the certification program at least provides a standardized set of requirements so that informed consumers understand what they are buying. Given the reduction in synthetic pesticide use, the program also allows consumers to support farming practices that are much better for soils and ecosystems and better for farmworkers, who face decreased risks from applying pesticides, than conventional farming practices.

### References

Pollan, M. 2001. Behind the organic-industrial complex. *New York Times,* May 13. http://www.nytimes.com/2001/05/13/magazine/13ORGANIC.html?pagewanted=1.

U.S. Department of Agriculture. *What is organic production?* http://www.nal.usda.gov/afsic/pubs/ofp/ofp.shtml.

# Nonrenewable Energy: Coal, Oil, Natural Gas, and Nuclear Fuels

## All Energy Use Has Consequences

A series of pivotal moments in the 1960s led to the first Earth Day in 1970. One of those events was an oil well explosion and rupture—called a *blowout*—off the coast of Santa Barbara, California, in January 1969. Before the resulting spill was contained, over 11.4 million liters (3 million gallons) of crude oil poured into the Santa Barbara Channel. Some of it washed ashore, coating sandy

Even after oil is safely extracted from underground and transported to a refinery, accidents can occur. In 2005, 15 workers died in an explosion at a BP oil refinery in Texas. Nor do the hazards of fossil fuel use end with production. After the oil that is refined into gasoline, jet fuel, or diesel reaches its end use location, and is burned to run a vehicle or heat a house, the combustion process emits pollutants, which cause a number of environmental problems.

> ## Despite repeated warnings that we are addicted to oil, our reliance on oil and other fossil fuels has only increased.

beaches and marine life with oil. The spill drew national attention—oil-soaked birds that were unable to fly were featured in newspapers and on the evening news. And yet, despite repeated warnings that we are addicted to oil, our reliance on oil and other fossil fuels has only increased.

Subsequent oil spills all over the world, on land and in water, have attracted national and international attention. Some spills have been caused by leaks or explosions where the oil was being extracted from the ground. Others have occurred while the oil was being transported by pipeline or tanker. In March 1989, the *Exxon Valdez,* a supertanker carrying 200 million liters (53 million gallons) of oil, crashed into a reef in Prince William Sound, Alaska. Roughly 42 million liters (11 million gallons) of oil spilled into the sound. Much of it washed up on shore, coating the coastline and killing hundreds of thousands of birds and thousands of marine mammals. This spill was the largest in U.S. waters for 21 years, until a blowout occurred at the BP Deepwater Horizon oil well in the Gulf of Mexico in April 2010. (BP used to be known as British Petroleum.) That accident killed 11 workers on the drilling platform, injured 17 others, and led to the release of well over 780 million liters (206 million gallons) of oil. This oil has dispersed in the Gulf of Mexico and washed up on the shores of Louisiana, Texas, Mississippi, Alabama, and Florida.

Other fossil fuels pose similar risks. In April 2010, an explosion in a coal mine in West Virginia killed 29 coal miners. This explosion was the worst coal mine disaster in the United States in 40 years, but only a century earlier, in the early 1900s, there were hundreds of accidental mining deaths per year in the United States. And long after they leave the mines, hundreds of thousands of coal miners develop black lung disease and other respiratory ailments that lead to disability or death.

Natural gas is considered to be the "clean" fossil fuel. Emissions of particulates, sulfur dioxide, and carbon dioxide are lower per unit energy obtained from natural gas than from oil or coal. But the extraction of natural gas has its consequences too. Exploration and extraction take place in many different locations on land and under water. "Thumper trucks," which generate seismic vibrations in order to identify natural gas deposits underground, can disturb soil and alter groundwater flow, causing certain areas to flood and wells to go dry. Drilling and ▶

An offshore oil drilling platform.

◀ An oil refinery in Antwerp, Belgium.

the use of water for gas extraction can cause contamination of groundwater. Once extracted, natural gas requires pipelines to transport it. Construction of pipelines is disruptive to the environment, and communities often oppose such construction.

The United States is dependent on fossil fuels for our energy supply. We are faced with constant reminders of that dependence and of the adverse consequences of using fossil fuels. Obviously, many of the benefits of our modern society—health care, comfortable living conditions, easy travel, plentiful food—come from our use of readily accessible and relatively affordable fossil fuels—but not without long-term costs. ■

*Sources:* J. Goodell, *Big Coal* (Mariner Books, 2007); L. Margonelli, *Oil on the Brain* (Broadway Books, 2008).

## UNDERSTAND THE KEY IDEAS

We use energy in all aspects of our daily lives: heating and cooling, cooking, lighting, communications, and travel. In these activities, humans convert energy resources such as natural gas and oil into useful forms of energy such as motion, heat, and electricity, with varying degrees of efficiency and environmental effects. We learned the fundamental energy concepts underlying these conversions in Chapter 2. This chapter focuses on conventional, nonrenewable energy resources—coal, oil, natural gas, and nuclear fuels.

After reading this chapter you should be able to

- describe how energy use and energy resources have varied over time, both in the United States and worldwide.
- compare the energy efficiencies of the extraction and conversion of different fuels.
- explain the various means of generating electricity.
- discuss the uses and consequences of using coal, oil, natural gas, and nuclear fuels.
- describe projections of future supplies of our conventional energy resources.

# Nonrenewable energy accounts for most of our energy use

Each energy choice we make has positive and negative consequences. In a society like the United States, where each person averages 10,000 watts of energy use continuously—24 hours per day, 365 days per year—this means there are a lot of consequences to understand, evaluate, and possibly try to change. In this chapter we will look at the fossil fuel and nuclear fuel supplies that we currently use. These types of energy resources are often called **nonrenewable** because, like the mineral resources we discussed in Chapter 8, once they are used up, they cannot be replenished. In other words, the supplies of these energy types are finite. Chapter 13 addresses the renewable energy resources that we use to some extent now and that many people expect we will use even more in the future.

The two primary categories of nonrenewable energy resources are *fossil fuels* and *nuclear fuels*. **Fossil fuels** are derived from biological material that became fossilized millions of years ago. Coal, oil, and natural gas are the three major fossil fuels. To access the ancient solar energy contained in the chemical bonds of fossil fuels, we burn those fuels and harness the heat energy from their combustion. **Nuclear fuels** are derived from radioactive materials that give off energy. We harness that energy by transferring heat as well.

We will begin our examination of conventional energy resources by looking at patterns of energy use in the world and in the United States.

## Worldwide Patterns of Energy Use

Every country in the world uses energy at different rates and relies on different energy resources. Factors that determine the rate at which energy is used include which resources are available and affordable. In the past few decades, environmental impacts have also come to play a part in some energy use decisions.

In order to talk about quantities of energy used, it is helpful to use specific measures. Recall from Chapter 2 that the basic unit of energy is the joule (J). A gigajoule (GJ) is 1 billion ($1 \times 10^9$) joules, or about as much energy as is contained in 30 L (8 gallons) of gasoline. An exajoule (EJ) is 1 billion ($1 \times 10^9$) gigajoules. In some figures, we also present the unit of energy that the U.S. government uses for reporting energy consumption. That unit, not used anywhere other than in the United States, is the quad, which is 1 quadrillion, or $1 \times 10^{15}$, Btu. One quad is equal to 1.055 EJ.

As FIGURE 12.1 shows, in 2008, total world energy consumption was approximately 495 EJ per year. This number amounts to roughly 75 GJ per person per year. Oil, coal, and natural gas were the three largest energy sources. Peat, a precursor to coal, is sometimes combined with coal for reporting purposes in certain countries, mostly in the developing world.

Energy use is not evenly distributed throughout the world, however, as FIGURE 12.2 shows. Energy consumption in the United States was 355 GJ per person per year in 2007, almost 5 times greater than the world average. In fact, although only 20 percent of the world's population lives in developed countries, those people use 70 percent of the world's energy each year. Note that of

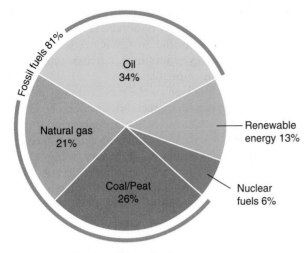

**Total = 495 exajoules
(469 quadrillion Btu, or "quads") per year**

**FIGURE 12.1** Annual energy consumption worldwide by resource. Oil, coal and peat, and natural gas are the major sources of energy for the world. [Data from the International Energy Agency, 2009.]

the countries shown in Figure 12.2, the United States has the greatest total energy consumption, whereas Canada has the greatest per capita energy consumption. At 0.12 EJ per year, Tanzania has the lowest annual energy consumption of the countries shown; annual per capita energy consumption in Tanzania is 2.85 GJ per person per year.

There are a variety of reasons for the patterns we see in Figure 12.2. In developed countries and in urban areas of some developing countries, individuals are likely to use fossil fuels such as coal, oil, and natural gas—either directly or indirectly through the use of electricity generated by burning those fuels. However, people living in rural areas of developing countries still primarily use fuels such as wood, charcoal, or animal waste, in addition to proportionately larger amounts of human and animal energy. Accordingly, we can distinguish between *commercial* and *subsistence* energy sources. **Commercial energy sources** are those that are bought and sold, such as coal, oil, and natural gas. Sometimes wood, charcoal, and animal waste are also sold commercially. **Subsistence energy sources** are those gathered by individuals for their own immediate needs. There is much greater use of subsistence energy sources in the developing world, especially in rural areas.

Changes in energy demand generally reflect the level of industrialization that is occurring in a country or region. As energy demand increases, societies change the types of fuels they use. Today, we see the same patterns of changing energy use in developing countries that have been observed historically in the United States. For example, as more people own automobiles, demand for gasoline and diesel increases. As industries develop and factories are built, demand for electricity

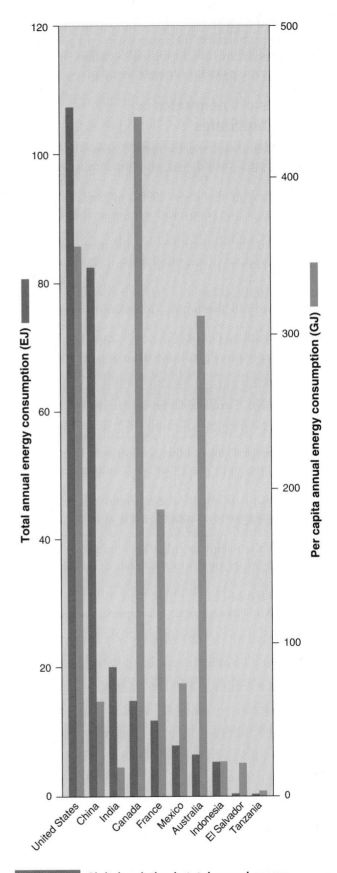

**FIGURE 12.2** Global variation in total annual energy consumption and per capita energy consumption. The 10 countries shown are among the largest and the smallest energy users in the world. [Data from the U.S. Department of Energy, Energy Information Administration, 2009.]

and nuclear fuels increases. Although worldwide energy use varies considerably, the United States is a particularly large energy consumer, so we will focus on energy use in the United States for much of this chapter.

## Patterns of Energy Use in the United States

FIGURE 12.3 shows the history of energy use in the United States. Wood was the predominant energy source until about 1875, when coal came into wider use. Starting in the early 1900s, oil and natural gas joined coal as the primary sources of energy. By 1950, electricity generated by nuclear energy became part of the mix, and hydroelectricity became more prominent. The 1970s saw a decline of oil and a resurgence of coal. These changes were the result of political, economic, and environmental factors that will continue to shape energy use into the future. Today, the three resources that supply the majority of the energy used in the United States and worldwide—in order of increasing importance—are oil, coal, and natural gas.

We can consider energy use in the United States today as the inputs and outputs of an enormous system (FIGURE 12.4). The boundaries of the system are social and technological as well as physical. For example, oil inputs enter the U.S. energy system both from domestic production and from other countries. Hydroelectric energy comes from water that flows within the physical boundaries of the country, as well as from neighboring Canada, but it is not an energy input until we move it into a technological system, such as a hydroelectric dam. One major output from the system is *work*—the end use of the energy, such as transportation or industry. The other major output is *waste:* heat, $CO_2$, and other pollutants that are released as energy is converted and entropy increases.

In 2008, U.S. energy input was approximately 105 exajoules ($1.05 \times 10^{18}$ J) per year. The energy mix of our input is 84 percent fossil fuel, 9 percent nuclear fuel, and 7 percent renewable energy resources. The United States produces 70 percent of the energy it needs. Nearly 30 percent of our energy comes from other countries, primarily in the form of petroleum imports. Industry uses the most energy, followed by the transportation sector.

Energy use varies regionally and seasonally. In the midwestern and southeastern states, coal is the primary fuel burned for electricity generation. The western and northeastern states generate electricity using a mix of nuclear fuels, natural gas, and hydroelectric dams. Highly populated areas tend to use less coal, which creates the most air pollution of any fuel. Northern areas consume more oil and natural gas during the winter months to meet the demand for heating. Southern areas consume more electricity in the summer months to meet air conditioning needs.

The type of energy that is used for a particular application is a function of many factors, including the features that each type of energy offers. Those features include ease of transportation and the amount of energy a given mass of fuel contains.

## Energy Types and Quality

Imagine running your car on coal or firewood. In order to run your car with one of these fuels and travel the same distance as a conventional car on one tank of gasoline, you would have to carry around a larger volume of material. Gasoline, the current fuel of choice for personal transportation, gets you a lot farther on a much smaller volume. This thought experiment highlights the importance of considering energy type and quality.

Certain forms of energy are best suited for particular purposes. For example, for transportation, we usually prefer gasoline or diesel—liquid energy sources that are relatively compact, meaning that they have a high energy-to-mass ratio. Energy-to-mass ratio is not the only consideration for fuel choice, however. A wood or

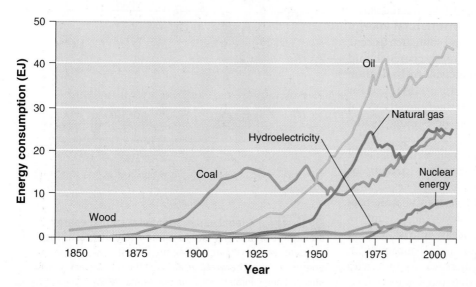

**FIGURE 12.3** Energy consumption in the United States since 1850. Wood and then coal used to dominate our energy supply. Today a mix of three fossil fuels accounts for much of our energy use. [After U.S. Department of Energy, Energy Information Administration, 2008.]

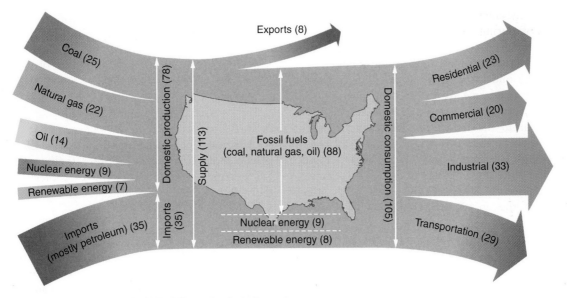

**(a)** Energy flowchart for the United States (exajoules)

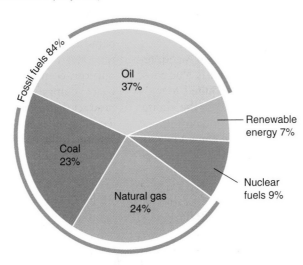

**Total = 105 EJ (99.3 quads) per year**

**(b)** U.S. energy consumption

**FIGURE 12.4** **Energy inputs and outputs in the United States.** (a) The types of fuels that are used are shown on the left, and their end uses are shown on the right. (Sums may differ by one whole unit due to rounding.) (b) Energy consumption by fuel type. [(a) After U.S. Department of Energy, Energy Information Administration, 2009; (b) data from U.S. Department of Energy, Energy Information Administration, 2009.]

coal fire starts relatively slowly. Gasoline and diesel are also ideal fuels for vehicles because of the speed with which they can provide or cut off the energy supply. Unfortunately, compared with other energy sources, such as natural gas or hydroelectricity, gasoline produces large amounts of air pollution per joule of energy released. In addition, gasoline requires a good deal of refining—chemical processing—to produce, unlike coal or wood.

**QUANTIFYING ENERGY EFFICIENCY** Although all conventional nonrenewable energy sources have environmental impacts, clearly some are better suited for particular jobs than others. One of the ways we can determine the best source to use is to consider energy efficiency: both the efficiency of the process of obtaining the fuel and the efficiency of the process that converts it into the work that is needed. Thus we can evaluate how effectively we use energy by quantifying both the energy expended to obtain a fuel and how efficiently we use it.

In Chapter 2 we discussed energy efficiency as well as energy quality, a measure of the ease with which stored energy can be converted into useful work. The second law of thermodynamics dictates that when energy is transformed, its ability to do work diminishes; some energy is lost during each conversion. In addition to these losses, for almost every fuel that we use, there is an energy expenditure involved in obtaining the fuel.

FIGURE 12.5 outlines the process of energy use from extraction of a resource to electricity generation and disposal of waste products from the power plant. As we can see from the red arrows in the figure, there are many opportunities for energy loss, each of which reduces energy efficiency. You may recall from Figure 2.15 that the efficiency of converting coal into electricity is approximately 35 percent. In other words, about two-thirds of the energy that enters a coal-burning electricity generation plant ends up as waste heat or other undesired outputs. If we included the energy used to extract the coal, and if we included the energy used to build the coal extraction machinery, to construct the power plant, and to remove and dispose of the waste material from the power plant, the efficiency of the process would be even lower. All of these other energy inputs are called *embodied energy*, and they will be discussed in greater detail in Chapter 16.

Every energy source, from coal to oil to wind, requires an expenditure of energy to obtain it. The most direct way to account for the energy required to produce a fuel, or energy source, is to calculate the energy return on energy investment (EROEI), or how much energy we get out of an energy source for every unit of energy expended on its production. EROEI is calculated as follows:

$$EROEI = \frac{\text{Energy obtained from the fuel}}{\text{Energy invested to obtain the fuel}}$$

For example, in order to obtain 100 J of coal from a surface coal mine, 5 J of energy is expended. Therefore,

$$EROEI = \frac{100\,J}{5\,J} = 20$$

As you might expect, a larger value for EROEI suggests a more efficient and more desirable process. The Science Applied section that follows Chapter 13, "Should Corn Become Fuel?" calculates the EROEI for ethanol, a fuel made from corn.

## FINDING THE RIGHT ENERGY SOURCE FOR THE JOB

When deciding between two energy sources for a given job, it is essential to consider the overall system efficiency. Sometimes the trade-offs are not immediately apparent. The home hot water heater is an excellent illustration of this principle.

Electric hot water heaters are often described as being highly efficient. Even though it is very difficult to convert an energy supply entirely to its intended purpose, converting electricity to hot water in a water heater comes very close. That's because heat, the waste product that usually makes an energy conversion system less efficient, is actually the intended product of the conversion. If the conversion from electricity to heat occurs inside the tank of water, which is usually the case with electric hot water heaters, very little energy is lost, and

FIGURE 12.5 **Inefficiencies in energy extraction and use.** This diagram uses energy generation from coal as an example. Energy is lost at each stage of the process, from extraction, processing, and transport of the fuel to the disposal of waste products.

we can say the efficiency is 99 percent. By contrast, a typical natural gas water heater, which transfers energy to water with a flame below the tank and vents waste heat and by-products of combustion to the outside, has an efficiency of about 80 percent. The overall efficiency, however, is actually lower because we have not factored in the energy expended to extract, process, and deliver natural gas to the home. But if a coal-fired power plant is the source of the electricity that fuels the electric water heater, we have to take into account the fact that conversion of fossil fuel into electricity is only about 35 percent efficient. This means that, even though an electric water heater has a higher direct efficiency than a natural gas water heater, the overall efficiency of the electric water heating system is lower—35 percent for the electric water heater versus something less than 80 percent—but not much less—for the gas water heater. There may be many situations in which it is a better choice to heat water with electricity rather than natural gas. But from an environmental perspective, it is important to look at the overall system efficiency when considering the pros and cons of an energy choice.

**EFFICIENCY AND TRANSPORTATION** Because nearly 30 percent of energy use in the United States is for transportation, this is an area in which efficiency is particularly important. *Transportation*—the movement of people and goods—is achieved primarily through the use of vehicles fueled by petroleum products, such as gasoline and diesel, and by electricity. These vehicles contribute to air pollution and greenhouse gas emissions. However, some modes of transportation are more efficient than others.

As you might expect, traveling by train or bus—that is, by public transportation—is much more efficient than traveling by car, especially when there is only

| TABLE 12.1 | Energy expended for different modes of transportation in the United States |
|---|---|
| Mode | MJ per passenger-kilometer |
| Air | 2.1 |
| Passenger car (driver alone) | 3.6 |
| Motorcycle | 1.1 |
| Train (Amtrak) | 1.1 |
| Bus | 1.7 |

*Source:* All data are from Bureau of Transportation Statistics, U.S. Department of Transportation, except for the passenger car, which was determined by assuming one occupant per vehicle obtaining average fuel efficiency of 22 mpg (9.4 km per liter).

one person in the car. And public ground transportation is usually more efficient than air travel. Table 12.1 shows the efficiencies of different modes of transportation. Note that the energy values report only energy consumed, in megajoules (MJ, $10^6$ J), per passenger-kilometer traveled and do not include the embodied energy used to build the different vehicles. Train and motorcycle are the most energy-efficient modes of transportation shown. If a car contained four passengers, we would divide the value in Table 12.1 for a lone driver by four, and thus the car would be the most efficient means of transportation. Cars traveling with four riders are relatively rare in the United States, however; single-occupant vehicles are the most common means of transportation. Crunch the Numbers "Efficiency of Travel" shows you how to calculate the efficiencies of different modes of transportation.

Transportation efficiency calculations do not take into account convenience, comfort, or style. Many people in

## CRUNCH THE NUMBERS

### Efficiency of Travel

Imagine that you had unlimited time and you needed to get from Washington, D.C., to Cleveland, Ohio. The distance is roughly 600 km (370 miles). For each mode of transportation, calculate how many megajoules of energy you would use.

Using the data in Table 12.1, we can determine the following:

Air     2.1 MJ/passenger-kilometer × 600 km/trip = 1,260 MJ/passenger-trip

Car     3.6 MJ/passenger-kilometer × 600 km/trip = 2,160 MJ/passenger-trip

Train    1.1 MJ/passenger-kilometer × 600 km/trip = 660 MJ/passenger-trip

Bus     1.7 MJ/passenger-kilometer × 600 km/trip = 1,020 MJ/passenger-trip

If a gallon of gasoline contains 120 MJ, how many gallons of gasoline does it take to make the trip by car?

$$\frac{2{,}160 \text{ MJ/passenger-trip}}{120 \text{ MJ/gallon}} = 18 \text{ gallons/passenger-trip}$$

Examining total joules expended makes it clear that the train is the most energy-efficient means of travel. Driving alone in a car is the least energy-efficient.

**Your Turn:** If you could carpool with three other people who needed to make the same trip, what would the energy expenditure be for each person?

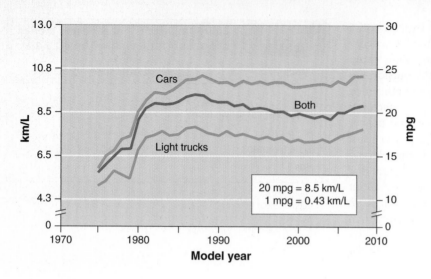

**FIGURE 12.6** **Overall fuel efficiency of U.S. automobiles.** As more buyers moved from cars to light trucks (a category that includes pickup trucks, minivans, and SUVs) for their personal vehicles, the fuel economy of the total U.S. fleet declined. Only recently has it begun to increase. [After U.S. Environmental Protection Agency.]

the developed world are quite particular about how they get from place to place and want the independence of a personal vehicle. They also tend to have strong feelings about what type of personal vehicle is most desirable. In the United States, light trucks—a category that usually includes sport-utility vehicles (SUVs), minivans, and pickup trucks—account for roughly one-half of total automobile sales, while hybrid electric vehicles account for between 2 and 3 percent of total sales.

Light trucks are comparatively heavy vehicles and so generally have mileage ratings of less than 8.5 km per liter, or 20 miles per gallon (mpg). Because they are exempt from certain vehicle emission standards, they emit more of certain air pollutants per liter of fuel combusted than passenger cars. Smaller cars with standard internal combustion engines can travel up to 19 km per liter (45 mpg) on the highway. Hybrid passenger cars, which use a gasoline engine, electric motors, and regenerative braking (a system that reclaims energy normally dissipated when car brakes are applied), obtain closer to 21 km per liter (50 mpg). Electric cars and plug-in hybrid electric cars, which are becoming more common in the United States, obtain even better fuel efficiency.

Despite the availability of fuel-efficient vehicles, many people drive vehicles that yield relatively low fuel efficiencies. As **FIGURE 12.6** shows, the overall fuel efficiency of the U.S. personal vehicle fleet declined from 1985 through 2005 as people chose light trucks and SUVs over cars. Only in the last few years have vehicle choice changed and vehicle efficiency slowly increased. Recently, legislation was passed to increase the average fuel efficiency of new cars and light trucks sold each year so as to deliver a combined fleet average of 15 km per liter (35 mpg) by 2016.

Energy efficiency is an important consideration when making fuel and technology choices, but it is not the only factor we must consider. Determining the best fuel for the job is not always easy, and it involves trade-offs among convenience, ease of use, safety, cost, and pollution.

**GAUGE YOUR PROGRESS**

✓ What are three examples of energy sources used by humans?

✓ Describe the difference between energy efficiency and energy quality.

✓ How do we determine the overall efficiency of energy use in a system?

# Electricity is a convenient form of energy

Because it can be generated from many different sources, from fossil fuels to the Sun, electricity is a form of energy in its own category. Coal, oil, and natural gas are *primary* sources of energy. Electricity is a *secondary* source of energy, meaning that we obtain it from the conversion of a primary source. By the nature of being a secondary source, electricity is an **energy carrier:** something that can move and deliver energy in a convenient, usable form to end users.

Approximately 40 percent of the energy used in the United States is used to generate electricity. But because of conversion losses during the electricity generation process, of that 40 percent, only 13 percent is available for end uses. In this section we will look at some of the basic concepts and issues related to generating electricity from fossil fuels. Chapter 13 discusses electricity generation from renewable energy sources.

## Electricity Generation

Electricity is produced by conversion of primary sources of energy such as coal, natural gas, or wind. Electricity is clean at the point of use; no pollutants are emitted in your home when you use a light bulb or computer.

When electricity is produced by burning fossil fuels, however, pollutants are released at the location of its production. In addition, as we have seen, the transfer of energy from a fuel to electricity is only about 35 percent efficient. Therefore, although electricity is highly convenient, from the standpoint of efficiency of the overall system and the total amount of pollution released, it is more desirable to transfer heat directly to a home with wood or oil combustion, for example, than via electricity generated from the same materials. The energy source that entails the fewest conversions from its original form to the end use is likely to be the most efficient.

Many types of fossil fuels, as well as nuclear fuels, can be used to generate electricity. Regardless of which fuel is used, all thermal power plants work in the same basic way: they convert the potential energy of a fuel into electricity.

FIGURE 12.7 illustrates the major features of a typical coal-burning power plant. Fuel—in this case, coal—is delivered to a boiler, where it is burned. The burning fuel transfers energy to water, which becomes steam. The kinetic energy contained within the steam is transferred to the blades of a **turbine,** a large device that resembles a fan or a jet engine. As the energy in the steam turns the turbine, the shaft in the center of the turbine turns the generator, which generates electricity. The electricity that is generated is then transported along a network of interconnected transmission lines known as the **electrical grid,** which connects power plants together and links them with end users of electricity. Once the electricity is on the grid, it is distributed to homes, businesses, factories, and other electricity consumers, where it may be converted into heat energy for cooking, kinetic energy in motors, or radiant energy in lights, or used to operate electronic and electrical devices. After the steam passes through the turbine, it is condensed back into water. Sometimes the water is cooled in a cooling tower or discharged into a nearby body of water. As we saw in Chapter 9, once-through use of water for thermal electricity generation is responsible for about one-half the water consumption in the United States.

**EFFICIENCY OF ELECTRICITY GENERATION** Whereas a typical coal-burning power plant has an efficiency of about 35 percent, newer coal-burning power plants may have slightly higher efficiencies. Power plants using other fossil fuels can be even more efficient. An improvement in gas combustion technology has led to the **combined cycle** natural gas–fired power plant, which has two turbines and generators. Natural gas is combusted, and the combustion products turn a gas turbine. In addition, the waste heat from this process boils water, which turns a conventional steam turbine. For this reason, a combined cycle plant can achieve efficiencies of up to 60 percent.

**FIGURE 12.7** **A coal-fired electricity generation plant.** Energy from coal combustion converts water into steam, which turns a turbine. The turbine turns a generator, which produces electricity.

A typical power plant in the United States might have a **capacity**—that is, a maximum electrical output—of 500 megawatts (MW). This means that when the plant is operating, it generates 500 MW of electricity. If the plant operated for one day, it would generate 500 MW × 24 hours = 12,000 megawatt-hours (abbreviated MWh). Most home electricity is measured in kilowatt-hours (abbreviated kWh). So the typical power plant we've just described would generate 12,000,000 kWh in a day. If it operated for 365 days per year, it would generate 365 times that daily amount.

Most power plants, however, do not operate every day of the year. They must be shut down for some time to allow for maintenance, refueling, or repairs. Therefore, it is useful to measure the amount of time a plant actually operates in a year. This number—the fraction of the time a plant is operating—is known as its **capacity factor.** Most thermal power plants have capacity factors of 0.9 or greater. As we will see in Chapter 13, power plants using some forms of renewable energy, such as wind, may have a capacity factor of only about 0.25. Crunch the Numbers "Calculating Electricity Supply" shows you how to calculate the amount of energy a power plant can supply.

When it is time to start up a power plant, nuclear and coal-fired plants may take a number of hours, or even a full day, to come up to full generating capacity. Because of the time it takes for them to become operational, electric companies tend to keep nuclear and coal-fired plants running at all times. As demand for electricity changes during the day or week, plants that are more easily powered up, such as those that use natural gas, oil, water, or wood, are used.

**COGENERATION** The use of a fuel to generate electricity *and* produce heat is known as **cogeneration.** Also called *combined heat and power,* cogeneration is a

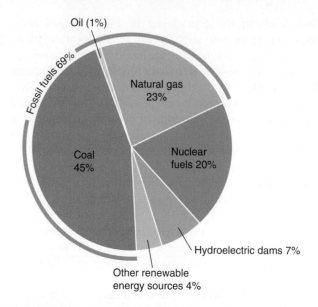

**FIGURE 12.8** **Fuels used for electricity generation in the United States.** Coal is the fuel most commonly used for electricity generation. [Data from U.S. Department of Energy, Energy Information Administration, 2009.]

method employed by certain users of steam for obtaining greater efficiencies. If steam used for industrial purposes or to heat buildings is diverted to turn a turbine first, the user will achieve greater overall efficiency than by generating heat and electricity separately. Cogeneration efficiencies can be as high as 90 percent, whereas steam heating alone might be 75 percent efficient, and electricity generation alone might be 35 percent efficient.

There are over 17,000 power plants in the United States. In 2009, they generated 3.9 billion MWh. **FIGURE 12.8** shows the fuels that were used to generate this electricity. As we can see, coal-fired power plants are

---

**CRUNCH THE NUMBERS**

**Calculating Electricity Supply**

A typical home in the United States uses approximately 900 kWh of electricity per month. On an annual basis, this is

900 kWh/month × 12 months/year = 10,800 kWh/year

How many homes can a 500 MW power plant with a 0.9 capacity factor support?

Begin by determining how much electricity the plant can provide per month:

500 MW × 24 hours/day × 30 days/month × 0.9
= 324,000 MWh/month

1 MWh equals 1,000 kWh, so to convert MWh per month to kWh per month, we multiply by 1,000:

324,000 MWh/month × 1,000 kWh/MWh
= 324,000,000 kWh/month

$$\frac{324,000,000 \text{ kWh/month}}{900 \text{ kWh/month/home}} = 360,000 \text{ homes}$$

On average, a 500 MW power plant can supply roughly 360,000 homes with electricity.

**Your Turn:** During summer months, in hot regions of the country, some homes run air conditioners continuously. How many homes can the same power plant support if average electricity usage increases to 1,200 kWh/month during summer months?

the backbone of electricity generation in the United States, responsible for 45 percent of all electricity produced. Natural gas, at 23 percent, and nuclear energy, at 20 percent, account for most of the remainder of our generating capacity. Together, these three sources, plus a very small amount of oil, account for 89 percent of electricity generation in the United States. Water and other renewable energy resources such as wind and solar energy account for the remainder of the electricity generated in the country.

**GAUGE YOUR PROGRESS**

✓ What is the basic process by which the energy in a fuel is converted into electricity?

✓ What are the major fuels that are used to generate electricity in the United States?

✓ What is a combined cycle plant, and what is cogeneration?

# Fossil fuels provide most of the world's energy

Fossil fuels provide most of the energy used in both developed and developing countries. The vast majority of the fossil fuels we use—coal, oil, and natural gas—come from deposits of organic matter that were formed 50 million to 350 million years ago. When organisms die, decomposers break down most of the dead biomass aerobically, and it quickly reenters the food web, as we saw in Chapter 3. However, in places such as swamps, river deltas, and the ocean floor, a large amount of detritus may build up quickly in an anaerobic environment.

Under these conditions, decomposers cannot break down all of the detritus. As this material is buried under succeeding layers of sediment and exposed to heat and pressure, the organic compounds within it are chemically transformed into high-energy solid, liquid, and gaseous components that are easily combusted. Because they come from ancient biomass, these components—coal, oil, and natural gas—are called fossil fuels. Figure 3.11, which shows the contemporary carbon cycle, illustrates the process of fossil fuel formation.

## Coal

We have seen that coal is the fuel most commonly used for electricity generation in the United States. In Chapter 8 we learned about the various methods of extracting coal from the ground. But how exactly is coal formed? **Coal** is a solid fuel formed primarily from the remains of trees, ferns, and other plant materials that were preserved 280 million to 360 million years ago. There are four types of coal: ranked from lesser to greater age, exposure to pressure, and energy content, they are *lignite, subbituminous, bituminous,* and *anthracite.* A precursor to coal, called *peat,* is made up of partly decomposed organic material, including mosses. **FIGURE 12.9** shows how the different types of coal are formed. Starting with an organic material such as peat, increasing time and pressure produce successively denser coal with more carbon molecules, and more potential energy, per kilogram.

The largest coal reserves are found in the United States, Russia, China, and India. The countries that are currently producing the greatest amounts of coal are China, the United States, India, and Australia.

**ADVANTAGES AND DISADVANTAGES OF COAL** Because it is energy-dense and plentiful, coal is used to generate electricity and for industrial processes such as making

Ancient forests cover much of Earth's surface.

Time and pressure

Peat

Lignite

(Sub)-bituminous coal

Anthracite

The vegetation dies and is buried under anaerobic conditions, forming peat (partially decomposed organic matter).

The peat is compressed between sediment layers to form lignite.

Further compression yields sub-bituminous and bituminous coal.

After yet more pressure and time, anthracite forms.

**FIGURE 12.9** **The coal formation process.** Peat is the raw material from which coal is formed. Over time and under increasing pressure, various types of coal are formed.

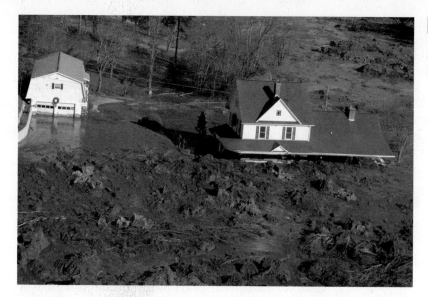

**FIGURE 12.10** **Results of a coal ash spill.** In 2008, this Tennessee home was buried in ash when an ash holding pond at a nearby power plant gave way.

steel. In many parts of the world, coal reserves are relatively easy to exploit by surface mining. The technological demands of surface mining are relatively small, and the economic costs are low. As surface coal is used up and becomes harder to find, however, subsurface mining becomes necessary. With subsurface mining, the technological demands and costs increase, and the human health consequences increase as well, as we saw in Chapter 8. Once it is extracted from the ground, coal is relatively easy to handle and needs little refining before it is burned. It can be transported to power plants and factories by train, barge, or truck. All of these factors make coal a relatively easy fuel for any country to use, regardless of its technological development and infrastructure.

Although coal is a relatively inexpensive fossil fuel, its use does have several disadvantages. Coal contains a number of impurities, including sulfur, that are released into the atmosphere when it is burned. The sulfur content of coal typically ranges from 0.4 to 4 percent by weight. Lignite and anthracite have a relatively low sulfur content, whereas bituminous coal often has a much higher sulfur content. Trace metals such as mercury, lead, and arsenic are also found in coal. Combustion of coal results in the release of these impurities, which leads to an increase of sulfur dioxide and other air pollutants, such as particulates, in the atmosphere, as we will see in Chapter 15. Compounds that are not released into the atmosphere remain behind in the resulting ash.

According to the U.S. Department of Energy, there are 1,450 coal mines in the United States, and they produced more than 1 billion metric tons of coal in 2008. Most of that coal is burned in the United States, and anywhere from 3 to 20 percent of it remains behind as ash. As a result, large deposits of ash are often stored near coal-burning power plants. One such ash deposit—a mixture of ash and water—was kept in a holding pond at a power plant near Knoxville, Tennessee. In December 2008, the retaining wall that contained the ash gave way and spilled 4.1 billion liters (1.1 billion gallons) of ash (**FIGURE 12.10**). Three houses were destroyed by the flow of muddy ash, which covered over 121 hectares (300 acres) of land. The event was the largest of its kind in U.S. history, and the cleanup is still ongoing.

Coal is 60 to 80 percent carbon. When it is burned, most of that carbon is converted into $CO_2$, and energy is released in the process. When it is combusted, coal produces far more $CO_2$ per unit energy released than either oil or natural gas. This $CO_2$ contributes to the increasing atmospheric concentrations of $CO_2$ discussed in Chapters 1, 3, and 19.

## Petroleum

**Petroleum,** another widely used fossil fuel, is a fluid mixture of hydrocarbons, water, and sulfur that occurs in underground deposits. While coal is an ideal fuel for stationary combustion applications, such as power plants and industry, the fluid nature of petroleum products such as oil and gasoline makes them ideal for mobile combustion applications, such as vehicles.

Petroleum is formed from the remains of ocean-dwelling phytoplankton (microscopic algae) that died 50 million to 150 million years ago. Deposits of phytoplankton are found in locations where porous sedimentary rocks, such as sandstone, are capped by nonporous rocks. Petroleum forms over millions of years and fills the pore spaces in the rock. Geologic events related to the tectonic cycle we discussed in Chapter 8 may deform the rock layers so that they form a dome. The petroleum is less dense than the rock, so over time, it migrates upward toward the highest point in the porous rock, where it is trapped by the nonporous rock, as **FIGURE 12.11** shows. In certain locations, petroleum flows out under pressure

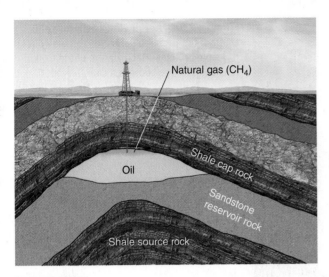

**FIGURE 12.11 Petroleum accumulation underground.**
Petroleum migrates to the highest point in a formation of porous rock and accumulates there. Such accumulations of petroleum can be removed by drilling a well.

the way water flows from an artesian well, as described in Chapter 9. But usually petroleum producers must drill wells into a deposit and extract the petroleum with pumps. After extraction, the petroleum must be transported by pipeline, if the well is on land, or by supertanker, if it is underwater, to a petroleum refinery.

Petroleum contains natural gas. As we can see from Figure 12.11, some of the gas separates out naturally. That's why you sometimes see a burning flame, known as a *gas flare,* in photographs of oil wells. The oil workers are *flaring,* or burning off, the natural gas under controlled conditions to prevent an explosion. Some of the gas is also extracted for use as fuel, as we will see in the following section.

Liquid petroleum that is removed from the ground is known as **crude oil.** The U.S. Department of Energy refers to oil, crude oil, and petroleum more or less as equivalent substances, and we will do the same in this chapter. Crude oil can be further refined into a variety of compounds. These compounds, including tar and asphalt, gasoline, diesel, and kerosene, are distinguished by the temperature at which they boil and can therefore be separated by heating the petroleum. This process takes place in an oil refinery, a large factory of the kind shown in this chapter's opening photograph. The refining process is complex and dangerous and requires a major financial investment. There are roughly 150 oil refineries in the United States; some of the larger ones can refine over 80 million liters (21 million gallons) per day. Oil production and sales are measured in *barrels* of oil; one barrel equals 160 liters (42 gallons).

As we saw in Figure 12.4, the United States uses more petroleum than any other fuel—roughly 3.1 billion liters (816 million gallons) of petroleum products

*per day.* Gasoline accounts for roughly one-half of that amount. The primary use of petroleum products is for transportation. Petroleum is also the raw material for petrochemicals, such as plastics, lubricants, raw materials for pharmaceuticals, and cleaning solvents. Worldwide petroleum consumption is almost 14 billion liters (3.7 billion gallons) per day. Thus the United States is responsible for about 22 percent of worldwide petroleum consumption.

The top petroleum-producing countries are Saudi Arabia, Russia, the United States, Iran, China, Canada, and Mexico, in that order. These seven countries account for roughly one-half of worldwide oil production.

### ADVANTAGES AND DISADVANTAGES OF PETROLEUM

Because petroleum is a liquid, it is extremely convenient to transport and use. It is relatively energy-dense and is cleaner-burning than coal. For these reasons, it is an ideal fuel for mobile combustion engines such as those found in automobiles, trucks, and airplanes. But because it is a fossil fuel, it releases $CO_2$ when burned. However, for every joule of energy released, oil produces only about 85 percent as much $CO_2$ as coal.

Oil, like coal, contains sulfur and trace metals such as mercury, lead, and arsenic, which are released into the atmosphere when it is burned. Some sulfur can be removed during the refining process, so it is possible, though more expensive, to obtain low-sulfur oil.

As we have seen, oil must be extracted from under the ground or beneath the ocean. Whenever oil is extracted and transported, there is the potential for oil to leak from the wellhead or be spilled from a pipeline or tanker. Some oil naturally escapes from the rock in which it was stored and seeps into water or out onto land. However, commercial oil extraction has greatly increased the number of leakage and spillage events and the amount of oil loss to land and water around the world. As we saw in this chapter's opening story, the largest oil spill in the United States until 2010 was the *Exxon Valdez* oil tanker accident in 1989. More recently, the blowout of the BP Deepwater Horizon oil well, drilled 81 km (50 miles) off the coast of Louisiana in 1,524 m (5,000 feet) of water, has led to a spill of well over 780 million liters (206 million gallons) of oil (**FIGURE 12.12**).

Larger oil spills have occurred elsewhere in the world. For example, during the 1991 Persian Gulf War, approximately 912 million liters (240 million gallons) of petroleum were spilled when wellheads were deliberately sabotaged or destroyed by the Iraqi army in Kuwait.

It is important to consider the various ways in which oil is spilled into the natural environment. A 2003 National Academy of Sciences study found that oil extraction and transportation were responsible for a relatively small fraction of the oil spilled into marine waters worldwide. Roughly 85 percent of the oil

**FIGURE 12.12** **Aerial view of an oil spill.** Oil from the BP Deepwater Horizon blowout is visible on the surface waters of the Gulf of Mexico, near the Louisiana coast, in 2010.

entering marine waterways came from runoff from land and rivers, airplanes, and small boats and personal watercraft, including deliberate and accidental releases of waste oil.

In the United States, debates continue over the trade-off between extracting oil domestically and the consequences for habitat and species living near oil wells or pipelines. For example, when a 1,300 km (800-mile) pipeline was constructed to transport oil overland from the North Slope of Alaska to tankers that would carry it to the continental United States, wildlife biologists predicted that the pipeline might melt permafrost and interfere with the calving grounds of caribou. Scientists continue to monitor the pipeline, but so far have come to no conclusions about its environmental impact. The debate about the environmental effects of land-based

oil extraction has continued with the proposal to allow oil exploration in the Arctic National Wildlife Refuge (ANWR), a 7.7 million hectare (19 million acre) tract of land in northeastern Alaska (**FIGURE 12.13**). Proponents of exploration suggest that ANWR might yield 95 billion liters (25 million gallons) to 1.4 trillion liters

**(a)**

**(b)**

**FIGURE 12.13** **The Arctic National Wildlife Refuge (ANWR).** (a) Caribou in ANWR. (b) Map of ANWR and adjacent areas on the North Slope of Alaska. The area in orange (the coastal plain, also called the "Ten-O-Two" area) is under consideration for petroleum exploration and extraction.

Point Barrow

Arctic Ocean

Prudhoe Bay

1002 area

National Petroleum Reserve

ANWR

Wilderness area

Northern margin of Brooks Range

Trans-Alaska Pipeline

Canada
United States

Known petroleum accumulation

150 km

(378 billion gallons) of oil and substantial quantities of natural gas. Opponents maintain that opening ANWR to exploration and petroleum extraction will harm pristine habitat for many species and affect people in the area as well.

Humans, as well as wildlife, have been harmed by oil extraction. In Nigeria and many other developing countries, oil fields are adjacent to villages. Thick, gelatinous crude oil covers the ground where people walk, sometimes in bare feet, and oil flaring—the burning off of excess natural gas—takes place close to homes. Concerns about the effects of oil extraction on health, human rights, and environmental justice have led to violent political protests against oil companies in Nigeria and elsewhere.

## Natural Gas

We have already mentioned natural gas in connection with petroleum, since it exists as a component of petroleum in the ground as well as in gaseous deposits separate from petroleum. Natural gas is 80 to 95 percent methane ($CH_4$) and 5 to 20 percent ethane, propane, and butane. Because natural gas is lighter than oil, it lies above oil in petroleum deposits (see Figure 12.11). Natural gas is generally extracted in association with petroleum; only recently has exploration specifically for natural gas been conducted.

The two largest uses of natural gas in the United States are electricity generation and industrial processes. Natural gas is also used to manufacture nitrogen fertilizer and in residences as an efficient fuel for cooking, heating, and operating clothes dryers and water heaters. Compressed natural gas can be used as a fuel for vehicles, but because it must be transported by pipeline, it is not accessible in all parts of the United States and is therefore unlikely to become an important fuel for cars. Liquefied petroleum gas (LPG)—which is similar to natural gas, but in a liquid form—is a slightly less energy-dense substitute. LPG can be transported via train or truck and stored at the point of use in tanks. This fuel is available practically everywhere in the United States and is used in place of natural gas and for portable barbecue grills and heaters. Overall, natural gas and LPG supply 24 percent of the energy used in the United States.

### ADVANTAGES AND DISADVANTAGES OF NATURAL GAS

Because of the extensive natural gas pipeline system in many parts of the United States, roughly one-half of homes use natural gas for heating. Compared with coal and oil, natural gas contains fewer impurities and therefore emits almost no sulfur dioxide or particulates during combustion. And for every joule of energy released during combustion, natural gas emits only 60 percent as much $CO_2$ as coal. On the other hand, unburned natural gas—methane—that escapes into the atmosphere is itself a potent greenhouse gas that is 25 times more efficient at absorbing infrared energy than $CO_2$. The leaking of natural gas after extraction is a suspected contributor to the steep rise in atmospheric methane concentrations that was observed in the 1990s.

While natural gas is referred to as the "clean" fossil fuel, its extraction and use still lead to environmental problems (FIGURE 12.14). The process of exploring for natural gas involves the "thumper trucks" mentioned in this chapter's opening story. The process of drilling and opening up the rock containing the natural gas, called host rock, in order to release natural gas is called hydraulic fracturing, or "fracking." This process entails drilling with water, sand, and proprietary chemicals whose names and effects do not have to be released to the public by the companies that use them. Large quantities of water are used during hydraulic fracturing; this water becomes contaminated with chemicals during the process and must be disposed of afterward. There is often groundwater contamination resulting from the drilling of natural gas wells.

**FIGURE 12.14** Natural gas field in Wyoming.

## Other Fossil Fuels: Oil Sands and Liquefied Coal

There are other types of fossil fuel deposits that contain a great deal of energy, but are not readily available. When we consider using these energy resources, the energy return on energy investment is always an important consideration.

Not all petroleum is easily extractable in conventional oil wells. **Oil sands** are slow-flowing, viscous deposits of *bitumen* mixed with sand, water, and clay. **Bitumen,** often called *tar* or *pitch,* is a degraded type of petroleum that forms when a petroleum deposit is not capped with nonporous rock. The petroleum migrates close to the surface, where bacteria metabolize some of the light hydrocarbons while others evaporate. The remaining mix no longer flows at ambient temperatures and pressures. It can be extracted by surface mining.

Although oil sand exploitation promises to extend our petroleum supply, it could have serious negative environmental impacts. The mining of bitumen is much more energy-intensive than conventional drilling for crude oil. As we saw in Chapter 8, surface mining creates large open pits. Extraction of the bitumen from the other material contaminates roughly 2 to 3 L of water for every liter of bitumen obtained, and many oil sands are located in areas where water is not an abundant resource. In addition, because oil sands require so much energy to process before they even get to a refinery, the overall system efficiency is lower, and the resulting $CO_2$ release greater, than for conventional oil production.

Whereas the availability of petroleum may become much more limited in this century as supplies diminish, the United States and China both have immense coal reserves. The technology to convert solid coal into a liquid fuel—a process known as **CTL,** short for "coal-to-liquid"—has been available for decades. CTL was widely used by the German military during World War II and has been used by other countries since then. But producing liquefied coal is relatively expensive and has many of the same drawbacks as exploiting oil sands. In terms of total energy content, there is over 1,000 times more energy in the world's coal reserves than in the world's petroleum reserves. Because there is so much coal in the United States, CTL has the potential to eliminate U.S. dependence on foreign oil. On the other hand, the U.S. Environmental Protection Agency estimates that total greenhouse gas emissions from liquefied coal are more than twice those from conventionally produced oil, and as we have seen, the environmental impacts of coal mining can be severe.

**GAUGE YOUR PROGRESS**

✓ How are the different types of coal formed?

✓ How is oil formed, and why does it need to be refined?

✓ What are the major advantages and disadvantages of using coal, oil, and natural gas?

# Fossil fuels are a finite resource

Although we know that the supply of fossil fuels is finite, there is some discussion within the environmental science community on whether or not that matters. Recall our discussion in Chapter 7 about the creativity of the human population. Many people believe that we will apply that creativity to the development of new energy sources. In the meantime, total energy use continues to increase, even though energy use per person has leveled off and energy use per unit of gross domestic product (GDP), known as **energy intensity,** has been steadily decreasing, as FIGURE 12.15 shows. In other words, we are using energy more efficiently in order to do what we need to do, but because there are more of us and we are doing more things that use energy, our overall energy use has increased. For example, think about how

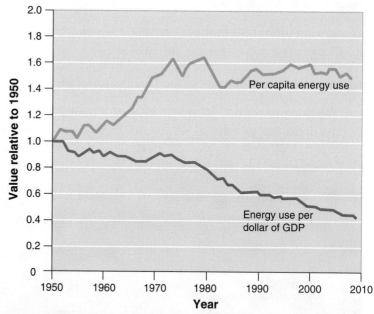

**FIGURE 12.15** **U.S. energy use per capita and energy intensity.** Our energy use per capita has been level while our energy intensity, or energy use per dollar of GDP, has been decreasing in recent years. However, because of our increasing population, our overall energy use continues to increase. [Data from U.S. Department of Energy, Energy Information Administration, 2009.]

many electronic devices you have, then ask your parents or grandparents how many electronic devices they had at your age.

Because fossil fuels take millions of years to form, they are nonrenewable resources. By definition, then, the use of fossil fuels is not sustainable because there is no way to limit our consumption to the rate at which they are being formed. Experts debate whether our economy will at some point be limited by the availability of this energy resource. In recent years, concerns have shifted away from the supply of fossil fuels to the consequences of fossil fuel combustion, particularly the release of $CO_2$ and its contribution to global warming. Many environmental scientists believe that these consequences will manifest themselves in adverse ways long before we run out of fossil fuels.

## The Hubbert Curve

In 1969, M. King Hubbert, a geophysicist and oil company employee, published a graph showing a bell-shaped curve representing oil use (FIGURE 12.16). The graph, which became known as the **Hubbert curve,** projected the point at which world oil production would reach a maximum and the point at which we would run out of oil. Hubbert used two estimates of total world petroleum reserves: an upper estimate and a lower estimate. He found that the total reserves did not greatly influence the time it would take to use up all of the oil in known reserves. Rather, he predicted that oil extraction and use would increase steadily until roughly half the supply had been used up. At that point, known as **peak oil,** extraction and use would begin to decline. Some oil experts

believe we have already reached peak oil, while others maintain that we may reach it very soon. Back in 1969, Hubbert predicted that 80 percent of the world's total oil supply would be used up in roughly 60 years.

Although there have been discoveries of large oil fields since Hubbert did his work, the conclusion he drew from his model still holds. Regardless of the exact amount of the total reserves, the total number of years we use petroleum will fall within a relatively narrow time window. That is, when we identify a fuel source, we tend to use it until we come upon a better fuel source. As a number of energy experts are fond of saying, "We did not move on from the Stone Age because we ran out of stones." Similarly, many people believe that ingenuity and technological advances in the renewable energy sector will one day render oil much less desirable.

## The Future of Fossil Fuel Use

If current global use patterns continue, we will run out of conventional oil supplies in less than 40 years. Natural gas supplies will last slightly longer. Coal supplies will last for at least 200 years, and probably much longer. While these predictions assume that we will continue our current use patterns, advances in technology, a shift to nonfossil fuels, or changes in social choices and population patterns could alter them.

In recent years, with greater acceptance of the theory that global climate change is resulting from anthropogenic increases in atmospheric greenhouse gas concentrations, a large number of researchers have suggested that the question we should be asking is not "When will we run out of oil?" but rather, "How can we transition

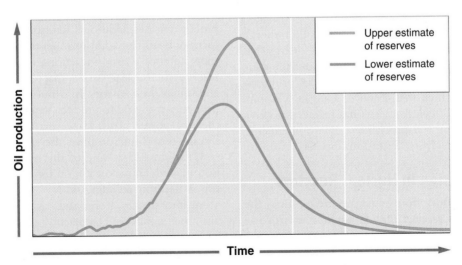

**FIGURE 12.16** **A generalized version of the Hubbert curve.** Whether an upper estimate or a lower estimate of total petroleum reserves is used, the date by which petroleum reserves will be depleted does not change substantially. [After M. K. Hubbert, The energy resources of the Earth, in *Energy and Power,* A Scientific American Book (W. H. Freeman, 1971).]

away from fossil fuels before their use causes further problems?"

Concerns about the scarcity, environmental impacts—especially the influence of $CO_2$ on global climate change—and rising costs of fossil fuels present many opportunities. Rising oil prices create a powerful incentive to invest in alternative energy resources and conservation. As we will see in Chapter 13, people have begun to explore the possibility of a new infrastructure to deliver wind energy, hydroelectricity, and solar energy in novel ways. On the other hand, the higher price of oil makes formerly unproductive mining and extraction methods cost-competitive, and so may result in greater total production of fossil fuels.

There is one other conventional, nonrenewable fuel that offers a very attractive feature: it does not contribute significantly to the addition of greenhouse gases to the atmosphere. That fuel is uranium.

---

**GAUGE YOUR PROGRESS**

✓ Explain the relationship between energy intensity and energy use per capita.

✓ Describe the Hubbert curve and its significance.

✓ What are the major considerations involved in the future of fossil fuels?

---

**FIGURE 12.17** **Nuclear fission.** Energy is released when a neutron strikes a large atomic nucleus, which then splits into two or more parts.

# Nuclear energy is getting a second look

Because the combustion of fossil fuels releases large quantities of $CO_2$ into the atmosphere, people have considered the advantages and disadvantages of many other energy sources. One of these alternatives, nuclear energy, has often been rejected because of concerns about the dangers of nuclear accidents, radioactivity, and the proliferation of radioactive fuels that could be used in weapons. Recently, however, nuclear energy has received positive attention, even from self-proclaimed environmentalists, because of its relatively low emissions of $CO_2$.

## The Use of Fission in Nuclear Reactors

Electricity generation from nuclear energy uses the same basic process as electricity generation from fossil fuels: steam turns a turbine that turns a generator that generates electricity. The difference is that a nuclear power plant uses a radioactive isotope, uranium-235 ($^{235}U$), as its fuel source. We presented the concepts of isotopes, radioactive decay, and half-lives in Chapter 2. Radioactive decay occurs when a parent radioactive isotope emits alpha or beta particles or gamma rays. Here,

we need to introduce one more concept before being able to fully describe a nuclear power plant. The naturally occurring isotope $^{235}U$, as well as other radioactive isotopes, undergoes a process called *fission*.

**Fission,** shown in **FIGURE 12.17**, is a nuclear reaction in which a neutron strikes a relatively large atomic nucleus, which then splits into two or more parts. This process releases additional neutrons and energy in the form of heat. The additional neutrons can, in turn, promote additional fission reactions, which leads to a chain reaction of nuclear fission that gives off an immense amount of heat energy. In a nuclear power plant, that heat energy is used to produce steam, just as in any other thermal power plant. However, 1 g of $^{235}U$ contains 2 million to 3 million times the energy of 1 g of coal.

Uranium-235 is one of the more easily fissionable isotopes, which makes it ideal for use in a nuclear reactor. A neutron colliding with $^{235}U$ splits the uranium atom into smaller atoms, such as barium and krypton, and three neutrons. The reaction is as follows:

$$1 \text{ neutron} + {}^{235}U \rightarrow {}^{142}Ba + {}^{91}Kr + 3 \text{ neutrons}$$
$$\text{in motion (kinetic energy)}$$

Many other radioactive daughter products are released as well. A properly designed nuclear reactor will harness the kinetic energy from the three neutrons in motion to produce a self-sustaining chain reaction of nuclear fission. The by-products of the nuclear reaction include radioactive waste that remains hazardous for many half-lives—that is, hundreds of thousands of years or longer.

FIGURE 12.18 shows how a nuclear reactor works. The containment structure encloses the nuclear fuel—which is contained within cylindrical tubes called **fuel rods**—and the steam generator. Uranium fuel is processed into pellets, which are then put into the fuel rods. A typical nuclear reactor might contain hundreds of bundles of fuel rods in the center, or reactor core, within the containment structure.

Within the containment structure, heat from nuclear fission is used to heat water, which circulates in a loop. This loop passes close to another loop of water, and heat is transferred from one loop to the other. In the process, steam is produced, which turns a turbine, which turns a generator, just as in most other thermal power plants. The nuclear power plant shown in Figure 12.18 is a light-water reactor, the only type of reactor used in the United States and the most common type used elsewhere in the world.

The objective of a nuclear power plant is to harness the heat energy from fission to make steam. But the plant must be able to slow the fission reaction to allow collisions to take place at the appropriate speed. To do this, nuclear reactors contain a moderator, such as water, to slow down the neutrons so that they can effectively trigger the next chain reaction. There is also a risk that the reaction will run out of control. Nuclear reactors contain **control rods,** cylindrical devices that can be inserted between the fuel rods to absorb excess neutrons, thus slowing or stopping the fission reaction. This is done routinely during the operation of the plant because nuclear fuel rods left uncontrolled will quickly become too hot and melt—an event called a *meltdown*—or cause a fire, either of which could lead to a catastrophic nuclear accident. Control rods are also inserted when the plant is being shut down during an emergency or for maintenance and repairs.

Depending on the ore, it may take up to 900 kg (2,000 pounds) of uranium ore to produce 3 kg (6.6 pounds) of nuclear fuel. In order to obtain uranium, miners remove large amounts of the host rock, extract and concentrate the uranium, and leave the remaining material in slag piles. Australia, the western United States, and parts of Canada have large commercial uranium mining operations for nuclear fuel. As we saw in Chapter 8, the mining of any material requires fossil fuel energy and results in mine tailings. This is also true for

FIGURE 12.18 **A nuclear reactor.** This schematic shows the basic features of the light-water reactor, the type of reactor found in the United States.

uranium, although, as we have noted, a much smaller volume and mass of uranium is needed to generate a quantity of electricity than coal.

Nuclear power plants rely on $^{235}$U as their fuel. However, most uranium ores contain as much as 99 percent $^{238}$U, another isotope of uranium that occurs with $^{235}$U but does not fission as easily. Therefore, when uranium ore is mined, it must be chemically enriched—a process to increase its concentration—in $^{235}$U to be useful as a fuel. Typically, suitable nuclear fuel contains 3 percent or greater $^{235}$U.

## Advantages and Disadvantages of Nuclear Energy

Nuclear power plants do not produce air pollution during their operation, so proponents of nuclear energy consider it "clean" energy. In countries with limited fossil fuel resources, nuclear energy is one way to achieve independence from imported oil. Nuclear energy generates 70 percent of electricity in France, and it is widely used in Lithuania, Germany, Spain, the United Kingdom, Japan, China, and South Korea, as well as other countries.

As Figure 12.8 shows, 20 percent of the electricity generated in the United States comes from nuclear energy. Early proponents of nuclear energy in the 1950s and 1960s claimed that it would be "too cheap to meter," meaning that it would be so inexpensive that there would be no point in trying to figure out how much each customer used. However, construction of new nuclear power plants became more and more expensive in the United States, in part because public protests, legal battles, and other delays increased the cost of construction. Public protests arose because of concerns that a nuclear accident would release radioactivity into the surrounding air and water. Other concerns included uncertainty about appropriate locations for radioactive waste disposal and fear that radioactive waste could fall into the hands of individuals seeking to make a nuclear weapon. By the 1980s, it had become prohibitively expensive—both monetarily and politically—to attempt to construct a new nuclear power plant.

There are currently 104 nuclear power plants in the United States—the same number as there were two decades ago. However, because of the relatively low $CO_2$ emissions associated with nuclear energy, there has been a resurgence of interest in constructing additional nuclear power plants. There are certainly $CO_2$ emissions related to mining, processing, and transporting nuclear fuel and constructing a nuclear power plant. However, these emissions are perhaps a few percent to 10 percent of those related to generating an equivalent amount of electricity from coal. Two major issues of environmental concern remain, however—the possibility of accidents and disposal of radioactive waste.

**THE POSSIBILITY OF ACCIDENTS** Two accidents contributed to the global protests against nuclear energy in the 1980s and 1990s. On March 28, 1979, at the Three Mile Island nuclear power plant in Pennsylvania, operators did not notice that a cooling water valve had been closed the previous day. This oversight led to a lack of cooling water around the reactor core, which overheated and suffered a partial meltdown. The reactor core was severely damaged, and a large part of the containment structure became highly radioactive. An unknown amount of radiation was released from the plant to the outside environment. A few thousand schoolchildren and pregnant women were evacuated from the area surrounding the plant by order of the governor of Pennsylvania. An estimated 200,000 other people chose to evacuate as well. A great deal of anxiety and fear were experienced by residents of the area, especially as reports of a potentially explosive gas bubble in the containment structure were evaluated in the days following the accident. Although there has been no documented increase in adverse health effects in the area of the plant as a result of this accident, several investigators maintain that infant mortality rates and cancer rates increased in the following years. This nuclear reactor, one of two at the Three Mile Island nuclear facility, has not been used since the accident. The Three Mile Island event, compounded by the coincidental release of the film *The China Syndrome*, about safety violations and a near-catastrophic accident at a nuclear power plant, led to widespread public fear and anger in Pennsylvania and elsewhere.

A much more serious accident occurred on April 26, 1986, at a nuclear power plant in Chernobyl, Ukraine. The accident occurred during a test of the plant when, in violation of safety regulations, operators deliberately disconnected emergency cooling systems and removed the control rods. With no control rods and no coolant, the nuclear reactions continued without control, and the plant overheated. These "runaway" reactions led to an explosion and fire that damaged the plant beyond use. At the time of the accident, 31 plant workers and firefighters died from acute radiation exposure and burns; many more died later of related causes.

After the accident, winds blew radiation from the plant across much of Europe, where it contaminated crops and milk from cows grazing on contaminated grass. More than a hundred thousand people were evacuated from the area around Chernobyl. Estimates of health effects vary widely, in part because of the paucity of information provided by the Soviet government, but a U.S. National Academy of Sciences panel estimated that 4,000 additional cancer deaths (over and above the average number of expected deaths) would occur over the next 50 years among people who lived near the plant or worked on the cleanup. There have been approximately 5,000 cases of thyroid cancer, most of them nonfatal, among children who were younger than 18 at the time of the accident and lived near the Chernobyl plant. Thyroid cancer may be caused by the absorption of radioactive iodine, one of the radioactive elements emitted during the accident.

**RADIOACTIVE WASTE** Long after nuclear fuel can produce enough heat to be useful in a power plant, it continues to emit radioactivity. At this point, it is considered **radioactive waste.** Because radioactivity can be extremely damaging to living organisms, radioactive materials must be stored in special, highly secure locations.

The use of nuclear fuels produces three kinds of radioactive waste: *high-level waste* in the form of used fuel rods; *low-level waste* in the form of contaminated protective clothing, tools, rags, and other items used in routine plant maintenance; and *uranium mine tailings,* the residue left after uranium ore is mined and enriched. Disposal of all three types is regulated by the government, but because it has the greatest potential impact on the environment, we will focus here on high-level radioactive waste.

After a period of time, nuclear fuel rods become "spent"—not sufficiently radioactive to generate electricity efficiently. As we discussed in Chapter 2, each radioactive isotope has a specific half-life: the time it takes for one-half of its radioactive atoms to decay. Uranium-235 has a half-life of 704,000,000 years, which means that 704 million years from today, a sample of $^{235}$U will be only half as radioactive as it is today. In another 704 million years it will lose another half of its radioactivity, so it will be one-fourth as radioactive as it is today.

Radiation can be measured with a variety of units. A **becquerel (Bq)** measures the rate at which a sample of radioactive material decays. 1 Bq is equal to the decay of one atom per second. A **curie,** another unit of measure for radiation, is 37 billion decays per second. If a material has a radioactivity level of 100 curies and has a half-life of 50 years, the radioactivity level in 200 years will be

$$\frac{200 \text{ years}}{50 \text{ years/half-life}} = 4 \text{ half-lives}$$

100 curies → 50 curies (one half-life) → 25 curies (two half-lives) → 12.5 curies (three half-lives) → 6.3 curies (four half-lives)

You can gain more experience in working with these calculations in Crunch the Numbers "Calculating Half-Lives" on the next page.

Spent fuel rods remain a threat to human health for 10 or more half-lives. For this reason, they must be stored until they are no longer dangerous. At present, nuclear power plants are required to store spent fuel rods at the plant itself. Initially, all fuel rods were stored in pools of water at least 6 meters (20 feet) deep. The water acts as a shield from radiation emitted by the rods. Currently, more than 100 sites around the country are storing spent fuel rods. However, some facilities have run out of pool storage and are storing them in lead-lined dry containers on land (**FIGURE 12.19**). Eventually, all of this material will need to be moved to a permanent radioactive waste disposal facility.

Disposing of radioactive waste is a challenge. It cannot be incinerated, safely destroyed using chemicals,

**FIGURE 12.19** **Storage of nuclear waste.** Nuclear waste is often stored in metal containers at the power plant where it was produced.

**Calculating Half-Lives**

Strontium-90 is a radioactive waste product from nuclear reactors. It has a half-life of 29 years. How many years will it take for a quantity of strontium-90 to decay to $\frac{1}{16}$ of its original mass?

It will take 29 years to decay to $\frac{1}{2}$ its original mass; another 29 years to $\frac{1}{4}$; another 29 years to $\frac{1}{8}$; another 29 years to $\frac{1}{16}$:

$$29 + 29 + 29 + 29 = 116 \text{ years}$$

**Your Turn:** You have 180 g of a radioactive substance. It has a half-life of 265 years. After 1,325 years, what mass remains?

shot into space, dumped on the ocean floor, or buried in an ocean trench because all of these options involve the potential for large amounts of radioactivity to enter the oceans or atmosphere. Therefore, at present, the only solution is to store it safely somewhere on Earth indefinitely. The physical nature of the storage site must ensure that the waste will not leach into the groundwater or otherwise escape into the environment. It must be far from human habitation in case of any accidents and secure against terrorist attack. In addition, the waste has to be transported to the storage site in a way that minimizes the risk of accidents or theft by terrorists.

In 1978, the U.S. Department of Energy began examining a site at Yucca Mountain, Nevada, 145 kilometers (90 miles) northwest of Las Vegas, as a possible long-term repository for the country's spent nuclear fuel. The Yucca Mountain proposal has generated enormous protest and controversy. In 2006, the Department of Energy released a report confirming the soundness of the research supporting the Yucca Mountain site. However, a few years later, after a change in presidential administrations, the Yucca Mountain project was cancelled.

Nuclear energy raises a unique question of sustainability. It is a source of electricity that releases much less $CO_2$ than coal, and none during electricity generation. But some argue that the use of a fuel that consistently generates large quantities of high-level radioactive waste is not a sustainable practice. Many critics of nuclear energy maintain that there will never be a safe enough place to store high-level radioactive waste.

## Fusion Power

As we have seen, nuclear fission occurs when the nuclei of radioactive atoms are broken apart into smaller, lighter nuclei. **Nuclear fusion,** the reaction that powers the Sun and other stars, occurs when lighter nuclei are forced together to produce heavier nuclei. In the process, a great deal of heat is generated. The nuclear fusion reaction that is most promising for electricity generation is that of two hydrogen isotopes fusing together into a

helium atom. As the reaction occurs, a small amount of mass is lost and an immense amount of energy is liberated.

Nuclear fusion seems to promise a seemingly unlimited source of energy that requires only hydrogen as an input and produces relatively small amounts of radioactive waste. Unfortunately, creating fusion on Earth requires a reactor that will heat material to temperatures 10 times those in the core of the Sun. These temperatures make containment extremely difficult. So far, the most promising techniques have involved suspending superhot material in a magnetic field, but the amount of energy required is greater than the energy output. Most experts believe that it will be several decades, or perhaps longer, before the promise of fusion power can be realized.

At present, our reliance on nuclear energy in the United States is subject to speculation and projections, but it is hard to know whether the positive aspects of nuclear energy will come to outweigh the risks of accidents and radioactive waste. Indeed, in 2006, applications for new nuclear power plants in the United States were filed with the Nuclear Regulatory Commission for the first time in more than two decades. Although no date for plant construction is yet certain, it does appear that some new nuclear power plants may come online in the United States before 2020.

Table 12.2 summarizes the major benefits and consequences of the conventional fuels we have discussed in this chapter.

### GAUGE YOUR PROGRESS

✓ How does a nuclear reactor work, and what makes it a desirable energy option?

✓ What are the two major concerns about nuclear energy?

✓ What are the promising aspects of and problems with nuclear fusion?

| TABLE 12.2 | Comparison of nonrenewable energy fuels | | | | |
|---|---|---|---|---|---|
| Energy type | Advantages | Disadvantages | Pollutant and greenhouse gas emissions | Electricity (cents/kWh) | Energy return on energy investment* |
| Oil/ gasoline | Ideal for mobile combustion (high energy/mass ratio) Quick ignition/turn-off capability Cleaner burning than coal | Significant refining required Oil spill potential effect on habitats near drilling sites Significant dust and emissions from fossil fuels used to power earth-moving equipment Human rights/ environmental justice issues in developing countries that export oil Will probably be much less available in the next 40 years or so | Second highest emitter of $CO_2$ among fossil fuels Hydrocarbons Hydrogen sulfide | Relatively little electricity is generated from oil | 4.0 (gasoline) 5.7 (diesel) |
| Coal | Energy-dense and abundant—U.S. resources will last at least 200 years No refining necessary Easy, safe to transport Economic backbone of some small towns | Mining practices frequently risk human lives and dramatically alter natural landscapes Coal power plants are slow to reach full operating capacity A large contributing factor to acid rain in the United States | Highest emitter of $CO_2$ among energy sources Sulfur Trace amounts of toxic metals such as mercury | 5 cents/ kWh | 14 |
| Natural gas | Cogeneration power plants can have efficiencies up to 60 percent Efficient for cooking, home heating, etc. Fewer impurities than coal or oil | Risk of leaks/explosions Twenty-five times more effective as a greenhouse gas than $CO_2$ Not available everywhere because it is transported by pipelines | Methane Hydrocarbons Hydrogen sulfide | 8–10 cents/ kWh | 8 |
| Nuclear energy | Emits no $CO_2$ once plant is operational Offers independence from imported oil High energy density, ample supply | Very unpopular; generates protests Plants are very expensive to build because of legal challenges Meltdown could be catastrophic Possible target for terrorist attacks | Radioactive waste is dangerous for hundreds of thousands of years No long-term plan currently in place to manage radioactive waste No air pollution during production | 12–15 cents/ kWh | 8 |

* Estimates vary widely.

## Meet TED: The Energy Detective

It is generally accepted that peer pressure influences behavior. When hotels placed small signs in the bathrooms telling their guests that *other* guests were reusing towels—rather than insisting on receiving fresh clean towels each day—the need for towel laundering dropped drastically. Many studies have shown that when people receive feedback about their electricity consumption, they will often respond just as the hotel guests did. A number of electric companies have experimented with mailings that show homeowners how much electricity they use in comparison to neighbors in similar-sized homes. However, if homeowners use less electricity than their neighbors, they may not be inclined to reduce their use further.

For decades, environmentalists have dreamed about a magical device that could allow homeowners to receive an instantaneous reading of actual electricity use in the home. Today, environmental scientists have produced such a device. The Energy Detective (TED) provides a readout on a small device that can sit on the kitchen table or be viewed on a laptop computer. And there are many other similar devices and software packages available, including Google Power Meter, which allows you to view your home electricity use on a laptop over the Internet (**FIGURE 12.20**).

TED and other such electricity monitoring devices provide an instantaneous readout of electricity use in your home. Suppose you are getting ready to head out for the evening. The refrigerator is plugged in, but is

quiet, meaning that the compressor is not running at that exact moment. Before you turn out the last light and leave, you glance at TED and see that your house is drawing 500 watts. Wait a minute, that's not right! Then you remember that you left your Xbox 360 video game and 42-inch plasma TV screen on in the other room. You run over and turn them both off. TED now reads that only 45 watts are being used in your home. But if everything is turned off, why is your house using 45 watts? It's probably due to the phantom load—unnecessary standby electricity—drawn by battery chargers, instant-on features on televisions, computers in sleep mode, and other electrical devices that are on even when you think they are off. Some TED owners have gone around their homes installing power strips that allow them to truly turn an appliance off and have seen their phantom loads drop as a result.

Since we know that all electricity use has environmental implications, a reduction in electricity use by any means is beneficial. The reductions that come from simple changes in behavior are some of the easiest to achieve.

### References

Ayres, I., S. Raseman, and A. Shih. 2009. *Evidence from Two Large Field Experiments That Peer Comparison Feedback Can Reduce Residential Energy Usage.* Working Paper 15386, National Bureau of Economic Research.

Schor, J. 2004. *Born to Buy.* Scribner.

**FIGURE 12.20** **The Energy Detective and Google Power Meter.** TED allows a user to instantaneously monitor electricity use in a home.

■ **Describe how energy use and energy resources have varied over time, both in the United States and worldwide.**

Energy use changes over time with the level of industrial development. The United States and the rest of the developed world have moved from a heavy reliance on wood and coal to fossil fuels and nuclear energy. The developing world still relies largely on wood, charcoal, and animal waste. Each source of energy is best suited for certain activities, and less well suited for others.

■ **Compare the energy efficiencies of the extraction and conversion of different fuels.**

Energy efficiency is an important consideration in determining the environmental impacts of energy use. In general, the energy source that entails the fewest conversions from its original form to the end use is likely to be the most efficient. Although multi-passenger transportation is the most energy-efficient way to travel, in the United States the single-passenger vehicle is the most popular.

■ **Explain the various means of generating electricity.**

Electricity generation plants convert the chemical energy of fuel into electricity. Coal, oil, natural gas, and nuclear fuels are the energy sources most commonly used for generating electricity. The electrical grid is a network of interconnected transmission lines that ties power plants together and links them with end users of electricity.

■ **Discuss the uses and consequences of using coal, oil, natural gas, and nuclear fuels.**

Coal is an energy-dense fossil fuel that is a common energy source for electricity generation. Coal combustion, however, is a major source of air pollution and greenhouse gas emissions. Petroleum includes both crude oil and natural gas. The United States uses more petroleum than any other fuel, primarily for transportation. Petroleum produces air pollution as well as greenhouse gas emissions. Oil spills are a major hazard to organisms and habitat. Natural gas is a relatively clean fossil fuel. Nuclear energy is a relatively clean means of electricity generation, though fossil fuels are used in constructing nuclear power plants and mining uranium. The major environmental hazards of nuclear energy are accidents and radioactive waste.

■ **Describe projections of future supplies of our conventional energy resources.**

Fossil fuels are a finite resource. Most observers believe that oil production will begin to decline some time in the next few decades. The transition away from oil will have important environmental consequences, depending upon how quickly it occurs and whether we make a transition to renewable energy resources or alternative fossil fuels.

## CHECK YOUR UNDERSTANDING

1. Which of the following is *not* a nonrenewable energy resource?
   (a) Oil
   (b) Coal
   (c) Natural gas
   (d) Wind
   (e) Nuclear fuels

2. The fact that global transfer of energy from fuels to electricity is about 35 percent efficient is mostly a consequence of
   (a) the Hubbert curve.
   (b) the law of conservation of matter.
   (c) the first law of thermodynamics.
   (d) the second law of thermodynamics.
   (e) the law of limiting factors.

3. Which of the following is the most fuel-efficient mode of transportation in terms of joules per passenger-kilometer?
   (a) Train
   (b) Bus
   (c) Airplane
   (d) Car with one passenger
   (e) Car with three passengers

4. Which of the following is *not* associated with the surface extraction of coal?
   (a) Low death rates among miners
   (b) Land subsidence and collapse
   (c) Large piles of tailings
   (d) Underground tunnels and shafts
   (e) Acid runoff into streams

**5.** Which of the following statements regarding petroleum is *correct?*

    I   It is formed from the decay of woody plants.

    II  It contains natural gas as well as oil.

    III It migrates through pore spaces in rocks.

(a) I, II, and III    (d) I and II
(b) I and III        (e) II and III
(c) II only

**6.** Nuclear power plants produce electricity using energy from the radioactive decay of
(a) uranium-235.    (d) plutonium-235.
(b) uranium-238.    (e) plutonium-238.
(c) uranium-239.

**7.** Currently, most high-level radioactive waste from nuclear reactors in the United States is
(a) stored in deep ocean trenches.
(b) buried in Yucca Mountain.
(c) reprocessed into new fuel pellets.
(d) chemically modified into safe materials.
(e) stored at the power plant that produced it.

**8.** A radioactive isotope has a half-life of 40 years and a radioactivity level of 4 curies. How many years will it take for the radioactivity level to become 0.25 curies?
(a) 80    (d) 200
(b) 120   (e) 240
(c) 160

**9.** Which of the following energy sources is responsible for the largest fraction of electricity generation in the United States?
(a) Natural gas    (d) Oil
(b) Coal         (e) Wood
(c) Uranium

**10.** In 1969, M. King Hubbert published a graph known as the Hubbert curve. This graph shows
(a) the amount of nuclear fuel available in North America.
(b) the amount of nuclear fuel available in the world.
(c) the point at which world oil production will reach a maximum and the point at which we will run out of oil.
(d) the point at which world oil production will increase.
(e) the coal reserves found in the United States, China, and Russia.

## APPLY THE CONCEPTS

**1.** Many college students have a mini fridge in their dorm room. A standard mini fridge costs roughly $100, uses about 100 watts of electricity, and can be expected to last for 5 years. The refrigerator is plugged into an electrical socket 24 hours a day, but is usually running only about 12 hours a day. Assume that electricity costs $0.10/kWh.
(a) Calculate the lifetime monetary cost of owning and operating the refrigerator.
(b) Assume that the electricity used to power the refrigerator comes from a coal-burning power plant. One metric ton of coal contains 29.3 GJ (8,140 kWh) of energy. Because of the inefficiency of electricity generation and transmission, only one-third of the energy in coal reaches the refrigerator. How many tons of coal are used to power the refrigerator during its lifetime?
(c) Assume that 15 percent of the mass of the coal burned in the power plant ends up as coal ash, a potentially toxic mixture that contains mercury and arsenic. How many tons of coal ash are produced as a result of the refrigerator's electricity use over its lifetime?
(d) What externalities does your answer from part (a) not include? Describe one social and one environmental cost associated with using this appliance.

(e) Describe two ways a college student could reduce the electricity use associated with having a mini fridge in his or her dorm room.

**2.** A number of U.S. electric companies have filed applications with the Nuclear Regulatory Commission for permits to build new nuclear power plants to meet future electricity demands.
(a) Explain the process by which electricity is generated by a nuclear power plant.
(b) Describe the two nuclear accidents that occurred in 1979 and 1986, respectively, that led to widespread concern about the safety of nuclear power plants.
(c) Discuss the environmental benefits of generating electricity from nuclear energy rather than coal.
(d) Describe the three types of radioactive waste produced by nuclear power plants and explain the threats they pose to humans.
(e) Discuss the problems associated with the disposal of radioactive waste and outline the U.S. Department of Energy's proposal for its long-term storage.

**Choosing a Car: Conventional or Hybrid?** One person buys a compact sedan that costs $15,000 and gets 20 miles per gallon. Another person pays $22,000 for the hybrid version of the same compact sedan, which gets 50 miles per gallon. Each owner drives 12,000 miles per year and plans on keeping the vehicle for 10 years.

(a) A gallon of gas emits 20 pounds of $CO_2$ when burned in an internal combustion engine. The average cost of a gallon of gas over the 10-year ownership period is $3.00.

   (i) Calculate how many gallons of gas each vehicle uses per year.

   (ii) Calculate the cost of the gas that each vehicle uses per year.

   (iii) Calculate the amount of $CO_2$ that each vehicle emits per year.

(b) Based on your answers to questions i–iii, complete the data table below.

| Year of operation | Sedan: total costs— purchase and gas ($) | Sedan: cumulative $CO_2$ emissions (pounds) | Hybrid: total costs— purchase and gas ($) | Hybrid: cumulative $CO_2$ emissions (pounds) |
|---|---|---|---|---|
| 1 | | | | |
| 2 | | | | |
| 3 | | | | |
| 4 | | | | |
| 5 | | | | |
| 6 | | | | |
| 7 | | | | |
| 8 | | | | |
| 9 | | | | |
| 10 | | | | |

(c) Use the data in the table to answer the following questions:

   (i) Estimate how many years it would take for the hybrid owner to recoup the extra cost of purchasing the vehicle based on savings in gas consumption.

   (ii) After the amount of time determined in (i), compare and comment on the total costs (purchase and gas) for each vehicle at that time.

   (iii) Over the 10-year ownership period, which vehicle is the more economically and environmentally costly to operate (in terms of dollars and $CO_2$ emissions), and by how much?

(d) Suggest ways that the owner of the conventional car could reduce the overall yearly $CO_2$ emissions from the vehicle.

(e) Suggest ways that the hybrid owner could become carbon-neutral in terms of operating the vehicle.

# ENGAGE YOUR ENVIRONMENT

Do you unintentionally waste energy in the ways you use electricity? In this Engage Your Environment exercise you will inventory your personal use of electricity, and identify ways to prevent energy waste, reduce environmental impacts, and save money. *See Engage Your Environment Activity 12 at the end of this book.*

# Renewable Energy: Innovative Uses of Earth, Sun, Wind, and Water

## Energy from the Moon

Moon power. On Roosevelt Island, a small residential island in the East River in New York City, the Moon is producing clean, renewable energy. A supermarket runs on it, and so does a parking garage.

Does "Moon power" sound far-fetched? Maybe the term "tidal energy" sounds less strange. Twice a day, variations in the Moon's gravitational pull interact with Earth's gravity and wind and the Sun, which generate electricity intermittently. In addition, tidal energy can be captured near major coastal population centers, where energy is needed most.

As the RITE project enters its third development phase, it will move forward with the installation of additional turbines and technology to improve the performance of the system. Eventually, the system will expand to operate 30 turbines. Even at full size, however, the system will be able to produce

> **On Roosevelt Island, a small residential island in the East River in New York City, the Moon is producing clean, renewable energy.**

rotation to make tides rise and fall along coastlines and near the mouths of rivers. The kinetic energy in the moving water can be captured and converted into a usable form as the water flows through turbines similar to those in "windmills" or large dams. Moon-driven tides are a continuous source of clean energy. What's more, since tidal currents move in two directions, tidal energy turbines can harness energy from both incoming and outgoing tides.

On Roosevelt Island, engineers have built the first commercial tidal energy plant in the United States. Called the Roosevelt Island Tidal Energy (RITE) project, the plant contains six underwater turbines. These turbines harness some of the kinetic energy of the water moving between the East River and Raritan Bay to produce 35 megawatt-hours (MWh) of electricity per year, enough electricity to supply approximately eight New York City residences for one year. The captured energy travels through transmission lines to the electrical grid and then on to local users.

Compared with some other electricity generation systems, the Roosevelt Island project has relatively little environmental impact. RITE does not divert water or store it behind dams, and mesh enclosures around the turbines protect fish. The system is invisible from shore and operates silently. The predictability of the tide gives this system an advantage compared with systems using other renewable energy resources such as

only 1 megawatt (MW) of energy, a small contribution when you consider that most fossil fuel power plants have capacities of 500 MW or more.

Despite its small size, the RITE project demonstrates that tidal energy—Moon power—could play an important role in a sustainable energy future. It also demonstrates that we must pursue a variety of possible ways to harness energy, even if some energy sources, such as the tides, can contribute only a small proportion of our energy needs. The RITE project is one example of an innovative approach to replacing conventional energy resources with renewable ones. In a limited way, tidal energy could be a part of a sustainable energy plan that would include energy conservation and increased energy efficiency as well as use of other renewable energy resources such as wind and the Sun. ■

*Sources:* University of Minnesota, St. Anthony Falls Laboratory, Tidal Power for New York City: The Wave of the Future? June 16, 2009, http://www .safl.umn.edu/tidal_power.html; Verdant Power, The RITE Project, East River— New York, NY, http://www.verdantpower .com/what-initiative

Underwater turbines capture tidal energy.

◄ We can capture the kinetic energy of the tides to produce electricity.

Throughout this book we have discussed sustainability as the foundation of our planet's environmental health. Sustainability is particularly important in energy use because energy is a resource humans cannot live without. In Chapter 12 we discussed the finite nature of traditional energy resources such as fossil fuels and the environmental consequences of their use. In this chapter we outline the components of a sustainable energy strategy, beginning with ways to reduce our use of energy through conservation and increased efficiency. We discuss an important carbon-based energy resource, biomass. We then describe innovations in obtaining energy from non-carbon-based resources such as moving water, the Sun, wind, Earth's internal heat, and hydrogen.

After reading this chapter you should be able to

- define renewable energy resources.
- describe strategies to conserve energy and increase energy efficiency.
- compare and contrast the various forms of biomass energy.
- explain the advantages and disadvantages of hydroelectricity, solar energy, geothermal energy, wind energy, and hydrogen as energy resources.
- describe the environmental and economic options we must assess in planning our energy future.

# What is renewable energy?

In Chapter 12 we learned that conventional energy resources, such as petroleum, natural gas, coal, and uranium ore, are in limited supply. Fossil fuels were formed over millions of years, by processes that occur so slowly that it would be impossible to replace the fuels we are currently extracting. From a systems analysis perspective, fossil fuels constitute an energy reservoir that is being depleted much faster than it can be replenished. Similarly, we have a finite amount of uranium ore available to use as fuel in nuclear reactors. These energy resources can be considered **nonrenewable.**

In contrast, some other sources of energy can be regenerated rapidly. Biomass energy resources are **potentially renewable** as long as we do not consume them more quickly than they can be replenished. There are still other energy resources that cannot be depleted no matter how much we use them. Solar, wind, geothermal, hydroelectric, and tidal energy are essentially **nondepletable** in the span of human time. For example, no matter how much wind energy we capture, there will always be more; the amount available tomorrow does not depend on how much we use today. In this book we refer to potentially renewable and nondepletable energy resources together as **renewable** energy resources. FIGURE 13.1 illustrates these categories of energy resources.

Many renewable energy resources have been used by humans for thousands of years. In fact, before humans began using fossil fuels, the only available energy sources were wood and

**FIGURE 13.1** **Renewable and nonrenewable energy resources.** Fossil fuels and nuclear fuels are nonrenewable energy resources. Renewable energy resources include potentially renewable energy sources such as biomass, which is renewable as long as humans do not use it faster than it can be replenished, and nondepletable energy sources, such as solar radiation and wind.

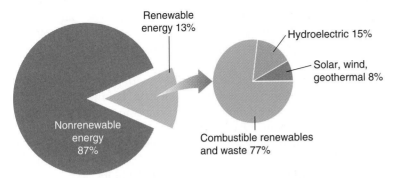

**FIGURE 13.2** **Global energy use, 2007.** Renewable energy resources provide about 13 percent of energy worldwide. [Data from International Energy Agency, World Energy Outlook, 2009.]

plants, animal manure, and fish or animal oils. Today, in parts of the developing world where there is little access to fossil fuels, people still rely on local biomass energy sources such as manure and wood for cooking and heating, sometimes to such an extent that they overuse the resource. For example, according to the U.S. Energy Information Administration, biomass is currently the source of 86 percent of the energy consumed in sub-Saharan Africa (excluding South Africa), and much of it is not harvested sustainably.

As **FIGURE 13.2** shows, renewable energy resources account for approximately 13 percent of the energy used worldwide. Most of that renewable energy is in the form of biomass. In the United States, which depends heavily on fossil fuels, renewable energy resources provide only about 7 percent of the energy used. That 7 percent, shown in detail in **FIGURE 13.3**, comes primarily from biomass and hydroelectricity.

Although renewable energy is a more sustainable energy choice than nonrenewable energy, using any form of energy has an impact on the environment. Biomass, for instance, is a renewable resource only if it is used sustainably. Overharvesting of wood leads to

deforestation and degradation of the land, as we saw in the discussion of Haiti in Chapter 3's opening story. Wind turbines can unintentionally kill birds and bats, and hydroelectric turbines kill millions of fish. Manufacturing photovoltaic solar panels requires heavy metals and a great deal of water. Because all energy choices have environmental consequences, the best approach to energy use is to minimize it through conservation and efficiency.

**GAUGE YOUR PROGRESS**

✓ What are some common kinds of nondepletable, potentially renewable, and nonrenewable energy resources?

✓ How do you know if an energy resource is nondepletable, potentially renewable, or nonrenewable?

✓ What are the most important differences between renewable and nonrenewable energy resources?

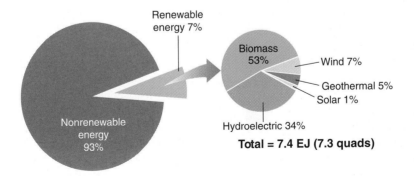

**FIGURE 13.3** **Energy use in the United States, 2008.** Only 7 percent of the energy used in the United States comes from renewable energy resources. [Data from U.S. Department of Energy, 2009.]

# How can we use less energy?

A truly sustainable approach to energy use must incorporate both energy conservation and energy efficiency. **Energy conservation** means finding ways to use less energy. FIGURE 13.4 lists some of the ways that an individual might conserve energy, including lowering the household thermostat during cold months, consolidating errands in order to drive fewer miles, or turning off a computer when it is not being used. On a larger scale, a government might implement energy conservation measures that encourage or require individuals to adopt strategies or habits that use less energy. One such top-down approach is to improve the availability of public transportation. Governments can also facilitate energy conservation by taxing electricity, oil, and natural gas, since higher taxes discourage their use. Alternatively, governments might offer rebates or tax credits for retrofitting a home or business to operate on less energy. Some electric companies bill customers with a **tiered rate system:** customers pay a low rate for the first increment of electricity they use and pay higher rates as their use goes up. All of these practices encourage people to reduce the amount of energy they use.

As we saw in Chapters 2 and 12, increasing energy efficiency means obtaining the same work from a smaller amount of energy. Energy conservation and energy efficiency are closely linked. One can conserve energy by not using an electrical appliance: doing so results in less energy consumption. But one can also conserve energy by using a more efficient appliance—one that does the same work with less energy.

Modern changes in electric lighting are a good example of how steadily increasing energy efficiency results in overall energy conservation. Compact fluorescent light bulbs use one-fourth as much energy to provide the same amount of light as incandescent bulbs. LED (light-emitting diode) light bulbs are even more efficient; they use one-sixth as much energy as incandescent bulbs. Over time, the widespread use of these efficient bulbs results in substantially less energy use to provide lighting.

Another way in which consumers can increase energy efficiency is to switch to Energy Star appliances, products that meet the efficiency standards of the U.S. Environmental Protection Agency's Energy Star program. For example, an Energy Star air conditioner may use 0.2 kilowatt-hours (kWh, or 200 watt-hours) less electricity per hour than a non–Energy Star unit. In terms of cost, a single consumer may save only 2 to 5 cents per hour by switching to an Energy Star unit. However, if 100,000 households in a city switched to Energy Star air conditioners, the city would reduce its energy use by 20 MW, or 4 percent of a typical power plant's output. Crunch the Numbers "Energy Star" on page 348 shows you how to calculate Energy Star savings.

## Benefits of Energy Conservation and Efficiency

As Amory Lovins, chief scientist at the Rocky Mountain Institute and an expert on energy efficiency, says, "The best energy choices are the ones you don't need." To achieve energy sustainability, we will need to rely as much, or more, on reducing the amount of energy we use as on developing renewable energy resources. Conservation and efficiency efforts save energy that can then be used later, just as you might save money in a bank account to use later when the need arises. In this sense, conservation and efficiency are sustainable energy "sources."

As we saw in Chapter 12, the demand for energy varies with time of day, season, and weather. When electricity-generating plants are unable to handle the demand during high-use periods, brownouts or black-outs may occur. To avoid this problem, electric companies must be able to provide enough energy to satisfy **peak demand,** the greatest quantity of energy used at any one time. Peak demand may be several times the overall average demand, which means that substantially more energy must be available than is needed under average conditions. To meet peak demand for electricity, electric companies often keep backup sources of energy available—primarily natural gas–fired generators.

---

**Home**

- Weatherize (insulate, seal gaps).
- Turn thermostat down in winter, up in summer.
- Reduce use of hot water (do laundry in cold water/take shorter showers).
- Replace incandescent bulbs with compact fluorescents or LEDs.

**Transportation**

- Walk or ride a bike.
- Take public transportation.
- Carpool.
- Consolidate trips.

**Electrical and electronic devices**

- Buy Energy Star devices and appliances.
- Unplug when possible or use a power strip.
- Use a laptop rather than a desktop computer.

FIGURE 13.4 **Reducing energy use.** There are many ways individuals can reduce their energy use in and outside the home.

Therefore, an important aspect of energy conservation is reducing peak demand, making it less likely that electric companies will have to build excess generating capacity that is used only sporadically. One way of reducing peak demand is to set up a variable price structure under which customers pay less to use electricity when demand is lowest (typically in the middle of the night and on weekends) and more when demand is highest. This approach helps even out the use of electricity, which reduces the burden on the company's generating capacity, and rewards the consumer at the same time.

The second law of thermodynamics tells us that whenever energy is converted from one form into another, some energy is lost as unusable heat. In a typical thermal fossil fuel or nuclear power plant, only about one-third of the energy consumed goes to its intended purpose; the rest is lost during energy conversions. In order to fully account for all energy conservation savings, we need to consider these losses. So, the amount of energy we save is the sum of the energy we did not use and the energy that would have been lost in converting

that energy into the form in which we would have used it. For example, if we can reduce our electricity use by 100 kWh, we may actually be conserving 300 kWh of an energy resource such as coal, since we save both the 100 kWh that we decide not to use and the 200 kWh that would have been lost during the conversion process to make the 100 kWh available to us.

Energy conservation and energy efficiency are the least expensive and most environmentally sound options for maximizing our energy resources. In many cases, they are also the easiest approaches to implement because they require fairly simple changes to existing systems, rather than a switch to a completely new technology.

## Sustainable Design

Sustainable design can improve the efficiency of the buildings and communities we live and work in. FIGURE 13.5 shows some key features of sustainable design applied to a single-family dwelling. As we saw in

**FIGURE 13.5** **An energy-efficient home.** A sustainable building design incorporates proper solar orientation and landscaping as well as insulated windows, walls, and floors. In the Northern Hemisphere, a southern exposure allows the house to receive more direct rays from the Sun in winter when the path of the Sun is in the southern sky.

**Energy Star**

Thinking clearly about energy efficiency and energy conservation can save you a lot of money in some surprising ways. If you are saving money, you are also saving energy and reducing the emission of pollutants. Consider the purchase of an air conditioner. Suppose you have a choice: an Energy Star unit for $300 or a standard unit for $200. The two units have the same cooling capacity. The Energy Star unit costs 5 cents per hour less to run. If you buy the Energy Star unit and run it 12 hours per day for 6 months of the year, how long would it take to recover the initial extra $100?

You would save

$$\$0.05/\text{hour} \times 12 \text{ hours/day} = \$0.60/\text{day}$$

Six months is about 180 days, so in the first year you would save

$$\$0.60/\text{day} \times 180 \text{ days} = \$108$$

Spending the extra $100 for the Energy Star unit would save you $8 in just 1 year of use. In 3 years of use, the savings would more than pay for the entire initial cost of the unit ($3 \times \$108 = \$324$). After that, you would pay only for the operating costs.

**Your Turn:** You are about to invest in a 66-inch flat screen TV. These TVs come in both Energy Star and non–Energy Star models. The cost of electricity is $0.15 per kilowatt-hour, and you expect to watch TV an average of 4 hours per day.

1. The non–Energy Star model uses 0.5 kW (half a kilowatt). How much will it cost you per year for electricity to run this model?

2. If the Energy Star model uses only 40 percent of the amount of electricity used by the non–Energy Star model, how much money would you save on your electric bill over 5 years by buying the efficient model?

---

Chapter 10, good community planning also conserves energy. Building houses close to where residents work reduces reliance on fossil fuels for transportation, which in turn reduces the amount of pollution and carbon dioxide released into the atmosphere.

Buildings consume a great deal of energy for cooling, heating, and lighting. Many sustainable building strategies rely on **passive solar design,** a technique that takes advantage of solar radiation to maintain a comfortable temperature in the building (**FIGURE 13.6**). Passive solar design stabilizes indoor temperatures without the need for pumps or other mechanical devices. For example, in the Northern Hemisphere, constructing a house with windows along a south-facing wall allows the Sun's rays to penetrate and warm the house, especially in winter when the Sun is more prominent in the southern sky. Double-paned windows insulate while still allowing incoming solar radiation to warm the house. Carefully placed windows also allow the maximum amount of light into a building, reducing the need for artificial lighting. Dark materials on the roof or exterior walls of a building absorb more solar energy than light-colored materials, further warming the structure. Conversely, using light-colored materials on a roof reflects heat away from the building, keeping it cooler. In summer, when the Sun is high in the sky for much of the day, an overhanging roof helps block out sunlight during the hottest period, keeping the indoor temperature cooler and thus reducing the need for ventilation fans or air conditioning. Window shades can also reduce solar energy entering the house.

To reduce demand for heating at night and for cooling during the day, builders can use construction materials that have high thermal inertia. **Thermal inertia** is the ability of a material to retain heat or cold. Materials with high thermal inertia stay hot once they have been heated and cool once they have been cooled. Stone and concrete have high thermal inertia, whereas

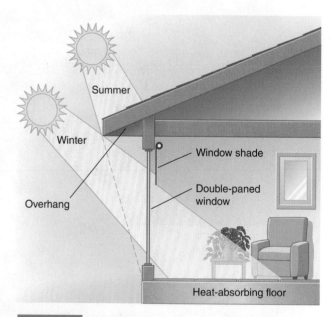

**FIGURE 13.6** **Passive solar design.** Roof overhangs make use of seasonal changes in the Sun's position to reduce energy demand for heating and cooling. In winter, when the Sun is low in the sky, it shines directly into the window and heats the house. In summer, when the Sun is higher in the sky, the overhang blocks incoming sunlight, and the room stays cool. High-efficiency windows and building materials with high thermal inertia are also components of passive solar design.

wood and glass do not; think of how a cement sidewalk stays warm longer than a wooden boardwalk after a hot day. A south-facing room with stone walls and a stone floor will heat up on sunny winter days and retain that heat long after the Sun has set.

Building a house into the side of a hill or roofing a building with soil and plants are less common approaches that also provide insulation and reduce the need for both heating and cooling. While "green roofs"—roofs with soil and growing plants—are somewhat unusual in the United States, many European cities, such as Berlin, have regulations regarding them on new or rebuilt structures—especially on high-rise buildings in downtown areas with little natural plant cover. These green roofs cool and shade the buildings themselves as well as the surrounding environment. Adding plants to an urban environment also improves overall air quality.

The use of recycled building materials is another method of energy conservation. It reduces the need for new construction materials, and thus reduces the amount of energy used to produce the components of the building. Many buildings now use recycled denim insulation in the walls and ceilings and fly ash (a by-product recovered from coal-fired power plants) in the foundation.

Many new homes now incorporate some or all of these sustainable design strategies. However, it is possible

to achieve energy efficiency even in very large buildings. The California Academy of Sciences in San Francisco's Golden Gate Park is a showcase for several of these sustainable design techniques (FIGURE 13.7). This structure, which incorporates a combination of passive solar design, radiant heating, solar panels, and skylights, actually uses 30 percent less energy than the quantity it is allowed to use under federal code requirements. Natural light fills 90 percent of the office space and many of the public areas. Windows, louvers, and skylights open as needed to allow air to circulate, capturing the ocean breezes and ventilating the building. Recycled denim insulation in the walls and a soil-covered rooftop garden provide insulation that reduces heating and cooling costs. As an added benefit, the living green roof grows native plants and captures 13.6 million liters (3.6 million gallons) of rainwater per year, which is then used to recharge groundwater stores.

In addition to these passive techniques, the building's designers incorporated active technologies that further reduce its energy use. An efficient radiant heating system carries warm water through tubes embedded in the concrete floor of the building, using a fraction of the energy required by a standard forced-air heating system. To produce some of the Academy's electricity directly, the designers added 60,000 photovoltaic solar cells to the roof of the building. These solar panels convert

**FIGURE 13.7** **The California Academy of Sciences.** The sustainable design of this San Francisco research institution maximizes the use of natural light and ventilation. The building generates much of its own electricity with solar panels on its roof and captures water in its rooftop garden.

energy from the Sun into 213,000 kWh of electricity per year and reduce greenhouse gas emissions by about 200 metric tons per year.

The California Academy of Sciences took an innovative approach to meeting its energy needs through a combination of energy efficiency and use of renewable energy resources. However, many, if not all, of these approaches will have to become commonplace if we are going to use energy in a sustainable way.

As fossil fuels become less available and more expensive, what will take their place? Probably it will be a mix of energy efficiency strategies, energy conservation, and new energy sources. In the rest of this chapter we will explore our renewable energy options, beginning with the renewable energy source that is most widely used today: biomass.

---

**GAUGE YOUR PROGRESS**

✓ How can we can use less energy as individuals and as societies?

✓ What is peak demand, and how does it relate to energy conservation?

✓ How can building design contribute to energy conservation and efficiency?

---

# Biomass is energy from the Sun

As we learned in Chapter 12, the Sun is the ultimate source of fossil fuels. Fossil fuels are created from dead plants and animals that are buried deep in sediments and slowly transformed into petroleum or coal. As **FIGURE 13.8** shows, most types of renewable energy are also derived from the Sun and cycles driven by the Sun, including solar, wind, and hydroelectric energy as well as plant biomass such as wood. In fact, the only important sources of energy that are not solar-based are nuclear, geothermal, and tidal energy.

Biomass energy resources encompass a large class of fuel types that include wood and charcoal, animal products and manure, plant remains, and municipal solid waste (MSW), as well as liquid fuels such as ethanol and biodiesel. Many forms of biomass used directly as fuel, such as wood and manure, are readily available all over the world. Because these materials are inexpensive and abundant, they account for more than 10 percent of world energy consumption (see Figure 13.2) and a much higher percentage in many developing countries. Biomass can also be processed or refined into liquid fuels such as ethanol and biodiesel, known collectively as **biofuels.** These fuels are used in more limited quantities due to the technological demands associated with

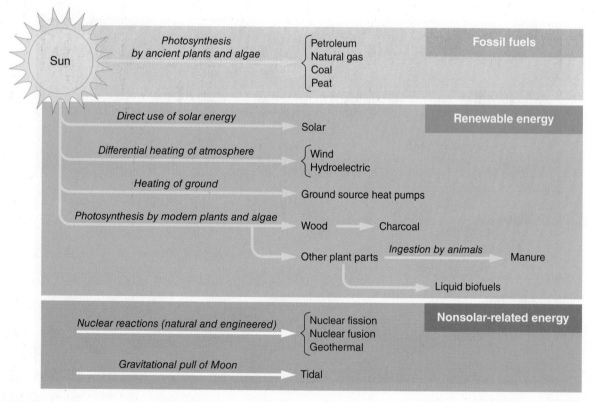

**FIGURE 13.8** **Sources of energy types.** The Sun is the ultimate source of almost all types of energy.

their use. For example, it is easier to burn a log in a fire than it is to develop the technology to produce a compound such as ethanol.

Biomass—including ethanol and biodiesel—accounts for more than one-half of the renewable energy and approximately 3.5 percent of all the energy consumed in the United States today (see Figure 13.3). However, the mix of biomass used in the United States differs from that in the developing world. Two-thirds of the biomass energy used in the United States comes from wood, with the remaining one-third divided evenly between MSW and biofuels. In the developing world, a larger percentage of biomass energy comes from wood and animal manure.

## Modern Carbon versus Fossil Carbon

Like fossil fuels, biomass contains a great deal of carbon, and burning it releases that carbon into the atmosphere. Given the fact that both fossil fuels and biomass raise atmospheric carbon concentrations, is it really better for the environment to replace fossil fuels with biomass? The answer depends on how the material is harvested and processed and on how the land is treated during and after harvest. It also depends on how long the carbon has been stored. The carbon found in plants growing today was in the atmosphere in the form of carbon dioxide until recently, when it was incorporated into the bodies of the plants through photosynthesis. Depending on the type of plant it comes from, the carbon in biomass fuels may have been captured through photosynthesis as recently as a few months ago, in the case of a corn plant, or perhaps up to several hundred years ago, in the case of the wood of a large tree. We call the carbon in biomass **modern carbon,** in contrast to the carbon in fossil fuels, which we call **fossil carbon.**

Unlike modern carbon, fossil carbon has been buried for millions of years. Fossil carbon is carbon that was out of circulation until humans discovered it and began to use it in increasing quantities. The burning of fossil fuels results in a rapid increase in atmospheric $CO_2$ concentrations because we are unlocking or releasing stored carbon that was last in the atmosphere millions of years ago. In theory, the burning of biomass (modern carbon) should not result in a net increase in atmospheric $CO_2$ concentrations because we are returning the carbon to the atmosphere where it had been until recently. And, if we allow vegetation to grow back in areas where biomass was recently harvested, that new vegetation will take up an amount of $CO_2$ more or less equal to the amount we released by burning the biomass. Over time, the net change in atmospheric $CO_2$ concentrations should be zero. An activity that does not change atmospheric $CO_2$ concentrations is referred to as **carbon neutral.**

Whether using biomass is truly carbon neutral, however, is an important question that is currently being discussed by scientists and policy makers. Sometimes the use of modern carbon ends up releasing $CO_2$ into the atmosphere that would otherwise have remained in the soil or in fossil fuels. And that is why we stated earlier that *in theory,* using modern carbon should not result in a net increase in atmospheric carbon dioxide. We discuss this issue further in the Science Applied section that follows this chapter ("Should Corn Become Fuel?").

## Solid Biomass: Wood, Charcoal, and Manure

Throughout the world, 2 billion to 3 billion people use wood for heating or cooking. In the United States, approximately 3 million homes rely on wood as the primary heating fuel, and more than 20 million homes use wood for energy at least some of the time. In addition, the pulp and paper industries, power plants, and other industries use wood waste and by-products for energy. In theory, cutting trees for fuel is sustainable if forest growth keeps up with forest use. Unfortunately, many forests, like those in Haiti, are cut intensively with little chance for regrowth.

Removing more timber than is replaced by growth, or **net removal** of forest, is an unsustainable practice that leads to deforestation. In addition, the combination of net removal of forest and the burning of wood results in a net increase in atmospheric $CO_2$: the $CO_2$ released from the burned wood is not balanced by photosynthetic carbon fixation in new tree growth. Harvesting the forest may also release carbon from the soil that would otherwise have remained buried deep in the A and B horizons. Although the mechanism is not entirely clear, it appears that some of this carbon release occurs due to soil disturbance by logging equipment.

Tree removal can be sustainable if we allow time for forests to regrow. In addition, in some heavily forested areas, extracting individual trees of abundant species and opening up spaces in the canopy to succession actually increases habitat diversity and may even increase total photosynthesis. More often, however, tree removal has the potential to result in soil erosion, increased water temperatures in nearby rivers and streams, and fragmentation of forest habitats when logging roads divide them. Tree removal may also harm species that are dependent on old-growth forest habitat.

Many people in the developing world use wood to make charcoal, which is a superior fuel for many reasons. Charcoal is lighter than wood and contains approximately twice as much energy per unit weight. A charcoal fire produces much less smoke than wood and does not need the constant tending of a wood fire. Although it is more expensive than wood, charcoal is a

**(b)**

**FIGURE 13.9** **Charcoal.** (a) A charcoal market in the Philippines. Many people in developing countries rely on charcoal for cooking and heating. (b) An area of land after it has been denuded for charcoal production.

**(a)**

fuel of choice in urban areas of the developing world and for families who can afford it (**FIGURE 13.9**).

In regions where wood is scarce, such as parts of Africa and India, people often use dried animal manure as a fuel for indoor heating and cooking. Burning manure can be beneficial because it removes harmful microorganisms from the surroundings, reducing the risk of disease transmission. However, as we will see in Chapter 15, burning manure also releases particulate matter and other pollutants into the air. These pollutants cause a variety of respiratory illnesses, from emphysema to cancer. The problem is exacerbated when the manure is burned indoors in poorly ventilated rooms, a common situation in many developing countries. The World Health Organization estimates that indoor air pollution is responsible for approximately 1.5 million deaths annually. Chapters 15 and 17 cover indoor air pollution and human health in more detail.

Whether indoors or out, burning biomass fuels produces a variety of air pollutants, including particulate matter, carbon monoxide (CO), and nitrogen oxides ($NO_x$), which are important components of photochemical smog (**FIGURE 13.10**). Again, as we have seen, there is no "good" choice of fuel; each has environmental costs.

## Biofuels: Ethanol and Biodiesel

The liquid biofuels—ethanol and biodiesel—can be used as substitutes for gasoline and diesel, respectively. **Ethanol** is an alcohol, the same one found in alcoholic beverages. It is made by converting starches and sugars from plant material into alcohol and carbon dioxide. More than 90 percent of the ethanol produced in the United States comes from corn and corn by-products, although ethanol can also be produced from sugarcane, wood chips, crop waste, or switchgrass. **Biodiesel** is a substitute for regular petroleum diesel. It is usually produced by extracting oil from algae and plants such as soybeans and palm. In the United States, many policy makers are encouraging the production of ethanol and biodiesel as a way to reduce the need to import foreign oil while also supporting U.S. farmers and declining rural economies.

**ETHANOL** The United States is the world leader in ethanol production, manufacturing approximately 34 billion liters (9 billion gallons) in 2008. In his State of the Union address in 2007, President George W. Bush proposed that the United States produce 133 billion liters (35 billion gallons) of ethanol by 2017. In May 2009, President Barack Obama issued a directive to accelerate U.S. investment in and production of biofuels, with the goal of reducing dependence on foreign oil by more than 47 billion liters (12.5 billion gallons) per year and reducing greenhouse gas emissions by an average of 160 million metric tons per year by 2022. Despite these actions, ethanol and biodiesel currently make up only about 0.6 percent of the U.S. energy supply.

Brazil, the world's second largest ethanol producer, is making biofuels a major part of its sustainable energy strategy. Brazil manufactures ethanol from sugarcane, which is easily grown in its tropical climate. Unlike corn, which must be replanted every year, sugarcane is replanted every 6 years and harvested by hand. Both factors reduce the amount of fossil fuel energy needed to harvest the crop.

Ethanol is usually mixed with gasoline, most commonly at a ratio of one part ethanol to nine parts gasoline. The result is *gasohol,* a fuel that is 10 percent ethanol.

**FIGURE 13.10** **Particulate emissions.** Smog and decreased visibility in Montreal, Canada, in January 2009 were reportedly caused by emissions of particulate matter due to the extensive use of woodstoves.

Gasohol has a higher oxygen content than gasoline alone and produces less of some air pollutants when combusted. In certain parts of the midwestern United States, especially corn-growing states, a fuel called E-85 (85 percent ethanol, 15 percent gasoline) is available. **Flex-fuel vehicles** can run on either gasoline or E-85. However, a study by General Motors revealed that most of the owners of the 7 million flex-fuel vehicles in use today did not know that their cars could run on E-85.

Proponents of ethanol claim that it is a more environmentally friendly fuel than gasoline, although opponents dispute that claim. Ethanol does have disadvantages. The carbon bonds in alcohol have a lower energy content than those in gasoline, which means that a 90 percent gasoline/10 percent ethanol mix reduces gas mileage by 2 to 3 percent compared with 100 percent gasoline fuel. As a result, a vehicle needs more gasohol to go the same distance it could go on gasoline alone. Furthermore, growing corn to produce ethanol uses a significant amount of fossil fuel energy, as well as land that could otherwise be devoted to growing food. Some scientists argue that using ethanol actually creates a net increase in atmospheric $CO_2$ concentrations. The benefits and drawbacks of using ethanol as a fuel are discussed in more detail in the Science Applied section that follows this chapter.

Research is under way to find viable alternatives to corn as sources for U.S. ethanol production. Switchgrass is one possibility. Switchgrass is a perennial crop, which means that farmers can harvest it without replanting, minimizing soil disturbance and erosion. Furthermore, switchgrass does not require as much fossil fuel input as corn to produce. However, crops such as corn and sugarcane produce ethanol more readily due to their high sugar content. Sugars are readily and rapidly converted into ethanol. In contrast, switchgrass and other alternative materials, such as wood chips, are composed primarily of cellulose—the material that constitutes plant cell walls—which must be broken down into sugars before it can be used in ethanol production. Scientists have not yet developed an efficient breakdown process for large-scale ethanol production from switchgrass. Developing such a process will increase the energy and carbon advantages of ethanol.

**BIODIESEL** Biodiesel is a direct substitute for petroleum-based diesel fuel. It is usually more expensive than petroleum diesel, although the difference varies depending on market conditions and the price of petroleum. Biodiesel is typically diluted to "B-20," a mixture of 80 percent petroleum diesel and 20 percent biodiesel. B-20 is available at some gas stations scattered around the United States and can be used in any diesel engine without modification.

Because biodiesel tends to solidify into a gel at low temperatures, higher concentrations of biodiesel work effectively only in modified engines. However, with a commercially sold kit, any diesel vehicle can be modified to run on 100 percent straight vegetable oil (SVO), typically obtained as a waste product from restaurants and filtered for use as fuel. Groups of students in the United States have driven buses around

**FIGURE 13.11** **A bus fueled by vegetable oil.** This converted school bus is fueled mostly by waste vegetable oil. Students from Dartmouth College drove and maintained the bus as part of a program to educate people around the United States about renewable fuels and energy conservation.

the country almost exclusively on SVO (**FIGURE 13.11**), and some municipalities, such as Portland, Maine, have community-based SVO recycling facilities. Although there is unlikely to be enough waste vegetable oil to significantly reduce fossil fuel consumption, SVO is nevertheless a potential transition fuel that may temporarily reduce our use of petroleum.

In the United States, most biodiesel comes from soybean oil or processed vegetable oil. However, scientists are working on ways to produce large quantities of biodiesel directly from wood or other forms of cellulose—especially waste wood from logging and sawmills. In addition, some species of algae appear to have great potential for producing biodiesel. Algae are photosynthetic microorganisms that can be grown almost anywhere, and of all biodiesel options, produce the greatest yield of fuel per hectare of land area per year and utilize the least amount of energy and fertilizer per quantity of fuel. One 2008 study reported that algae produce 15 to 300 times more fuel per area used than conventional crops. Algae can be grown on marginal lands, in brackish water, on rooftops, and in other places that are not traditionally thought of as agricultural space. Algae can even be harvested from ships: the city of Venice, Italy, has a plan to gather algae growing on ships that enter its waterways and convert that algae into oils that could fuel a power plant and provide 50 percent of the seaport's electricity needs.

Emissions of carbon monoxide from combustion of biodiesel are lower than those from petroleum diesel. In addition, since it contains modern carbon rather than fossil carbon, biodiesel should be carbon neutral. However, as with ethanol, some critics question whether biodiesel is truly carbon neutral. For instance, producing biodiesel from soybeans requires less fossil fuel input per liter of fuel than producing ethanol from corn, but it requires more cropland, and therefore may transfer more carbon from the soil to the atmosphere. In contrast, producing biodiesel from wood waste or algae may require very little or no cropland and a minimal amount of other land.

> **GAUGE YOUR PROGRESS**
>
> ✓ What are the major forms of biomass energy?
>
> ✓ Why is energy from modern carbon potentially carbon neutral?
>
> ✓ Why is it important to find abundant sources of biomass energy? What are the advantages and disadvantages of different forms of biomass energy?

# The kinetic energy of water can generate electricity

**Hydroelectricity** is electricity generated by the kinetic energy of moving water. It is the second most common form of renewable energy in the United States and in the world, and it is the form most widely used for electricity generation. As we saw in Chapter 12 (Figure 12.8), hydroelectricity accounts for approximately 7 percent of the electricity generated in the United States. More than one-half of that hydroelectricity is generated in three states: Washington, California, and Oregon. Worldwide, nearly 20 percent of all electricity comes from hydroelectric power plants. China is the world's leading producer of hydroelectricity, followed by Brazil and the United States.

Although impounding water from rivers and streams behind dams is the most common approach to generating hydroelectricity, *run-of-the-river* and tidal energy systems, like the one installed at New York's Roosevelt Island, also capture the energy of moving water. These systems will be discussed in the following section.

## How Hydroelectricity Generation Works

Moving water, either falling over a vertical distance or flowing with a river or tide, contains kinetic energy. A hydroelectric power plant captures this kinetic energy and uses it to turn a turbine, just as the kinetic energy of steam turns a turbine in a coal-fired power plant. The turbine, in turn, transforms the kinetic energy of water or steam into electricity, which is then exported to the electrical grid via transmission lines.

The amount of electricity that can be generated at any particular hydroelectric power plant depends on the flow rate, the vertical distance the water falls, or both. Where falling water is the source of the energy, the amount of electricity that can be generated depends on the vertical distance the water falls; the greater the distance, the more potential energy the water has, and the more electricity it can generate (see Figure 2.12). The amount of electricity a hydroelectric power plant can generate also depends on the flow rate: the amount of water that flows past a certain point per unit time. The higher the flow rate, the more kinetic energy is present, and the more electricity can be generated.

**RUN-OF-THE-RIVER SYSTEMS** In **run-of-the-river** hydroelectricity generation, water is retained behind a low dam and runs through a channel before returning to the river. Run-of-the-river systems do not store water in a reservoir. These systems have several advantages that reduce their environmental impact: relatively little flooding occurs upstream, and seasonal changes in river flow are not disrupted. However, run-of-the-river systems are generally small, and because they rely on natural water flows, electricity generation can be intermittent. Heavy spring runoff from rains or snowmelt cannot be stored, and the system cannot generate any electricity in hot, dry periods when the flow of water is low.

**WATER IMPOUNDMENT SYSTEMS** Storing water in a reservoir behind a dam is known as **water impoundment**. FIGURE 13.12 illustrates the various features of a water impoundment system. Water impoundment is the most common method of hydroelectricity generation because it usually allows for electricity generation on demand. By controlling the opening and closing of the gates, the dam operators determine the flow rate of water and therefore the amount of electricity generated. The largest hydroelectric water impoundment dam in the United States is the Grand Coulee Dam in Washington State, which generates 6,800 MW at peak capacity. The Three Gorges Dam on the Yangtze River in China, depicted in Figure 9.10, is the largest dam in the world. It has a capacity of 18,000 MW and can generate almost 85 billion kilowatt-hours per year,

**FIGURE 13.12** A water impoundment hydroelectric dam. Arrows indicate the path of water flow.

approximately 11 percent of China's total electricity demand.

**TIDAL SYSTEMS  Tidal energy** also comes from the movement of water, although the movement in this case is driven by the gravitational pull of the Moon. Tidal energy systems, like the one on Roosevelt Island described in this chapter's opening story, use gates and turbines similar to those used in run-of-the-river and water impoundment systems to capture the kinetic energy of water flowing through estuaries, rivers, and bays and convert it into electricity.

Although tidal power plants are operating in many parts of the world, including France, Korea, and Canada, tidal energy does not have the potential to become a major energy source. In many places the difference in water level between high and low tides is not great enough to provide sufficient kinetic energy to generate a large amount of electricity. In addition, to transfer the energy generated, transmission lines must be constructed on or near a coastline or estuary. This infrastructure may have a disruptive effect on coastal, shoreline, and marine ecology as well as on tourism that relies on the aesthetics of coastal regions.

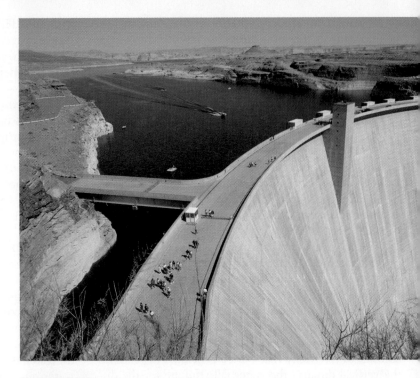

**FIGURE 13.13**  **Glen Canyon National Recreation Area and Lake Powell.** Lake Powell is a popular recreation area created by water impoundment at the hydroelectric Glen Canyon Dam.

## Is Hydroelectricity Sustainable?

Major hydroelectric dam projects have brought renewable energy to large numbers of rural residents in many countries, including the United States, Canada, India, China, Brazil, and Egypt. Although hydroelectric dams are expensive to build, once built, they require a minimal amount of fossil fuel. In general, the benefits of water impoundment hydroelectric systems are great: they generate large quantities of electricity without creating air pollution, waste products, or $CO_2$ emissions. Electricity from hydroelectric power plants is usually less expensive for the consumer than electricity generated using nuclear fuels or natural gas. In the United States, the price of hydroelectricity ranges from 5 cents to 11 cents per kilowatt-hour.

In addition, the reservoir behind a hydroelectric dam can provide recreational and economic opportunities as well as downstream flood control for flood-prone areas. For example, Lake Powell, the reservoir impounded by the hydroelectric Glen Canyon Dam, draws more than 2.5 million visitors to the Glen Canyon National Recreation Area each year, generating more than $400 million annually for the local and regional economies in Arizona and Utah (**FIGURE 13.13**). In China, the Three Gorges Dam provides flood control and protection to many millions of people.

Water impoundment does have negative environmental consequences, however. In order to form an impoundment, a free-flowing river must be held back. The resulting reservoir may flood hundreds or thousands of hectares of prime valley agricultural land or canyons that have great aesthetic or archeological value. It may also force people to relocate. As we saw in Chapter 9, the construction of the Three Gorges Dam displaced more than 1.3 million people. It also submerged ancient cultural and archaeological sites as well as large areas of farmland. In addition, impounding a river in this way may make it unsuitable for organisms or recreational activities that depend on a free-flowing river. Large reservoirs of standing water hold more heat and contain less oxygen than do free-flowing rivers, affecting which species can survive in the waters. Certain human parasites become more abundant in impounded waters in tropical regions.

By regulating water flow and flooding, dams also alter the dynamics of the river ecosystem downstream. In some rivers, for example, sandbars are created during periods of low flow following periods of flooding. Some plant species, such as cottonwood trees, cannot reproduce in the absence of these sandbars. The life cycles of certain aquatic species, such as salmon, certain trout species, and freshwater clams and mussels, also depend on seasonal variations in water flow. By controlling the flow

**(a)** Before removal

**(b)** After removal

 **FIGURE 13.14** **Dam removal.** (a) Oregon's Marmot Dam, on the Sandy River, viewed from downstream of the dam. (b) The same stretch of the Sandy River after the dam was removed in 2007.

of water to make it consistently plentiful for hydroelectricity generation, impoundment systems disrupt these life cycles.

It is possible to address some of these problems. For example, as we saw in Chapter 9, the installation of a fish ladder (see Figure 9.13) may allow fish to travel upstream around a dam. Such solutions are not always optimal, however; some species fail to utilize them, and some predators learn to monitor the fish ladders for their prey.

Other environmental consequences of water impoundment systems include the release of greenhouse gases to the atmosphere, both during dam construction and after filling the reservoir. Production of cement—a major component of dams—is responsible for approximately 5 percent of global anthropogenic $CO_2$ emissions to the atmosphere. Once the dam is completed, the impounded water covers forests or grasslands. The dead plants and organic materials in the flooded soils decompose anaerobically and release methane, a potent greenhouse gas. Some researchers assert that in tropical regions, the methane released from a hydroelectric water impoundment contributes more to climate warming than a coal-fired power plant with about the same generating capacity.

The accumulation of sediments in reservoirs has negative consequences not only for the environment, but also for the electricity-generating capacity of hydroelectric dams. A fast-moving river carries sediments that settle out when the river feeds into the reservoir. The accumulation of these sediments on the bottom of the reservoir is known as **siltation.** Over time, as the reservoir fills up with sediments, the amount of water that can be impounded, and thus the generating capacity and life span of the dam, is reduced. This process may take hundreds of years or only decades, depending on the geology of the

area. The only way to reverse the siltation process is by dredging: removal of the sediment, usually with machinery that runs on fossil fuels.

Because of either environmental concerns or heavy siltation, a number of hydroelectric dams are being dismantled, as we saw in Chapter 9's opening story. In 1999, the Edwards Dam was removed from the Kennebec River in Maine. A decade later, native fishes such as bass and alewives have returned to the waters and are flourishing. In 2007, the Marmot Dam on Oregon's Sandy River was removed using explosives. The restored river now hosts migrating salmon and steelhead trout (*Oncorhynchus mykiss*) for the first time since 1912 (**FIGURE 13.14**).

---

**GAUGE YOUR PROGRESS**

✓ How can water be used as a renewable energy resource?

✓ What are the types of hydroelectricity generation systems?

✓ What are the trade-offs associated with using hydroelectricity compared with biomass energy?

---

# The Sun's energy can be captured directly

In addition to driving the natural cycles of water and air movement that we can tap as energy resources, the Sun also provides energy directly. Every day, Earth is bathed in solar radiation, an almost limitless source of energy. The amount of solar energy available in a particular

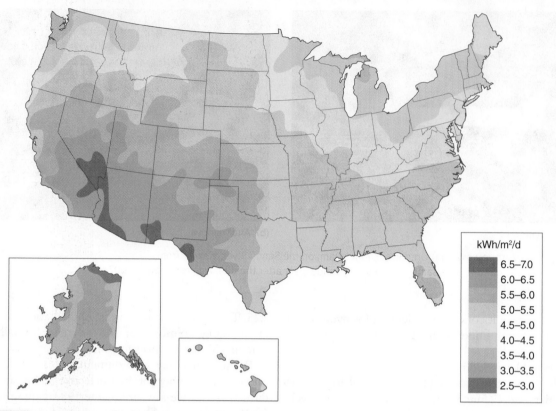

| kWh/m²/d |
|---|
| 6.5–7.0 |
| 6.0–6.5 |
| 5.5–6.0 |
| 5.0–5.5 |
| 4.5–5.0 |
| 4.0–4.5 |
| 3.5–4.0 |
| 3.0–3.5 |
| 2.5–3.0 |

**FIGURE 13.15** **Geographic variation in solar radiation in the United States.** This map shows the amount of solar energy available to a flat photovoltaic solar panel in kilowatt-hours per square meter per day, averaged over a year. [Data from National Renewable Energy Laboratory, U.S. Department of Energy.]

place varies with cloudiness, time of day, and season. The average amount of solar energy available varies geographically. In the continental United States, average daily solar radiation ranges from 3 kWh of energy per square meter in the Pacific Northwest to almost 7 kWh per square meter in parts of the Southwest, as **FIGURE 13.15** shows.

## Passive Solar Heating

We have already seen several applications of passive solar heating, including positioning windows on south-facing walls to admit solar radiation in winter, covering buildings with dark roofing material in order to absorb the maximum amount of heat, and building homes into the side of a hill. None of these strategies relies on intermediate pumps or technology to supply heat. Solar ovens are another practical application of passive solar heating. For instance, a simple "box cooker" concentrates sunlight as it strikes a reflector on the top of the oven. Inside the box, the solar energy is absorbed by a dark base and a cooking pot and is converted into heat energy. The heat is distributed throughout the box by reflective material lining the interior walls and is kept from escaping by a glass top. On sunny days, such box cookers can maintain temperatures of 175°C (350°F), heat several liters of

water to boiling in under an hour, or cook traditional dishes of rice, beans, or chicken in 2 to 5 hours.

Solar ovens have both environmental and social benefits. Using solar ovens in place of firewood reduces deforestation. In areas that are unsafe for travel, having a solar oven means not having to leave the relative safety of home to seek firewood. For example, over 10,000 solar ovens have been distributed in refugee camps in the Darfur region of western Sudan, Africa, where leaving the camps to find cooking fuel would put women at risk of attack (**FIGURE 13.16**).

## Active Solar Energy Technologies

In contrast to passive solar design, **active solar energy** technologies capture the energy of sunlight with the use of technologies. These technologies include small-scale solar water heating systems, photovoltaic solar cells, and large-scale concentrating solar thermal systems for electricity generation.

**SOLAR WATER HEATING SYSTEMS** Solar water heating applications range from providing domestic hot water and heating swimming pools to a variety of business and home heating purposes. In the United States, heating swimming pools is the most common application

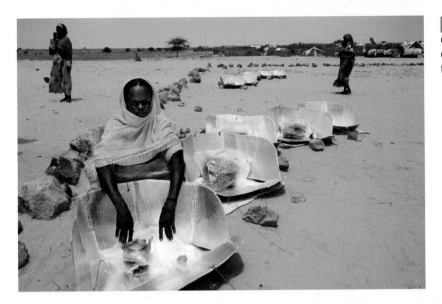

**FIGURE 13.16 Solar cookers.** Residents of this refugee camp in Chad use solar cookers to conserve firewood and reduce the need to travel outside the camp.

of solar water heating, and is also the one that pays for itself the most quickly.

A household solar water heating system, like the domestic hot water system shown in FIGURE 13.17, allows heat energy from the Sun to be transferred directly to water or another liquid, which is then circulated to a hot water or heating system. The circulation of the liquid is driven either by a pump (in active systems) or by natural convection (in passive systems). In both cases, cold liquid is heated as it moves through a solar collector mounted on the roof, wall, or ground. The simplest systems pump cold water directly to the collector to be heated; the heated water then flows back to an insulated storage tank. In areas that are sunny but have temperatures below freezing, the water is kept in the storage tank, and a "working" liquid containing nontoxic antifreeze circulates in pipes between the storage tank and the solar collector. The nonfreezing circulating liquid is heated by the Sun in the solar collector, then returned to the storage tank, where it flows through a heat exchanger that transfers its heat to the water. The energy needed to run the pump is usually much less then the energy gain from using the system, especially if the pump runs on electricity from the Sun. Solar water heating systems typically include a backup energy source, such as an electric heating element or a connection to a fossil fuel–based central heating system, so that hot water is available even when it is cloudy or cold.

**PHOTOVOLTAIC SYSTEMS** In contrast to solar water heating systems, **photovoltaic solar cells** capture energy from the Sun as light, not heat, and convert it directly into electricity. FIGURE 13.18 shows how a photovoltaic system, also referred to as PV, delivers electricity to a house. A photovoltaic solar cell makes use of the fact that certain semiconductors—very thin, ultraclean layers of material—generate a low-voltage electric current when they are exposed to direct sunlight. That low-voltage direct current is usually converted into higher-voltage alternating current for use in homes or businesses. Typically, photovoltaic solar cells are 12 to 20 percent efficient in converting the energy of sunlight into electricity.

**FIGURE 13.17 A solar domestic hot water system.** A nonfreezing liquid is circulated by an electric pump through a closed loop of pipes. This circulating liquid moves from a water storage tank to a solar collector on the roof, where it is heated, and back to the tank, where a heat exchanger transfers the heat to water.

**(a)** Schematic of photovoltaic (PV) system

**(b)** California home with PV panels on roof

**FIGURE 13.18** **Photovoltaic solar energy.** (a) In this domestic photovoltaic system, photovoltaic solar panels convert sunlight into direct current (DC). An inverter converts DC into alternating current (AC), which supplies electricity to the house. Any electricity not used in the house is exported to the electrical grid. (b) Photovoltaic panels on the roof of this house in California provide 4,200 kWh of electricity per year—nearly all of the electricity this family uses.

Electricity produced by photovoltaic systems can be used in several ways. Solar panels—arrays of photovoltaic solar cells—on a roof can be used to supply electricity to appliances or lights directly, or they can be used to charge batteries for later use. The vast majority of photovoltaic systems are tied to the electrical grid, meaning that any extra electricity generated is sent to the electric company, which buys it or gives the customer credit toward the cost of future electricity use. Homes that are "off the grid" may rely on photovoltaic solar cells as their only source of electricity, using

batteries to store the electricity until it is needed. Photovoltaic solar cells have other uses in locations far from the grid where a small amount of electricity is needed on a regular basis. For example, small photovoltaic solar cells charge the batteries that keep highway emergency telephones working. In several U.S. cities, photovoltaic solar cells provide electricity for new "smart" parking meter systems that have replaced aging urban coin-operated parking meters.

**CONCENTRATING SOLAR THERMAL ELECTRICITY GENERATION** Concentrating solar thermal (CST) systems are a large-scale application of solar energy to electricity generation. CST systems use lenses or mirrors and tracking systems to focus the sunlight falling on a large area into a small beam, in the same way you might use a magnifying glass to focus the Sun's energy and perhaps burn a hole in a piece of paper. In this case, however, the heat of the concentrated beam is used to evaporate water and produce steam that turns a turbine to generate electricity. CST power plants operate much like conventional thermal power plants; the only difference is that the energy to produce the steam comes directly from the Sun, rather than from fossil fuels. The arrays of lenses and mirrors required are large, so CST power plants are best constructed in expansive desert areas where there is consistent sunshine and plenty of open space (**FIGURE 13.19**).

Although CST systems have been in place for 10 years or more, they are becoming more common. In the United States, several plants are under development in California and the Southwest. One 35 MW plant being planned in California calls for reflectors to cover 65 ha (160 acres) of land. The drawbacks of these plants include the large amount of land required and their inability to generate electricity at night.

## Benefits and Drawbacks of Active Solar Energy Systems

Active solar energy systems offer many benefits. They can generate hot water or electricity without producing air pollution, water pollution, or $CO_2$ in the process. In addition, photovoltaic solar cells and CST power plants can produce electricity when it is needed most: on hot, sunny days when demand for electricity is high, primarily for air conditioning. By producing electricity during peak demand hours, these systems can help reduce the need to build new fossil fuel power plants.

**FIGURE 13.19** **A concentrating solar thermal power plant.**
Mirrors and reflectors concentrate the Sun's energy onto a
"power tower," which uses the sunlight to heat water and
make steam for electricity generation.

In many areas, small-scale solar energy systems are
economically feasible. For a new home located miles
away from the grid, installing a photovoltaic system may
be much less expensive than running electrical transmission lines to the home site. When a house is near the
grid, the initial cost of a photovoltaic system may take
5 to 20 years to pay back; once the initial cost is paid
back, however, the electricity it generates is almost free.

Despite these advantages, a number of drawbacks
have inhibited the growth of solar energy use in the
United States. Photovoltaic solar panels are expensive
to manufacture and install. Although the technology is
changing rapidly as industrial engineers and scientists
seek better, cheaper photovoltaic materials and systems,
the initial cost to install a photovoltaic system can be
daunting, and the payback period is a long one. In parts
of the country where the average daily solar radiation
is low, the payback period is even longer. Some countries, such as Germany, have made solar energy a part
of their sustainable energy agenda by subsidizing their
solar industry. In the United States, recent tax breaks,
rebates, and funding packages instituted by various states
and the federal government have made solar electricity
and water heating more affordable for consumers and
businesses.

The use of photovoltaic solar cells has environmental
as well as financial costs. Manufacturing photovoltaic
solar cells requires a great deal of energy and water and
involves a variety of toxic metals and industrial chemicals that can be released into the environment during
the manufacturing process. Newer types of photovoltaic solar cells may reduce their reliance on toxic
materials. For systems that rely on batteries for energy
storage, there are environmental costs associated with
manufacturing, disposing of, or recycling the batteries, as well as energy losses during charging, storage,
and recovery of electricity in batteries. The end-of-
life reclamation and recycling of photovoltaic solar cells
is another potential source of environmental contamination, particularly if the cells are not recycled properly.
However, solar energy advocates, and even most critics, agree that the energy expended to manufacture
photovoltaic solar cells is usually recovered within a
few years of their operation, and that if the life span
of photovoltaic solar cells can be increased to 30 to
50 years, they will be a very promising source of renewable energy.

---

### GAUGE YOUR PROGRESS

✓ In what ways do humans capture solar energy
for their use?

✓ Why are active solar energy systems not feasible
everywhere?

✓ How do active and passive solar systems work?
What are the advantages of each?

---

# Earth's internal heat is a source of nondepletable energy

Unlike most forms of renewable energy, geothermal
energy does not come from the Sun. **Geothermal
energy** is heat that comes from the natural radioactive decay of elements deep within Earth. As we saw in
Chapter 8, convection currents in Earth's mantle bring
hot magma toward Earth's surface. Wherever magma
comes close enough to groundwater, that groundwater
is heated. The pressure of the hot groundwater sometimes drives it to the surface, where it visibly manifests
itself as geysers and hot springs, like those in Yellowstone
National Park. Where hot groundwater does not naturally rise to the surface, humans may be able to reach it
by drilling.

Many countries obtain clean, renewable energy from
geothermal resources. The United States, China, and
Iceland, which have substantial geothermal resources,
are the largest geothermal energy producers.

## Harvesting Geothermal Energy

Geothermal energy can be used directly as a source
of heat. Hot groundwater can be piped directly into

household radiators for heating a home. In other cases, heat exchangers can collect heat by circulating cool liquid underground, where it is heated, and returning it to the surface. Iceland, a small nation with vast geothermal resources, heats 87 percent of its homes this way.

Geothermal energy can also be used to generate electricity. The electricity-generating process is much the same as that in a conventional thermal power plant; in this case, however, the steam to run the turbine comes from water evaporated by Earth's internal heat instead of by burning fossil fuels.

The heat released by decaying radioactive elements deep within Earth is essentially nondepletable in the span of human time. However, the groundwater that so often carries that heat to Earth's surface is not. As we learned in Chapter 9, groundwater, if used sustainably, is a renewable resource. Unfortunately, long periods of harvesting groundwater from a site may deplete it to the point at which the site is no longer a viable source of geothermal energy. Returning the water to the ground to be reheated is one way to use geothermal energy sustainably.

Iceland currently produces about 20 percent of its electricity using geothermal energy. In the United States, geothermal energy is the third most commonly used renewable energy resource, after hydroelectricity and biomass, constituting about 5 percent of the renewable energy used. Geothermal power plants are currently in operation in California, Nevada, Hawaii, and Utah. Geothermal energy has less growth potential than wind or solar energy, however, simply because it is not easily accessible everywhere. Another drawback of geothermal energy is the hazardous gases and steam that may be emitted from geothermal power plants.

## Ground Source Heat Pumps

Another approach to tapping Earth's thermal resources is the use of **ground source heat pumps,** which take advantage of the high thermal inertia of the ground. Earth's temperature about 3 m (10 feet) underground remains fairly constant year-round, at 10°C to 15°C (50°F–60°F), because the ground retains the Sun's heat more effectively than does the ambient air. We can take advantage of this fact to heat and cool residential and commercial buildings. Although the heat tapped by ground source heat pumps is often referred to informally as "geothermal," it comes not from geothermal energy, but from solar energy.

FIGURE 13.20 shows how a ground source heat pump transfers heat from the ground to a house. In contrast to the geothermal systems described above, ground source heat pumps do not remove steam or hot water from the ground. In much the same way that a solar water heating system works, a ground source heat pump cycles fluid through pipes buried underground. In winter, this fluid absorbs heat from underground. The slightly warmed fluid is compressed in the heat pump to increase its temperature even more, and the heat is distributed throughout the house. The fluid is then allowed to expand, which causes it to cool, and run through the cycle again, picking up more heat from the ground. In summer, when the underground temperature is lower than the ambient air temperature, the fluid is cooled underground and pulls heat from the house as it circulates, resulting in a cooler house and heat transferred underground.

Ground source heat pumps can be installed anywhere in the world, regardless of whether there is geothermal

1. At the compressor, the circulating fluid is compressed to form a hot gas. Heat is given off into the house from the heat exchanger as the gas cools to form a liquid.

2. As the fluid expands and cools, it becomes a gas that is much cooler than the ground through which it will move.

3. The warmer ground heats the cool fluid, which cycles through buried tubing. Thus heat from the ground is essentially "pumped" into the building.

Heat exchanger

Expansion valve

Heat

Compressor

Heat

Heat

Heat

**FIGURE 13.20** **A ground source heat pump.** By exchanging heat with the ground, a ground source heat pump can heat and cool a building using 30 to 70 percent less energy than traditional furnaces and air conditioners.

energy accessible in the vicinity. The operation of the pump requires some energy, but in most cases the system uses 30 to 70 percent less energy to heat and cool a building than a standard furnace or air conditioner.

---

# Wind energy is the most rapidly growing source of electricity

The wind is another important source of nondepletable, renewable energy. Winds are the result of the unequal heating of Earth's surface by the Sun (see Chapter 4). Warmer air rises and cooler, denser air sinks, creating circulation patterns similar to those in a pot of boiling water. Ultimately, the Sun is the source of all winds—it is solar radiation and ground surface heating that drives air circulation.

Before the electrical grid reached rural areas of the United States in the 1920s, windmills dotted the landscape. Today, **wind energy** is widely used in some countries and is making a comeback in others. FIGURE 13.21 shows installed wind energy generating capacity and the percentage of electricity generated by wind for a number of countries. The United States has the largest wind energy generating capacity in the world, followed by Germany, Spain, China, India, and Italy. Despite its large generating capacity, the United States obtains less than 1 percent of its electricity from wind. The largest amounts are generated in California and Texas, but more than one-half of U.S. states produce at least some wind-generated electricity. Denmark, a country of 5.5 million people, obtains more than 21 percent of its electricity from wind—the greatest percentage anywhere in the world.

## Generating Electricity from Wind

In much the same way that a hydroelectric turbine harnesses the kinetic energy of moving water, a **wind turbine** converts the kinetic energy of moving air into electricity. A modern wind turbine like the one shown in FIGURE 13.22 may sit on a tower as tall as 100 m (330 feet) and have blades 40 to 75 m (130–250 feet) long. Under average wind conditions, a wind turbine on land might produce electricity 25 percent of the time. While it is spinning, it might generate between 2,000 and 3,000 kW (2–3 MW), and in a year it might produce more than 4.4 million kilowatt-hours of electricity, enough to supply more than 400 homes. Offshore wind conditions are even more desirable for electricity

**(a)** Wind capacity

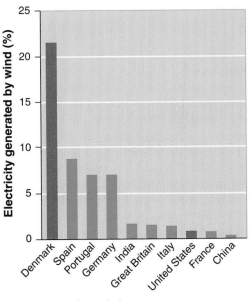

**(b)** Percentage from wind

FIGURE 13.21 **Installed wind energy capacity by country.** (a) The United States generates more electricity from wind energy than any other country. (b) However, some relatively small countries, such as Denmark, generate a much higher percentage of their electricity from wind. [After Global Wind Energy Council.]

**Blade**
(turned by wind)

**Wind direction**

**Gearbox**
transfers mechanical energy to generator.

**Motorized drive**
turns the turbine to face into the wind, which maximizes efficiency.

**Generator**
converts mechanical energy into electricity.

**FIGURE 13.22** **How a wind turbine generates electricity.** The wind turns the blade, which is connected to the generator, which generates electricity.

**FIGURE 13.23** **An offshore wind park in Denmark.** Capacity factors at near-offshore locations are generally higher than on land, which makes the location shown here highly desirable for wind generation.

generation, and turbines can be made even larger in an offshore environment.

Wind turbines on land are typically installed in rural locations, away from buildings and population centers. However, they must also be close to electrical transmission lines with enough capacity to transport the electricity they generate to users. For these reasons, as well as for political and regulatory reasons and to facilitate servicing of the equipment, it is practical to group wind turbines into wind farms or wind parks.

Some of the most rapidly growing wind energy projects are offshore wind parks, clusters of wind turbines located in the ocean within a few miles of the coastline (**FIGURE 13.23**). Offshore wind parks are operating in Denmark, the Netherlands, the United Kingdom, Sweden, and elsewhere. The proposed Cape Wind project may become the first offshore wind farm in the United States. Located off Cape Cod, Massachusetts, in Nantucket Sound, the project would feature 130 wind turbines with the potential to produce up to 420 MW

of electricity, or up to 75 percent of the electricity used by Cape Cod and the nearby islands.

## A Nondepletable Resource

Wind energy is the fastest-growing major source of electricity in the world. As **FIGURE 13.24** shows, global installed wind energy capacity has risen from less than 10 gigawatts in 1996 to more than 160 gigawatts today. Wind energy offers many advantages over other energy resources. Like sunlight, wind is a nondepletable, clean, and free energy resource; the amount available tomorrow does not depend on how much we use today. Furthermore, once a wind turbine has been manufactured and installed, the only significant energy input comes from the wind. The only substantial fossil fuel input required, once the turbines are installed, is the fuel needed to travel to the wind farm to maintain the equipment. Thus, wind-generated electricity produces no pollution and no greenhouse gases. Finally, unlike

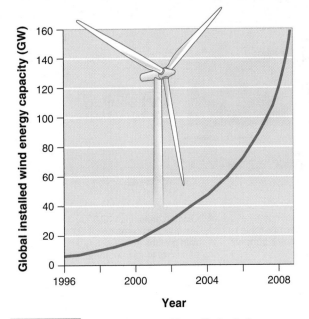

**FIGURE 13.24** **Global growth of installed wind energy capacity.** Worldwide, installed wind energy capacity is now almost 160 gigawatts (GW). [After Global Wind Energy Council.]

hydroelectric, CST, and conventional thermal power plants, wind farms can share the land with other uses. For example, wind turbines may share land with grazing cattle.

Wind-generated electricity does have some disadvantages, however. Currently, most off-grid residential wind energy systems rely on batteries to store electricity. As we have discussed, batteries are expensive to produce and hard to dispose of or recycle. Furthermore, some people object to the noise produced by large wind farms. Some people find them ugly, although probably just as many find them aesthetically pleasing. Controversies have erupted over the placement of wind farms where people can see or hear them. In addition, birds and bats may be killed by collisions with wind turbine blades. According to the National Academy of Sciences, as many as 40,000 birds may be killed by wind turbine blades in the United States each year—approximately 4 deaths per turbine. Bat deaths are not as well quantified. New turbine designs and location of wind farms away from migration paths have reduced these deaths to some extent, as well as some of the noise and aesthetic disadvantages.

**GAUGE YOUR PROGRESS**

✓ What are the most practical and feasible locations for wind farms?

✓ How is wind used to generate electricity?

✓ How do near-offshore and land-based wind farms differ? How are they the same?

# Hydrogen fuel cells have many potential applications

We will conclude our coverage of sustainable energy types with another energy technology that has received a great deal of attention for many years: hydrogen fuel cells.

## How Fuel Cells Work

A **fuel cell** is a device that operates much like a common battery, but with one key difference. In a battery, electricity is generated by a reaction between two chemical reactants, such as nickel and cadmium. This reaction happens in a closed container to which no additional materials can be added. Eventually the reactants are used up, and the battery goes dead. In a fuel cell, however, the reactants are added continuously to the cell, so the cell produces electricity for as long as it is supplied with fuel.

**FIGURE 13.25** shows how hydrogen functions as one of the reactants in a hydrogen fuel cell. Electricity is generated by the reaction of hydrogen with oxygen, which forms water:

$$2 \, H_2 + O_2 \rightarrow energy + 2 \, H_2O$$

Although there are many types of hydrogen fuel cells, the basic process forces protons from hydrogen gas through a membrane, while the electrons take a different pathway. The movement of protons in one direction and electrons in another direction generates an electric current.

Using a hydrogen fuel cell to generate electricity requires a supply of hydrogen. Supplying hydrogen is a challenge, however, because free hydrogen gas is relatively rare in nature and because it is explosive. Hydrogen tends to bond with other molecules, forming compounds such as water ($H_2O$) or natural gas ($CH_4$). Producing hydrogen gas requires separating it from these compounds using heat or electricity. Currently, most commercially available hydrogen is produced by an energy-intensive process of burning natural gas in order to extract its hydrogen; carbon dioxide is a waste product of this combustion. In an alternative process, known as **electrolysis,** an electric current is applied to water to "split" it into hydrogen and oxygen. Energy scientists are looking for other ways to obtain hydrogen; for example, under certain conditions, some photosynthetic algae and bacteria, using sunlight as their energy source, can give off hydrogen gas.

Although it may seem counterintuitive to *use* electricity to create electricity, the advantage of hydrogen is that it can act as an energy carrier. Renewable energy sources such as wind and the Sun cannot produce electricity constantly, but the electricity they produce can

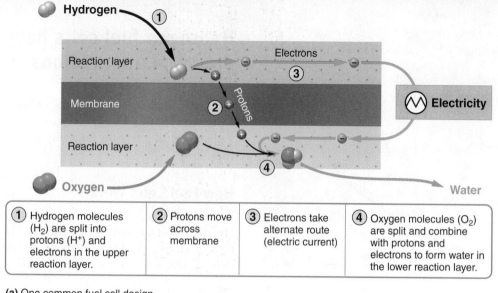

Reaction layer — Electrons

Membrane — **Electricity**

Reaction layer — Protons

**Hydrogen** ①

② ③ ④

**Oxygen** — Water

| ① Hydrogen molecules ($H_2$) are split into protons ($H^+$) and electrons in the upper reaction layer. | ② Protons move across membrane | ③ Electrons take alternate route (electric current) | ④ Oxygen molecules ($O_2$) are split and combine with protons and electrons to form water in the lower reaction layer. |

**(a)** One common fuel cell design

Oxygen

Fuel cell stack    Battery    Hydrogen cylinders

**Hydrogen**

Electric current

Water

**(b)** Fuel cell vehicle    Electric motor

**FIGURE 13.25** **How a hydrogen fuel cell works.** (a) Hydrogen gas enters the cell from an external source. Protons from the hydrogen molecules pass through a membrane, while electrons flow around it, producing an electric current. Water is the only waste product of the reaction. (b) In a fuel cell vehicle, hydrogen is the fuel that reacts with oxygen to provide electricity to run the motor.

be used to generate hydrogen, which can be stored until it is needed. Thus, if we could generate electricity for electrolysis using a clean, nondepletable energy resource such as wind or solar energy, hydrogen could potentially be a sustainable energy carrier.

## Is Hydrogen a Viable Energy Alternative?

Hydrogen fuel cells are considered by some policy makers to be the future of energy and the solution to many of the world's energy problems. They are 80 percent efficient in converting the potential energy of hydrogen and oxygen into electricity, and their only by-product is water. In contrast, thermal fossil fuel power plants are only 35 to 50 percent efficient, and they produce a wide range of pollutants as by-products. However, there are some who believe that hydrogen fuel cells will not be a solution to our energy problems.

Despite the many advantages of hydrogen as a fuel, it also has a number of disadvantages. First, scientists must learn how to obtain hydrogen without expending more fossil fuel energy than its use would save. This means that the energy for the hydrogen generation process must come from a renewable resource such as wind or solar energy instead of fossil fuels. Second, suppliers will need a distribution network to safely deliver hydrogen to consumers—something similar to our current system of gasoline delivery trucks and gasoline stations. Hydrogen can be stored as a liquid or as a gas; each storage medium has its limitations. In a fuel cell vehicle, hydrogen would probably be stored in the form of a gas in a large tank under very high pressure. Vehicles would have to be redesigned with fuel tanks much larger than current gasoline tanks to achieve an equivalent travel distance per tank. There is also the risk of a tank rupture, in which case the hydrogen might catch fire or explode.

Given these obstacles, why is hydrogen considered a viable energy alternative? Ultimately, hydrogen-fueled vehicles could be a sustainable means of transportation because a hydrogen-fueled car would use an electric motor. Electric motors are more efficient than internal combustion engines: while an internal combustion engine converts about 20 percent of the fuel's energy into the motion of the drive train, an electric motor can convert 60 percent of its energy into motion. So if we generated electricity from hydrogen at 80 percent efficiency, and used an electric motor to convert that electricity into vehicular motion at 60 percent efficiency, we would have a vehicle that is much more efficient than one with an internal combustion engine. Thus, even if we obtained hydrogen by burning natural gas, the total amount of energy used to move an electric vehicle using hydrogen might still be substantially less than the total amount needed to move a gasoline-fueled car. Using solar or wind energy to produce the hydrogen would lower the environmental cost even more, and the energy supply would be renewable. Under those circumstances, an automobile could be fueled by a truly renewable source of energy that is both carbon neutral and pollution free.

---

### GAUGE YOUR PROGRESS

✓ How do we obtain hydrogen for use in fuel cells?

✓ How does a fuel cell work?

✓ Why is hydrogen useful as an alternative to fossil fuels?

---

# How can we plan our energy future?

- - - - - - - - - - - - - - - - - - - - - - - - - - -

Each of the renewable energy resources we have discussed in this chapter has unique advantages. None of these resources, however, is a perfect solution to our energy needs. Table 13.1 lists some of the advantages and limitations of each. In short, no single energy resource that we are currently aware of can replace nonrenewable energy resources in a way that is completely renewable, nonpolluting, and free of impacts on the environment. A sustainable energy strategy, therefore, must combine energy efficiency, energy conservation, and the development of renewable and nonrenewable energy resources, taking into account the costs, benefits, and limitations of each one.

## The Challenge of a Renewable Energy Strategy

Energy expert Amory Lovins suggests that innovation and technological advances, not the depletion of a resource, provide the driving force for moving from one energy technology to the next. Extending this concept to the present, one can argue that we will develop new energy technologies before we run out of the fuels on which we currently depend.

Despite their tremendous potential, however, renewable energy resources are unlikely to replace fossil fuels completely in the immediate future unless nations commit to supporting their development and use through direct funding and financial incentives such as tax cuts and consumer rebates. In fact, the U.S. Department of Energy predicts that fossil fuel consumption will continue to increase in the United States well into the middle of the twenty-first century. In spite of their extremely rapid growth, wind and solar energy still account for far less than 1 percent of the energy produced in the United States. Government funding or other sources of capital are needed to support research to overcome the current limitations of many renewable energy resources.

## Improving the Electrical Grid

In addition to investing in *new* energy sources, the United States will have to upgrade its *existing* electrical infrastructure—its power plants, storage capacity, and distribution networks. Approximately 40 percent of the energy used in the United States is used to generate electricity. The U.S. electricity distribution system is outdated and subject to overloads and outages, which cost the U.S. economy over $100 billion per year. There are regions of the country that cannot supply enough generating capacity to meet their needs. There are other locations where the electrical infrastructure cannot accommodate all the electricity that could be generated. Furthermore, the current system requires that electricity be moved long distances from the power plant to the consumer. Approximately 5 to 10 percent of the electricity generated is lost as it is transported along electrical transmission lines, and the farther it is transported, the more is lost. While the storage capacity of batteries improves each year, batteries are not a sustainable solution for this problem of energy loss. Many people are focusing their attention on improving the electrical grid to make it as efficient as possible at moving electricity from one location to another, thereby reducing the need for storage capacity.

An energy economy based on nondepletable energy sources requires reliable electricity storage and affordable (or at least more effective and efficient) distribution networks. U.S. energy scientists maintain that because we currently do not have a cost-effective, reliable means of storing energy, depending on intermittent sources such as wind and solar energy for more than about 20 percent of our total electricity production could lead to risky instability in the grid.

One solution currently in development may be the **smart grid,** an efficient, self-regulating electricity

| TABLE 13.1 | Comparison of renewable energy resources | |
|---|---|---|
| **Energy resource** | **Advantages** | **Disadvantages** |
| Liquid biofuels | Potentially renewable<br>Can reduce our dependence on fossil fuels<br>Reduces trade deficit<br>Possibly more environmentally friendly than fossil fuels | Loss of agricultural land<br>Higher food costs<br>Lower gas mileage<br>Possible net increase in greenhouse gas emissions |
| Solid biomass | Potentially renewable<br>Eliminates waste from environment<br>Available to everyone<br>Minimal technology required | Deforestation<br>Erosion<br>Indoor and outdoor air pollution<br>Possible net increase in greenhouse gas emissions |
| Photovoltaic solar cells | Nondepletable resource<br>After initial investment, no cost to harvest energy | Manufacturing materials requires high input of metals and water<br>No plan in place to recycle solar panels<br>Geographically limited<br>High initial costs<br>Storage batteries required for off-grid systems |
| Solar water heating systems | Nondepletable resource<br>After initial investment, no cost to harvest energy | Manufacturing materials requires high input of metals and water<br>No plan in place to recycle solar panels<br>Geographically limited<br>High initial costs |
| Hydroelectricity | Nondepletable resource<br>Low cost to run<br>Flood control<br>Recreation | Limited amount can be installed in any given area<br>High construction costs<br>Threats to river ecosystems<br>Loss of habitat, agricultural land, and cultural heritage; displacement of people<br>Siltation |
| Tidal energy | Nondepletable resource<br>After initial investment, no cost to harvest energy | Potential disruptive effect on some marine organisms<br>Geographically limited |
| Geothermal energy | Nondepletable resource<br>After initial investment, no cost to harvest energy<br>Can be installed anywhere (ground source heat pump) | Emits hazardous gases and steam<br>Geographically limited |
| Wind energy | Nondepletable resource<br>After initial investment, no cost to harvest energy<br>Low up-front cost | Turbine noise<br>Deaths of birds and bats<br>Geographically limited to windy areas near transmission line<br>Aesthetically displeasing to some<br>Storage batteries required for off-grid systems |
| Hydrogen fuel cell | Efficient<br>Zero pollution | Producing hydrogen is an energy-intensive process<br>Lack of distribution network<br>Hydrogen storage challenges |

*Sources:* I. Kubiszewski and C. J. Cleveland, Energy return on investment (EROI) for wind energy, in *Encyclopedia of Earth,* ed. C. J. Cleveland, 2008; C. Hall, Why EROI matters (Part 1 of 6), *The Oil Drum,* April 1, 2008, http://www.theoildrum.com/node/3786.

| Emissions (pollutants and greenhouse gases) | Electricity cost ($/kWh) | Energy return on energy investment |
| --- | --- | --- |
| $CO_2$ and methane | | 1.3 (from corn) 8 (from sugarcane) |
| Carbon monoxide Particulate matter Nitrogen oxides Possible toxic metals from MSW Danger of indoor air pollution | | |
| None during operation Some pollution generated during manufacturing of panels | 0.2 | 8 |
| None during operation Some pollution generated during manufacturing of panels | | |
| Methane from decaying flooded vegetation | 0.05–0.11 | 12 |
| None during operation | | 15 |
| None during operation | 0.05–0.30 | 8 |
| None during operation | 0.04–0.06 | 18 |
| None during operation | | 8 |

distribution network that accepts any source of electricity and distributes it automatically to end users. A smart grid uses computer programs and the Internet to tell electricity generators when electricity is needed and electricity users when there is excess capacity on the grid. In this way, it coordinates energy use with energy availability. In late 2009, President Obama announced a plan to invest $3.4 billion in smart grid technology.

How does a smart grid work? FIGURE 13.26 shows one example. With "smart" appliances plugged into a smart grid, at bedtime, a consumer could set an appliance such as a dishwasher to operate before he wakes up. A computer on the dishwasher would be programmed to run it anytime between midnight and 5:00 AM, depending on when there is a surplus of electricity. The dishwasher's computer would query the smart grid and determine the optimal time, in terms of electricity availability, to turn the appliance on. The smart grid could also help manage electricity demand so that peak loads do not become too great. We cannot control the timing of all electricity demand, but by improving consumer awareness of electricity abundance and shortages, using smart appliances, and setting variable pricing for electricity, we can make electricity use much more regular, and thus more sustainable.

Our current electrical infrastructure relies on a system of large energy producers—regional electricity generation plants. When one plant goes off-line (shuts down), the reduction in available generating capacity puts greater demands on the rest of the system. Some energy experts maintain that a better system would consist of a large number of small-scale electricity generation "parks" that rely on a mix of fossil fuel and renewable energy sources. These experts maintain that a system of decentralized energy parks would be the least expensive and most reliable electrical infrastructure for our future needs. Small, local energy parks would save money and energy by transporting electricity a shorter distance. In addition, decentralized generators would be safest from both breakdowns and sabotage: each small energy park might serve only a few thousand people, so widespread outages would be much less likely.

## Addressing Energy Cost and Storage

The major impediments to widespread use of wind, solar, and tidal energy—the forms of renewable energy with the least environmental impact—are cost and the limitations of energy storage technology. Fortunately, the cost of renewable energy has been falling, to the point at which wind energy is cost-competitive with natural gas and coal in some markets. Throughout this book we have seen that the efficiency of making something improves with technological advances and experience. In general, as we produce more of something—as we get experience making it—we learn to produce it less expensively. Production processes become dramatically

FIGURE 13.26 **A smart house.**
A smart grid optimizes the use of energy in a home by continuously coordinating energy use with energy availability.

Computer

Smart thermostat

Electricity

Smart meter

High-speed Internet connection

Smart appliances

more efficient, more companies enter the market, and developing new technologies has a clear payoff. For the consumer, this technological advancement also has the benefit of lowering prices: solar, wind, and natural gas electricity generation demonstrate that costs tend to decline in a fairly regular way as installed capacity grows.

What are the implications of this relationship between experience and efficiency? Any technology in widespread use has an advantage over a new technology because the less expensive something is, the more people will buy it, leading to further reductions in its price. State and federal subsidies and tax incentives also help to lower the price of a technology. Tax credits and rebates have been instrumental in reducing the cost of solar and wind energy systems for consumers.

Similarly, in time, researchers will develop solutions to the problem of creating efficient energy storage systems, which might reduce the need to transport electricity over long distances. One very simple and effective approach is to use excess capacity (during off-peak hours) to pump water uphill to a reservoir. Then, during hours of peak demand, the water can be released through a turbine to generate the necessary electricity—cleanly and efficiently. Research into battery technology and hydrogen fuel cell technology continues.

Progress on these and other technologies may accelerate with government intervention, taxes on industries that emit carbon dioxide, or a market in which consumers are willing to pay more for technologies with minimal environmental impacts. In the immediate future, we are more likely to move toward a sustainable energy mix if nonrenewable energy becomes more expensive. Consumers have shown more willingness to convert in large numbers to renewable energy sources, or to engage in further energy conservation, when fossil fuel prices increase. We have already seen instances of this shift in behavior. In 2008, energy conservation increased when oil prices rose rapidly to almost $150 per barrel and gasoline in most of the United States cost more than $4.00 per gallon. People used public transportation more often, drove more fuel-efficient vehicles, and carpooled more than they did before the price spike.

Other ways to spur conservation are initiatives that regulate the energy mix itself—for example, by requiring that a certain fraction of electricity be generated using renewable energy sources. One such initiative is the Regional Greenhouse Gas Initiative (RGGI), whereby 10 eastern states have committed to reducing greenhouse gas emissions from electricity generation plants by 10 percent by 2018.

**GAUGE YOUR PROGRESS**

✓ What are the barriers to increasing our use of renewable energy sources?

✓ What are some of the ways that we are working to overcome these barriers?

✓ Consider the idea of a sustainable energy strategy for the United States. What factors must be considered for a successful strategy? How achievable do you think it is in the next decade? Explain your answer.

The people of Iceland use more energy per capita than the people of any other nation, including the United States. However, the energy Icelanders use is almost all in the form of local, renewable resources that do not pollute or contribute greenhouse gases to the environment.

## Building an Alternative Energy Society in Iceland

This isolated European island nation has had to learn to be self-sufficient or suffer the high cost of importing fuel. When the Vikings first came to Iceland over a thousand years ago, they relied on biomass, in the form of birch wood and peat, for fuel. The resulting deforestation, and the slow regrowth of forests in Iceland's cold temperatures and limited sunlight, restricted human population growth and economic development for the next thousand years. With the beginning of the Industrial Revolution in the eighteenth century, Iceland began to supplement its biomass fuel with imported coal, but the expense of importing coal also limited its economic growth.

In the late nineteenth and early twentieth centuries, Iceland began to look to its own resources for energy. Soon it had tapped its abundant freshwater resources to generate hydroelectricity—the country's major energy source for residential and industrial use. This transition led to the general electrification and economic modernization of the country and greatly reduced its dependence on imported fossil fuels. Iceland did not stop with hydroelectricity, however, but sought ways to utilize its other major renewable energy source: the thousands of geysers and hot springs on this volcanic island that provide ready access to geothermal energy.

Geothermal energy is now the primary energy source for home heating in Iceland, and geothermal and hydroelectric resources provide energy for nearly all of the country's electricity generation. Even so, the potential of these resources remains relatively underutilized. Iceland has harvested only about 17 percent of its hydroelectric potential, and there is even more potential in its geothermal resources.

Despite its commitment to renewable energy, Iceland is still dependent on imported fossil fuels to run its cars, trucks, buses, and fishing vessels. In a further push for energy independence, Iceland is now developing new technologies that could harness another vast and clean source of energy: hydrogen. In 2000, in partnership with the automobile company DaimlerChrysler, the European energy giant Norsk Hydro, and the oil company Royal Dutch Shell, Iceland embarked on an ambitious project that could wean the country from fossil fuels by 2050. The plan is to use its sustainably generated electricity to split water and obtain hydrogen, then use that hydrogen as fuel.

In April 2003, Iceland opened one of the world's first filling stations for hydrogen-fueled vehicles. In August of that year, hydrogen-fueled buses for public transportation were introduced in the country's capital, Reykjavik (**FIGURE 13.27**). Hydrogen-fueled rental cars are also available. Although the global banking crisis of 2008 changed near-term projections significantly, Iceland could have the world's first hydrogen-fueled ocean fishing fleet by 2030. Iceland could also become a net exporter of clean hydrogen energy. Iceland is an example of an ongoing experiment in taking advantage of one's resources to develop clean, sustainable energy sources.

### References

Blanchette, S. 2008. A hydrogen economy and its impact on the world as we know it. *Energy Policy* 36:522–530.

Maack, M. H., and J. B. Skulason. 2006. Implementing the hydrogen economy. *Journal of Cleaner Production* 14:52–64.

**FIGURE 13.27** **A hydrogen-fueled bus in Reykjavik, Iceland.** All 80 buses in the Reykjavik fleet were converted to hydrogen fuel cells in 2003.

■ **Define renewable energy resources.**

Renewable energy resources include nondepletable energy resources, such as the Sun, wind, and moving water, and potentially renewable energy resources, such as biomass. Potentially renewable energy resources will be available to us as long as we use them sustainably.

■ **Describe strategies to conserve energy and increase energy efficiency.**

Turning down the thermostat and driving fewer miles are examples of steps individuals can take to conserve energy. Buying appliances that use less energy and switching to compact fluorescent light bulbs are examples of steps individuals can take to increase energy efficiency. Buildings that are carefully designed for energy efficiency can save both energy resources and money. Reducing the demand for energy can be an equally effective or more effective means of achieving energy sustainability than developing additional sources of energy.

■ **Compare and contrast the various forms of biomass energy.**

Biomass is one of the most common sources of energy in the developing world, but biomass energy is also used in developed countries. In theory, biomass energy is carbon neutral; that is, the carbon produced by combustion of biomass should not add to atmospheric carbon concentrations because it comes from modern, rather than fossil, carbon sources. Wood is a potentially renewable resource because, if harvests are managed correctly, it can be a continuous source of biomass energy. Ethanol and biodiesel have the potential to supply large amounts of

renewable energy, but their growing and processing make demands on land and energy resources.

■ **Explain the advantages and disadvantages of hydroelectricity, solar energy, geothermal energy, wind energy, and hydrogen as energy resources.**

Most hydroelectric systems use the energy of water impounded behind a dam to generate electricity. Run-of-the-river hydroelectric systems impound little or no water and have fewer environmental impacts, although they often produce less electricity. Active solar technologies have high initial costs but can potentially supply large amounts of energy. Geothermal energy can heat buildings directly or be used to generate electricity. However, geothermal power plants can be located only in places where geothermal energy is accessible. Wind is a clean, nondepletable energy resource, but people may object to wind farms on the basis of aesthetics or the hazards they pose to birds and bats. The only waste product from a hydrogen fuel cell is water, but obtaining hydrogen gas for use in fuel cells is an energy-intensive process. If hydrogen could be obtained using renewable energy sources, it could become a truly renewable source of energy.

■ **Describe the environmental and economic options we must assess in planning our energy future.**

Although many scenarios have been predicted for the world's energy future, conserving energy, increasing energy efficiency, a greater reliance on renewable energy sources, and new technologies to improve energy distribution and storage will all be necessary for achieving energy sustainability.

1. Which of the following is *not* an example of a potentially renewable or nondepletable energy source?
   (a) Hydroelectricity
   (b) Solar energy
   (c) Nuclear energy
   (d) Wind energy
   (e) Geothermal energy

2. Renewable energy sources are *best* described as
   (a) those that are the most cost-effective and support the largest job market.
   (b) those that are, or can be, perpetually available.
   (c) those that are dependent on increasing public demand and decreasing supply.
   (d) those that are being depleted at a faster rate than they are being replenished.
   (e) those that are reusable, and therefore eliminate waste energy released into the environment.

3. An energy-efficient building might include all of the following *except*
   (a) building materials with low thermal inertia.
   (b) a green roof.
   (c) southern exposure with large double-paned windows.
   (d) reused or recycled construction materials.
   (e) photovoltaic solar cells as a source of electricity.

4. Which of the following sources of energy is *not* (ultimately) solar-based?
   (a) Wind
   (b) Biomass
   (c) Tides
   (d) Coal
   (e) Hydroelectricity

5. Which of the following demonstrate(s) the use of passive solar energy?
   I   A south-facing room with stone walls and floors
   II  Photovoltaic solar cells for the generation of electricity
   III A solar oven

   (a) I only
   (b) II only
   (c) III only
   (d) I and III
   (e) II and III

6. A study of small wind turbines in the Netherlands tested the energy output of several models. The results for two models are shown in the graphs that follow.

Which of the following statements can be inferred from these data?
   (a) As wind speed increases, energy output decreases.
   (b) The annual energy output of model 1 can exceed 6,000 kWh.
   (c) As energy output surpasses 50 kWh per week, noise pollution increases.
   (d) Model 2 is likely to cause more bird and bat deaths.
   (e) Model 1 is more cost-effective.

**(a)** Weekly energy output

**(b)** Annual energy output

7. The primary sources of renewable energy in the United States are
   (a) solar and wind energy.
   (b) hydroelectricity and tidal energy.
   (c) biomass and hydroelectricity.
   (d) geothermal and tidal energy.
   (e) wind and geothermal energy.

8. The environmental impacts of cutting down a forest to obtain wood as fuel for heating and cooking could include
   I   deforestation and subsequent soil erosion.
   II   release of particulate matter into the air.
   III   a large net rise in atmospheric concentrations of sulfur dioxide.

   (a) I only
   (b) II only
   (c) III only
   (d) I and II
   (e) II and III

Use the diagram below, which represents annual U.S. energy consumption by source and sector for 2007, to respond to questions 9 and 10.

9. Which of the following statements *best* describes the *sources* of energy in U.S. energy consumption patterns?
   (a) Most of the renewable energy is used in the industrial, residential, and commercial sectors.
   (b) Most of the electricity generated in the United States comes from nuclear energy.
   (c) The industrial sector is heavily dependent on coal and renewable energy.
   (d) Fossil fuels continue to be the major energy source for all sectors.
   (e) The transportation sector uses the greatest amount of energy.

10. Which of the following *best* describes U.S. energy *use?*
    (a) Transportation is the largest end use of energy in the United States.
    (b) Transportation is fueled mainly by coal.
    (c) Electricity generation is the largest end use of energy in the United States.
    (d) Electricity generation is powered mainly by nuclear energy.
    (e) Industry is the largest end use of energy in the United States.

11. Which of the following statements *best* describes the role of renewable energy in the United States?
    (a) It is the dominant source of energy.
    (b) It is the largest contributor of greenhouse gases.
    (c) It is a large contributor to the transportation sector.
    (d) Its largest contribution is to the electricity generation sector.
    (e) It is never sustainable.

12. In order to best achieve energy sustainability, humans must consider which of the following strategies?
    I   Building large, centralized power plants
    II   Improving energy efficiency
    III   Developing new energy technologies

    (a) I only          (d) I and II
    (b) II only         (e) II and III
    (c) III only

## APPLY THE CONCEPTS

1. Hydroelectricity provides about 7 percent of the electricity generated in the United States.
   (a) Explain how a hydroelectric power plant converts energy stored in water into electricity.
   (b) Identify *two* factors that determine the amount of electricity that can be generated by an individual hydroelectric power plant.
   (c) Describe the *two* main types of land-based hydroelectric power plants.
   (d) Describe *two* economic advantages and *two* environmental disadvantages of hydroelectricity.

2. The following table shows the amounts of electricity generated by photovoltaic solar cells and by wind in the United States from 2002 to 2007 (in thousands of megawatt-hours):

| Energy source | Year | | | | | |
|---|---|---|---|---|---|---|
| | 2002 | 2003 | 2004 | 2005 | 2006 | 2007 |
| PV solar cells | 555 | 534 | 575 | 550 | 508 | 606 |
| Wind | 10,354 | 11,187 | 14,143 | 17,810 | 26,589 | 32,143 |

(a) Describe the trend in electricity generation by photovoltaic solar cells from 2002 to 2007. Calculate the approximate percentage change between 2002 and 2007.
(b) Describe the trend in electricity generation by wind from 2002 to 2007. Calculate the approximate percentage change between 2002 and 2007.
(c) Identify and explain any difference between the two trends you described in (a) and (b).
(d) A homeowner wants to install either photovoltaic solar cells or wind turbines to provide electricity for her home in Nevada, which gets ample sunlight and wind. Provide two arguments in favor of installing one of these technologies, and explain two reasons for not choosing the other.
e) Would the installation of either PV solar cells or wind turbines be considered the application of energy conservation or energy efficiency? Explain.

## MEASURE YOUR IMPACT

**Choosing a Light Bulb**  The following table provides information on two types of light bulbs available at a local store.

| Type of bulb | Estimated lifetime (hours) | Cost per bulb ($) |
|---|---|---|
| 75 W incandescent | 750 | 0.50 |
| 18 W compact fluorescent | 7,500 | 5.00 |

Assume that
   (i) each bulb lasts exactly its estimated lifetime.
   (ii) each bulb provides the same amount of usable light.
   (iii) the cost of electricity is $0.10 per kilowatt-hour.

(a) Calculate the total lifetime cost for using each type of bulb.
(b) Calculate the total cost of using the incandescent bulb and of using the compact fluorescent bulb for 7,500 hours (the lifetime of the compact fluorescent bulb).
(c) Most electricity in the United States is generated by fossil fuel–burning power plants that produce atmospheric pollution. Given that knowledge, explain which type of light bulb has the smaller ecological footprint.

## ENGAGE YOUR ENVIRONMENT

How energy efficient is your home? Explore the interior and exterior of your home for energy efficiency, and determine ways to address energy losses. *See Engage Your Environment Activity 13 at the end of this book.*

# science applied

## Should Corn Become Fuel?

Corn-based ethanol is big business—so big, in fact, that to offset demand for petroleum, U.S. policy calls for an increase in annual ethanol production from 34 billion liters (9 billion gallons) in 2008 to 57 billion liters (15 billion gallons) by 2015.

Ethanol proponents maintain that substituting ethanol for gasoline decreases air pollution, greenhouse gas emissions, and our dependence on foreign oil. Opponents counter that growing corn and converting it into ethanol uses more energy than we recover when we burn the ethanol for fuel. Perhaps more importantly, those energy inputs release additional carbon into the atmosphere. Are the policies that encourage corn ethanol production firmly grounded in science, or are they a concession to politically powerful farming interests?

### Does ethanol reduce air pollution?

Ethanol ($C_2H_6O$) and gasoline (a mixture of several compounds, including heptane: $C_7H_{16}$) are both hydrocarbons. Under ideal conditions, in the presence of enough oxygen, burning hydrocarbons produces only water and carbon dioxide. In reality, however, gasoline-only vehicles always produce some carbon monoxide (CO) because there can be insufficient oxygen present at the time of combustion. Carbon monoxide has direct effects on human health and also contributes to the formation of photochemical smog (see Chapter 15 for more on CO and air pollution). Modern car engines regulate the fuel/oxygen mix to maximize the combustion of gasoline and minimize CO production. However, these regulatory systems are not fully operational until a car has "warmed up." As a result, whenever you start your engine, you release some CO. Many older cars do not have these regulators at all.

Because ethanol is an **oxygenated fuel**—a fuel with oxygen as part of the molecule—adding ethanol to a car's fuel mix should ensure that more oxygen is present, and thus that combustion is more complete, reducing CO production. However, an ethanol/gasoline blend evaporates more readily than pure gasoline. In the atmosphere, evaporated fuel produces photochemical smog in the same way that CO does. In other words, the fuel may burn more cleanly, but the extra evaporation may counteract some of this benefit.

### Does ethanol reduce greenhouse gas emissions?

Biofuels are modern carbon, not fossil carbon. In theory, burning biofuels should not introduce additional carbon into the atmospheric reservoir because the carbon captured in growing the crops and the carbon released in burning the fuel should cancel each other out. However, agriculture in the United States depends heavily on inputs of fossil fuels to grow and process crops. Does the argument that ethanol use is carbon neutral hold up when we examine the entire ethanol production cycle from farm to filling station?

To analyze this question further, we need to examine ethanol's energy return on energy investment (EROEI), or how much energy we get out of ethanol for every unit of energy we put in.

Scientists at the U.S. Department of Agriculture have analyzed this problem, examining the energy it takes to grow corn and convert it into ethanol (the inputs) and the return on this energy investment (the outputs). Energy to run farm machinery, to produce chemicals (especially nitrogen fertilizer), and to dry the corn are inputs. Additional energy is required to transport the corn, convert it into ethanol, and ship it. Ethanol is the primary output, but in the course of growing and processing the corn, several by-products are produced, including distiller's grains, corn gluten,

Outputs

Producing by-products from scratch would require 5,390,000 kJ, so credit is given for these.

Inputs

Growing 0.4 ha (1 acre) of corn requires 8,545,000 kilojoules (kJ).

Break-even point

1,420 L (375 gallons) of ethanol are produced, containing 33,217,000 kJ of energy (a surplus of 3,036 kJ).

Converting corn into ethanol and distributing it requires 21,636,000 kJ.

30,181,000 kJ        38,607,000 kJ

Net energy return on energy investment = 1.3

$$\left( \frac{38,607,000}{30,181,000} \right)$$

**FIGURE SA6.1 Producing ethanol requires energy.** An analysis of the energy costs of growing and converting 0.4 hectares (1 acre) of corn into ethanol shows a slight gain of usable energy when corn is converted into ethanol.

and corn oil. Each by-product would have required energy to produce, had they been manufactured independently of the ethanol manufacturing process, and so energy "credit" is assigned to these by-products. As FIGURE SA6.1 shows, there is a slight gain of usable energy when corn is converted into ethanol: for every unit of fossil fuel energy we put in, we produce about 1.3 units of ethanol. A land area of 0.4 hectares (1 acre) produces enough net energy to displace about 236 liters (62 gallons) of gasoline.

Our analysis indicates that replacing part of our gasoline needs with ethanol will displace some greenhouse gases. There will be a climate "benefit," but perhaps not to the extent that one would imagine, because we rely so heavily on fossil fuels to support agriculture. Moreover, in the United States, the ethanol production process currently uses more coal than natural gas. Because coal emits nearly twice as much $CO_2$ per joule of energy as natural gas (see Chapter 12), producing the ethanol may reverse many of the benefits of replacing the fossil carbon in gasoline with the modern carbon in ethanol. Quite possibly, producing ethanol with coal releases as much carbon into the atmosphere as simply burning gasoline in the first place.

Finally, we have to take into account the corn-growing process itself. Various aspects of corn production, such as plowing and tilling, may release

additional $CO_2$ into the atmosphere from organic matter that otherwise would have remained undisturbed in the A and B horizons of the soil. Furthermore, greater demand for corn will increase pressure to convert land that is forest, grassland, or pasture into cropland. There is increasing evidence from recent studies that these conversions result in a net transfer of carbon from the soil to the atmosphere. This transfer leads to additional increases in atmospheric $CO_2$ concentrations.

## Does ethanol reduce our dependence on gasoline?

If there is a positive energy return on energy investment, then using ethanol should reduce the amount of gasoline we use and therefore the amount of foreign oil we must import. However, if in order to use less gasoline we use more coal to produce ethanol, we might increase the release of greenhouse gases, so the overall benefit is questionable. Furthermore, if we create greater demand for a crop that until now has been primarily a food source, there are other implications as well.

Replacing a significant fraction of U.S. gasoline consumption with corn-based ethanol would involve the large-scale conversion of cropland from food to fuel production. Even if we converted every acre of potential cropland to ethanol production, we could not produce

**FIGURE SA6.2** **Rising food prices.** Increased demand for corn-based ethanol accounted for about 10 to 15 percent of the rise in food prices from April 2007 to April 2008, according to the Congressional Budget Office.

enough ethanol to displace more than 20 percent of U.S. annual gasoline consumption. Furthermore, all agricultural products destined for the dinner table would have to be imported from other countries. Clearly this is not a practical solution.

What seems more likely is that we will be able to replace some smaller fraction of gasoline consumption with biofuels. Lester Brown of the Earth Policy Institute points out, however, that the 10 bushels of corn that it takes to produce enough ethanol to fill a 95-liter (25-gallon) SUV fuel tank contain the number of calories needed to feed a person for about a year. He argues that higher ethanol demand would increase corn and other grain prices and would thus make it harder for lower-income people around the world to afford food (FIGURE SA6.2).

Indeed, in the summer of 2007, corn prices in the United States rose to $4 per bushel, roughly double the price in prior years, primarily because of the increased demand for ethanol. Since then, prices have stayed above $3 per bushel. People in numerous countries have had difficulty obtaining food because of higher grain prices. In 2008, there were food riots in Afghanistan, Bangladesh, Egypt, and Haiti, among other places around the globe.

Our analysis indicates that corn-based ethanol has the potential to displace a small amount of total U.S. gasoline consumption, but that increasing corn ethanol consumption to the levels suggested by some politicians may require troublesome trade-offs between driving vehicles and feeding the world.

## Are there alternatives to corn ethanol?

Stimulating demand for corn ethanol may spur the development of another ethanol technology: **cellulosic ethanol.** Cellulose is the material that makes up plant cell walls: grasses, trees, and plant stalks are made primarily of cellulose. If we were able to produce large quantities of ethanol from cellulose, we could replace fossil fuels with fuel made from a number of sources. Ethanol could be manufactured from fast-growing grasses such as switchgrass or *Miscanthus* (FIGURE SA6.3), tree species that require minimal energy input, and many waste products, including discarded paper and agricultural waste. It is also possible that algae could be used as the primary material for ethanol.

Producing cellulosic ethanol requires breaking cellulose into its component sugars before distillation. This is a difficult and expensive task because the bonds between the sugar molecules are very strong. One method of breaking down cellulose is to mix it with enzymes that sever these bonds. In 2007, the first commercial cellulosic ethanol plant was built in Iowa. At the moment, however, cellulosic ethanol is more expensive to produce than corn ethanol.

How much land would it take to produce significant amounts of cellulosic ethanol? Some scientists suggest that the impact of extensive cellulosic ethanol production would be very large, while others have calculated that, with foreseeable technological improvements, we could replace all of our current gasoline consumption without large increases in land under cultivation or significant losses in food production. Because the technology is so new, it is not yet clear who is correct. There will still be other considerations, such as the impact on biodiversity whenever land is dedicated to growing biofuels.

The good news is that many of the raw materials for cellulosic ethanol are perennial crops such as grasses. These crops do not require the high energy, fertilizer, and water inputs commonly used to grow annuals such as corn. Furthermore, the land used to grow grass

**FIGURE SA6.3** **A potential source of cellulosic ethanol.** *Miscanthus,* a fast-growing tall grass, may be a source of ethanol in the future.

would not need to be plowed every year. Fertilizers and pesticides would also be unnecessary, eliminating the large inputs of energy needed to produce and apply them.

Algae may be an even more attractive raw material for cellulosic ethanol because its production would not need to utilize land that could otherwise be used for growing food crops or serve as a repository for carbon.

Many forms of ethanol are now being produced or are very close to being produced commercially. Production methods that use biomass grown with fewer fossil fuel inputs, at a lower level of intensity on agricultural land, or without the use of land at all are the most desirable from virtually all environmental perspectives.

## Summarizing the science

Adding ethanol to gasoline reduces carbon monoxide formation and may reduce photochemical smog if evaporation can be controlled. When we look at the entire "life cycle" of ethanol, from farm to tank, it's not clear whether ethanol will reduce greenhouse gas emissions. Because its EROEI is fairly low, and because the amount of available agricultural land is limited, corn ethanol will not replace more than a small fraction of gasoline consumption. Cellulosic ethanol shows the potential to have a significant effect on fossil fuel use, at least in part because of lower energy inputs to obtain the raw material and convert it into a fuel. Corn should not be used for fuel to any great extent, but it may serve as a stepping-stone to other biofuels.

## References

Farrell, A. E., R. J. Plevin, B. T. Turner, A. Jones, M. O'Hare, and D. M. Kammen. 2006. Ethanol can contribute to energy and environmental goals. *Science* 311:506–508.

Searchinger, T. D., et al. 2009. Fixing a critical climate accounting error. *Science* 326:527–528.

# Water Pollution: Causes and Effects

## The Chesapeake Bay

The Chesapeake Bay is the largest estuary in the United States and it faces some of the biggest environmental challenges. Located between Virginia, Maryland, and Delaware, the Chesapeake Bay receives fresh water from numerous rivers and streams that mixes with the salt water of the ocean to produce an extremely productive estuary. Many of the rivers and streams that dump into the bay travel long distances and

Increased sediments are also an issue in the Chesapeake Bay. Sediments are soils washed away from fields and forests as well as soils washed away from the banks of streams and the ocean shoreline. The current estimate is that 8.2 billion kg (18.7 billion pounds) of sediments come into the bay each year. The tiniest soil particles stay suspended in the water, make the water cloudy, and prevent sunlight from reaching the grasses that have been historically abundant in the bay.

> Anthropogenic chemicals appeared to be responsible for the discovery that 82 percent of male smallmouth bass developed into hermaphrodites with male sex organs that grow female eggs.

drain water from a large watershed of urban, suburban, and agricultural areas. Indeed, the watershed that supplies the bay extends from northern Virginia all the way up into central New York State. One of the consequences of receiving water from such a large watershed is that the water coming into the bay contains an abundance of nutrients, sediments, and chemicals.

Among the major nutrients that enter the bay are 272 million kg (600 million pounds) of nitrogen and 14 million kg (30 million pounds) of phosphorus. The nitrogen and phosphorus come from three major sources. The first source is water discharged from sewage treatment facilities that carries high amounts of nutrients from human waste. The second source is animal waste produced by concentrated animal feeding operations generating large amounts of manure that can make its way into nearby streams and rivers. The third source is fertilizer that is spread on both agricultural fields and suburban lawns. Much of this fertilizer leaches out of the soil and into local streams, eventually making its way to the Chesapeake Bay. When these nutrients reach the bay, the algae in the bay experience an explosive population growth that is known as an algal bloom.

These grasses are important because they serve as a habitat for fish and blue crabs (*Callinectes sapidus*).

The Chesapeake Bay also receives inputs of anthropogenic chemicals. Many of these are pesticides that are sprayed throughout the watershed for growing crops and controlling pests. They can arrive in the bay by direct application over water, by running off the surface of the land when it rains, or by being carried by the wind immediately after application. The bay also contains pharmaceutical drugs that pass through the human body and enter sewage treatment plants, eventually discharging into the streams and rivers that feed the bay. In 2009, the U.S. Fish and Wildlife Service announced that anthropogenic chemicals appeared to be responsible for the discovery that 23 percent of male largemouth bass and 82 percent of male smallmouth bass developed into hermaphrodites with male sex organs that grow female eggs. This impact is not only a concern for fish, ▶

The Chesapeake Bay watershed extends from New York down to Virginia.

◀ **The Chesapeake Bay watershed.** The Chesapeake Bay receives water from a very large geographic area that carries sediments, nutrients, and chemicals from a variety of urban, suburban, and agricultural areas.

but also a concern for humans who share many similarities in their endocrine systems.

The enormous size of the Chesapeake Bay watershed means that cleaning it up will require a monumental effort. In 2000 the surrounding states formed a partnership along with multiple federal departments to develop the Chesapeake Bay Action Plan. This plan outlines a series of goals to reduce the impacts of nutrients, sediments, and chemicals coming into the bay. In 2010 the governors of the surrounding states, along with many local leaders, announced that many of the Action Plan's goals were being met, including a reduction in nitrogen, an increase in water clarity, and an increase in the number of blue crabs. In fact, the improved water quality combined with earlier reductions in the number of blue crabs that could be harvested allowed the blue crab population to increase 60 percent from 2009 to 2010. This represents the largest crab population in the bay in 13 years. The story of Chesapeake Bay serves as an excellent example of the wide variety of pollutants that can impact aquatic ecosystems and of effective and substantial efforts that can only be made when all parties work together toward a common goal. ■

Sources: Chesapeake Bay Program: A watershed partnership, http://www.chesapeakebay.net/index.aspx?menuitem=13853; U.S. Fish and Wildlife Service, News Release, April 22, 2009, http://www.fws.gov/ChesapeakeBay/pdf/IntersexPR.pdf.

## UNDERSTAND THE KEY IDEAS

Water is a key resource for life on Earth. All organisms, including humans, require water to live, but growing human populations combined with industrialization have led to the contamination of water supplies. Water contamination has a wide variety of causes. The consequences for people and ecosystems can be severe.
After reading this chapter you should be able to

- distinguish between point and nonpoint sources of pollution.
- identify the ways in which human wastewater can cause water pollution.
- evaluate the different technologies that humans have developed for treating wastewater.
- identify the major types of heavy metals and other substances that pose serious hazards to humans and the environment.
- discuss the impacts of oil spills and how such spills can be remediated.
- identify contaminants that are nonchemical pollutants.
- explain the connections among industrialization, affluence, and water-pollution legislation.

# Pollution can come from specific sites or broad areas

**Water pollution** is generally defined as the contamination of streams, rivers, lakes, oceans, or groundwater with substances produced through human activities and that negatively affect organisms. As we will see, this broad definition encompasses a wide range of substances that vary in their sources, prevalence, and impact. Although this chapter focuses on the contaminants that exist in water and their effects on aquatic organisms and humans, it is important to remember that there are many ecological connections between aquatic and terrestrial ecosystems. As a result, water pollution has the potential to impact both aquatic and terrestrial organisms.

Regardless of the specific contaminant, pollution can come from either *point sources* or *nonpoint sources* (**FIGURE 14.1**). **Point sources** are distinct locations such as a particular factory that pumps its waste into a nearby stream or a sewage treatment plant that discharges its wastewater from a pipe into the ocean. **Nonpoint sources** are more diffuse areas such as an entire farming region, a suburban community with many lawns and septic systems, or storm runoff from parking lots. It is important to differentiate between the two types of pollution sources because the distinction can help in controlling pollutant inputs to waterways. For example, if a municipality determines that the bulk of water pollution is coming from one or two point sources, it can target those specific sources to reduce their pollution output. It can be more difficult to control pollution from nonpoint sources. As we discussed in Chapter 10, pollution represents an externality because the cost to the environment is not reflected in the cost of making the products that generate pollution.

The broad range of pollutants that can be found in water includes human and animal waste, inorganic substances, organic compounds, synthetic organic compounds, and nonchemical pollutants. For each of these pollutant groups, we need to consider where the pollutant comes from, its negative effects on humans and the environment, and what can be done to reduce these effects.

**(a)**

**(b)**

**FIGURE 14.1** **Two types of pollution sources.** Pollution can enter water bodies either as (a) point sources, as in sewage pipes, or as (b) nonpoint sources, as in rainwater that runs off hundreds of square kilometers of agricultural fields and into streams.

**GAUGE YOUR PROGRESS**

✓ What is water pollution? Why is it important to learn about water pollution?

✓ What are point and nonpoint sources? How do they differ?

✓ What are the most common types of pollutants in the water?

# Human wastewater is a common pollutant

Human **wastewater** is the water produced by human activities including human sewage from toilets and gray water from bathing and washing clothes and dishes. For centuries, one of the biggest challenges has been to keep human wastewater from contaminating human drinking water. This can be difficult because throughout the world many people routinely use the same water source for drinking, bathing, washing, and disposing of sewage (**FIGURE 14.2**). Environmental scientists are concerned about human wastewater as a pollutant for three major reasons. First, wastewater dumped into bodies of water naturally undergoes decomposition by bacteria, which creates a large demand for oxygen in the water. Second, the nutrients that are released from

**FIGURE 14.2** **Washing clothes in the Tuo River of China.** Using water for bathing, washing, and disposing of sewage without contaminating sources of drinking water is a long-standing challenge.

wastewater decomposition can make the water more fertile. Third, wastewater can carry a wide variety of disease-causing organisms.

## Oxygen demand

**Oxygen-demanding waste** is organic matter that enters a body of water and feeds the growth of the microbes that are decomposers. Because these microbes require oxygen to decompose the waste, the more waste that enters the water, the more the microbes grow and the more oxygen they demand. Oxygen-demanding waste is measured in terms of **biochemical oxygen demand,** or **BOD**—the amount of oxygen a quantity of water uses over a period of time at a specific temperature. Lower BOD values indicate that a water body is less polluted by wastewater, whereas higher BOD values indicate that a water body is more polluted by wastewater. If we were to test the BOD of natural waters over a five-day period in a liter of water, for example, we might find a BOD of 5 to 20 mg of oxygen coming from the decomposition of leaves, twigs, and perhaps a few dead organisms. In contrast, domestic wastewater might have a BOD of 200 mg of oxygen.

When bodies of water have a high oxygen demand due to microbial decomposition, the amount of oxygen remaining for other organisms can be very low. Low oxygen concentrations are lethal to many organisms including fish. Low oxygen can also be lethal to organisms that cannot move, such as many plants and shellfish. In some areas, there is so little oxygen, and therefore so little life, that we call such areas **dead zones.** Such dead zones can be self-perpetuating, with the dying organisms subsequently decomposing and causing continued oxygen demand by microbes.

## Nutrient release

The oxygen required to decompose large amounts of wastewater is clearly an important factor in water pollution. We also have to think about the products of decomposition, which include nitrogen and phosphorus. Additional sources of these nutrients include soaps and detergents. As you may recall from Chapter 3, nitrogen and phosphorus are generally the two most important elements for limiting the abundance of producers in aquatic ecosystems. The decomposition of wastewater—because it adds these elements—can provide an abundance of fertility to a water body, a phenomenon known as **eutrophication.** As noted in the chapter opener, the Chesapeake Bay experiences this problem of nutrients from wastewater decomposition as well as from nutrients that are leached from agricultural lands during periods of precipitation. When a body of water experiences an increase in fertility due to anthropogenic inputs of nutrients, it is called **cultural eutrophication.**

Eutrophication initially causes a rapid growth of algae, known as an algal bloom. As we discussed in Chapter 3, this enormous amount of algae eventually dies, microbes rapidly begin digesting the dead algae, and the increase in microbes consumes most of the oxygen in the water (see Figure 3.14 on page 73). In short, the release of nutrients from wastewater initiates a chain of events that eventually leads to a lack of oxygen and the creation of dead zones once again. One of the most impressive dead zones in the world occurs where the Mississippi River dumps into the Gulf of Mexico. The Mississippi River receives water from 41 percent of the land of the continental United States. Each summer there is an influx of wastewater and fertilizer that causes large algal blooms followed by substantial decreases in

**(a)** Oxygen concentrations in Gulf Coast waters

**(b)** A fish die-off in Florida

**FIGURE 14.3** **Dead zone.** When raw sewage is dumped directly into bodies of water, subsequent decomposition by microbes can consume nearly all of the oxygen in the water and cause a dead zone to develop. Shown here are (a) oxygen concentrations in Gulf Coast waters and (b) a massive fish die-off as a result of low oxygen conditions in Lake Trafford, Florida.

oxygenated water and massive die-offs of fish (**FIGURE 14.3**). In 2006, the United Nations estimated that there were nearly 200 dead zones caused by pollution around the world.

## Disease-causing organisms

Human wastewater can carry a variety of illness-causing viruses, bacteria, and parasites that we collectively call pathogens. Pathogens in wastewater are responsible for a number of diseases that can be contracted by humans or other organisms that come in contact with or ingest the water. These pathogens cause cholera, typhoid fever, various types of stomach flu, and diarrhea. Worldwide, the major waterborne diseases are cholera and hepatitis. Cholera, which claims thousands of lives annually in developing countries (**FIGURE 14.4**), is not common in the United States. However, hepatitis A is appearing more frequently, usually originating in restaurants that lack adequate sanitation practices. The bacterium *Cryptosporidium* has caused a number of outbreaks of gastrointestinal illness in this country. Large-scale disease outbreaks from municipal water systems are relatively rare in the United States, but they do occasionally occur. They are relatively common in many parts of the developing world.

The World Health Organization estimates that 1.1 billion people, nearly one-sixth of the world's population, do not have access to sufficient supplies of safe drinking water. In addition, half of the 3.1 million annual deaths from diarrheal diseases and malaria could be prevented with safe drinking water, proper sanitation, and proper hygiene. Approximately 42 percent of the world's population lacks access to proper sanitation and over half of these people live in China and India. In sub-Saharan Africa, only 36 percent of the people have access to improved sanitation.

Given the risk that so many pathogens pose, we need to be able to easily test whether pathogens are in our drinking water. It is not feasible to test for all of the many different pathogens that can exist in drinking water. Instead, scientists have settled on using an **indicator species,** an organism that indicates whether or not disease-causing pathogens are likely to be present. The best indicators for potentially harmful water are **fecal coliform bacteria,** a group of generally harmless microorganisms that live in the intestines of human beings and other animals. One of the most common species of fecal coliform bacteria is *Escherichia coli,* abbreviated *E. coli.* Most strains of *E. coli* live naturally in humans and are not harmful, although there are strains that can be deadly to people who are very young,

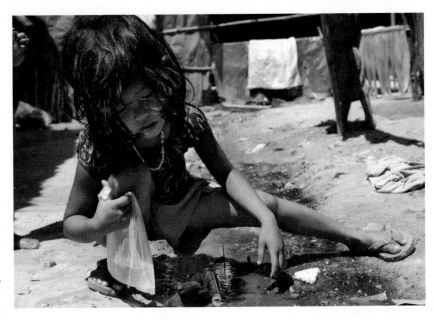

**FIGURE 14.4** **Cholera is prevalent in raw sewage.** Children who play in water contaminated by raw sewage, such as this girl in Cambodia, face a high risk of contracting the cholera pathogen.

very old, or possess weak immune systems. Given that *E. coli* is commonly found in human intestines, detecting *E. coli* in a body of water indicates that human waste has entered the water. This does not necessarily mean that the water is harmful to drink, but the presence of *E. coli* does indicate that there is an increased risk of other wastewater pathogens being in the water.

Public water supplies, such as drinking water sources and swimming pools, are routinely tested. Homeowners with a single-family well might test their water less frequently or not at all. Public health authorities recommend declaring water unsuitable for human consumption if any bacteria are present. For safe swimming and fishing, the acceptable level of *E. coli* is higher; for example, swimming at a public beach or in a river is considered safe as long as the fecal coliform bacteria levels are less than 500 to 10,000 colonies per 100 mL of water. A pool, beach, or campground with contaminated water likely would be posted with a sign such as "The Department of Health has closed this water supply because of the presence of fecal coliform bacteria."

### GAUGE YOUR PROGRESS

✓ How does high BOD influence water quality?

✓ What is a "dead zone"? How do nitrogen and phosphorus contribute to dead zones?

✓ What pathogens are common in poorly treated wastewater? Where are they the biggest threat?

# We have technologies to treat wastewater from humans and livestock

Given the importance of proper sanitation, we need ways of treating human wastewater to reduce the risk of waterborne pathogens. Humans have devised a number of ways to handle wastewater. The various solutions all have the same basic approach—bacteria break down the organic matter into carbon dioxide and inorganic compounds such as nitrate and phosphate. The two most widespread systems for treating human sewage are *septic systems* and *sewage treatment plants*. The most prevalent system to treat waste from large livestock operations is a *manure lagoon*.

## Septic systems

In rural areas of low population density, houses often have their own sewage treatment system called a **septic system.** As shown in FIGURE 14.5, this is a relatively small and simple system with two components: a *septic tank* and a *leach field*.

The **septic tank** is a large container that receives wastewater from the house. Having a capacity of 1,900 to 4,700 liters (500–1,250 gallons), the septic tank is buried underground adjacent to the house. Wastewater from the house flows into the tank at one end and leaves the tank at the other. After the tank has been operating for some time, three layers develop. Anything that will float rises to the top of the tank and forms a scum layer. Anything heavier than water sinks to form the **sludge** layer. In the middle is a fairly clear water layer called

FIGURE 14.5 **A septic system.** Wastewater from a house is held in a large septic tank where solids settle to the bottom and bacteria break down the sewage. The liquid moves through a pipe at the top of the tank and passes through perforated pipes that distribute the water through a leach field.

**septage.** The septage contains large quantities of bacteria and also may contain pathogenic organisms and inorganic nutrients such as nitrogen and phosphorus.

The septage moves out of the septic tank by gravity into several underground pipes laid out across a lawn below the surface. The combination of pipes and lawn makes up the **leach field.** The pipes contain small perforations so the water can slowly seep out and spread across the leach field. The septage that seeps out of the pipes is slowly absorbed and filtered by the surrounding soil. The harmful pathogens can settle and become part of the sludge, be outcompeted by other microorganisms in the septic tank and therefore diminish in abundance, or be degraded by soil microorganisms in the leach field. The organic matter is broken down into carbon dioxide and inorganic nutrients. Eventually, the water and nutrients are taken up by plants or enter a nearby stream or aquifer.

There are significant environmental advantages to septic systems. Because most septic systems rely on gravity—water from the house flows downhill to the septic tank, and water from the septic tank flows downhill to the leach field—no electricity is needed to run a septic system. However, sludge from the septic tank must be pumped out periodically (every five to ten years) and taken to a sewage treatment plant.

## Sewage treatment plants

Household septic systems work well for rural areas in which each house has sufficient land for a leach field. This is not a feasible solution for more developed areas with greater population densities and little open land. In developed countries, municipalities use centralized sewage treatment plants that receive the wastewater from hundreds or thousands of households via a network of underground pipes. In a traditional sewage treatment plant, wastewater is handled using a *primary treatment* followed by a *secondary treatment*. FIGURE 14.6 gives an overview of the process.

The goal of the primary treatment in a sewage plant is for the solid waste material to settle out of the wastewater. This solid material is then dried and classified as sludge. To reduce the volume of material and help remove many of the pathogens, sludge is typically exposed to bacteria that can digest it. Most of the water is then removed from the sludge to reduce its volume and weight prior to transporting it away from the sewage treatment plant. This final form of sludge can be placed into a landfill, burned, or converted into fertilizer pellets for agricultural fields, lawns, and gardens.

After the sludge has settled out of the wastewater and been treated, the remaining wastewater undergoes a secondary treatment. The goal of the secondary treatment is to use bacteria to break down 85 to 90 percent of the organic matter in the water and convert it to carbon dioxide and inorganic nutrients such as nitrogen and phosphorus. The secondary treatment typically includes

**1** Underground pipes carry waste to treatment plant.

**2** Large debris filtered out by screens and sent to landfill.

**7** Exposure to chemicals or ultraviolet light kills pathogens.

**8** Treated water released into river or lake.

Wastewater
Sludge

Primary treatment

Secondary treatment

**3** Solid waste (sludge) settles to bottom of tanks.

**4** Bacteria break down organic material to $CO_2$ and inorganic nutrients. Settled particles added to sludge.

**5** Sludge thickened by removing water.

**6** Thickened sludge taken to landfill, burned, or used for fertilizer.

**FIGURE 14.6** **A sewage treatment plant** In large municipalities, great volumes of wastewater are handled by separating the sludge from the water and then using bacteria to break down both components.

aeration of the water to promote the growth of aerobic bacteria, which emit less offensive odors than anaerobic bacteria. This treated water sits for several days to allow particles to settle out. These settled particles are added to the sludge from a primary treatment. The remaining water is disinfected, using chlorine, ozone, or ultraviolet light to kill any remaining pathogens. The treated water is then released into a nearby river or lake, where it is once again part of the global water cycle.

Sewage treatment plants are very effective at breaking down the organic matter into carbon dioxide and inorganic nutrients. Unfortunately, these inorganic nutrients can still have undesirable effects on the waterways into which they are released. Although nitrogen and phosphorus are important nutrients for increasing primary productivity, when high concentrations of these nutrients are released into bodies of water from sewage treatment plants, they fertilize the water, which can lead to large increases in the abundance of algae and aquatic plants. In response to this problem, large sewage treatment plants are now developing tertiary treatments that remove nitrogen and phosphorus from the wastewater. The ultimate goal is to release wastewater similar in quality to the waterway receiving it.

**LEGAL SEWAGE DUMPING** Sewage treatment plants are critical to human health because they remove a great

deal of harmful organic matter and associated pathogens that cause human illness. It might surprise you to know that even in the most developed countries, raw sewage can sometimes be directly pumped into rivers and lakes. Sewage treatment plants are typically built to handle wastewater from local households and industries. However, many older sewage treatment plants also receive water from storm-water drainage systems. During periods of heavy rain, the combined volume of storm water and wastewater overwhelms the capacity of the plants. When this happens, the treatment plants are allowed to bypass their normal treatment protocol and pump vast amounts of water directly into an adjacent body of water.

How big a problem is this? According to the U.S. Environmental Protection Agency, overflows of raw sewage occur approximately 40,000 times per year in the United States. In Indianapolis, for example, more than 3.8 billion liters (1 billion gallons) of raw sewage are dumped into surrounding water bodies each year during periods of high rain. Around the country, such incidents result in the contamination of drinking water, beaches, fish, and shellfish, and can lead to human illness. Between 1.8 million and 3.5 million illnesses are associated with swimming in sewage-contaminated water in the United States each year, and 500,000 illnesses are linked to drinking sewage-contaminated

**FIGURE 14.7** **Feedlot and manure lagoon.** Large-scale agricultural operations, such as this one in Georgia, produce great amounts of waste that can be held in lagoons until pumped out and removed.

water. Illnesses caused by eating contaminated shellfish are estimated to cost $2.5 million to $22 million annually. The solution to this problem is straightforward but quite expensive. Municipalities facing sewage overflow will need to modernize their sewage treatment systems at considerable expense to prevent the influx of storm water from overwhelming their capacity to treat human wastewater.

## Animal feed lots and manure lagoons

Although the impact of human waste on the environment and human health is generally well appreciated, less well known is the issue of animal waste. The problem of animal waste is actually quite similar to the problem of human waste, since it is only a major problem when large numbers of animals live in one place. On a small scale, animal manure can contaminate waters when farm animals are allowed access to streams for food and water. Inevitably, they will defecate in the stream. As you may recall from Chapter 11, large-scale manure issues arise with concentrated animal feeding operations that raise thousands of cattle, hogs, and poultry. The manure from these operations not only contains digested animal food,

but also can contain a variety of hormones and antibiotics given to animals to improve their growth and health when living under crowded conditions.

Raising thousands of animals in a single location poses the interesting challenge of how to handle the tremendous amount of manure that is produced. Rather than dumping the manure into bodies of water, many of these farms have built *manure lagoons* (**FIGURE 14.7**). **Manure lagoons** are large, human-made ponds lined with rubber to prevent the manure from leaking into the groundwater. After the manure has been broken down by bacteria—the same process that occurs in sewage treatment plants—the manure can be spread onto farm fields to serve as a fertilizer. A major risk of manure lagoons is the possibility of developing a leak in the liner. A leak would allow the waste to seep into the underlying groundwater, contaminating it. A possible overflow into adjacent water bodies is another danger. Like human wastewater, overflow of animal waste into rivers can lead to disease outbreaks in humans and wildlife. Finally, the application of manure as fertilizer can create runoff that moves into nearby water bodies. Crunch the Numbers "Building a Manure Lagoon" shows you how to calculate the appropriate size for a manure lagoon.

---

## CRUNCH THE NUMBERS

### Building a Manure Lagoon

Concentrated animal feeding operations typically use manure lagoons to hold the manure produced by the cattle that are being held.

If an individual animal produces 53 L of manure each day and the average concentrated animal feeding operation holds 900 cattle on any given day, how much manure is produced each day?

Daily manure production =
53 L/animal/day × 900 animals = 47,700 L/day

### Your Turn

1. If a manure lagoon needs to hold 30 days' worth of manure production, what is the minimum capacity that the lagoon needs to hold?

2. After the manure has broken down, you need to spread the manure onto farm fields. A modern manure spreader can hold 40,000 L of liquid manure. How many trips will it take for the manure spreader to remove the 30 days' worth of manure that is held in the manure lagoon?

✓ What problems are associated with sewage?

✓ Describe and contrast the two most common ways to treat wastewater.

✓ Describe the advantages and disadvantages of both septic systems and sewage plants.

✓ What is the role of bacteria in the treatment of human and animal waste?

# Heavy metals and other substances can pose serious threats to human health and the environment

Some compounds such as nitrogen and phosphorus cause environmental problems by overfertilizing the water. Other inorganic compounds, including heavy metals (lead, arsenic, and mercury), acids, and synthetic compounds (pesticides, pharmaceuticals, and hormones), can directly harm humans and other organisms.

## Lead

Lead is a heavy metal that poses a serious health threat. Lead is rarely found in natural sources of drinking water. Instead, it contaminates water when the water passes through the pipes of older homes that contain lead-lined pipes, brass fittings containing lead, and lead-containing materials such as solder used to fasten pipes together. Fetuses and infants are the most sensitive to lead, and exposure can damage the brain, nervous system, and kidneys. A series of federal guidelines for building construction implemented during the past three decades now requires the installation of lead-free pipes, pipe fittings, and pipe solders. These changes and the increased use of water filtration systems in many homes are gradually reducing the problem of lead in drinking water.

## Arsenic

Arsenic is a compound that occurs naturally in Earth's crust and can dissolve into groundwater. As a result, naturally occurring arsenic in rocks can lead to high concentrations of arsenic in groundwater and drinking water. Human activity also contributes to higher arsenic concentrations in groundwater. For example, mining breaks up rocks deep underground, and industrial uses of arsenic for items such as wood preservatives can add to the amount of arsenic found in drinking water. Fortunately, arsenic can be removed from water via fine membrane filtration, distillation, and reverse osmosis (see Figure 9.14 on page 245).

FIGURE 14.8 indicates where high concentrations of arsenic have been found in well water throughout the United States. As you can see, concentrations are highest in the Midwest and West. Arsenic in drinking water is associated with cancers of the skin, lungs, kidneys, and bladder. These illnesses can take 10 years or more after exposure to develop. Even very low concentrations of arsenic, such as those found in many wells around the world and measured in micrograms (µg, or $10^{-6}$ grams) per liter, can cause severe health problems. In fact, though cancers can develop with less than 50 µg/L of arsenic in drinking water, 50 µg/L was set as the upper "safe" limit for U.S. drinking water from 1942 to 1999. In 1999, the EPA lowered the upper limit to 10 µg/L, a compromise reached after much debate between environmental groups pushing for 5 µg/L and water, mining, and wood preservative industries arguing that even the 10 µg/L limit would be too expensive to implement. In 2001, the EPA delayed the implementation of the 10 µg/L standard until more data could be evaluated. Later that year, the U.S. National Academy of Sciences recommended a standard of 5 µg/L and the EPA decided to implement a standard of 10 µg/L. As a result, after substantial investment required to improve many water treatment facilities, people in the United States

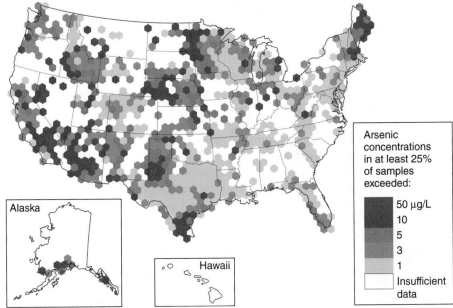

Alaska

Hawaii

Arsenic concentrations in at least 25% of samples exceeded:

50 µg/L
10
5
3
1
Insufficient data

**FIGURE 14.8** **Arsenic in U.S. well water.** The highest concentrations of arsenic are generally found in the upper Midwest and the West.

will now have lower amounts of cancer-causing arsenic in their drinking water.

The problem of arsenic in drinking water is a world-wide issue. During the 1980s and 1990s, for example, water engineers in regions of Bangladesh and eastern India drilled millions of very deep wells in an attempt to find sources of groundwater that were not contaminated by local pollution. While this had the desired effect of obtaining less-polluted water, officials were not aware that this deeper groundwater was contaminated by naturally occurring arsenic. Currently, 140 million people in this region drink arsenic-contaminated water and thousands of individuals have been diagnosed with arsenic poisoning. Current efforts to solve this problem include plans to collect uncontaminated rainwater as a source of drinking water and research to develop inexpensive filters that can remove arsenic from the well water.

## Mercury

Mercury is another naturally occurring heavy metal found in increased concentrations in water as a result of human activities. FIGURE 14.9 shows mercury releases from different regions of the world as the result of activities such as burning coal (see Chapter 12). Among regions of the world, 7 percent of human-produced mercury comes from North America and more than half comes from Asia, including 28 percent from China.

Approximately two-thirds of all mercury produced by human activities comes from the burning of fossil fuels, especially coal. Other important sources of mercury include the incineration of garbage, hazardous waste, medical supplies, and dental supplies. One of the less well-known sources of mercury comes from the raw materials that go into the manufacturing of cement for construction. The limestone used to make cement can contain mercury that is released during the heating process. Moreover, the source of heat is often coal

that also releases mercury when it is burned. Petroleum exploration is a source of both mercury and lead pollution. Each petroleum well produces roughly 180,000 gallons of contaminated wastewater and mud over its lifetime. This water is usually dumped at the drilling site and, depending on the soils and topography, can either run into nearby waterways or infiltrate the soil and contaminate the underlying groundwater.

The mercury emitted by these activities eventually finds its way into water. Inorganic mercury (Hg) is not particularly harmful, but its release into the environment can be hazardous because of a chemical transformation it undergoes. In wetlands and lakes, bacteria convert inorganic mercury to methylmercury, which is highly toxic to humans. Methylmercury damages the central nervous system, particularly in young children and in the developing embryos of pregnant women. The result is impairment of coordination and the senses of touch, taste, and sight.

Human exposure to methylmercury occurs mostly through eating fish and shellfish. Methylmercury can move up the food chain in aquatic ecosystems, which results in the top consumers containing the highest concentrations of mercury in their bodies. Given that oceans are contaminated with mercury and that tuna are top predators, it is not surprising that these fish contain high concentrations of mercury. In 2008, a reporter for the *New York Times* purchased tuna sushi from 20 locations in New York City and, after analysis by a private laboratory, found that, at most restaurants, a diet of six pieces of sushi per week would exceed the EPA standard for human consumption of mercury.

What can be done about mercury pollution? In 2009 the United States announced a new effort to work with other governments around the world to reduce global mercury production. In one of the first steps toward this goal, the EPA has proposed that cement manufacturing plants reduce mercury emissions by 81 percent. New EPA rules on reducing mercury emissions from other major sources, including coal-burning power plants, are expected in the near future.

## Acid deposition and acid mine drainage

About 40 years ago, people throughout the northeastern United States, northern Europe, China, and Russia began to notice that the forests, lakes, and streams were becoming more and more acidic. As a consequence, some trees were killed and some bodies of water became too acidic to sustain fish. After much debate, it became clear that the source of the lower pH of the water was the presence of very tall smokestacks of industrial plants that were burning coal and releasing sulfur dioxide and nitrogen dioxide into the air. These tall smokestacks kept the emissions away from local populations, but sent the chemicals into the atmosphere where they were converted to sulfuric acid and nitric acid. These acids returned to Earth hundreds of

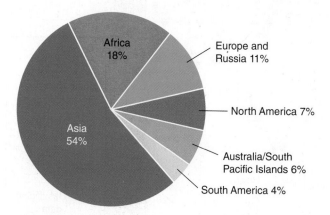

**FIGURE 14.9** **World mercury production.** Mercury production from human activities varies greatly among regions of the world. [Data from E. G. Pacyna, J. M. Pacyna, F. Steenhuisen, and S. Wilson. 2006. Global anthropogenic mercury emission inventory for 2000. *Atmospheric Environment* 40: 4048–4063.]

kilometers away as **acid deposition.** Wet-acid deposition occurs in the form of rain and snow (also known as acid precipitation or acid rain, described in Chapter 8), whereas dry-acid deposition occurs as gases and particles that attach to the surfaces of plants, soil, and water. Acid deposition reduced the pH of water bodies from 5.5 or 6 to below 5, which can be lethal to many aquatic organisms, leaving these water bodies devoid of many species. To combat the problem of acids being released into the atmosphere, many coal-burning facilities have installed *coal scrubbers.* Coal scrubbers pass the hot gases through a limestone mixture. The limestone reacts with the acidic gases and removes them from the hot gases that subsequently leave the smokestack. We will look at the issues raised by acid deposition in greater detail in Chapter 15 on air pollution.

Low pH in water bodies also occurs when very acidic water comes from belowground. This problem begins with the development of underground mines that, once abandoned, flood with groundwater. The combination of water and air allows a type of rock, named *pyrite,* to break down and produce iron and hydrogen ions. This chemical reaction produces acidic water with a low pH. Water in the mine containing these ions can find its way up to the surface in the form of springs that feed into streams. As we discussed in Chapter 8, a similar effect occurs during mountaintop mining operations in which the tops of mountains are removed and the soil dumped into stream valleys. In either case, low-pH water can also cause many other harmful metal ions to become soluble, including zinc, copper, aluminum, and manganese. The streams that are fed by springs from these mines are infiltrated with water that can have a pH close to zero. Although some water bodies such as bogs are naturally acidic and contain species that are adapted to acidic conditions, the combination of very low pH and toxic metals can produce environments that are too harsh for most organisms to survive. Because much of the dissolved iron precipitates out of solution as the low-pH water of the mine mixes with the high-pH water of the stream, these streams often have a striking red or yellow color (FIGURE 14.10).

A number of strategies are being examined to counteract the acidity of these streams, including passing stream water through limestone treatment facilities that raise the pH of the water and remove toxic metals to levels tolerable to stream organisms. Unfortunately, many mining companies responsible for making streams uninhabitable for fish and other organisms are no longer in business. As a result, they are often not held accountable for the environmental damage they caused, which presents a major challenge for dealing with the problem of acidic water from mines and the toxic metals that are produced.

## Synthetic organic compounds

Synthetic, or human-made, compounds can enter the water supply either from industrial point sources where

**FIGURE 14.10** **Acid mine drainage.** The low pH of water emerging from abandoned mines mixes with stream water to lower the pH of streams, causing iron to precipitate out of solution and form a rusty red oxidized iron. This problem occurs around the world, including this stream in Colline Metallifere, Italy.

they are manufactured or from nonpoint sources when they are applied over very large areas. These organic (carbon-containing) compounds include pesticides, pharmaceuticals, and industrial cleaners. Synthetic organic compounds have a variety of effects on organisms. They can be toxic, cause genetic defects, and, in the case of compounds that resemble animal hormones, interfere with growth and sexual development (see Chapter 17).

**PESTICIDES AND INERT INGREDIENTS** Pesticides serve an important role in helping to control pest organisms that pose a threat to crop production and human health (FIGURE 14.11). Although natural pesticides such

**FIGURE 14.11** **Applying pesticides.** Pesticides provide benefits to humans, but they also can have unexpected effects on humans and other nonpest organisms that are not fully understood and have not been adequately investigated. These airplanes are spraying insecticides over a lake to help control mosquito populations.

as arsenic have been used for centuries, the first generation of synthetic pesticides was developed during World War II. As we discussed in Chapter 11, these chemicals proved to be very effective in killing a variety of undesired plants (herbicides), fungi (fungicides), and insects (insecticides). In the decades that have followed, however, environmental scientists have identified a number of concerns about the unintended effects of pesticides.

The first concern is that most pesticides do not target particular species of organisms, but generally kill a wide variety of related organisms. For example, an insecticide that is sprayed to kill mosquitoes is typically lethal to many other species of invertebrates, including insects that might be desirable as predators of the pest. Some pesticides are lethal to unrelated species. It has recently been discovered that the insecticide endosulfan, a chemical designed to kill insects, is highly lethal to amphibians even at very low concentrations. Even a pesticide that is not directly lethal to a species can indirectly affect organisms by altering the species composition of the community.

The second concern is that although synthetic pesticides are generally designed specifically to target particular aspects of a pest species' physiology, they can also alter other physiological functions. Conceptually, the unintended side effects of pesticides are similar to the unintended side effects of pharmaceutical chemicals. Designed to target one physiological function, they often have other significant side effects. For example, most insecticides target the nervous system of insects, yet they can have unintended impacts on pests as well as on many other nonpest species. The insecticide DDT (dichlorodiphenyltrichloroethane) is a prime example. While DDT was designed to target nerve transmissions in insects, the chemical can move up an aquatic food chain all the way to eagles that consume fish. Eagles that consumed DDT-contaminated fish produced eggs with thinner shells that would prematurely break during incubation. After the United States banned the spraying of DDT in 1972, the bald eagle and other birds of prey increased in numbers. However, DDT is still manufactured in developed nations and sprayed in developing countries as a preferred way to control the mosquitoes that carry the deadly malaria virus.

The third concern is the role of *inert ingredients* added to commercial formulations of pesticides. Inert ingredients are additives that make a pesticide more effective, allowing it to dissolve in water for spraying or to penetrate inside a pest species. Although the term "inert" may suggest that these chemicals are harmless, this is often not true. The popular herbicide Roundup, for example, is composed of a chemical that is highly effective at killing plants but has difficulty getting past the waxy outer layer of leaves without the help of an added inert ingredient. Since inert ingredients are legally classified as trade secrets and most are not required to be tested for safety, their effects are not

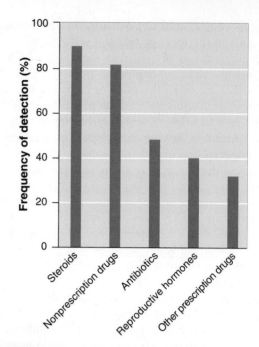

**FIGURE 14.12** **Contaminants in streams.** Streams contain a wide variety of chemicals including pharmaceutical drugs and hormones. These come from a combination of wastewater inputs, agriculture, forestry, and industry. [After Pharmaceuticals, hormones, and other organic wastewater contaminants in U.S. streams, 1999–2000: A national reconnaissance, *Environmental Science & Technology* 36: 1202–1211.]

always known before a product comes to market. In the case of Roundup, recent research has discovered that the herbicide is highly toxic to amphibians, not because of the plant-killing chemical but because of this inert ingredient. It appears that the same properties that allow the penetration of leaves also allow the penetration of tadpole gill cells. The gills burst and the tadpoles suffocate. Unintended consequences such as these have roused interest in both Europe and North America to require that inert ingredients be tested for potentially harmful effects.

**PHARMACEUTICALS AND HORMONES** While most people know that pesticides are commonly found in the environment, they often are surprised to learn that pharmaceutical drugs are also common. For example, the U.S. Geological Survey tested 139 streams across the United States for a variety of chemical contaminants. FIGURE 14.12 shows data for the frequency of detection. Among the different types of chemicals that were present at detectable levels, approximately 50 percent of all streams contained antibiotics and reproductive hormones, 80 percent contained nonprescription drugs, and 90 percent contained steroids. In most cases the concentrations of these chemicals are quite low and currently are not thought to pose a risk to environmental or human health. Some chemicals such as hormones, however, operate at very low concentrations inside the tissues of organisms and we have a poor understanding of their effects. As noted in the Chesapeake Bay example at the beginning of this chapter, it is now clear that low concentrations of pharmaceutical drugs that mimic estrogen are connected to male fish growing female eggs in their testes. The extent of hormone effects on humans and wildlife around the world is currently unknown but is receiving increased attention by environmental scientists.

**MILITARY COMPOUNDS** In regions of the world where military rockets are manufactured, tested, or dismantled, a group of harmful chemicals known as *perchlorates* sometimes contaminate the soil. Used for rocket fuel, perchlorates come in many forms. The United States Space Shuttle, for example, uses a booster rocket that contains 70 percent ammonium perchlorate. Perchlorates are easily leached from contaminated soil into the groundwater where they can persist for many years. Human exposure to perchlorates comes primarily through consumption of contaminated food and water. In the human body, perchlorates can affect the thyroid gland and reduce the production of hormones necessary for proper functioning of the human body.

**INDUSTRIAL COMPOUNDS** Industrial compounds are chemicals that are used in manufacturing. Unfortunately, some of these compounds have been dumped directly into bodies of water as a method of disposal. Over the years in the United States, this was a common practice. One of the most widely publicized consequences of this occurred in the Cuyahoga River of Ohio. For more than 100 years, industries along the river had dumped industrial wastes that formed a slick of pollution along the surface, killing virtually all animal life. The problem became so bad that the river actually caught fire and burned several times over the decades (**FIGURE 14.13**). The fire on the river in 1969 garnered

national attention and led to a movement to clean up America's waterways. Today, the Cuyahoga River and most other rivers in the United States are much cleaner because of legislation that substantially reduced the amount of industrial and other waste that can be legally dumped into waterways.

Polychlorinated biphenyls, or **PCBs,** represent one group of industrial compounds that has caused many environmental problems. PCBs were used in manufacturing plastics and insulating electrical transformers until 1979. Although they are no longer manufactured or used in the United States, because of their long-term persistence PCBs are still present in the environment. Ingested PCBs are lethal and *carcinogenic,* or cancer-causing. One particularly high-profile case revolves around two General Electric manufacturing plants in New York State that dumped 590,000 kg (1.3 million pounds) of PCBs into the Hudson River from 1947 to 1977. In 2002, the EPA ruled that General Electric must pay for the dredging and removal of approximately 2.03 million cubic meters (2.65 million cubic yards) of PCB-contaminated sediment from a 64-km (40-mile) stretch of the upper Hudson River in New York State. General Electric argued that the dredging should not occur because it would stir up the sediments and resuspend the PCBs in the river water, causing further problems. The courts and EPA scientists disagreed and in 2009 the dredging of the PCBs finally began. As this case demonstrates, though it can take decades of scientific research and debate, in the end we can partially reverse the contamination of our water bodies.

While PCBs have long been a concern, there is a growing uneasiness over compounds known as **PBDEs** (polybrominated diphenyl ethers). PBDEs are most commonly known as flame retardants added to a wide variety of items including construction materials, furniture, electrical components, and clothing. They make

**FIGURE 14.13** **A river on fire.** In 1952, the polluted Cuyahoga River in Ohio caught fire after a spark ignited the film of industrial pollution that was floating on the surface of the water.

buildings and their contents considerably less flammable than they would be otherwise. Since the 1990s, however, scientists have been detecting PBDEs in some unexpected places, including fish, aquatic birds, and human breast milk. Exposure to some types of PBDEs can lead to brain damage, especially in children. As a result, the European Union and several states including Washington and California have banned the manufacture of several types of PBDEs.

**GAUGE YOUR PROGRESS**

✓ Describe the primary dangers associated with heavy metals in water.

✓ Explain the role of acid deposition in water pollution.

✓ Name examples of synthetic compounds that have been found in our water supply and explain why they are a concern.

# Oil pollution can have catastrophic environmental impacts

In Chapter 12 we noted that the pollution of Earth's oceans and shorelines from crude oil and other petroleum products can be an environmental catastrophe as well as an ongoing problem. Petroleum products are highly toxic to many marine organisms, including birds, mammals, and fish, as well as to the algae and microorganisms that form the base of the aquatic food chain. Oil is a persistent substance that can spread below and across the surface of the water for hundreds of kilometers and leave a thick, viscous covering on shorelines that is extremely difficult to remove.

One source of oil in the water comes from drilling for undersea oil using offshore platforms. There are approximately 5,000 offshore oil platforms in North America and another 3,000 in other parts of the world. These oil platforms often experience leaks. The best estimate for the amount of petroleum leaking into North American waters is 146,000 kg (322,000 pounds) per year. In other parts of the world, antipollution regulations are often less stringent. Estimates of the amount of petroleum leaking into the ocean annually from foreign oil platforms range from 0.3 million to 1.4 million kg (0.6 million–3.1 million pounds).

One of the most famous oil leaks from an offshore platform occurred in 2010 on a BP operation in the Gulf of Mexico. In this case, an explosion on the Deepwater Horizon platform caused a pipe to break on the ocean floor nearly 1.6 km (1 mile) below the surface

of the ocean. From the time of the explosion in April until the well was sealed in August 2010, the broken pipe released an estimated 780 million liters (206 million gallons) of crude oil into the Gulf of Mexico. Given the magnitude of the oil spill and its potential to contaminate beaches, wildlife, and the estuaries that serve as habitats for the reproduction of commercially important fish and shellfish, this accident has the potential to be one of the largest environmental disasters in history.

Oil and other petroleum products can also enter the oceans as spills from oil tankers. One of the best-known spills involved the tanker *Exxon Valdez* that ran aground off the coast of Alaska in 1989 (FIGURE 14.14). As we discussed in Chapter 12, the ship spilled 41 million liters (11 million gallons) of crude oil that spread across the surface of several kilometers of ocean and coastline. The spill killed 250,000 seabirds, 2,800 sea otters, 300 harbor seals, and 22 killer whales. Cleanup efforts have been going on for two decades.

**FIGURE 14.14** The *Exxon Valdez* oil spill. In 1989, the oil tanker ran aground and spilled millions of liters of crude oil onto the shores of Alaska, where the oil killed thousands of animals.

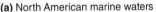

**(a)** North American marine waters

**(b)** Worldwide marine waters

**FIGURE 14.15** **Sources of oil in the ocean.** Oil contamination in the ocean, both (a) in North America and (b) worldwide, comes from a variety of sources including natural seeps, extraction of oil from underneath the ocean, transport of oil by tanker or pipeline, and consumption of petroleum-based products. [After http://oceanworld.tamu.edu/resources/oceanography-book/contents.htm.]

In 2009, 20 years after the spill, scientists evaluated the state of the contaminated Alaskan ecosystem. They concluded that the harmed populations of many species have rebounded, including bald eagles and salmon. Those that have not include killer whales (*Orcinus orca*) and sea otters (*Enhydra lutris*). Nor has the oil been completely removed from the environment. Pits dug into the shoreline suggest that approximately 55,000 liters (14,500 gallons) of oil remain. It is estimated that this oil will take more than 100 years to break down and the long-lasting effects will only become apparent over the coming decades. For its part, Exxon has paid $1 billion for the cleanup and $500 million in damages. The company also changed the ship's name, although the ship has been banned from carrying oil in North America. The *Valdez* accident sparked new rules that oil tankers in North America must have a double-hull design, requiring the ship's body to be constructed of two steel walls that would help contain leaking oil if there were an accident.

Though it is an often underappreciated aspect of oil pollution, a large fraction of oil in the ocean occurs naturally. In fact, the U.S. National Academy of Sciences recently estimated that natural releases of oil from *seeps* in the bottom of the ocean account for 60 percent of all oil in the waters surrounding North America and 45 percent of all oil in water worldwide. FIGURE 14.15 shows the proportion of different sources of oil in water for both North American and worldwide marine waters. In the waters controlled by the United States, the ocean seeps more than 270,000 liters (70,000 gallons) of oil every day. This means that when we assess

the environmental impact of oil in our oceans, we must consider the combination of both natural and anthropogenic releases of oil.

## Ways to remediate oil pollution

Since the 1989 *Exxon Valdez* oil spill, experiments have been underway to determine how best to remediate oil spills. For contaminated mammals and waterfowl, there is little debate about what to do; they must be cleaned by hand. Bird feathers that are covered with oil, for example, become heavy and lose their ability to insulate. The best approach to cleaning up the spilled oil, however, is not always clear.

Oil spilled in the ocean can either float on the surface or remain far below in the form of *underwater plumes*. For oil floating on the surface of the open ocean, a common approach is to contain the oil within an area and then suck it off the surface of the water. Containment occurs by laying out oil containment booms that consist of plastic barriers floating on the surface of the water and extending down into the water for several meters. These plastic walls keep the floating oil from spreading further. Once the oil is contained, boats equipped with giant oil vacuums suck up as much oil as possible (FIGURE 14.16). In shallow areas and along the coastline, absorbent materials are used to suck up the spilled oil.

A second approach to treating oil floating on the surface is to apply chemicals that help break up the oil, making it disperse before it hits the shoreline and causes damage to the coastal ecosystems. Although the dispersants can be effective, they can also be toxic to

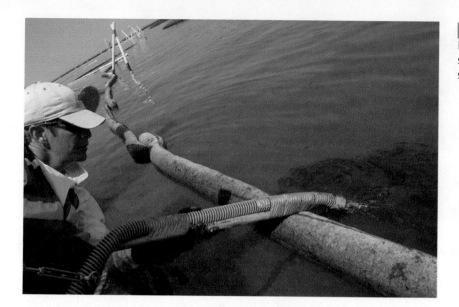

**FIGURE 14.16** **Oil-spill containment.** Floating plastic walls can contain oil spills while the oil is sucked off the surface of the water.

marine life. Current research is examining ways to make chemical dispersants more environmentally friendly.

A third approach to cleaning up oil uses genetically engineered bacteria. Several years ago scientists discovered a naturally occurring bacterium that obtained its energy by consuming oil emerging from natural seeps. These bacteria were typically rare in the ocean, but very abundant in areas where oil spills or seeps occurred. Scientists, currently trying to determine the genes that confer the bacteria's ability to consume oil, hope to insert copies of these genes into genetically modified bacteria to consume oil spills even faster.

Oil in underwater plumes persists as a mixture of water and oil, similar to the mixture of vinegar and oil in a salad dressing. In the case of the BP platform explosion in the Gulf of Mexico, scientists reported observing an oil plume moving approximately 1,000 meters (3,000 feet) below the surface of the ocean. The plume was approximately 24 to 32 km (15–20 miles) long, 8 km (5 miles) wide, and hundreds of meters thick. There is currently no agreed-upon method of removing underwater plumes from the water.

There is some debate over the best way to treat rocky coastlines after an oil spill. Scientists have been monitoring parts of Prince William Sound that were treated in different ways after the *Exxon Valdez* spill. Workers cleaned some areas with high-pressure hot water to remove the oil. Unfortunately, this also removed the plants and animals that inhabited the rocks and, in some cases, removed the fine-grained sediments containing nutrients. The hot water sprayers not only removed the oil, but also removed most of the marine life. Without the fine-grained sediment, many organisms were unable to recolonize the coast.

Other parts of the coastline received no human intervention. Over the years since the spill, the repeated action of waves and tides slowly removed much of the oil. However, the remaining oil existing in crevices of the rocky shoreline continues to have a negative effect on organisms that live among the rocks. Thus, leaving the beaches uncleaned also poses problems. At present, there is no clear consensus on the best way to respond to oil spills on coastlines.

---

**GAUGE YOUR PROGRESS**

✓ Name several ways in which oil gets into the ocean.

✓ Describe the effects of an oil spill.

✓ What are three ways to remediate an oil spill?

---

# Not all water pollutants are chemicals

When we think about water pollution, we most commonly envision scenes of dirty water contaminated with toxic chemicals. Though such scenarios certainly receive a great deal of public attention, there are other, less familiar types of water pollution that include solid waste, sediment, heat, and noise.

## Solid waste pollution

Solid waste consists of discarded materials from households and industries that do not pose a toxic hazard to humans and other organisms. Much solid waste is what we call garbage. In the United States, solid waste is generally disposed of in landfills, but in some cases it is dumped into bodies of water and can later wash up

Environmental regulations have greatly reduced the amount of solid waste that is dumped in U.S. waters, but other parts of the world such as the Citarum River in Indonesia still face a major environmental challenge.

on coastal beaches. In 1997, scientists discovered a large area of solid waste, composed mostly of discarded plastics, floating in the North Pacific gyre (see Figure 4.11 on page 96). This area, named the *Great Pacific Garbage Patch,* appears to collect much of the solid waste that is dumped into waters and to concentrate it in the middle of the rotating currents in an area the size of Texas. The other ocean gyres appear to have the same ability to collect garbage. Given the vastness of these areas, no one is certain how much solid waste is floating but current estimates are in the range of hundreds of millions of kilograms.

Garbage on beaches and in the ocean is not only unsightly, but also dangerous to both marine organisms and people. Plastic rings from beverage six-packs, for instance, can strangle many animals. Medical waste poses a threat to people on the beach, particularly children. During the 1970s and 1980s, the dumping of garbage off the coasts of the United States by both municipalities and cruise ships received increased attention and public outcry, especially when it was found to include medical waste such as used hypodermic needles. As a result, this practice was curtailed in the early 1980s. Today, garbage being dumped in U.S. waters is not nearly the problem that it is in many developing countries where there is often a lack of political mechanism and economic ability to manage proper garbage disposal (FIGURE 14.17).

Another major source of solid waste pollution is the coal ash and coal slag that remain behind when coal is burned. As we saw in Chapter 12, such waste contains a number of harmful chemicals including mercury, arsenic, and lead. The solid waste products from burning coal can contaminate groundwater since they are typically dumped into landfills, ponds, or abandoned mines. In the United States, the Environmental Protection Agency considers the waste from burning coal and other fossil fuels to be "special waste" and exempt from federal regulations for the disposal of hazardous waste. In addition, most states have either no regulations or weak regulations governing this waste.

## Sediment pollution

As noted in the chapter opener, sediments are particles of sand, silt, and clay carried by moving water in streams and rivers that eventually settle out in another location where water movement is slowed, such as where streams empty into lakes and rivers empty into oceans forming deltas. The transport of sediments by streams and rivers is a natural phenomenon, but sediment pollution is the result of human activities that can substantially increase the amount of sediment entering natural waterways (FIGURE 14.18).

Numerous human activities lead to increased sediments. Construction of houses and shopping malls, for example, requires digging up the soil, and the plant-free landscape created can lose some of its soil to erosion. As we saw in Chapters 8 and 10, plowed agricultural fields are susceptible to erosion from rain and wind. Sediments can also enter streams and rivers when natural vegetation is removed from the water's edge and replaced with either crops or domesticated animals that continually disturb the soil when they come to drink. Recent estimates are that, of all sediments in our waterways, 30 percent come from natural sources while 70 percent come from human activities.

What is the effect of increased sediment from human activities? Perhaps the most noticeable is waterways that become brown and cloudy due to the suspension of soil particles in the water. Increased sediment in the water column reduces the infiltration of sunlight, which can reduce the productivity of aquatic plants and algae. It

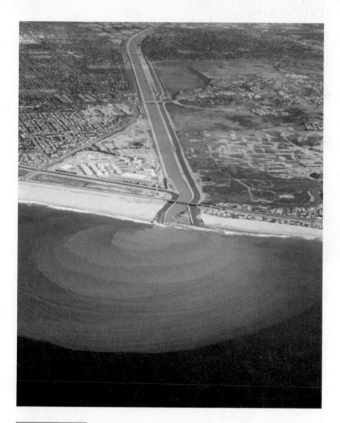

**FIGURE 14.18** **Sediments carried by rivers.** Some rivers, such as the Santa Ana River in California, which empties into the Pacific Ocean, carry a large amount of sediment that gets emptied into lakes and oceans to form deltas.

can also clog fish gills and hinder their ability to obtain oxygen. In locations where water moves slowly, the sediments settle out and accumulate on the bottom of the water body. This accumulation can clog the gills

of bottom dwellers such as oysters or clams. Since the sediments can contain nutrients, this pollution may also contribute to increased nutrients coming into the ecosystem. In total, the U.S. Environmental Protection Agency estimates that the annual cost of sediment pollution is $16 billion in environmental damage.

## Thermal pollution

A third type of nonchemical water pollution, **thermal pollution,** occurs when human activities cause a substantial change in the temperature of water. Although temperature can become either warmer or cooler, the most common cause of thermal pollution occurs when cold water is removed from a natural supply, used to absorb heat as part of some industrial process, and then returned as heated water back to the natural supply. A variety of industries make use of water for cooling, including steel mills that need to cool their machines and electric power plants. For example, as described in Chapter 12, electric power plants use nearly half of all water extracted for cooling purposes. The plant must cool the steam converted from water back into water. Thus, power plants bring in cold water from rivers, lakes, or oceans, use it to cool the steam, and then return it back to nature at significantly warmer temperatures—from 10°C to 15°C (18°F–27°F) warmer.

Since species in a given community are generally adapted to a particular natural range of temperatures, a dramatic change in temperature can kill many species, a phenomenon called **thermal shock.** High temperatures also cause organisms to increase their respiration rate, yet warmer water does not contain as much dissolved oxygen as cold water. When both effects are present, many animals will simply suffocate. In recent years,

**FIGURE 14.19** **Thermal pollution.** Nuclear reactors, such as the Svartsengi power plant in Iceland, use water to generate steam. To cool this steam, they either use cooling towers or empty the water into holding ponds. In both cases, the water must be cooled before it is returned to the natural source of water such as a lake or river.

steps have been taken to help reduce thermal pollution, including pumping the heated water into outdoor holding ponds where it can further cool before being pumped back into natural water bodies (FIGURE 14.19).

The EPA regulates how much heated water can be returned to natural water bodies. Compliance is measured by the number of degrees the temperature of the water body can be increased. This becomes a real challenge in the summer when high demand for electricity causes an increased demand for cooling water even though the water in rivers and lakes will probably be at its lowest volume and at its highest temperature. One common solution to this problem has been the construction of cooling towers that release the excess heat into the atmosphere instead of into the water. A cooling tower relies on the cooling power of evaporation to reduce the temperature of the water, much as we depend on our own sweat and a breeze to cool ourselves on a hot day. Some industries have built closed systems in which they cool the hot water in a cooling tower and then recycle the water to be heated again. In this way, the industries do not extract water from natural water bodies, nor do they release any heated water back into nature.

## Noise pollution

It may strike you as odd to think of noise as a type of water pollution. Indeed, noise pollution has received the least amount of attention from environmental scientists and, as a result, we know the least about it. Sounds emitted by ships and submarines that interfere with animal communication are the major concern. Especially loud sonar could negatively affect species such as whales that rely on low-frequency, long-distance communication. Several instances of beached whales in the Bahamas, the Canary Islands, and the Gulf of California have been connected to the use of military sonar and loud, underwater air guns. As a result, in 2003 a federal judge rejected the U.S. Navy's request to install a network of long-range sonar systems across the ocean floor to detect incoming submarines, because of suspected negative impacts on endangered whales and other species of marine animals. In 2008, however, the U.S. Supreme Court ruled that the president of the United States could exempt the navy from environmental laws that were related to potential sonar effects on ocean life.

### GAUGE YOUR PROGRESS

✓ Name three types of nonchemical water pollution.

✓ How can nonchemical water pollution be addressed?

✓ What are some examples of noise pollution as it relates to water?

# A nation's water quality is a reflection of the nation's water laws and their enforcement

Around the world, water quality improves when citizens demand it and nations can afford it. In the United States, the most important water-pollution laws are the Clean Water Act and Safe Drinking Water Act. Similar laws can be found around the world, although they are more common in developed countries than in developing ones.

## Clean Water Act

Water quality in much of the United States was quite bad in the 1960s but growing awareness of the problem encouraged a series of laws to fight water pollution. The Federal Water Pollution Control Act of 1948 was the first major piece of legislation affecting water quality, and it was substantially expanded in 1972 into what is now known as the Clean Water Act.

The Clean Water Act supports the "protection and propagation of fish, shellfish, and wildlife and recreation in and on the water" by maintaining and, when necessary, restoring the chemical, physical, and biological properties of natural waters. Note that this objective does not include the protection of groundwater. The Clean Water Act originally focused mostly on the chemical properties of surface waters. More recently, there has been increased focus on ensuring that the biological component, including the abundance and diversity of various species, also receives attention. Most importantly, the Clean Water Act issued water quality standards that defined acceptable limits of various pollutants in U.S. waterways. Associated with these limits, the act allowed the EPA and state governments to issue permits to control how much pollution industries can discharge into the water. Over time, more and more categories of pollutants have been brought under the jurisdiction of the Clean Water Act, including animal feedlots and storm runoff from municipal sewer systems.

## Safe Drinking Water Act

In addition to the Clean Water Act, other legislation has been passed to regulate water pollution, including the Safe Drinking Water Act (1974, 1986, 1996), which sets the national standards for safe drinking water. Under the Safe Drinking Water Act, the EPA is responsible for establishing **maximum contaminant levels (MCL)** for 77 different elements or substances in both surface water and groundwater. This list includes some well-known microorganisms, disinfectants, organic chemicals, and inorganic chemicals (Table 14.1). These maximum concentrations consider both the concentration of each

| TABLE 14.1 | The maximum contaminant levels (MCL) for a variety of contaminants in drinking water as determined by the U.S. Environmental Protection Agency, in parts per billion (ppb) | |
|---|---|---|
| Contaminant category | Contaminant | Maximum contaminant level (ppb) |
| Microorganism | Giardia | 0 |
| Microorganism | Fecal coliform | 0 |
| Inorganic chemical | Arsenic | 10 |
| Inorganic chemical | Mercury | 2 |
| Organic chemical | Benzene | 5 |
| Organic chemical | Atrazine | 3 |

Source: U.S. Environmental Protection Agency, http://www.epa.gov/safewater/contaminants/index.html.

compound that can cause harm as well as the feasibility and cost of reducing the compound to such a concentration.

MCLs are somewhat subjective and are subject to political pressures. For example, the MCL for arsenic was kept at 50 ppb for many years because it was argued that despite the evidence that 50 ppb caused harm in humans, reducing arsenic in drinking water to 10 ppb was too expensive for many communities to afford.

What has been the impact of these water-pollution laws? In general, they have been very successful. The EPA defines bodies of water in terms of their designated uses, including aesthetics, recreation, protection of fish, and as a source of safe drinking water. The EPA then determines if a particular waterway fully supports all of the designated uses. In 2004 (the most recent data), the EPA determined that 56 percent of all streams, 35 percent of lakes and ponds, and 70 percent of bays and estuaries in the United States now fully support their designated uses. This is a large improvement from

decades past but, as Table 14.2 shows, we still have a lot of work to do to improve the remaining waterways. Today, the water in municipal water systems in the United States is generally safe. Water regulations have greatly reduced contamination of waters and nearly eliminated major point sources of water pollution. But nonpoint sources such as oil from parking lots and nutrients and pesticides from suburban lawns are not covered under existing regulations.

## Legislation in the developing world

If we look at water-pollution legislation around the world, there is a clear difference between developed and developing countries. Developed countries, including those in North America and Europe, experienced tremendous industrialization many decades ago and widely polluted their air and water. More recently they have turned their attention to cleaning up their polluted areas and to addressing the problem of pollution

| TABLE 14.2 | The current leading causes and sources of impaired waterways in the United States | |
|---|---|---|
| | Causes of impairment | Sources of impairment |
| Streams and rivers | Bacterial pathogens, habitat alteration, oxygen depletion | Agriculture, water diversions, dam construction |
| Lakes, ponds, and reservoirs | Mercury, PCBs, nutrients | Atmospheric deposition, agriculture |
| Bays and estuaries | Bacterial pathogens, oxygen depletion, mercury | Atmospheric deposition, municipal discharges including sewage |

Source: Data from U.S. Environmental Protection Agency. 2004. *National Water Quality Inventory: Report to Congress.*

**FIGURE 14.20** **The Tietê River in Brazil.** The Tietê River, which passes through the large city of São Paolo, was badly polluted in the 1950s but is much cleaner today.

with legislation. Developing countries are still in the process of industrializing, have less restrictive environmental laws, and have less money to fund water-quality improvements such as wastewater treatment plants. In some cases, contaminating industries move from developed countries to developing countries. Although the developing countries suffer from the additional pollution, they benefit economically from the additional jobs that the new industries bring with them.

Water-pollution problems are prevalent in many of the developing nations of Africa, Asia, Latin America, and eastern Europe. China and India, for example, have undergone rapid industrialization and have many areas of dense human population—a recipe for major water-pollution problems. However, as a nation becomes more affluent, there is often more interest in the environment with resources made available to address environmental issues. In Brazil, for example, industrialization began to take off in the 1950s. By the 1990s, the Tietê River, which passes through the large city of São Paulo, was badly polluted. More than a million Brazilians signed a petition in 1992 requesting that the government regulate the industrial and municipal pollution being dumped into the river. Today the Tietê River is a much cleaner place (**FIGURE 14.20**).

**GAUGE YOUR PROGRESS**

✓ What is an MCL?

✓ What is the relationship between economic development and clean water?

✓ Describe some of the legislative actions taken to protect clean water and why they are significant.

 # WORKING TOWARD SUSTAINABILITY

A s we have seen, one of the major sources of water pollution is human wastewater from toilets, showers, washing machines, and sinks. While the EPA has set regulations for towns and cities on the proper treatment of wastewater, many municipalities find it difficult to afford modern wastewater treatment plants. What alternative solutions can be used to allow for the safe treatment of sewage?

The small town of Ashfield, Massachusetts, had a hard time affording a traditional sewage treatment facility. In 1972, the town was notified that it was in violation of the Clean Water Act because much of its

### Building "Green" Solutions to Wastewater Treatment

wastewater was being dumped directly into a nearby stream. For the next 25 years, the residents of the town debated how they could address this federal violation and its potential health risks in an affordable way. If they could find a way to do so in an environmentally friendly manner, they could obtain grants from the state and federal governments that would cover about 70 percent of the construction costs.

On the other side of the state, in Ipswich, New England BioLabs was facing a similar problem. As a major manufacturer of biological chemicals, the company was large enough to produce a substantial volume of sewage, but too far away from the nearest central sewage treatment plant.

The town of Ashfield and New England BioLabs each needed to build sewage treatment facilities. Each wanted a facility that was aesthetically pleasing, odor-free, and used environmentally friendly technologies. In each town, a proposal was made to use a system of greenhouses and wetlands that would help purify wastewater. This solution would be effective and affordable, with no unpleasant odor, and it would be aesthetically pleasing as well.

The greenhouse technology contains many of the same processes found in traditional wastewater treatment plants, but executes these processes in some unique ways (**FIGURE 14.21**). Like a traditional system, wastewater from homes and businesses flows through pipes to a holding tank at the facility, but then the liquid is pumped into a series of 3,785 L (1,000-gallon) transparent tanks that sit in a greenhouse where communities of microbes, algae, plants, and snails grow and consume the excess nutrients present in the wastewater. Air is bubbled into tanks so that the tanks experience aerobic bacterial decomposers—as opposed to the anaerobic decomposers that produce the offensive odor in many treatment plants. The liquid coming out of these tanks is held to allow the solids to settle and then the water is filtered through sand and pumped into a human-made wetland inside a greenhouse. Here denitrifying bacteria remove most of the nitrogen in the water by converting it to nitrogen gas. This water is then exposed to ultraviolet light to kill any harmful viruses or bacteria and released into a leach field where the liquid can slowly be filtered through the soil.

Because these alternative sewage treatment plants are essentially giant greenhouses, they fit in well with the aesthetics of many towns. To make them even more appealing, designers planted bananas, figs, and flowers throughout the greenhouses, often with walkways built to allow educational tours. The fruits that are produced are not eaten by people since the plants take up sewage water during growth that may contain harmful chemicals. Instead, fruits and dead plants are removed to a compost pile.

In northern latitudes the greenhouses require heat to maintain proper temperatures during the winter months, and this is a major challenge in using this more environmentally friendly technology. As the price of heating fuels climbed sharply during the past decade, the annual costs of running sewage greenhouses has also increased. A decade ago the annual cost of maintaining the greenhouse sewage facility was similar to traditional facilities. Today, the high cost of heating oil and natural gas makes this type of alternative sewage treatment better suited for municipalities in warmer climates. For some communities and industries, however, the higher costs of operating the greenhouses are worth the benefit of having a local, environmentally friendly sewage treatment plant.

**FIGURE 14.21** A "green" wastewater treatment plant. The citizens of Ashfield, Massachusetts, voted to build a more environmentally friendly wastewater plant. The plant includes greenhouses that decompose the organic material and remove excess nitrogen.

### References

Leue, T. 1998. Ashfield, Massachusetts, showcases "green" wastewater treatment plant. *Government Finance Review,* August 1.

Wacker, T. 2005. Garden grows from waste: Natural process treats raw sewage. *Boston Globe,* December 8. http://www.boston.com/news/local/articles/2005/12/08/garden_grows_from_waste/.

■ **Distinguish between point and nonpoint sources of pollution.**

Point sources of pollution have distinct locations, such as a pipe from a factory that discharges toxic chemicals into a stream. In contrast, nonpoint sources of pollution are more diffuse and cover very large areas, such as agricultural fields that leach fertilizer into a nearby stream.

■ **Identify the ways in which human wastewater can cause water pollution.**

Human wastewater can have a number of effects on natural water bodies. Wastewater adds organic matter that increases the biochemical oxygen demand, nutrients that cause eutrophication and algal blooms, and disease-causing pathogens that can harm both humans and wildlife.

■ **Evaluate the different technologies that humans have developed for treating wastewater.**

Single residences in rural areas with sufficient land space use simple septic systems that consist of a holding tank and leach field. In large communities with denser human populations and less open land, wastewater treatment plants are needed. These systems are much more complex and can be constructed either in a traditional way or by using newer, more environmentally friendly technologies.

■ **Identify the major types of heavy metals and other substances that pose serious hazards to humans and the environment.**

The major inorganic compounds that are of concern for water pollution are mercury, arsenic, and acids. Most arsenic occurs in well water through natural processes, but pollution from mercury, acid precipitation, and acid mine drainage largely occur as a result of human industrial activities. The major organic compounds composing water pollution are pesticides and their inert ingredients; pharmaceuticals, including hormones; and industrial compounds, including PCBs.

■ **Discuss the impacts of oil spills and how such spills can be remediated.**

Oil spills occur both from tankers that transport oil as well as from offshore drilling platforms that leak during oil extraction. There is general agreement about containing and removing the oil slicks that float on the surface of the water. However, scientists still debate whether oil spills that hit the coastline should be remediated by washing the coastline with hot water or leaving it to recover without human intervention.

■ **Identify contaminants that are nonchemical pollutants.**

Though nonchemical pollutants receive much less attention, they can be very harmful. These pollutants include sediments, heat, noise, and solid waste such as garbage.

■ **Explain the connections among industrialization, affluence, and water-pollution legislation.**

Most modern nations have experienced periods of industrialization and widespread pollution followed by greater affluence that allows an improvement in the quality of their waterways. Developed countries have the resources to address pollution issues. Many developing countries are still in the phase of rapid industrial growth and consequently have poor water quality.

1. Which of the following statements about nonpoint source (NPS) pollution is *not* correct?
   (a) NPS results from rain or snowmelt moving over or permeating through the ground.
   (b) NPS is a form of water pollution that is more difficult to control, measure, and regulate.
   (c) NPS includes sediment from improperly managed construction sites as a pollutant.
   (d) NPS is water pollution that originates from a distinct source such as a pipe or tank.
   (e) NPS disperses pollutants over a large area, such as oil and grease in a parking lot.

2. Human wastewater results in which of the following water-pollution problems?
   I   The organic matter decomposes and reduces dissolved oxygen levels.
   II  Decomposition of organic matter releases great quantities of nutrients.
   III Pathogenic organisms are carried into surface waters.

   (a) I only
   (b) II only
   (c) III only
   (d) I and III
   (e) I, II, and III

3. Which of the following would be an indication that a body of water was contaminated by human wastewater?
   (a) Low BOD and a fecal coliform bacteria count of zero
   (b) High levels of nutrients, such as nitrogen and phosphorus, and high BOD
   (c) Low BOD and low levels of nutrients, such as nitrogen and phosphorus
   (d) Low levels of nutrients, such as nitrogen and phosphorus, and a fecal coliform bacteria count of zero
   (e) A lack of dead zones

4. Both septic systems and sewage treatment plants utilize bacteria to break down organic matter. Where in each system does this process occur?
   (a) Septic tank and leach field; primary treatment and secondary treatment
   (b) Septic tank only; primary treatment and chlorination
   (c) Leach field only; secondary treatment only
   (d) Septic tank and leach field; secondary treatment only
   (e) Leach field only; secondary treatment and chlorination

5. Under which of the following circumstances is a sewage treatment plant legally permitted to bypass normal treatment protocol and discharge large amounts of sewage directly into a lake or river?
   (a) When the receiving surface water is designated for swimming only
   (b) When the population of the surrounding community surpasses the plant's capacity
   (c) When combined volumes of storm water and wastewater exceed the capacity of an older plant
   (d) When a permit to modernize the plant is denied by the Environmental Protection Agency
   (e) When an extended period of drought restricts water flow in a lake or river

6. Which of the following inorganic substances is naturally occurring in rocks, soluble in groundwater, and toxic at low concentrations?
   (a) Mercury
   (b) Lead
   (c) PCBs
   (d) Copper
   (e) Arsenic

7. Acid mine drainage results from acidic water formed belowground that makes its way to the surface; the acidic water is formed as a result of the flooding of abandoned mines where the underground water
   (a) reacts with a type of rock, pyrite, which releases iron and hydrogen ions.
   (b) reacts with sulfur dioxide and nitrogen dioxide to form sulfuric and nitric acids.
   (c) flushes out the chemicals used in the mining process.
   (d) permeates a limestone layer that lowers the pH.
   (e) reacts with copper and aluminum to form pyrite rock and hydrogen ions.

8. All of the following are problems that result from the use of pesticides *except*
   (a) most pesticides are not target-specific and kill other related and nonrelated species.
   (b) pesticide runoff enters surface waters and increases the solubility of heavy metals.
   (c) pesticides affect nontarget organisms by changing community relationships.
   (d) pesticides target specific physiological functions, but also disrupt other functions.
   (e) most inert ingredients are not tested for safety and may pose unacceptable risks.

9. Based on Figure 14.12, which of the following pharmaceuticals was detected *most often* in the streams studied by the U.S. Geological Survey?

   (a) Nonprescription drugs
   (b) Antibiotics
   (c) Steroids
   (d) Reproductive hormones
   (e) Other prescription drugs

10. All of the following are the result of the *Exxon Valdez* oil tanker running aground off the coast of Alaska in 1989 *except*

    (a) millions of gallons of crude oil spread over the ocean and shoreline.
    (b) thousands of animals were killed; some species have recovered, others have not.
    (c) the oil that contaminated the shoreline is now completely broken down.
    (d) Exxon paid $500 million in damages and changed the name of the ship.
    (e) all oil tankers in North America must have a double-hull design to resist leaking.

11. Successful remediation of oil spills floating on the surface of the ocean from leaking oil tankers or offshore platforms includes which of the following?

    I   Containment and vacuuming up the oil
    II  High-pressure hot water
    III Chemicals that break down oil

    (a) I only
    (b) II only
    (c) III only
    (d) I and III
    (e) II and III

12. All of the following are nonchemical forms of water pollution *except*

    (a) industrial waste.
    (b) solid waste or garbage.
    (c) sediments.
    (d) noise.
    (e) thermal pollution.

13. Which of the following provisions is defined in the Safe Drinking Water Act?

    (a) Support the protection and propagation of fish, shellfish, wildlife, and recreation in and on the water.
    (b) Maintain and restore the chemical, physical, and biological properties of natural waters.
    (c) Protect the abundance and diversity of various species in and on the water.
    (d) Issue permits to control how much pollution industries can discharge into the water.
    (e) Establish maximum contaminant levels (MCLs) for specific substances in both surface water and groundwater.

14. The passage of water-pollution legislation is *most likely* to occur when

    (a) a developing country is still in the process of industrialization and rapid population growth.
    (b) a developed country has experienced industrialization and can afford the control measures demanded by its citizens.
    (c) a developing country has not yet industrialized and water quality is still good.
    (d) contaminating industries relocate in a developing country and begin to abuse the less-restrictive environmental laws.
    (e) nonpoint sources of water pollution outnumber point sources.

1. Based on the data collected by the Maryland Department of Natural Resources and the Chesapeake Bay Program (see the figure below), answer the following questions.

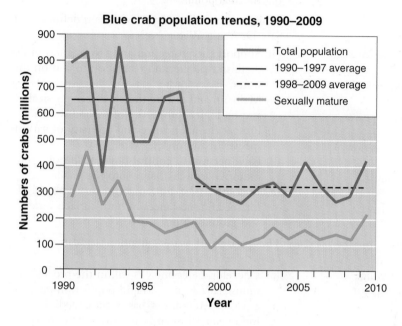

**Blue crab population trends, 1990–2009**

Legend:
- Total population
- 1990–1997 average
- 1998–2009 average
- Sexually mature

(x-axis) Year
(y-axis) Numbers of crabs (millions)

(a) Calculate the difference between the 1990–1997 average blue crab population and the 1998–2009 average blue crab population. Predict the average blue crab population for 2010–2020 and explain your answer.

(b) Identify and explain *three* possible factors *related to water pollution* that could have contributed to the decline in the total blue crab population in the Chesapeake Bay.

(c) Select *one* factor stated in (b) and describe how that source of water pollution could be managed and controlled.

(d) What federal legislation would apply to the Chesapeake Bay and the blue crabs?

2. The Food and Drug Administration (FDA) has developed guidelines for the consumption of canned tuna fish. These guidelines were developed particularly for children, pregnant women, or women who were planning to become pregnant, because mercury poses the most serious threat to these segments of society. However, the guidelines can be useful for everyone.

(a) Identify *two* major sources of mercury pollution and *one* means of controlling mercury pollution.

(b) Explain how mercury is altered and finds its way into albacore tuna fish.

(c) Identify *two* health effects of methylmercury on humans.

# MEASURE YOUR IMPACT

## Gaining Access to Safe Water and Proper Sanitation

One of the main causes of diarrheal disease is the transmission of pathogenic microorganisms through contaminated fresh water. One way to compare countries is to assess the percentage of a country's population that has access to technologies that ensure safe water and sanitation (defined by the World Health Organization as *improved water sources* and *improved sanitation*). Based on the data in the table below, answer the following questions.

| Country | 2000 %Total population with sustainable access to improved drinking water sources | 2006 %Total population with sustainable access to improved drinking water sources | 2000 %Total population with sustainable access to improved sanitation | 2006 %Total population with sustainable access to improved sanitation | 2000 Deaths among children under five years of age due to diarrheal diseases (%) | Water footprint* ($m^3$/capita/year) Global average = 1,240 ($m^3$/capita/year) | % Water derived from outside the country |
|---|---|---|---|---|---|---|---|
| United States | 99.0 | 99.0 | 100.0 | 100.0 | 0.1 | 2,483 | 19 |
| China | 80.0 | 88.0 | 59.0 | 65.0 | 11.8 | 702 | 7 |
| India | 82.0 | 89.0 | 23.0 | 28.0 | 20.3 | 980 | 2 |
| Japan | 100.0 | 100.0 | 100.0 | 100.0 | 0.4 | 1,153 | 64 |

*Water footprint is defined by the Water Footprint Network as "the volume of water needed for the production of goods and services by the inhabitants of the country."

*Sources:* Water Footprint Network, http://www.waterfootprint.org/?page=files/home; World Health Organization Core Health Indicators, http://apps.who.int/whosis/database/core/core_select_process.cfm,

(a)
   (i) List the countries in order from the highest to lowest percentage of deaths among children under five due to diarrheal diseases.
   (ii) How does this compare with access to improved drinking water sources and improved sanitation for the year 2000?
   (iii) For each country, predict how the 2006 data will affect the deaths among children under five due to diarrheal diseases.

(b) Based on your answers to (a), how could each country reduce the death rate due to diarrheal diseases of children under five years of age?

(c) For each country, calculate the ratio of its water footprint to the global average. Based on the definition of water footprint, state a relationship between the ratios calculated and water pollution.

(d) Even if each of these countries was able to achieve zero water pollution, discuss two reasons why poor water quality could still be a problem.

# ENGAGE YOUR ENVIRONMENT

How can your community minimize its storm water runoff? Explore ways to reduce the volume and velocity of runoff reaching local waterways. *See Engage Your Environment Activity 14 at the end of this book.*

# Air Pollution: Causes, Effects, and Stratospheric Ozone Depletion

## Cleaning Up in Chattanooga

Chattanooga, Tennessee, sits along the Tennessee River in a natural basin formed by the Appalachian Mountains, one of which—Lookout Mountain—rises 600 meters (1,970 feet) over the city. After the Civil War, foundries, textile mills, and other industrial plants were quickly built and Chattanooga soon became one of the leading manufacturing centers in the nation.

> By 1957, Chattanooga had the third-worst particulate pollution in the country and respiratory diseases were well above the national average.

The economic boom in Chattanooga came with an environmental cost. Like Los Angeles and many other highly polluted cities, Chattanooga is located in a bowl formed by the surrounding mountains and this geography trapped pollutants, which hovered above the city. By 1957, Chattanooga had the third-worst particulate pollution in the country and rates of respiratory diseases were well above the national average. Over the next decade conditions worsened and by the 1960s, people were often unable to see Lookout Mountain from even a quarter mile away.

In 1969, a U.S. survey of the nation's air quality confirmed what many Chattanooga residents suspected: their city topped the list of the worst cities in the United States for particulate air pollution. Obviously the poor quality of the air needed to be addressed. In 1969, before passage of the federal 1970 Clean Air Act, Chattanooga, in conjunction with Hamilton County, created its own air pollution legislation by enacting the Air Pollution Control Ordinance. It controlled the emissions of sulfur oxides, allowed open burning by permit only, placed regulations on odors and dust, outlawed visible automobile emissions, capped the sulfur content of fuel at 4 percent, and limited visible emissions from industry. At the

same time, the city and county governments put in place new pollution monitoring techniques to make sure the ordinance was being followed.

The city and county governments, along with private industry, poured approximately $40 million into the cleanup effort. Actions to improve air quality did not hinder business, but rather, created new industrial opportunities related to the cleanup effort, such as the establishment of a local manufacturer of smokestack scrubbers. As a result of all these measures, in 1972—just 3 years after passage of the city ordinance and 2 years after the Clean Air Act—Chattanooga achieved compliance with the Clean Air Act air-quality standards.

The people of Chattanooga and the local government recognized that keeping their air clean and maintaining economic sustainability would be an ongoing effort. To maintain their newly improved air quality, the city government and local businesses began several programs. One was a comprehensive recycling program, chosen as an alternative to a waste incinerator that would have added particles to the air. The public and private sectors successfully partnered to achieve both environmental and economic sustainability in creating the largest municipal fleet of electric buses in the United States, manufactured by another new local business, Advanced Vehicle Services. ▶

An electric bus in downtown Chattanooga.

◀ Recent aerial view of Chattanooga, Tennessee, with Lookout Mountain clearly visible in the background.

Unfortunately, while Chattanooga's efforts dramatically reduced the levels of particulate pollutants, the concentration of ozone, mostly from automotive precursors within and beyond the city limits, continued to climb. Ozone concentrations exceeded the 1997 standard of 0.08 parts per million by volume set by the Environmental Protection Agency. Chattanooga has responded to the new air pollution problem in the same way it faced the particulate pollutant problem of the 1960s—through a combined effort of government, the public, and local industries. The city and county governments formed an Early Action Compact with the EPA, agreeing to improve ozone concentrations ahead of EPA requirements in return for not being labeled a "nonattainment area," a designation that can result in the loss of federal highway funds and create a negative image that makes industrial recruitment and economic development more difficult. Like the 1969 Air Pollution Control Ordinance, the new Early Action Compact calls for a concerted effort by private and public sectors, and includes educating people on how they can limit ozone production on high-ozone days.

Chattanooga attained the 0.08 parts per million standard in 2007, 2 years ahead of schedule. However, national legislation has since lowered the ozone standard to 0.075 parts per million. At present, it is too early to determine whether Chattanooga will meet this new, lower ozone standard. However, residents know that to achieve their goals of an economically vibrant city with clean air, they must continue to encourage cooperation among government, people, and business. ■

*Sources:* Chattanooga Area Chamber of Commerce, *Summary of the Chattanooga Area Chamber of Commerce's Position on Strengthening the National Ambient Air Quality Standard for Ozone,* 2010, www.chattanoogachamber.com; National Ambient Air Quality Standards: www.epa.gov/ttn/naaqs.

## UNDERSTAND THE KEY IDEAS

Because air is a common resource on Earth, air pollution crosses many system boundaries. Human activity contributes to both outdoor and indoor air pollution. To understand air pollution and its effects, we need to look at all air pollutants, where they come from, and what happens to them after they are released into the atmosphere.

After reading this chapter you should be able to

■ identify the major air pollutants and where they come from.

■ explain how photochemical smog and acid deposition are formed and describe the effects of each.

■ examine various approaches to the control and prevention of outdoor air pollution.

■ explain the causes and effects of stratospheric ozone depletion.

■ discuss the hazards of indoor air pollution, especially in developing countries.

# Air pollutants are found throughout the entire global system

**Air pollution** is defined as the introduction of chemicals, particulate matter, or microorganisms into the atmosphere at concentrations high enough to harm plants, animals, and materials such as buildings, or to alter ecosystems. Generally, the term *air pollution* refers to pollution in the troposphere, the first 16 km (10 miles) of the atmosphere above the surface of Earth. (See Chapter 4 for a discussion of the atmosphere.) Tropospheric pollution is sometimes called *ground-level pollution.*

Air pollution can occur naturally, from sources such as volcanoes and fires, or it can be anthropogenic, from sources such as automobiles and factories. Although many urban regions in North America experience air pollution today, in general the air is much cleaner than it was just 40 years ago. In recent years, developing countries in Asia have had the worst outdoor air quality (FIGURE 15.1). Indoor air pollution is also a human health issue, primarily in Asia, Africa, and South America.

Since one of the major repositories for air pollutants is the atmosphere, which envelops the entire globe, we must think of the air pollution system as a global system. In fact, evidence appears to link air pollution across long distances. For example, in recent years, air pollution in Asia has been responsible for acidic rainfall on the West Coast of the United States.

The air pollution system has many inputs, which are the sources of pollution. It also has many outputs, which are components of the atmosphere and biosphere that remove air pollutants. It is difficult to conceptualize this system because the inputs do not originate from just one location. Air pollution inputs can come from automobiles on the ground, airplanes in the sky, or vegetation (tree leaves) 100 feet in the air. Similarly, air pollution can be removed or altered by vegetation, soil, and components of the atmosphere such as clouds, particles, or gases.

**FIGURE 15.1** **Particulate pollution and visibility.** Particulates and sulfate aerosols are most responsible for reducing visibility in cities, such as this location in China.

To understand the global air pollution system, we need to identify the major pollutants and determine where they come from.

## Major Air Pollutants

Even though air pollution has been with us for millennia, both the definition of pollution and the classification of a substance as a pollutant are still in transition. The atmosphere is a public resource—in effect, a global commons—and consequently the science of air pollution is closely intertwined with legislation and social perspectives. In formulating the U.S. Clean Air Act of 1970 and subsequent amendments, legislators used information from environmental scientists and others on the most important air pollutants to monitor and control. The act ultimately identified six pollutants that significantly threaten human well-being, ecosystems, and/or structures: sulfur dioxide, nitrogen oxides, carbon monoxide, particulate matter, tropospheric ozone, and lead. These were called *criteria* air pollutants because under the Clean Air Act, the EPA must specify allowable concentrations of each pollutant.

The definition of pollution has continued to undergo rapid change. Although carbon dioxide was not included among the major air pollutants in the 1970s, today it is widely accepted that carbon dioxide is altering ecosystems in a substantial way. In 2007, the U.S. Supreme Court ruled that carbon dioxide should be considered an air pollutant under the Clean Air Act, and in 2009, the EPA proposed that carbon dioxide should be considered an air pollutant at some point in the future. In addition, volatile organic compounds and mercury, though not officially listed in the Clean Air Act, are commonly measured air pollutants that have the potential to be harmful.

The sources and effects of the major air pollutants including the six criteria air pollutants are summarized in Table 15.1.

**SULFUR DIOXIDE** Sulfur dioxide ($SO_2$) is a corrosive gas that comes primarily from combustion of fuels such as coal and oil. It is a respiratory irritant and can adversely affect plant tissue as well. Because all plants and animals contain sulfur in varying amounts, the fossil fuels derived from their remains contain sulfur. When these fuels are combusted, the sulfur combines with oxygen to form sulfur dioxide. Sulfur dioxide is also released in large quantities during volcanic eruptions and can be released, though in much smaller quantities, during forest fires.

**NITROGEN OXIDES** Nitrogen oxides are generically designated $NO_X$, with the X indicating that there may be either one or two oxygen atoms per nitrogen: nitrogen oxide (NO), a colorless, odorless gas, and nitrogen dioxide ($NO_2$), a pungent, reddish-brown gas, respectively. When we use the term nitrogen oxides in our discussion, we will be referring to either nitrogen oxide or nitrogen dioxide since in the atmosphere they easily transform from one to the other. The atmosphere is 78 percent nitrogen gas ($N_2$), and all combustion in the atmosphere leads to the formation of some nitrogen oxides. Motor vehicles and stationary fossil fuel combustion are the primary anthropogenic sources of nitrogen oxides. Natural sources include forest fires, lightning, and microbial action in soils. Atmospheric nitrogen oxides play a role in forming tropospheric ozone and other components of photochemical smog.

**CARBON OXIDES** Carbon monoxide (CO) is a colorless, odorless gas that is formed during incomplete combustion of most matter, and therefore is a common emission in vehicle exhaust and most other combustion processes. Carbon monoxide can be a significant component of air pollution in urban areas. It also can be a dangerous indoor air pollutant when exhaust systems on natural gas heaters malfunction and, primarily in developing countries, where there may be poor ventilation when cooking with manure, charcoal, or kerosene.

Carbon dioxide ($CO_2$) is a colorless, odorless gas that is formed during the complete combustion of most matter, including fossil fuels and biomass. It is absorbed by plants during photosynthesis. It is also released during respiration. In general, the complete combustion of matter that produces carbon dioxide is more desirable than the incomplete combustion that produces carbon monoxide and other pollutants. However, burning fossil fuels has contributed additional carbon dioxide to the atmosphere and led to its becoming a major pollutant.

| TABLE 15.1 | Major air pollutants | | |
|---|---|---|---|
| Compound | Symbol | Human-derived sources | Effects/impacts |
| **Criteria air pollutants** | | | |
| Sulfur dioxide | $SO_2$ | Combustion of fuels that contain sulfur, including coal, oil, gasoline. | Respiratory irritant, can exacerbate asthma and other respiratory ailments. $SO_2$ gas can harm stomates and other plant tissue. Converts to sulfuric acid in atmosphere, which is harmful to aquatic life and some vegetation. |
| Nitrogen oxides | $NO_x$ | All combustion in the atmosphere including fossil fuel combustion, wood, and other biomass burning. | Respiratory irritant, increases susceptibility to respiratory infection. An ozone precursor, leads to formation of photochemical smog. Converts to nitric acid in atmosphere, which is harmful to aquatic life and some vegetation. Also contributes to overfertilizing terrestrial and aquatic systems (as discussed in Chapter 3). |
| Carbon monoxide | CO | Incomplete combustion of any kind, malfunctioning exhaust systems, and poorly ventilated cooking fires | Bonds to hemoglobin thereby interfering with oxygen transport in the bloodstream. Causes headaches in humans at low concentrations; can cause death with prolonged exposure at high concentrations. |
| Particulate matter | $PM_{10}$ (smaller than 10 micrometers) $PM_{2.5}$ (2.5 micrometers and less) | Combustion of coal, oil, and diesel, and of biofuels such as manure and wood. Agriculture, road construction, and other activities that mobilize soil, soot, and dust. | Can exacerbate respiratory and cardiovascular disease and reduce lung function. May lead to premature death. Reduces visibility, and contributes to haze and smog. |
| Lead | Pb | Gasoline additive, oil and gasoline, coal, old paint. | Impairs central nervous system. At low concentrations, can have measurable effects on learning and ability to concentrate. |
| Ozone | $O_3$ | A secondary pollutant formed by the combination of sunlight, water, oxygen, VOCs, and $NO_x$. | Reduces lung function and exacerbates respiratory symptoms. A degrading agent to plant surfaces. Damages materials such as rubber and plastic. |
| **Other air pollutants** | | | |
| Volatile organic compounds | VOC | Evaporation of fuels, solvents, paints; improper combustion of fuels such as gasoline. | A precursor to ozone formation. |
| Mercury | Hg | Coal, oil, gold mining. | Impairs central nervous system. Bioaccumulates in the food chain. |
| Carbon dioxide | $CO_2$ | Combustion of fossil fuels and clearing of land. | Affects climate and alters ecosystems by increasing greenhouse gas concentrations. |

This topic will be covered in more detail in Chapter 19 where we discuss the issue of climate change.

**PARTICULATE MATTER Particulate matter (PM),** also called **particulates** or **particles,** is solid or liquid particles suspended in air. FIGURE 15.2 outlines both the sources and the physical characteristics of particulate matter. Particulate matter comes from the combustion of wood, animal manure and other biofuels, coal, oil, and gasoline. It is most commonly known as a class of pollutants released from the combustion of fuels such as coal and oil. Diesel-powered vehicles give off more particulate matter, in the form of black smoke, than do gasoline-powered vehicles. Particulate matter can also come from road dust and rock-crushing operations. Volcanoes, forest fires, and dust storms are important natural sources of particulate matter.

Particulate matter ranges in size from 0.01 micrometer ($\mu$m) to 100 $\mu$m (1 micrometer = 0.000001 m). For comparison, a human hair has a diameter of roughly 50 to 100 $\mu$m. Particulate matter larger than 10 $\mu$m is usually filtered out by the nose and throat and is not regulated by the EPA. Particles smaller than 10 $\mu$m are called Particulate Matter-10, written as $PM_{10}$, and are of

Particulate matter in the atmosphere can either absorb or scatter sunlight, creating haze and reducing light reaching Earth's surface.

These dots, representing particulate matter, have been magnified 500 times. Even at this magnification, it would be impossible to see a dot that represents 0.01 μm without using a microscope.

$0.01\ \mu m < PM_{2.5} < 2.5\ \mu m$

$2.5\ \mu m < PM_{10} < 10\ \mu m$

Volcanoes

Fossil fuel–burning power plant smoke

Wood fires

Vehicle exhaust

Road dust

10 μm

The largest particulate matter ranges from 10 to 100 μm. For scale, human hair is about 50 to 100 μm in diameter.

100 μm

**FIGURE 15.2** **Particulate matter.** Particulate matter can be natural or anthropogenic. It ranges considerably in size and can absorb or scatter light.

concern to air pollution scientists because they are not filtered out by the nose and throat and can be deposited deep within the respiratory tract. Particles of 2.5 μm and smaller, called $PM_{2.5}$, are an even greater health concern because they deposit deeply within the respiratory tract and they tend to be composed of more toxic substances than particles in larger size ranges.

Particulate matter also scatters and absorbs sunlight. If the atmospheric concentration of particulate matter is high enough, as it would be immediately following a large forest fire or a volcanic eruption, the reduction of incoming solar radiation in the region will affect photosynthesis. This happened in 1816, a year after a large volcanic eruption in Java released more than 150 million metric tons of particles that slowly spread around the globe. That year is commonly referred to as the year without a summer.

Reduced visibility, also known as **haze,** is caused primarily when particulate matter from air pollution scatters light. But as we will see in the next section, ozone and photochemical oxidants can also play an important indirect role in the formation of haze.

**PHOTOCHEMICAL OXIDANTS INCLUDING TROPOSPHERIC OZONE** Oxides are reactive compounds that remove electrons from other substances. **Photochemical oxidants** are a class of air pollutants formed as a result of sunlight acting on compounds such as nitrogen oxides and sulfur dioxide. Though there are many photochemical oxidants, and they are generally harmful to

plant tissue, human respiratory tissue, and construction materials, attention is frequently focused on ozone ($O_3$), three oxygen atoms bound together.

In Science Applied 1 "Were We Successful in Halting the Growth of the Ozone Hole?" we discussed ozone that forms naturally in the stratosphere. We will return to this topic later in this chapter. Ozone is probably better known as the most abundant and most frequently measured photochemical oxidant in the troposphere. Tropospheric ozone is harmful to both plants and animals and causes respiratory inflammations such as asthma and emphysema. In the presence of nitrogen oxides and *volatile organic compounds,* ozone reacts to form even more harmful oxidants. In the presence of sulfur and nitrogen oxides, ozone and other photochemical oxidants can also enhance the formation of certain particulate matter, which contributes to scattering light. The resulting mixture of oxidants and particulate matter is commonly referred to as **smog,** a word derived by combining the words *smoke* and *fog.* Smog is partly responsible for the hazy view and reduced sunlight observed in many cities. Smog can be divided into two categories: **photochemical smog,** which is dominated by oxidants such as ozone and is sometimes called **Los Angeles-type smog** or **brown smog,** and **sulfurous smog,** which is dominated by sulfur dioxide and sulfate compounds and is sometimes called **London-type smog** or **gray smog.** *Atmospheric brown cloud* is a relatively new descriptive term that has been given to the combination of particulate matter and ozone. Derived primarily from combustion of fossil fuel and burning biomass, atmospheric brown clouds have been observed in cities and throughout entire regions, especially in Asia. The brownish tint that characterizes these clouds of pollution is typically caused by the presence of black or brown light-absorbing carbon particles and/or nitrogen dioxide. In addition to human health problems, particulate matter and photochemical oxidants also cause economic harm, since poor visibility in popular vacation destinations can reduce tourism revenues for recreation areas, and for businesses such as hotels and restaurants in these areas.

**LEAD AND OTHER METALS** Lead (Pb) is a trace metal that occurs naturally in rocks and soils. Present in small concentrations in fuels including oil and coal, lead compounds were added to gasoline for many years to improve vehicle performance. During that time, lead compounds were released into the air, traveled with the prevailing winds, and were deposited on the ground by rain or snow, becoming pervasive around the globe, including polar regions far from combustion sources.

In the United States, lead was phased out as a gasoline additive between 1975 and 1996, and since then its concentration in the air has dropped considerably. A campaign to phase out lead use in gasoline globally is still underway. Another persistent source of lead is lead-based paint in older buildings; when the paint peels off, the resulting dust or chips can be toxic to the central nervous system and can affect learning and intelligence, particularly for young children who may ingest the dust or chips, attracted by their sweet taste.

*Mercury* (Hg), another trace metal, is also found in coal and oil and, like lead, is toxic to the central nervous system of humans and other organisms. The EPA regulates mercury through its hazardous air pollutants program. As a result of the release of mercury into the air, primarily from the combustion of fossil fuels, especially coal, the concentrations of mercury in both air and water have increased dramatically in recent years. As a result of *bioaccumulation*—which will be discussed in greater detail in Chapter 17—mercury concentrations

in some fish have also increased. People who eat these fish increase their own mercury concentrations—an example of the interconnectedness of air pollution, air, water, aquatic health, and human health. Over the past 20 years, mercury emissions in the United States from waste incinerators have been reduced substantially. Because coal-fired electricity generation plants remain the largest uncontrolled source of mercury, emissions standards for coal plants will likely be the focus of future regulations.

**VOLATILE ORGANIC COMPOUNDS** Organic compounds that become vapors at typical atmospheric temperatures are called **volatile organic compounds (VOCs).** Many VOCs are hydrocarbons—compounds that contain carbon-hydrogen bonds, such as gasoline, lighter fluid, dry-cleaning fluid, oil-based paints, and perfumes. Compounds that give off a strong aroma are often VOCs since the chemicals are easily released into the air. VOCs play an important role in the formation of photochemical oxidants such as ozone. VOCs are not necessarily hazardous; many, such as VOCs given off by conifer trees, cause no direct harm. VOCs are not currently considered a criteria air pollutant, but because they can lead to the formation of photochemical oxidants, they have the potential to be harmful and are therefore of concern to air pollution scientists.

## Primary and Secondary Pollutants

Pollutants in the air can be categorized as *primary* or *secondary*. FIGURE 15.3 outlines the relationship between the two. We will look at each in turn.

**PRIMARY POLLUTANTS** **Primary pollutants** are polluting compounds that come directly out of the smokestack, exhaust pipe, or natural emission source. They include CO, $CO_2$, $SO_2$, $NO_X$, and most suspended particulate matter. Many VOCs are also primary pollutants. For example, as gasoline is burned in a car, it volatilizes from a liquid to a vapor, some of which is emitted from the exhaust pipe in an uncombusted form. The effect is

**FIGURE 15.3** **Primary and secondary air pollutants.** The transformation from primary to secondary pollutant requires a number of factors including sunlight, water (clouds), and the appropriate temperature.

more pronounced if the car is not operating efficiently. The resulting VOC becomes a primary air pollutant.

**SECONDARY POLLUTANTS Secondary pollutants** are primary pollutants that have undergone transformation in the presence of sunlight, water, oxygen, or other compounds. Because solar radiation provides energy for many of these transformations, and because water is usually involved, the conversion to secondary pollutants does not occur as rapidly at night as during the day or in dry environments as in wet environments.

Ozone is an example of a secondary pollutant. Ozone is not emitted from a smokestack or automobile tailpipe but is formed in the atmosphere as a result of the emission of the primary air pollutants $NO_x$ and VOCs in the presence of sunlight. The main components of acid deposition—sulfate ($SO_4^{2-}$) and nitrate ($NO_3^-$)—are also secondary pollutants. Both of these secondary pollutants will be discussed more fully below.

When trying to control secondary pollutants, it is necessary to consider the primary pollutants that create them, as well as factors that may lead to the breakdown or reduction in the secondary pollutants themselves. For example, when municipalities such as Chattanooga, described in the chapter opener, try to reduce ozone concentrations in the air, they focus on reducing the compounds that lead to ozone formation—$NO_x$ and VOCs—rather than on the ozone itself.

# Air pollution comes from both natural and human sources

A great deal of attention is directed toward air pollution that comes from human activity. However it is important to recognize that natural processes also cause air pollution. In this section we will examine both natural and human sources of air pollution.

## Natural Emissions

Volcanoes, lightning, forest fires, and plants, both living and dead, all release compounds that can be classified as pollutants. Volcanoes release sulfur dioxide, particulate matter, carbon monoxide, and nitrogen oxides. Lightning strikes create nitrogen oxides from atmospheric nitrogen. Forest fires release particulate matter, nitrogen oxides, and carbon monoxide. Living plants release a variety of VOCs, including ethylene and terpenes (**FIGURE 15.4**). The fragrant smell from conifer trees such as pine and fir and the smell from citrus fruits are mostly from terpenes; though we enjoy their fragrance, they can be precursors to photochemical smog. Long before anthropogenic pollution was common, the natural VOCs from plants gave rise to smog and photochemical oxidant pollution—hence the names of the forested mountain ranges in the southeastern United States, the Blue Ridge and the Smoky Mountains. Large nonindustrial areas such as agricultural fields can give rise to particulate matter when they are plowed, as happened in the dustbowl of the 1930s. *The Encyclopedia of Earth,* using data in part from the Intergovernmental Panel on Climate Change, estimates that across the globe, sulfur dioxide emissions are 30 percent natural, nitrogen oxide emissions are 44 percent natural, and volatile organic compound emissions are 89 percent natural. However,

(a)

(b)

**FIGURE 15.4** **Natural sources of air pollution.** There are many natural sources of air pollution, including volcanoes, lightning strikes, forest fires, and plants. (a) A forest fire in Ojai, California, produces air pollution. (b) The Great Smoky Mountains were named for the natural air pollutants that reduce visibility and give the landscape a smoky appearance.

in certain regions, such as North America, the anthropogenic contribution is much greater, perhaps as much as 95 percent for nitrogen oxides and sulfur dioxide.

The effects of these various compounds, especially when major natural events occur, depend in part on natural conditions such as wind direction. In May 1980, volcanic emissions from Mount St. Helens in Washington State were carried by the prevailing westerly winds and distributed particulate matter and sulfur oxides across the United States. The rainfall pH was noticeably lower in the eastern United States that summer.

## Anthropogenic Emissions

In contrast to natural emissions, emissions from human activity are monitored, regulated, and in many cases controlled. The EPA reports periodically on the emission sources of the criteria air pollutants for the entire United States, listing pollution sources in a variety of categories, such as on-road vehicles, power plants, industrial processes, and waste disposal. FIGURE 15.5 shows some of the most recent data. On-road vehicles, also referred to as the general category of *transportation,* are the largest source of carbon monoxide and nitrogen oxides. Electricity generation, which as we know from Chapter 12 is almost 50 percent fueled by coal, is the major source of anthropogenic sulfur dioxide. Particulate matter comes from a variety of sources including natural and human-made fires, road dust, and the generation of electricity.

The Clean Air Act and its various amendments require that the EPA establish standards to control pollutants that are harmful to "human health and welfare." The term *human health* means the health of the human population and includes the elderly, children, and sensitive populations such as those with asthma. The term *welfare* refers to visibility, the status of crops, natural vegetation, animals, ecosystems, and buildings. Through the National Ambient Air Quality Standards (NAAQS), the EPA periodically specifies concentration limits for each air pollutant. For each pollutant the NAAQS note a concentration that should not be exceeded over a specified time period. The chapter opener described the standard for ozone: for each locality in the United States, the average ozone concentration for any 8-hour period should not exceed 0.075 parts of ozone per million parts of air by volume more than 4 days per

**FIGURE 15.5** **Emission sources of criteria air pollutants for the United States.** Recent EPA data show that on-road vehicles, categorized as "transportation," are the largest source of (a) carbon monoxide and (b) nitrogen oxides. The major source of (c) anthropogenic sulfur dioxide is the generation of electricity primarily from coal. Among the sources of (d) particulate matter are road dust, industrial processes, electricity generation, and natural and human-made fires.

**(a)** Carbon monoxide

**(b)** Nitrogen oxides

**(c)** Sulfur dioxide

**(d)** Particulate matter (PM$_{2.5}$)

year, averaged over a 3-year period. If a locality violates the ozone air-quality standard and does not make an attempt to improve air quality, it is subject to penalties.

Each year, the EPA issues a report that shows the national level of the six criteria air pollutants relative to the published standards. FIGURE 15.6 shows that all criteria air pollutants have decreased in the United States over the last two decades. Only ozone and particulate matter concentrations exceed NAAQS on a regular basis. The cases where particulate matter exceeds national standards are not evident from Figure 15.6. Lead has decreased most significantly because it is no longer added to gasoline.

The situation is less positive in other parts of the world. A large area of Germany, Poland, and the Czech Republic contains a great deal of "brown" coal or lignite, which provides fuel for nearby coal-fired power plants and other industries. Emissions from combustion of this high-sulfur-content coal have caused this so-called *Black Triangle* to become one of the most polluted areas in the world. In addition to human health problems such as respiratory illnesses, many types of forest ecosystem damage have become apparent in this region in the last 25 years. In many parts of Asia, air quality has been so severely impaired by particulate matter and sulfates that visibility has been reduced by as much as 20 percent. A variety of nongovernmental and environmental organizations have prepared lists indicating the top 10 or top 20 most-polluted cities in the world. Cities in China usually dominate the list.

### GAUGE YOUR PROGRESS

✓ What are the major natural sources of air pollution?

✓ What are the major sources of anthropogenic air pollution?

✓ How does the Clean Air Act regulate anthropogenic emissions?

# Photochemical smog is still an environmental problem in the United States

"EPA Says Half of the United States Is Breathing Excessive Levels of Smog." You might think this headline was from a newspaper in the 1970s, before the Clean Air Act was fully in effect. But headlines like these have appeared in recent years. In June 2010, the EPA reported that 50 regions within the United States did not comply with the maximum allowable ozone concentration in the air of 0.075 parts of ozone per million parts of air over 8 hours, as described in the previous section. Although sulfur, nitrogen, and carbon monoxide pollution have been reduced well below the specified standards since the Clean Air Act was implemented, photochemical smog and ozone present especially difficult challenges. The reason lies in the chemistry of smog formation.

## The Chemistry of Smog Formation

As we mentioned earlier, the term *smog* was originally used to describe the combination of smoke, fog, and sometimes sulfur dioxide that used to occur in cities that burned a lot of coal. Today, Los Angeles-type brown, photochemical smog is still a problem in many U.S. cities. The formation of this photochemical smog is complex and still not well understood. A number of pollutants are involved and they undergo a series of complex transformations in the atmosphere.

FIGURE 15.7 shows a portion of the chemical process that creates photochemical smog. The first part of the process, shown in the upper part of the figure, takes place during the day, in the presence of sunlight. If there is an abundance of nitrogen oxides in the atmosphere, but very few VOCs, ozone ($O_3$) forms. A few hours later, when sunlight intensity decreases, nitrogen oxide is still present in the atmosphere and the ozone recombines with the NO, and reforms into $O_2 + NO_2$. When petrochemicals or volatile organic compounds from human activity are absent or limited, the cycle of ozone

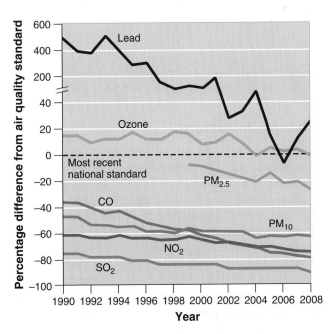

**FIGURE 15.6 Criteria and other air pollutant trends.** Trends in the criteria air pollutants in the United States between 1990 and 2008. All criteria air pollutants have decreased during this time period. The decrease for lead is the greatest. [After U.S. EPA.http://www.epa.gov/air/airpollutants.html.]

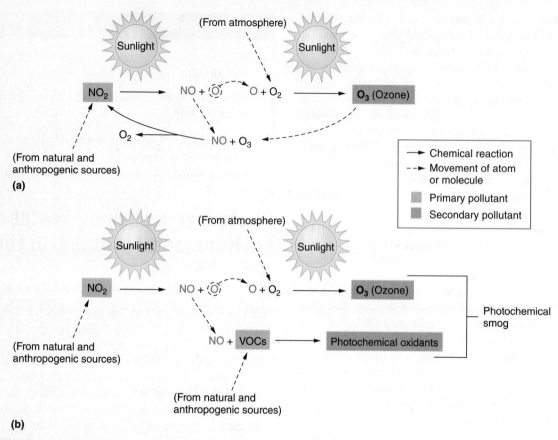

**FIGURE 15.7** **Tropospheric ozone and photochemical smog formation.** (a) In the absence of VOCs, ozone will form during the daylight hours and break down after sunset. (b) In the presence of VOCs, ozone will form during the daylight hours. The VOCs combine with nitrogen oxides to form photochemical oxidants, which reduce the amount of ozone that will break down later and contribute to prolonged periods of photochemical smog.

formation and destruction generally takes place on a daily basis with relatively small amounts of photochemical smog formation. However, as shown in the lower half of Figure 15.7, when VOCs are present, they combine with nitrogen oxide. This means the nitrogen oxide is not available to break down ozone by recombining with it and a larger amount of ozone accumulates. This explains, in part, the daytime accumulation of ozone in urban areas with an abundance of VOCs and $NO_x$.

Smog is usually thought of as an urban problem, but it is not limited to urban areas. Trees and shrubs in rural areas produce VOCs that can contribute to the formation of photochemical smog, as do forest fires that begin naturally in rural areas.

Atmospheric temperature influences the formation of smog in several important ways. Emissions of VOCs from vegetation such as trees, as well as from evaporation of volatile liquids like gasoline, increase as the temperature increases. $NO_x$ emissions from electric utilities also increase as air-conditioning demands for electricity increase on the hottest days. Many of the chemical reactions that form ozone and other photochemical oxidants also proceed more rapidly at higher temperatures. These and other factors combine to lead to greater smog concentrations at higher temperatures.

## Thermal Inversions

Another factor links temperature with air pollution conditions. Normally, temperature decreases as altitude increases. But sometimes a relatively warm layer of air at mid-altitude covers a layer of cold, dense air below. In this situation, known as a **thermal inversion,** the warm **inversion layer** traps emissions that then accumulate beneath it. The trapped emissions often cause a severe pollution event. **FIGURE 15.8** compares normal conditions with a thermal inversion. Thermal inversions that create pollution events are particularly common in some cities, where high concentrations of vehicle exhaust and industrial emissions are easily trapped by the inversion layer.

Thermal inversions can also lead to other forms of pollution. A striking example occurred in spring 1998 in the northern Chinese city of Tianjin. A cold spell that occurred after the city had shut off its central heating system for the season led many households to use

**Day**
Cooler
Cool
Warm

Altitude
Temperature

**(a)** Normal conditions

Air pollution trapped near surface
**Night**
Cooler
Relatively warm
Cool

Altitude
Temperature

**(b)** Thermal inversion

**FIGURE 15.8** **A thermal inversion.** (a) Under normal conditions, where temperatures decrease with increasing altitude, emissions rise into the atmosphere. (b) When a mid-altitude, relatively warm inversion layer blankets a cooler layer, emissions are trapped and accumulate.

individual coal-burning stoves for heat. A temperature inversion trapped the carbon monoxide and particulate matter from the coal, and caused over 1,000 people to suffer carbon monoxide poisoning or respiratory ailments from the polluted air. Eleven people died.

**GAUGE YOUR PROGRESS**

✓ What is photochemical smog?

✓ How does photochemical smog form?

✓ How does an inversion layer influence air pollution events?

# Acid deposition is much less of a problem than it used to be

As we discussed in Chapter 2 and elsewhere, all rain is naturally somewhat acidic; the reaction between water and atmospheric carbon dioxide lowers the pH of precipitation from neutral 7.0 to 5.6. Acid deposition refers to deposition with a pH lower than 5.6. Acid deposition is largely the result of human activity, although natural processes, such as volcanoes, may also contribute.

## How Acid Deposition Forms and Travels

FIGURE 15.9 shows how acid deposition forms. Nitrogen oxides and sulfur dioxide are released into the atmosphere as a result of numerous natural and anthropogenic combustion processes. In the presence of atmospheric oxygen and water, these primary pollutants are transformed, through a series of reactions, into the secondary pollutants nitric acid and sulfuric acid. The latter compounds break down further, producing nitrate, sulfate—inorganic pollutants that we have discussed earlier—and hydrogen ions ($H+$) that generate the acidity in acid deposition. These transformations occur over a number of days, and during this time, the

Primary pollutants

$SO_2$ Sulfur dioxide

$NO_x$ Nitrogen oxides

Oxidants

Secondary pollutants

$H_2SO_4$ Sulfuric acid → $2H^+ + SO_4^{2-}$

$HNO_3$ Nitric acid → $H^+ + NO_3^-$

Acid precipitation

**FIGURE 15.9** **Formation of acid deposition.** The primary pollutants sulfur dioxide and nitrogen oxides are precursors to acid deposition. After transformation to the secondary pollutants—sulfuric and nitric acid—dissociation occurs in the presence of water. The resulting ions—hydrogen, sulfate, and nitrate—cause the adverse ecosystem effects of acid deposition.

pollutants may travel a thousand kilometers (600 miles) or more. Eventually, these secondary acidifying pollutants are washed out of the air in precipitation and in dry form and are deposited on vegetation, soil, or water. This process was also discussed in Chapter 14 in relation to water pollution.

Acid deposition has been reduced in the United States as a result of lower sulfur dioxide and nitrogen oxide emissions that we showed in Figure 15.6. Much of this improvement is a result of the Clean Air Act Amendments that were passed in 1990 and implemented in 1990 and 1995.

Studies have documented regional acid deposition in West Africa, South America, Japan, China, and many areas in eastern and central Europe. Acid deposition crosses the border between the United States and Canada and is carried from England, Germany, and the Netherlands to Scandinavia. One implication of this mobility is that the precursors to acid deposition emitted in one region may have a significant impact on another region or another country. For example, over the years, there have been legislative and legal attempts to restrict emissions from coal-burning power plants in the midwestern United States that fall as acid deposition in Canada.

## Effects of Acid Deposition

Acid deposition in the United States increased substantially from the 1940s through the 1990s due to human activity. It had a variety of effects on materials, on agricultural lands, and on both aquatic and terrestrial natural habitats. Newspaper headlines in the United States and Europe in the 1980s contained frequent reports about adverse effects of acid deposition on forests, lakes, and streams.

Effects of acid deposition may be direct, such as lowering of the pH of lake water, or indirect. It is often difficult to determine whether an effect is direct or indirect, making remediation challenging.

As we saw in Chapter 14, the greatest effects of acid deposition have been on aquatic ecosystems. Lower pH of lakes and streams in areas of northeastern North America, Scandinavia, and the United Kingdom has caused decreased species diversity of aquatic organisms. As we saw in Chapter 6, many species are able to survive and reproduce only within a narrow range of environmental conditions. Many amphibians, for instance, will survive when the pH of a lake is 6.5, but when the lake acidifies to pH 6.0 or 5.5 the same organism will begin to have developmental or reproductive problems. In water below pH 5.0, most salamander species won't survive.

Lower pH can also lead to mobilization of metals, an indirect effect. When this happens, metals bound in organic or inorganic compounds in soils and sediments are released into surface water. Because metals such as aluminum and mercury can impair the physiological

functioning of aquatic organisms, exposure can lead to species loss. Decreased pH can also affect the food sources of aquatic organisms, creating indirect effects at several trophic levels. On land, at least one species of tree, the red spruce (*Picea rubens*), at high elevations of the northeastern United States was shown to have been harmed by acid deposition. It is likely that these trees have been harmed by both the acidity of the deposition as well as the nitrate and sulfate ions.

People are not harmed by direct contact with precipitation at the acidities commonly experienced in the United States or elsewhere in the world, because the human skin is a sufficiently robust barrier to this irritant. Human health is more affected by the precursors to acid deposition such as sulfur dioxide and nitrogen oxides.

Acid deposition can, however, harm human-built structures. Statues, monuments, and buildings—even some of the buildings from ancient Greece that have stood for around 2,000 years—have been seriously eroded over the last half century by this form of air pollution (FIGURE 15.10). The damage happens because acid deposition reacts with building materials. When the hydrogen ion in acid deposition interacts with

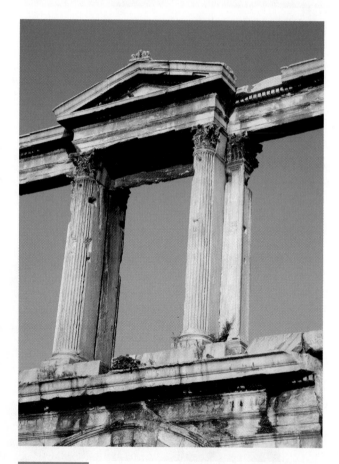

**FIGURE 15.10** **Hadrian's Arch.** This monument, near the Acropolis in Athens, Greece, has been damaged by acid deposition. It is made of marble, which contains calcium carbonate, and is susceptible to deterioration from acid deposition and acids in the air.

limestone or marble, the calcium carbonate reacts with $H^+$ and gives off $Ca^{2+}$. In the process, the calcium carbonate material is partially dissolved. The more acidic the precipitation, the more hydrogen ions there are to interact with the calcium carbonate. In the case of the Acropolis and some other stone structures, gaseous sulfur dioxide ($SO_2$) or sulfuric acid vapor (other components of acidic deposition) have contributed to the deterioration. Acid deposition also erodes many exposed painted surfaces, including automobile finishes.

> **GAUGE YOUR PROGRESS**
>
> ✓ What is acid deposition?
>
> ✓ What are the two primary pollutants that lead to the formation of acid deposition?
>
> ✓ What are the major effects of acid deposition?

# Pollution control includes prevention, technology, and innovation

As with other types of pollution, the best way to decrease air pollution emissions is to avoid them in the first place. This can be achieved through the use of fuels that contain fewer impurities. Coal and oil, for example, both occur naturally with different sulfur concentrations and are available for purchase at a variety of sulfur concentrations. In addition, during refining and processing, the concentration of sulfur can be reduced in both fuels.

Use of a low-sulfur coal or oil is certainly one of the best means of controlling air pollution, although typically low-sulfur coal or oil is more expensive to purchase than coal or oil with higher sulfur concentrations. As we discussed in Chapters 12 and 13, other ways to reduce air pollution include increased efficiency and conservation: use less fuel and you will produce less air pollution. However, switching to lower-impurity fuel and reduced use of fuel can only reduce emissions by a certain amount. Where fuel is combusted, pollution will be emitted. So ultimately, most attempts to reduce air pollution will rely to some extent on the control of pollutants after combustion, or, as we discussed in Chapter 13, on developing alternative energy sources that are not dependent on combustion.

## Control of Sulfur and Nitrogen Oxide Emissions

Removing sulfur dioxide from coal exhaust during combustion by *fluidized bed combustion,* in which granulated coal is burned in close proximity to calcium carbonate, reduces sulfur dioxide emissions. The heated calcium carbonate absorbs sulfur dioxide and produces calcium sulfate, which can be used in the production of gypsum wallboard, also known as sheetrock, for houses. Some of the sulfur oxide that does escape the combustion process can be captured by other methods after combustion.

Nitrogen oxides are produced in virtually all combustion processes in the atmosphere of Earth, which is 78 percent nitrogen gas. Hotter burning conditions and the presence of oxygen allow proportionally more nitrogen oxide to be generated per unit of fuel burned. In order to reduce nitrogen oxide emissions, burn temperatures must be reduced and the amount of oxygen must be controlled—a procedure that is sometimes utilized in factories and power plants with certain air pollution control technology. However, lowering temperatures and oxygen supply can result in less-complete combustion, reducing the efficiency of the process and increasing the amount of particulates and carbon monoxide. Finding the exact mix of air, temperature, oxygen, and other factors is a significant challenge

Nitrogen oxide emissions from automobiles have also been reduced significantly in the United States over the last 35 years. Beginning in 1975, all new automobiles sold in the United States were required to include a catalytic converter, which reduced the nitrogen oxide and carbon monoxide emissions. In order to operate properly, the precious metals in the catalytic converter (mostly platinum and palladium) could not be exposed to lead. This, in turn, meant that gasoline could no longer contain lead. Lead in gasoline was a major source of atmospheric lead and a hazard to human health. As Figure 15.6 illustrates, the change in gasoline formulation caused a significant reduction in emissions of lead from automobiles and in lead concentrations in the atmosphere. At the same time, improvements in the combustion technology of power plants and factories reduced emissions of nitrogen oxides from those sources.

## Control of Particulate Matter

The removal of particulate matter is the most common means of pollution control. Sometimes, in the process of removing particulate matter, sulfur is removed as well. There are a variety of methods used to remove particulate matter. The simplest removal method is gravitational settling, which relies on gravity to remove some of the particles as the exhaust travels through the smokestack. Ash residue that accumulates must be disposed of in a landfill. Depending on the fuel that was burned, the ash may contain sufficiently high concentrations of metals that require special disposal. This subject will be covered in more detail in Chapter 16 on solid waste.

Pollution control devices remove particulate matter and other compounds after combustion. Each has its advantages and disadvantages and all of them use energy—most commonly electricity—and as a result, generate additional pollution. Fabric filters, shown in

FIGURE 15.11 is labeled in the figure with:

Clean air outlet ③

Shaker mechanism

② 

Filter bag

Dirty air inlet ①

Trapped particles on inner bag surface

④

Collection hopper

① Dirty air enters housing.

② Combustion exhaust stream moves through and dust particles are trapped in a series of filter bags.

③ Cleaner, filtered air moves out of unit.

④ A shaker mechanism is activated periodically to dislodge trapped particles, which can then be collected from beneath the unit.

**FIGURE 15.11** **The baghouse filter.** In this air pollution control device, particles are removed by a series of filter bags that physically filter out the particles.

FIGURE 15.11, are a type of filtration device that allow gases to pass through them but remove particulate matter. Often called baghouse filters, certain fabric filters can remove almost 100 percent of the particulate matter emissions. Electrostatic precipitators, shown in FIGURE 15.12, use an electrical charge to make particles coalesce so they can be removed. Polluted air enters the precipitator and the electrically charged particles within are attracted to negative or positive charges on the sides of the precipitator. The particles collect and relatively clean gas exits the precipitator. A scrubber, shown in FIGURE 15.13, uses a combination of water and air that actually separates and removes particles. Particles are removed in the scrubber in a liquid or sludge form and clean gas exits. Borrowing from the concept utilized in the electrostatic precipitator, particles are sometimes ionized before entering the scrubber to increase its efficiency. Scrubbers are also used to reduce the emissions of sulfur dioxide. All three types of pollution control devices, because they use additional energy and increase resistance to air flow in the factory or power plant, require the use of more fuel and result in increased carbon dioxide emissions.

Devices such as the electrostatic precipitator and the scrubber have helped to reduce pollution significantly before it is released into the atmosphere. It is much

harder—if not impossible—to remove pollutants from the environment after they have been dispersed over a wide area.

## Smog Reduction

As the examples cited earlier make clear, many cities in the United States and around the world continue to have smog problems. Smog is particularly challenging to overcome because the main component of photochemical smog, ozone, is a secondary pollutant. Because of this, control efforts must be directed toward reducing the precursors, or primary pollutants, that contribute to smog. Historically, most local smog reduction measures have been directed primarily at reducing emissions of VOCs in urban areas. As we discussed earlier and illustrated in Figure 15.7, with fewer VOCs in the air there are fewer compounds to interact with nitrogen oxides, and thus more nitrogen oxide will be available to recombine with ozone. More recently, regional efforts to control ozone have focused on reducing nitrogen oxide emissions, which appears to be a more effective method of controlling smog in areas away from urban centers.

② High-voltage negative electrode

③ Positive collection electrode

① Dirty air

④ Cleaner air

⑤ Collection hopper

① Dirty air enters precipitator unit.

② Particles in combustion exhaust stream pass by negatively charged plates, which gives them a negative charge.

③ The negatively charged particles are attracted to positively charged collection plates.

④ Cleaner air moves out of the unit.

⑤ The positive collection plates are periodically discharged, which causes the particles to fall off so that they can be removed from the system.

**FIGURE 15.12** **The electrostatic precipitator.** In this air pollution control device, particles are given a negative charge. This causes them to be attracted to a positively charged plate, where they are held. Periodically, they are removed from the plate and collected for disposal.

**8** Cleaner air

**7** Mist eliminator

**6** Scrubbing water

**1** Dirty air

**2**

**3**

**4**

**5** Sludge to disposal

Sludge removal system

**1** Dirty air enters scrubber.

**2** Combustion exhaust stream moves upward in shower of water mist.

**3** Mist collects particles ("scrubs" the air) and brings them down to bottom of unit.

**4** Dirty water moves to a sludge removal system.

**5** Sludge is separated from water and disposed of.

**6** Water moves back to scrubber for reuse.

**7** Excess mist collects on screen.

**8** Cleaner air exits through stack.

**FIGURE 15.13** **The scrubber.** In this air pollution control device, particles are "scrubbed" from the exhaust stream by water droplets. A water-particle "sludge" is collected and processed for disposal.

## Innovative Pollution Control

A number of cities around the world, including those in China, Mexico, and England, have taken innovative and often controversial measures to reduce smog levels. Municipalities have passed measures, for example, to reduce the amount of gasoline spilled at gasoline stations, restrict the evaporation of dry-cleaning fluids, or restrict the use of lighter fluid (a VOC) for starting charcoal barbecues. Suburban areas have taken additional actions such as calling for a reduction in the use of wood-burning stoves or fireplaces that would reduce emissions of not only nitrogen oxide but also particulate matter, VOCs, and carbon monoxide. A number of California municipalities even discussed reducing the number of bakeries within certain areas, as the emissions from rising bread contain VOCs. This proposal was not very popular, as you can imagine, but emissions from bakeries along with many other businesses are sometimes regulated by local air-quality ordinances.

Since cars are responsible for large emissions of nitrogen oxides and VOCs in urban areas and these two compounds are the major contributors to smog

formation, some municipalities have tried to achieve lower smog concentrations by restricting automobile use. A number of cities, including Mexico City, have instituted plans permitting automobiles to be driven only every other day—for example, those with license plates ending in odd numbers may be used on one day and those with even-numbered license plates on alternate days. In China, during the 2008 Beijing Olympics, the government successfully expanded public transportation networks, imposed motor vehicle restrictions, and temporarily shut down a number of industries as a way to reduce photochemical smog and improve visibility.

Limiting automobile use has also helped to reduce other air pollutants. Carpool lanes, available in many municipalities, reduce the number of cars on the road by encouraging two or more people to share one vehicle. Improving the quality and accessibility of public transportation encourages people to leave their cars at home. A number of cities in England, including London, have been experimenting with charging individual user fees (tolls) to use roads at certain times of the day or within certain parts of a city as a way to reduce automobile traffic. Road user fees have been proposed for cities in

## Calculating Annual Sulfur Reductions

Given the data presented below for total $SO_2$ emission reductions in the United States, calculate the total percentage reduction and the annual percentage reduction of $SO_2$ emissions.

23.5 million metric tons – 10.3 million metric tons
= 13.2 million metric tons total reduction

Divide the reduction by the original amount and multiply by 100 to obtain a percent reduction:

$$\frac{13.2 \text{ million metric tons}}{23.5 \text{ million metric tons}} \times 100 = 56\%$$

The total reduction was 56 percent. To calculate the reduction per year divide 56 percent by the number of years from beginning to end:

$$2008 - 1982 = 26 \text{ years}$$

$$\frac{56\%}{26 \text{ years}} = 2.2\%/\text{year}$$

**Your Turn:** In the United States, nitrogen oxide emissions decreased from 22.6 million metric tons (25 million U.S. tons) in 1990 to 17.2 million metric tons (19 million U.S. tons) in 2005. Calculate the total percent reduction and the annual percent reduction for nitrogen oxides

---

the United States, including New York City, but none has yet been implemented.

In 1990 and again in 1995, scientists, policy makers, and academics collaborated on amendments to the Clean Air Act that would allow the free market to determine the least expensive ways to reduce emissions of sulfur dioxide. The free-market program was implemented in two phases between 1995 and 2000, and approximately 3,000 power plants are now covered under the Acid Rain Program of the act. So far, each phase has led to significant reductions in sulfur emissions.

One of the most innovative aspects of the Clean Air Act amendments was the provision for the buying and selling of allowances that authorized the owner to release a certain quantity of sulfur. Such an allowance authorizes a power plant or industrial source to emit one ton of $SO_2$ during a given year. Sulfur allowances are awarded annually to existing sulfur emitters proportional to the amounts of sulfur they were emitting before 1990, and the emitters are not allowed to emit more sulfur than the amount for which they have permits. At the end of a given year, the emitter must possess a number of allowances at least equal to its annual emissions. In other words, a facility that emits 1,000 tons of $SO_2$ must possess at least 1,000 allowances that are usable in that year. Facilities that emit quantities of $SO_2$ above their allowances must pay a financial penalty.

Sulfur allowances can be bought and sold on the open market by anyone. If emitters wanted to exceed their allowance level, say because they intended to increase their industrial output, they would be required to purchase more allowances from another source. If, on the other hand, a company decreased its sulfur emissions more than it needed to in order to comply with its allowance amount, it could sell any unused sulfur emission allowances. Over time, the number of allowances distributed each year has been gradually reduced, such that the total $SO_2$ emissions from all sources in the United States has declined from 23.5 million metric tons (26 million U.S. tons) in 1982 to 10.3 metric tons (11.4 million U.S. tons) in 2008. Crunch the Numbers "Calculating Annual Sulfur Reductions" shows you how to calculate these decreases as percentages. The overall economic cost for achieving these reductions has been about one-quarter of the original cost estimate. Global change researchers have used the sulfur allowance example as a model for the more recent experiments with buying and selling carbon dioxide allowances. We discuss this further in the Science Applied that follows Chapter 20.

### GAUGE YOUR PROGRESS

✓ Describe pollution control methods for sulfur dioxide, nitrogen oxides, and particulates.

✓ What are some approaches to smog reduction?

✓ Explain the purpose of sulfur allowances and how they work.

# The stratospheric ozone layer provides protection from ultraviolet solar radiation

Earlier in this chapter, we discussed tropospheric or ground-level pollution, which has been shown to contribute to a number of problems in the natural world, to exacerbate asthma and breathing difficulties in humans, and to cause cancer. Now we turn to the effects of certain pollutants in the stratosphere that have a substantial impact on the health of humans and ecosystems. In the troposphere, ozone is an oxidant that can harm respiratory systems in animals and damage a number of structures in plants. However, in the stratosphere ozone forms a necessary, protective shield against radiation from the Sun. In Science Applied 1, "Were We Successful in Halting the Growth of the Ozone Hole?" (pages 52–55), we examined the reductions in stratospheric ozone. As we saw there, ozone in the stratosphere absorbs ultraviolet light and thereby prevents harmful ultraviolet radiation from reaching Earth.

## The Benefit of Stratospheric Ozone

As we saw in Figure 2.11 on page 37, energy from the Sun occurs at many wavelengths including harmful high-energy ultraviolet waves, medium-energy waves that we see as visible light, and lower-energy infrared heat waves. For much of Earth's history, organisms could exist only in water because ultraviolet radiation from the Sun would cause mutations or death to almost any organism that was not protected by the filtration water provided. The ultraviolet (UV) spectrum is made up of three increasingly energetic ranges: UV-A, UV-B, and UV-C, each of which has different properties. UV-A passes through the atmosphere without being absorbed and contributes to, and possibly initiates, skin cancer. UV-B and UV-C have enough energy to cause potentially significant damage to the tissues and DNA of living organisms. However, a protective layer of oxygen and ozone in the stratosphere absorbs over 99 percent of all incoming UV-B and UV-C radiation, allowing life to exist on land in its current form.

It is easy to confuse stratospheric ozone with tropospheric, or ground-level, ozone that we discussed earlier in this chapter because it *is* the same gas, $O_3$. However, stratospheric ozone occurs in a different region of the atmosphere and therefore has a very different function. Tropospheric ozone acts as an air pollutant that damages lung tissue and plants. Because living organisms are not found high in the atmosphere, stratospheric ozone is not harmful to them. In fact, its ability to absorb ultraviolet radiation and thereby shield the surface below makes stratospheric ozone critically important to life on Earth.

## Formation and Breakdown of Stratospheric Ozone

Stratospheric ozone forms and breaks down naturally in a closed-loop cycle. (For an illustration, turn to Figure SA1.1 on page 53.)

First, UV-C radiation breaks the bonds holding together the oxygen molecule ($O_2$), leaving two free oxygen atoms:

$$O_2 + UV\text{-}C \rightarrow 2O \quad (1)$$

This only happens to a small fraction of oxygen molecules at any given time, so most of the molecular oxygen $O_2$ remains unaffected in the atmosphere. But when it reacts with the free oxygen atoms (O) formed in reaction 1, the result is ozone:

$$O_2 + O \rightarrow O_3 \quad (2)$$

The net result of these two reactions is that, in the presence of ultraviolet radiation, oxygen is converted to ozone.

Ozone is broken down into $O_2$ and free oxygen atoms when it absorbs both UV-C and UV-B ultraviolet light:

$$O_3 + UV\text{-}B \text{ or } UV\text{-}C \rightarrow O_2 + O \quad (3)$$

The free oxygen atoms and molecular oxygen $O_2$ may again react to produce ozone molecules; thus, ozone is continuously formed and continuously broken down in the presence of sunlight, maintaining a steady state concentration of ozone. Each cycle absorbs UV energy, but note that only ozone absorbs in the UV-B range. In fact, ozone is the only molecule in the atmosphere that absorbs strongly in this range. Without ozone, much more UV-B would reach the surface than it does now.

## Anthropogenic Contributions to Ozone Destruction

Just as the ultraviolet radiation in sunlight can break down ozone, so can certain chemical catalysts, the most important of which is chlorine. Chlorine atoms enter the atmosphere in many ways. The major source of chlorine in the stratosphere is a class of anthropogenic compounds known as **chlorofluorocarbons (CFCs),** a family of organic compounds whose properties make them ideal for use in refrigeration and air conditioning, as propellants in aerosol cans to deliver ingredients such as deodorant and insect repellant, and as "blowing agents" to inject air into foam products like Styrofoam cups and foam insulation. In the past, as a result of these applications, CFCs were released into the atmosphere or escaped from leaky or broken appliances.

Some of the features that made CFCs ideal refrigerants—they are extremely stable, inert, nontoxic, and nonflammable—are the same features that enable them to harm the stratosphere. After release into the troposphere, a CFC molecule does not degrade, dissolve in water, or undergo any significant chemical change, but slowly circulates in the atmosphere. When it reaches the stratosphere, the ultraviolet radiation present is energetic enough to break the bond connecting chlorine to the CFC molecule, which can then undergo the following reactions with ozone and free oxygen atoms (available from reaction 1). First, chlorine breaks ozone's bonds and pulls off one atom of oxygen, forming a chlorine monoxide molecule (ClO) and $O_2$:

$$O_3 + Cl \rightarrow ClO + O_2 \quad (4)$$

Next, a free oxygen atom pulls the oxygen atom from ClO, liberating the chlorine and creating one oxygen molecule. The free chlorine molecule is ready to break down more ozone.

$$ClO + O \rightarrow Cl + O_2 \quad (5)$$

Notice that ClO gets made from Cl in the first reaction, and reconverted into Cl in the second, so after reactions 4 and 5, the only things that really change are the oxygen compounds. Thus we can leave the Cl compounds out of the "net reaction," which is just a simpler way to write equations 4 and 5:

$$O_3 + O \rightarrow 2 O_2 \quad (6)$$

Chlorine is acting here as a catalyst, bringing about the reaction without getting used up itself. One chlorine atom can catalyze the breakdown of as many as 100,000 ozone molecules before it leaves the stratosphere. Reaction 6 converts ozone to molecular oxygen $O_2$, just like reaction 3, but reaction 6 does not involve the absorption of any UV light; the ozone that chlorine breaks down is no longer available to absorb incoming UV-B radiation.

In addition to CFCs, many other anthropogenic compounds, including nitrogen oxides, bromines (used as fumigants for soil pests such as termites), and carbon tetrachloride (formerly used as a cleaning solvent), can contribute to the destruction of stratospheric ozone, although these do not reach the stratosphere as efficiently as CFCs because they are more reactive in the troposphere.

## Depletion of the Ozone Layer

In the mid-1980s, atmospheric researchers noticed that stratospheric ozone in Antarctica had been decreasing each year, beginning in about 1979. Since the late 1970s, global ozone concentrations had decreased by more than 10 percent. Depletion was greatest at the poles, but occurred worldwide. One study from Switzerland showed an erratic but clearly decreasing trend of ozone concentrations since 1970. The graph in FIGURE 15.14 shows the results of this study.

Researchers also determined that, in the Antarctic, ozone depletion was seasonal: each year depletion occurred from roughly August through November (late winter through early spring in the Southern Hemisphere). The depletion caused an area of severely reduced ozone concentrations over most of Antarctica, which has come to be called the "ozone hole." A depletion of ozone also occurs over the Arctic in January through April, but it is not as severe, varies more from year to year, and does not cause a "hole" as in the Antarctic.

The cause of the formation of the ozone hole, which has received a great deal of media attention and has been studied intensively, is complex. It appears that extremely cold weather conditions during the polar winter cause a buildup of ice crystals mixed with nitrogen oxide. This in turn provides the perfect surface for the formation of the stable molecule $Cl_2$, which accumulates as atmospheric chlorine interacts with the ice crystals. When the Sun reappears in the spring, UV radiation breaks down this molecule into Cl again, which in turn catalyzes the destruction of ozone as shown above. Because almost no ozone had been formed in the dark of the polar winter, a large "hole" occurs. Only after the temperatures warm up and the chlorine gets diluted by air from outside of the polar region does the hole diminish. In contrast, the overall global trend of decreasing stratospheric ozone concentration is not related to temperature, but is caused by the breakdown reactions described earlier that result from increased concentrations of chlorine in the atmosphere.

Decreased stratospheric ozone has increased the amount of UV-B radiation that reaches the surface of

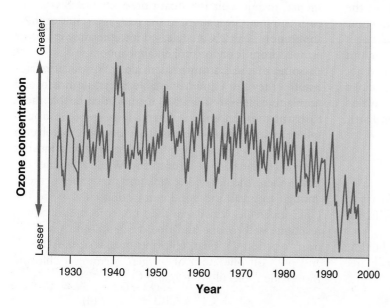

FIGURE 15.14 Stratospheric ozone concentration. This data for one area of Switzerland shows a generally decreasing trend from 1970 to 2000.

Earth. A United Nations study showed that in mid-latitudes in North America, UV radiation at the surface of Earth increased about 4 percent between 1979 and 1992. The increase is greater closer to the poles. For plants, both on land and in water, increasing exposure to UV-B radiation can be harmful to cells and can reduce photosynthetic activity, which could have an adverse impact on ecosystem productivity, among other things. In humans, particularly those with light skin, increasing exposure to UV-B radiation is correlated with increased risks of skin cancer, cataracts and other eye problems, and a suppressed immune system. Significant increases in skin cancers have already been recorded, especially in countries near the Antarctic ozone hole such as Chile and Australia.

## Efforts to Reduce Ozone Depletion

In 1986, when nations first began meeting in Montreal to consider ways to reduce ozone destruction, success seemed unlikely. CFCs were considered ideal compounds for numerous aspects of modern life, and many argued that the economic costs of a reduction would be staggering. While most scientists agreed that CFCs caused reduced ozone levels, confusing reports in the popular press on the role of volcanoes provided political cover for those who felt that a CFC reduction was unwarranted. Moreover, because chlorine in the stratosphere lasts for tens to hundreds of years, it appeared that a reduction in CFCs would not lead to changes in human exposure to UV radiation for a significant period of time, making immediate benefits seem small.

It was therefore a pleasant surprise when 24 nations in 1987 signed an agreement called the Montreal Protocol on Substances That Deplete the Ozone Layer, committing to concrete steps toward a solution and resolving to reduce CFC production 50 percent by the year 2000. This was the most far-reaching environmental treaty to date, in which global CFC exporters like the United States appeared in some ways to prioritize the protection of the global biosphere over their short-term economic self-interest. Moreover, a series of increasingly stringent amendments was eventually signed by more than 180 countries, requiring the elimination of CFC production and use in the developed world by 1996, although some developing nations were exempted from this strict deadline. In total, the protocol addressed 96 ozone-depleting compounds.

Because of these efforts, the concentration of chlorine in the stratosphere has stabilized at about 5 ppb (parts per billion) and should fall to about 1 ppb by 2100. The chlorine concentration reduction process is slow because CFCs are not easily removed from the stratosphere and in some recent years ozone depletion has continued to reach record levels. However, with the leveling off of chlorine concentrations, stratospheric ozone depletion should decrease in subsequent decades.

The number of additional skin cancers should eventually decrease as well, although this effect will take some time due to the long time it takes for cancer to develop.

---

**GAUGE YOUR PROGRESS**

✓ How does the stratospheric ozone layer form and why is it beneficial?

✓ What has caused the depletion of the ozone layer?

✓ What steps are being taken to reduce ozone depletion?

---

# Indoor air pollution is a significant hazard, particularly in developing countries

When we think of air pollution, we usually don't associate it with air inside our buildings, but indoor air pollution causes more deaths each year than outdoor air pollution. The amount of time one spends indoors depends on culture, climate, and economic situation. The quality of indoor air is highly variable and when polluting activities take place indoors, exposure to pollutants in a confined space can be a significant health risk.

## Developing Countries

In Chapter 13 we saw that around the world, more than three billion people use wood, animal manure, or coal indoors for heat and cooking. Biomass and coal are usually burned in open-pit fires that lack the proper mix of fuel and air to allow complete combustion. Usually, there is no exhaust system and little or no ventilation in the home, which makes indoor air pollution from carbon monoxide and particulates a particular hazard in developing countries (FIGURE 15.15). Exposure to indoor air pollution from cooking and heating increases the risk of acute respiratory infections, pneumonia, bronchitis, and even cancer. The World Health Organization estimates that indoor air pollution is responsible for more than 1.6 million deaths annually worldwide, and that 56 percent of those deaths occur among children less than 5 years of age. Ninety percent of deaths attributable to indoor air pollution are in developing countries.

## Developed Countries

There are a number of factors that have caused the quality of air in homes in developed countries to take on greater importance in recent decades. First of all, people in much of the developed world have begun to spend more and more time indoors. Homes have become

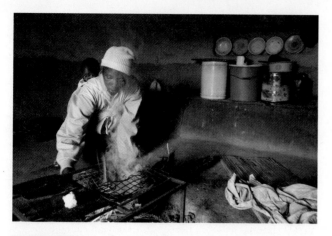

FIGURE 15.15 **Indoor air pollution in the developing world.** A woman and her child in their home in Zimbabwe.

more tightly insulated, which allows existing air to remain in contact with inhabitants for greater amounts of time. This is in part a result of improved insulation and tightly sealed building envelopes, which have been implemented in order to reduce energy consumption.

Finally, there are more materials in the home that are made from plastics and other petroleum-based materials that can give off chemical vapors. As FIGURE 15.16 shows, all of these factors combine to allow many possible sources of indoor air pollution to impact occupants.

**ASBESTOS Asbestos** is a long, thin, fibrous silicate mineral with insulating properties. For many years it was used as an insulator on steam and hot-water pipes and in shingles for the siding of buildings. The greatest health risk from asbestos has been respiratory diseases such as asbestosis and lung cancer found in very high rates among those who have mined asbestos. In manufactured form, asbestos is relatively stable and not dangerous until it is disturbed. When insulating materials become old or are damaged or disrupted, however, the fine fibers can become airborne and can enter the respiratory tract. In the United States, asbestos is no longer used as an insulating material, but it can still be found in older buildings, including schools. Removal of asbestos insulation must be done under tightly controlled conditions so that the fibers, typically less than 10 microns in diameter, cannot enter the air inside the building. Some studies

**Furniture; carpets; foam insulation; pressed wood**
Pollutant: VOCs

**Tobacco smoke**
Pollutants: Many toxic or carcinogenic compounds

**Household products: Pesticides; paints; cleaning fluids**
Pollutant: VOCs and others

**Old paint**
Pollutant: Lead

**Fireplaces; wood stoves**
Pollutant: Particulate matter

**Floor and ceiling tiles; pipe insulation**
Pollutant: Asbestos

**Leaky or unvented gas and wood stoves and furnaces; car left running in garage**
Pollutant: Carbon monoxide

**Rocks and soil beneath house**
Pollutant: Radon

FIGURE 15.16 **Some sources of indoor air pollution in the developed world.** A typical home in the United States may contain a variety of chemical compounds that could, under certain circumstances, be considered indoor air pollutants. [After U.S. EPA http://www.epa.gov/iaq/.]

have shown that when asbestos removal is complete, the concentration of asbestos in the air of the remediated building can be greater in the year after removal than during the year before removal. For this reason, it is absolutely necessary that asbestos removal be carefully done by qualified asbestos abatement personnel.

**CARBON MONOXIDE** Carbon monoxide has already been described as an outdoor air pollutant. It can be an even more dangerous indoor air pollutant, present as a result of malfunctioning exhaust systems on household heaters, most typically natural gas heaters. When the exhaust system malfunctions, exhaust air escapes into the living space of the house. Because natural gas burns relatively cleanly with little odor, a malfunctioning natural gas burner allows the colorless, odorless carbon monoxide to build up in a house without the occupants noticing, particularly if they are asleep. In the body, carbon monoxide binds with hemoglobin more efficiently than oxygen, thereby interfering with oxygen transport in the blood. Extended exposure to high concentrations of carbon monoxide in air can lead to oxygen deprivation in the brain and ultimately death.

**RADON** Radon-222, a radioactive gas that occurs naturally from the decay of uranium, exists in granitic and some other rocks and soils in many parts of the world. The map in FIGURE 15.17 shows areas for potential radon exposure in the United States. Humans can receive significant exposure to radon if it seeps into a home through cracks in the foundation or soil, or from drinking the water from underlying rock, soil, or groundwater. Radon-222 decays within four days to a radioactive daughter product, Polonium-210. Either the radon or the polonium can attach to dust and other particles in the air and then be inhaled. The EPA, the federal agency most responsible for identifying, measuring, and addressing environmental risks, estimates that about 21,000 people die each year from radon-induced lung cancer. This is 15 percent of yearly lung cancer deaths, and makes radon the second leading cause of lung cancer, after smoking. The EPA suggests that people test their homes for airborne radon. If radon levels are determined to be high, it is important to increase ventilation in the home. Other relatively inexpensive actions, such as sealing cracks in the basement, can be beneficial if radon is coming from underlying soil and bedrock.

**VOCS IN HOME PRODUCTS** Many volatile organic compounds are used in building materials, furniture, and other home products such as glues and paints. One of the most toxic is formaldehyde, which is used widely to manufacture a variety of building products such as particle board and carpeting glue. Formaldehyde is common in new homes and new products made from pressed wood, such as cabinets. The pungent smell that you may have noticed in a new home or one with new carpeting comes from formaldehyde, which is volatile and degasses over time. A high enough concentration in a confined space can cause a burning sensation in the eyes and throat, and breathing difficulties and asthma in some people. There is evidence that people develop a sensitivity to formaldehyde over time; though not very sensitive at first, with continued exposure they can experience irritation from ever smaller exposures. Formaldehyde has been shown to cause cancer in laboratory animals and has recently been suspected of being a human carcinogen.

Many other consumer products such as detergents, dry-cleaning fluids, deodorizers, and solvents may

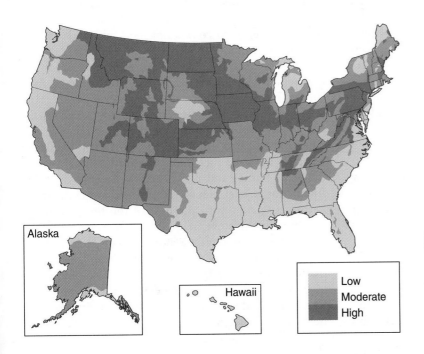

**FIGURE 15.17** **Potential radon exposure in the United States.** Depending on the underlying bedrock and soils, the potential for exposure to radon exists in houses in certain parts of the United States. [After U.S. Geological Survey and http://www.epa.gov/radon/zonemap.html.]

Low
Moderate
High

contain VOCs and can be harmful if inhaled. Plastics, fabrics, construction materials, and carpets may also release VOCs over time.

**SICK BUILDING SYNDROME** In newer buildings in developed countries in the temperate zone, more and more attention is given to insulation and prevention of air leaks in order to reduce the amount of heating or cooling necessary for a comfortable existence. This reduces energy use but may have the unintended side effect of allowing the buildup of toxic compounds and pollutants in an airtight space. In fact, such a phenomenon has been observed often enough in new or renovated buildings to be given a name: **sick building syndrome.** Because new buildings contain many products made with synthetic materials and glues that may not have fully dried out, a significant amount of off-gassing occurs. Usually, this means that the indoor levels of VOCs, hydrocarbons, and other potentially toxic materials are quite high. Sick building syndrome has been observed particularly in office buildings, where large numbers of workers have reported a variety of maladies such as headaches, nausea, throat or eye irritations, and fatigue.

The EPA has identified four specific reasons for sick building syndrome: inadequate or faulty ventilation; chemical contamination from indoor sources such as glues, carpeting, furniture, cleaning agents, and copy machines; chemical contamination in the building from outdoor sources such as vehicle exhaust transferred through the air intakes for the building; and biological contamination from inside or outside, such as molds and pollen.

> **GAUGE YOUR PROGRESS**
>
> ✓ What are the main sources of indoor pollution in the developing world?
>
> ✓ List common sources of indoor air pollution in the developed world.
>
> ✓ What is sick building syndrome?

 ## WORKING TOWARD SUSTAINABILITY

### A New Cook Stove Design

In China, India, and sub-Saharan Africa, people in 80 to 90 percent of households cook food using wood, animal manure, and crop residues as their fuel. Since women do most of the cooking, and young children are with the women of the household for much of the time, it is the women and young children who receive the greatest exposure to carbon monoxide and particulate matter. When biomass is used for cooking, concentrations of particulate matter in the home can be 200 times higher than EPA-recommended exposure limits. A wide range of diseases have been associated with exposure to smoke from cooking. Earlier in the chapter, we described that indoor air pollution is responsible for 1.6 million deaths annually around the world, and indoor cooking is a major source of indoor air pollution.

There are hundreds of projects underway around the world to enable women to use more efficient cooking stoves, ventilate cooking areas, cook outside whenever possible, and change customs and practices that will reduce indoor air pollution exposure. The use of an efficient cook stove will have the added benefit of consuming less fuel. This improves air quality and reduces the amount of fuel needed, which has environmental benefits and also reduces the amount of time that a woman must spend searching for fuel.

Increasing the efficiency of the combustion process requires the proper mix of fuel and oxygen. One effective method of ensuring a cleaner burn is the use of a small fan to facilitate greater oxygen delivery. However, most homes in developing countries with significant indoor air pollution problems do not have access to electricity. Therefore, some sort of internal source of energy for the fan is needed.

Two innovators from the United States developed a cook stove for backpackers and other outdoor enthusiasts who needed to cook a hot meal with little impact on the environment. They described their stove as needing no gasoline and no batteries, both desirable features for people carrying all their belongings on their backs. They soon realized that their stove, which could burn wood, animal manure, or crop residue, could make an important contribution in the developing world. This stove, called BioLite, (FIGURE 15.18) physically separates the solid fuel from the gases that form when the fuel is burned and allows the stove to burn the gases. In addition, a small electric fan, located inside the stove, harnesses energy from the heat of the fire and moves air through the stove at a rate that ensures complete combustion. The result is a more efficient burn, less fuel use, and less release of carbon monoxide and particulate matter. The stove weighs 0.7 kilograms (1.6 pounds).

How do they generate the electricity? They added a small semiconductor that generates electricity from the heat of the stove. All components of the stove except

**FIGURE 15.18** **BioLite cookstove.** This small stove, and others like it, has the potential to reduce the amount of firewood needed to cook a meal, and lower the amount of indoor air pollution emitted as well.

the semiconductor could be manufactured or repaired in a developing country. The BioLite stove won an international competition in early 2009 for the lowest emission stove. It was also the only stove in the competition that required no additional electricity inputs to operate. The company hopes to have the BioLite stove commercially available in 2011. There are many possible hurdles for those who are trying to introduce cleaner, more efficient cooking apparatus to the developing world. Manufacturing costs might make the stove difficult to afford. There has been some concern about possible reluctance to accept a different kind of cooking appliance. However, a number of studies in the developing world suggest that most households are quite receptive to using efficient stoves because of the benefits of improved air quality and reduced time spent obtaining fuel. Other promising ways to reduce fuel use and improve indoor air quality include the solar cooker shown in Figure 13.16 on page 359.

**References**

Bilger, B. 2009. Annals of Invention, Hearth Surgery, *The New Yorker*, December 21, p. 84; http://www.newyorker.com/reporting/2009/12/21/091221fa_fact_bilger#ixzz0sMCnDR00.

www.biolitestove.com, homepage of BioLite stove.

## REVISIT THE KEY IDEAS

- **Identify the major air pollutants and where they come from.**

Sulfur oxides, nitrogen oxides, carbon monoxide, carbon dioxide, particulates, and ozone are some of the major ground-level air pollutants. They come from both natural and human sources. Human activities that release these pollutants or their precursors include transportation, generation of electricity, space heating, and industrial processes.

- **Explain how photochemical smog and acid deposition are formed and describe the effects of each.**

Smog forms when sunlight, nitrogen oxides, and volatile organic compounds are present. The secondary pollutant ozone is a major component of photochemical smog. Sulfur is the dominant ingredient in sulfurous smog. Smog impairs respiratory function in human beings. Acidic deposition, which is comprised of hydrogen, sulfate, and nitrate ions, forms from both sulfur dioxide and nitrogen oxides. Acid deposition is harmful to aquatic organisms and can reduce forest productivity in sensitive ecosystems.

- **Examine various approaches to the control and prevention of outdoor air pollution.**

Air pollution is best controlled by increasing efficiency of processes that pollute, thereby reducing emissions, or by removing pollutants from fuel before combustion. After combustion occurs, filters and scrubbers remove pollutants from the exhaust stream before they can be released into the environment. The use of filters and scurbbers is preferable to trying to remove pollutants after they have been distributed throughout the environment.

- **Explain the causes and effects of stratospheric ozone depletion.**

Although ozone is the same gas that is a component of photochemical smog in the troposphere, in the stratosphere it is an important gas that absorbs harmful ultraviolet radiation. Chlorine-containing compounds such as CFCs lead to a reduction in stratospheric ozone. International efforts to reduce CFC emissions have helped stratospheric ozone levels to recover.

- **Discuss the hazards of indoor air pollution, especially in developing countries.**

Indoor air pollution is a more pervasive problem in developing than in developed countries. Carbon monoxide and particulates are the most harmful aspects of indoor air pollution and account for 1.6 million deaths annually worldwide. Fifty-six percent of those deaths occur among children less than 5 years of age.

# CHECK YOUR UNDERSTANDING

Questions 1–4 refer to the selections (a)–(e) below. Match the lettered item with the correct numbered statement.

(a) CO
(b) $CH_4$
(c) $NO_2$
(d) $SO_2$
(e) $PM_{10}$

1. A pungent reddish-brown gas often associated with photochemical smog

2. A corrosive gas from burning coal often associated with industrial smog

3. A dangerous indoor air pollutant

4. Emitted from both diesel and burning wood

5. All of the following are examples of primary air pollutants *except*
   (a) sulfur dioxide.
   (b) carbon dioxide.
   (c) tropospheric ozone.
   (d) nitrogen oxide.
   (e) particulates.

6. The greatest emission of sulfur dioxide comes from
   (a) on-road vehicles.
   (b) biofuels.
   (c) industrial processes.
   (d) electricity generation.
   (e) fires.

7. The least amount of nitrogen oxide emissions comes from
   (a) on-road vehicles.
   (b) fossil fuel combustion.
   (c) industrial processes.
   (d) electricity generation.
   (e) fires.

8. The accumulation of tropospheric ozone at night depends mainly upon the atmospheric concentration of
   (a) nitrogen dioxide.
   (b) volatile organics.
   (c) chlorofluorocarbons.
   (d) sulfates and nitrates.
   (e) nitric acid.

9. Under natural conditions the pH of rainfall is closest to
   (a) 8.5.
   (b) 7.1.
   (c) 5.6.
   (d) 4.5.
   (e) 3.1.

10. The effects of acid deposition include all of the following *except*
    (a) mobilization of metal ions from the soil into surface water.
    (b) increased numbers of salamanders in ponds and streams.
    (c) lower food sources for aquatic organisms.
    (d) erosion of marble buildings and statues.
    (e) erosion of painted automobile finishes and metals.

11. The World Health Organization estimates that over half of the deaths worldwide due to indoor air pollution occur among
    (a) children less than 5 years old.
    (b) elderly people over 65 years of age.
    (c) people who work in office buildings.
    (d) workers in the smelting industry.
    (e) workers who manufacture asbestos.

12. Two major factors involved in the conversion of primary pollutants into secondary pollutants are
    (a) sunlight and water.
    (b) sulfates and sunlight.
    (c) water and volatile organics.
    (d) nitrogen oxides and sulfates.
    (e) sulfur dioxide and sulfuric acid.

13. The pollutant *least* likely to be emitted from a smokestack would be
    (a) carbon monoxide.
    (b) carbon dioxide.
    (c) ozone.
    (d) sulfur dioxide.
    (e) particulates.

14. The EPA identifies all of the following as reasons for sick building syndrome *except*
    (a) faulty ventilation systems.
    (b) emissions from carpets and furniture.
    (c) contamination from outdoor air.
    (d) contamination from molds and pollen.
    (e) high levels of radon in the basement.

15. Which statement regarding the decreased levels of stratospheric ozone is *correct*?
    (a) Increased photosynthetic activity has been measured in the phytoplankton around Antarctica.
    (b) The largest decrease in the level of stratospheric ozone over the Arctic region occurs between January and April.

(c) Although the Montreal Protocol led to a reduction in the use of CFCs, it will have little effect on stratospheric ozone levels in the long term.

(d) There is no correlation between the incidence of cataracts and skin cancers and the lower levels of stratospheric ozone.

(e) The global crop yields of wheat, rice, and corn have increased since the reduction in CFC use.

16. The type(s) of ultraviolet radiation most strongly absorbed by ozone in the stratosphere is/are

(a) UV-A.

(b) UV-B.

(c) UV-C.

(d) UV-B and UV-C.

(e) UV-A and UV-B.

17. Natural sources of air pollution include all of the following *except*

(a) forest fires.

(b) volcanic eruptions.

(c) waterfalls.

(d) dust storms.

(e) conifer trees.

18. A thermal inversion

(a) rarely occurs in cities but is common in rural areas.

(b) helps remove pollutants from the atmosphere.

(c) leads to decreased amounts of ground-level smog.

(d) occurs when a warm air layer overlies a cooler layer.

(e) occurs when a cool air layer overlies a warmer layer.

# APPLY THE CONCEPTS

1. Examine the following ambient air data collected for Pittsburgh, Pennsylvania, and answer the following questions.

**2008 Monthly Ambient Air Monitoring Report, Pittsburgh, PA**

| Month | Monthly maximum ozone levels (ppb) | Monthly average ozone levels (ppb) | Monthly average solar radiation (watts/m²) |
|---|---|---|---|
| January | 37 | 14 | 65 |
| February | 49 | 15 | 63 |
| March | 56 | 23 | 86 |
| April | 76 | 31 | 81 |
| May | 75 | 27 | 152 |
| June | 77 | 32 | 208 |
| July | 95 | 31 | 215 |
| August | 92 | 27 | 204 |
| September | 89 | 20 | 153 |
| October | 48 | 14 | 109 |
| November | 57 | 12 | 64 |
| December | 30 | 14 | 45 |

*Source:* http://www.ahs2.dep.state.pa.us/aq_apps/aadata/default .asp?varAction=form&varform=2.

(a) Based on the National Ambient Air Quality Standards (NAAQS), the 2008 standard for ozone states that the average ozone levels are not to exceed 0.075 ppm (75 ppb) in any 8-hour period. Is Pittsburgh in compliance with this standard? Discuss how this NAAQS may not truly reflect the overall air quality.

(b) Ozone is classified as a secondary pollutant. Identify the primary pollutants necessary for its formation and describe how tropospheric ozone is formed.

(c) Identify two relationships between the data presented. Apply these relationships to your answer in (b) to explain the pattern of ozone levels in Pittsburgh.

(d) Explain how the same ozone that is harmful in the troposphere is beneficial in the stratosphere.

2. The following appeared in *Medical News Today* on November 28, 2003.

**Ultraviolet light in ventilation system cures sick building syndrome**

So-called sick building syndrome (SBS) is caused by the increasing use of air conditioning in modern offices. Bacteria can build up in the ventilation systems and leave office workers suffering from a range of problems, such as breathing difficulties, headaches, sore throats, stuffy noses and itchy eyes....

Researchers from the Montreal Chest Institute in Canada tested three offices in the city where people suffered from SBS. They installed UVGI (ultraviolet germicidal irradiation) in the ventilation systems of the buildings and compared reports of illness when the lights were turned on with sickness rates when they were off.

The lights resulted in a 20 percent reduction in all symptoms. There were 40 percent fewer complaints about respiratory problems and a 30 percent reduction in people complaining of stuffy noses. Muscular complaints also halved, according to the study.

Scientists said the UVGI killed the bacteria and molds in ventilation systems. Installing the systems would be relatively cheap and could save companies millions by reducing rates of sickness...

Excerpted from this article URL: http://www.medicalnewstoday .com/articles/4757.php.

(a) According to the article, what is the cause of sick building syndrome and what are the advantages of this cure?

(b) What other sources of indoor pollution contribute to sick building syndrome? Will ultraviolet light be effective against these sources? Suggest *one* additional control measure that would be effective against SBS.

(c) Will the cure mentioned in the article be effective against indoor pollution in developing countries? Discuss why or why not.

## MEASURE YOUR IMPACT

**Mercury Release from Coal** Coal-burning power plants are the largest source of mercury emissions in the United States.

| Coal Type | Mercury concentration ($\mu$g/g) | Amount used per year (million metric tons) | Total mercury emissions (tons/year) |
|---|---|---|---|
| U.S. average | 0.1 | 1.5 | _____ |
| High-mercury coal | 0.5 | 1.5 | _____ |
| Low-mercury coal | 0.04 | 1.5 | _____ |

Coal contains impurities at a wide range of concentrations. The average mercury concentration in typical coal found in the United States is approximately 0.1 $\mu$g/g. A concentration of 0.5 $\mu$g/g is toward the higher end of the range. A concentration of 0.04 $\mu$g/g is at the lower end of the range

(a) Fill in the table to calculate the amount of mercury released by a typical coal-burning power plant that burns 1.5 million metric tons of coal per year. Observe the difference in mercury released depending on the mercury concentration of the coal burned.

(b) Assuming that the power plant burns coal with average mercury content and serves 100,000 households, each with roughly equal energy consumption, how much mercury (in grams) can be attributed to each household per year?

(c) Identify three different ways you could reduce mercury emissions from coal-burning power plants.

How can you use air quality data to help avoid respiratory injury from air pollution? This Engage Your Environment exercise examines practical ways to reduce your contribution to air pollution and to stay healthy when participating in outdoor activities. *See Engage Your Environment Activity 15 at the end of this book.*

## CHAPTER 16

# Waste: Solid Waste Generation and Disposal Systems

## Paper or Plastic?

Polystyrene is a plastic polymer that has high insulation value. More commonly known by its trade name, Styrofoam, it is particularly useful for food packaging because it minimizes temperature changes in both food and beverages. Polystyrene is lighter, insulates better, and is less expensive than the alternatives. However, a number of years ago, polystyrene was deemed harmful to the environment because, like all plastics, and bark, a renewable material. A polystyrene cup requires 3 grams of petroleum, a nonrenewable material, but no wood or bark. About twice as much energy, and much more water, is needed to make the paper cup. A paper cup of the exact same size as a polystyrene cup is substantially heavier, requiring more energy to transport a paper cup to the location where it will be used. Air emissions are different in the manufacturing of each cup and it is difficult to say which are more harmful to

> Today, there is still no definitive answer as to whether the paper cup or the polystyrene cup causes less harm to the environment.

it is made from petroleum and because it does not decompose in landfills. In response to public sentiment, most food businesses greatly reduced or eliminated their use of polystyrene. All over the country, schools, businesses, and public institutions have purged their cafeterias of polystyrene cups and most have replaced them with disposable paper cups.

Was the elimination of polystyrene the environmental victory many thought it to be? It is hard to quantify the exact environmental benefits and costs of using a paper cup versus using a polystyrene cup. For example, because a paper cup does not insulate as well as a Styrofoam cup, paper cups filled with hot drinks are too hot to hold and vendors often wrap them in a cardboard band that becomes additional waste. To fully quantify the environmental costs and benefits of each type of cup, one must create a list of inputs and outputs related to their manufacture, use, and disposal. This input-output analysis of all energy and materials is also called a *cradle-to-grave,* or *life-cycle, analysis.* When we make a list of all the materials and all the energy required to produce and then dispose of each type of cup, we find that it is not easy to determine which choice is better for the environment.

One study found that making a paper cup requires approximately 2 grams of petroleum along with 33 grams of wood

the environment. Since more energy is needed to make and transport a paper cup, it is reasonable to assume that using it generates more air pollution. A paper cup is normally used once or at most a few times while the polystyrene cup can, in theory, be reused many times. It is possible that the paper cup could be recycled or composted, but, in reality, both are usually thrown away after one use. There has been concern among some scientists—but no consensus—that a polystyrene cup might leach chemicals from the plastic into the coffee; if this is true, using a paper cup could pose less risk to human health. However, without proper disposal, the bleach used to make paper cups in a paper mill, and small amounts of the associated by-product, dioxin, can cause harm to aquatic life when the water is discharged into rivers and streams. Incineration of both cups could yield a small amount of energy. In a landfill, the paper cup will degrade and eventually produce methane gas, while the polystyrene cup, because it is made of an inert material, will remain there for a very long time.

Weighing these and other factors, one study concluded ▶

Polystyrene cup.

Paper cup.

that a polystyrene cup is more desirable than a paper cup for one-time use. Critics of that study felt that the author did not consider the toxicity of emissions from making polystyrene, the exposure of workers to those emissions, the impact of both cups on global carbon dioxide emissions, or the possibility of making the cup from materials other than petroleum or paper. Today, there is still no definitive answer as to whether the paper cup or the polystyrene cup causes less harm to the environment.

These types of studies illustrate that analyzing the environmental effects of the products we use is complex since it involves the synthesis of many aspects of environmental studies. Not only does it include science, ethics, and social judgments, it also necessitates a systems-based understanding of waste generation, waste reduction, and waste disposal. There is widespread agreement that paper and Styrofoam are not the only alternatives. Reusable mugs are a possibility but they would require consideration of a host of different life-cycle issues such as greater inputs for manufacturing and energy consumption, as well as the water needed to clean them after each use. ■

*Sources:* M. B. Hocking, Paper versus polystyrene: A complex choice, *Science* 251 (1991): 504–505. DOI: 10.1126/science.251.4993.504; M. Brower and L. Warren, *The Consumer's Guide to Effective Environmental Choices* (Three Rivers Press, 1999).

## UNDERSTAND THE KEY IDEAS

As life in many countries has become increasingly dependent on disposable items, the generation of solid waste has become more of a problem for both the natural and human environments. This chapter examines solid waste generation and disposal systems.

After reading this chapter you should be able to

- define waste generation from an ecological and systems perspective.

- describe how each of the three Rs—Reduce, Reuse, and Recycle—as well as composting can avoid waste generation.

- explain the implications of landfills and incineration.

- understand the problems associated with the generation and disposal of hazardous waste.

- present a holistic approach to avoiding waste generation and to treating solid waste.

# Humans generate waste that other organisms cannot use

Throughout this book, we've described systems in terms of inputs, outputs, and internal changes. In an ecological system, plant materials, nutrients, water, and energy are the inputs. In human systems, inputs are very similar but may contain materials manufactured by humans as well. Within the system, humans use these inputs and materials to produce goods. Outputs include anything not useful or consumed, and nonuseful products generated within the system; these outputs we call **waste**. FIGURE 16.1 shows a diagram of the relationship between inputs and outputs in a human system.

Notice that we are defining waste as the nonuseful products of a system. But how do we determine what is useful? The detritivores we described in Chapter 3 recycle the waste from animals and plants, using the energy and nourishment they obtain from them and turning the remainder into compost or topsoil that nourishes other organisms. Dung beetles, for example, live on the energy and nutrients contained within elephant and other dung; in the natural world, this is not waste, it is food (FIGURE 16.2). Even humans make use of animal waste—for fertilizer, heat, and cooking fuel. In most situations, the waste of one organism becomes a source of energy for another. Humans are the only organisms that produce waste others cannot use. To explore this further, we need to learn why materials generated by humans become waste, and what that waste contains.

**Inputs**

Raw materials, energy

**Outputs**

Material that can be recycled or disposed of

Waste energy

Use (and reuse) of a product

FIGURE 16.1 **The solid waste system.** Waste is a component of a human-dominated system in which products are manufactured, used, and eventually disposed of (arrows are not proportional). At least some of the waste of this system may become the input of another system.

## The Throw-Away Society

Until a society becomes relatively wealthy, it generates little waste. Every object that no longer has value for its

**FIGURE 16.2** **A dung beetle.** This dung beetle is using elephant waste as a resource. The waste of most organisms in the natural world end up being a resource for other organisms.

original purpose becomes useful for something else. In 1900 in the United States, virtually all metal, wood, and glass materials were recycled, although no one called it recycling back then. Those who collected recyclables were called junk dealers, or scrap metal dealers. For example, if a wooden bookcase broke and was unable to be repaired, the pieces could be used to make a step stool. When the step stool broke, the wood was burned in a wood stove to heat the house. After World War II and the rapid population growth that occurred in the United States, consumption patterns changed. The increasing industrialization and wealth of the United States, as well as cultural changes, made it possible for people to purchase household conveniences that could be used and thrown away. Families were large, and people were urged to buy "labor-saving" household appliances and also to dispose of them as soon as a new model was available. *Planned obsolescence,* the design of

a product so that it will need to be replaced within a few years, became a typical characteristic of everything from toasters to cars. TV dinners, throw-away napkins, and disposable plates and forks became common. In the 1960s disposable diapers became widely available and eventually replaced reusable cloth diapers. The components of household materials also changed. Objects frequently contained mixtures of different materials, making them harder to use for another purpose or to recycle. The United States became the leader of what came to be known as the "throw-away society."

Refuse collected by municipalities from households, small businesses, and institutions such as schools, prisons, municipal buildings, and hospitals is known as **municipal solid waste (MSW).** The Environmental Protection Agency (EPA) estimates that approximately 60 percent of MSW comes from residences and 40 percent from commercial and institutional facilities. FIGURE 16.3 shows the trend toward greater generation of MSW both overall and on a per capita basis from 1960 to 2008. In the first 47 years of this period, the total amount of MSW generated in the United States increased from 80 million metric tons (88 million U.S. tons) to 232 million metric tons (255 million U.S. tons) per year. In the last year for which there are data (2008), the total amount of MSW actually decreased by a small amount. The increase for all but the last year can be explained in part by a growth in population and in part because individuals have been generating increasing amounts of MSW. In the year 2008, average waste generation was 2.0 kg (4.5 pounds) of MSW per person per day. Waste generation varies by season of the year, socioeconomic status of the individual generating the waste, and even geographic location within the country. In addition, there are many other kinds of waste that are generated in the United States: agricultural waste, mining waste,

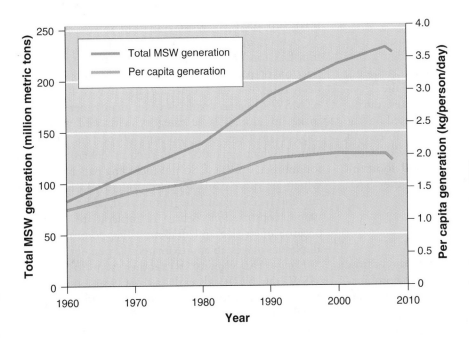

**FIGURE 16.3** **Municipal solid waste generation in the United States, 1960–2008.** Total MSW generation and per capita MSW generation had been increasing from 1960 through 2008. It has recently started to decrease. [After U.S. Environmental Protection Agency, *MSW Generation, Recycling and Disposal in the United States: Facts and Figures for 2008.*]

and industrial waste are just three examples. For most of these other kinds of waste, material is deposited and processed on-site rather than being transferred to a different location for disposal. Although some of these other categories generate a much greater percentage of yearly total solid waste, this chapter focuses on MSW.

Waste generation in much of the rest of the world stands in contrast to the United States. In Japan, for example, each person generates an average of 1.1 kg (2.4 pounds) of MSW each day. The 2010 UN-HABITAT estimate for the developing world is 0.55 kg (1.2 pounds) per person per day. The estimate for the developed world ranges from 0.8 to 2.2 kg (1.8–4.8 pounds) per person per day. Some indigenous people create virtually no waste per day, with as much as 98 percent of MSW being used for something by someone. The remaining 2 percent ends up in a landfill or waste pile. Even there, impoverished people scavenge and reuse some of the discarded material (**FIGURE 16.4**).

Developing countries have become responsible for a greater portion of global MSW because of their growing populations and because developing countries, since they are producing more of the goods used in the developed world, are left with the waste generated during production. For example, computers, invented in the United States and assembled from parts made in Taiwan, Singapore, and China, are discarded in many developing countries—evidence that the ecological footprint of both the manufacturing process and the user has a global spread.

## Content of the Solid Waste Stream

MSW is comprised of the things we use and then throw away. The goods that we use are generally a combination of organic items, fibers, metals, and plastics, made from petroleum. A certain amount of waste is generated during any manufacturing process. Waste is also generated from packaging and transporting goods.

Consumers use the materials, products, and goods they possess. Depending on what these are and how they are used, they can remain in the consumer use system for a long time. For example, a ceramic plate or drinking mug might last for 5 to 10 years. In most cases, a disposable paper cup leaves the system within minutes or hours after it is used. Ultimately, all products wear out, lose their value, or are discarded. At this point they enter the **waste stream**—the flow of solid waste that is recycled, incinerated, placed in a solid waste landfill, or disposed of in another way.

**COMPOSITION OF MUNICIPAL SOLID WASTE** FIGURE 16.5a shows the data for MSW composition in the United States in 2008 by category. The category "paper," which includes newsprint, office paper, cardboard, and boxboard such as cereal and food boxes, comprised 31 percent of the 231 million metric tons (254 million U.S. tons) of waste generated before recycling. The fraction of paper in the solid waste stream has been decreasing; less than a decade ago it was 40 percent of MSW. Organic materials other than paper products make up another large category, with yard waste and food scraps together comprising 26 percent of MSW. Wood, which includes construction debris, accounts for another 7 percent. So, not including paper products, which are more easily recycled, roughly 33 percent of current MSW could be composted, although some wood construction debris is difficult to compost because of its size and thickness. The combination of all plastics makes up approximately 12 percent of MSW.

Long-term viability is another way to consider MSW: durable goods will last for years, nondurable goods are disposable, and compostable goods are those largely made up of organic material that can decompose

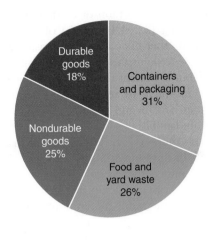

**(a)** Breakdown of MSW by composition

**(b)** Breakdown of MSW by source

**FIGURE 16.5** **Composition and sources of municipal solid waste (MSW) in the United States.** (a) The composition, by weight, of MSW in the United States in 2008 before recycling. Paper, food, and yard waste comprise more than half of the MSW by weight. (b) The major sources, by weight, of MSW in the United States. Containers and packaging comprise almost one-third of MSW. [After U.S. Environmental Protection Agency, *MSW Generation, Recycling and Disposal in the United States: Facts and Figures for 2008.*]

under proper conditions. As **FIGURE 16.5b** shows, containers and packaging make up 31 percent of MSW and are typically intended for one use. Food and yard waste are 26 percent, and nondurable goods such as newspaper, white paper, printed products like telephone books, clothing, and plastic items like utensils and cups are 25 percent of the solid waste stream. Durable goods such as appliances, tires, and other manufactured products make up 18 percent of the waste stream. In addition to considering waste by weight, there is sometimes merit in considering waste by volume, especially when con-

sidering how much can be transported per truckload and how much will fit in a particular landfill.

**E-WASTE** Electronic waste, or e-waste, is one component of MSW that is small by weight but very important and rapidly increasing. Consumer electronics that include televisions, computers, portable music players, and cell phones account for roughly 2 percent of the waste stream. This may not sound like a large amount but the environmental effect of these discarded objects is far greater than represented by their weight. The older-style

**FIGURE 16.6** **Electronic waste recycling in China.** Much of the recycling is done without protective gear and respirators that would typically be used in the United States. In addition, children are sometimes part of the recycling workforce in China.

cathode-ray tube (CRT) television or computer monitor contains 1 to 2 kg (2.2–4.4 pounds) of the heavy metal lead as well as other toxic metals such as mercury and cadmium. These toxic metals and other components can be extracted, but at present there is little formalized infrastructure or incentive to recycle them. However, many communities have begun voluntary programs to divert e-waste from landfills.

It generally costs more to recycle a computer than to put it in a landfill. In the United States, most electronic devices are not designed to be easily dismantled after they are discarded. The EPA estimates that approximately 18 percent of televisions and computer products discarded in 2007 were sent to recycling facilities. Unfortunately, much e-waste from the United States is exported to China where adults as well as some children separate valuable metals from other materials using fire and acids in open spaces with no protective clothing and no respiratory gear (FIGURE 16.6). So even when consumers do send electronic products to be recycled, there is a good chance that it will not be done properly.

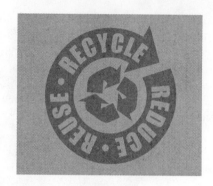

**FIGURE 16.7** **Reduce, Reuse, Recycle.** This is a popular slogan because it emphasizes the actions to take in the proper order.

### GAUGE YOUR PROGRESS

✓ What are the main sources of waste?

✓ What is the relationship between availability of and access to resources and the production of waste?

✓ How does the solid waste stream differ between a developed and a developing country?

# The three Rs and composting divert materials from the waste stream

Starting in the 1990s, people in the United States began to promote the idea of diverting materials from the waste stream with a popular phrase "**Reduce, Reuse, Recycle**," also known as the **three Rs.** The phrase incorporates a practical approach to the subject of solid waste management, with each technique presented in the order of benefit to the environment, from the most desirable to the least (FIGURE 16.7).

## Reduce

"Reduce" is the first choice among the three Rs because reducing inputs is the optimal way to achieve a reduction in solid waste generation. This strategy is also known as *waste minimization* and *waste prevention*. If the input of materials to a system is reduced, the outputs will also be reduced, in this case, the amount of material

that must be discarded. One approach, known as **source reduction,** seeks to reduce waste by reducing, in the early stages of design and manufacture, the use of materials—toxic and otherwise—destined to become MSW. In many cases, source reduction will also increase energy efficiency because it produces less waste to begin with, avoiding disposal processes. Since fewer resources are being expended, source reduction also provides economic benefits.

Source reduction can be implemented both on individual and on corporate or institutional levels. For example, if an instructor has two pages of handout material for a class, she could reduce her paper use by 50 percent if she provided her students with double-sided photocopies. A copy machine that can automatically make copies on both sides of the page might use more energy and require more materials in manufacturing than a copy machine that only prints on one side, but with up to half the number of copies needed, the overall energy used to produce them over time will probably be less. Further source reduction could be achieved if the instructor did not hand out any sheets of paper at all but sent copies to the class electronically, with class members refraining from printing out the documents.

Source reduction in manufacturing will result from reducing the materials that go into packaging. If the new packaging can provide the same amount of protection to the product with less material, successful source reduction has occurred. Consider the incremental source reduction that occurred with purchasing music: Music compact discs used to be packaged in large plastic sleeves that were three times the size of the CD. Today, most CDs are wrapped with a small amount of plastic material that just covers the CD case. Many people no longer purchase CDs and instead download their music from the Web. Less wrapping on CDs and avoiding the purchase of CDs are both examples of source reduction.

Source reduction can also be achieved by material substitution. In an office where workers drink water and coffee from paper cups, providing every worker with a reusable mug will reduce MSW. In some categorization schemes, this could be considered reuse rather than source reduction. Nevertheless, cleaning the mugs will require water, energy to heat the water, soap, and processing of wastewater. The break-even point, beyond which there are gains achieved by using a ceramic mug, will depend on a variety of factors, but it might be at 50 uses. The break-even point is shorter for a reusable plastic mug, in part because less energy is required to manufacture the plastic mug and, because it is lighter, less energy is used to transport it. Source reduction may also involve substituting less toxic materials or products in situations where manufacturing utilizes or generates toxic substances. For example, switching from an oil-based paint that contains toxic petroleum derivatives to a relatively nontoxic latex paint is a form of source reduction.

Car manufacturer Subaru of America utilizes all aspects of the Reduce, Reuse, Recycle strategy in its zero-waste manufacturing plant. Subaru manufactured over 200,000 automobiles in the United States in 2009 and claims to send absolutely no waste to landfills. The company reuses materials such as shipping boxes and packaging material and reclaims solvents and chemicals after they are no longer useful. The plant diligently recycles all materials that would otherwise go to landfills. The remaining 1 percent of material that cannot be diverted in any other way is converted into energy, in a waste-to-energy plant, a process we will discuss later in this chapter.

## Reuse

**Reuse** of a soon-to-be-discarded product or material, rather than disposal, allows a material to cycle within a system longer before becoming an output. In other words, its mean residence time in the system is greater. Optimally, no additional energy or resources are needed for the object to be reused. For example, a mailing envelope can be reused by covering the first address with a label and writing the new address over it. Here we are increasing the residence time of the envelope in the system and reducing the waste disposal rate. Or we could reuse a disposable polystyrene cup more than once, though reuse might involve cleaning the cup, adding some energy cost and generating some wastewater. Sometimes reuse may involve repairing an existing object, costing time, labor, energy, and materials.

Energy may also be required to prepare or transport an object for reuse by someone other than the original user. For example, certain companies reuse beverage containers by shipping them to the bottling factory where they are washed, sterilized, and refilled. Although energy is involved in the transport and preparation of the containers, it is still less than the energy that would be required for recycling or disposal.

Reuse is still common in many countries. It was common practice in the United States before we became a "throw-away society," and it is still practiced in many ways that we might think of as reuse. For example, people often reuse newspapers for animal bedding or art projects. Many businesses and universities have surplus-equipment agents who help find a home for items no longer needed. Flea markets, swap meets, and even popular Web sites such as eBay, craigslist, and Freecycle are all agents of reuse.

## Recycle

**Recycling** is the process by which materials destined to become MSW are collected and converted into raw materials that are then used to produce new objects. We divide recycling into two categories: *closed-loop* and *open-loop*. FIGURE 16.8 shows the process for each. **Closed-loop recycling** is the recycling of a product into the same product. Aluminum cans are a familiar example; they are collected, brought to an aluminum plant, melted down, and made into new aluminum cans. This process is called a closed loop because in theory it is possible to keep making aluminum cans from only old aluminum cans almost indefinitely; the process is thus similar to a closed system. In **open-loop recycling,** one product, such as plastic soda bottles, is recycled into another product, such as polar fleece jackets. Although recycling plastic bottles into other materials avoids sending the plastic bottles to a landfill, it does not reduce demand for the raw material, in this case petroleum, to make plastic for new bottles.

Recycling is not new in the United States, but over the past 25 years it has been embraced enthusiastically by both individuals and municipalities in the belief that it measurably improves environmental quality. The graph in FIGURE 16.9 shows both the increase in the weight of MSW in the United States from 1960 to 2008 and the increase in the percent of waste that was recycled over the same period of time. Recycling rates have increased in the United States since 1975, and today we recycle roughly one-third of MSW. In Japan, recycling rates are closer to 50 percent. Some colleges and universities in the United States report recycling rates of 60 percent for their campuses.

Extracting resources from Earth requires energy, time, and usually a considerable financial investment. As we have seen, these processes generate pollution. Therefore on many levels, it makes sense for manufacturers to utilize resources that have already been extracted. Today, many communities are adopting zero-sort recycling programs. These programs allow residents to mix all types of recyclables in one container that is deposited on the curb outside the home or brought to a transfer station. This saves time for residents who were once required to sort materials. At the sorting facility, workers sort the materials destined for recycling into whatever categories

**FIGURE 16.8** **Closed- and open-loop recycling.** (a) In closed-loop recycling, a discarded carpet can be recycled into a new carpet, although some additional energy and raw material is needed. (b) In open-loop recycling, a material such as a beverage container is used once and then recycled into something else, such as a fleece jacket.

**(a)** Closed-loop recycling

**(b)** Open-loop recycling

are in greatest demand at a given time and offer the greatest economic return (**FIGURE 16.10**). The markets for glass and paper are highly volatile. There is always a demand for materials such as aluminum and copper, but at times there is little demand for recycled newspapers. In such periods, the single-stream sorting facility might pull out newsprint to sell or give to local stables for use as horse bedding. At other times, when demand for paper is higher, the newsprint might be kept with other paper to be recycled into new paper products.

Nevertheless, because recycling requires time, processing, cleaning, transporting, and possible modification

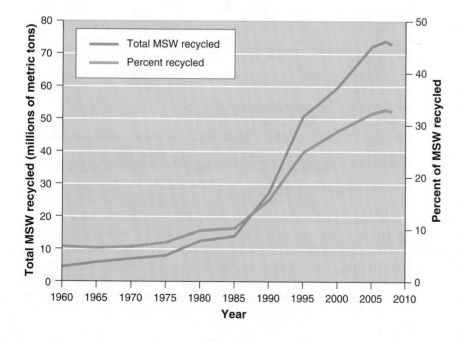

**FIGURE 16.9** **Total weight of Municipal Solid Waste recycled and percent of MSW recycled in the United States over time.** Both the total weight of MSW that is recycled and the percentage of MSW that is recycled have increased over time. [After U.S. Environmental Protection Agency, *MSW Generation, Recycling and Disposal in the United States: Facts and Figures for 2008.*]

FIGURE 16.10 A mixed single-stream solid waste recycling facility in San Francisco, California. With single-stream recycling, also called no-sort or zero-sort recycling, consumers no longer have to worry about separating materials.

before the waste is usable as raw material, it does require more energy than reducing or reusing materials. Such costs caused New York City to make a controversial decision in 2002 that suspended glass and plastic recycling. This was a major shift in policy for the city, which had been encouraging as much recycling as possible, including collection of mixed recyclables—glass, plastic, newspapers, magazines, and boxboard—at the same time as waste materials. After collection, the recyclables were sent to a facility where they were sorted. According to city officials, the entire process of sorting glass and plastic from other recyclables and selling the material was not cost-effective. In 2004, the recycling of all materials was reinstated in New York City. Today, the goal in most recycling programs is to maximize diversion from the landfill, even if that means collecting materials that have less economic value. However, given that the price paid for glass and plastic fluctuates widely, many communities periodically have difficulty finding buyers for them.

The New York City case is just one example of why recycling is the last choice among the three Rs. This doesn't mean that we should abandon recycling programs. Not only does it work well for materials such as paper and aluminum, it also encourages people to be more aware of the consequences of their consumption patterns. Nevertheless, in terms of the environment, source reduction and reuse are preferable. The environmental implications of recycling are considered further in Science Applied "Is Recycling Always Good for the Environment?" on page 490.

## Compost

While diversion from the landfill is usually referred to as the three Rs, there is one more diversion pathway that is equally, if not more, important—composting. Organic materials such as food and yard waste that end up in landfills cause two problems. Like any material, they take up space. But unlike glass and plastic that are chemically inert, organic materials are also unstable. As we will see later in the chapter, the absence of oxygen in landfills causes organic material to decompose anaerobically, which produces methane gas, a much more potent greenhouse gas than carbon dioxide.

An alternative way to treat organic waste is through *composting*. **Compost** is organic matter that has decomposed under controlled conditions to produce an organic-rich material that enhances soil structure, cation exchange capacity, and fertility. Vegetables and vegetable by-products such as cornstalks, grass, animal manure, yard wastes such as grass clippings, leaves, and branches, and paper fiber not destined for recycling are suitable for composting. Normally, meat and dairy products are not composted because they do not decompose as easily, produce foul orders, and are more likely to attract unwanted visitors such as rats, skunks, and raccoons.

Outdoor compost systems can be as simple as a pile of food and yard waste in the corner of a yard, or as sophisticated as compost boxes and drums that can be rotated to ensure mixing and aeration. From the decomposition process described in Chapter 3, we are already familiar with the process that takes place during composting. In order to encourage rapid decomposition, it is important to have the ratio of carbon to nitrogen (C:N) that will best support microbial activity— about 30:1. While it is possible to calculate the carbon and nitrogen content of each material you put into a compost pile, most compost experts recommend layering dry material such as leaves or dried cut grass—normally brown material—with wet material such as kitchen vegetables—normally green material. This will provide the correct carbon to nitrogen ratio for optimal composting. Frequent turning or agitation is usually necessary to ensure that decomposition processes are aerobic and to maintain appropriate moisture levels; otherwise the compost pile will produce methane and associated gases, emitting a foul odor. If the pile becomes particularly dry, water needs to be added. Although many people assume that

**FIGURE 16.11 Composting.** Good compost has a pleasant smell and will enhance soil quality by adding nutrients to the soil and by improving moisture and nutrient retention. In a compost pile that is turned frequently, compost can be ready to use in a month or so.

a compost pile must smell bad, with proper aeration and not too much moisture, the only odor will be that of fresh compost in 2 to 3 months' time (**FIGURE 16.11**).

Large-scale composting facilities currently operate in many municipalities in the United States. Some facilities are indoors, but most employ the same basic process we have described, though on a much larger scale. **FIGURE 16.12** shows the process. Organic material is piled up in long, narrow rows of compost. The material is turned frequently, exposing it to a combination of air and water that will speed natural aerobic decomposition. As with household composting, the organic material must include the correct combination of green (fresh) and brown (dried) organic material so that the ratios of carbon and nitrogen are optimal for bacteria. Various techniques are used to turn the organic material over periodically, including rotating blades that move through the piles of organic material or a front loader that turns over the piles. The respiration activity of the microbes generates enough heat to kill any pathogenic bacteria that may be contained in food scraps, which is typically a concern only in large municipal composting systems. If the pile becomes too hot, it should be turned more frequently. If the pile doesn't become hot enough, operators should check to make sure their C:N ratio is optimal, or they should slow the turnover rate. Within a matter of weeks, the organic waste becomes compost. Large-scale municipal composting systems with relatively little mechanization and labor may take up to a year to create a finished compost.

It is not necessary to have an outdoor space to compost household waste; composting is possible even in a city apartment or a dorm room. It is even possible to set up a composting system in a kitchen or basement.

① Waste is dumped in tipping area.

② Compostable and noncompostable materials are separated.

③ Noncompostable material is removed to landfill.

Screen

To landfill

Mixing drums

Organic waste: Newspaper, leaves and grass, food scraps, woody materials, etc.

④ Compostable material is aerated and turned one or more times (to speed up aerobic respiration) for a period of 30 days to 1 year.

Finished compost

⑥ Finished compost is transported for use.

⑤ Composted material is allowed to cure.

**FIGURE 16.12 A municipal composting facility.** A typical facility collects almost 100,000 metric tons of food scraps and paper per year and turns it into usable compost. Most facilities have some kind of mechanized system to allow mixing and aeration of the organic material, which speeds conversion to compost.

The very popular book *Worms Eat My Garbage: How to Set Up and Maintain a Worm Composting System* by Mary Appelhof has encouraged thousands of individuals across the country to compost kitchen waste using red wiggler worms. A small household recycling bin is large enough to serve as a worm box. As with an outdoor compost pile, a properly maintained worm box does not give off bad odors.

The composting process does take time and space. Source separation can be an inconvenience or, in some situations, not possible. Also, in certain environments, storing materials before they are added to the compost pile can attract flies or vermin. Finally, the compost pile itself can attract unwanted animals such as rats, skunks, raccoons, and even bears. But because compost is high in organic matter, which has a high cation exchange capacity and contains nutrients, it enhances soil quality when added to agricultural fields, gardens, and lawns.

---

**GAUGE YOUR PROGRESS**

✓ What are the three Rs? What are the benefits and disadvantages of each?

✓ What is the difference between open- and closed-loop recycling?

✓ Why is composting an important activity in waste management?

---

# Currently, most solid waste is buried in landfills or incinerated

Historically, people deposited their waste in open dumps, where it attracted pests and polluted the air and water. Though open dumps are now rare in developed countries, they still exist in the developing world, where they pose a considerable health hazard.

Beginning in the 1930s in the United States, with growing public opposition to open dumps, the most convenient locations for disposal of MSW became holes in the ground created by the removal of soil, sand, or other earth material used for construction purposes. When people filled those holes with waste, the sites became known as *landfills*. Initially, there were few concerns about what material went into a landfill. Those responsible for collecting and disposing of solid waste in landfills did not realize that components of the MSW could generate harmful runoff and **leachate**—the water that leaches through the solid waste and removes various chemical compounds with which it comes into contact. Nor did they recognize the harm a landfill could cause when located near sensitive features of the landscape such as aquifers, rivers, streams, drinking water supplies, and human habitation.

Today, some environmental scientists would say that we should not use landfills at all, and in later sections of this chapter we will discuss alternative means of waste disposal. Since landfills are still a component of solid waste management in the United States today, we can make them much less harmful than the ones utilized in the past. In this section we will examine landfill basics, how a landfill is sited, the problems of using landfills, and incineration as an option for the disposal of MSW.

## Landfills

When certain materials reach the end of their useful lives, or are no longer wanted or needed, they end up destined for disposal. As **FIGURE 16.13** shows, in the United States a third of our waste is recovered through reuse and recycling, while more than half is discarded. The remainder is converted into energy through incineration. Over the past few decades, the amount of material in the United States that has been reused, recycled, and composted has been increasing.

**LANDFILL BASICS** In the United States today, repositories for MSW, known as **sanitary landfills,** are engineered ground facilities designed to hold MSW with as little contamination of the surrounding environment as possible. These facilities, like the one illustrated in **FIGURE 16.14**, generally utilize a variety of technologies that safeguard against the problems of traditional dumps.

Sanitary landfills are constructed with a clay or plastic lining at the bottom. Clay is often used because it can impede water flow and retain positively charged ions, such as metals. A system of pipes is constructed below the landfill to collect leachate, which is sometimes recycled back into the landfill. Finally, a cover of soil and clay, called a cap, is installed when the landfill reaches capacity.

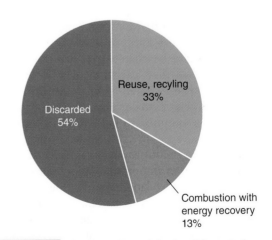

**FIGURE 16.13** **The fate of municipal solid waste in the United States.** The majority is disposed of in landfills. [After U.S. Environmental Protection Agency, *MSW Generation, Recycling and Disposal in the United States: Facts and Figures for 2008.*]

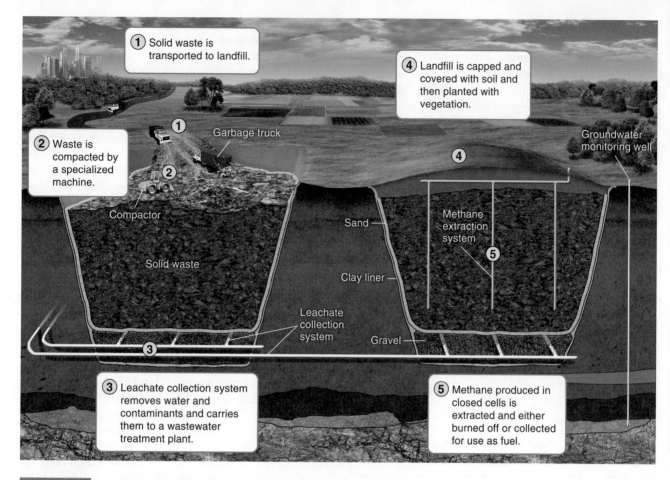

**FIGURE 16.14** **A modern sanitary landfill.** A landfill constructed today has many features to keep components of the solid waste from entering the soil, water table, or nearby streams. Some of the most important environmental features are the clay liner, the leachate collection system, the cap—which prevents additional water from entering the landfill—and, if present, the methane extraction system.

Rainfall and other water inputs are minimized because excess water in the landfill causes a greater rate of anaerobic decomposition and consequent methane release. In addition, with a large amount of water entering the landfill from both MSW and rainfall, there is a greater likelihood that some of that water will leave the landfill as leachate. Leachate that is not captured by the collection system may leach into nearby soils and groundwater. Leachate is tested regularly for its toxicity and if it exceeds certain toxicity standards, the landfill operators could be required to collect it and treat it as a toxic waste.

Once the landfill is constructed, it is ready to accept MSW. Perhaps the most important component of operating a safe modern-day landfill is controlling inputs. The materials destined for a landfill are those least likely to cause environmental damage through leaching or generating methane. Composite materials made of plastic and paper, such as juice boxes for children, are good candidates for a landfill because they are difficult to recycle. Aluminum and other metals such as copper may contribute to leaching. Because of this, and

because they are valuable as recyclables, aluminum and copper should never go into a landfill. Glass and plastics are both chemically inert, making them suitable for a landfill when reuse or recycling is not possible. Toxic materials, such as household cleaners, oil-based paints, automotive additives such as motor oil and antifreeze, consumer electronics, appliances, batteries, and anything that contains substantial quantities of metals, should not be deposited in landfills. All organic materials, such as food and garden scraps and yard waste, are potential sources of methane and should not be placed in landfills.

The MSW added to a landfill is periodically compacted into "cells," which reduces the volume of solid waste, thereby increasing the capacity of the landfill. The cells are covered with soil minimizing the amount of water that enters them and reducing odor. When a landfill is full, it must be closed off from the surrounding environment, so that the input and output of water are reduced or eliminated. Once a landfill is closed and capped, the MSW within it is entombed. Some air and water may enter from the outside environment, but this should be minimal if the landfill was well designed and properly sealed. The design and topography of the landfill cap, which is a combination of soil, clay, and sometimes plastic, encourages water to flow off to the sides, rather than into the landfill. Closed landfills can be *reclaimed,* meaning that some sort of herbaceous, shallow-rooted vegetation is planted on the topsoil layer, both for aesthetic reasons and to reduce soil erosion. Construction on the landfill is normally restricted for many years, although parks, playgrounds, and even golf courses have been built on reclaimed landfills (FIGURE 16.15).

A municipality or private enterprise constructs a landfill at a tremendous cost. These costs are recovered by charging a fee, called a **tipping fee** because each truck-load is put on a scale, and after the MSW is weighed it is tipped into the landfill. Tipping fees at solid waste land-fills average $35 per ton in the United States, although in certain regions, such as the Northeast, fees can be twice that much. These fees create an economic incentive to reduce the amount of waste that goes to the landfill. Many localities accept recyclables at no cost but charge for disposal of material destined for a landfill. This practice encourages individuals to separate recyclables. Some localities mandate that recyclable material be removed from the waste stream and disposed of separately. However, if tipping fees become too high, and regulations too stringent, a locality may inadvertently encourage illegal dumping of waste materials in locations other than the landfill and recycling center.

**CHOOSING A SITE FOR A SANITARY LANDFILL** A landfill should be located in a soil rich in clay to reduce the migration of contaminants. It should be located away from rivers, streams, and other bodies of water and drinking water supplies. A landfill should also be sufficiently far from population centers so that trucks transporting the waste and animal scavengers such as seagulls and rats present a minimal risk to people. However, the energy needed to transport MSW must also be considered in siting; as distance from a population center increases, so does the amount of energy required to move MSW to the landfill. Regional landfills are becoming more common because sending all waste to a single location often offers the greatest economic advantage.

The **siting,** or designation of a location, is always highly controversial and sometimes politically charged. Due to their unsightliness and odor, landfills are not considered a desirable neighbor. Landfill siting has been the source of considerable environmental injustice. People with financial resources or political influence often adopt what has been popularly called a "not-in-my-backyard," or NIMBY, attitude about landfill sites. Because of this, a site may be chosen not because it meets the safety criteria better than other options but because its neighbors lack the resources to mount a convincing opposition.

In Fort Wayne, Indiana, for example, the Adams Center Landfill was located in a densely populated, low-income, and predominantly minority neighborhood. A University of Michigan environmental justice study quotes Darrell Leap, a hydrogeologist and professor at Purdue University, as saying that on a scale of 1 to 10, with 10 being a geologically ideal site, the Adams Center Landfill would rate a "3, possibly 4" because the site held a substantial risk of water contamination. When the communities surrounding the Adams Center Landfill learned of the report and this danger, they protested the renewal of the federal permit, and fought expansion of the landfill at both the state and local levels. Ultimately, a 1997 decision by the Indiana Department of Environmental Management closed the landfill.

**SOME PROBLEMS WITH LANDFILLS** Though sanitary landfills are an improvement over open dumps, we have already seen they present many problems. Locating landfills near populations that do not have the resources to object is a global problem. No matter how careful the design and engineering, there is always the possibility that leachate from a landfill will contaminate underlying and adjacent waterways. The EPA estimates that virtually all landfills in the United States have had some leaching. Even after a landfill is closed, the potential to harm adjacent waterways remains. The amount of leaching, the substances that have leached out, and how far they will travel are impossible to know

**FIGURE 16.15** **Reclamation of a landfill.** This playground and athletic field were constructed on a closed and capped landfill in the 1970s in Anoka, Minnesota.

in advance. To get an idea of how much leachate is generated from a landfill and how much might be collected, see Crunch the Numbers "How Much Leachate Might Be Collected?"

The risk to humans and ecosystems from leachate is uncertain. Public perception is that landfill contaminants pose a great threat to human health, though the EPA has ranked this risk as fairly low compared to other risks such as global climate change and air pollution. But methane and other organic gases generated from decomposing organic material in landfills do release greenhouse gases.

When solid waste is first placed in a landfill, some aerobic decomposition may take place, but shortly after the waste is compacted into cells and covered with soil, most of the oxygen is used up. At this stage, anaerobic decomposition begins, a process that generates methane and carbon dioxide—both greenhouse gases—and other gaseous compounds. The methane also creates an explosion hazard. For this reason, landfills are vented so that methane does not accumulate in highly explosive quantities. In recent years, more and more landfill operators are collecting the methane the landfill produces and using it to generate heat or electricity. An even more desirable environmental choice would be to keep organic material out of landfills and to use it to make compost.

Professor William Rathje, from the University of Arizona, is well known for using archaeological tools to examine modern-day garbage. Using a bucket auger, a type of very large drill, Rathje and others have obtained information on the decomposition rates of MSW in landfills. Prior to Professor Rathje's work, most people conceived of landfills as places with a great deal of biological and chemical activity breaking down organic matter, implying that landfills would shrink over time as the material in them was converted into carbon dioxide or methane and released to the atmosphere. However, Rathje and colleagues have found newspapers with headlines still legible 40 years after being deposited in landfills, proving that little decomposition had taken place. Today it is widely accepted that decomposition takes place only in those areas of a landfill where the correct mixtures of air, moisture, and organic material are present. Because most areas do not contain this necessary mixture, the landfills will probably remain the sizes they were when capped.

## Incineration

Given all the problems of landfills, people have turned to a number of other means of solid waste disposal, including *incineration*. More than three-quarters of the material that constitutes municipal solid waste is easily combustible. Because paper, plastic, and food and yard waste are composed largely of carbon, hydrogen, and oxygen, they are excellent candidates for **incineration,** the process of burning waste materials to reduce their volume and mass and sometimes to generate electricity or heat. An efficient incinerator operating under ideal conditions may reduce the volume of solid waste by up to 90 percent and the weight of the waste by approximately 75 percent, although the reductions vary greatly depending on the incinerator and the composition of the MSW.

**INCINERATION BASICS** FIGURE 16.16 shows a mass-burn municipal solid waste incinerator. Typically at this incinerator, MSW is sorted and certain recyclables are diverted to recycling centers. The remaining material is dumped or "tipped" from a refuse truck onto a platform where certain materials such as metals are identified and removed. A moving grate or other delivery system transfers the waste to a furnace. Heat is released as combustion rapidly converts much of the waste into carbon dioxide and water, which are released into the atmosphere.

Particulates, more commonly known as *ash* in the solid waste industry, are an end product of combustion. **Ash** is the residual nonorganic material that does not combust during incineration. Residue collected underneath the furnace is known as **bottom ash** and residue collected beyond the furnace is called **fly ash.**

① Waste is dumped into refuse bunker.

② Crane moves material from bunker to hopper.

③ Waste is burned in incineration chamber.

④ Ash is collected and removed from plant.

⑤ A baghouse filter helps clean air before it is released through chimney.

⑥ Heat energy can be used to create steam and generate electricity (not illustrated).

Chimney

Crane

Hopper

Baghouse filter

Incineration chamber

Refuse bunker

Ash bunker

**FIGURE 16.16** **A municipal mass-burn waste-to-energy incinerator.** In this plant, MSW is combusted and the exhaust is filtered. Remaining ash is disposed of in a landfill. The resulting heat energy is used to make steam, which turns a generator that generates electricity in the same manner as was illustrated in Figure 12.7 on page 323.

Because incineration often does not operate under ideal conditions, ash typically fills roughly one-quarter the volume of the precombustion material. Disposal of this ash is determined by its concentration of toxic metals. The ash is tested for toxicity by leaching it with a weak acid. If the leachate is relatively low in concentration of contaminants such as lead and cadmium, the ash can be disposed of in a landfill. Ash deemed safe can also be used for other purposes such as fill in road construction or as an ingredient in cement blocks and cement flooring. If deemed toxic, the ash goes to a special ash landfill designed specifically for toxic substances.

Metals and other toxins in the MSW may be released to the atmosphere or may remain in the ash, depending on the pollutant, the specific incineration process, and the type of technology used. Exhaust gases from the combustion process, such as sulfur dioxide and nitrogen oxides, move through collectors and other devices that reduce their emission to the atmosphere. These collectors are similar in design to those described in Chapter 15 on air pollution. Acidic gases such as hydrogen chloride (HCl), which results from the incineration of certain materials including plastic, are recovered in a scrubber, neutralized, and sometimes treated further before disposal in a regular landfill or ash landfill.

Incineration also releases a great deal of heat energy that is often used in a boiler immediately adjacent to the furnace either to heat the incinerator building or to generate electricity, using a process similar to that of a coal, natural gas, or nuclear power plant. When heat generated by incineration is used rather than released to the atmosphere, it is known as a **waste-to-energy** system. Although energy generation is a positive benefit of incineration, as we shall see, there are a number of environmental problems with incineration as a method of waste disposal.

**SOME PROBLEMS WITH INCINERATION** Though incinerators address some of the problems of landfills, they have problems of their own. In order to cover the costs of construction and operation of an incinerator, tipping fees are charged, just as they are charged for disposal of waste in a MSW landfill. Generally, tipping fees are higher at incinerators than at landfills; national averages are around $70 per U.S. ton. The siting of an incinerator raises NIMBY and environmental justice issues similar to those of landfill siting. An incinerator may release air pollutants such as organic compounds from the incomplete combustion of plastics and metals contained in the solid waste that was burned. Some environmental scientists believe that incinerators are a poor solution to solid waste disposal because they produce ash that is more concentrated and thus more toxic than the original MSW. In addition, because incinerators are generally quite large and expensive to build, they require large quantities of MSW on a daily basis in order to burn efficiently and to be profitable. In order to support these costs, communities that use incinerators may be less likely to encourage recycling. Rate structures and other programs can be designed to encourage MSW reduction and diversion, with the goal of using incineration only as a last resort.

Incinerators may not completely burn all the waste deposited in them. Plant operators can monitor and modify the oxygen content and temperature of the burn, but because the contents of MSW are extremely variable and lumped all together, it is difficult to have a uniform burn. Consider a truckload of MSW from your neighborhood. The same load may contain food waste with high moisture content and, right next to that, packaging and other dry, easily burnable material. It is challenging for any incinerator—even a state-of-the-art modern facility—to burn all of these materials uniformly.

Inevitably, MSW contains some toxic material. The concentration of toxics in MSW is generally quite low relative to all the paper, plastic, glass, and organics in the waste. However, rather than being dissipated to the atmosphere, most metals remain in the bottom ash or are captured in the fly ash. As we have already mentioned, incinerator ash that is deemed toxic must be disposed of in a special landfill for toxic materials. As with other topics, there is no good choice for waste disposal. Sometimes the decision between a landfill and an incinerator is in part a decision about the kind of pollution a community prefers. Again, this emphasizes that the best choice is the production of less material for either the landfill or the incinerator.

---

### GAUGE YOUR PROGRESS

✓ What are the features of a modern sanitary landfill? How does a modern landfill compare to the older practice of putting MSW in holes in the ground?

✓ When or why might incineration be used instead of a landfill?

✓ What are the advantages and disadvantages of landfills and incineration?

---

# Hazardous waste requires special means of disposal

**Hazardous waste** is liquid, solid, gaseous, or sludge waste material that is harmful to humans or ecosystems. According to the EPA, over 20,000 hazardous waste generators in the United States produce about 36 million metric tons (40 million U.S. tons) of hazardous waste each year. Only about 5 percent of that waste is recycled.

The majority of hazardous waste is the by-product of industrial processes such as textile production, cleaning of machinery, and manufacturing of computer equipment, but it is also generated by small businesses such as dry cleaners, automobile service stations, and small farms. Even individual households generate hazardous waste—1.5 million metric tons (1.6 million U.S. tons) per year in the United States, including materials such as oven cleaners, batteries, and lawn fertilizers. All of these materials have a much greater likelihood of causing harm to humans or ecosystems than do materials such as newspapers or plastic beverage containers and should not be disposed of in regular landfills.

## Handling and Treatment of Hazardous Waste

Most municipalities do not have regular collection sites for hazardous waste or household hazardous waste. Rather, homeowners and small businesses are asked to keep their hazardous waste in a safe location until periodic collections are held (FIGURE 16.17). Every aspect of the treatment and disposal of hazardous waste is more expensive and more difficult than the disposition of ordinary MSW. A can of oven cleaner in your kitchen is just that—a can of oven cleaner that cost perhaps $5 to purchase. But as soon as a municipality collects oven cleaners or any other substances used in the household such as oil-based paints, motor oil, or chemical cleaners, it becomes regulated hazardous waste. Collection sites are designated as hazardous waste collection facilities that must be staffed with specially trained personnel. Sometimes the materials gathered are unlabeled and unknown and must be treated with extreme caution. Ultimately, the wastes may be sorted into a number of categories, such as fuels, solvents, and lubricants, for example. Some items, such as paint, may be reused, while others may be sent to a special facility for treatment.

Hazardous waste must be treated before disposal. Treatment, according to the EPA definition, means making it less environmentally harmful. To accomplish this, the waste must usually be altered through a series of chemical procedures.

As with other waste, there are no truly good options for disposing of hazardous waste. The most beneficial and one of the least expensive ways is source reduction: don't create the waste in the first place. In the case of household hazardous waste, community groups and municipalities encourage consumers to substitute products that are less toxic or to use as little of the toxic substances as possible.

## Legislative Response

Because of the dangers that hazardous waste presents, disposal has received much public attention. Regulation

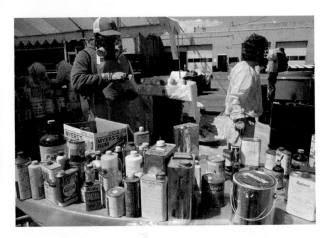

**FIGURE 16.17** A typical household hazardous waste collection site in Seattle, Washington. Residents are encouraged to keep hazardous household waste separate from their regular household waste. Collections are held periodically.

and oversight of the handling of hazardous waste falls under two pieces of federal legislation, the U.S. Resource Conservation and Recovery Act (RCRA) and the Comprehensive Environmental Response, Compensation, and Liability Act (CERCLA), usually referred to as the *Superfund* act.

In 1976, RCRA expanded previous solid waste laws. Its main goal was to protect human health and the natural environment by reducing or eliminating the generation of hazardous waste. Under RCRA's provision for "cradle-to-grave" tracking, the EPA maintains lists of hazardous wastes and works with businesses and state and local authorities both to minimize the generation of hazardous waste and to make sure that it is tracked until proper disposal. In 1984, RCRA was modified with the federal Hazardous and Solid Waste Amendments (HSWA) that encouraged waste minimization and phased out the disposal of hazardous wastes on land. The amendments also increased law enforcement authority in order to punish violators.

CERCLA, or the **Superfund** act, is well known because of a number of sensational cases that have fallen under its jurisdiction. Originally passed in 1980 and amended in 1986, this legislation has several parts. First, it imposes a tax on the chemical and petroleum industries. The revenue from this tax is used to fund the cleanup of abandoned and nonoperating hazardous waste sites where a responsible party cannot be established. The name *Superfund* came from this provision. CERCLA also authorizes the federal government to respond directly to the release or threatened release of substances that may pose a threat to human health or the environment.

Under Superfund, the EPA maintains the National Priorities List (NPL) of contaminated sites that are eligible for cleanup funds. For a long time, very little Superfund money was disbursed, and few sites underwent remediation. The map in **FIGURE 16.18** shows the

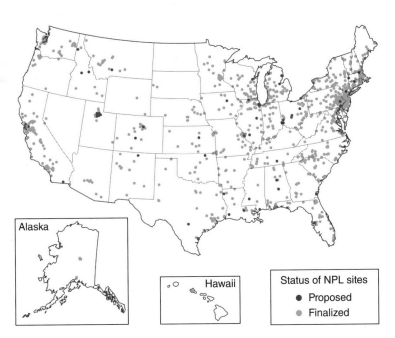

**FIGURE 16.18** Distribution of NPL (Superfund) sites in the United States. Under Superfund, the EPA maintains the National Priorities List (NPL) of contaminated sites that are eligible for cleanup funds. [After http://www.epa.gov/superfund/sites/products/nplmap.pdf.]

**FIGURE 16.19** **Love Canal, New York.** Love Canal became a symbol of hazardous chemical pollution in the United States in the 1970s. Lois Gibbs, with no prior experience in environmental science or activism, became a spokesperson for the neighborhood of Love Canal and is widely credited with bringing national attention to her community, including the elementary school which had been constructed on top of large quantities of hazardous chemical waste. Gibbs is currently the executive director of the Center for Health, Environment and Justice.

location of the NPL sites. As of October 2010, there were 1,282 Superfund sites—at least one in every state except North Dakota. New Jersey, with 114 Superfund sites, has the most. California and Pennsylvania have the next highest number, with 94 Superfund sites each. New York has 85 sites.

Perhaps the best known site is Love Canal, New York (**FIGURE 16.19**). Originally a hazardous waste landfill, Love Canal was covered with fill and topsoil and used as a site for a school and a housing development. In 1978 and 1980, known cancer-causing wastes such as benzene (a solvent) and trichloroethylene (a degreasing agent) were found in the basements of homes in the area. When it became clear that a disproportionately large number of illnesses, possibly connected to the chemical waste, had been diagnosed in the local population, the situation attracted national attention. The contamination was so bad that in 1983 Love Canal was listed as a Superfund site and the inhabitants of the area were evacuated. In 1994, the EPA removed Love Canal from the National Priorities List because the physical cleanup had been completed and the site was no longer deemed a threat to human health.

Since its inception, many have observed that CERCLA has not had enough funding to clean up the numerous hazardous waste sites around the country. The listing of NPL sites in a state does not necessarily include all contaminated sites within that state. For example, the Department of Environmental Protection in New Jersey, which has 114 listed Superfund sites, believes that there are 9,000 sites where soil or groundwater is contaminated with hazardous chemicals.

**BROWNFIELDS** The Superfund designation is reserved strictly for those locations with the highest risk to public health, and Superfund sites are managed solely by the federal government. In 1995, the EPA created the *Brownfields* Program to assist state and local governments in cleaning up contaminated industrial and commercial land that did not achieve conditions necessary to be in the Superfund category. **Brownfields,** like Superfund sites, are contaminated industrial or commercial sites that may require environmental cleanup before they can be redeveloped or expanded. Old factories, industrial areas and waterfronts, dry cleaners, gas stations, landfills, and rail yards are common examples of brownfield sites. Brownfields legislation has prompted the revitalization of several sites throughout the country. One notable instance is Seattle's Gasworks Park. Previously used as a coal and oil gasification plant, the city of Seattle purchased the land in 1962 to rehabilitate the site into a park. After undergoing chemical abatement and environmental cleanup, the park has become a distinctive landmark for the city and is the site of many public events throughout the year.

The Brownfields Program has been criticized as an inadequate solution to the estimated 450,000 contaminated locations throughout the country. Since the cleanup is managed entirely by state and local governments, brownfields management can vary widely from region to region. Furthermore, the brownfields legislation lacks legal liability controls to compel polluters to rehabilitate their properties. Without legal recourse, many brownfields sites remain unused and contaminated, posing a continued risk to public health.

## International Consequences

Because of the difficulties of disposing of hazardous waste, municipalities and industries sometimes try to send the waste to countries with less stringent regulations. This has been illustrated by the many reports of garbage and ash barges that travel the oceans looking for a developing country willing to accept hazardous waste from the United States in exchange for a cash payment. There are many examples of hazardous wastes being accepted and disposed of in developing nations around the world. Perhaps the most famous is the story of the cargo vessel *Khian Sea,* which left Philadelphia in 1986 with almost 13,000 metric tons of hazardous ash from an incinerator. It traveled to a number of countries in the Caribbean in search of a dumping place. Some of the ash was dumped in Haiti, and some was dumped in the ocean. In 1996, the United States ordered that the ash dumped in Haiti be retrieved and returned to the United States. After being held at a dock in Florida, the ash was deemed nonhazardous by the EPA and other agencies and in 2002 the ash was placed in a landfill in Franklin County, Pennsylvania, not very far from its source of origin.

In 2003, a notable instance of a waste transfer in the other direction occurred. A Pennsylvania company that specializes in recovering mercury from a wide variety of products agreed to accept 270 metric tons of mercury waste generated by a company in the state of Tamil Nadu, India, during the manufacture of thermometers. India had no facilities for recycling mercury waste, and most of the thermometers manufactured at the plant were shipped to the United States and Europe, so the transfer of material seemed appropriate. The mercury was shipped from India to the United States in May 2003. The mercury was concentrated, purified, and then sold to industrial users of the metal.

---

### GAUGE YOUR PROGRESS

✓ What is the definition of hazardous waste, and what are its main sources?

✓ Why is disposal of hazardous waste a challenge?

✓ Which acts authorize which agencies to regulate and oversee hazardous waste?

---

# There are newer ways of thinking about solid waste

Throughout this chapter we have described a variety of ways of managing solid waste, from creating less of it to burying it in a landfill to burning it. Each method has both benefits and drawbacks. Because there is no obvious best method and because waste is a pervasive fact of contemporary life, the problem of waste disposal seems overwhelming. How can an individual, a small business, an institution, or a municipality best deal with its solid waste? The answer is highly specific to each case and varies by region. Every method of waste disposal will have adverse environmental effects; the challenge is to find the option least detrimental. How do we decide which choices are best? Life-cycle analysis and a holistic approach are two useful approaches to gaining insights to the question of what we should do with our solid waste.

## Life-Cycle Analysis

In the chapter opening story about polystyrene, we attempted to make an objective assessment of solid waste disposal options. This process is known as *life-cycle analysis*. **Life-cycle analysis** is an important systems tool that looks at the materials used and released throughout the lifetime of a product—from the procurement of raw materials through their manufacture, use, and disposal. That's why it is also called a cradle-to-grave analysis. As we saw in the small-scale example of comparing a paper cup to a polystyrene cup, the full inventory a life-cycle analysis requires will sometimes yield surprising results.

In theory, conducting a life-cycle analysis should help a community determine whether incineration is more or less desirable than using a landfill. However, there are problems in determining the overall environmental impact of a specific material. For example, it is not possible to determine whether the particulates and nitrogen oxides released from incinerating food waste are better or worse for the environment than the amount of methane that might be released if the same food waste was placed in a landfill. So in the case of waste that contains food matter, it is not possible to directly compare the full environmental impact of disposal in a landfill versus incineration. This is the same problem we encountered when we compared the types of pollution generated by producing the paper cup versus the plastic cup. Although life-cycle analysis can be used, there are limitations to the comparisons that can be made with it. Ultimately, the best choice for disposing of food waste might be to compost it.

Although life-cycle analysis may not be able to determine absolute environmental impact, it can be very helpful in assessing other, especially economic and energy, considerations. For example, a glass manufacturing plant might pay $5 per ton for green glass that it will recycle into new glass. For a municipality, it might be better to receive $5 for a ton of glass than to pay a $65 per ton tipping fee to throw the material in a landfill. But the municipality must also consider the lower cost of transporting the glass to a relatively close landfill rather than to a distant glass plant.

A life-cycle analysis should also consider the energy content of gasoline or diesel fuel used and the pollution generated in trucking the material to each destination, as well as the monetary, energy, and pollution savings achieved if the new glass was made from old glass rather than from raw materials (sand, potash, lime). Reconciling all these competing factors is very challenging and the ultimate decisions based on such analyses are often debatable.

As we have seen, waste disposal choices are influenced by economic as well as scientific choices. In some parts of the country, the cost of waste disposal is covered by local taxes; in other locations, municipalities, businesses, or households may have to pay directly for disposal of their solid waste. Whether direct or indirect, there is always a cost to waste disposal. Normally, disposal of recyclables costs less than material destined for the landfill, because the landfill always involves a tipping fee while recyclables either incur a lower tipping fee or generate revenue. However, costs change depending on many factors, including market conditions. For example, in a particular year, there may not be a market for recycled newspapers in the United States, but in the following year a huge increase in Japanese demand for newsprint may cause the price to go up in the United States. It is therefore essential for municipalities to have many choices and to be able to modify these choices as market environments change.

## Alternative Ways to Handle Waste

A more holistic method seeks to develop as many options as possible, emphasizing reduced environmental harm and cost. This is **Integrated Waste Management,** which employs several waste reduction, management, and disposal strategies in order to reduce the environmental impact of MSW. Such options include a major emphasis on source reduction and include any combination of recycling, composting, use of landfills, incineration, and whatever additional methods are appropriate to the particular situation. FIGURE 16.20 shows how a nation or a community could consider a series of steps, starting with source reduction during manufacturing and procurement of items. After that, behavior related to use and disposal can be considered and possibly altered in order to obtain the desired outcome: less generation of MSW. According to this approach, no community should be forced into any one method of waste disposal. If a region makes a large investment in an incinerator, for example, there is a risk that it would then need to attract large quantities

of waste to pay for that incinerator, thereby reducing the incentive to recycle or use a landfill. Landfill space may be abundant or scarce, depending on the location of the community. If the municipality is free to consider all options, it can make the choice or choices that are efficient, cost effective, and least harmful to the environment.

The architect and former University of Virginia dean William McDonough looks at holistic waste management from a more far-reaching perspective. In the book *Cradle to Cradle,* McDonough and coauthor Michael Braungart propose a new approach to the manufacturing process. They argue that it is first necessary to assess existing practices in order to minimize waste generation before, during, and after manufacturing. Beyond that, manufacturers of durable goods such as automobiles, computers, appliances, and furniture should develop plans for disassembling the goods when they are no longer useful so that parts or materials can be recycled with as little as possible becoming part of the waste stream. This is already being done in some industries. Volkswagen, for example, manufactures some of its cars so that they can be easily taken apart and materials of different composition easily separated to allow recycling. Certain carpet manufacturers design their carpets so that when worn out they can be easily recycled into new carpeting (FIGURE 16.21). This is typically done by making a base that is extremely durable with a top portion of the carpet that can be changed when the color fades, is worn out, or is no longer desired. Finally, McDonough and Braungart point out that many organisms in the natural world, such as the turtle, produce very hard, impact-resistant materials, such as a shell, without producing any toxic waste. They suggest that humans should examine how a tortoise creates such a hard shell without the production of toxic wastes. Humans can use this example as a goal for other kinds of production where no toxic wastes are produced.

**FIGURE 16.20** **A holistic approach to waste management.** Depending on the kind of waste and the geographic location, much less time and money can be spent on reducing waste than disposing of it. Horizontal arrows indicate the waste stream from manufacture to disposal and curved arrows indicate ways in which waste can either be reduced or removed from the stream, thereby reducing the amount of waste incinerated or placed in landfills.

**FIGURE 16.21** **A recyclable carpet.**
FLOR carpet tiles are designed to be easily replaced and easily recycled when the carpet wears out.

Although there are many ways to improve our handling of solid waste, there is some evidence that the nation as a whole has taken source reduction seriously because per capita waste generation has been level since 1990. It appears that recycling has been taken seriously too, since recycling rates have increased since 1985. However, considering both the amount of waste generated in this country and the recycling rates found in other countries, we have a long way to go.

---

**GAUGE YOUR PROGRESS**

✓ What is a life-cycle analysis and how is it useful?

✓ How is holistic waste management different from other approaches to waste management?

✓ What are some of the economic issues to consider when making waste disposal decisions?

---

 **WORKING TOWARD SUSTAINABILITY**

## Recycling E-Waste in Chile

Electronic waste is a small part of the waste stream but it contains a large fraction of the hazardous waste that ends up in landfills, dumps, and other locations. Toxic metals such as lead, mercury, and cadmium as well as carcinogenic organic compounds are common in electronic waste. Recycling rates for electronic waste vary around the world. In Switzerland, more than 80 percent of electronic waste is recycled. In the United States, the recycling rate for e-waste is about 20 percent, with about 70 percent of the recycled U.S. e-waste exported to China. While China currently receives the largest percentage of e-waste in the world, India and African countries also receive e-waste. In much of the world, electronic waste is mixed in with all other solid waste. For e-waste that is collected separately, fully assembled electronic devices are usually exported for disassembly and recycling elsewhere. Not only does this practice export the pollution burden, it also uses unnecessary energy and space to export the entire device rather than just the components that need recycling.

Until the year 2000, the South American nation of Chile recycled less than 1 percent of its electronic waste. Businessman and entrepreneur Fernando Nilo decided he wanted to increase the e-waste recycling rate in Chile but he did not want to follow the typical e-waste model of exporting unsorted, completely assembled e-waste to other countries. He devised a business plan that won a local competition and subsequently obtained funding to launch his plan. In 2005, the company Recycla Chile of Santiago opened the first recycling facility in Latin America (**FIGURE 16.22**), which received electronic devices such as computers, cell phones, scanners, and televisions. While in China and other countries, workers, including children, pick apart e-waste without proper face and respiratory protection, at Recycla the workers are trained to dismantle and separate electronic

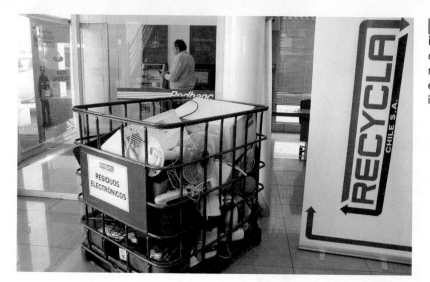

**FIGURE 16.22** **A Recycla collection site in Chile.** In this facility, electronic waste is deposited for recycling, thereby keeping metals and other harmful components of electronic products from the landfill or incinerator.

components safely. Certain separated materials are sold to reputable dealers within Chile, while other materials are compacted to minimize shipping and energy costs and sent to environmentally certified metal smelters around the globe. These metal smelting companies safely recover valuable metals for recycling and use in new electronic devices. One of the innovations introduced by Recycla is the first "Green Seal" in Latin America. The manufacturer typically passes along the responsibility for recycling a product to the new owner. For example, when a customer purchases a printer, he or she is then responsible for proper disposal of that printer. Recycla introduced a program where the manufacturer contracted with Recycla to be responsible for the disposal of a product whenever the customer decides it is no longer needed. The firm that made the electronic device is financially or physically responsible for recycling that object. With this guarantee at the time of purchase, the object is much more likely to end up in a proper location at the end of its useful life, with no additional cost to the consumer.

Nilo and coworkers began a national advertising campaign both to generate revenue for their company and to educate leaders in their country and around the world about the importance of recycling electronic waste. Recycla developed slogans such as, "If you don't pay to recycle today, our children will pay tomorrow." Today Recycla promotes environmental sustainability in Chile and around the world, and provides jobs for marginalized people including prisoners and former prisoners. The company is profitable and has received seven awards, including the Technology Pioneer Award from the World Economic Forum in Davos, Switzerland, in 2009. When Recycla started, Chile was recycling less than 1 percent of its electronic waste. Their goal is to recycle 10 percent within a few years. In 2009, they reported that they had achieved a 5 percent recycling rate in Chile, well on the way to their goal and more than five times higher than the recycling rate when they started.

### References

Information Development Incubator Support Center. *Environmental Impact of IT Solutions: Recycling E-Waste in Chile.* http://www.idisc.net/en/Article.38520.html.

Recycla Chile, SA. *Solving the E-Waste Problem* (company report).

■ **Define waste generation from an ecological and systems perspective.**

From an ecological and systems perspective, waste is composed of the nonuseful products of a system. Much of the solid waste problem in the United States stems from the attitudes of the "throw-away society" adopted after World War II. While the United States may be a major generator of solid waste, the lifestyle and goods disseminated around the world have made solid waste a global problem.

■ **Describe how each of the three Rs—Reduce, Reuse, and Recycle—as well as composting can avoid waste generation.**

The three Rs—Reduce, Reuse, and Recycle—divert materials from the waste stream. Composting, source reduction, and reuse generally have lower energy and financial costs than recycling, but all are important ways to minimize solid waste production.

■ **Explain the implications of landfills and incineration.**

Currently, most solid waste in the United States is buried in landfills. Contemporary landfills entomb the garbage and keep water and air from entering and leachate from escaping. The potential for toxic leachate to contaminate surrounding waterways is one major problem in landfills; the generation of methane gas is another. Siting of landfills often raises issues of environmental justice. Incineration is an alternative to landfills. Its main benefit is that it reduces the waste material to roughly one-quarter its original volume. In addition, waste-to-energy incineration often uses the excess heat produced to generate electricity. However, incineration generates air pollution and ash, which can sometimes contain a high concentration of toxic substances and require disposal in a special ash landfill.

■ **Understand the problems associated with the generation and disposal of hazardous waste.**

Hazardous waste is a special category of material that is especially toxic to humans and the environment. It includes industrial by-products and some household items such as batteries and oil-based paints, all requiring special means of disposal. Though legislation has been passed to deal with hazardous waste, many problems remain.

■ **Present a holistic approach to avoiding waste generation and to treating solid waste.**

By using life-cycle analysis, which tracks material "from cradle to grave," and integrated waste management, which draws on all the available treatment methods, we can make optimal decisions regarding our solid waste. The best solution is to design products with a strategy for their ultimate reuse or dismantling and recycling.

## CHECK YOUR UNDERSTANDING

1. It is important to keep household batteries out of landfills because of all of the following *except*
   (a) they can leach toxic metals.
   (b) their decomposition can contribute to greenhouse gas emissions.
   (c) they can be recycled, which would reduce the need for new raw materials.
   (d) they can be recycled, which would reduce the need for additional energy.
   (e) they take up space in landfills and we have a finite supply of landfill space.

2. All of the following are desired outcomes of MSW incineration *except*
   (a) extracting energy.
   (b) reducing volume.
   (c) prolonging life of landfills.
   (d) increasing air pollution.
   (e) generating electricity.

3. In the last 15 years, MSW per capita in the United States has
   (a) decreased drastically.
   (b) decreased, then increased drastically.
   (c) increased drastically.
   (d) increased moderately, and then decreased drastically.
   (e) stayed the same.

4. The EPA estimates that approximately _____ percent of municipal solid waste comes from residences and _____ percent comes from commercial and institutional facilities.
   (a) 30:70
   (b) 40:60
   (c) 50:50
   (d) 60:40
   (e) 70:30

5. In 2008, which of the following materials comprised the largest component of municipal solid waste?
   (a) Metals
   (b) Yard waste
   (c) Food scraps
   (d) Discarded electronic devices
   (e) Paper

6. From an environmental waste perspective, which of the following is the *most* desirable?
   (a) Reduce
   (b) Reuse
   (c) Recycle
   (d) Compost
   (e) Incinerate

7. In the United States, how much of generated waste ends up being recycled?
   (a) < 23 percent
   (b) < 33 percent
   (c) < 43 percent
   (d) > 53 percent
   (e) > 60 percent

8. The legislation that imposes a tax on the chemical and petroleum industries to generate funds to pay for the cleanup of hazardous substances is
   (a) RCRA.
   (b) Cradle-to-Grave Act.
   (c) the National Priorities List.
   (d) HSWA.
   (e) CERCLA.

9. Of the following, which contributes most to the weight of MSW?
   (a) Packaging
   (b) E-waste
   (c) Compact discs
   (d) Tires
   (e) Ores such as gold and silver

For questions 10–12, refer to the following lettered choices and choose the compound that is most associated with each numbered statement.
   (a) Benzene
   (b) Dioxin
   (c) Methane
   (d) Hydrochloric acid

10. It may be present in the emissions from waste incinerators.

11. It contaminated the land and water near the housing development Love Canal in New York.

12. It is produced by anaerobic decomposition in landfills.

## APPLY THE CONCEPTS

1. The total amount of municipal solid waste (MSW) generated in the United States has increased from 80 million metric tons (88 million U.S. tons) in 1960 to 232 million metric tons (255 million U.S. tons) in 2007.
   (a) Describe reasons for this increase and explain how the United States became the leader of the "throw-away society."
   (b) Explain why reducing is more favorable than reusing, which in turn is more favorable than recycling.
   (c) Describe the process of composting and compare a home composting system with that of a large-scale municipal facility.

2. In many developing countries garbage is still deposited in open dumps. Developed countries have phased out open dumps in favor of landfills or incinerators for handling garbage in the waste stream.
   (a) Explain why open dumps have been phased out in developed countries.
   (b) Describe how a modern-day sanitary landfill is constructed and explain how it is an improvement on the earlier landfills.
   (c) Discuss *one* benefit and *one* problem of landfilling waste.
   (d) Describe how a waste incinerator works and how it could be used to generate electricity.
   (e) Discuss *one* benefit and *one* problem of incinerating waste.

## MEASURE YOUR IMPACT

**Understanding Household Solid Waste** The table below shows the percentage composition of typical municipal solid waste (MSW) generated by a household in the United States according to a 2007 report by the EPA.

| Material | Percentage (%) |
|---|---|
| Food scraps | 12.5 |
| Paper and paperboard | 32.7 |
| Plastic | 12.1 |
| Metals | 8.2 |
| Glass | 5.3 |
| Yard trimmings | 12.8 |

Consider a household that has four residents and generates 4 pounds of MSW per day per person. The tipping fee in their location is $60 per U.S. ton. The household has already reduced its amount of garbage through reducing and reusing materials, but now wants to lower the amount further by recycling. The major recyclables in the waste stream of this household are paper, plastic, glass, and metals.

Use the preceding table and information to answer the following questions:
(a) How much MSW does the household generate in a month (31 days)?
(b) How much MSW does the household generate in a year (365 days)?
(c) How much does the household pay for tipping fees each year?
(d) Approximately how long does it take the household to generate a ton of MSW?
(e) If the household decided to compost all of its food scraps, paper, and yard trimmings, how much would it save on tipping fees each year?
(f) If the household decided to recycle all of its plastic and glass, how much would it save on tipping fees each year?
(g) How much would the household save if it did both composting and recycling as in e and f, and how much MSW would it generate a year as a result?

## ENGAGE YOUR ENVIRONMENT

How does your production of municipal solid waste compare to that of the average person in the United States? In this activity you will tally all of the waste you discard in a single day and then compare your discards to that of the average person in the United States. *See Engage Your Environment Activity 16 at the end of this book.*

# Human Health and Toxicology: Environmental Sources of Health Risk

## Citizen Scientists

The neighborhood of Old Diamond in Norco, Louisiana, is composed of four city blocks located between a chemical plant and an oil refinery, both owned by the Shell Oil Company. There are approximately 1,500 residents in the neighborhood, largely lower-income African Americans. In 1973, a pipeline explosion blew a house off its foundation and killed two residents. In 1988, an accident at the refinery killed seven workers and sent more than 70 million kg (159 million pounds) of potentially toxic chemicals into the air. Nearly one-third of the children in Old Diamond suffered from asthma and there were many cases of cancer and birth defects. The unusually high rates of disease raised suspicions that the residents were being affected by the two nearby industrial facilities.

By 1989, local resident and middle school teacher Margie Richard had seen enough. Richard organized the Concerned Citizens of Norco. The primary goal of the group was to get Shell to buy the residents' properties at a fair price so they could move away from the industries that were putting their health at risk. Richard contacted environmental scientists and quickly learned that to make a solid case to the company and to the U.S. Environmental Protection Agency (EPA), she needed to be more than an organizer; she also needed to be a scientist.

The residents all knew that the local air had a foul smell, but they had no way of knowing which chemicals were present or their concentrations. To determine whether the air they were breathing exposed the residents to chemical concentrations that posed a health risk, the air had to be tested. Richard learned about specially built buckets that could collect air samples. She organized a "Bucket Brigade" of volunteers and slowly collected the data she and her collaborators needed. As a result of these efforts, scientists were able to document that the Shell refinery was releasing more than 0.9 million kg (2 million pounds) of toxic chemicals into the air each year.

The fight against Shell met strong resistance from company officials and went on for 13 years. But in the end, Margie Richard won her battle. In 2002, Shell agreed to purchase the homes of the Old Diamond neighborhood. The company also agreed to pay an additional $5 million for community development and it committed to reducing air emissions from the refinery by 30 percent to help improve the air quality for those residents who remained in the area. In 2007, Shell agreed that it had violated air pollution regulations in several of its Louisiana plants and paid the state of Louisiana $6.5 million in penalties.

For her tremendous efforts in winning the battle in Norco, Margie Richard was the North American recipient of the Goldman Environmental Prize, which honors grassroots environmentalists. Since then, Richard has brought her message to many other minority communities located near large polluting industries. She teaches people that success requires a combination of organizing people to take action to protect their environment and learning how to be a citizen scientist. ■

> ## The unusually high rates of disease raised suspicions that the residents were being affected by two nearby industrial facilities.

Sources: The Goldman Environmental Prize: Margie Richard. http://www.goldmanprize.org/node/100; M. Scallan, Shell, DEQ settle emission charges, *Times-Picayune* (New Orleans), March 15, 2007. http://www.nola.com/news/t-p/riverparishes/index.ssf?/base/news-3/1173941825153360.xml&coll=1.

Margie Richard became a citizen scientist to help document the health risk of nearby chemical plants.

◀ The citizens of Norco, Louisiana, live in the shadows of chemical plants and oil refineries.

Health risks come from a variety of environmental sources including diseases and harmful chemicals. Some diseases have existed for millennia and others have emerged during the past few decades. In many cases, the likelihood of contracting a specific disease is associated with economic status. Harmful chemicals also pose a risk to humans and other organisms.

After reading this chapter you should be able to

- identify the three major categories of human health risk.

- list the major historical and emerging infectious diseases.
- name the five major types of toxic chemicals.
- distinguish between dose-response studies, retrospective studies, and prospective studies.
- describe the factors that help determine the chemical concentrations that organisms experience.
- explain the factors that go into a risk analysis and distinguish between the two major philosophies of chemical regulation.

# Human health is affected by a large number of risk factors

The number of health risks that we face in our lives can feel overwhelming. Sometimes it seems that we hear new warnings every day. How do we evaluate and manage these risks? We can begin to answer this question by determining which risks are common and the current state of our understanding about each of them. After that, we will look at how to assess and manage these risks. As we will see, although many health risks do exist in both the developed and developing worlds, we can do a great deal to manage these risks and improve our lives.

## Categories of Human Health Risk

The first step in understanding health risks is to consider the three major categories of risks that can

harm human health: physical, biological, and chemical. Physical risks include environmental factors, such as natural disasters, that can cause injury and loss of life. Physical risks also include less dramatic factors such as excessive exposure to ultraviolet radiation from the Sun, which causes sunburn, and exposure to radioactive substances such as radon, which we discussed in Chapter 15. Biological risks are those associated with *diseases*. A **disease** is any impaired function of the body with a characteristic set of symptoms. Chemical risks are associated with exposure to chemicals ranging from naturally occurring arsenic to synthetic chemicals and pesticides. In this chapter, we will focus on biological and chemical risks.

## Types of disease

Given these three categories of human health risk, biological risks cause the most human deaths. As you can see in **FIGURE 17.1**, approximately three-quarters

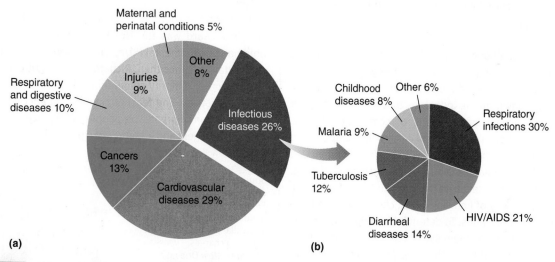

**(a)**

**(b)**

**FIGURE 17.1** **Leading causes of death in the world.** (a) More than three-quarters of all world deaths are caused by diseases, including respiratory and digestive diseases, various cancers, cardiovascular diseases, and infectious diseases. (b) Among the world's deaths caused by infectious diseases, 94 percent are caused by only six types of diseases. [Data from World Health Organization, 2004.]

of human deaths worldwide stem from various types of diseases. **Infectious diseases** are those caused by infectious agents, known as *pathogens*. Examples include pneumonia and sexually transmitted diseases. Diseases not caused by pathogens include cardiovascular diseases, respiratory and digestive diseases, and most cancers.

The pathogens that cause most infectious diseases include viruses, bacteria, fungi, protists, and a group of parasitic worms called helminths. Only six types of illnesses account for 94 percent of all deaths caused by infectious disease. The three top types of infectious diseases are those caused by respiratory infections (such as pneumonia), those caused by the virus that causes Acquired Immune Deficiency Syndrome (or *AIDS*), and the variety of pathogens that cause diarrhea. The next three are tuberculosis, malaria, and childhood diseases such as measles and tetanus. We will discuss many of these important infectious diseases later in the chapter.

All diseases fall into two categories—*acute* and *chronic*. **Chronic diseases** slowly impair the functioning of a person's body. Heart disease and most cancers, for example, are chronic diseases that develop over several decades. In contrast, **acute diseases** rapidly impair the functioning of a person's body. In some cases, such as a disease called *Ebola hemorrhagic fever* that we will discuss later in this chapter, death can come in a matter of days or weeks.

## RISK FACTORS FOR CHRONIC DISEASE IN HUMANS

Numerous factors cause people to be at a greater risk of chronic diseases such as cancer, cardiovascular diseases, diabetes, and chronic infectious diseases. The World Health Organization (WHO) has found that these risk factors differ substantially between low- and high-income countries. **FIGURE 17.2** shows the WHO data for a variety of risks.

In low-income countries, the top 10 risk factors leading to chronic disease are associated with poverty, including unsafe drinking water, poor sanitation, and malnutrition. In Chapter 11 we looked at some of the factors that contribute to malnutrition, including reduced government support for agriculture and higher costs of fossil fuels leading to increased food prices. As an example of poverty leading to chronic disease, nearly half of the children under the age of 5 who die from pneumonia succumb to the disease because they suffer from poor nutrition. Similarly, nearly three-quarters of children who die from diarrhea are simultaneously malnourished. With improved nutrition, these children would be better able to fight infectious diseases and many would survive.

In contrast, malnutrition and poor sanitation are not prevalent risk factors for chronic disease in high-income countries. Because people in these countries can afford better nutrition and proper sanitation, fewer die young

**(a)**

Percentage of total

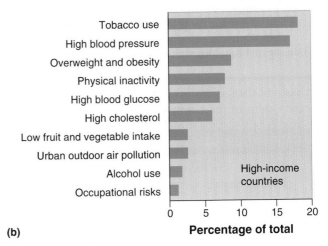

**(b)**

Percentage of total

**FIGURE 17.2** **Leading health risks in the world.** If we consider all deaths that occur and separate them into different causes, we can examine which categories cause the highest percentage of all deaths. (a) The leading health risks for low-income countries include issues related to low nutrition and poor sanitation. (b) The leading risks for high-income countries include issues related to tobacco use, inactivity, obesity, and urban air pollution. [After World Health Organization, 2009.]

from diseases such as pneumonia and diarrhea. Risk factors for people in high-income countries include an increased availability of tobacco, and a combination of less active lifestyles, poor nutrition, and overeating that leads to high blood pressure and obesity. In short, being affluent changes the major health risk factors for chronic disease.

The change in risk factors between low- and high-income countries occurs over time as a given country becomes more affluent. The graph in **FIGURE 17.3** illustrates how this transition in economic development affects health risk. A poor country initially faces the challenge of supplying food and proper sanitation to its citizens. As it begins to accumulate wealth, the health

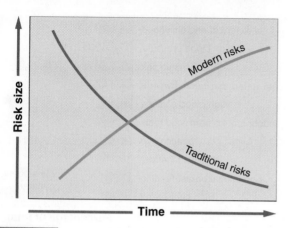

**FIGURE 17.3** **The transition of risk.** As a nation becomes more developed over time and attains higher income levels, the risks of inadequate nutrition and sanitation decline while the risks of tobacco, obesity, and poor urban air quality rise. [After World Health Organization, 2009.]

risks will change in a predictable fashion and the health care system of the country must change as well.

### GAUGE YOUR PROGRESS

✓ What are the three major categories of risk for human health? Give an example of each.

✓ What is the difference between an acute and a chronic disease?

✓ How is the economic development level of a country related to disease?

Water

Air

Food

Other humans

Wild animals (insects, rats, etc.)

Domesticated animals (livestock, pets)

**FIGURE 17.4** **Pathways of transmitting pathogens.** Pathogens have evolved a wide variety of ways to infect humans.

# Infectious diseases have killed large numbers of people

Although diseases can have genetic causes, environmental scientists are generally interested in diseases that have environmental causes, especially those caused by pathogens such as fungi, bacteria, and viruses. These pathogens have evolved a wide variety of pathways for infecting humans including transmission of pathogens from other humans, other animals, the food we eat, and the water we drink. **FIGURE 17.4** illustrates some of these relationships.

Throughout human history, disease-causing pathogens have taken a large toll on human health and mortality. When a pathogen causes a rapid increase in disease, we call it an **epidemic.** When an epidemic occurs over a large geographic region such as an entire continent, we call it a **pandemic.** Among the diversity of human diseases that have caused epidemics and pandemics, we will consider both those that have been historically important and those that have emerged recently.

## Historically Important Infectious Diseases

In Chapters 7 and 14 we looked at several diseases associated with poor sanitation and unsafe drinking water. These include cholera, hepatitis, and diarrheal diseases. All of these diseases are considered historical. Historical diseases that are passed between hosts include *plague, malaria,* and *tuberculosis.*

**PLAGUE** **Plague** is one of the most familiar diseases of human history. Also known by several historical names including *bubonic plague* and *Black Death,* plague is caused by an infection from a bacterium (*Yersinia pestis*) that is carried by fleas. Fleas attach to rodents such as mice and rats, giving the fleas tremendous mobility. When humans live in close contact with mice and rats, the bacterium can be transmitted either by flea bites or by handling the rodents. Individuals who become infected often experience swollen glands, black spots on their skin, and extreme pain. Plague is estimated to have killed hundreds of millions of people throughout history, including nearly one-fourth of the European population in the 1300s (**FIGURE 17.5**). The last major pandemic of plague occurred in Asia in the early 1900s. Today there are still occasional small outbreaks of plague around the world, including in the southwestern United States, but modern antibiotics are

**FIGURE 17.5** **The Black Death in Europe.** As depicted in Carlo Coppola's *The Marketplace in Naples During the Plague of 1656*, plague pandemics repeatedly swept through Europe from the 1300s through the 1800s, and killed millions of people. Because the disease caused black sores on people's bodies, it also had the name *Black Death*.

highly effective at killing the bacterium and preventing human death.

**MALARIA Malaria** is another widespread disease that has killed millions of people over the centuries. Malaria is caused by an infection from any one of several species of protists in the genus *Plasmodium*. The parasite spends one stage of its life inside a mosquito and another stage of its life inside a human. Infections cause recurrent flulike symptoms. Each year, 350 to 500 million people in the world contract the disease and 1 million people, mostly children under 5 years of age, die from it. The regions hardest hit include sub-Saharan Africa, Asia, the Middle East, and Central and South America. Since 1951, the malaria parasite has been eradicated from the United States by mosquito eradication programs. Although more than 1,000 cases of malaria are diagnosed in the United States each year, these are found in people who have returned from regions of the world where the malaria parasite lives.

The traditional approach to combating malaria was widespread spraying of insecticides such as DDT to eradicate the mosquitoes. Eradication efforts have proven to be ineffective in many parts of the world. Moreover, as we will see later in this chapter, the widespread use of many insecticides can create new problems. At the end of this chapter, Working Toward Sustainability "The Global Fight Against Malaria" examines the latest approaches toward combating malaria.

**TUBERCULOSIS Tuberculosis** is a highly contagious disease caused by a bacterium (*Mycobacterium tuberculosis*) that primarily infects the lungs. Tuberculosis is spread when a person coughs and expels the bacteria into the air. The bacteria can persist in the air for several hours and infect a person who inhales them. Symptoms of an infection include weakness, night sweats, and coughing up blood. As is the case with many pathogens, a person can be infected but not develop the tuberculosis disease. Indeed, it is estimated that one-third of the world's population is infected with tuberculosis. Each year 9 million people develop the disease and 2 million of them die.

Most tuberculosis infections can be easily treated by taking antibiotics for a year. In countries such as the United States, where the medicines are readily available, there has been a dramatic fall in both the number of new cases and the number of deaths from tuberculosis. **FIGURE 17.6** shows the decline of tuberculosis in the United States since the mid-1950s. In other parts of the world, especially in developing countries, the medicines are not as available or affordable and those who receive the medicine sometimes do not take the prescribed dose for the full duration of time. When a patient stops taking antibiotics before the last bacteria have been killed, there are two consequences. First, the pathogen can quickly rebuild its population inside the person's body. Second, because the last few bacteria are generally the most drug-resistant, stopping the antibiotics before the bacteria are eradicated selects for drug-resistant strains. Drug-resistant strains of tuberculosis are becoming a major concern, particularly in parts of Africa and Russia, where up to 20 percent of the people infected with tuberculosis carry a drug-resistant strain. Such strains are much harder to kill and therefore require newer antibiotics that can cost 100 times more than the traditional drugs.

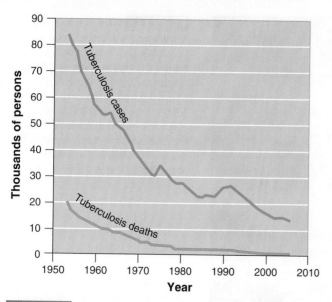

**FIGURE 17.6** **The decline of tuberculosis in the United States.** Due to effective and available medicines, tuberculosis has gone from being one of the most deadly diseases in the United States to a disease that rarely kills today.

## Emergent Infectious Diseases

In recent decades, we have witnessed the appearance of many **emergent infectious diseases,** which are defined as infectious diseases that were previously not described or have not been common for at least the prior 20 years. FIGURE 17.7 locates some of the best-known emergent infectious diseases. Since the 1970s, the world has observed an average of one emergent disease each year. Many of these new diseases have come from pathogens that normally infect animal hosts but then unexpectedly jump to human hosts. This typically occurs because the diseases can mutate rapidly, eventually producing a genotype that can infect humans. Some of the most high-profile diseases that have jumped from animals to humans include HIV/AIDS, Ebola, mad cow disease, bird flu, and West Nile virus. These new diseases are of particular concern in the world today because the rapid movement of people and cargo can spread them to nearly any place on Earth within 24 hours.

**HIV/AIDS** In the late 1970s, rare types of pneumonia and cancer began appearing in individuals with weak immune systems, a condition that was soon named **Acquired Immune Deficiency Syndrome (AIDS).** In 1983, scientists discovered that the weak immune system was caused by a previously unknown virus that they named the **Human Immunodeficiency Virus (HIV).** This virus was spread both through sexual contact and by drug users who were sharing needles that had not been sanitized between uses.

The origin of this new virus remained a mystery until 2006 when researchers found a genetically similar virus in a wild population of chimpanzees living in the African nation of Cameroon (FIGURE 17.8). The researchers hypothesized that local hunters were exposed to the virus when butchering or eating the chimps (a common practice in this part of the world). With this exposure, the virus was able to infect a new host, humans. Today, more than 33 million people in the world are infected with HIV and 25 million people have died from AIDS-related illnesses.

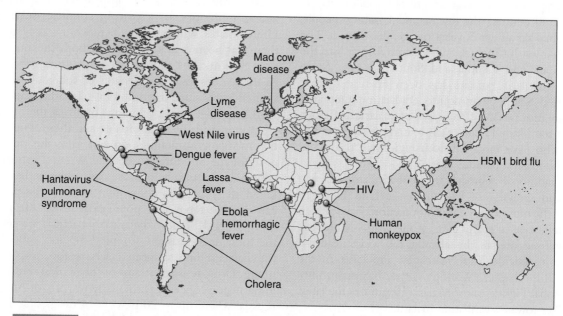

**FIGURE 17.7** **The emergence of new diseases.** Since the 1970s, new diseases, or diseases that have been rare for more than 20 years, have been appearing throughout the world at a rate of approximately one per year. [After http://ww3.niaid.nih.gov/NR/rdonlyres/DF9CBA79-F005-4550-857A-63D62FEFBCA8/0/emerging_diseases1.gif.]

**FIGURE 17.8** **The source of HIV.** In 2006, researchers found that chimpanzees in Cameroon carried a virus that was genetically very similar to HIV. Thus, these chimps are the most likely source of this emerging human disease.

Fortunately, new antiviral drugs are able to maintain low HIV populations inside the human body and thereby substantially extend life. From the lessons learned in combating other diseases such as tuberculosis, combinations of antiviral drugs are being used to reduce the risk that the virus will evolve resistance to any single drug. Unfortunately, many of these drugs are quite expensive and not affordable to most people living in low-income countries. This is changing, however, and both the distribution and the availability of these drugs to those who cannot afford them have greatly improved.

**EBOLA HEMORRHAGIC FEVER** In 1976, researchers first discovered **Ebola hemorrhagic fever,** a disease caused by the Ebola virus. First discovered in the Democratic Republic of Congo near the Ebola River, the virus has infected several hundred humans and a variety of other primates from several countries in central Africa. Although infections of humans have been sporadic and have not reached epidemic proportions, the Ebola virus is of particular concern because it kills a large percentage of those infected. Infected individuals have suffered a 50 to 89 percent death rate from different outbreaks of the disease. Those infected quickly begin to experience fever, vomiting, and sometimes internal and external bleeding (**FIGURE 17.9**). Death occurs within 2 weeks, and there are no drugs available to fight the virus. Unlike the progress that has been made with identifying the origin of HIV, the natural source of the Ebola virus remains unknown. Because the virus also kills other primates at high rates, leaving no primate hosts for the virus, primates are not a likely long-term source of the virus.

**MAD COW DISEASE** In the 1980s, scientists first described a neurological disease, later known as **mad cow disease,** in which a pathogen slowly damages a cow's nervous system. The cow loses coordination of its body (a condition compared to a person going mad), and then dies (**FIGURE 17.10**). Scientists now know that small, beneficial proteins in brains of cattle, called **prions,** occasionally mutate into deadly proteins that act as pathogens and subsequently cause mad cow disease. Prions are not well understood and represent a new category of pathogen.

In 1996, scientists in Great Britain announced that mad cow disease, also known as bovine spongiform encephalopathy (BSE), could be transmitted to humans who ate meat from infected cattle. Unlike harmful bacteria that can be killed with proper cooking, prions are difficult to destroy by cooking. Infected humans

**FIGURE 17.9** **Ebola hemorrhagic fever.** The Ebola virus is highly lethal to humans and there are no drugs for treatment. When treating a person infected with the virus, researchers and medical workers have to exercise extreme caution.

**FIGURE 17.10** **Mad cow disease.** Cows that have been fed the remains of dead cows and sheep can become infected with harmful prions. These prions cause the cows to develop glazed eyes, body tremors, and a loss of coordination, eventually leading to death. Humans who consume the beef from infected cows can become infected and suffer a similar fate.

developed variant Creutzfeldt-Jakob disease (vCJD) and suffered a fate similar to the infected cattle.

Mutant prions cannot be transmitted among cattle living together. Transmission requires an uninfected cow to consume the nervous system of an infected cow. As a result, when cattle feed on grass together in a pasture, a rare mutation in a prion would be restricted to a single cow and not spread to other cattle. In the 1980s, however, cattle diets in Europe commonly included the ground-up remains of dead cattle as a source of additional protein. If these remains happened to contain a mutant prion, the prions spread rapidly through the entire cattle population and, in turn, infected humans

who ate the beef. In Britain, a total of 180,000 cattle were infected and 166 people had died as of 2009. It is estimated that several thousand people are currently infected, but the prions can exist in the human body for many years before they begin to cause symptoms of the disease. In response, the European Union temporarily banned British beef imports in 1996 and the British government destroyed tens of thousands of cattle. Since that time, the disease has been found in several other countries including Canada and the United States, but only in a few cattle and no humans. Today, new rules exist that forbid the feeding of animal remains to cattle. As a result, the current risk of mad cow disease to humans has been greatly reduced.

**BIRD FLU** Humans commonly contract many types of flu viruses. As we saw in Chapter 5, the *Spanish flu* of 1918 killed up to 100 million people. Spanish flu was an avian influenza, also known as **bird flu,** caused by the H1N1 virus. This virus is similar to a flu virus that humans normally contract, but H1N1 normally infects only birds. Infections are rarely deadly to wild birds but can frequently cause domesticated birds such as ducks, chickens, and turkeys to become very sick and die. In 2006, reports emerged from Asia that a related virus, known as H5N1, had jumped from birds to people, primarily to people who were in close contact with birds (**FIGURE 17.11**). Although humans often contract a variety of flu viruses, they have no evolutionary history with the H5N1 virus and, as a result, have few defenses against it. As of 2009, more than 400 people had become infected by H5N1 and more than half of them died. Governments responded to this risk by destroying large numbers of infected birds. Currently the H5N1 virus is not easily passed among people, but if a future mutation makes transmission easier, scientists estimate that H5N1 has the potential to kill 150 million people.

**FIGURE 17.11** **Bird flu.** A farm worker feeds a large number of chickens on a farm in the Republic of Niger in western Africa. The virus that causes bird flu normally infects only birds. In 2006, however, the virus began jumping to human hosts where people and birds were in close contact.

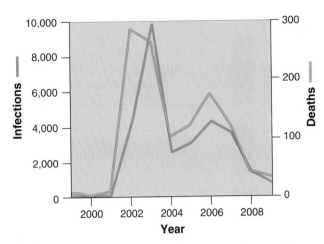

FIGURE 17.12 **FIGURE 17.12** **West Nile virus in the United States.** Following the first appearance of West Nile virus in the United States in 1999, the number of human infections and deaths rapidly increased. Efforts to control populations of mosquitoes that carry the virus are helping to reduce the prevalence of the disease. [Data from http://www.cdc.gov/ncidod/dvbid/westnile/surv&control.htm.]

**WEST NILE VIRUS** The **West Nile virus** lives in hundreds of species of birds and is transmitted among birds by mosquitoes. Although the virus can be highly lethal to some species of birds, including blue jays (*Cyanocitta cristata*), American crows (*Corvus brachyrhynchos*), and American robins (*Turdus migratorius*), most species of birds survive the infection. During the latter half of the twentieth century there were increasing reports that the virus could sometimes infect horses and humans who had been bitten by mosquitoes. The first human case was identified in 1937 in the West Nile region of Uganda, thus giving the virus its name. In humans, the virus causes an inflammation of the brain leading to illness and sometimes death. In 1999, the virus appeared in New York and quickly spread throughout much of the United States. **FIGURE 17.12** shows the history of the West Nile virus in the United States. The highest numbers of infections and deaths from the virus occurred in 2002 and 2003. Increased efforts to combat mosquito populations and protect against mosquito bites are causing a decline in the disease.

## The Future of Human Health

Humans face a large number of health risks, but we have an excellent understanding of the risk factors that are important and the ways to combat many historical and emerging infectious diseases. To combat diseases in low-income countries, the primary needs are improvements in nutrition, wider availability of clean drinking water, and proper sanitation. In high-income countries, public health efforts should promote healthier lifestyle choices such as increased physical activity, a balanced diet, and limiting excess food consumption and tobacco

use. In all countries, continued education is needed to reduce the spread of diseases such as HIV and tuberculosis. Though many historical diseases are either currently under control or likely to be soon if the financial resources become available, emerging infectious diseases may present a greater challenge. New diseases often arise from new pathogens with which we have no experience. Since we cannot predict which diseases will emerge next, public health officials throughout the world must develop rapid response plans when a particular disease does appear. These include rapid worldwide notification of newly identified diseases and strategies to isolate infected persons, which will slow the spread of the disease and provide time for researchers to develop appropriate tactics to combat the threat.

### GAUGE YOUR PROGRESS

✓ What is the difference between historical and emergent infectious diseases?

✓ Which diseases that affect humans are transmitted from animals to humans?

✓ What is the outlook for disease in both developing and developed nations? How is each different?

# Toxicology is the study of chemical risks

The complexity of the biological risks humans face is matched by the complexity of chemical risks. Our modern society has developed an incredible array of chemicals to improve human life around the world, including pharmaceuticals, insecticides, herbicides, and fungicides. In addition to these beneficial chemicals manufactured to improve human health and food production, we have seen other chemicals that are often part of the by-products from industry and the generation of energy, some of which have proven harmful to humans and the environment.

The large number of chemicals released into the environment naturally raises questions about potential effects these chemicals have on humans and other organisms. Many pharmaceuticals, for example, have unexpected consequences when released into the environment. In this section we will look at the types of chemicals that can have harmful effects, how scientists study these chemicals, and what kinds of effects they have.

## Types of Harmful Chemicals

Chemicals can have many different effects on organisms, and some of the most harmful are common in our environment; Table 17.1 lists those of current concern. They can be grouped into five categories: *neurotoxins, carcinogens, teratogens, allergens,* and *endocrine disruptors.*

**NEUROTOXINS** **Neurotoxins** are chemicals that disrupt the nervous systems of animals. Many insecticides, for example, are neurotoxins that interfere with an insect's ability to control its nerve transmissions. Insects and other invertebrates are highly sensitive to neurotoxin insecticides. These animals can become completely paralyzed, cannot obtain oxygen, and quickly die. Other important neurotoxins include lead and mercury. As we discussed in Chapters 14 and 15, lead and mercury are very harmful heavy metals that can damage a person's kidneys, brain, and nervous system. Mercury remains a major problem. Since the federal government required a gradual elimination of lead in gasoline and paint in the 1970s, lead exposure in the United States has declined sharply (**FIGURE 17.13**). However, lead contamination in children remains a serious problem in low-income neighborhoods due to the presence of old lead paint in buildings.

**CARCINOGENS** **Carcinogens** are chemicals that cause cancer. Carcinogens cause cell damage and lead to uncontrolled growth of these cells either by interfering with the normal metabolic processes of the cell or by damaging the genetic material of the cell. Carcinogens that cause damage to the genetic material of a cell are

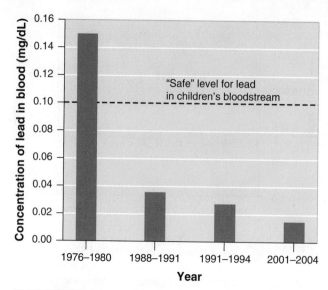

**FIGURE 17.13** **The decline of lead in children over time.** Lead poses a particular risk to childhood development. Lead was gradually phased out of gasoline and paint in the 1970s. Since that time, the average concentration of lead in the bloodstream of children between 1 and 5 years old has declined dramatically. [Data from CDC NHANES survey of blood lead levels in children.]

called **mutagens** (although not all mutagens are carcinogens). Some of the most well-known carcinogens include asbestos, radon, formaldehyde, and the chemicals found in tobacco.

| TABLE 17.1 | Some chemicals of major concern | | |
|---|---|---|---|
| **Chemical** | **Sources** | **Type** | **Effects** |
| Lead | Paint, gasoline | Neurotoxin | Impaired learning, nervous system disorders, death |
| Mercury | Coal burning, fish consumption | Neurotoxin | Damaged brain, kidneys, liver, and immune system |
| Arsenic | Mining, groundwater | Carcinogen | Cancer |
| Asbestos | Building materials | Carcinogen | Impaired breathing, lung cancer |
| Polychlorinated biphenyls (PCBs) | Industry | Carcinogen | Cancer, impaired learning, liver damage |
| Radon | Soil, water | Carcinogen | Lung cancer |
| Vinyl chloride | Industry, water from vinyl chloride pipes | Carcinogen | Cancer |
| Alcohol | Alcoholic beverages | Teratogen | Fetuses with reduced fetal growth, brain and nervous system damage |
| Atrazine | Herbicide | Endocrine disruptor | Feminization of males, low sperm counts |
| DDT | Insecticide | Endocrine disruptor | Feminization of males, thin eggshells of birds |
| Phthalates | Plastics, cosmetics | Endocrine disruptor | Feminization of males |

**FIGURE 17.14** **The effects of thalidomide.** Thalidomide was widely prescribed to pregnant women in the late 1950s to alleviate the symptoms of morning sickness, but it had the unanticipated effect of causing birth defects in tens of thousands of newborn children.

**TERATOGENS** **Teratogens** are chemicals that interfere with the normal development of embryos or fetuses. One of the most infamous teratogens was the drug thalidomide, prescribed to pregnant women during the late 1950s and early 1960s to combat morning sickness. Sadly, tens of thousands of these mothers around the world gave birth to children with defects before the drug was taken off the market in 1961 (**FIGURE 17.14**). One of the most common modern teratogens is alcohol.

Excessive alcohol consumption reduces the growth of the fetus and damages the brain and nervous system, a condition known as fetal alcohol syndrome. This is why physicians recommend that women not consume alcoholic beverages while they are pregnant.

**ALLERGENS** **Allergens** are chemicals that cause allergic reactions. Although allergens are not pathogens, allergens are capable of causing an abnormally high response from the immune system. In some cases, this response can cause breathing difficulties and even death. Typically, a given allergen only causes allergic reactions in a small fraction of people. Some common chemicals that cause allergic reactions include the chemicals naturally found in peanuts and milk and several drugs including penicillin and codeine.

**ENDOCRINE DISRUPTORS** **Endocrine disruptors** are chemicals that interfere with the normal functioning of hormones in an animal's body. Hormones are normally manufactured in the endocrine system and released into the bloodstream in very low concentrations. As the hormones move through the body, they bind to specific cells. Binding stimulates the cell to respond in a way that regulates the functioning of the body.

As noted in our discussion of water pollution in Chapter 14, wastewater may contain hormones from animal-rearing facilities, hormones from human birth control pills found in residential sewage, and pesticides that mimic animal hormones. In waterways containing this wastewater, scientists are increasingly finding that male fish, reptiles, and amphibians are becoming feminized, with males possessing testes that have low sperm counts and, in some cases, testes that produce both eggs and sperm. This change occurs because males normally convert the female hormone *estrogen* into the male chemical *testosterone*. As **FIGURE 17.15** shows, some

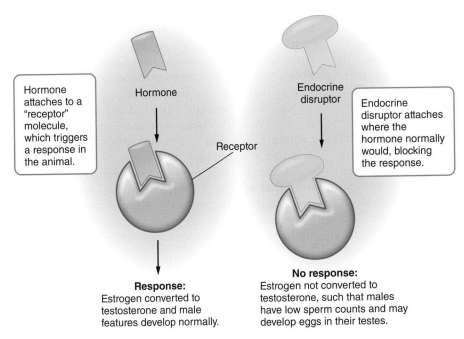

Hormone attaches to a "receptor" molecule, which triggers a response in the animal.

Hormone

Receptor

**Response:** Estrogen converted to testosterone and male features develop normally.

Endocrine disruptor

Endocrine disruptor attaches where the hormone normally would, blocking the response.

**No response:** Estrogen not converted to testosterone, such that males have low sperm counts and may develop eggs in their testes.

**FIGURE 17.15** **Endocrine disruption.** Male animals normally make estrogen and then convert it to testosterone, an important hormone for male reproduction. Some endocrine disrupting hormones are thought to interfere with this process, causing less testosterone to be produced. As a result, these males can have low sperm counts and eggs may develop inside their testes.

endocrine disruptors appear to interfere with the production of testosterone, resulting in males having higher concentrations of estrogen and lower concentrations of testosterone in their bodies. Such discoveries raise serious concerns about whether endocrine disruptors might affect the normal functioning of human hormones. These effects include low sperm counts in men and increased risks of breast cancer in women.

**FIGURE 17.16** **LD50 studies.** To determine the dose of a chemical that causes a 50 percent death rate, scientists expose animals to different doses of a chemical and determine what proportion of the animals die at each dose. Such an experiment typically produces an S-shaped curve.

---

**GAUGE YOUR PROGRESS**

✓ What are some of the beneficial ways humans use chemicals?

✓ What is the impact on humans of each of the five major types of chemicals?

✓ How are chemical and biological risks related?

---

# Scientists can determine the concentrations of chemicals that harm organisms

To assess the risk a chemical poses to any organism, we need to determine the concentrations that cause harm. To learn this, scientists use *dose-response studies, prospective studies,* and *retrospective studies.*

## Dose-Response Studies

**Dose-response studies** expose animals or plants to different amounts of a chemical and then observe a variety of possible responses including mortality or changes in behavior or reproduction. These chemical amounts can be measured as the *concentration* of a chemical in the air, water, or food. They can also be measured as the *dose* of a chemical, which is the amount of chemical that is absorbed or consumed by an organism. For reasons of efficiency, most dose–response studies only last for 1 to 4 days. Because of their short duration, such experiments are called **acute studies.**

Dose-response studies most commonly measure mortality as a response. At the end of a dose-response experiment, scientists count how many individuals die after exposure to each concentration. When the data are graphed, the data generally follow an S-shaped curve, like the one in **FIGURE 17.16.** If you examine the purple curve, you will see that at the lowest dose no individuals die. At slightly higher doses, a few individuals die. The dose at which an effect can be detected is called the *threshold.* These individuals generally are in poorer health or genetically are not very tolerant to the chemical. As the dose is further increased, many more individuals begin to die. At the highest concentrations all individuals die.

A helpful measurement for comparing the harmful effects of different chemicals is the **LD50,** which is an abbreviation for the *lethal dose* that kills 50 percent of the individuals. The LD50 value is important for assessing the relative toxicity of a chemical. By quantifying the LD50 value for a new chemical, for example, scientists can compare the value to thousands of previous tests. In doing so, they can determine whether a new chemical is more or less lethal in comparison to other chemicals that are being used.

The amount of death that a chemical causes can differ among species and among different groups of species, including mammals, birds, fish, and invertebrates. Since conducting LD50 studies on humans would be unethical, studies are conducted on animals such as mice and rats and the results are extrapolated to humans. To understand the effects of chemicals on nonhuman animals, test results from mice and rats are used to represent all mammals, birds such as pigeons and quail are used to represent all birds, fish such as trout are used to represent all fish, and common invertebrates such as water fleas are used to represent all invertebrates.

Not all dose-response experiments measure death as a response. In many cases, scientists are interested in other harmful effects that a chemical might have, including its acting as a teratogen, carcinogen, or neurotoxin that could alter the behavior of an individual. We call these **sublethal effects.** In these cases, the experiments are conducted to determine the *effective dose* that causes 50 percent of the individuals to display the harmful, but nonlethal, effect. This dose is known as the **ED50.**

**TESTING STANDARDS** In the United States, the effects of chemicals on humans and wildlife are regulated by the Environmental Protection Agency (EPA). The Toxic Substances Control Act of 1976 gives the EPA the authority to regulate many chemicals, though excluding food, cosmetics, and pesticides. Pesticides are regulated under a separate law—the Federal Insecticide, Fungicide, and Rodenticide Act of 1996. Under this act, a manufacturer must demonstrate that a pesticide "will not generally cause unreasonable adverse effects on the environment."

With approximately 10 million species of organisms on Earth, no chemical can be tested on every organism. As a result, scientists have devised a system of testing a few species—a bird, mammal, fish, and invertebrate—that are thought to be among the most sensitive in the world. The particular species tested from each of the four animal groups can vary, depending on which species is thought to be the most sensitive to a particular chemical. The logic is that if we know the sensitivity of the most sensitive species in a group, then any regulations that are devised to protect it would automatically protect all other species in that group.

You might have noticed that the four groups of animals required to be tested do not include amphibians or reptiles. Unfortunately, the standards for testing chemicals were set up before there was much interest in protecting amphibians and reptiles. Currently, test results from fish are used to represent aquatic amphibians and reptiles, whereas test results from birds are used to represent terrestrial amphibians and reptiles. Because amphibians and reptiles are now experiencing population declines throughout the world, there is increased interest in requiring tests of these two groups as well.

Using the LD50 and ED50 values from dose-response experiments, regulatory agencies such as the EPA can determine the concentrations in the environment that should cause no harm. For most animals, a safe concentration is obtained by taking the LD50 value and dividing it by 10. The logic is that if the LD50 value causes 50 percent of the animals to die, then 10 percent of the LD50 value should cause few or no animals to die.

For humans, however, the regulatory agencies are much more conservative in setting concentrations. As mentioned earlier, dose-response tests cannot be conducted on humans. Therefore scientists conduct dose-response experiments on rats and mice and then extrapolate the results to humans. The LD50 or ED50 values are then divided by 10 to determine a safe concentration for rats and mice. This value is divided by 10 again to reflect that rats and mice may be less sensitive to a chemical than humans. Finally, this value is often divided by 10 again to ensure an extra level of caution. In short, the LD50 and ED50 values obtained from rats and mice are divided by 1,000 to set the safe values for humans. Crunch the Numbers "Estimating LD50 Values and Safe Exposures" shows you how to make this calculation.

**CHRONIC STUDIES** Although the vast majority of toxicology studies are only conducted for a few days, some studies are conducted for longer periods of time. These experiments of longer duration are called **chronic studies.** Chronic studies will often last from the time

## CRUNCH THE NUMBERS

### Estimating LD50 Values and Safe Exposures

Using our knowledge of how scientists conduct LD50 studies, we can consider an example. Let's imagine that you are a scientist charged with determining the safe levels for mammals of a pesticide in the environment. Using lab rats, you feed them a diet that contains different amounts of the pesticide, ranging from 0 to 4 mg of pesticide per kg of the rat's mass. After feeding them these diets for 4 days, you count how many rats are still alive. When you plot the data, you obtain the graph in **FIGURE 17.17**.

What is the LD50 value for lab rats? To determine this, we can draw a horizontal line at the point of 50 percent mortality on the y axis. Where this line intersects the purple line, we can draw another line straight down to the x axis. This second line crosses the x axis at 2 mg/kg of mass.

Based on this LD50 study, what amount of pesticide would be considered safe for mammals to ingest? Recall that we can calculate this number by taking the LD50 value and dividing it by 10. Thus the safe amount of pesticide for a rat is:

$$\frac{2 \text{ mg/kg of mass}}{10} = 0.2 \text{ mg/kg of mass}$$

**FIGURE 17.17** **Calculating safe doses.** When a rat is fed increased amounts of a chemical, the amount of mortality increases.

**Your Turn:** Using the same LD50 study, what amount of pesticide would be considered safe for a human to ingest?

(a)                                        (b)

**FIGURE 17.18** **Conducting dose-response experiments.** (a) Researchers determine how chemicals affect the mortality, behavior, and reproduction of animals using both acute and chronic dose-response experiments. (b) In the experiment shown, researchers are examining the effects of different insecticide concentrations on the survival of tadpoles.

an organism is very young to when it is old enough to reproduce. For some species such as fish, chronic experiments can take several months. The goal of chronic studies is to examine the long-term effects of chemicals, including their effects on survival and their impacts on reproduction (**FIGURE 17.18**).

## Retrospective versus Prospective Studies

Estimating the effects of chemicals on humans is a major challenge. One approach that we have discussed conducts dose-response experiments on rats and mice and extrapolates the results to humans. An alternative approach examines large populations of humans or animals who are exposed to chemicals in their everyday lives and then determines whether these exposures are associated with any health problems. Such investigations fall within the study of **epidemiology**, a field of science that strives to understand the causes of illness and disease in human and wildlife populations. There are two ways of conducting this type of research: *retrospective studies* and *prospective studies.*

**Retrospective studies** monitor people who have been exposed to a chemical at some time in the past. In such studies, scientists identify a group of people who have been exposed to a potentially harmful chemical and a second group of people who have not been exposed. Both groups are then monitored for many years to see if the exposed group experiences greater health problems than the unexposed group. In 1984, for example, there was an accidental release of methyl isocyanate gas from a Union Carbide pesticide factory in Bhopal, India. More than 36,000 kg (80,000 pounds) of hazardous gas spread through the city of 500,000 inhabitants. An estimated 2,000 people died that night and another 15,000 died later from effects related to the

exposure. Scientists have now been monitoring many citizens of Bhopal for more than two decades to determine if survivors of the accident have developed any additional health problems over time. The retrospective studies have found that approximately 100,000 people are still suffering illness from the accident. As shown in **FIGURE 17.19**, the survivors have higher rates of respiratory symptoms and still births; they also have higher rates of genetic abnormalities, infant mortality, kidney failure, and learning disabilities.

In contrast to retrospective studies, **prospective studies** monitor people who might become exposed to harmful chemicals in the future. In this case, scientists might select a group of 1,000 participants and ask them to keep track of the food they eat, the tobacco they use, and the alcohol they drink for the next 40 years. As time passes, the researchers can determine if the habits of the participants have any association with future health problems. Prospective studies can be quite challenging because a participant's habits, such as tobacco use, can also be associated with many other risk factors including socioeconomic status. Of particular concern is when multiple risks cause **synergistic interactions**, in which two risks together cause more harm than one would expect based on their individual risks. For example, the health impact of a carcinogen such as asbestos can be much higher if an individual also smokes tobacco.

Studies of lead in children are often conducted using prospective studies. In one study conducted by researchers at Harvard University on the effects of lead on children's intelligence, 276 children in Rochester, New York, were followed from 6 months to 5 years of age. At the age of 5, children can take reliable IQ tests. In addition to lead exposure, the researchers also accounted for other factors that might affect childhood IQ including the mother's IQ, exposure to tobacco, and the

(a)

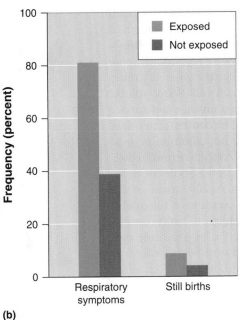

(b)

**FIGURE 17.19** The chemical disaster in Bhopal, India. (a) In 1984, a massive release of methyl isocyanate gas killed and injured thousands of people. (b) Retrospective studies that have followed the survivors of the accident have identified a large number of longer-term health effects from the accident. [Data from P. Cullinan, S. D. Acquilla, and V. R. Dhara, Long-term morbidity in survivors of the 1984 Bhopal gas leak, *National Medical Journal of India* (January–February 1996): 5–10.]

intellectual environment of their homes. After controlling for these other factors, the researchers found that among children who had been exposed to lead in the environment, primarily from lead dust and consumed lead paint chips, those with higher lead exposures scored lower on the subsequent IQ tests. Such prospective studies can be very helpful in helping regulators determine acceptable levels of chemical exposure.

## Factors That Determine Concentrations of Chemicals Organisms Experience

Knowing the concentrations of chemicals that can harm humans or other animals is important, but it is only useful when combined with information about the concentrations that an individual might actually experience in the environment. If a chemical is quite harmful at some moderate concentration but individuals only experience lower concentrations, we might not be particularly concerned. Therefore, to identify and understand the effects of chemical concentrations that organisms experience, we need to know something about how the chemicals behave in the environment.

**ROUTES OF EXPOSURE** The ways in which an individual might come into contact with a chemical are known as **routes of exposure.** As FIGURE 17.20 illustrates, the full range of possibilities is complex because it includes potential exposures from the air, from water used for drinking, bathing, or swimming, from food, and from the environments of places where people live, work, or visit. For any particular chemical, however, the major routes of exposure are usually limited to just a few of the many possible routes. For example, bisphenol A is a chemical used in the manufacturing of hard plastic items such as toys, food containers, and baby bottles.

Recent research has raised concerns that bisphenol A may be responsible for early puberty and increased rates of cancer. While these effects are being debated and investigated, it is clear that a child's exposure to bisphenol A is limited to toys, food containers, and baby bottles. Knowing this, scientists can then determine the chemical's *solubility* and its potential for bioaccumulation and *biomagnification*.

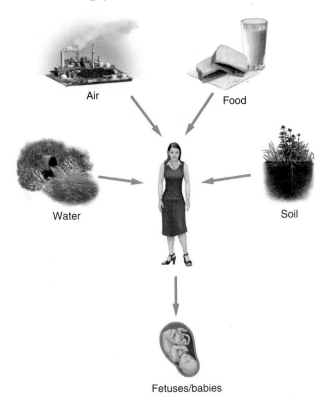

**FIGURE 17.20** **Routes of exposure.** Despite a multitude of potential routes of exposure to chemicals, most chemicals have a limited number of major routes.

**SOLUBILITY OF CHEMICALS, BIOACCUMULATION, AND BIOMAGNIFICATIONS** The movement of a chemical in the environment depends in part on its **solubility**—how well a chemical can dissolve in a liquid. Some chemicals are readily soluble in water whereas others are much more soluble in fats and oils. Water-soluble chemicals can be pervasive in groundwater and surface waters including rivers and lakes. In contrast, chemicals that are soluble in fats and oils are not commonly found in water, but are found in higher concentrations in soils, including the benthic soils that underlie bodies of water.

Oil-soluble chemicals are also readily stored in the fat tissues of animals. Continued exposure to oil-soluble chemicals can cause bioaccumulation, an increased concentration of a chemical within an organism over time. The process of bioaccumulation begins when an individual incorporates small amounts of a chemical from the environment into its body. As the chemical continually accumulates over time, often in fat tissues, the concentration of the chemical inside the organism increases. Fish, for example, are exposed to low concentrations of methyl mercury when they drink water, pass water over their gills to breathe, and consume food that contains mercury. A fish stores mercury in its fat tissues and, over time, the mercury accumulates. The rate of accumulation for any animal will depend on the concentration of the chemical in the environment, the rate that the animal takes up each source of the chemical, the rate at which the chemical breaks down inside the animal, and the rate at which it is excreted by the animal.

**Biomagnification** is the increase in chemical concentration in animal tissues as the chemical moves up the food chain. For example, primary consumers can obtain an oil-soluble chemical from the environment and the chemical bioaccumulates in their fat tissues. Secondary consumers then consume the primary consumers and the chemical that they contain. The secondary consumers bioaccumulate the chemical they have ingested by consuming the chemical that is stored in the fat tissues of the primary consumers. The chemical is now stored in the secondary consumer's fat tissues. As we continue to move up the food chain, each trophic level is exposed to higher concentrations of chemicals from the food it consumes. In this way, the original concentration in the environment is magnified to occur at a much higher concentration in the top predator of the community.

The classic example of biomagnifications is the case of DDT, an insecticide that has been widely used to kill insect pests in agriculture and to kill the mosquitoes that carry malaria and other diseases. DDT is not soluble in water, so when sprayed over water it quickly binds to particulates in the water and the underlying soil or is quickly taken up by the tiny zooplankton that act as primary consumers on algae. As we see in **FIGURE 17.21**, the very low concentration of DDT in the water bioaccumulates in the bodies of the zooplankton where

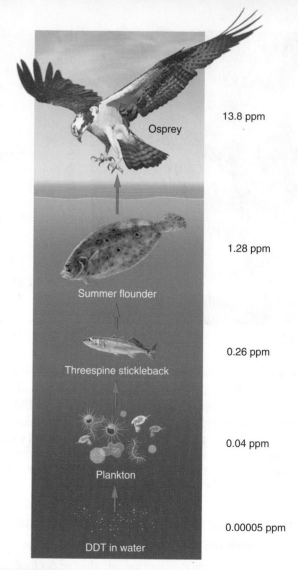

Osprey — 13.8 ppm

Summer flounder — 1.28 ppm

Threespine stickleback — 0.26 ppm

Plankton — 0.04 ppm

DDT in water — 0.00005 ppm

**FIGURE 17.21** **The biomagnification of DDT.** The initial exposure is primarily in a low trophic group such as the plankton in a lake. Consumption causes the upward movement of the chemical where it is accumulated in the bodies at each trophic level. The combination of bioaccumulation at each trophic level and upward movement by consumption allows the concentration to magnify to the point where it can be substantially more concentrated in the top predator than it was in the water. [Data from G. M. Woodwell, C. F. Wurster, Jr., and Peter A. Isaacson, DDT Residues in an East Coast Estuary: A Case of Biological Concentration of a Persistent Insecticide, *Science,* New Series, 156 (3776) (May 12, 1967): 821–824. http://www.jstor.org/stable/1722018.]

it becomes approximately 10,000 times more concentrated. Small fish eat the zooplankton for many weeks or months and the DDT is further concentrated approximately 10-fold. Large fish spend their lives eating the contaminated smaller fish and the DDT in the large fish is further concentrated approximately fourfold. Finally, fish-eating birds such as pelicans and eagles spend years eating the large fish and further magnify the DDT in

| TABLE 17.2 | The persistence of various chemicals in the environment, measured in terms of their half-life | |
| --- | --- | --- |
| **Chemical** | **Half-life** | |
| Malathion insecticide | 1 day | |
| Radon | 4 days in air | |
| Vinyl chloride | 4.5 days in air | |
| Phthalates | 4.5 days in water | |
| Roundup herbicide | 7 to 70 days in water | |
| Atrazine herbicide | 224 days in wetland soils | |
| Polychlorinated biphenyls (PCBs) | 8 to 15 years in water | |
| DDT | 30 years in soil | |

Source: Hazardous Substances Data Bank, http://toxnet.nlm.nih.gov/cgi-bin/sis/htmlgen?HSDB/.

organisms may become even larger risks when they persist for many years. For this reason, many modern chemicals are designed to break down much more rapidly so that any unintended effects will be short-lived.

**GAUGE YOUR PROGRESS**

✓ Why is it difficult to test the potential effects of chemicals on humans?

✓ How does route of exposure influence toxicity? How does solubility affect exposure to chemicals?

✓ In what ways are persistence and bioaccumulation similar concepts? How are they related to one another?

their own bodies. Because of biomagnification along the food chain, the concentration of DDT in the birds is nearly 8 million times higher than the concentration in the water. The concentrated DDT caused fish-eating birds to produce thin-shelled eggs that often broke when the parent birds incubated the eggs. This was a primary cause in the decline of these birds in the 1960s. Since DDT was banned in the United States in 1972, the populations of these birds have dramatically increased. Working Toward Sustainability "The Global Fight Against Malaria" discusses the continued use of DDT in other parts of the world.

**PERSISTENCE** The **persistence** of a chemical refers to how long the chemical remains in the environment. Persistence depends on a number of factors including temperature, pH, whether the chemical is in water or soil, whether the chemical can be degraded by sunlight, and whether the chemical can be broken down by microbes. Scientists often measure persistence by observing the time needed for a chemical to degrade to half its original concentration, or the *half-life* of the chemical (Table 17.2). DDT for example, has a half-life in soil of up to 30 years. Thus, even after DDT is no longer sprayed in an area, half of the chemical that was absorbed in the soil would still be present after 30 years, and one-fourth would be present after 60 years. Chemicals that cause harmful effects on humans and other

# Risk analysis helps us assess, accept, and manage risk

Many of our actions involve some amount of risk from environmental hazards. For our purposes, an **environmental hazard** is anything in our environment that can potentially cause harm. Environmental hazards include substances such as pollutants or other chemical contaminants, human activities such as draining swamps or logging forests, or natural catastrophes such as volcanoes and earthquakes. The hazards we face may be voluntary, as when we make a decision to smoke tobacco, or they may be involuntary, as when we are exposed to air pollution.

When assessing the risk of different environmental hazards, regulatory agencies, environmental scientists, and policy makers usually follow the three steps listed in **FIGURE 17.22**. These are risk assessment, risk acceptance, and risk management.

## Risk Assessment

Risk assessment seeks to identify a potential hazard and determine the magnitude of the potential harm. There are two types of risk assessment—*qualitative* and *quantitative*.

| Risk assessment | Risk acceptance | Risk management |
| --- | --- | --- |
| 1. Identify the hazard.<br>2. Characterize toxicity (dose/response).<br>3. Determine extent of exposure. | Determine acceptable level of risk (balanced against social, economic, political considerations). | Determine policy with input from private citizens, industry, interest groups. |

**FIGURE 17.22** **The process of risk analysis.** Risk analysis involves risk assessment, risk acceptance, and risk management.

**QUALITATIVE RISK ASSESSMENT** Each of us has some idea of the risk associated with different environmental hazards. We generally make qualitative judgments in which we might categorize our decisions as having low, medium, or high risks. When we choose to slow down on a wet highway or to buy a more expensive car because it is safer, for example, we are making qualitative judgments of the relative risks of various decisions. That is, we make judgments that are based on our perceptions but that are not based on actual data. It would be unusual for us to consider the actual probability—that is, the statistical likelihood—of an event occurring and the probability of that event causing us harm. Because our personal risk assessments are not quantitative, they often do not match the actual risk.

**PERCEIVED RISK VERSUS ACTUAL RISK** The perception that certain behaviors or activities carry a high risk does not always match the reality. For example, some people find air travel very stressful because they are afraid the plane might crash. These same people often prefer riding in a car, which they perceive to be much safer.

Or, a person may be very cautious about safety while walking in an area with heavy traffic but never consider the health dangers of smoking or a lack of exercise. To manage our risk effectively, we need to ask how closely our perceptions of risk match the reality of actual risk.

In the United States, the actual risks of various hazards are well known based on death statistics that are kept by the government. By knowing the total number of people who die in a year and their different causes of death, researchers can calculate the probability of an individual dying from each cause. FIGURE 17.23 provides current data on causes of death in the United States. Because these risk estimates are based on real data, they are quantitative rather than qualitative. If we examine this figure, we see that the probability of dying in an automobile is far greater than the probability of dying in an airplane. Similarly, the probability of dying from heart disease is monumentally greater than the risk of dying in a pedestrian accident. These numbers underscore the fact that our perceptions of risk can often be very different from the actual risk. Because a catastrophic event, such as a nuclear plant meltdown or a plane crash, can

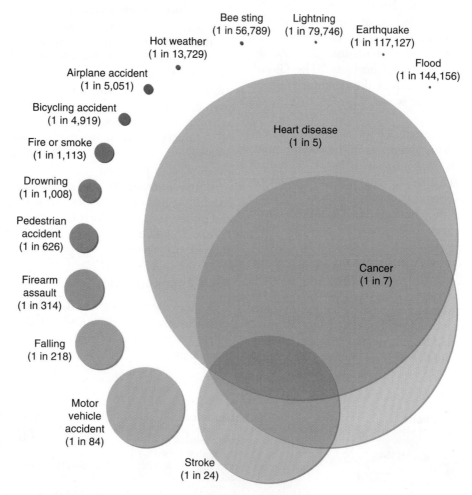

**FIGURE 17.23** **The probabilities of death in the United States.** Some factors that people perceive as having high risks of death, such as dying in an airplane crash, actually are quite low. In contrast, some factors that people rate as low risk, such as dying from heart disease, actually pose the greatest risk. [After National Geographic Society (2006); data from National Safety Council (2005).]

do a great deal of harm and receives great media attention, people believe that it is very risky to use nuclear reactors or to fly in airplanes. However, as these events rarely occur the risk of harm is low. In contrast, we tend to downplay the risk of activities that provide us with cultural, political, or economic advantages such as drinking alcohol or working in a coal mine.

**QUANTITATIVE RISK ASSESSMENT** The most common approach to conducting a quantitative risk assessment can be expressed with a simple equation:

$$\text{Risk} = \text{probability of being exposed to a hazard} \times \\ \text{probability of being harmed if exposed}$$

Using this equation, we could ask whether it is riskier to fly on commercial airlines for 1,000 miles per year or to eat 40 tablespoons of peanut butter per year (which contains tiny amounts of a carcinogenic chemical produced naturally by a fungus that sometimes occurs in peanut butter). The risk of dying in a plane crash depends on the probability of your plane crashing (which is very low) multiplied by your probability of dying if the plane does crash (which approaches 100 percent). The risk of dying of cancer from consuming peanut butter depends upon your probability of eating peanut butter (which is near 100 percent) multiplied by the probability that consuming peanut butter will cause you to develop lethal cancer (which is very small). It turns out that both behaviors produce a 1 in 1 million chance of dying. This example demonstrates a fundamental rule of risk assessment: the risk of a rare event that has a high likelihood of causing harm can be equal to the risk of a common event that has a low likelihood of causing harm.

Quantitative risk assessments bring together tremendous amounts of data. The estimates of harm can come from acute and chronic dose-response experiments, retrospective studies, and prospective studies. The estimates of which concentrations of a chemical an organism will experience in the environment incorporates the concentrations found in nature, routes of exposure, solubility, persistence, and the potential for the chemical to bioaccumulate or biomagnify. Together, these two groups of data are integrated to estimate the probability of harm.

**A CASE STUDY IN RISK ASSESSMENT** As we saw in Chapter 14 on water pollution, from the 1940s to the 1970s some companies manufacturing electrical components dumped PCBs (polychlorinated biphenyls) into rivers. Beginning in the 1960s, there was increasing evidence that PCBs might have harmful health effects on organisms that came into contact with them, including liver damage in animals and impaired learning in human infants.

Once the EPA identified PCBs as a potential hazard, it began a risk assessment. The agency brought together a range of data. First, scientists had to determine which

concentrations of PCBs might cause cancer. To accomplish this objective, they examined dose-response studies on laboratory rats exposed to different concentrations of PCBs. They also examined retrospective studies of cancer cases in workers employed by industries that used PCBs. Next, they had to determine what concentrations people might experience. To accomplish this, scientists examined data on current concentrations in the air, soil, and water and considered the half-life of the chemical. Because PCBs were found throughout the environment and because they are very persistent, the probability of coming into contact with PCBs was considered relatively high. They also considered the potential routes of exposure: eating contaminated fish, drinking contaminated water, and breathing contaminated air.

The final result of the risk assessment on PCBs showed that the risk from eating contaminated fish is higher than the risk from drinking contaminated water and much higher than breathing contaminated air. As a result, signs were posted on the Hudson River instructing anglers not to consume the fish that they had caught (**FIGURE 17.24**). With limited fish consumption, the EPA concluded that the absolute risk of an individual developing cancer from PCB exposure was low. As discussed in Chapter 14, however, the risk was high enough to cause the EPA to recommend a dredging of the Hudson River to remove a large fraction of the PCBs that had settled at the bottom of the river.

## Risk Acceptance

Once the risk assessment is completed, the second step is determining risk acceptance— the level of risk that

**FIGURE 17.24** **The outcome of a risk assessment of PCBs.** Based on a risk assessment of humans consuming fish, the EPA determined that the fish living in the Hudson River in New York State and in Silver Lake in Massachusetts had unacceptably high concentrations of PCBs due to illegal dumping of PCBs by General Electric. As a result, anglers were not allowed to keep and consume the fish that they caught.

can be tolerated. Risk acceptance may be the most difficult of the three steps in the risk-analysis process. No amount of information on the extent of the risk will overcome the conflict between those who are willing to live with some amount of risk and those who are not. Even among those people who are willing to accept some risk, the precise amount of acceptable risk is open to heated disagreement. For example, according to the EPA, a 1 in 1 million risk is acceptable for most environmental hazards. Some people believe this is too high. Others feel that a 1 in 1 million chance of death from radiation leaks is a small price to pay for the electricity generated by nuclear energy. While personal preferences will always complicate the determination of risk acceptance, environmental scientists, economists, and others can help us weigh the options as objectively as possible by providing accurate estimates of the costs and benefits of activities that affect the environment.

## Risk Management

Risk management, the third part of the risk-analysis process, seeks to balance possible harm against other considerations. Risk management integrates the scientific data on risk assessment and the analysis of acceptable levels of risk with a number of additional factors including economic, social, ethical, and political issues. Whereas risk assessment is the job of environmental scientists, risk management is a regulatory activity that is typically carried out by local, national, or international government agencies.

The regulation of arsenic in drinking water provides an excellent example of the difference between risk assessment and risk management. As we saw in Chapter 14, despite the fact that scientists knew that 50 $\mu$g/L of arsenic could cause cancer in people, from 1942 to 1999 the federal government set the acceptable concentration of arsenic at 50 $\mu$g/L. In 1999, the EPA announced it was lowering the maximum concentration of arsenic in drinking water to 10 $\mu$g/L, which matched the standards set by the European Union and the World Health Organization. This new regulation threatened to place a new economic burden on mining companies that produced arsenic as a by-product of mining and on several municipalities in western states with naturally high concentrations of arsenic in their drinking water. Both groups lobbied hard against the lower arsenic limits because the lower limits would require a large financial investment to remove the arsenic from drinking water. In 2001, weeks before the new lower limits were to go into effect, the EPA announced that it would return to the 50 $\mu$g/L limit. The agency argued that further risk assessments needed to be conducted and any risk assessment had to be balanced by economic interests. Later in 2001, the National Academy of Sciences concluded that the acceptable amount of arsenic was actually 5 $\mu$g/L, which was lower than some previous estimates. This new risk assessment played a key role in striking

a balance between the scientific data and economic interests and the EPA revised its ruling, setting the safe arsenic concentration at 10 $\mu$g/L.

## Worldwide Standards of Risk

There are currently about 80,000 registered chemicals in the world but they are not regulated the same way around the globe. A key factor determining the type of chemical regulation is whether the regulations are guided by the *innocent-until-proven-guilty principle* or the *precautionary principle,* outlined in FIGURE 17.25. The **innocent-until-proven-guilty principle** is based on the philosophy that a potential hazard should not be considered a hazard until the scientific data can definitively demonstrate that a potential hazard actually causes harm. This philosophy allows the introduction of beneficial chemicals more quickly. The downside of this philosophy is that harmful chemicals can affect humans or wildlife for decades before sufficient scientific evidence accumulates to confirm that they are harmful.

In contrast, the **precautionary principle** is based on the philosophy that when a hazard is plausible but not yet certain, we should take actions to reduce or remove the hazard. The plausibility of the risk must have a scientific basis, rather than simple speculation. In addition, the intervention should be in proportion to the potential harm that might be caused by the hazard. The benefit of this philosophy is that fewer harmful chemicals will enter the environment. However, the introduction of beneficial chemicals that are ultimately found not to pose harm could be delayed for many years if the initial assessment indicates a plausible risk. Moreover, the slower pace of approval can reduce the financial motivation of manufacturers to invest in research for new chemicals. In short, there is a trade-off between greater safety with slower introduction of beneficial chemicals versus greater potential risk with a greater rate of discovery of helpful chemicals. Use of the precautionary principle has been growing throughout many parts of the world and was instituted by the European Union in 2000. The United States, however, continues to use the innocent-until-proven-guilty principle.

The potential benefit of the precautionary principle can be illustrated using the case of asbestos. Asbestos is a white, fibrous mineral that is very resistant to burning. This made asbestos a popular building material throughout much of the twentieth century. It is now widely accepted that dust from asbestos can cause a number of deadly diseases including asbestosis (a painful inflammation of the lungs) and several types of cancer. When asbestos was first mined in 1879, there was no evidence that it harmed humans. The first report of deaths in humans was in 1906 and the first experiment showing harmful effects in rats was conducted in 1911. In 1930, it was reported that 66 percent of workers in an asbestos factory suffered from asbestosis. In 1955, researchers found that asbestos workers had a higher

**FIGURE 17.25** **The two different philosophies for managing risk.** The innocent-until-proven-guilty principle requires that researchers prove harm before the chemical is restricted or banned. The precautionary principle requires that when there is scientific evidence that demonstrates a plausible risk, the chemical must then be further tested to demonstrate it is safe before it can continue to be used.

risk of lung cancer than other groups. In 1965, a study linked a rare form of cancer with workers who were exposed to asbestos dust. Despite all of the growing scientific evidence that asbestos was harming human health, little was done to reduce the exposure of workers and the public. Indeed, it was not until 1998 that the European Union banned asbestos. Today, workers go to great lengths not to be exposed to asbestos dust (**FIGURE 17.26**). A study in the Netherlands estimated that had asbestos been banned in 1965 when the harm to health became clear, the country would have had 34,000 fewer deaths from asbestos and would have saved approximately $25 billion in cleanup and compensation costs. Because the effects of asbestos can take several decades to harm a person's health, the European Union estimates that from 2005 to 2040 they will have 250,000 to 400,000 more people die as a result of past exposures to asbestos. Had the European Union been using the precautionary principle decades earlier, the number of deaths would have been considerably less.

**INTERNATIONAL AGREEMENTS ON HAZARDOUS CHEMICALS** In 2001 a group of 127 nations gathered in Stockholm, Sweden, to reach an agreement on restricting the global use of some chemicals. The agreement, known as the **Stockholm Convention**, produced a list of 12 chemicals to be banned, phased out, or reduced. These 12 chemicals came to be known as the "dirty dozen" and included pesticides such as DDT, industrial chemicals such as PCBs, and certain chemicals that are by-products of manufacturing processes. All of the chemicals were known to be endocrine disruptors, and

**FIGURE 17.26** **The risks of asbestos dust.** Despite nearly a century of studies on the risks of asbestos dust to human health, only recently have workers been required to go to great lengths to prevent exposure. Today, they dress in chemical suits and respirators when removing asbestos from a building. Applying the precautionary principle would have required protection of workers many decades earlier and saved hundreds of thousands of lives.

a number of them had already been banned or were experiencing declining use in many countries. However, bringing countries together in a forum to discuss controlling the most harmful chemicals was the great achievement of the Stockholm Convention. In 2009, 9 additional chemicals were added to the original list of 12 and several more have been suggested for future listing.

In 2007, the 27 nations of the European Union put into effect an agreement on how chemicals should be regulated within the European Union. Known as **REACH,** an acronym for *r*egistration, *e*valuation, *a*uthorisation, and restriction of *ch*emicals, the agreement embraces the precautionary principle by putting more responsibility on chemical manufacturers to confirm that chemicals used in the environment pose no risk to people or the environment. This regulation was

enacted because many chemicals used for decades in the European Union had not been subjected to rigorous risk analyses. The new regulations are being phased in over several years to permit sufficient time for chemical manufacturers to complete the required testing.

---

### GAUGE YOUR PROGRESS

✓ Why is risk acceptance the most complex and difficult of the three steps in risk analysis?

✓ What is the difference between the innocent-until-proven-guilty principle and the precautionary principle?

✓ Why might the United States be resistant to the precautionary principle approach?

---

# WORKING TOWARD SUSTAINABILITY

## The Global Fight Against Malaria

Bill Gates is best known as the founder of Microsoft, the computer software company, but he is also an active philanthropist. In 2007 he stood up in front of a large group of scientists in Seattle, Washington, and declared that the world needed to eradicate malaria. In challenging the scientists of the world, he asked, "Why would anyone want to follow a long line of failures by becoming the umpteenth person to declare the goal of eradicating malaria?"

Bill Gates knew the history of malaria. People have died from this disease for thousands of years. In modern times, 350 million to 500 million people are infected each year and 1 million of them die. Most malaria cases are in Africa and most of those who die are children. Several eradication efforts have been attempted over the past six decades, mostly focused on eradicating the mosquitoes that carry the malaria pathogen. In the United States, eradication was achieved in 1951 through widespread spraying of the insecticide DDT as well as the application of numerous other public health measures. This spraying program became controversial in the 1960s and 1970s because DDT was found to be widely distributed around the globe and it was linked to the thinning of egg shells in large birds of prey due to its biomagnification. DDT is still sprayed in many parts of the world to assist in the eradication of malaria, but malaria has not been eradicated.

Malaria is difficult to eliminate for a number of reasons. First, mosquito populations that are impacted by spraying insecticides can rebound quickly. In Sri Lanka, for example, consistent spraying to kill mosquitoes

reduced the number of malaria cases from 1 million to a mere 18. Because of this great success, the spraying program was stopped there, but within a few years, malaria cases rapidly increased to a half million. In short, the spraying program was ended before the job was done. Moreover, if one country is spraying to kill mosquitoes and neighboring countries are not, mosquitoes will continue to enter from the neighboring countries. Second, mosquitoes can rapidly evolve resistance to insecticides such as DDT. Third, the malaria pathogen can rapidly evolve resistance to antimalaria drugs. Finally, eradicating malaria is expensive. Typically, countries use multiple strategies including insecticide spraying, antimalarial drugs, and the distribution of mosquito tents in which people can sleep and avoid being bitten during the night (FIGURE 17.27). Collectively, these strategies can carry a price tag that many poor countries cannot afford. Additionally, other social and economic priorities of these countries, as well as social disruption, have precluded or curtailed malaria control programs.

So in 2007, after so many past failures, why did Bill Gates think there was now a possibility of eradicating malaria? Earlier that year, scientists had reported that a new drug, combined with a new style of mosquito net, produced large reductions in malaria cases—as much as a 97 percent reduction in Uganda. The new nets, which were impregnated with more modern insecticides, could last 3 to 5 years. This was a big improvement from the earlier nets that only lasted up to 3 months. In addition to having a new drug and longer-lasting nets, the key to the success of the Ugandan program was to

**FIGURE 17.27** **Combating malaria.** By distributing medicine and nets impregnated with insecticide to households in Africa, childhood death from malaria has declined by as much as 60 percent.

pay for and distribute the drug and nets to everyone who needed them. Employing this strategy around the world is an expensive endeavor, and that's where Bill Gates comes in.

The Bill and Melinda Gates Foundation funds projects that have been historically underfunded. Equally important, by throwing its prominent name and financial resources behind a cause like malaria eradication, the foundation can rally significant financial support from other foundations and from developed countries. Western governments joined the movement and increased malaria funding from $50 million to $1.1 billion. This gave new hope to the declared goal of eradicating malaria from the globe within 50 years.

Many challenges remain. One of the largest is simply organizing distribution systems to hand out the drugs and millions of mosquito tents. In some regions, there are no roads into the villages and the items must be brought in by foot or by boat. There is also the challenge to continue research into new strategies against the pathogen and the mosquito. Currently, a new antimalaria drug from China has proven very effective against the pathogen and is quite inexpensive to manufacture. The manufacturer of this drug agreed to sell it at less than the cost of manufacturing it, making the

drug a very attractive option for low-income countries. The company estimates that this decision cost it $253 million in profits but saved 550,000 lives and brought the company very positive publicity. Another possibility is the development of a vaccine that would provide immunity to malaria infections. Research is ongoing.

Today there is tremendous hope that Bill Gates's dream of eradicating malaria is gaining ground. Most experts agree that malaria cases could be reduced by at least 85 percent in most African countries. The reduction in illness and death would also be highly beneficial to the economies of these low-income countries by realizing reduced health costs and a healthier, and therefore more productive, workforce. The success of the global fight against malaria critically depends on sustained financial support from foundations and governments, continued discovery of new drugs and vaccines, and the recognition that we cannot stop fighting malaria until the job is done.

### References

Kingsbury, K. 2009. A better deal on malaria. *Time,* February 26.

McNeil, D., Jr. 2008. Nets and new drug make inroads against malaria. *New York Times,* February 1.

McNeil, D., Jr. 2008. Eradicate malaria? Doubters fuel debate. *New York Times,*, March 4.

■ **Identify the three major categories of human health risk.**

The major categories of human health risk are physical risks such as natural catastrophes, biological risks such as diseases, and chemical risks such as pesticides.

■ **List the major historical and emerging infectious diseases.**

Some of the major historical infectious diseases include plague, malaria, and tuberculosis. Some of the emerging infectious diseases include HIV/AIDS, Ebola hemorrhagic fever, mad cow disease, bird flu, and West Nile virus.

■ **Name the five major types of toxic chemicals.**

The five major types of toxic chemicals are neurotoxins, carcinogens, teratogens, allergens, and endocrine disruptors.

■ **Distinguish between dose-response studies, retrospective studies, and prospective studies.**

Dose-response studies expose animals to a range of chemical concentrations to determine which concentrations cause harmful effects. Retrospective studies identify a group of people who have been exposed to a chemical in the past and follow them through time to determine if they suffer any harmful effects. Prospective studies identify a group of people and determine whether future exposures to chemicals are associated with any harmful effects.

■ **Describe the factors that help determine the chemical concentrations that organisms experience.**

The amount of exposure to a chemical depends on the potential routes of exposure, how soluble the chemical is in the environment and in the human body, whether the chemical can bioaccumulate with an organism, and whether the chemical can biomagnify up a food chain.

■ **Explain the factors that go into a risk analysis and distinguish between the two major philosophies of chemical regulation.**

A risk analysis starts by identifying a potential hazard and assessing the risk that concentrations in the environment pose to humans or other organisms. The analysis then determines an acceptable level of risk. Finally, social, economic, political, and ethical considerations are weighed to ultimately manage the risk. The innocent-until-proven-guilty principle requires scientists to demonstrate that a chemical causes harm to humans or the environment before any restrictions are imposed. According to the precautionary principle, if scientists have made a plausible association between a chemical and harm to humans or the environment, then the use of the chemical should be restricted until scientists can demonstrate that the chemical is safe.

## CHECK YOUR UNDERSTANDING

1. Which statement is true regarding human health risks?
   (a) More people die from infectious diseases than from noninfectious diseases.
   (b) More people die from accidents than from any other cause.
   (c) More people die from chemical risks than from physical or biological risks.
   (d) More people die from cancer than from any other cause.
   (e) More people die from heart disease than from any other cause.

2. Which statement is true regarding the relationship between health risks and income?
   (a) A major risk in high-income countries is a lack of food.
   (b) A major risk in high-income countries is poor sanitation.
   (c) A major risk in low-income countries is obesity.
   (d) A major risk in low-income countries is a lack of food.
   (e) The major risks in high- and low-income countries are similar.

3. Which statement about historical infectious diseases is *not* true?
   (a) Plague is a disease that is carried by fleas attached to rodents.
   (b) Malaria is a disease that is carried by mosquitoes.
   (c) Tuberculosis is a disease that is transmitted through the air.
   (d) The pathogen that causes tuberculosis can become drug-resistant.
   (e) Historically important infectious diseases no longer pose a health risk.

4. Which statement about emerging infectious diseases is *not* true?
   (a) HIV is a virus that most likely came from chimps.
   (b) Ebola hemorrhagic fever causes a high rate of death.
   (c) Mad cow disease is spread by feeding grass to cows.
   (d) Bird flu is a virus that jumps from birds to people.
   (e) West Nile virus is a virus that comes from birds.

5. Which statement about toxins is *correct?*
   (a) Neurotoxins impair the nervous system.
   (b) Carcinogens cause birth defects.
   (c) Teratogens cause cancer.
   (d) Allergens mimic naturally occurring hormones.
   (e) Endocrine disruptors cause allergic reactions.

6. Which statement about dose-response studies is *not* true?
   (a) Dose-response studies test chemicals across a range of concentrations.
   (b) Dose-response studies only test for lethal effects.
   (c) Dose-response studies can last for days or months.
   (d) LD50 values are divided by 10 to determine safe concentrations for wildlife.
   (e) LD50 values are divided by 1,000 to determine safe concentrations for humans.

7. Which statement about retrospective and prospective toxicity studies is *correct?*
   (a) Retrospective studies are not conducted on humans.
   (b) Prospective studies are only conducted on wildlife.
   (c) Retrospective studies monitor health effects from future chemical exposures.
   (d) Prospective studies monitor health effects from future chemical exposures.
   (e) Prospective studies monitor health effects from past chemical exposures.

8. The concentration of chemical exposure does not depend on
   (a) the persistence of the chemical.
   (b) the solubility of the chemical.
   (c) the ability of the chemical to bioaccumulate.
   (d) the ability of the chemical to biomagnify.
   (e) the LD50 value of the chemical.

9. Which statement is *not* correct?
   (a) Risk assessment quantifies the potential harm that a chemical poses.
   (b) Risk assessment does not include social, political, and economic considerations.
   (c) Risk acceptance determines the amount of permissible risk.
   (d) Risk management includes social, political, and economic considerations.
   (e) Risk management does not consider the potential harm that a chemical poses.

10. What is *not* true about the two philosophies of regulating chemicals?
    (a) The innocent-until-proven-guilty principle assumes chemicals are safe unless harm can be demonstrated.
    (b) The precautionary principle is used in the United States.
    (c) The precautionary principle assumes chemicals are harmful unless safety can be demonstrated.
    (d) The innocent-until-proven-guilty principle allows rapid approval of chemicals by regulatory agencies but increases the risk of harmful chemicals being approved.
    (e) The precautionary principle can cause delays in the use of beneficial chemicals but reduces the risk of harmful chemicals being approved.

## APPLY THE CONCEPTS

1. You are an employee of the Environmental Protection Agency. You are given the task of conducting risk management for the spraying of insecticides to kill the mosquitoes that carry West Nile virus.
   (a) How might you determine the proper concentration needed to kill mosquitoes?
   (b) How might you determine whether the concentration used to kill mosquitoes might also kill other species of insects?
   (c) If you knew the insecticide's LD50 value for humans, what concentration would be the safe upper limit for humans?
   (d) Given the information you have accumulated as part of your risk assessment, describe the factors that might be important in the risk management of spraying insecticides to kill the mosquitoes that carry West Nile virus.

2. Given the differences in health risks that exist between low- and high-income countries, consider the following issues.
   (a) What strategies might you use to reduce the health risks of low-income countries?
   (b) What strategies might you use to reduce the health risks of high-income countries?
   (c) Suppose a low-income country discovers oil and is projected to become a high-income country within a decade. What changes in the country's healthcare system might you suggest?

## MEASURE YOUR IMPACT

**How Does Risk Affect Your Life Expectancy?** An interesting way of examining risky behaviors is to determine how different behaviors reduce your life expectancy. Using U.S. government statistics, we know that the life expectancy for men is 75.6 years and the life expectancy for women is 80.8 years.

(a) If you choose to smoke, the loss of life expectancy will be 6.6 years for the average man and 3.9 years for the average woman. What is the life expectancy for men and women who smoke?
(b) Alcoholism leads to a 12-year decline in life expectancy in both sexes. What would your life expectancy be if you were an alcoholic man who also smoked?

(c) Being overweight causes a loss of 36 days of life expectancy for every pound that you are overweight. If you become 20 pounds overweight, by how many years will your life expectancy be reduced?
(d) Based on the above numbers, what is the life expectancy of an alcoholic woman who smokes and is 20 pounds overweight?

How can you reduce risks associated with Sun exposure when more UV radiation from the Sun is reaching the surface of Earth today than 40 years ago? This Engage Your Environment exercise examines how you can reduce health risks associated with exposure to UV radiation. *See Engage Your Environment Activity 17 at the end of this book.*

# science applied

## Is Recycling Always Good for the Environment?

One of the three ways to reduce solid waste is to recycle. As we discussed in Chapter 16, when we recycle items such as paper, plastic, bottles, and cans, less material ends up in landfills and fewer natural resources need to be extracted to produce these items in the future. At first glance, recycling appears to make a lot of sense both economically and environmentally. Indeed, many state and local governments have encouraged or required recycling programs and the public generally associates recycling as being good for the environment (FIGURE SA7.1). But what do the data

tell us? When we decide to recycle, what are the measurable benefits for the environment? How do these benefits compare to benefits from other decisions we make, such as the type of car we drive? The answers to these questions may surprise you.

### How do we begin to assess the benefits of recycling?

To determine the overall effect of recycling any type of waste, we need to consider the full range of costs and benefits of recycling and then compare these to the costs

**FIGURE SA7.1** **Recycling.** There is increasing interest in recycling many materials. From a perspective of energy savings, some items are more important to recycle than others.

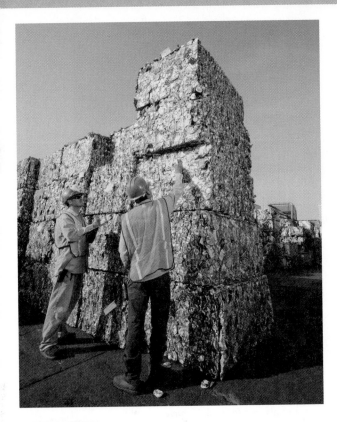

**Recycling aluminum cans.** Converting old aluminum cans into new aluminum cans requires only 6 percent of the energy used to convert aluminum ore from a mine into new aluminum cans.

and benefits of manufacturing the same item from raw materials. For example, to assess the benefits of recycling paper, we need to compare the cost of recycling old paper into new paper products versus the cost of manufacturing new paper products from trees.

As we saw in Chapter 16, the best way to answer these questions is to complete a life-cycle analysis. Let's look at two examples: aluminum cans and plastic containers. To compare the environmental and economic costs of transporting and manufacturing these items from recycled materials versus raw materials, we begin at the manufacturing facility.

The recycling of aluminum, primarily from aluminum cans, is widespread in the United States. According to the Aluminum Association, 51 billion cans are recycled in the United States each year, representing more than 50 percent of all aluminum cans that are manufactured (FIGURE SA7.2). To manufacture aluminum cans from raw materials, aluminum ore or bauxite must be mined and processed into pure aluminum. Not only does mining have environmental impacts as discussed in Chapter 8, but this processing of aluminum from ore also takes a substantial amount of energy while manufacturing aluminum cans from recycled cans requires

only 6 percent of this energy. In short, making new cans from recycled cans saves a large amount of energy and therefore saves manufacturers a lot of money. When this energy comes from burning fossil fuels, it also means that manufacturing recycled cans reduces the amount of carbon dioxide and other pollutants in the atmosphere.

The recycling of plastic containers is also widely practiced. According to the American Chemistry Council and the Association of Postconsumer Plastic Recyclers, the recycling of plastic continues to grow each year with 1.1 billion kg (2.4 billion pounds) of bottles recycled annually, representing 27 percent of all plastic bottles that are manufactured. However, in contrast to aluminum cans, the cost of energy required to make new plastic bottles from raw material—in this case petroleum—is substantially less. As a result, manufacturing plastic bottles from recycled plastic bottles results in much smaller energy savings; it requires nearly 50 percent of the energy required in manufacturing plastic bottles from oil. This means that the economic and environmental benefits of using recycled plastic are much smaller than the economic benefits of using recycled aluminum.

## What other costs of recycling do we need to consider?

Regardless of the type of material that is being recycled, we have to remember that there are several additional costs of recycling beyond the cost of energy. To understand these costs, let's start at your house. If you rinse out your cans and bottles before recycling, energy is needed to get the water to your sink, particularly if you use hot water. If you use hot water to rinse out the peanut butter from a plastic peanut butter jar, for example, you are very likely using more energy to clean the jar than is saved when you recycle the jar.

After they are cleaned, the recycled materials must be transported to a central recycling facility. Depending on location, the homeowner must either set out recycled items on the curb for pickup by a collection truck (FIGURE SA7.3) or bring them to a central facility. Both scenarios require burning fossil fuels for transportation. Additional fossil fuels must be consumed to transport the recycled items from the collection facility to the manufacturing facility. Although transportation costs will vary among different towns and cities, in terms of energy consumed and pollutants produced, they reduce the benefits of recycling. However, we can easily compare the cost of transporting recycled materials to manufacturers against the cost of transporting raw materials from their source, such as an aluminum mine. In addition, when homeowners pay for transporting the items to the collection center through taxes or trash collection fees, they will avoid the costs of putting the waste in a landfill.

**FIGURE SA7.3** **Transportation costs.** Although early recycling programs had garbage trucks make separate trips to pick up trash and recycling materials, modern trucks have separate compartments that allow both garbage and recycled items to be picked up at the curb in a single trip. A single trip saves time and money, and reduces consumption of fossil fuels as well as the production of air pollutants.

## What other benefits of recycling do we need to consider?

The primary argument for recycling is that it reduces the need for raw materials and keeps solid waste out of landfills. During the 1990s, there was a growing concern that the United States was running out of landfill space and that recycling was critical to extending the life of existing landfills. While it is true that many landfills are nearing capacity, particularly in the northeastern United States, there is still a large amount of land throughout the country that could serve as landfill space if people in those areas agreed to the construction of new landfills.

Reducing the amount of solid waste going into landfills allows existing landfills to operate longer. This, in turn, reduces the costs of closing and monitoring existing landfills. It also reduces the costs of building more landfills in the future and the costs of trucking the waste to new landfills likely to be farther away. Increased trucking raises both the economic cost and the environmental impact.

## How do the benefits of recycling compare to other choices we could make to help our environment?

Public support of recycling has grown tremendously in recent years. But how does recycling compare to other actions that affect the environment? For example, based only on the energy used in the manufacturing processes, the energy saved by recycling aluminum cans is 0.15 percent of all energy used in the United States. The energy saved by recycling plastics is 0.008 percent of all energy used in the United States. How do these values compare to other choices we can make to save energy?

One easy comparison is the decision about the type of vehicle to purchase. According to the Bureau of Transportation Statistics, in 1989 cars comprised 69 percent of all passenger vehicles whereas light trucks and SUVs comprised 31 percent. By 2006, the fraction of light trucks and SUVs climbed to 42 percent. As we discussed in Chapter 12, such a shift in vehicle choice affects the use of fossil fuels; the average car built in 2009 achieved approximately 10.1 km per liter (24 miles per gallon) whereas the average light truck or SUV achieved 7.6 km per liter (18 mpg). A recent analysis of such data demonstrated that if consumers had not increased their purchases of light trucks and SUVs beyond the 1989 levels, the energy savings would be 0.6 percent of all energy used in the United States. In other words, deciding to replace your car with another car instead of switching to a light truck or SUV saves 4 times more energy than deciding to recycle aluminum cans and 75 times more energy than the decision to recycle plastic.

Of course, a person could decide to drive a more fuel-efficient car *and* to recycle plastic and aluminum. But when we consider how we can improve the environment it is often helpful to gather the scientific data to make objective comparisons rather than simply make decisions based on perceptions. In the case of recycling, the analysis of the data makes it clear that recycling certain materials will have much greater environmental and economic benefit than recycling other materials. This helps us understand why manufacturers might be much more inclined to promote the recycling of certain items such as aluminum cans. Identifying the full range of costs and benefits also helps us identify the complexity of the question. The energy costs of recycling, for example, are wide-ranging and include homeowner costs, transportation costs, manufacturing costs, and landfill costs. By identifying all of the costs and benefits, we can strive to design more efficient recycling programs. Finally, examining the costs and benefits of a particular effort to help our environment versus other efforts gives some perspective on which provides the greatest benefits. In many cases, of course, we do not have to choose between two actions. We can simultaneously recycle and drive a more fuel-efficient vehicle.

### References

Friedland, A., et al. 2003. Personal decisions and their impacts on energy use and the environment. *Environmental Science and Policy* 6: 175–179.

*Municipal Solid Waste Generation, Recycling, and Disposal in the United States: Facts and Figures for 2007.* www.epa.gov/epawaste/nonhaz/municipal/pubs/msw07-fs.pdf.

# Conservation of Biodiversity: Protection of Earth's Species and Ecosystems

## Modern Conservation Legacies

The biodiversity of the world is currently declining at such a rapid rate that many scientists have declared that we are in the midst of a sixth mass *extinction*. There are many causes of this decline, but all are related to human activities ranging from habitat destruction to overharvesting plant and animal populations. In response to this crisis, there is growing interest in conserving biodiversity by setting aside areas that are protected from many human activities.

Islands, 36 million ha (90 million acres) of these marine habitats were set aside as the Papahānaumokuākea Marine National Monument. This protected region is immense, covering an area about the size of California.

The marine ecosystem that surrounds the Hawaiian Islands contains a great deal of biodiversity—more than 7,000 marine species, approximately one-fourth of which are found nowhere else in the world. Unfortunately, in recent decades human activity has caused a decline in this diversity. The

> There is growing interest in conserving biodiversity by setting aside areas that are protected from many human activities.

The conservation of biodiversity has a long history. The United States, for example, has been protecting habitats as national parks, national monuments, national forests, and wilderness areas for more than a century. Yellowstone National Park was the first national park in the United States, designated in 1872 by President Ulysses Grant. During the presidency of Theodore Roosevelt (1901–1909), nearly 93 million ha (230 million acres) received federal protection. This included the creation of more than a hundred national forests, although much of this land was set aside to ensure a future supply of trees for lumber and therefore lacked complete protection. Efforts to protect critical wildlife habitats continue today. In 2009, the Obama administration set aside 58 million ha (more than 200,000 square miles) of Alaska coastline and waters as critical habitat for polar bears. This does not prevent activities such as gas and oil drilling, but it does mean that potential impacts on polar bears must now be considered when such activities are proposed in this area. From 2006 to 2009, President George W. Bush designated a total of 95 million ha (215 million acres) of marine habitats as protected around the northwestern Hawaiian Islands and other U.S. Pacific islands. In the northwestern Hawaiian

anthropogenic causes of declining diversity are wide ranging. Although Hawaii has only 1.3 million residents, 7 million tourists visit each year. Both individuals and commercial operations have exploited marine life, including coral and fish. In addition to this exploitation, there are thousands of kilograms of old fishing equipment lying at the bottom of the ocean that sometimes wash up on shore, entangling wildlife in old fishing lines. Invasive species of algae also dominate some areas. ▶

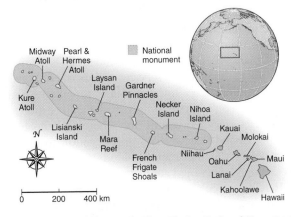

Map of the Papahānaumokuākea Marine National Monument.

◀ The Papahānaumokuākea Marine National Monument, designated in 2006, surrounds the northwestern Hawaiian Islands and protects more than 7,000 species of marine organisms, including these Hawaiian squirrel fish (*Sargocentron xantherythrum*).

The Papahānaumokuākea monument presents an opportunity for improving the Hawaiian marine environment. As a national monument, the area is protected from fishing, harvesting of coral, and the extraction of fossil fuels. Large amounts of solid waste debris are to be removed from the shorelines and coral reefs, and efforts are also under way to clean out much of the invasive algae. It is expected that the biodiversity of the area will quickly respond to these efforts. As the populations of organisms increase in the protected areas, individuals will disperse and add to the populations in the larger surrounding area. In this way, the protected area can serve as a constant supply of individuals to help neighboring areas maintain their diversity of species.

In the United States and the rest of the world, conserving the biodiversity of marine areas through the creation of marine reserves is a relatively new activity for governments, but the idea is gaining ground. In the Galápagos Islands, where, as we discussed in Chapter 5, different species of finch inspired Charles Darwin, the nation of Ecuador recently designated a marine reserve that extends 64 km (40 miles) into the ocean from the islands and allows only limited fishing. Marine reserves have also been designated by Russia, the United Kingdom, Australia, Canada, and Belize. As more countries develop marine reserves, we have to make sure these areas are large enough to ensure long-term protection of local species and consider how each new reserve is positioned relative to other reserves so that individuals may be able to move among them. Furthermore, countries must decide what human activities will be allowed in each reserve, perhaps protecting a core area and allowing tourism, fishing, or extraction of fossil fuels to occur in more distant areas. These are exciting times that demonstrate that there is a great potential for conserving biodiversity in the twenty-first century. ■

*Sources:* Bush creates world's biggest ocean preserve, MSNBC, June 16, 2006, http://www.msnbc.msn.com/id/13300363/; P. Thomas, President Bush to add marine reserves; not all are applauding, *Los Angeles Times,* January 6, 2009, http://latimesblogs.latimes.com/outposts/2009/01/news-flash-pres.html.

## UNDERSTAND THE KEY IDEAS

The conservation of biodiversity is a continuous challenge in a world that is increasingly affected by human population growth and development.

After reading this chapter you should be able to

- understand how genetic diversity, species diversity, and ecosystem function are changing over time.
- identify the causes of declining biodiversity.
- describe the single-species approach to conserving biodiversity including the major laws that protect species.
- explain the ecosystem approach to conserving biodiversity and how size, shape, and connectedness affect the number of species that will be protected.

# We are in the midst of a sixth mass extinction

As we have seen, protecting the biodiversity of individuals, species, and ecosystems is important because the biodiversity of the world provides a number of instrumental and intrinsic values to humans. Instrumental values include provisions such as food, medicine, and building materials; regulating services such as the ability of plants to remove human-added $CO_2$ from the atmosphere; and support services such as the pollination of agricultural crops. Intrinsic values provide no direct benefit to people but are simply the belief that individuals, species, and ecosystems are inherently valuable in themselves and that we have an obligation to preserve them. Given the importance of biodiversity, scientists are increasingly concerned about its rapid decline.

In Chapter 5, we noted that the world has experienced five major extinctions during the past 500 million years and that we are currently experiencing a sixth major extinction event. In scientific terms, an **extinction** occurs when the last member of a species dies. Scientists estimate that the world is currently experiencing approximately 50,000 species extinctions per year, amounting to 0.5 percent of the world's species each year. One unique aspect of this sixth mass extinction is that it is happening over a relatively short period of time and is the first to occur since humans have been present on Earth. Indeed, the rate of decline has been 100 to 1,000 times faster during the past 50 years than at any other time in human history and rivals the rates observed during the mass extinction event that eliminated the dinosaurs 65 million years ago. Another unique feature is that the current mass extinction has a human cause. In this section, we will examine the patterns of declining genetic diversity, species diversity, and ecosystem function.

## Global Declines in the Genetic Diversity of Wild Organisms

At the lowest level of complexity, environmental scientists are concerned about conserving genetic diversity. One reason for this is that populations with low genetic diversity are not well suited to surviving environmental change. In addition, populations with low genetic diversity are prone to *inbreeding depression*. **Inbreeding**

**depression** occurs when individuals with similar genotypes—typically relatives—breed with each other and produce offspring that have an impaired ability to survive and reproduce. This impaired ability occurs when each parent carries one copy of a harmful mutation in his or her genome. When the parents breed, some of their offspring receive two copies of the harmful mutation and, as a result, have poor chances of survival and later reproduction. High genetic diversity ensures that a wider range of genotypes are present, which reduces the probability that an offspring will receive a harmful mutation from both parents. In addition, high genetic diversity improves the probability of surviving future change in the environment. This occurs because high genetic diversity produces a wide range of phenotypes that survive and reproduce under different environmental conditions.

Some declines in genetic diversity have natural causes. Cheetahs (*Acinonyx jubatus*), for example, possess very low genetic diversity. Researchers have determined that this condition is the result of a population bottleneck that occurred approximately 10,000 years ago (see Chapter 6). Other declines in genetic diversity have human causes. For example, the panther, also known as a mountain lion or cougar, once ranged over much of North America. Because this predator could be a threat to humans and domestic livestock, efforts to eradicate it became intense as human populations grew. At the same time, much of the panther's habitat disappeared as humans settled the land and increased agricultural activities. One subspecies, known as the Florida panther, once roamed throughout the southeastern United States (**FIGURE 18.1**). Because of hunting and

habitat destruction, the population of the Florida panther shrank to only a small group in south Florida. This led to inbreeding since this group was too far away from other panther populations, such as a group in Texas, with which it could breed. This inbreeding caused a number of harmful effects, including heart defects and a high proportion of morphologically abnormal sperm. When this lack of genetic diversity caused the Florida panther population to decline even further—down to a total of 20 to 30 animals in 1995—scientists released 8 panthers from Texas into the remaining Florida habitat. This effort to add genetic diversity to the Florida subspecies and reduce the problems associated with inbreeding was a success. Today, the Florida panther population is estimated at 80 to 100 individuals.

## Global Declines in the Genetic Diversity of Crops and Livestock

Although declining genetic variation of plants and animals in the wild is of great concern to scientists, there are also major concerns about declining genetic variation in the species of crops and livestock on which humans depend. The United Nations notes that the majority of livestock species comes from seven species of mammals (donkeys, buffalo, cattle, goats, horses, pigs, and sheep) and four species of birds (chickens, ducks, geese, and turkeys). In different parts of the world, these species have been bred by humans for a variety of characteristics including adaptations to survive local climates. This wide variety of characteristics, which is underlain by a great deal of genetic variation, could be used for adapting to changing environmental conditions in the future or resisting new diseases. Unfortunately, livestock producers have concentrated their efforts on the breeds that are most productive and much of this genetic variation is being lost. In Europe, for example, half of the breeds of livestock that existed in 1900 are now extinct. Of those that remain, 43 percent are currently **endangered,** meaning they are at serious risk of extinction. Of the 200 breeds of domesticated animals that have been evaluated in North America, 80 percent of these breeds are either declining or are already facing extinction (**FIGURE 18.2**).

A similar story exists for crop plants. A century ago, most of the crops that humans consumed were composed of hundreds or thousands of unique genetic varieties. Each variety grew well under specific environmental conditions and was usually resistant to local pests. In addition, each variety often had its own unique flavor. As we saw in Chapter 11, the green revolution in agriculture focused on techniques that increased productivity. Farmers planted fewer varieties, concentrating on those with higher yields. Fertilizers and irrigation helped humans control many of the abiotic conditions, allowing fewer but higher yielding varieties to be grown across large regions of the world. For example, at the turn of the twentieth century, farmers

**FIGURE 18.1** **Decline in genetic diversity.** The Florida panther was reduced to such a small population that it suffered severe effects of inbreeding. In recent years the introduction of new genoytpes from a Texas population has allowed the Florida panther to rebound.

Holstein

Boran

Texas Longhorn

Ankole

Highland

**FIGURE 18.2** **The genetic diversity of livestock.** Over thousands of years, humans have selected for numerous breeds of domesticated animals to thrive in local climatic conditions and to resist diseases common in their local environments. Modern breeding, which focuses on productivity, has caused the decline or extinction of many of these animal breeds.

grew approximately 8,000 varieties of apples. Today, that number has been reduced to about 100, and considerably fewer are available in your local grocery store.

Planting only a few varieties leaves us open to crop loss if the abiotic or biotic environment changes. For example, in the 1970s, a fungus spread through cornfields of the southern United States and killed half the crop. Although the fungus was uncommon, the high-yielding variety of corn that most farmers planted turned out to be susceptible to it. Following this crisis, scientists modified this high-yielding corn by adding a gene from a variety that is resistant to the fungus. Had the resistant variety not been preserved, this gene would not have been available.

The nations of the world have recognized the problem of declining seed diversity and have responded by storing seed varieties in specially designed warehouses to preserve genetic diversity. In fact, there are currently more than 1,400 such storage facilities around the world. However, many of these facilities are at risk from war and natural disasters. In the past decade, nations and philanthropists have funded an international storage facility known as the Svalbard Global Seed Vault (**FIGURE 18.3**). This facility was built into the side of a frozen mountain in the Arctic region of northern Norway. It was designed to resist a wide range of possible calamities, including natural disasters and global warming. Should the environment change in future years, either in terms of abiotic conditions or emergent diseases, the seed bank will be available to help scientists

address the challenge. The Svalbard facility opened in 2008 with a capacity of 14.5 million seed varieties. As of 2010, more than 430,000 seed varieties had already been sent to Svalbard for long-term storage.

## Global Declines in Species Diversity

To understand the current loss of biodiversity, we will begin by looking at how particular groups of species are declining. When considering the status of a species, we use one of five categories defined by the International Union for Conservation of Nature (IUCN). *Data-deficient* species have no reliable data to assess their status; they may be increasing, decreasing, or stable. Species for which we have reliable data are placed in one of four categories. *Extinct* species are those that were known to exist as recently as the year 1500 but no longer exist today. *Threatened* species have a high risk of extinction in the future and *near-threatened* species are very likely to become threatened in the future. *Least concern* species are widespread and abundant. These categories provide a mechanism for comparing the status of different groups of species.

Evaluating the status of different plant and animal groups presents several challenges. Many species fall under the category of data-deficient. At the same time, we are still discovering many new species, particularly in remote areas of the world. Since the number of species known to science constantly increases, it is not possible to evaluate every species and our estimates of what

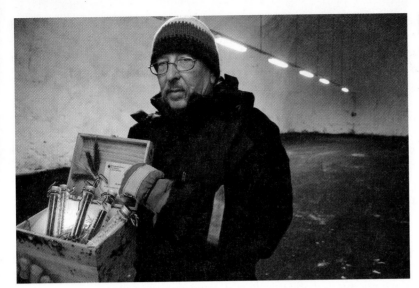

**FIGURE 18.3** **A global seed bank.**
The Svalbard Global Seed Vault in northern Norway is an international storage area for many varieties of crop seeds from throughout the world.

fraction of species are declining will constantly change. Finally, the work is expensive. Making an assessment for even one group of species, such as birds or mammals, requires thousands of scientists and millions of dollars.

Of the estimated 10 million species that currently live on Earth, ranging from bacteria to whales, only about 50,000 have been assessed to determine whether their populations are increasing, stable, or declining. Across all groups of organisms that have been assessed, nearly one-third are threatened with extinction. Given that some of the best data are for birds, mammals, and amphibians, we will examine these groups in more detail. FIGURE 18.4 shows the most recent data for those species that are not yet extinct.

Since the year 1500, nearly 10,000 bird species have existed and 133 have become extinct. Today, 21 percent are threatened or near-threatened. Among the 800 species of birds living in the United States, nearly one-third of them are experiencing declining populations. These include 40 percent of bird species that live in grasslands and 30 percent of bird species that live in arid regions.

Multiple threats, including reduced habitat and rising sea levels, have caused a growing concern for all species of birds that live on coastlines or on islands.

A similar pattern exists for mammals. Of the nearly 5,500 species of mammals known to have existed after 1500, 79 are extinct. Among the approximately 4,600 species for which there are reliable data, 25 percent are threatened and 32 percent are either threatened or near-threatened. This means that more than 1,400 species of mammals may be at risk of extinction.

Amphibians are experiencing the greatest global declines. Of the more than 6,200 species of amphibians, 39 species are extinct. However, a recent assessment of amphibian populations suggests that the number of extinctions may accelerate soon in the coming decades. Among the approximately 4,000 species for which reliable data exist, 49 percent are either threatened or near-threatened. This means that nearly 2,000 species of amphibians are declining around the world.

Many other groups of organisms are also experiencing large declines, but complete assessments have not

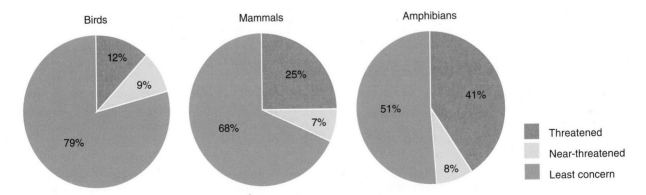

**FIGURE 18.4** **The decline of birds, mammals, and amphibians.** Based on those species for which scientists have reliable data, 21 percent of birds, 32 percent of mammals, and 49 percent of amphibians are currently classified as threatened or near-threatened with extinction. [After International Union for Conservation of Nature, 2009.]

yet been conducted because of the time and money required for each assessment. However, from the sample of species that have been assessed in each group, we see an emerging picture that is far from positive. For example, of the species that have been assessed so far, approximately one-third of all reptiles, fish, and invertebrates are threatened with extinction. Similarly, one-fourth of plant species are threatened. These results suggest that when the assessments are complete, the news will most likely not be good.

## Global Declines in Ecosystem Function

As we discussed in Chapter 3, ecosystems provide valuable services to humans, including both instrumental values and intrinsic values. These services include items that humans use directly such as lumber, food crops, and prescription drugs. They also include support systems such as pollinating many crops and filtering drinking water to reduce contaminants and pathogens. These services are critically important to humans.

Because species help determine the services that ecosystems can provide, we would expect declines in species diversity to be associated with declines in ecosystem function. In the Millennium Ecosystem Assessment conducted in 2006, scientists from around the world examined the current state of various ecosystem functions, including food, clean water, pollination, water purification, and nutrient cycling. Of 24 different ecosystem functions, 15 were found to be in decline. If we want to improve ecosystem functions, we need to improve the fate of the species and ecosystems that provide these services.

> **GAUGE YOUR PROGRESS**
>
> ✓ Why should we be concerned about inbreeding?
>
> ✓ What are the reasons for the declining genetic diversity of domesticated plants and animals?
>
> ✓ What are some of the challenges associated with understanding which species are threatened with extinction?

# Declining biodiversity has many causes

Declines in biodiversity are happening around the globe. In Science Applied "How Should We Prioritize the Protection of Biodiversity?" on page 144 we discussed biodiversity hotspots, threatened areas that are rich in biodiversity. The need to protect biodiversity, however, exists far beyond these hotspots because the threats to biodiversity exist throughout the world. In

Chapter 5, we discussed the basics of population and community ecology. In this chapter we build on these basics to understand how habitat loss, intrusion of *alien species,* overharvesting, pollution, and climate change all affect our efforts to conserve biodiversity.

## Habitat Loss

For most species, the greatest cause of decline and extinction is habitat loss. In modern times, the primary cause of habitat loss is human development that removes natural habitats and replaces them with homes, industries, agricultural fields, shopping malls, and roads. Many species can only thrive in a particular habitat within a narrow range of abiotic and biotic conditions. Species requiring such specialized habitats are particularly prone to population declines, especially when their favored habitat is limited, restricting their distribution to a limited geographic area suitable only for a small population.

The alteration of a habitat, such as the removal of trees or the damming of streams that give the habitat its distinctive characteristics, has an effect on the organisms that live in that habitat. For example, for thousands of years the northern spotted owl (*Strix occidentalis caurina*) lived in old-growth forests—those dominated by trees that are hundreds of years old—in the Pacific northwestern region of the United States and British Columbia (see Figure 10.1 on page 263). This habitat provided the ideal sites for nesting, roosting, and catching small mammals to eat. The removal of the old trees, for lumber and housing developments, has transformed much of the former old-growth forest into a different habitat (**FIGURE 18.5**). This habitat alteration reduces the number of northern spotted owls because they have fewer trees in which to nest and less forest in which to find food.

The map in **FIGURE 18.6** shows the changing face of forest habitats over the past few decades. As we saw in Chapter 6, much of the forest in the United States was logged during the 1700s and 1800s for lumber and was cleared for agriculture. In recent decades, forested land has been increasing, although the new forests have often been planted by humans and have a lower diversity of species than the original forest. At the same time, developing countries in South America, sub-Saharan Africa, and Southeast Asia are clearing their forests much as the United States and Europe did in years past. As a result, large declines in forest cover are occurring in developing countries that were once forested. It is currently unclear whether these countries will follow the pattern of Europe and North America and allow their forested areas to increase.

Although deforestation receives a lot of attention, many other habitats are also being lost. According to the Millennium Ecosystem Assessment, approximately 70 percent of the woodland/shrubland ecosystem that borders the Mediterranean Sea has been lost. Similarly, across the globe we have lost nearly 50 percent

(a)

(b)

**FIGURE 18.5** | **Habitat loss.** Habitat loss caused by humans is the largest threat to biodiversity. (a) Old-growth forests, such as this one in the Mount Rainier National Park in Washington State, serve as habitat for spotted owls. (b) After clear-cutting an old-growth forest for timber, such as this site in the Olympic National Forest in Washington State, the forest provides a very different habitat. It may take hundreds of years before the habitat again becomes suitable for the spotted owl.

of grassland habitats and 30 percent of desert habitats. Wetlands exhibit a mixed picture: although the amount of wetland habitat is less than half of what existed in the United States during the 1600s, from 1998 to 2004 the amount of freshwater wetland habitat increased. This overall growth occurred due to large increases in the Great Lakes region despite a decline in coastal wetlands in the eastern United States and the Gulf of Mexico because of growing human populations that require more roads, homes, and businesses.

In marine systems, there has been a sharp decline in the amount of living coral in the Caribbean Sea, as

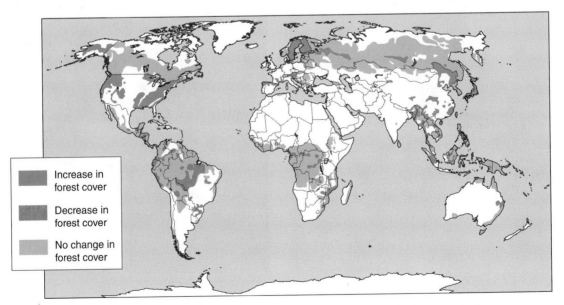

Increase in forest cover

Decrease in forest cover

No change in forest cover

**FIGURE 18.6** | **Changing forests.** Some regions of the world experienced large declines in the amount of forested land from 1980 to 2000 while other regions have shown little change or have seen increases in forest cover. [After *Global Biodiversity Outlook 2,* Convention on Biological Diversity, 2006.]

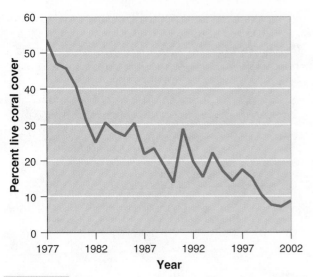

**FIGURE 18.7** **Changing coral reefs.** The percentage of coral that remains alive in coral reefs has declined sharply in the Caribbean since 1977. [After *Global Biodiversity Outlook 2,* Convention on Biological Diversity, 2006.]

shown in **FIGURE 18.7**, from a high of 50 percent live coral in the 1970s to a mere 10 percent by 2002. Living coral provide habitat for thousands of other species, which makes them particularly vital to the persistence of marine habitats. The decline in coral is the result of human impacts including the warming of oceans (associated with global warming), increased pollution,

and the removal of coral by collectors. This loss of coral habitat is occurring at a rapid rate throughout the world.

A species may decline in abundance or become extinct even without complete habitat destruction. A reduction in the size of critical habitat also can lead to extinctions through a variety of processes. As we saw in the case of the Florida panther, a smaller habitat supports a smaller population, reducing genetic diversity. Less habitat also reduces the variety of physical and climatic options available to individuals during periods of extreme conditions. The presence of cooler, high-altitude areas in a habitat, for example, allows animals a place to move during periods of hot weather. Also, loss of habitat can restrict the movement of migratory or highly mobile species. While many species can thrive in small habitats, other species, such as mountain lions, wolves, and tigers, require large tracts of relatively unin-habited, undisturbed land.

Smaller habitats can also cause increased interactions with other harmful species. For example, many song-birds in North America live in forests. When these birds make their nests near the edge of the forests—where the forest meets a field—they often have to contend with the brown-headed cowbird (*Molothrus ater*). The cowbird is a nest parasite—it does not build its own nest, but lays its eggs in the nests of several other species of birds (**FIGURE 18.8**). In this way the cowbird tricks forest birds into raising its offspring, which takes food away from the other birds' own offspring. In some cases,

(a)

(b)

**FIGURE 18.8** **The brown-headed cowbird.** (a) Increased fragmentation of forests has caused forest songbirds to come into increasing contact with the brown-headed cowbird. (b) The cowbird does not make its own nest. Instead, it lays its brown, spotted eggs in the nests of other species, such as this nest containing four blue eggs of the chipping sparrow (*Spizella passerina*).

the host bird will simply abandon the nest. As forests are broken up into smaller fragments, the proportion of forest near the edge increases and, therefore, the number of bird nests that are susceptible to brown-headed cowbirds increases. Over time, increased fragmentation has allowed brown-headed cowbirds to cause declines in many species of North American songbirds.

## Alien Species

**Native species** are species that live in their historical range, typically where they have lived for thousands or millions of years. In contrast, **alien species,** also known as **exotic species,** are species that live outside their historical range. For example, honeybees (*Apis mellifera*) were introduced to North America in the 1600s to provide a source of honey for European colonists. Red foxes (*Vulpes vulpes*), now abundant in Australia, were introduced there in the 1800s for the purpose of fox hunts, which were popular in Europe at the time.

During the past several centuries, humans have frequently moved animals, plants, and pathogens around the world. Some species are also moved accidentally, such as rats stowing away in shipping containers and often ending up far from their original port, sometimes on oceanic islands. Since many oceanic islands have never had rats or other ground predators, there had never been any natural selection against nesting on the ground. As a result, numerous island bird species evolved to nest on the ground. When the rats arrived, they found the eggs and hatchlings from ground nests an easy source of food, resulting in a high rate of extinction in ground-nesting birds in places such as Hawaii. Similar accidental movements have occurred for many pathogens, including fungi that have killed nearly all American elm (*Ulmus americana*) and American chestnut (*Castanea dentata*) trees in eastern North America and a protist that has caused avian malaria and driven many species of Hawaiian birds to extinction. Other species are moved intentionally, such as exotic plants that are sold in greenhouses for houseplants and outdoor landscape plants, or animals that are sold as pets or to game ranches that raise exotic species of large mammals for hunting.

Not all alien species are a threat to biodiversity. In many cases, the alien species live in their new surroundings and have no negative effect on the native species. In other cases, the alien species rapidly increase in population size and cause harmful effects on native species. When alien species spread rapidly across large areas, we call them **invasive species.** Rapid spread of invasive species is possible because invasive species, which have natural enemies in their native regions that control their population, often have no natural enemies in the regions where they are introduced. Two of the best known examples of invasive alien species in North America are the kudzu vine (*Pueraria lobata*) and the zebra mussel (*Dreissena polymorpha*).

The kudzu vine is native to Japan and southeast China but was introduced to the United States in 1876. Throughout the early 1900s, farmers in the southeastern states were encouraged to plant kudzu to help reduce erosion in their fields. By the 1950s, it became apparent that the southeastern climate was ideal for kudzu, with growth rates of the vine approaching 0.3 m (1 foot) *per day*. Because herbivores in the region do not eat kudzu, the species has no enemies and can spread rapidly. The vine grows up over most wildflowers and trees and shades them from the sunlight, causing the plants to die. Indeed, the vine grows over just about anything that does not move (**FIGURE 18.9a**). Kudzu currently covers approximately 2.5 million ha (7 million acres) in the United States.

The zebra mussel is native to the Black Sea and the Caspian Sea in Asia. Over the years, large cargo ships that traveled in these seas unloaded their cargo in Asian ports and then pumped seawater into the holding tanks to ensure that the ship sat low enough in the water to remain stable. This water that is pumped into the ship is called *ballast water*. When the ships arrived in the St. Lawrence River and the Great Lakes, they loaded on new cargo and no longer needed the weight of the ballast water, so they pumped the ballast water out of the ship into local waters. One consequence of transporting ballast water from Asia to North America is that many aquatic species from Asia, including zebra mussels, have been introduced into the aquatic ecosystems of North America. Because the St. Lawrence River and the Great Lakes provided an ideal ecosystem for the zebra mussel, and because a single zebra mussel can produce up to 30,000 eggs, the mussel spread rapidly through the Great Lakes ecosystem. On the positive side, because the mussels feed by filtering the water, they remove large amounts of algae and some contaminants, which, to some degree, counteracts the cultural eutrophication that has occurred in the Great Lakes ecosystem. On the negative side, the zebra mussels physically crowd out many native mussel species and the zebra mussels can consume so much algae that they negatively affect native species that also need to consume the algae. Moreover, the invasive mussels can achieve such high densities that they can clog intake pipes and impede the flow of water on which industries and communities rely (**FIGURE 18.9b**).

A new threat to the Great Lakes is the silver carp (*Hypophthalmichthys molitrix*), a fish that is native to Asia but has been transported around the world to consume excess algae that accumulates in aquaculture operations and the holding ponds of sewage treatment plants. After being brought to the United States, some of the fish escaped and rapidly spread through many of the major river systems, including the Mississippi River. Over the years, the carp population has expanded northward, and in 2010 it was rapidly approaching a canal where the Mississippi River connects to Lake Michigan. There

(a)

(b)

(c)

**FIGURE 18.9** **Invasive alien species.** (a) The fast-growing kudzu vine is native to Asia but was introduced to the United States to control erosion. It has since spread rapidly, growing over the top of nearly anything that does not move. (b) The zebra mussel was accidentally introduced to the Great Lakes and has covered all hard surfaces, including the water intake and outlet pipes of industries that use the water. (c) The silver carp has been introduced into the Mississippi River and is quickly heading toward an invasion of the Great Lakes.

are two major concerns about this invading fish. First, scientists worry that it will outcompete native species of fish that also consume algae. Second, the silver carp has an unusual behavior; it jumps out of the water when startled by passing boats (**FIGURE 18.9c**). Given that the carp can grow to 18 kg (40 pounds) and jump up to 3 m (10 feet) into the air, this poses a major safety issue to boaters.

Around the world, invasive alien species pose a serious threat to biodiversity by acting as predators, pathogens, or superior competitors to native species. Some of the most complete data exist in the Nordic countries of Iceland, Sweden, Finland, Norway, and Denmark. As **FIGURE 18.10** shows, during the past one hundred years, these countries have experienced a steady increase of nearly 1,700 alien species in terrestrial, freshwater, and marine ecosystems combined. A number of efforts are currently being used to reduce the introduction of invasive species, including the inspection of goods coming into a country and the prohibition of wooden packing crates made of untreated wood that could contain insect pests.

## Overharvesting

Hunting, fishing, and other forms of harvesting are the most direct human influences on wild populations of plants and animals. Most populations can be harvested to some degree, but a species is overharvested when individuals are removed at a rate faster than the population can replace them. In the extreme, overharvesting of a species can cause extinction. In the seventeenth century, for example, ships sailing from Europe stopped for food and water at Mauritius, an uninhabited island in the Indian Ocean. On Mauritius, the sailors would hunt the dodo (*Raphus cucullatus*), a large flightless bird that had no innate fear of humans because it had never seen humans during its evolutionary history (**FIGURE 18.11**). The dodo, unable to protect itself from human hunters and the rats (introduced by humans) that consumed dodo eggs and hatchlings, became extinct in just 80 years. This same scenario appears to have taken place with many other large animal species as well. These animals include the giant ground sloths, mammoths, American camels of North and South America, and the 3.7 m (12 feet) tall moa birds of New Zealand. Each species became extinct soon after humans arrived. This timing suggests that the animals' demise may have been due to overharvesting.

Overharvesting has also occurred in the more recent past. In the 1800s and early 1900s, for example, market hunters slaughtered wild animals to sell their parts on such a scale that many species declined dramatically,

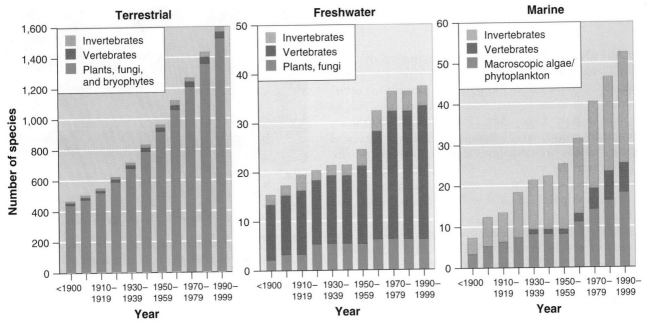

**FIGURE 18.10** **Exotic species.** Over the decades, there has been a steady increase in the number of alien species. This example shows the number of alien species recorded in three environments from the Nordic countries of Iceland, Sweden, Finland, Norway, and Denmark. [After *Global Biodiversity Outlook 2*, Convention on Biological Diversity, 2006.]

including the American bison (*Bison bison*). Bison were once abundant on the western plains, with estimates ranging from 60 to 75 million individuals. By the late 1800s fewer than 1,000 were left. Following enactment of legal protections, the bison population today has increased to more than 500,000, including both wild bison and bison raised commercially for meat.

**FIGURE 18.11** **Overharvesting.** The dodo was a large flightless bird that served as an easy source of meat for sailors and settlers on the island of Mauritius. Because it evolved on an island with no large predators or humans, the dodo had no instinct to fear humans.

Not all species harvested by market hunters fared as well as the American bison. The passenger pigeon (*Ectopistes migratorius*) was once one of the most abundant species of birds in North America. Population estimates range from 3 to 5 million birds in the nineteenth century. In fact, during annual migrations, people observed continuous flocks of pigeons flying overhead for 3 days straight in densities that blocked out most of the sun. Breeding flocks could cover 40,000 ha (100,000 acres) with 100 nests built into each tree in these areas. With such high densities, market hunters could shoot or net the birds in very large numbers and fill train cars with pigeons to be sold in eastern cities. This overharvesting, combined with the effects of forest clearing for agriculture, caused the passenger pigeon to decline quickly. The last passenger pigeon died in 1914 at the Cincinnati Zoo.

During the past century, regulations have been passed to prevent the overharvesting of plants and animals. In the United States, for example, state and federal regulations restrict hunting and fishing of game animals to particular times of the year and limit the number of animals that can be harvested. Similar agreements have been reached among countries through international treaties. In general, these regulations have proven very successful in preventing species declines from overharvesting. In some regions of the world, however, harvest regulations are not enforced and illegal poaching, especially of large, rare animals that include tigers, rhinoceroses, and apes, continues to threaten species with extinction. Harvesting of rare plants, birds, and coral reef dwellers for private collections has also jeopardized these species.

**PLANT AND ANIMAL TRADE** For some species, the legal and illegal trade in plants and animals represents a serious threat to their ability to persist in nature. One of the earliest laws in the United States to control the trade of wildlife was the **Lacey Act.** First passed in 1900, the act originally prohibited the transport of illegally harvested game animals, primarily birds and mammals, across state lines. Over the years, a number of amendments have been added so that the Lacey Act today forbids the interstate shipping of all illegally harvested plants and animals.

At the international level, the United Nations **Convention on International Trade in Endangered Species of Wild Fauna and Flora,** also known as **CITES,** was developed in 1973 to control the international trade of threatened plants and animals. Today, CITES is an international agreement among 175 countries of the world. The IUCN maintains a list of threatened species known as the **Red List.** Each member country assigns a specific agency to monitor and regulate the import and export of animals on the list. For example, in the United States, oversight is conducted by the U.S. Fish and Wildlife Service.

Despite such international agreements, much illegal plant and animal trade still occurs throughout the world. In 2008, a report by the Congressional Research Service estimated that illegal trade in wildlife was worth $5 billion to $20 billion annually. In some cases, animals are sold for fur or for body parts that are thought to have medicinal value. In other cases, rare animals are in demand as pets. For example, in 2001 a population of the Philippine forest turtle (*Siebenrockiella leytensis*), once thought to be extinct, was discovered on a single island in the Philippines (**FIGURE 18.12**). This animal, one of the most endangered species in the world, cannot be traded legally, but demand for it as a pet has caused it to be sold illegally and the last remaining population has declined sharply in only a few years. A single turtle sells for $50 to $75 in the Philippines and up to $2,500 in the United States and Europe. Similar cases of illegal trade occur in rare species of trees for lumber such as big-leaf mahogany (*Swietenia macrophylla*), rare species of plants for medicine such as goldenseal (*Hydrastis canadensis*), and many rare species of orchids for their beautiful flowers.

Sometimes even when trade in a particular species is legal, it can pose a potential long-term threat to species persistence. In the American southwest, for example, there is a growing movement to reduce water use by replacing grassy lawns with desert landscapes. One of the unintended consequences is the increased demand for cacti and other desert plants that are collected from the wild. Sales are currently estimated to be $1 million annually. Given the slow growth of desert plants, this increased demand is causing heightened concern for these plant populations in the wild.

**FIGURE 18.12** **Philippine forest turtle.** The only remaining population of this turtle lives on a single island in the Philippines. Although protected by law, illegal trade has caused a rapid decline of this species in the wild.

## Pollution

In Chapters 14 and 15 we saw how water and air pollution harm ecosystems. Threats to biodiversity come from toxic contaminants such as pesticides, heavy metals, acids, and oil spills. Other contaminants, such as endocrine disrupters, can have nonlethal effects that prevent or inhibit reproduction. Pollution sources that cause declines in biodiversity also include the release of nutrients that cause algal blooms and dead zones as well as thermal pollution that can make water bodies too warm for species to survive.

In 2010, for example, an oil platform in the Gulf of Mexico named the Deepwater Horizon exploded, causing a massive release of oil that lasted for several months. This release of oil, from a platform owned by BP, caused a tremendous amount of death across a wide range of animal species including sea turtles, pelicans, fish, and shellfish. In response to the massive oil spill, BP released hundreds of thousands of liters of oil dispersant, a chemical designed to break up large areas of oil into tiny droplets that can be consumed by specialized species of bacteria. However the dispersant is also toxic to many species of animals. The total impact of the spilled oil and applied dispersants on the wildlife of the Gulf of Mexico may not be known for many years.

## Climate Change

We have mentioned climate change in previous chapters and will discuss it in detail in Chapter 19. As a threat to biodiversity, the primary concern about climate change is how it will affect patterns of temperature and precipitation in different regions of the world. In some regions, a species may be able to respond to warming temperatures and changes in precipitation by migrating to a place where the climate is well suited to the species niche. In other cases, this is not possible. For example, in southwestern Australia, a small woodland/shrubland peninsula exists on the edge of the continent with a much larger area of subtropical desert farther inland. (You can see this biome on the map in Figure 4.17 on page 100.) Scientists expect conditions on the peninsula to become drier during the next 70 years. If this occurs, many species of plants in this small ecosystem will not have a nearby hospitable environment to which they can migrate, since the surrounding desert ecosystem is already too dry for them. An examination of 100 species of plants in the area (all from the genus *Banksia*) has led scientists to project that 66 percent of the species will decline in abundance and up to 25 percent will become extinct. As we will see in Chapter 19, many species in the world are expected to be affected by climate change.

# The conservation of biodiversity often focuses on single species

Given the large number of factors that can cause a reduction in biodiversity, it is important that we consider how to protect and increase biodiversity. There are two general approaches to conserving biodiversity: the single-species approach and the ecosystem approach.

The single-species approach to conserving biodiversity focuses our efforts on one species at a time. When a species declines to a status of threatened or endangered,

the natural response is to encourage a population rebound by improving the conditions in which it exists. This might be accomplished by providing additional habitat or reducing the presence of a contaminant that is causing impaired reproduction. When the population of a species has declined to extremely low numbers, sometimes the remaining few individuals will be captured and brought into captivity. Captive animals are bred with the intention of returning the species to the wild. A well-known example of captive breeding occurred with the California condor. As we discussed in Chapter 6, the condor had declined to a mere 22 birds in 1987. Thanks to captive breeding and several improvements in the condor's habitat, the population today numbers more than 300 birds. Programs such as these are a major function of zoos and aquariums around the world.

## Conservation Legislation

The single-species approach to conservation formed the foundation for the passage of the *Marine Mammal Protection Act* in 1972 and the *Endangered Species Act* in 1973 in the United States. The Marine Mammal Protection Act was passed in response to declining populations of many marine mammals, including polar bears, sea otters (*Enhydra lutris*), manatees (*Trichechus manatus*), and California sea lions (*Zalophus californianus*) (FIGURE 18.13). The act prohibits the killing of all marine mammals in the United States and prohibits the import or export of any marine mammal body parts. Only the U.S. Fish and Wildlife Service and the National Marine Fisheries Service are allowed to approve any exceptions to the act.

The Endangered Species Act, which has been amended several times since its initial passage in 1973, implements the international CITES agreement. The act authorizes the U.S. Fish and Wildlife Service to determine which species can be listed as threatened or endangered and prohibits the harming of such species, including prohibitions on the trade of listed species,

(a)

(b)

**FIGURE 18.13** **Protected marine mammals.** The Marine Mammal Protection Act protects marine mammals in the United States from being killed. (a) Sea otter, (b) California sea lion.

their fur, or their body parts. The act also authorizes the government to purchase habitat that is critical to the conservation of threatened and endangered species and to develop recovery plans to increase the population of threatened and endangered species. This is often one of the most important steps in allowing endangered species to persist. Today, the species that have been listed as threatened or endangered include 201 invertebrate animals, 381 vertebrate animals, and 795 plants. An additional 245 species are currently being considered for listing, a process that can take several years. Once listed, however, many threatened and endangered species have experienced stable or increasing populations. Indeed, some species have experienced sufficient increases in numbers that they have been taken off the endangered species list, including the bald eagle (FIGURE 18.14), the American alligator (*Alligator mississippiensis*), and the eastern Pacific population of the gray whale (*Eschrichtius robustus*). Other species are currently increasing in number and may be taken off the list in the future. The gray wolf, for example, was reintroduced into Yellowstone National Park to help improve the species' abundance in the United States and it is now no longer endangered.

The Endangered Species Act has sparked a great deal of controversy in recent years because it restricts certain human activities in areas where listed species live. For example, it has prevented or altered some construction projects to accommodate threatened or endangered species. Organizations whose activities are restricted often pit the protection of listed species against the jobs of people in the region. In the 1990s, for example, logging companies wanted to continue logging the old-growth forest of the Pacific Northwest. As we discussed earlier in this chapter, these forests are home

to the threatened northern spotted owl. Although automation had caused a decline in the number of logging jobs in the preceding several decades, many loggers perceived the Endangered Species Act as a further threat to their livelihood. They denounced the act for placing more value on the spotted owl than on the humans who depended on logging. In the end, a compromise allowed continued logging on some of the old-growth forest while the rest became protected habitat for the spotted owl, as well as the many other species depending on old-growth forest.

During the past decade, several politicians have attempted to weaken the Endangered Species Act. However, strong support from the public and scientists has allowed it to retain much of its original power to protect threatened and endangered species. The biggest current challenge is a lack of sufficient funds and personnel required to implement the law.

**THE CONVENTION ON BIOLOGICAL DIVERSITY** Protection of biodiversity is an international concern. In 1992, world nations came together and created the **Convention on Biological Diversity,** an international treaty to help protect biodiversity. The treaty had three objectives: conserve biodiversity, sustainably use biodiversity, and equitably share the benefits that emerge from the commercial use of genetic resources such as pharmaceutical drugs.

In 2002, the convention developed a strategic plan to achieve a substantial reduction in the worldwide rate of biodiversity loss by 2010. The nations that signed this agreement recognized both the instrumental and intrinsic values of biodiversity. In 2010, the convention evaluated the current trends in biodiversity around the world and concluded that the goal had not been met.

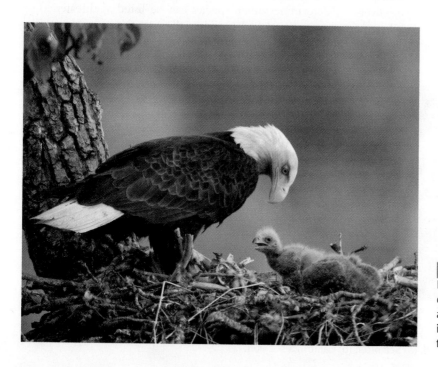

**FIGURE 18.14 Bald eagle hatchlings.** Habitat protection and reduced contaminants in the environment have allowed bald eagle populations to increase to the point where they could be taken off the Endangered Species List.

They identified the following trends from 2002 to 2010:

- On average, species at risk of extinction have moved closer to extinction.
- One-quarter of all plant species are still threatened with extinction.
- Natural habitats are becoming smaller and more fragmented.
- The genetic diversity of crops and livestock is still declining.
- There is a widespread loss of ecosystem function.
- The causes of biodiversity loss have either stayed the same or increased in intensity.
- The ecological footprint of humans has increased.

Collectively, the message emerging from the convention is not very positive. From the perspectives of genetic diversity, species diversity, and ecosystem services, all of the trends during the 8-year period continue to move in the wrong direction.

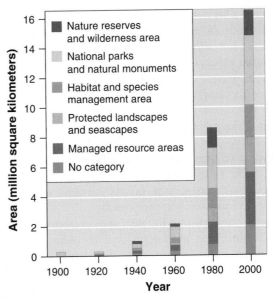

**FIGURE 18.15** **Changes in protected land.** Since the 1960s, there has been a large increase in the amount of land that is under various types of protection throughout the world. [After *Global Biodiversity Outlook 2,* Convention on Biological Diversity, 2006.]

**GAUGE YOUR PROGRESS**

✓ What is the single-species approach to conserving biodiversity?

✓ What makes the Endangered Species Act controversial? Why is it important to legislate species protection?

✓ In what ways is biodiversity protection an international issue?

# The conservation of biodiversity sometimes focuses on protecting entire ecosystems

Awareness of a sixth mass extinction in which humans have played a major role has brought a growing interest in the ecosystem approach to conserving biodiversity, which recognizes the benefit of preserving particular regions of the world, such as biodiversity hotspots. Protecting entire ecosystems has been one of the major motivating factors in setting aside national parks and marine reserves. In some cases, these areas were originally protected for their aesthetic beauty, but today they are also valued for their communities of organisms. The amount of protected land has increased dramatically worldwide since 1960. **FIGURE 18.15** shows changes in the amount of protected land worldwide since 1900.

When protecting ecosystems to conserve biodiversity, a number of factors must be taken into consideration including the size and shape of the protected area. We must also consider the amount of connectedness to other protected areas and how best to incorporate conservation while recognizing the need for sustainable habitat use for human needs.

## The Size, Shape, and Connectedness of Protected Areas

A number of questions arise when we consider protecting areas of land or water. For example, how large should the designated area be? Should we protect a single large area or several smaller areas? Does it matter whether protected areas are isolated or if they are near other protected areas? To help us answer these questions, we can return to our discussion of the theory of island biogeography from Chapter 6.

As you may recall, the theory of island biogeography looks at how the size of islands and the distance between islands and the mainland affect the number of species that are present on different islands. Larger islands generally contain more species because they support larger populations of each species, which makes them less susceptible to extinction. Larger islands also contain more species because they typically contain more habitats and, therefore, provide a wider range of niches for different species to occupy. The distance between an island and the mainland, or between one island and another, is another crucial factor, since more species are capable of dispersing to close islands than to islands farther away.

Although the theory of island biogeography was originally applied to oceanic islands, it has since been applied to islands of protected areas in the midst of less hospitable environments. For example, we can think of all the state and national parks, natural areas, and wilderness areas as islands surrounded by environments subject to high levels of human activity including agricultural fields, logged forests, housing developments, and industrial parks (FIGURE 18.16). Applying the theory of island biogeography from this perspective gives us some idea of the best ways to design and manage protected areas. For example, when protected areas are far apart, it is less likely that species can travel among them. This means that when a species has been lost from one ecosystem, it will be harder for individuals of that species from other ecosystems to recolonize it. So when we create smaller areas, they should be close enough for species to move among them easily.

Decisions regarding the design of protected areas can also be informed by the concept of metapopulations. As we learned in Chapter 6, a metapopulation is a collection of smaller populations connected by occasional dispersal of individuals along habitat corridors. Each population fluctuates somewhat independently of the other populations and a population that declines or goes extinct, due to a disease, for example, can be rescued by dispersers from a neighboring population. So if we set aside multiple protected areas, and recognize the need for connecting habitat corridors, a species is more likely to be protected from extinction by a decimating event such as a disease or natural disaster that could eliminate all individuals in a single protected area.

The concepts of island biogeography and metapopulations raise an interesting dilemma for conservation efforts. If we have limited resources to protect the biodiversity of a region, should we protect a single large area or several small areas? A single large area would support larger populations, but a species is more likely to survive a disease or natural disaster if it occupies several different areas. The debate over the best approach is known as SLOSS, which is an acronym for "*single large or several small*." While both approaches have their merits, in reality, human development and other factors often mean

that only one of the two strategies is available. For example, due to human development of a region, there may simply not be a single large area available to protect, so the only available strategy is to protect several small areas.

A final consideration regarding the size and shape of protected areas is the amount of **edge habitat** that an area contains. Edge habitat occurs where two different communities come together, typically forming an abrupt transition, such as where a grassy field meets a forest. While some species will live in either field or forest, others, like the brown-headed cowbird, specialize in living at the forest edge. So another challenge of protecting many small areas is the comparatively larger amount of edge habitat. When we protect several small forests, for example, the proliferation of species such as the cowbird in this larger amount of edge habitat can have a detrimental effect on songbirds that typically live farther inside a forest.

## Biosphere Reserves

In Chapter 10 we saw that managing national parks and other protected areas so they serve multiple users can be a challenge. On the one hand we want to make places of great natural beauty available to everyone. Yet when large numbers of people use an area for recreation, at least some degradation is very likely. To address this problem, the United Nations Educational, Scientific and Cultural Organization (UNESCO) developed the innovative concept of *biosphere reserves*. **Biosphere reserves** are protected areas consisting of zones that vary in the amount of permissible human impact. These reserves protect biodiversity without excluding all human activity. FIGURE 18.17 shows the different zones in a model biosphere reserve. The central core is an area that receives minimal human impact and is therefore the best location for preserving biodiversity. A buffer zone encircles the core area. Here, modest amounts of human activity are permitted, including tourism, environmental education, and scientific research facilities. Farther out is a transition area containing sustainable logging, sustainable agriculture, and residences for the local human population.

Designing reserves with these three zones represents an ideal scenario. In reality, biosphere reserves can take many forms depending on their location, though all attempt to maintain

FIGURE 18.16 **Islands of protected areas.** Central Park in New York City is an extreme example of an island of hospitable habitat surrounded by an urban environment that is not hospitable to most species.

**Zonation system of biosphere reserves**

| | Human settlements | | Monitoring |
| --- | --- | --- | --- |
| | Research station | | Tourism and recreation |
| | Education and training | | |

Core area

Buffer zone

Transition area

**FIGURE 18.17** Biosphere reserves. Biosphere reserves ideally consist of core areas that have minimal human impact and outer zones that have increasing levels of human impacts.

low-impact core areas. Currently there are 564 biosphere reserves worldwide—47 in the United States—with a total of 109 nations participating.

One well-known biosphere reserve is Big Bend National Park in Texas. The park itself serves as the core area and receives relatively little human impact, although hikers are permitted to walk through the beautiful desert landscapes and tree-covered mountain peaks. The park contains several dozen threatened and endangered species and more than 400 species of birds, many of which pass through the Big Bend region during their annual spring and fall migrations. Outside the boundaries of the park is a region of increased human impact including tourist facilities, human settlements, and agriculture (FIGURE 18.18).

**GAUGE YOUR PROGRESS**

✓ How does island biogeography help us decide how to protect species?

✓ What are the different ways that reserves can be designed?

✓ What is a biosphere reserve and how does it help preserve biodiversity?

**FIGURE 18.18** Big Bend National Park. The entire park, located in southwest Texas, serves as a low-impact core area of the Big Bend biosphere reserve.

## Swapping Debt for Nature

Preserving biodiversity is expensive. A case in point is the money required to set aside terrestrial or aquatic areas for protection. For example, if the land is privately owned, it must be purchased. Indirect costs, such as the income lost from not using the land, water, or other natural resources such as wood materials, metals, and fossil fuels, can be high. Finally, the costs of maintaining the protected area, ranging from monitoring the biodiversity to hiring guards to prevent illegal activities such as poaching, can be prohibitive. Given the fact that preserving biodiversity is expensive, how can the developing nations of the world, which contain much biodiversity but little wealth, afford it?

In 1984, Thomas Lovejoy from the World Wildlife Fund came up with an idea to help protect large areas of land and at the same time improve the economic conditions of developing countries. Lovejoy observed that developing nations possessed much biodiversity but were often deep in debt to wealthier, developed countries. Developing countries borrowed large amounts for the purpose of improving economic conditions and political stability. The developing countries were slowly repaying their loans with interest, but some countries had fallen so far behind on these payments that debtor nations doubted that the loans would ever be repaid in full. These debtor countries had little money left over for investment in an improved environment after they had paid their loans to developed countries. Lovejoy considered the possibility that debtor nations could use their position to motivate investment in biodiversity conservation.

The "debt-for-nature" swap has been used several times in Central and South America. In these swaps, the United States government and prominent environmental organizations provide cash to pay down a portion of a country's debt to the United States. The debt is then transferred to environmental organizations within the country. The debtor government then makes payments to the environmental organizations rather than to the United States. This does not mean that the country is out of debt, just that it now sends its annual loan payments to the environmental organizations for the purpose of protecting the country's biodiversity. In short, the indebted country switches from sending its money out of the country to investing in its own environmental conservation.

One of the largest debt-for-nature swaps recently happened in the Central American country of Guatemala. The United States government paired with two conservation organizations to provide $17 million to Guatemala. Over a period of 15 years, this amount, with interest, would have grown to more than $24 million, or about 20 percent of Guatemala's debt to the United States. In exchange, Guatemala agreed to pay $24 million over 15 years to improve conservation efforts in four areas of the country, including the purchase of land, the prevention of illegal logging, and future grants to conservation organizations helping to document and preserve the local biodiversity (**FIGURE 18.19**). The four areas include

**FIGURE 18.19 Swapping debt for nature in Guatemala.** The Maya Biosphere Reserve is one of four areas of Guatemala that will be better protected under an agreement between the governments of the United States and Guatemala as well as several conservation organizations. More than twice the size of Yellowstone National Park in the United States, this reserve offers important protection to biodiversity while also preserving historic Mayan temples that are part of Guatemala's cultural heritage and allowing sustainable use of some of the forest by local people.

two ecosystems—mangrove forests and tropical forests. Each forms a core area within a biosphere reserve that contains a large number of rare and endangered species including the jaguar (*Panthera onca*).

To date, the United States has used the debt-for-nature swap in 12 countries to protect tropical forests from Central America to the Philippines. To take part in the swap program, the countries are required to have a democratically elected government, a plan for improving their economies, and an agreement to cooperate with the United States on issues related to combating drug trafficking and terrorism. The results of these agreements have been encouraging. In Belize, for example, a debt-for-nature swap allowed 9,300 ha (23,000 acres) to be protected and an additional 109,000 ha (270,000 acres) to be managed for conservation. In Peru, a $10.6 million debt-for-nature swap led to the protection of more than 11 million ha (27 million acres) of tropical forest. Although these arrangements are only currently being applied to tropical forests, there is no inherent reason that this unique, modern-day conservation strategy would not also work in many other developing countries around the world.

**References**

How debt-for-nature swaps protect tropical forests. The Nature Conservancy. http://www.nature.org/wherewework/centralamerica/tropicalforests/news/news2113.html.

Lacey, M. 2006. U.S. to cut Guatemala's debt for not cutting trees. *New York Times,* October 2. http://www.nytimes.com/2006/10/02/world/americas/02conserve.html?ex=1317441600&en=a415721a6d698962&ei=5088&partner=rssnyt&emc=rss.

## REVISIT THE KEY IDEAS

- **Understand how genetic diversity, species diversity, and ecosystem function are changing over time.**

The genetic diversity of both wild and domesticated species is declining, as is the species diversity for all groups of organisms that have been assessed so far. Because species help to determine the function of ecosystems, declines in species diversity have, in turn, caused losses in ecosystem services.

- **Identify the causes of declining biodiversity.**

Because each species relies on a particular habitat, a major cause of declining biodiversity is the loss of habitat. Additional causes include overharvesting, legal and illegal trading in plants and animals, introductions of alien species, pollution, and climate change.

- **Describe the single-species approach to conserving biodiversity including the major laws that protect species.**

The single-species approach to conserving biodiversity focuses on saving one species at a time, often by using laws such as the Marine Mammal Protection Act and the Endangered Species Act.

- **Explain the ecosystem approach to conserving biodiversity and how size, shape, and connectedness affect the number of species that will be protected.**

The ecosystem approach to conserving biodiversity focuses on protecting not just a particular species of interest, but entire communities of organisms that live in an ecosystem. This is the goal of the biosphere reserve program of the United Nations. In protecting areas, we have to consider the theories of island biogeography and metapopulations because we can protect more biodiversity if we save larger areas of habitat that are close enough to each other to allow organisms to disperse.

# CHECK YOUR UNDERSTANDING

1. Which is a cause of declining global biodiversity?
   - I Pollution
   - II Habitat loss
   - III Overharvesting
   - (a) I
   - (b) I and II
   - (c) I and III
   - (d) II and III
   - (e) I, II, and III

2. Which statement about global biodiversity is *correct?*
   - (a) Species diversity is decreasing but genetic diversity is increasing.
   - (b) Species diversity is decreasing and genetic diversity is decreasing.
   - (c) Species diversity is increasing but genetic diversity is decreasing.
   - (d) Declines in genetic diversity are occurring in wild plants but not in crop plants.
   - (e) Declines in genetic diversity are occurring in crop plants but not in wild plants.

3. Which group of animals is declining in species diversity around the world?
   - I Fish and amphibians
   - II Birds and reptiles
   - III Mammals
   - (a) I
   - (b) I and II
   - (c) I and III
   - (d) II and III
   - (e) I, II, and III

4. Which of the following species was historically overharvested?
   - (a) Brown-headed cowbird
   - (b) Honeybees
   - (c) Kudzu vine
   - (d) Dodo bird
   - (e) Zebra mussel

5. Which statement is *incorrect* regarding the genetic diversity of livestock?
   - (a) The use of only the most productive breeds improves genetic diversity.
   - (b) Livestock come from very few species.
   - (c) The genetic diversity of livestock has declined during the past century.
   - (d) Different breeds are adapted to different climatic conditions.
   - (e) Different breeds are adapted to different diseases.

6. Which statement is *incorrect* about invasive alien species?
   - (a) Their populations grow rapidly.
   - (b) They often have no major predators or herbivores.
   - (c) They are often competitively inferior.
   - (d) A well-known invasive alien plant is the kudzu vine.
   - (e) A well-known invasive alien animal is the zebra mussel.

7. Which is an example of the single-species approach to conservation?
   - I The Endangered Species Act
   - II The Marine Mammal Protection Act
   - III The Biosphere Reserve
   - (a) I
   - (b) I and II
   - (c) I and III
   - (d) II and III
   - (e) I, II, and III

8. Which principle of island biogeography is *incorrectly* applied to protecting areas of land or water?
   - (a) A larger protected area should contain more species.
   - (b) Protected areas that are closer to each other should contain more species.
   - (c) National parks can be thought of as islands of biodiversity.
   - (d) A larger protected area will have fewer habitats.
   - (e) Marine reserves can be thought of as islands of biodiversity.

9. Which statement *correctly* reflects the idea of a biosphere reserve?
   - (a) Sustainable agriculture and tourism are permitted in different zones.
   - (b) Human activities are allowed throughout the reserve.
   - (c) Human activities are restricted to the central core of the reserve.
   - (d) No human activities are permitted in a biosphere reserve.
   - (e) Sustainable agriculture is permitted, but tourism is not.

10. Which statement is *correct* regarding swapping debt for nature?
    - (a) Protecting land and water is typically not expensive.
    - (b) Developing countries can pay part of their debt by investing in their own environment.
    - (c) Developing countries pay their debt to the United States by investing in U.S. national parks.
    - (d) Having a plan to improve the economy of a developing country is not important.
    - (e) The only expense of protecting biodiversity is the purchase of an area.

## APPLY THE CONCEPTS

1. The conservation of biodiversity is an international problem.
   (a) Name and describe one U.S. law that is intended to prevent the extinction of species.
   (b) Name and describe one international treaty that is intended to prevent the extinction of species.
   (c) Explain the benefits of taking an ecosystem approach, as opposed to a single-species approach, to conserving biodiversity.

2. Tropical rainforests are home to a tremendous diversity of species. As a result, you need to develop a plan to protect this diversity.
   (a) Describe the advantages and disadvantages of protecting a single large area versus several small areas.
   (b) How might increasing the amount of edge habitat affect species that typically live deep in the forest?
   (c) Discuss the merits of preserving individual species that are threatened and endangered versus preserving the function of the ecosystem.
   (d) Describe three characteristics of organisms that would make them particularly vulnerable to extinction.

## MEASURE YOUR IMPACT

**How Large Is Your Home?** One of the major worldwide threats to biodiversity is habitat loss, including the loss of forests as the result of logging. Given that the demand for lumber drives much of the market for logging, consider how you and your family might influence the demand for lumber.

(a) From 1970 to 2010, the average size of a house in the United States has doubled while the average size of a family has been reduced by half. Based on this information, how much more space per person does a modern house have?

(b) The average house today uses the lumber from 50 trees. If homes were built to be half the size, and there are approximately 400,000 new homes built each year, how many trees could be saved?

(c) Rather than demolishing an older house and building a new one, many homeowners have chosen to move an older house to a new location. This effectively recycles the older house. There are currently 50,000 homes moved annually. Assuming that the average house uses the lumber from 50 trees, how many trees are saved when houses are moved rather than demolished?

## ENGAGE YOUR ENVIRONMENT

What are ways to conserve biodiversity? In this activity you will evaluate actions that benefit animal and plant life and determine ways that you can help conserve bio- diversity. *See Engage Your Environment Activity 18 at the end of this book.*

# Global Change: Climate Alteration and Global Warming

## Walking on Thin Ice

The polar bear (*Ursus maritimus*) is one of the best-known species of the Arctic. With a geographic distribution surrounding the North Pole, the polar bear plays an important role in this frozen ecosystem. Polar bears are voracious predators that specialize in eating several species of seals. The seals live under large expanses of ocean ice and come up for air it becomes smaller during the summer and larger during the winter. Because of the rise in Arctic temperatures, the amount of summer shrinking of the ice has increased. To measure the extent of this shrinkage, scientists compare yearly measurements taken at the end of each summer. Compared to the average amount of ice present from 1979 to 2000, scientists found that there was 25 to 39 percent less ice each year from

> If Arctic summer sea ice disappears by the end of this century, as some scientists predict, polar bears could decline and could even become extinct.

where there are holes in this ice. The bears roam the ice in search of these holes and pounce on seals that come to the surface. In many cases, the bears only consume seal blubber because it contains a great deal of energy critical for an organism living in such a cold environment. The portion of the seal carcass not consumed by the polar bear serves as a significant food source for other animals, including the Arctic fox (*Vulpes lagopus*). The polar bear is also important for the indigenous people who have long depended on its meat for food and its fur for clothing.

But over the last few decades, temperatures in the Arctic have risen much faster than in other parts of the world. Warmer air and ocean water have caused the polar ice cap to melt, causing concern among both scientists and the general public because of the popular appeal of the polar bear and its importance to the ecosystem and native peoples.

Scientists have been taking satellite photos of the polar ice cap for over 30 years. In 1978, photos revealed that the ice cap extended from Russia and Norway to Greenland, Canada, and Alaska. Over the years, however, the ice cap has shrunk and retreated away from the land. During each year, a natural cycle of shrinking and expansion of the polar ice cap occurs;

2006 to 2009 — a total area twice the size of Texas. In addition, the remaining ice is considerably thinner, making it more vulnerable to future melting.

As the polar ice cap melts, the polar bears have been losing habitat. During the summer, when the ice cap becomes unusually small and ice retreats far away from the land, polar bears can no longer reach it to hunt for seals. Today, the sea ice melts 3 weeks earlier than it did 30 years ago. Because of the shorter time that they are able to hunt for seals, male polar bears near Hudson Bay in Canada currently weigh 67 kg (150 pounds) less than they weighed 30 years ago. If Arctic summer sea ice disappears by the end of this century, as some scientists predict, polar bears could decline and could even become extinct.

A sharp decline or extinction of polar bears would have a wide range of effects on the ecosystem and on the indigenous people in the area. Seal populations could increase with the demise of their major predator, while other species, such as the Arctic fox, ▶

Seals that swim up through ice holes to breathe are susceptible to predation by polar bears.

◀ The polar ice cap currently melts 3 weeks earlier than it did 30 years ago, leaving polar bears less time to hunt for seals.

would decline because there would no longer be seal carcasses to feed on. These changes in population sizes could, in turn, cause reverberations throughout the Arctic food web. The indigenous people would also be affected, not only in terms of the food and clothing that the polar bear provides, but also from the perspective of their cultural and social identity. As the ecosystem changes with the decline of the polar bear, people of the Arctic may find their lives and livelihoods altered.

In 2008, the United States classified polar bears as a threatened species because of the decline in their ice habitat. If current conditions persist, the species will continue to decline. In 2009, the five nations with polar bear populations (Canada, Greenland, Norway, Russia, and the United States) agreed that the polar bear should be classified as threatened throughout its entire global range. While acknowledging that pollution and hunting contributed to the bears' bleak future,

the nations agreed that the effect of global warming on the ice cap posed the greatest threat to polar bears. This means that the solution to the problem is not simply in the hands of those few nations that contain polar bear habitats, but is in the hands of the wider world community.

While the plight of the polar bear has drawn attention to the effects of global warming, it is only one of many indicators that our world is rapidly changing because of human activity. In this chapter, we will try to understand the human role in altering the natural world, what kinds of effects these changes might have, and how we can address or mitigate them. ■

*Sources:* A. Revkin, Nations near Arctic declare polar bears threatened by climate change, *New York Times,* March 20, 2009, http://www.nytimes.com/2009/03/20/science/earth/20bears.html; B. Harden, Experts predict polar bear decline, *Washington Post,* July 7, 2005, http://www.washingtonpost.com/wp-dyn/content/article/2005/07/06/AR2005070601899.html.

## UNDERSTAND THE KEY IDEAS

Over the past two centuries, the globe has experienced a vast array of changes, many of them the result of increases in the human population. One of the global changes currently facing the world is that of global climate change. In this chapter, we will consider both the causes and the consequences of this change. We will also discuss the steps that are being taken to slow global warming on several fronts.

After reading this chapter you should be able to

■ distinguish among global change, global climate change, and global warming.

■ explain how solar radiation and greenhouse gases warm our planet.

■ discuss how $CO_2$ concentrations and temperatures have changed over time.

■ describe the importance of feedback loops in the process of global warming.

■ identify how global warming is affecting people and the environment.

■ discuss how the Kyoto Protocol aims to reduce global warming.

# Global change includes global climate change and global warming

The melting of the polar ice cap is just one example of the changes taking place around the globe. Throughout this book, we have highlighted a wide variety of ways in which the world has changed as a result of a rapidly growing human population. Human activity has placed increasing demands on natural resources such as water, trees, minerals, and fossil fuels. We have also emitted greater amounts of carbon dioxide, nitrogen compounds, and sulfur compounds into the atmosphere than we have historically. Our agricultural methods depend on chemicals, including fertilizers and pesticides. Finally, a growing population faces challenges of waste disposal, sanitation, and the spread of human diseases.

Change that occurs in the chemical, biological, and physical properties of the planet is referred to as **global change** (FIGURE 19.1). Some types of global change are natural and have been occurring for millions of years. Global temperatures, for example, have fluctuated over millions of years, with periods of cold temperatures causing ice ages. In modern times, however, the rates of change have often been much higher than those that have occurred historically. Many of these changes are the result of human activities, and can have significant, sometimes cascading, effects. For example, as we saw in Chapter 17, emissions from coal-fired power plants and waste incinerators have increased the amount of mercury in the air and water, with concentrations roughly tripling over preindustrial levels. This mercury bioaccumulates in fish caught thousands of kilometers away from the sources of pollution. Because mercury has harmful effects on the nervous system of children, women who might become pregnant and children are

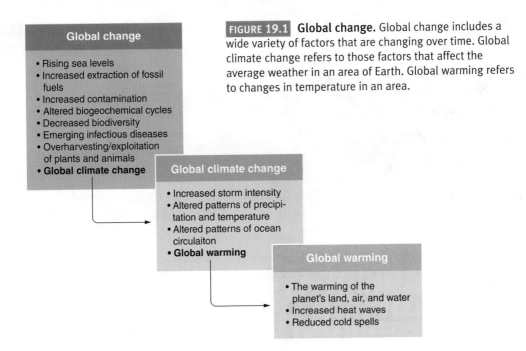

**FIGURE 19.1** **Global change.** Global change includes a wide variety of factors that are changing over time. Global climate change refers to those factors that affect the average weather in an area of Earth. Global warming refers to changes in temperature in an area.

**Global change**

- Rising sea levels
- Increased extraction of fossil fuels
- Increased contamination
- Altered biogeochemical cycles
- Decreased biodiversity
- Emerging infectious diseases
- Overharvesting/exploitation of plants and animals
- **Global climate change**

**Global climate change**

- Increased storm intensity
- Altered patterns of precipitation and temperature
- Altered patterns of ocean circulaiton
- **Global warming**

**Global warming**

- The warming of the planet's land, air, and water
- Increased heat waves
- Reduced cold spells

advised to avoid eating top predators such as swordfish and tuna. Far-reaching effects on this scale were unimaginable just 50 years ago.

One type of global change of particular concern to contemporary scientists is *global climate change*. **Global climate change** refers to changes in the climate of Earth—the average weather that occurs in an area over a period of years or decades. Changes in climate can be categorized as either natural climate change or anthropogenic climate change. For example, you might recall from Chapter 4 that El Niño events, which occur every 3 to 7 years, alter global patterns in temperature and precipitation (see Figure 4.13 on page 98). Anthropogenic activities such as fossil fuel combustion and deforestation also have major effects on global climates. **Global warming** refers specifically to one aspect of climate change: the warming of the oceans, landmasses, and atmosphere of Earth. In this chapter, we will focus on the phenomenon of global climate change. This chapter ties together many of the major themes that we have developed throughout the book: the interconnectedness of the systems of Earth, the environmental indicators that enable us to measure and evaluate the status of Earth, and the interaction of environmental science and policy.

**GAUGE YOUR PROGRESS**

✓ In what ways are humans involved in global change?

✓ How is current global change different from historic global change?

✓ How is climate change similar to or different from global change?

# Solar radiation and greenhouse gases make our planet warm

The physical and biogeochemical systems that regulate temperature at the surface of Earth—the concentrations of gases, distribution of clouds, atmospheric currents, and ocean currents—are essential to life on our planet. It is therefore critical that we understand how exactly the planet is warmed by the Sun, including how sources of radiation are converted into other sources of radiation and how the greenhouse effect, which we first mentioned in Chapter 1, contributes to the warming of Earth.

## The Sun-Earth Heating System

The ultimate source of almost all energy on Earth is the Sun. In the most basic sense, the Sun emits solar radiation that strikes Earth. As the planet warms, it emits radiation back toward the atmosphere. However, the types of energy radiated from the Sun and Earth are different (see Figure 2.11 on page 37). Because the Sun is very hot, most of its radiated energy is in the form of high-energy visible radiation and ultraviolet radiation—also known as visible light and ultraviolet light. When this radiation strikes Earth, the planet warms and radiates energy. Earth is not nearly as hot as the Sun, so it emits most of its energy as infrared radiation—also known as infrared light. We cannot see infrared radiation, but we can feel it being emitted from warm surfaces, like the heat that radiates from an asphalt road on a hot day.

Differences in the types of radiation emitted by the Sun and Earth, in combination with the greenhouse

**① Incoming solar radiation consists primarily of UV and visible light.**

**② About one-third of this solar radiation is reflected—from the atmosphere, clouds, and the surface of the planet—back into space.**

**④ Much of the emitted infrared radiation from Earth is absorbed by greenhouse gases in the atmosphere. The remainder is emitted into space.**

Incoming solar radiation

Reflected by atmosphere and clouds

Reflected from surface

Absorbed by clouds

Outgoing infrared radiation

Absorbed by surface

Greenhouse gases in atmosphere

**③ The remaining solar radiation is absorbed by clouds and the surface of the planet. Both become warmer and then emit infrared radiation.**

**⑤ As the greenhouse gases absorb infrared radiation, they warm and emit infrared radiation, with much of it going back toward Earth. The greater the concentration of greenhouse gases, the more infrared radiation is absorbed and emitted back toward Earth.**

**FIGURE 19.2** **The greenhouse effect.** When the high-energy radiation from the Sun strikes the atmosphere, about one-third is reflected from the atmosphere, clouds, and the surface of the planet. Much of the high-energy ultraviolet radiation is absorbed by the ozone layer, where it is converted to low-energy infrared radiation. The remaining ultraviolet radiation and visible light strike the land and water of Earth where they are also converted into low-energy infrared radiation. The infrared radiation radiates back toward the atmosphere, where it is absorbed by greenhouse gases that radiate much of it back toward the surface of Earth. Collectively, these processes cause warming on the planet.

effect, cause the planet to warm. Using **FIGURE 19.2**, we can walk through each step of this process. As radiation from the Sun travels toward Earth, about one-third of the radiation is reflected back into space. Although some ultraviolet radiation is absorbed by the ozone layer in the stratosphere, the remaining ultraviolet radiation, as well as visible light, easily passes through the atmosphere. Once it has passed through the atmosphere, this solar radiation strikes clouds and the surface of Earth. Some of this radiation is reflected from the surface of the planet back into space. The remaining radiation is absorbed by clouds and the surface of Earth, which become warmer and begin to emit lower-energy infrared radiation back toward the atmosphere. Unlike ultraviolet and visible radiation, infrared radiation does not easily pass through the atmosphere. It is absorbed by gases, which then warm. The warmed gases emit

infrared radiation, some of which goes out into space. The rest of the infrared radiation goes back toward the surface of Earth, causing the surface to become even warmer. This absorption of infrared radiation by atmospheric gases and reradiation of the energy back toward Earth is the greenhouse effect.

The greenhouse effect gets its name from the idea that solar radiation causes a gardener's greenhouse to become very warm. However, the process by which actual greenhouses are warmed by the Sun involves glass windows holding in heat, whereas the process by which Earth is warmed involves greenhouse gases radiating infrared energy back toward the surface of the planet.

In the Sun–Earth heating system, the net flux of energy is zero; the inputs of energy to Earth equal the outputs from Earth. Over the long term—thousands or millions of years—the system has been in a steady state.

However, in the shorter term—over years or decades—inputs can be slightly higher or lower than outputs. Factors that influence short-term fluctuations include changes in incoming solar radiation from increased solar activity and changes in outgoing radiation from an increase in atmospheric gases that absorb infrared radiation. If incoming solar energy is greater than the sum of reflected solar energy and radiated infrared energy from Earth, then the energy accumulates faster than it is dispersed and the planet becomes warmer. If incoming solar energy is less than the sum of the two outputs, the planet becomes cooler. Such natural changes in inputs and outputs cause natural changes in the temperature of Earth over time.

## The Greenhouse Effect

As we have seen throughout this book, certain gases in the atmosphere can absorb infrared radiation emitted by the surface of the planet and radiate much of it back toward the surface. The gases in the atmosphere that absorb infrared radiation are known as *greenhouse gases.*

The two most common gases in the atmosphere, $N_2$ and $O_2$, comprising 99 percent of the atmosphere, do not absorb infrared radiation. Therefore $N_2$ and $O_2$ are not greenhouse gases and do not contribute to the warming of Earth. In fact, greenhouse gases make up a very small fraction of the atmosphere. Perhaps surprisingly, the most common greenhouse gas is water vapor ($H_2O$). Water vapor absorbs more infrared radiation from Earth than any other compound, although a molecule of water vapor does not persist nearly as long as other greenhouse gases. Other important greenhouse gases include carbon dioxide ($CO_2$), methane ($CH_4$), nitrous oxide ($N_2O$), and ozone ($O_3$). In the case of ozone, we have seen that its effects on Earth are diverse. Ozone in the stratosphere is beneficial because it filters out harmful ultraviolet radiation. In contrast, ozone in the lower troposphere acts as a greenhouse gas and can cause increased warming of Earth. It also is an air pollutant in the lower troposphere, where it can cause damage to plants and human respiratory systems. All of these gases have been a part of the atmosphere for millions of years, and have kept Earth warm enough to be habitable. There is one other type of greenhouse gas, chlorofluorocarbons (CFCs), which does not exist naturally and occurs in the atmosphere exclusively due to synthesis by humans. Although we commonly think of the greenhouse effect as detrimental to our environment, without any greenhouse gases the average temperature on Earth would be approximately −18°C (0°F) instead of its current average temperature of 14°C (57°F). Concern about the danger of greenhouse gases is part of a growing awareness that an increase in the concentration of these gases may cause the planet to warm even more.

The contribution of each gas to global warming depends in part on its *greenhouse warming potential.* The **greenhouse warming potential** of a gas estimates how much a molecule of any compound can contribute to global warming over a period of 100 years relative to a molecule of $CO_2$. In calculating this potential, scientists consider the amount of infrared energy that a given gas can absorb and how long a molecule of the gas can persist in the atmosphere. Because greenhouse gases can differ a great deal in these two factors, greenhouse warming potentials span a wide range of values. For example, water vapor has a lower potential compared to carbon dioxide. The remaining greenhouse gases have much higher values, either because they absorb more infrared radiation or because they persist for very long periods of time. Table 19.1 shows the global warming potential for five common greenhouse gases. Compared to $CO_2$, the greenhouse warming potential is 25 times higher for methane ($CH_4$), nearly 300 times higher for nitrous oxide ($N_2O$), and up to 13,000 times higher for chlorofluorocarbons (CFCs), which we discussed in Science Applied "Were We Successful in Halting the Growth of the Ozone Hole?"

| TABLE 19.1 | The major greenhouse gases | | |
|---|---|---|---|
| The major greenhouse gases differ in their ability to absorb infrared radiation and the duration of time that they stay in the atmosphere. The units "ppm" are parts per million. | | | |
| Greenhouse gas | Concentration in 2010 | Global warming potential (over 100 years) | Duration in the atmosphere |
| Water vapor | Variable with temperature | <1 | 9 days |
| Carbon dioxide | 390 ppm | 1 | Highly variable (ranging from years to hundreds of years) |
| Methane | 1.8 ppm | 25 | 12 years |
| Nitrous oxide | 0.3 ppm | 300 | 114 years |
| Chlorofluorocarbons | 0.9 ppm | 1,600 to 13,000 | 55 to >500 years |

*Source:* Data on concentration are from the National Oceanic and Atmospheric Administration. www.esrl.noaa.gov/gmd/aggi.
Data on global warming potential are from the United Nations Framework Convention on Climate Change.

The effect of each greenhouse gas depends on both its warming potential and its concentration in the atmosphere. Although carbon dioxide has a relatively low warming potential, it is much more abundant than most other greenhouse gases. One exception is water vapor, which can have a concentration similar to carbon dioxide. However, human activity appears to have little effect on the amount of water vapor in the atmosphere. In contrast, human activity has caused substantial increases in the amount of the other greenhouse gases. Among these, carbon dioxide remains the greatest contributor to the greenhouse effect because its concentration is so much higher than any of the others. As a result, scientists and policy makers focus their efforts on ways of reducing carbon dioxide in the atmosphere.

Given what we now know about how greenhouse gases work, the concentrations of each gas, and which gases absorb the most energy, we can understand how changes in the composition and concentrations of greenhouse gases can contribute to global warming. Increasing the concentration of any historically present greenhouse gas should cause more infrared radiation to be absorbed in the atmosphere, which will then radiate more energy back toward the surface of the planet. Temperatures on Earth will become warmer. Likewise, producing new greenhouse gases that can make their way into the atmosphere, such as CFCs, should also cause increased absorption of infrared radiation in the atmosphere and cause the temperatures on Earth to become warmer.

---

### GAUGE YOUR PROGRESS

✓ How does the energy of the Sun cause Earth to heat?

✓ What is a greenhouse gas? Which greenhouse gases are the most common on Earth?

✓ What determines the effect of a greenhouse gas? Which greenhouse gas has the strongest effect?

---

# Sources of greenhouse gases are both natural and anthropogenic

As we have seen, greenhouse gases include a variety of compounds including water vapor, carbon dioxide, methane, nitrous oxide, and CFCs. These gases have natural and anthropogenic sources. After reviewing the different sources of greenhouse gases, we will discuss the relative ranks of the different anthropogenic sources.

## Natural Sources of Greenhouse Gases

Natural sources of greenhouse gases include volcanic eruptions, decomposition, digestion, denitrification, evaporation, and evapotranspiration.

**VOLCANIC ERUPTIONS** Over the scale of geologic time, volcanic eruptions can add a significant amount of carbon dioxide to the atmosphere. Other gases and the large quantities of ash released during volcanic eruptions can also have important, short-term climatic effects. In particular, the large quantity of ash emitted in the atmosphere from a volcanic eruption can have a major effect on global temperatures by reflecting incoming solar radiation back out into space, thereby cooling Earth. In 1991, for example, Mount Pinatubo in the Philippines erupted and spewed millions of tons of ash into the atmosphere, as far as 20 km (12 miles) high (**FIGURE 19.3**). This reduced the amount of radiation striking Earth, which caused a 0.5°C (0.9°F) decline in the temperature on the surface of the planet. Such effects usually last only for a few years, after which the ash and small particles settle out of the atmosphere.

**METHANE** When decomposition occurs under high-oxygen conditions, the dead organic matter is ultimately converted into carbon dioxide. As we saw in Chapter 16 when discussing landfills, methane is created when there is not enough oxygen available to produce carbon dioxide. This is a common occurrence at the bottom of wetlands where plants and animals decompose and oxygen is in low supply. Wetlands are the largest natural source of methane.

A similar situation occurs when certain animals digest plant matter. Animals that consume significant quantities of wood or grass, including termites and grazing antelopes, require gut bacteria to digest the plant material. Because the digestion occurs in the animal's gut, the bacteria do not have access to oxygen and methane is produced as a by-product. Because a single termite colony can contain more than a million termites and because termites are abundant

**FIGURE 19.3** Volcanic eruptions. Volcanic eruptions, such as the eruption of Mount Pinatubo in the Philippines in 1991, send millions of tons of ash into the atmosphere where it can absorb incoming solar radiation, reradiate it back to space, and cause a cooling of Earth.

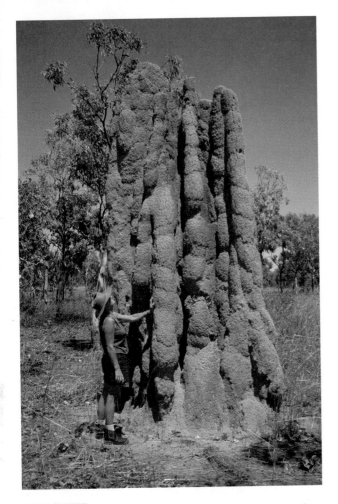

**FIGURE 19.4** **Termite mounds.** The bacteria that live in the anaerobic gut environment of herbivores such as termites produce methane as a by-product of their digestive activities. Because termite colonies, such as this one in Australia, can achieve population sizes of more than one million, they collectively can produce large amounts of methane.

throughout the world, especially in the tropics, they represent the second largest natural source of methane (FIGURE 19.4).

**NITROUS OXIDE** As we learned in Chapter 3, nitrous oxide ($N_2O$) is a natural component of the nitrogen cycle that is produced through the process of denitrification. Denitrification occurs in the low-oxygen environments of wet soils and at the bottoms of wetlands, lakes, and oceans. (Figure 3.12 on page 70 shows the nitrogen cycle.) In these environments, nitrate is converted to nitrous oxide gas, which then enters the atmosphere as a powerful greenhouse gas.

**WATER VAPOR** As we stated earlier, water vapor is the most abundant greenhouse gas in the atmosphere and the greatest natural contributor to global warming. In Chapter 3 we examined the role of water vapor in the hydrologic cycle. (Figure 3.10 on page 67 shows the hydrologic cycle.) Water vapor is produced when liquid water from land and water bodies evaporates and by the evapotranspiration process of plants. Because the amount of evaporation into water vapor varies with climate, the amount of water vapor in the atmosphere can vary regionally.

## Anthropogenic Sources of Greenhouse Gases

There are many anthropogenic sources of greenhouse gases. The most significant of these are the burning of fossil fuels, agriculture, deforestation, landfills, and industrial production of chemicals (FIGURE 19.5).

**USE OF FOSSIL FUELS** Tens to hundreds of millions of years ago, organisms were sometimes buried without

**FIGURE 19.5** **Anthropogenic sources of greenhouse gases.** Human activities are major contributors of greenhouse gases including $CO_2$, methane, and nitrous oxide. These activities include the use of fossil fuels, agricultural practices, the creation of landfills, and the industrial production of new greenhouse gases.

first decomposing into carbon dioxide. In Figure 3.11 on page 68 we outlined the process by which the carbon contained in these organisms, called fossil carbon, is slowly converted to fossil fuels deep underground. When humans burn fossil fuels, we produce $CO_2$ that goes into the atmosphere. Because of the long time required to convert carbon into fossil fuels, the rate of putting carbon into the atmosphere by burning fossil fuels is much greater than the rate at which producers take $CO_2$ out of the air and both the producers and consumers contribute to the pool of buried fossil carbon.

Because fossil fuels differ in how they store energy, each type of fossil fuel produces different amounts of carbon dioxide. For a given amount of energy, coal produces the most $CO_2$. In comparison, oil produces 85 percent as much $CO_2$ as coal, and natural gas produces 56 percent as much. As we saw in Chapter 12, this is why, from the perspective of $CO_2$ emissions, natural gas is considered better for the environment than coal. The production of fossil fuels, such as the mining of coal, and the combustion of fossil fuels can also release methane and, in some cases, nitrous oxide.

Particulate matter, while not a greenhouse gas, may also play an important role in global warming. Although particulate matter, also known as black soot, may reflect solar radiation under some conditions, recent findings suggest that it may be responsible for up to one-quarter of observed global warming during the past century. Particulates that fall on ice and snow in the higher latitudes absorb more of the Sun's energy by lowering the albedo. As the snow and ice begin to melt, the particulates become more concentrated on the surface. The increased concentration raises the amount of solar radiation absorbed, which increases melting. This positive feedback system might help explain warming that occurred early in the last century, when atmospheric concentrations of greenhouse gases had only begun to increase but soot from the burning of coal was widespread.

**AGRICULTURAL PRACTICES** Agricultural practices can produce a variety of greenhouse gases. Agricultural fields that are overirrigated, or those that are deliberately flooded for cultivating crops such as rice, create low-oxygen environments similar to wetlands and therefore can produce methane and nitrous oxide. Synthetic fertilizers, manures, and crops that naturally fix atmospheric nitrogen—for example, alfalfa—can create an excess of nitrates in the soil that are converted to nitrous oxide by the process of denitrification.

Raising livestock can also produce large quantities of methane. Many livestock such as cattle and sheep consume large quantities of plant matter and rely on gut bacteria to digest this cellulose. As we saw in the case of termites, gut bacteria live in a low-oxygen environment and digestion in this environment produces methane as a by-product. Manure from livestock operations will decompose to $CO_2$ under high-oxygen conditions, but in low-oxygen conditions, for example in manure lagoons that are not aerated, it will decompose to methane.

**DEFORESTATION** Each day, living trees remove $CO_2$ from the atmosphere during photosynthesis, and decomposing trees add $CO_2$ to the atmosphere. This part of the carbon cycle does not change the net atmospheric carbon because the inputs and outputs are approximately equal. However when forests are destroyed by burning or decomposition and not replaced, as can happen during deforestation, the *net* destruction of vegetation will contribute to the increase in atmospheric $CO_2$. This is because the mass of carbon that made up the trees is added to the atmosphere by combustion or decomposition. The shifting agriculture described in Chapter 11, which involves clearing forests and burning the vegetation to make room for crops, is a major source of both particulates and a number of greenhouse gases, including carbon dioxide, methane, and nitrous oxide.

**LANDFILLS** As we saw in Chapter 16, landfills receive a great deal of household waste that slowly decomposes under layers of soil. When the landfills are not aerated properly, they create a low-oxygen environment, like wetlands, in which decomposition causes the production of methane as a by-product.

**INDUSTRIAL PRODUCTION OF NEW GREENHOUSE CHEMICALS** The creation of new industrial chemicals often has unintended effects on the atmosphere. In both Science Applied "Were We Successful in Halting the Growth of the Ozone Hole?" and Chapter 15 we looked at CFCs, the family of chemicals that serve as refrigerants used in air conditioners, freezers, and refrigerators. CFCs were used in the past until scientists discovered that they were damaging the protective ozone layer. CFCs are also potent greenhouse gases with very high greenhouse warming potentials. As we discussed in Chapter 15, the nations of the world joined together to sign the Montreal Protocol, which phased out the production and use of CFCs by 1996. Unfortunately, many of the alternative refrigerants that are less harmful to the ozone layer, including a group of gases known as hydrochlorofluorocarbons (HCFCs), still have very high greenhouse warming potentials. As a result, developed countries will phase out the use of HCFCs by 2030.

**RANKING THE ANTHROPOGENIC SOURCES OF GREENHOUSE GASES** We have seen that there are multiple anthropogenic sources of greenhouse gases. What is the relative contribution of each source? FIGURE 19.6 shows the major anthropogenic sources of greenhouse gases in the United States. Figure 19.6a shows that the three major contributors of methane in the atmosphere

**FIGURE 19.6** **Anthropogenic sources of greenhouse gases in the United States.** (a) The largest contributions of methane in the atmosphere are gut bacteria that help many livestock species digest plant matter, landfills that experience decomposition in low-oxygen environments, and the production, storage, and transport of natural gas and petroleum products from which methane escapes. (b) The largest contributions of nitrous oxide in the atmosphere arise from the agricultural soils that obtain nitrogen from applied fertilizers, applied manures, and the planting of nitrogen-fixing crops. This nitrogen is converted to nitrous oxide through the process of denitrification. (c) Approximately 94 percent of all $CO_2$ emissions come from the burning of fossil fuels. [Data from http://www.epa.gov/methane/sources.html, http://www.epa.gov/nitrousoxide/sources.html, http://www.epa.gov/climatechange/emissions/co2_human.html.]

are the digestive processes of livestock, landfills, and the production of natural gas and petroleum products. The major contributor of nitrous oxide, shown in Figure 19.6b, is agricultural soil that receives nitrogen from synthetic fertilizers, applications of manure as an organic fertilizer, and nitrogen-fixing crops such as alfalfa. Finally, looking at the numbers for carbon dioxide in Figure 19.6c, we see that approximately 94 percent of all $CO_2$ emissions come from the burning of fossil fuels.

**GAUGE YOUR PROGRESS**

✓ What are the main natural and anthropogenic sources of greenhouse gases?

✓ Which of the anthropogenic sources are the easiest to reduce? Why?

✓ Why do we rank the sources of greenhouse gases?

# Changes in $CO_2$ and global temperatures have been linked for millennia

Now that we know something about greenhouse gases and their role in global warming, we can explore the evidence that an increased concentration of greenhouse gases is causing Earth to become warmer. If Earth is, in fact, becoming warmer, we can begin to evaluate whether or not warming is caused by human activities that have released greenhouse gases. One way to make these assessments is to determine gas concentrations and temperatures from the past and compare them to gas concentrations and temperatures in the present day. We can also use information about changes in gas concentrations and temperatures to predict future climate conditions.

In 1988 the United Nations and the World Meteorological Organization created the Intergovernmental Panel on Climate Change (IPCC), a group of more than 3,000 scientists from around the world working together to assess climate change. Their mission is to understand the details of the global warming system, the effects of climate change on biodiversity and energy fluxes in ecosystems, and the economic and social effects of climate change. The IPCC enables scientists to assess and communicate the state of our knowledge and to suggest research directions for improving our understanding in the future. This effort has produced an excellent understanding of how $CO_2$ and temperatures are linked.

## Increasing $CO_2$ Concentrations

Through the work of the IPCC, we now understand that $CO_2$ is an important greenhouse gas that can contribute to global warming, but we didn't always realize this. In the first half of the twentieth century, most scientists believed that if any excess $CO_2$ were being produced, it would be absorbed by the oceans and vegetation. In addition, because the concentration of atmospheric $CO_2$ was low compared to gases such as oxygen and nitrogen, it was difficult to measure accurately.

Charles David Keeling was the first to test prevailing assumptions rigorously and to overcome the technical difficulties in $CO_2$ measurement. When Keeling set out to measure the precise level of $CO_2$ in the atmosphere, most atmospheric scientists believed that two measurements several years apart would be sufficient to answer the question of whether human activities were causing increased concentrations of $CO_2$ in the atmosphere. Keeling did not agree, and in 1958 he began collecting data throughout the year at the Mauna Loa Observatory in Hawaii. After just 1 year of work, Keeling found that $CO_2$ levels varied seasonally and that the concentration of $CO_2$ increased from year to year. His results prompted him to take measurements for several more years, and he and his students have continued this work into the twentieth-first century. The results, shown in **FIGURE 19.7**, confirm Keeling's early findings: although $CO_2$ concentrations vary between seasons, there is a clear trend of rising $CO_2$ concentrations across the years. This increase over time is correlated to increased human emissions of carbon from the combustion of fossil fuels and net destruction of vegetation. Crunch the Numbers "Projecting Future Increases in $CO_2$" gives you an opportunity to estimate how much $CO_2$ concentrations will increase by the end of the century.

What causes the seasonal variation? Each spring, as deciduous trees, grasslands, and farmlands in the Northern Hemisphere turn green, they increase their absorption rates of $CO_2$ to carry out photosynthesis. At the same time, bodies of water begin to warm and the algae and plants also begin to photosynthesize. In doing so, these producers take up some of the $CO_2$ in the

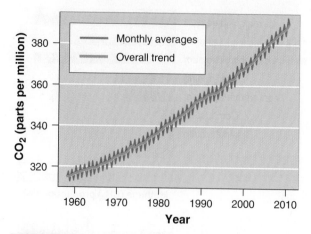

**FIGURE 19.7** **Changes in atmospheric $CO_2$ over time.** Carbon dioxide levels have risen steadily since measurement began in 1958. [After http://www.esrl.noaa.gov/gmd/ccgg/trends/.]

atmosphere. Conversely, in the fall, as leaves drop, crops are harvested, and bodies of water cool, the uptake of atmospheric $CO_2$ by algae and plants declines and the amount of $CO_2$ in the atmosphere increases.

## Emissions from the Developed versus Developing World

Throughout this book we have seen that per capita consumption of fossil fuel and materials is greatest in developed countries. It is not surprising, then, that the production of carbon dioxide has also been greatest in the developed world. For many decades, the 20 percent of the population living in the developed world—roughly 1 billion people—produced about 75 percent of the carbon dioxide. However, as some developing nations industrialize and acquire more vehicles that burn fossil fuels these percentages are changing. In 2009,

---

## CRUNCH THE NUMBERS

### Projecting Future Increases in $CO_2$

Since Charles David Keeling and his colleagues began measuring $CO_2$ in 1958, we have an excellent record of how $CO_2$ concentrations have changed in the atmosphere over time. From 1960 to 2010, the concentration of $CO_2$ in the atmosphere increased from 320 to 390 ppm (parts per million).

Based on these two points in time, what has been the average annual increase of $CO_2$ in the atmosphere?

Time = 2010–1960 = 50 years
Increase in $CO_2$ = 390 ppm–320 ppm = 70 ppm

Average annual increase in

$$CO_2 = \frac{70 \text{ ppm}}{50 \text{ years}} = 1.4 \text{ ppm/year}$$

**Your Turn**

1. If the annual rate of $CO_2$ increase is 1.4 ppm, what concentration of $CO_2$ do you predict for the year 2100?

2. From 2000 to 2010, the rate of increase was faster than in previous decades, achieving a rate of 1.9 ppm per year. Based on this faster rate, what concentration of $CO_2$ do you predict for the year 2100?

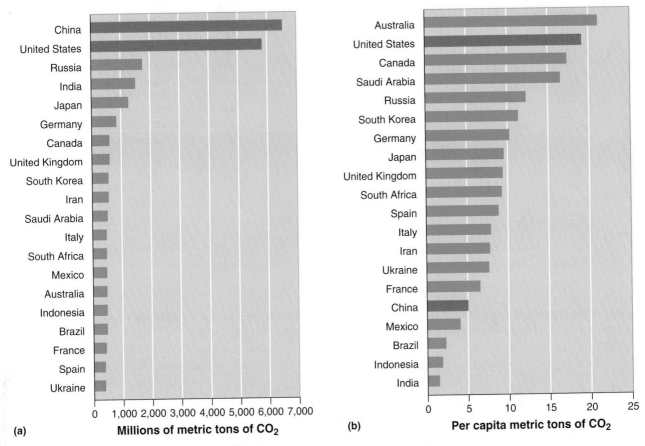

**FIGURE 19.8** $CO_2$ **emissions by country.** (a) When we consider the total amount of $CO_2$ produced by a country, we see that the largest contributors are the developed and rapidly developing countries of the world. (b) On a per capita basis, some major $CO_2$ emitters have relatively low per capita $CO_2$ emissions. [Data from http://www.ucsusa.org/global_warming/science_and_impacts/science/each-countrys-share-of-co2.html.]

developing countries surpassed developed countries in the production of $CO_2$.

Development has been especially rapid in China and India, which together contain one-third of the world's population. From 2001 to 2008, emissions of carbon dioxide from China nearly doubled as the country built many new coal-powered electrical plants that increased its ability to burn coal. **FIGURE 19.8a** shows that today China is the leading emitter of $CO_2$. China emits more than 6,500 million metric tons of $CO_2$, representing 21 percent of all global $CO_2$ emissions. The United States is in second place, emitting more than 5,800 million metric tons. This represents 19 percent of all global $CO_2$ emissions, yet the United States contains only 5 percent of the world's people. If we consider the amount of per capita $CO_2$ emissions, shown in **FIGURE 19.8b**, we obtain a very different picture of which countries produce the most $CO_2$. Australia recently emerged as the leading per capita emitter of $CO_2$, followed by the United States and Canada. Because the populations of China and India are so large, on a per capita basis they rank as the sixteenth and twentieth leading emitters of $CO_2$, respectively.

## Global Temperatures Since 1880

Before we can determine if global temperature increases are a recent phenomenon and if these increases are unusual, we must establish what the temperatures of Earth were in the past. Since about 1880, there have been enough direct measurements of land and ocean temperatures that the NASA Institute for Space Studies has been able to generate a graph of global temperature change over time. This graph, updated monthly, is shown in **FIGURE 19.9**. Comprising thousands of measurements from around the world, the graph shows global temperatures have increased 0.8°C (1.4°F) from 1880 through 2009. In fact, of the 10 warmest years on record, 9 occurred from 2000 to 2009, making this decade the warmest on record.

An increase in average global temperature of 0.8°C (1.4°F) may not sound very substantial. This temperature increase, however, is not evenly distributed around the globe. As the map in **FIGURE 19.10** shows, some regions, including parts of Antarctica, have experienced temperatures that are cooler than experienced in earlier

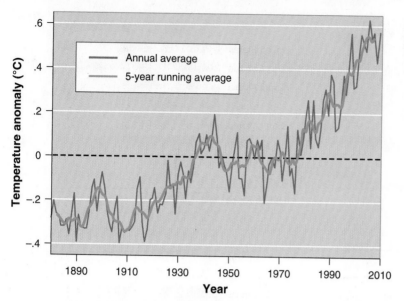

**FIGURE 19.9** **Changes in mean global temperatures over time.** Although annual mean temperatures can vary from year to year, temperatures have exhibited a slow increase from 1880 to today. This pattern becomes much clearer when scientists compute the average temperature each year for the past 5 years. [After http://data.giss.nasa.gov/gistemp/graphs/Fig.A2.lrg.gif.]

years. Some regions, including areas of the oceans, have experienced no change in temperature. Finally, some regions, such as those in the extreme northern latitudes, have experienced increases of 1°C to 4°C (1.8°F–7.2°F). The substantial increases in temperatures in the northern latitudes have caused, among other problems, nearly 40 percent of the northern ice cap to melt, threatening polar bears and their ecosystem, as discussed in the chapter opener.

The data collected by the NASA Institute for Space Studies clearly demonstrate that the globe has been slowly warming during the past 120 years. However, it is possible that such changes in temperature are simply a natural phenomenon. If we want to know whether

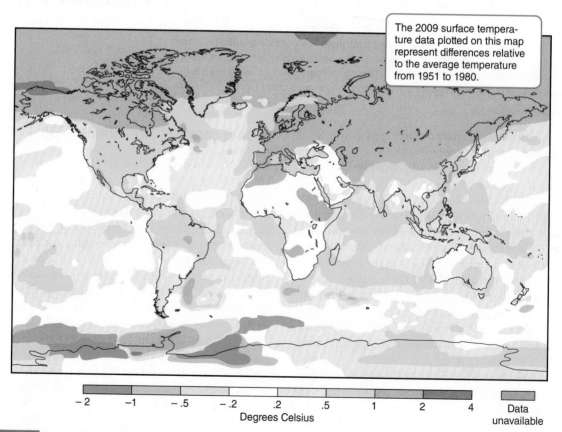

The 2009 surface temperature data plotted on this map represent differences relative to the average temperature from 1951 to 1980.

**FIGURE 19.10** **Changes in mean annual temperature in different regions of the world.** During the decade of 2000–2009, some regions have become cooler, some regions have had no temperature change, and the northern latitudes have become substantially warmer. [After http://climate.nasa.gov/BrowseImage.cfm?ImgID=59.]

**FIGURE 19.11** **Ancient shells of foraminifera.** The tiny shells of the protists become buried in layers of ocean sediments. By knowing the age of different ocean sediments and the preferred temperature of different species of foraminifera, scientists can indirectly estimate ocean temperature changes over time based on which species are found in each sediment layer.

these changes are typical, we must examine a much longer span of time.

## Global Temperatures and Greenhouse Gas Concentrations During the Past 400,000 Years

Since no one was measuring temperatures thousands of years ago, we must use indirect measurements. Common indirect measurements include changes in the species composition of organisms that have been preserved over millions of years and chemical analyses of ice that was formed long ago.

One commonly used biological measurement is the change in species composition of a group of small protists, called foraminifera. Foraminifera are tiny, marine organisms having hard shells that resist decay after death (**FIGURE 19.11**). In some regions of the ocean floor, the tiny shells have been building up in sediments for millions of years. The youngest sediment layers are near the top of the ocean floor, whereas the oldest sediment layers are much deeper. Fortunately, different species of foraminifera prefer different water temperatures. As a result, when scientists identify the predominant species of foraminifera in a layer of sediment, they can infer the likely temperature of the ocean at the time the layer of sediment was deposited. By examining thousands of sediment layer samples, we can gain insights into temperature changes over millions of years.

Scientists can also determine temperature change over long periods of time by examining ancient ice. In cold areas such as Antarctica and at the top of the Himalayas, the snowfall each year eventually compresses to become ice. Similar to marine sediments, the youngest ice is near the surface and the oldest ice is much deeper. During the process of compression, the ice captures small air bubbles. These bubbles contain tiny samples of the atmosphere that existed at the time the ice was formed. Scientists have traveled to these frozen regions of the world to drill deep into the ice and extract long tubes of ice called *ice cores* (**FIGURE 19.12**). Samples of ice cores can span up to 500,000 years of ice formation. Scientists determine the age of different layers in the ice core and then melt the ice from a piece associated with a particular time period. When the piece of ice melts, air bubbles are released and scientists measure the concentration of greenhouse gases that are trapped within the ice.

(a)

(b)

**FIGURE 19.12** **Ice cores.** (a) Ice cores can be extracted from very cold regions of the world such as this glacier on Mount Sajama in Bolivia. (b) Ice cores have tiny trapped air bubbles of ancient air that can provide indirect estimates of greenhouse gas concentrations and global temperatures.

In addition, oxygen atoms in melted ice cores can be used to determine temperatures from the distant past. Oxygen atoms occur in two forms, or *isotopes:* light oxygen, also known as oxygen-16 ($O^{16}$), contains 8 protons plus 8 neutrons. In contrast, heavy oxygen, also known as oxygen-18 ($O^{18}$), contains 8 protons plus 10 neutrons. Ice formed during a period of warmer temperatures contains a higher percentage of heavy oxygen than ice formed during colder temperatures. By examining changes in the percentage of heavy oxygen atoms from different layers of the ice core, we can indirectly estimate temperatures from hundreds of thousands of years ago.

Combining data from different biological and physical measurements, researchers have created a picture of how the atmosphere and temperature of Earth have changed over hundreds of thousands of years. FIGURE 19.13 shows the pattern of atmospheric $CO_2$. Notice that for over 400,000 years, the atmosphere never contained more than 300 ppm of $CO_2$. In contrast, from 1958 to 2009 the concentration of $CO_2$ in the atmosphere has rapidly climbed from 310 to 390 ppm. This means that the rise of $CO_2$ in the atmosphere during the past 50 years is unprecedented.

During the past 10,000 years, $CO_2$ is not the only greenhouse gas whose concentration has increased. As you can see in FIGURE 19.14, methane and nitrous oxide show a pattern of increase that is similar to the pattern we saw for $CO_2$. For all three gases, there was little change in concentration for most of the previous 10,000 years. After 1800, however, concentrations of the three gases all rose dramatically. Given what we now know about the anthropogenic sources of greenhouse gases, this increase in greenhouse gases

makes sense because this time marks the start of the Industrial Revolution when humans began burning large amounts of fossil fuel and producing a variety of greenhouse gases.

FIGURE 19.15 charts historic temperatures and $CO_2$ concentrations. Looking at the blue line, we see that temperatures have changed dramatically over the past 400,000 years. Most of these rapid shifts occurred during the onset of an ice age or during the transition from an ice age to a period of warm temperatures after an ice age. Because these changes occurred before humans could have had an appreciable effect on global systems, scientists suspect the changes were caused by small, regular shifts in the orbit of Earth. The path of the orbit, the amount of tilt on its axis, and the position relative to the Sun all change regularly over hundreds of thousands of years. These changes alter the amount of sunlight that hits high northern latitudes in the winter, the amount of snow that can accumulate, and the way the albedo effect keeps energy from being absorbed and converted to heat. These changes could give rise to fairly regular shifts in temperature over a long period of time.

The more important insight from Figure 19.15 is the close correspondence between historic temperatures and $CO_2$ concentrations. But the graph does not tell us the nature of this relationship. Did periods of increased $CO_2$ cause increased temperature; did periods of increased temperature cause increased production of $CO_2$; or is another factor at work? Scientists believe that the relationship between fluctuating levels of $CO_2$ and the temperature is complex and that both factors play a role. As we know, the increase of $CO_2$ in the atmosphere causes a greater capacity for warming through

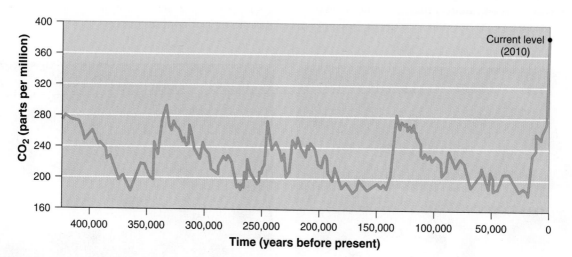

FIGURE 19.13 **Historic $CO_2$ concentrations.** Using a variety of indirect indicators including air bubbles trapped in ancient ice cores, scientists have found that for more than 400,000 years $CO_2$ concentrations never exceeded 300 ppm. After 1950, $CO_2$ concentrations have sharply increased to their current level of 390 ppm. [After http://climate.nasa.gov/evidence/.]

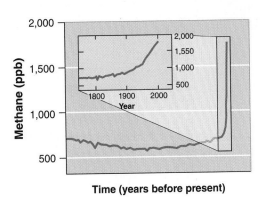

**FIGURE 19.14** Historic concentrations of $CO_2$, methane, and nitrous oxide. Using samples from ice cores and modern measurements of the atmosphere, scientists have demonstrated that the concentrations of all three greenhouse gases have increased dramatically during the past 200 years. [After IPCC, 2007: Summary for Policymakers. In *Climate Change 2007: The Physical Science Basis.* Contribution of Working Group I to the Fourth Assessment Report of the Intergovernmental Panel on Climate Change, ed. S. D. Solomon et al. (Cambridge University Press).]

the greenhouse effect. However, when Earth experiences higher temperatures, the oceans warm and cannot contain as much $CO_2$ gas and, as a result, they release $CO_2$ into the atmosphere. What ultimately matters is the net movement of $CO_2$ between the atmosphere and the oceans and how these different feedback loops work together to affect global temperatures.

We can also examine temperatures over somewhat shorter time periods. For example, scientists have used indirect and direct measures to determine global temperatures during the past 2,000 years. They then selected the average temperature from 1901 to 2000 as a baseline against which all other years can be compared. **FIGURE 19.16** shows that average past temperatures have

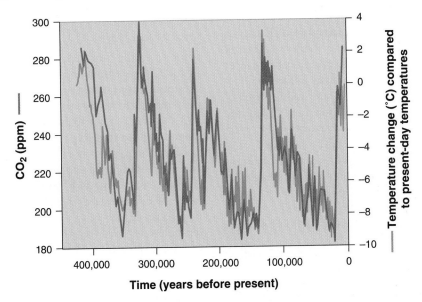

**FIGURE 19.15** Historic temperature and $CO_2$ concentrations. Ice cores used to estimate historic temperatures and $CO_2$ concentrations indicate that the two factors vary together. [After http://www.ncdc.noaa.gov/paleo/globalwarming/temperature-change.html.]

**FIGURE 19.16** **Temperature changes in the Northern Hemisphere for the past 2,000 years.** By combining indirect measures of temperatures with direct measures after 1880, we can gain a longer-range perspective on how current global warming compares to past global warming. The average temperature from 1901 to 2000 serves as a baseline against which all other years are compared. As a baseline, the average temperature from 1901 to 2000 is set at "zero temperature change." [After A. Moberg et al., Highly variable Northern Hemisphere temperatures reconstructed from low- and high-resolution proxy data, *Nature* 433 (2005): 613–617.]

remained cooler than the average during the twentieth century. Temperatures tended to be warmer during 1000 CE to 1200 CE and then cooler from 1500 to 1800. It is particularly striking that the change in temperature during the past century has increased rapidly and that the average temperatures that we are experiencing today are higher than those experienced in the past 2,000 years.

## Recent Temperature Increases

We have seen that the surface temperature of Earth has increased roughly 0.8°C (1.4°F) over the past 120 years. But larger changes in temperature have occurred over the past 400,000 years without human influence. How can we tell if the recent changes are anthropogenic? One explanation for warming temperatures during the past century is an increase in solar radiation. Another possibility is that warming is caused by increased $CO_2$ in addition to warming caused by natural fluctuations in solar radiation. In other words, both factors might be important. Unfortunately, simply looking at temperature and $CO_2$ data averages around the globe will not allow us to determine if either of these two possibilities is correct..

One way to approach the problem is to look for more detailed patterns in temperature changes. For example, if increased $CO_2$ concentrations caused global warming by preventing heat loss, then periods of elevated $CO_2$ would be associated with higher temperatures more commonly in winter than in summer, at night rather than doing the day, and in the Arctic rather than in

warmer latitudes. These three scenarios are all associated with colder temperatures, so reducing heat loss would have a bigger impact on temperature than in scenarios in which the temperatures were already quite warm. In fact, we already observed this when we examined the changes in temperature around the world in Figure 19.10—the Arctic regions are experiencing the greatest amount of warming.

On the other hand, if increased solar radiation were the cause of global warming, periods of elevated solar radiation would be associated with higher temperatures more commonly when the Sun is shining more— namely, in the summer, during the day, and at low latitudes. These times and locations on Earth receive the greatest amount of sunlight, so an increased intensity of solar radiation would cause these times and places on Earth to warm more than other times and places. When scientists make these types of detailed comparisons, they find that the patterns in temperature change are strongly consistent with increased greenhouse gases such as $CO_2$ and not consistent with increased solar radiation. This body of evidence led the IPCC to conclude in 2007 that most of the observed increase in global average temperatures since the mid-twentieth century has been the result of increased concentrations of anthropogenic greenhouse gas.

## Climate Models and Future Conditions

Just as indirect indicators can help us get a picture of what the temperature has been in the distant past, computer models can help us predict future climate conditions. Researchers can determine how well a model approximates real-world processes by applying it to a time in the past for which we have accurate data on conditions such as air and ocean temperatures, $CO_2$ concentration, extent of vegetation, and sea ice coverage at the poles. Modern models reproduce recent temperature fluctuations well over large spatial scales. From this work modelers are fairly confident that climate models capture the most significant features of today's climate.

Although climate models cannot forecast future climates with total accuracy, as the models improve scientists have been able to place more confidence in their predictions of temperature change. In contrast, they have had a more difficult time predicting changes in precipitation. Because assumptions vary among different climate models, when multiple models predict similar changes, we can have increased confidence that the predictions are robust. FIGURE 19.17 shows recent predictions based on the data from climate models. Scientists

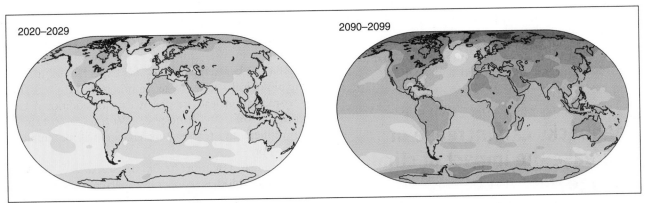

**(a)** Low increase in $CO_2$

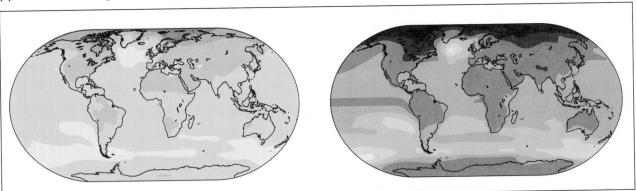

**(b)** Moderate increase in $CO_2$

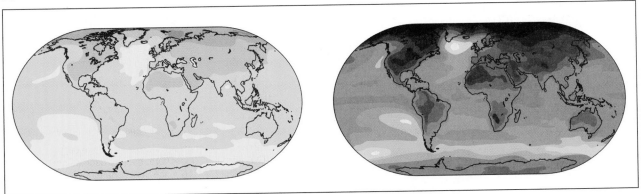

**(c)** High increase in $CO_2$

**Surface temperature change (°C)**

0    1    2    3    4    5    6    7    8

FIGURE 19.17 **Predicted increase in global temperatures by 2100.** The predictions depend on whether we expect (a) low, (b) moderate, or (c) high increases in how much $CO_2$ the world emits during the current century. [After http://www.epa.gov/climatechange/science/futuretc.html.]

generally agree that average global temperatures will rise by 1.8°C to 4°C (3.2°F–7.2°F) by the year 2100, depending on whether $CO_2$ emissions experience slow, moderate, or high growth over time.

# Feedbacks can increase or decrease the impact of climate change

The global greenhouse system is made up of several interconnected subsystems with many potential positive and negative feedbacks. As we saw in Chapter 2 with population systems, positive feedbacks amplify changes. Because of this, positive feedback often leads to an unstable situation in which small fluctuations in inputs lead to large observed effects. On the other hand, negative feedbacks dampen changes. When we think about how anthropogenic greenhouse gases will affect Earth, we must ask whether positive or negative feedbacks will predominate. We do not currently have enough evidence to settle this question conclusively, but we can examine some of the feedback cycles and the way they influence temperatures on Earth.

## Positive Feedbacks

There are many ways that a rise in temperatures could create a positive feedback. For example, global soils contain more than twice as much carbon as the amount currently in the atmosphere. As shown in **FIGURE 19.18a**, higher temperatures are expected to increase the biological activity of decomposers in these soils. This decomposition leads to the release of additional $CO_2$ from the soil to the atmosphere. With more $CO_2$ in the atmosphere, the temperature change will be amplified even more.

A similar, but more troubling, scenario is expected in tundra biomes containing permafrost. As atmospheric concentrations of $CO_2$ from anthropogenic sources increase, the Arctic regions become substantially warmer and the frozen tundra begins to thaw. As it thaws, the tundra develops areas of standing water with little oxygen available under the water as the thick organic layers of the tundra begin to decompose. As a result, the organic material experiences anaerobic decomposition. This decomposition produces methane,

**(a)** Positive feedback system

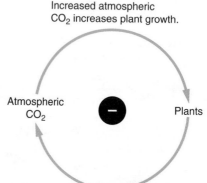

**(b)** Negative feedback system

**FIGURE 19.18 Global change feedback systems.** (a) Temperature and $CO_2$ represent a positive feedback system. When the concentration of $CO_2$ increases in the atmosphere, it can cause global temperatures to increase. This in turn can cause more rapid decomposition, thereby releasing even more $CO_2$ into the atmosphere. (b) Carbon dioxide and producers represent a negative feedback system. Increased $CO_2$ in the atmosphere from anthropogenic sources can be partially removed by increased photosynthesis by producers.

a stronger greenhouse gas than $CO_2$. This should lead to even more global warming.

## Negative Feedbacks

One of the most important negative feedbacks occurs as plants respond to increases in atmospheric carbon. FIGURE 19.18b shows this cycle. Because carbon dioxide is required for photosynthesis, an increase in $CO_2$ can stimulate plant growth. More plants will cause more $CO_2$ to be removed from the atmosphere. This negative feedback causes carbon dioxide and temperature increases to be smaller than they otherwise would have been. This negative feedback appears to be one of the reasons why only about half of the $CO_2$ emitted into the atmosphere by human activities has remained in the atmosphere.

## The Limitation of Feedbacks

Most of the feedbacks we have discussed are limited by features of the systems in which they take place. For example, the soil-carbon feedback is limited by the amount of carbon in soils. While warming soils could add large amounts of $CO_2$ to the atmosphere for a time, eventually soil stocks will become so low that biological activity falls back to earlier rates. The enhanced $CO_2$ uptake by plants is also limited: studies indicate that only some plants benefit from $CO_2$ fertilization, and often the growth is enhanced only until another factor becomes limiting, such as water or nutrients.

The magnitude and direction of many feedbacks are complex. For example, water vapor has both positive and negative feedbacks and there are limits to each. As temperatures increase, water can evaporate into the atmosphere more easily. Because water vapor is an important greenhouse gas, this increased evaporation will lead to further warming. There is a limit, however, to the amount of water vapor that can exist in the air. As we discussed in Chapter 4, air can become saturated with water vapor and the amount of saturation changes with temperature. Increased water vapor in the atmosphere can lead to the formation of clouds and clouds can shield the surface of Earth from solar radiation, leading to a negative feedback. In short, the net effect of water vapor on global temperatures depends on several simultaneously occurring processes that make predictions difficult.

---

### GAUGE YOUR PROGRESS

✓ What are positive and negative feedbacks?

✓ What is an example of a positive and a negative feedback in climate change?

✓ What is the benefit of identifying positive and negative feedbacks?

---

# Global warming has serious consequences for the environment and organisms

A wide range of environmental indicators demonstrates that global warming is affecting global processes and contributing to overall global change. In many cases, we already have clear evidence of how global warming is having an effect. In other cases, we can use climate models to make predictions about future changes. As with all predictions of the future, there is a fair degree of uncertainty regarding the future effects of global warming. In this section, we will discuss how global warming is expected to affect the environment and organisms living on Earth.

## Consequences to the Environment

Warming temperatures are expected to have a wide range of impacts on the environment. Many of these effects are already happening, including melting of polar ice caps, glaciers, and permafrost and rising sea levels. Other effects are predicted to occur in the future, including an increased frequency of heat waves, reduced cold spells, altered precipitation patterns and storm intensity, and shifting ocean currents, but it is less clear whether they will actually occur.

**POLAR ICE CAPS** As we have seen, the Arctic has already warmed by 1°C to 4°C (1.8°F–7.2°F). FIGURE 19.19 illustrates the extent of the reduction in the size of the ice cap that surrounds the North Pole. Over the next 70 years, the Arctic is predicted to warm by an additional 4°C to 7°C (7°F–13°F) compared to the temperatures experienced from 1980 to 1990. If this prediction is accurate, large openings in sea ice will continue to expand and the ecosystem of the Arctic region will be negatively affected. At the same time, though, there may also be benefits to humans. For example, the opening in the polar ice cap, could create new shipping lanes that would reduce by thousands of kilometers the distance some ships have to travel. Also, it is estimated that nearly one-fourth of all undiscovered oil and natural gas lies under the polar ice cap, and a melted polar ice cap might make these fossil fuels more easily obtainable. However, the combustion of these fossil fuels would further facilitate global warming, representing another example of a positive feedback.

In addition to the polar ice cap, Greenland and Antarctica have experienced melting. From 2000 to 2008, Greenland lost approximately 1,360 trillion kg (3,000 trillion pounds) of ice. From 2002 to 2009, Antarctica lost approximately 1,320 trillion kg (2,900 trillion pounds) of ice. Current evidence shows that the melting rate of these ice-covered regions is continuing to

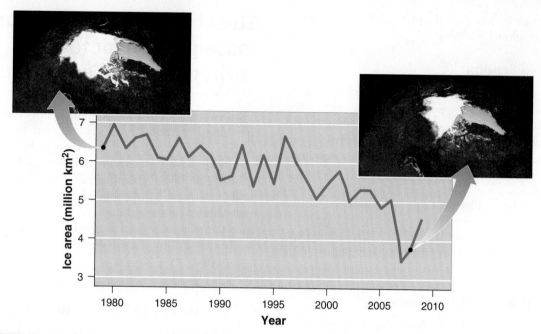

**FIGURE 19.19** **The melting polar ice cap.** Because northern latitudes have experienced the greatest amount of global warming, the ice cap near the North Pole is currently up to 39 percent smaller during the summer than it was from 1979 to 2000. [After http://svs.gsfc.nasa.gov/vis/a000000/a003500/a003563/index.html.]

increase. As we will see, such large amounts of melted ice cause sea levels to rise.

**GLACIERS** As we discussed in Chapter 9, global warming has caused the melting of many glaciers around the world. Glacier National Park in northwest Montana, for example, had 150 individual glaciers in 1850 but has only 25 glaciers today. It is estimated that by 2030 the park will no longer have any glaciers. The loss of glaciers is not simply a loss of an aesthetic natural wonder. In many parts of the world the spring melting of glaciers provides a critical source of water for many communities. Historically, these glaciers would partially melt during the spring and summer and then grow back to their full size during the winter. However, as summers become warmer, glaciers are melting faster than they can grow back in the winter, leaving some people without a reliable water supply.

**PERMAFROST** With warmer temperatures causing ice caps and glaciers to melt, it is perhaps not surprising that areas of permafrost are also melting. You may recall from the discussion on biomes in Chapter 4 that permafrost is permanently frozen ground that exists in the cold regions of high altitudes and high latitudes, which include the tundra and boreal forest biomes. About 20 percent of land on Earth contains permafrost and in some places it can be as much as 1,600 m (1 mile) thick. Melting of the permafrost causes overlying lakes to become smaller as the lake water drains deeper down into the ground. Melting can also cause substantial

problems with human-built structures that are anchored into the permafrost, including houses and oil pipelines. As the frozen ground melts, the ground can subside and slide away.

Melting permafrost also means that the massive amounts of organic matter contained in the tundra will begin to decompose. Because this decomposition would occur in wet soils under low-oxygen conditions, it would release substantial amounts of methane, increasing the concentration of this potent greenhouse gas. This chain of events could produce a positive feedback in which the warming of Earth melts the permafrost, releasing more methane and causing further global warming.

**SEA LEVELS** The rise in global temperatures affects sea levels in two ways. First, the water from melting glaciers and ice sheets on land adds to the total volume of ocean water. Second, as the water of the oceans becomes warmer, it expands. **FIGURE 19.20** shows that, as a result of both effects, sea levels have risen 220 mm (9 inches) since 1870. Scientists predict that by the end of the twenty-first century, sea levels could rise an additional 180 to 590 mm (7–23 inches) above 1999 levels. This could endanger coastal cities and low-lying island nations by making them more vulnerable to flooding, especially during storms, with more saltwater intrusion into aquifers and increased soil erosion. Currently, 100 million people live within 1 m (3 feet) of sea level. The actual impact on these areas of the world will depend on the steps taken to mitigate these effects. For example, as we saw in Chapter 9, some countries may

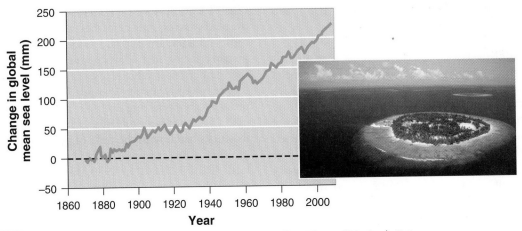

**FIGURE 19.20** **Rising sea levels.** Since 1870, sea levels have risen by 220 mm (9 inches). Future sea level increases are predicted to be 180 to 590 mm (7–23 inches) above 1999 levels by the end of this century. Nearly 100 million people live within 1 m (3 feet) of sea level, such as on this island in the Maldives in the Indian Ocean. [After http://www.cmar.csiro.au/sealevel/index.html.]

be able to build up their shorelines with dikes to prevent inundation from rising sea levels. Countries with less wealth are not expected to be able to respond as effectively to global changes such as coastal flooding.

**HEAT WAVES** As temperatures increase, long periods of hot weather—known as heat waves—are likely to become more frequent. Heat waves cause an increased energy demand for cooling the areas where people live and work. For people who lack air conditioning in their homes, especially the elderly, heat waves increase the risk of death. Heat waves also cause heat and drought damage to crops, prompting the need for greater amounts of irrigation. The increased energy required for irrigation would raise the cost of food production.

**COLD SPELLS** With global temperatures rising, minimum temperatures are expected to increase over most land areas, with fewer extremely cold days and fewer days below freezing. Such conditions would have two major positive effects for humans: fewer deaths due to cold temperatures and a decrease in the risk of crop damage from freezing temperatures. It may also make new areas available for agriculture that are currently too cold to grow crops. In addition, warmer temperatures would decrease the energy needed to heat buildings in the winter. However, a decrease in freezing temperatures that normally would cause the death of some pest species might allow these pest species to expand their range. The hemlock woolly adelgid (*Adelges tsugae*), for example, is an invasive insect from Asia that is causing the death of hemlock trees in North America by feeding on the trees' sap. Researchers have found that the range of the species is limited by cold temperatures. Warmer conditions in future decades are expected to allow this pest to expand its range and kill hemlock trees over a much larger area.

**PRECIPITATION PATTERNS** Because warmer temperatures should drive increased evaporation from the surface of Earth as part of the hydrologic cycle (see Figure 3.10 on page 67), global warming is projected to alter precipitation patterns. As mentioned earlier, climate models have a difficult time making solid predictions about precipitation. Current models predict that some areas will experience increased rainfall, but the models differ in predicting which regions of the world would be affected. Regions receiving increased precipitation would benefit from an increased recharge to aquifers and higher crop yields, but they could also experience more flooding, landslides, and soil erosion. In contrast, other regions of the world are predicted to receive less precipitation. In these regions, it will be more difficult to grow crops and will require greater efforts to supply water.

**STORM INTENSITY** Although it is impossible to link any particular weather event to climate change because of the multiple factors that are always involved, ocean warming does appear to have increased the intensity of Atlantic storms. Hurricanes Katrina and Rita, which in 2005 devastated coastal areas in Texas, Louisiana, and Mississippi, appear to have become as powerful as they did because waters in the Gulf of Mexico were unusually warm, with scientists at the National Center for Atmospheric Research concluding that climate change was responsible for at least half of this warming. As temperatures increase, such conditions should become more frequent, and hurricanes are likely to become more common farther north. The devastation in New Orleans did not come as a surprise; for many years scientists had warned that a strong hurricane could flood the city because of its position below sea level. Unfortunately, scientists caution that New Orleans is not the only American city that could be devastated by a powerful storm. Other cities, including New York, Miami, and Tampa, are also at risk.

**OCEAN CURRENTS** Global ocean currents may shift as a result of more fresh water being released from melting ice. In Chapter 4 we saw that ocean currents have major effects on the climate of nearby continents. If the currents change, the distribution of heat on the planet could be disrupted. Scientists are particularly concerned about the thermohaline circulation, which, as we saw in Chapter 4, is a deep-ocean circulation driven by water that comes out of the Gulf of Mexico and moves up to Greenland, where it becomes colder and saltier and sinks to the ocean floor. This sinking water mixes with the deep waters of the ocean basin, resurfaces near the equator, and eventually makes its way back to the Gulf of Mexico (see Figure 4.12 on page 97). This circulating water moves the warm water from the Gulf of Mexico up toward Europe and moves cold water from the North Atlantic down to the equator. However, increased melting from Greenland and the northern polar ice cap could dilute the salty ocean water sufficiently to stop the water from sinking near Greenland and thereby shut off the thermohaline circulation. If this occurred, it would cause much of Europe to experience significantly colder temperatures.

## Consequences to Living Organisms

The warming of the planet is not only affecting polar ice caps and sea levels but is also affecting living organisms. These effects range from temperature-induced changes in the timing of plant flowering and animal behavior to the ability of plants and animals to disperse to more hospitable habitats. Global warming also affects humans in a wide variety of areas including habitation, health, and even tourism.

**WILD PLANTS AND ANIMALS** During the last decade, the IPCC reviewed approximately 2,500 scientific papers that reported the effects of warmer temperatures on plants and animals. The panel concluded that over the preceding 40 years, the growing season for plants had lengthened by 4 to 16 days in the Northern Hemisphere, with the greatest increases occurring in higher latitudes. Indeed, scientists are finding that many species of plants now flower earlier, birds arrive at their breeding grounds earlier, and insects emerge earlier in the Northern Hemisphere. At the same time, the ranges occupied by different species of plants, birds, insects, and fish have shifted toward both poles.

Rapid temperature changes have the potential to cause harm if organisms do not have the option of moving to more hospitable climates and do not have sufficient time to evolve adaptations. Historically, organisms have migrated in response to climatic changes. This ability to migrate is one reason that temperature shifts have not been catastrophic over the past few million years. Today, however, fragmentation of certain habitats by roads, farms, and cities has made movement much more difficult. In fact, this fragmentation may be the primary factor that allows a warming climate to cause the extinction of species.

The pied flycatcher (*Ficedula hypoleuca*) is a bird that provides an interesting example. In the Netherlands, the pied flycatcher has evolved to synchronize the time that its chicks hatch with the time of peak abundance of caterpillars, a major source of food for the newly hatched birds (**FIGURE 19.21**). In 1980, the date of hatching preceded the peak in caterpillar abundance by a few days, so there was plenty of food for the new chicks. Twenty years later, warmer spring temperatures have caused trees to produce leaves earlier in the spring. Because the caterpillars feed on tree leaves, the peak abundance of caterpillars now occurs about 2 weeks earlier than it did in 1980. However, the hatching date of the pied flycatcher has not been affected by temperature change. Thus, by the time the chicks hatch, the caterpillars are no longer abundant and the hatchlings lack a major source of food. As a result, flycatcher populations in these areas have declined by 90 percent. There are other similar examples of the effects of rapid temperature change throughout the natural world.

Corals are one group of organisms that are particularly sensitive to global warming because their range of temperature tolerance is quite small. Many corals

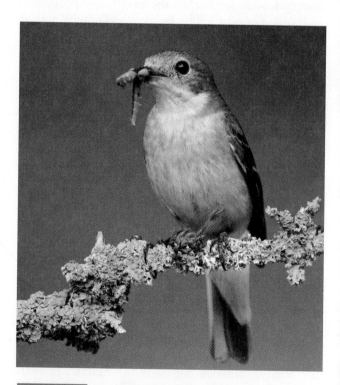

**FIGURE 19.21 Pied flycatcher.** Due to global warming in the Netherlands, the bird's main food source, a caterpillar, now becomes abundant two weeks earlier. However, the time when eggs hatch for the flycatcher has not changed. As a result, the birds hatch after the caterpillar population has begun to decline.

worldwide are currently undergoing "bleaching," which occurs when stressed corals eject the algae providing their energy and turn white. The underlying causes of coral bleaching appear to be a combination of warming oceans, pollution, and sedimentation. Bleaching is sometimes temporary, but if it lasts for more than a short time, the corals die. While new corals should colonize regions at higher latitudes, it will take centuries before major new reef systems can be built. More coral bleaching is expected from global warming even if the climate changes are kept relatively small.

**HUMANS** Global warming and climate change could also affect many aspects of our lives. For example, some people will have to relocate from such vulnerable areas as coastal communities and some ocean islands. Poorer communities close to or along coastlines might not have the resources to rebuild on higher ground. If these communities do not obtain financial assistance from others, they will face severe consequences from flooding and saltwater intrusion. On the other hand, certain areas that have not been suitable for human habitation might become more hospitable if they become warmer, although other factors, such as water availability, might still limit their habitability.

Climate change has the potential to affect human health. Continued warming of the planet could affect the geographic range of temperature-limited disease vectors. For example, the mosquitoes that carry West Nile virus and malaria could spread beyond their current geographic range and bring health threats to regions that were once relatively untouched. As the climate changes, heat waves could cause more deaths to the very young, the very old, and those without access to air conditioning. Infectious diseases and bacterial and fungal illnesses might extend over a wider range than they do at present.

Climate changes will also have economic consequences. In northern locations, for example, warmer temperatures and shorter winters may sound appealing at first, leading to, for example, reduced heating bills. However, they would drastically alter the character of northern communities that depend on snow for tourism. In the Alps, for example, many ski resorts are already adjusting to reduced snow on the mountains by catering to new groups of tourists who are more interested in spending time biking in the mountains than skiing (**FIGURE 19.22**). In warmer regions, the damage to corals reefs would negatively affect tourism as well. The economic impact on these types of tourist attractions depends on the rate of climate change and the ability of the tourism industry to adjust.

## The Controversy of Climate Change

How much controversy is there regarding climate change? This question is the subject of much discussion today, and it has been portrayed in many ways by various

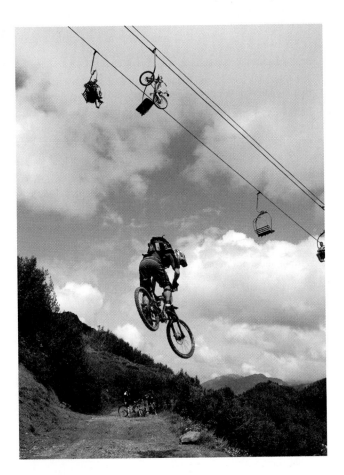

**FIGURE 19.22** **Adjusting tourism to climate change.** In the French Alps, global warming has caused many ski resorts to receive less snow in the winter. Some ski resorts, particularly those at lower altitudes, are responding by offering longer off-season activities such as mountain biking. They are also building new facilities, including year-round resorts and shopping centers, which can attract customers in the absence of snow.

interest groups. Advocates in the environmental community talk of a "scientific consensus" on the topic of global warming, while opponents of government regulation often speak of the global warming "controversy."

The fundamental basis of climate change—that greenhouse gas concentrations are increasing and that this will lead to global warming—is not in dispute among the vast majority of scientists. The first contention has been documented with real data and the second contention is a simple application of physics. Furthermore, the fact that the globe is warming is not in dispute because the data have clearly demonstrated this trend during the past 120 years. However, what is unclear is exactly by how much world temperatures will increase for a given change in greenhouse gases, because that depends upon the various feedbacks that we have discussed.

In its 2007 report, the IPCC attempted to address some of the uncertainty by listing the likelihood that various types of climate changes are already occurring, the likelihood that humans contributed to the changes,

## TABLE 19.2　The 2007 assessment of global change by the Intergovernmental Panel on Climate Change (IPCC)

The scientists considered the likelihood that specific changes have occurred, the likelihood that humans contributed to the change, and the likelihood that current trends will continue.

Definitions: More likely than not = more than 50% certain; Likely = more than 60% certain; Very likely = more than 90% certain; Virtually certain = more than 99% certain.

| Phenomenon and direction of trend | Likelihood that trend occurred in late 20th century (typically post-1960) | Likelihood of a human contribution to observed trend | Likelihood of future trends based on projections for 21st century from *Special Report on Emissions Scenarios* |
|---|---|---|---|
| Warmer and fewer cold days and nights over most land areas | Very likely | Likely | Virtually certain |
| Warmer and more frequent hot days and nights over most land areas | Very likely | Likely (nights) | Virtually certain |
| Warm spells/heat waves. Frequency increases over most land areas | Likely | More likely than not | Very likely |
| Heavy precipitation events. Frequency (or proportion of total rainfall from heavy falls) increases over most areas | Likely | More likely than not | Very likely |
| Area affected by droughts increases | Likely in many regions since 1970s | More likely than not | Likely |
| Intense tropical cyclone activity increases | Likely in some regions since 1970 | More likely than not | Likely |
| Increased incidence of extreme high sea level (excludes tsunamis) | Likely | More likely than not | Likely |

and the likelihood that these trends will continue through the twenty-first century. Their results are shown in Table 19.2. For example, the panel concluded that a decline in cold days and an increase in warm days very likely occurred in most land areas, that these changes likely had a human contribution, and that it was virtually certain that these trends would continue through the twenty-first century. They also concluded that heat waves, droughts, heavy precipitation, and hurricanes have likely increased since 1960, that these changes probably were influenced by human activities, and that these trends were likely to continue through the twenty-first century.

### GAUGE YOUR PROGRESS

✓ What is the evidence that global warming is affecting Earth?

✓ What changes are predicted to occur as temperature increases?

✓ How will climate change affect humans? What are some examples of direct and indirect effects?

# The Kyoto Protocol addresses climate change at the international level

Awareness of global change is a relatively recent phenomenon. In the past, most environmental issues could be dealt with at the national, state, or even local level. Global change is different because the scale of impact is so much larger and because the people and ecosystems affected can be extremely distant from the cause. In the case of climate change, many of the adverse effects are expected to be in the developing world, which has received disproportionately fewer benefits from the use of fossil fuels that led to the change in the first place. It would be impossible for one nation to pass legislation allowing it to avoid the impacts of climate change. To address the problem of global warming, the nations of the world must work together.

In 1997, representatives of the nations of the world convened in Kyoto, Japan, to discuss how best to control the emissions contributing to global warming. Under the agreement known as the **Kyoto Protocol,** global emissions of greenhouse gases from all industrialized countries will be reduced to 5.2 percent below their 1990 levels by 2012. Due to special circumstances and political pressures, countries agreed to different levels of emission restrictions, including a 7 percent reduction for the United States, an 8 percent reduction for the countries of the European Union, and a 0 percent reduction for Russia. Developing nations, including China and India, did not have emission limits imposed by the protocol. These nations argued that different restrictions on developed and developing countries are justified because developing countries are unfairly exposed to the consequences of global warming that in large part come from the developed nations. Indeed, the poorest countries in the world have only contributed to 1 percent of historic carbon emissions but are still affected by global warming. Thus, having the countries who emit the most $CO_2$ pay most of the costs of reducing $CO_2$ seemed appropriate.

The main argument for the Kyoto Protocol is grounded in the precautionary principle, which, as we saw in Chapter 17, states that in the face of scientific uncertainty we should behave cautiously. In the case of climate change, this means that since there is sufficient evidence to suggest human activities are altering the global climate, we should take measures to stabilize greenhouse gas concentrations either by reducing emissions or by removing the gases from the atmosphere. The first option includes trying to increase fuel efficiency or switching from coal and oil to energy sources that emit less or no $CO_2$ such as natural gas, solar energy, wind-powered energy, or nuclear energy. The second option includes **carbon sequestration**—an approach that involves taking $CO_2$ out of the atmosphere. Methods of carbon sequestration might include storing carbon in agricultural soils or retiring agricultural land and allowing it to become pasture or forest, either of which would return atmospheric carbon to longer-term storage in the form of plant biomass and soil carbon. Researchers are also working on cost-effective ways of capturing $CO_2$ from the air, from coal-burning power stations, and from other emission sources. This captured $CO_2$ is then compressed and pumped into abandoned oil wells or the deep ocean (**FIGURE 19.23**). Such technologies

**FIGURE 19.23** **Carbon sequestration.** To help reduce the amount of $CO_2$ that ends up in the atmosphere, efforts are being made to capture carbon at its source and either to convert it into plant biomass and soil carbon or to pump it down into the ground or the deep sea.

are still being developed, so their economic feasibility and potential environmental impacts are not yet known.

In developed countries, reductions in $CO_2$ emissions would require major changes to manufacturing, agriculture, or infrastructure at significant expense and economic impact. In 1997, before the Kyoto Protocol was finalized, the U.S. Senate voted unanimously (95−0) that the United States should not sign any international agreement that lacked restrictions on developing countries or any agreement that would harm the economy of the United States. Despite this vote, the vice president at the time, Al Gore, was instructed by President Bill Clinton (1993–2001) to sign the protocol on behalf of the United States. President Clinton, however, never sent the agreement to the Senate for ratification since it was clear the Senate would not vote in favor. Since it was never ratified, the protocol is not legally binding on the United States.

In 2001, the Kyoto Protocol was modified to convince more developed nations to ratify it. At that time, the United States, under the administration of President George W. Bush (2001–2009), argued that there was too much uncertainty in global warming predictions to justify ratification of the protocol. The administration also argued that the costs of controlling carbon dioxide emissions would unfairly disadvantage businesses in the United States while businesses in China and India—two developing countries that were not significant emitters of greenhouse gases prior to 1990 but are significant emitters today—have no reduction requirements. Ultimately, the United States argued that efforts to limit greenhouse gases should wait until there is more scientific evidence for global warming and its effects before accepting the costs that the protocol would entail. Furthermore, the United States argued that *all* countries should be subject to emissions limits. Otherwise polluting factories could simply relocate to the developing world, and little good would have been accomplished. Proponents of reducing greenhouse gas emissions argue that profits gained from manufacturing new pollution-control technologies and savings in fuel costs through greater efficiency will offset any costs or decrease in short-term profits.

More recently, the U.S. government has taken stronger steps to regulate $CO_2$ emissions. In 2007, the U.S. Supreme Court ruled that the U.S. Environmental Protection Agency not only had the authority to regulate greenhouse gases as part of the Clean Air Act,

but was required to do so. As a result, in 2009 the EPA announced it would begin regulating greenhouse gases for the first time. In 2010, under the administration of President Barack Obama, the EPA began to look more closely at possible ways to regulate emissions of carbon dioxide. A proposed increase in fuel efficiency requirements for automobiles, which received strong support from automobile manufacturers, would allow a 30 percent reduction in $CO_2$ and other greenhouse gases by 2016. The more fuel-efficient cars are expected to cost an extra $1,000, but the reduced consumption of gasoline is expected to save the average driver $3,000 over the lifetime of the vehicle. This would allow the United States to reduce greenhouse gases, invest in new automotive technology, reduce its consumption of fossil fuels, and save money.

As of 2010, 190 countries have ratified the Kyoto Protocol, including most developed and developing countries. The United States is the only developed country that has not yet ratified the agreement. However, changes in $CO_2$ emissions in various countries have been mixed. For example, the total emissions of greenhouse gases from 1990 to 2007 have increased by 119 percent for Australia, 57 percent for Canada, and 17 percent for the United States. In contrast, greenhouse gases have declined by 21 percent for Germany and 4 percent for Russia. Across all 41 developed countries, the reduction in greenhouse gases for this period was 5.2 percent.

In December 2009, the United Nations convened a gathering of its member nations in Copenhagen, Denmark, to further address climate change. Unfortunately, they could not agree to any legally binding $CO_2$ reductions. The meeting simply produced an agreement stating that climate change was a major challenge and that the nations of the world should try to prevent mean global temperature increases from exceeding 2°C.

## GAUGE YOUR PROGRESS

✓ What is the Kyoto Protocol?

✓ How is the Kyoto Protocol an example of the precautionary principle?

✓ How will the Kyoto Protocol affect developing countries? How will it affect developed countries?

Although the United States signed the original Kyoto Protocol, the U.S. Congress never ratified the agreement and the protocol has never been legally binding on the United States. The administration of President George W. Bush argued that there was no scientific consensus on global warming and that the costs of reducing greenhouse gases were simply too high. However, many state and local governments felt they had waited long enough for change at the federal level and in 2005, mayors from 141 cities and both major political parties gathered in San Francisco to organize their own efforts to reduce the causes and consequences of global warming. Their goal was to reduce greenhouse emissions in their own cities by the same 7 percent that the United States had agreed to in the Kyoto Protocol.

By 2009, more than 1,000 out of 1,139 city mayors had signed the U.S. Congress of Mayors Climate Protection Agreement. Among the reasons they cited for supporting this agreement were concerns in their communities over increasing droughts, reduced supplies of fresh water due to melting glaciers, and rising sea levels in coastal cities. "The United States inevitably will have to join this effort," Seattle Mayor Greg Nickels said. "Ultimately we will make it impossible for the federal government to say no. They will see that it can be done without huge economic disruption and that there's support throughout the country to do this."

In 2005, the governor of California, Arnold Schwarzenegger, stated at a press conference, "The debate is

## Local Governments and Businesses Lead the Way on Reducing Greenhouse Gases

over . . . and we know the time for action is now" (FIGURE 19.24). In 2006, Governor Schwarzenegger signed the California Global Warming Solutions Act of 2006. The goal of the act was to bring California into compliance with the Kyoto Protocol by 2020, an effort that would require a 25 percent reduction in greenhouse gases for a state that, if a country, would be the tenth largest producer of greenhouse gases in the world. At the signing ceremony, the governor stated, "I say unquestionably it is good for businesses. Not only large, well-established businesses, but small businesses that will harness their entrepreneurial spirit to help us achieve our climate goals." Indeed, a cost analysis by the California Air Resources Board in 2008 indicated that the law would add $27 billion to the economy of the state and add 100,000 jobs.

The California effort is gaining popularity around the country. In the Northeast, for example, 11 states and the District of Columbia have joined together collectively to control regional production of greenhouse gases. A similar group has emerged in the western region of the country in which seven western states and four Canadian provinces have joined together. These states and provinces understand the causes and consequences of climate change and hope to find solutions at the local level rather than waiting for legislation at the federal level.

A number of large businesses are also joining in efforts to reduce greenhouse gases. General Electric, for example, announced in 2005 that it would voluntarily reduce

**FIGURE 19.24** **Leading the way.** California governor Arnold Schwarzenegger announces his plan for California to lead the country in reducing the greenhouse gases that cause global warming.

its emissions. In addition, the company has invested billions of dollars in research and development designing technologies that can reduce greenhouse gases. In 2009, General Electric announced that its revenues from this technology were $20 billion in 2009 and were expected to be $25 billion in 2010. Similarly, the retailer Wal-Mart announced that it would invest in more energy-efficient technologies in its stores and vehicles. In 2010, Wal-Mart further announced that it would push its 100,000 suppliers to reduce their carbon emissions by 20 million metric tons (22 million U.S. tons) by 2015. This is the equivalent of taking 3.8 million cars off the road for one year.

From these stories, it is clear that progress on reducing greenhouse gases that cause global warming does not have to wait for national and international agreements. The public overwhelmingly understands that the globe is warming, states and cities are pushing forward with solutions that save money, and large corporations understand that reducing emissions can reduce costs and improve profits over the long term. In short, curbing greenhouse gases and global warming is not only good for humans and the environment, it is often good for business as well.

### References

Peer, M. 2009. GE's green goals. *Forbes,* June 11. http://www.forbes.com/2009/06/11/general-electric-energy-markets-equities-alternative.html.

Sappenfield, M. 2005. American mayors target global-warming. *USA Today,* June 5. http://www.usatoday.com/news/nation/2005-06-05-mayors-kyoto-csm_x.htm?csp=34.

U.S. mayors abide by Kyoto treaty. *News Hour with Jim Lehrer,* PBS, August 8, 2005. http://www.pbs.org/newshour/extra/features/july-dec05/kyoto_8-08.html.

## REVISIT THE KEY IDEAS

- **Distinguish among global change, global climate change, and global warming.**

Global change includes changes in the chemistry, biology, and physical properties of the planet. One type of global change is climate change, which refers to changes in the climate of Earth. One aspect of global climate change is global warming, which refers to the warming of oceans, landmasses, and the atmosphere of Earth.

- **Explain how solar radiation and greenhouse gases warm our planet.**

Solar energy is largely composed of radiation in the ultraviolet and visible range. Nearly half of this radiation is reflected from the atmosphere. Of the remaining half, much of the ultraviolet radiation is absorbed by the ozone layer and the remaining solar radiation penetrates the atmosphere to strike the land and water. On land and in the water, the radiation is converted to infrared radiation and is emitted back toward the atmosphere. Greenhouse gases in the atmosphere absorb the infrared radiation and emit some of it back toward Earth. This absorption and emission of infrared radiation by the greenhouse gases cause the planet to warm. Natural sources of greenhouse gases include volcanoes, decomposition and digestion under low-oxygen conditions, denitrification under low-oxygen conditions, and the evaporation and evapotranspiration of water vapor. Anthropogenic sources include the burning of fossil fuels, certain agricultural practices, deforestation, landfills, and industries.

- **Discuss how $CO_2$ concentrations and temperatures have changed over time.**

Measurements of $CO_2$ during the past 50 years have demonstrated that $CO_2$ in the atmosphere has increased by nearly 30 percent. The major source of $CO_2$ emissions is from developed countries including the United States and the European Union and from rapidly developing countries including China and India. Indirect measurements indicate that $CO_2$ and temperatures have fluctuated widely over hundreds of thousands of years and changes in $CO_2$ are closely associated with changes in temperature. During the past 120 years, temperatures have risen an average of 0.8°C (1.4°F) with northern latitudes experiencing increases of 1°C to 4°C (1.8°F–7.2°F).

- **Describe the importance of feedback loops in the process of global warming.**

Predicting global warming and global climate change is complex because of positive and negative feedback loops. Positive feedback loops include warmer soils increasing their rates of decomposition and thereby increasing their emission of $CO_2$. Negative feedback loops include plants being able to increase their growth under elevated $CO_2$ environments, thereby reducing some of the $CO_2$ in the atmosphere.

- **Identify how global warming is affecting people and the environment.**

Many consequences of global warming are already occurring while others are predicted to occur in the future. Current effects of global warming include melting ice caps, glaciers, and permafrost, rising sea levels, and changes that impact the lives of many species in nature. Future impacts include more frequent heat waves, rarer cold spells, changes in precipitation patterns, disrupted ocean currents, and effects on human health and economics.

## CHECK YOUR UNDERSTANDING

1. Which of the following activities causes a cooling of Earth?
   (a) Volcanic eruptions
   (b) Emissions of anthropogenic greenhouse gases
   (c) Evaporation of water vapor
   (d) Combustion of fossil fuels
   (e) Deforestation

2. In regard to the greenhouse effect, which statement is *not* true?
   (a) Ultraviolet and visible radiation are converted to infrared radiation at the surface of Earth.
   (b) Approximately one-third of the radiation of the Sun does not enter the atmosphere of Earth.
   (c) Infrared radiation is absorbed by greenhouse gases.
   (d) Greenhouse gases were not historically present in the atmosphere.
   (e) Ultraviolet radiation is absorbed by ozone.

3. Which of the following is *not* a greenhouse gas?
   (a) Carbon dioxide
   (b) Water vapor
   (c) Methane
   (d) Nitrous oxide
   (e) Nitrogen

4. Of the following factors, which ones are important when considering the effect of a greenhouse gas on global warming?
   I   How much infrared radiation the gas can absorb
   II  How long the gas remains in the atmosphere
   III The concentration of the gas in the atmosphere
   (a) I
   (b) I and II
   (c) I and III
   (d) II and III
   (e) I, II, and III

5. Which greenhouse gas is *not* correctly paired with one of its sources?
   (a) Nitrous oxide : landfills
   (b) Methane : termites
   (c) Water vapor : evaporation
   (d) Nitrous oxide : automobiles
   (e) $CO_2$ : deforestation

6. Which statement about global warming is *true?*
   (a) The planet is not warming.
   (b) The planet is warming, but humans have not played a role.
   (c) The planet has had many periods of warming and cooling in the past.
   (d) Greenhouse gases compose only a small fraction of the atmosphere, so they cannot be important in causing global warming.
   (e) Such small increases in average global temperatures could not cause any important effects on polar bears.

7. Which sources of data have been used to assess changes in global $CO_2$ and temperature?
   I   Air bubbles in ice cores from glaciers
   II  Thermometers placed around the globe
   III $CO_2$ sensors placed around the globe
   (a) I
   (b) I and II
   (c) I and III
   (d) II and III
   (e) I, II, and III

8. Which statement about feedback loops that occur with climate change is *true?*
   (a) All feedback loops are positive.
   (b) All feedback loops are negative.
   (c) Increased soil decomposition under warmer temperatures represents a positive feedback loop.
   (d) Increased evaporation under warmer temperatures represents a negative feedback loop.
   (e) Increased plant growth under higher $CO_2$ concentrations represents a positive feedback loop.

9. Which predicted consequence of global warming has *not* yet occurred?
   (a) Melting ice caps
   (b) Rising sea levels
   (c) Melting permafrost
   (d) Shutting down the thermohaline circulation of the ocean
   (e) Altered breeding times and flowering times of animals and plants

10. Which statement regarding the Kyoto Protocol is *true?*
    (a) Developed and developing nations all agreed to reduce their emission of greenhouse gases.
    (b) All nations agreed to stop their emission of greenhouse gases.
    (c) The developed nations agreed to different levels of emission reductions.
    (d) Developing nations agreed to reduce their emission of greenhouse gases.
    (e) Developing nations agreed to stop their emission of greenhouse gases.

## APPLY THE CONCEPTS

1. During a debate on climate change legislation in 2009, a U.S. congressman declared that human-induced global warming is a "hoax" and that "there is no scientific consensus."
   (a) If you were a member of Congress, what points might you raise in the debate to demonstrate that global warming is real?
   (b) What points might you raise to demonstrate that global warming has been influenced by humans?
   (c) What are some human health and economic effects that could occur because of global warming?

2. Given what you have learned about global warming and global climate change,
   (a) what actions might you propose in the United States to reduce $CO_2$ emissions?
   (b) what actions might you propose in the United States to reduce methane emissions?
   (c) what actions might you propose in the United States to reduce nitrous oxide emissions?
   (d) what evidence have scientists used to support that global warming is happening?

## MEASURE YOUR IMPACT

### Carbon Produced by Different Modes of Travel

Approximately 6.3 billion metric tons of carbon—$6.3 \times 10^{12}$ kg—are emitted worldwide each year. Approximately 20 percent comes from the United States. In this exercise, you can calculate the carbon emission you contribute to this total through your use of transportation in a typical year. Next, you will calculate what the world's carbon emissions from transportation would be if everyone's carbon emissions were the same as yours.

1. Estimate your personal transportation carbon emissions using the information provided below. Don't worry that your numbers will not be exact.

   **Personal transportation per year by automobile**

   (a) Miles traveled per year: _____
   (b) Miles per gallon: _____
   (c) Number of gallons used per year: _____
   (d) Convert to liters of gasoline used per year (1 gallon = 3.875 liters): _____
   (e) Amount of carbon produced (1,577 liters of gasoline produces one metric ton of carbon): _____

   **Public transportation per year by bus**

   (a) Miles traveled per year: _____
   (b) If an average bus travels 4 miles per gallon of diesel fuel, number of gallons used per year: _____
   (c) If an average bus carries 40 people, number of gallons used by *you* each year: _____
   (d) Convert to liters of gasoline used per year (1 gallon = 3.875 liters): _____
   (e) Amount of carbon produced (1,382 liters of diesel fuel produces one metric ton of carbon): _____

   **Public transportation per year by airplane**

   (a) Miles traveled: _____
   (b) Metric tons of carbon released (1 mile traveled by air releases 0.0002 metric tons of carbon): _____

2. Determine the total metric tons of carbon you emitted by summing all the individual values:

Total annual metric tons of carbon you released = metric tons from personal transportation + metric tons from bus travel + metric tons from air travel: _____

3. Multiply your personal annual carbon release from transportation by 310 million to see what the U.S. release from transportation would be if everyone had a carbon emission similar to yours.

4. Multiply your personal annual carbon release by 6.9 billion to see what the global release would be if everyone in the world consumed carbon at the same rate you do.

5. Discuss briefly if this is a good estimate of global carbon emissions. Why or why not? Consider these questions as you compose your answer: Is transportation the only source of carbon emissions (you can find the fraction of total emissions from transportation on the Web)? Would your personal estimate be different if you lived in a different country?

6. If your main means of transportation is a car, what would happen to your carbon emissions if you were to increase your reliance on public transportation?

## ENGAGE YOUR ENVIRONMENT

How do your greenhouse gas emissions compare to those of the average person in the United States? Explore how your daily activities create carbon dioxide emissions, and think of ways to decrease your carbon dioxide contributions. *See Engage Your Environment Activity 19 at the end of this book.*

# Working Toward Sustainability: Environmental Economics, Equity, and Policy

## Assembly Plants, Free Trade, and Sustainable Systems

Although citizens of Ciudad Juárez, Mexico, call the border with the United States *la línea,* or "the line," this nearly 3,200 km (2,000 mile) stretch is more a network of passageways than a division. Trade among people and cultures across the national boundary clearly affects both countries. The link between the Mexican and American economies, strengthened by globalization and increased free trade, is exemplified by the *maquiladora,* or assembly plant, industry.

Environmental regulations are often lenient or nonexistent, and the majority of companies do not comply with mandates that maquiladora waste be shipped to the company's home country. Disposal of toxic chemicals and heavy metals into the local environment causes groundwater and surface water pollution and significant harm to human health. Many maquiladora employees are women of reproductive age, a population that is particularly vulnerable to exposure to toxic chemicals.

> If we could give equal attention to economic profit, environmental integrity, and human welfare, could we ultimately create more sustainable development?

Established in the 1960s, this industry allows international companies to import materials and equipment tariff-free to Mexican maquiladoras, and then to export the finished product to markets in other countries. In 1994, the United States, Canada, and Mexico passed the North American Free Trade Agreement (NAFTA) that was intended to increase trade among the three countries by reducing tariffs and other taxes, and regulations. After NAFTA, the use of maquiladoras increased significantly; export of assembled products tripled between 1995 and 2000. Maquiladoras, which export 90 percent of their products to the United States, currently constitute 80 percent of the economy in the northern border region and a quarter of Mexico's total GDP. And while the jobs have been welcomed in these economically depressed areas, there have been many negative consequences as well, including industrial pollution, poor working conditions, and discrimination. In addition, the maquiladoras raise questions of social justice because much of the profit is sent to other countries.

In terms of the environment, maquiladora operations often contaminate the border region with toxic industrial waste.

In addition to pollution from the manufacturing processes, an increase in the human population in maquiladora areas has added greatly to other environmental problems. Many municipalities do not have sewage treatment facilities or trash collection capabilities. The solid waste pollutes water sources, and seasonal floods spread garbage throughout the cities.

Social abuses also occur in this system. Women are often tested for pregnancy before hiring, and those who become pregnant may be illegally fired. Managers employ underage workers. Factory conditions are hazardous, and employees are often unaware of risks because of the lack of "right to know" laws and an absence of warning labels in Spanish. Companies exploit the poverty of the region by offering wages that barely ▶

Rusted barrels and other waste disposed of near a residential community.

support employee needs. An average maquiladora worker earns the equivalent of a few dollars per day, and these wages have remained stagnant for years even as living costs have risen.

Sometimes the profits from these factories do not enter the Mexican economy, but rather go to the home countries of the companies that run the plants. Many observers believe that northern Mexico pays the social and environmental prices for the maquiladora industry, while foreign corporations reap the benefits.

Free trade and globalization agreements like NAFTA are designed to enhance developing economies by facilitating international business. However, in northern Mexico, increased free trade has stimulated an industry that in some cases may sacrifice social *well-being* and environmental health. Nevertheless, there are many people who are employed in the maquiladora industry and therefore money enters the local economy and helps individuals. Environmental scientists interested in human social welfare and the well-being of the environment look at situations such as these and ask: If we could give equal attention to economic profit, environmental integrity, and human welfare, could we ultimately create more sustainable development? ■

*Sources:* J. Carrillo and R. Zarate, The evolution of maquiladora best practices: 1965–2008, *Journal of Business Ethics* (2009) 88: 335–348; J. G. Samstad and S. Pipkin, Bringing the firm back in: Local decision making and human capital development in Mexico's maquiladora sector, *World Development* (2005) 33: 805–822.

## UNDERSTAND THE KEY IDEAS

Throughout this book we have seen that economic development, social justice, and sustainable environmental practices are often in conflict. In recent years, environmental science has begun to apply the tools from economics and a few other fields to help find ways in which we can achieve a more equitable existence for all inhabitants on Earth.

After reading this chapter you should be able to

■ discuss sustainability in a variety of environmental contexts including human well-being.

■ evaluate ways in which the use of economic analysis can do a better job of including the costs of economic activities on the environment and on people.

■ understand that economic systems are based on three forms of capital—natural, human, and manufactured.

■ explain the role of laws and regulations in attempting to protect our natural and human capital.

■ define and discuss the relationship among sustainability, poverty, personal action, and stewardship.

# Sustainability is the ultimate goal of sound environmental science and policy

*Sustainability* is a relatively new and evolving concept in contemporary environmental science. We have used it in a variety of contexts throughout this text. In the most comprehensive definition of the term, we say that something is sustainable when it meets the needs of the present generation without compromising the ability of future generations to meet their own needs. Although human needs can be defined in various ways, for our purposes we identify the basic necessities as access to food, water, shelter, education, and a healthy, disease-free existence. In order for these five necessities to be available, there must be functioning environmental systems that provide us with breathable air, drinkable water, and productive land for growing food, fiber, and other raw materials: the ecosystem services that we have described in this book.

As we saw in Chapter 7, the quest to obtain resources and increase **well-being**—the status of being healthy, happy, and prosperous—has caused individuals and nations to exploit and degrade natural resources such as air, land, water, wildlife, minerals, and even entire ecosystems. To address questions of sustainability, we need to be able to understand where human well-being and the condition of environmental systems are in conflict. To do this we will consider economic theory, ecological economics and ecosystem services, and the role of regulatory agencies in bringing about environmental regulation and protection.

### GAUGE YOUR PROGRESS

✓ What is sustainability?

✓ What are some of the variables associated with well-being? Which can be measured directly? Which are harder to measure?

# Economics studies how scarce resources are allocated

In an attempt to reduce environmental harm, researchers and policy makers have experimented with a variety of techniques to encourage consumers to change their behavior in ways that would be beneficial to the environment. We explored some of these techniques in Chapter 10 where we discussed externalities and in Chapters 15 and 19 where we discussed the buying and selling of air pollution allowances, and charging a fee or tax for the use of certain resources or for the emission of certain pollutants. **Economics** examines how humans either as individuals or as companies allocate scarce resources in the production, distribution, and consumption of goods and services.

Throughout this text we have already applied many concepts from the field of economics. When we looked at the problem of externalities and pollution, we were using economic theory. Life-cycle analysis is very similar to the cost-benefit analysis that economic policy makers use. In this section we will look at some basic economic concepts and learn how they can be applied to environmental issues.

## Supply, Demand, and the Market

In today's world, most economies are *market* economies. In the simplest sense, a market occurs wherever people engage in trade. In a market economy, the cost of a good is determined by supply and demand. When a good is in great demand and wanted by many people, producers are typically unable to provide an unlimited supply. Price is the way that producers and consumers communicate the value of an item and allocate the scarce item.

The graph shown in **FIGURE 20.1** illustrates the relationship between supply, demand, and prices. The supply curve (S) shows how many units suppliers of a given product or service, for example, T-shirts, are willing to provide at a particular price. Factors that influence supply include input prices (the cost of the resources used to produce the item), technology, expectations about future prices, and the number of people selling the product. For example, if you are the only person selling T-shirts and many people want them, you are likely to be willing to make the investment required to produce many T-shirts. However if a new T-shirt seller comes along, you might be concerned that you will not sell as many, so you will decrease your production because you now must share the market with other suppliers.

The demand curve (D) shows how much of a good consumers want to buy. Factors that influence demand include income, prices of related goods, tastes, expectations, and the number of people who also want the good. For example, if your boss gives you a raise, you may feel like you can afford that T-shirt you've been wanting to buy.

Notice that the demand curve is downward-sloping. In other words, as the price of T-shirts rises, the demand for them declines. This illustrates the law of demand, which states that when the price of a good rises, the quantity demanded falls and when the price falls, demand rises. Conversely, the supply curve is upward-sloping.

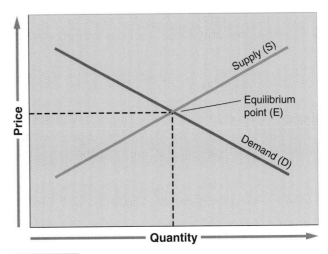

**FIGURE 20.1** **Supply and demand.** A manufacturer will supply a certain number of units of an item based on the revenue that will be received. A consumer will demand a certain number of units of that item based on the price paid. The intersection of the supply and demand curves determines the market equilibrium point for that item.

This reflects the law of supply, which states that when the price of a good rises, the quantity supplied of that good will rise and when the price of a good falls, the quantity of the good supplied will also fall.

The laws of supply and demand make intuitive sense. After all, if you are selling T-shirts and you find that your profits have shrunk, you are more likely to use your resources to produce something more popular, and more profitable. If you are a consumer of T-shirts, the less expensive they are, the more you are inclined to buy.

With these different interests, how do demand and supply ever meet? In a market system, without any restrictions such as taxes or other regulations, the price of a good will come to an equilibrium point (E) where the two curves on the graph intersect. Here the quantity demanded and the quantity supplied are exactly equal. At this price, suppliers find it worthwhile to supply exactly as many T-shirts as consumers are willing to buy.

Unfortunately, markets—composed of many buyers and sellers—do not always take all costs of production into account. We have already seen that this is the case in situations of land degradation where people, organizations, or even governments deplete or damage a natural resource because they do not bear any direct costs for doing so. What happens to supply and demand if we account for the costs of *externalities,* the costs or impact of a good or service on people and the environment not included in the economic price of that good or service? As we saw in Chapter 10, externalities include the costs for using common resources such as water, air, land, or the oceans and the costs of air and water pollution or solid waste products.

The dollar cost of coal-generated electricity includes the cost of the coal, the cost of paying people to operate the power plant, and the cost of distribution to

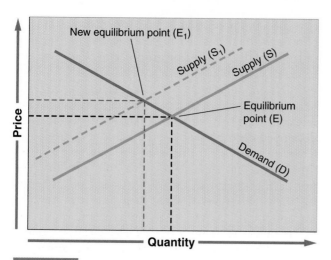

**FIGURE 20.2** **Supply and demand with externalities.**
When the cost of emitting pollutants is included in the price of a good, for any given quantity of items, the price increases. This causes the supply curve to shift to the left, from S to $S_1$. Since the law of demand states that when the price of a good goes up, demand falls, the amount demanded falls, and the market reaches a new equilibrium, $E_1$.

customers. However, the cost to the environment of emitting sulfur dioxide, carbon dioxide, and other waste products, all of which are negative externalities, is largely missing from the price customers pay; but these negative externalities certainly add costs, both financially and in terms of the well-being of people living downwind from the power plant. For example, someone with a respiratory ailment could incur greater medical expenses because of increased sulfur dioxide and particulates in the air. There may be provisions requiring polluters to pay some of the costs related to these emissions, but often these payments are not sufficient to cover the total cost of the pollution. In addition, they often do not reach the affected individuals or groups.

If the dollar cost of a good included externalities such as the expenses incurred by emitting pollutants into the air, or the expenses related to removing the pollutants before they were emitted, then the cost for most items produced would be greater. This could only occur if a tax were imposed by a regulatory agency. When the cost of production rises due to this tax, the supply curve shifts to the left, from S to $S_1$ as shown in **FIGURE 20.2**. The new market equilibrium $(E_1)$ is at a higher cost and, as a result, fewer items are manufactured and purchased. In other words, including the externalities raises the price and lowers demand. Therefore the price including externalities is more reflective of the true cost of the item.

## Wealth and Productivity

There are a variety of national economic measurements that gauge the economic wealth of a country in terms of its productivity and consumption. Most of them do not take externalities into account. The most common

is the gross domestic product (GDP), which refers to the value of all products and services produced in a year in a given country. GDP includes four types of spending: consumer spending, investments, government spending, and exports minus imports. As a measure of well-being, GDP has been criticized for a number of reasons. Costs for health care contribute to a higher GDP, so a society that has a great deal of illness would have a higher GDP than an equivalent society without a great deal of illness. This does not appear to be an accurate reflection of the "wealth" or "well-being" of a society. Because externalities such as pollution and land degradation are not included in GDP, measurement of GDP does not reflect the true cost of production.

Some social scientists maintain that the best way to improve the global environment is to increase the GDP in the less developed world. In Chapter 7 we examined the relationship between rising income and falling birth rates; as GDP increases, population growth slows. This, in turn, should lead to a reduction in anthropogenic environmental degradation. Wealthier, developed countries are able to purchase goods and services that will lead to environmental improvements—for example, pollution control devices such as catalytic converters—and to use their resources more efficiently. On the other hand, as we have seen, developed countries use many more resources than developing countries, which leads to more environmental degradation.

**THE GPI** We have seen that GDP is an incomplete measurement of the economic status of a country because it only considers production. The **genuine progress indicator (GPI)** attempts to address this shortcoming by including measures of personal consumption, income distribution, levels of higher education, resource depletion, pollution, and the health of the population. While GDP in the United States has been steadily rising for the last 60 years or more, GPI has been level since about 1970 (**FIGURE 20.3**). A number of countries, including England, Germany, and Sweden, have recalculated their GDP using the GPI. They have found that their overall wealth, when human and environmental welfare are included, has steadily declined over the last three decades.

## The Kuznets Curve

To address some of the shortcomings of GDP as a measurement of wealth, some environmental economists and scientists advocate using a model known as the Kuznets curve. The Kuznets curve, shown in **FIGURE 20.4**, suggests that as per capita income in a country increases, environmental degradation first increases and then decreases. The model is controversial because it is not easily applicable to all situations. For example, despite the increasing affluence of developed countries, carbon dioxide emissions and municipal solid waste (MSW) generation have both *continued* to increase. It is possible that these developed

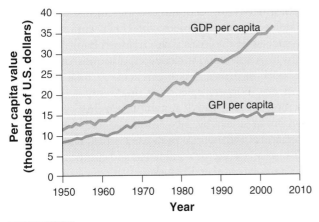

**FIGURE 20.3** Genuine progress indicator versus gross domestic product, per capita, for the United States from 1950 to 2004. While gross domestic product measures the value of all products and services a country produces, the genuine progress indicator attempts to include the level of education, personal consumption, income distribution, resource depletion, pollution, and the health of the population.

countries are not yet wealthy enough to deal with these problems effectively, but it is also possible that there are certain problems that cannot be solved with greater wealth. For example, as countries become wealthier, residents tend to use more fossil fuel for travel, to consume more resources, and to generate more waste.

Sometimes less developed countries experience technological leaps without going through each phase of technological development. These kinds of changes may influence the shape of the Kuznets curve or influence how well it characterizes a given situation. **Technology transfer** happens when less developed countries adopt technological innovations developed in wealthy countries. For example, in many less developed countries

a significant proportion of the population uses cell phones without ever having had access to a network of landlines. Such **leapfrogging** occurs whenever new technology develops in such a way that makes the older technology unnecessary or obsolete. In this way, the developing nations can take advantage of the expensive research, development, and experience of the more developed nations.

Solar energy is a particularly good example of leapfrogging. In industrialized nations, solar electricity has not been cost-competitive with gas- or coal-generated electricity. However, it has been very successful in nations in Africa, Asia, and South America that lack the resources to build a reliable electrical distribution grid. Solar energy is a small-scale energy source not dependent on outside connections to an electrical grid (see Chapter 13). In fact, it is possible that many less developed countries will continue to increase their use of solar energy and skip the step of building a nationwide electrical grid, much like what has happened with telephone service and cell phones versus landlines. Solar energy allows countries to produce and distribute their own electricity without investment in the massive infrastructure of an electrical distribution grid needed in a developed country (**FIGURE 20.5**).

**MICROLENDING** The transfer of technology can sometimes help developing nations avoid the environmental degradation that often comes with improving the productivity of a nation. Another important effort to improve the economic well-being of people while also trying to minimize adverse effects on the environment is *microlending*. **Microlending** is the practice of loaning small amounts of money to people who intend to start a small business in less developed countries. It is not a model of charity, but a financial approach that allows an

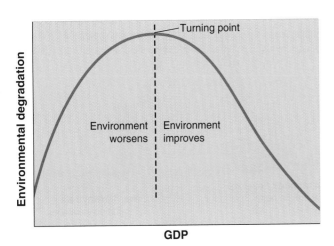

**FIGURE 20.4** **The Kuznets curve.** This model suggests that as per capita income in a country increases, environmental degradation first increases and then decreases. In many respects, China is on the first part of this curve while the United States is on the second part of the curve.

**FIGURE 20.5** **Solar panels in Africa.** In areas where the electrical grid is not established and electricity supply lines are not present, the installation and use of photovoltaic solar cells may be less expensive and less environmentally disruptive than a traditional electrical infrastructure.

initial sum of money to be reinvested many times over to the benefit of local economies. The recipients of a microloan are typically too poor to obtain traditional loans from banks. Microlending helps foster sustainable development by encouraging people to be economically solvent and to care for the environment, goals adhered to by the microlending program and the representatives of that program.

Microlending began in 1983, when Muhammad Yunus founded the Grameen Bank in Bangladesh, and it has expanded rapidly since then. Through the Grameen Bank, some of the poorest people in less developed countries have received microloans, usually between $50 and $500. FIGURE 20.6 shows how microloans have resulted in a higher quality of life. The loan recipients can use the money to buy items that will help them earn an income. For example, a woman growing and selling vegetables might be able to bring more vegetables to market if she had a wheelbarrow. With a microloan, the woman is able to buy the wheelbarrow, thereby increasing her income, and improving her well-being and that of her family. The expectation is that she will be able to pay off the loan in a reasonable amount of time and continue with the business that the microloan initiated.

In 2006 Dr. Yunus received the Nobel Peace Prize for pioneering the concept of microlending. Two insights helped Dr. Yunus to make the Grameen Bank a success. First, many poor people have good ideas and can generate a significant profit out of a very small investment.

Second, providing economic opportunities to women and allowing them to meet with other women improve the success of the loans. The Grameen Bank focuses on loans to women and includes stipulations that maintain the goal of helping women achieve financial independence. For example, many loans stipulate that the borrower must attend a training seminar with other women in a nearby city. Attending the seminar and interacting with other motivated small entrepreneurs can inspire a sense of confidence and motivation.

The environmental benefits of this type of development can be large. The Grameen Bank asks borrowers to pledge to keep their families small, which can have immediate economic benefits to the family as well as broader environmental benefits because smaller families have less impact on the environment. Since the loans help develop local markets for services and goods, less fuel is used to bring in distant goods. Many microlending institutions strongly encourage a sustainable approach to the natural world as an ethical responsibility and foster sustainable business practices that do not deplete natural resources.

There has been criticism of microlending in recent years. In some cases, the interest rates being charged are extremely high, and some critics have questioned exactly how much environmental benefit comes from microlending. Nevertheless, globally, microlending programs are considered a success. Estimates are difficult to obtain and vary widely, but there may be over 10,000 micro-

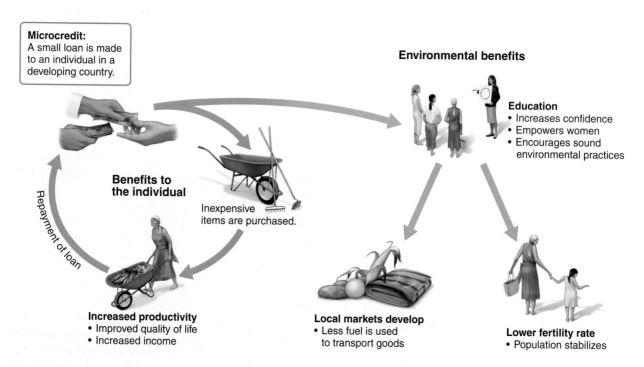

**Microcredit:** A small loan is made to an individual in a developing country.

Repayment of loan

**Benefits to the individual**

Inexpensive items are purchased.

**Increased productivity**
• Improved quality of life
• Increased income

**Environmental benefits**

**Education**
• Increases confidence
• Empowers women
• Encourages sound environmental practices

**Local markets develop**
• Less fuel is used to transport goods

**Lower fertility rate**
• Population stabilizes

**FIGURE 20.6** **The microcredit cycle.** In a small business such as farming and selling produce in nearby markets, the availability of a wheelbarrow and farming tools may greatly increase revenues and allow the farmer to rise from poverty. Microcredit allows the loan of tens or hundreds of dollars that will allow the purchase of the items needed to increase revenues.

finance institutions worldwide, lending to over 100 million people. The average loan is typically less than $300. According to the World Bank, more than 95 percent of borrowers in Bangladesh report improved quality of life, better and more food, and better clothing and housing as a result of their loans. And in many cases, these improvements have led to improved environmental quality as well.

---

**GAUGE YOUR PROGRESS**

✓ How do the laws of supply and demand influence the price of a good or service?

✓ How do GPI and the Kuznets curve address the shortcomings of GDP as a measurement of a country's well-being?

✓ What is microlending and how does it improve the lives of people and the environment in developing nations?

---

# Economic health depends on the availability of natural capital and basic human welfare

Capital, or the totality of our economic assets, is typically divided into three categories: natural, human, and manufactured. **Natural capital** refers to the resources of the planet, such as air, water, and minerals. **Human capital** refers to human knowledge and abilities. **Manufactured capital** refers to all goods and services that humans produce. Economists usually base their assessment of national wealth on productivity and consumption. Environmental scientists point out that all economic systems require a foundation of natural capital. Without natural capital, humans would not be able to produce very much and would probably not survive.

## Environmental and Ecological Economics

Some advocates of a purely free-market system believe that as long as market forces are left alone, human work and creativity will find solutions to problems of natural resource degradation and depletion. But as we have seen, externalities are not assessed appropriately if the cost of environmental degradation is not charged to the individuals responsible for that degradation. When the economic system does not appropriately account for all costs, it is referred to as a **market failure.** Among those economic thinkers who have sought ways to respond to market failures, many have become part of the discussion in the fields of *environmental economics* and *ecological economics.* **Environmental economics** is a

subfield of economics that examines the costs and benefits of various policies and regulations that seek to regulate or limit air and water pollution and other causes of environmental degradation. **Ecological economics** treats the field of economics as a component of ecological systems rather than as a distinctly separate field of study. Ecological economics is a method of understanding and managing the economy as a subsystem of both natural and human systems. It has as a goal the preservation of natural capital, the goods and services related to the natural world.

Both environmental and ecological economists attempt to assign monetary value to intangible benefits and natural capital. This practice is called **valuation.** For example, they have developed methods for assessing the monetary value of a pristine nature preserve, a spotted owl, or a scenic view. One method is to calculate the revenue generated by people who pay for the benefit—for example, the amount tourists pay to visit a nature preserve would represent the dollar value of the preserve. Another method is to use surveys. They might ask a number of people how much they are willing to pay just to know that spotted owls exist, even if they are unlikely ever to see one. The most extensive assessments have attempted to determine the value of ecosystem services such as oxygen that plants produce or pollination by insects. Although estimates vary, global ecosystem services might have a dollar value of approximately $30 trillion per year. The 2004 Millennium Ecosystem Assessment categorized the variety of services that ecosystems provide for the benefit of humans. In many cases, it is possible to estimate the cost of a particular service if it were provided using technology rather than naturally. For example, as we discussed in Chapter 3, New York City could have constructed a massive water purification system at a known cost. Instead, it chose to protect watersheds in the Catskill Mountains, a region north of New York City that supplies the city with water, so the water would not need expensive purification. Accordingly, the ecosystem service of water purification has a known value, which can be used to help calculate the total dollar value of ecosystem services.

Given all of the natural capital and ecological services distributed around the world, it is quite likely that human activities will generate multiple negative externalities. Economic tools can be used to incorporate the dollar cost of the externalities in the price of goods and services. We have seen examples of this with our discussions of charging for allowances to emit sulfur dioxide and carbon dioxide. These economic tools can be used in many other ways as well. Typically, a tax or regulation calls for reducing externalities through a market-driven system. This system calls for the incorporation of negative environmental impacts of a commodity or service in its cost of production. For example, a car manufacturer would include in the cost of production for each car not only the cost of labor and natural resources, such as steel and water, but also the cost of the air pollution caused by the manufacturing process. Viewed this way, the cost of production of a car will immediately increase. Typically, the manufacturer would want to distribute at least part of this additional cost to the consumer by raising the price of the car. Calculating the full costs of a commodity or service by internalizing externalities will likely cause consumers to buy fewer items with high negative impacts, since those impacts will be reflected in higher prices. The most obvious way for the costs of externalities to be included is by requiring the producer to pay them. This could be achieved through regulation, imposition of a tax, or some sort of public action mandating reparation for externalities or making it difficult for the company to produce its product in a way that pollutes. Much of the debate in environmental and ecological economics revolves around how to impose the dollar cost on the producer.

## Sustainable Economic Systems

Critics of our current economic system maintain that it is based on maximizing the utilization of resources, energy, and human labor. This encourages the extraction of large amounts of natural resources and does not provide any incentives that would reduce the amount of waste generated. A system analysis of the current economic situation, shown in **FIGURE 20.7a**, suggests that continuing with such a system is not sustainable. In the current system, large amounts of extracted resources and energy and relatively small amounts of ecosystem services are the inputs and large amounts of waste are the outputs.

A sustainable economic system will rely more on ecosystem services and reuse of existing manufactured materials and less on resource extraction (**FIGURE 20.7b**). In this system, there is greater reliance on ecosystem services and less reliance on resource extraction that requires energy. The system in Figure 20.7b also takes some of the waste stream and reuses it in the production and consumption cycle, as indicated by the arrow labeled "Waste stream recycling." Therefore, the cycle in Figure 20.7b would use more renewable energy, lessen negative externalities, and reuse more of the products that were destined for the waste stream. This model has led architects, environmental scientists, and engineers to a collaborative discussion of the optimal way to design, manufacture, use, and dispose of objects such as automobiles, houses, and consumer goods. Currently, a consumer purchases an object, such as an automobile or computer, and when that object has reached the end of its useful life, the consumer is responsible for its disposal. As we pointed out in Chapter 16, because the responsibility for the object rests with the consumer, there is no incentive for the manufacturer to make it easy to reuse or recycle an object. Some observers of the current situation have noted the irony that a can of a chemical oven cleaner purchased for $5 may cost $20 for disposal. These kinds of discussions have led to the cradle-to-grave and cradle-to-cradle analysis presented in Chapter 16.

Because the cradle-to-cradle system includes human capital, resource, and energy inputs as well as a redirection of the waste stream, it gains even greater importance when we consider the entire economic system (**FIGURE 20.8**). The ultimate goal is to produce a good that at the

**(a)** Less sustainable economy

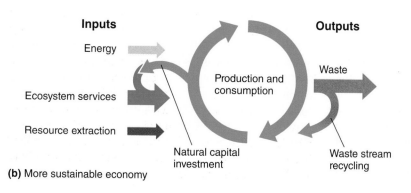

**(b)** More sustainable economy

**FIGURE 20.7** **Systems diagrams of two economic systems.** (a) A less sustainable system, like our current economy, is based on maximizing the utilization of resources and results in a fairly large waste stream. (b) A more sustainable system is based on greater use of ecosystem services, less resource extraction, and minimizing the waste stream.

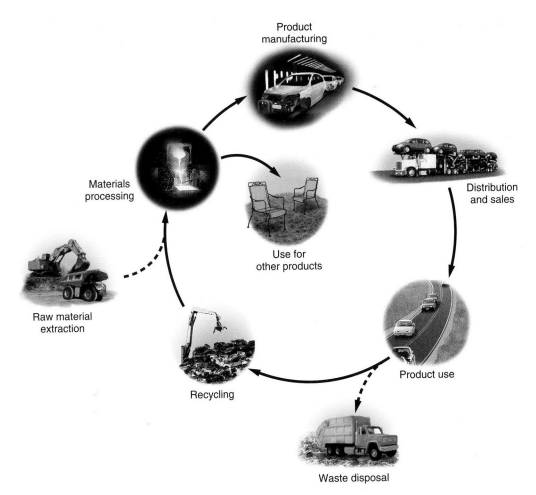

Product
manufacturing

Materials
processing

Raw material
extraction

Use for
other products

Distribution
and sales

Product use

Recycling

Waste disposal

**FIGURE 20.8** **A cradle-to-cradle system for material use and waste recycling.** The manufacture of automobiles serves as one example. Products made at a factory are made from recycled materials whenever possible. Product design and manufacturing are done with the goal of being able to recycle as much of the automobile as possible when its useful life is over. Energy costs in manufacturing, distribution, and use are all taken into consideration when designing the automobile and the distribution network.

end of its useful life—as it approaches its "grave"—can be easily reused to make a new product. That is, most or all of the parts will become the "cradle"—the beginning of life—for the same or other products. An automobile would be ideal for a cradle-to-cradle system because each individual car contains a ton or more of steel and other metals as well as rubber, plastic, and a host of other materials that could be reused.

| GAUGE YOUR PROGRESS |
| --- |
| ✓ What is capital? Name three types of capital. |
| ✓ What do environmental and ecological economics add to the study of economics? |
| ✓ What are the features of a sustainable economic system? |

# Agencies, laws, and regulations are designed to protect our natural and human capital

We have looked at the economic system and the roles of natural capital and human capital. This understanding provides us with the tools we need to evaluate different options for monitoring and managing human systems in a way that will result in the least amount of harm to the natural environment. Many different techniques and approaches are used with others currently under consideration. Before examining some of them, we must familiarize ourselves with an important factor that shapes the way nations approach policy making—how people look at the world.

## Environmental Worldviews

An **environmental worldview** has three dimensions: how you think the world works; how you view your role in it; and what you believe to be proper environmental behavior. Three types of environmental worldviews dominate: human-centered, life-centered, and Earth-centered.

The **anthropocentric,** or human-centered, worldview considers that human beings have intrinsic value. In other words, nature has an instrumental value to provide for our needs. There are variations on the human-centered worldview. For example, those who favor a free-market approach to economics are optimistic about the results of unlimited competition and minimal government intervention. The planetary management school, while optimistic that we can solve resource depletion issues with technological innovations, believes that nature requires protection and that government intervention is at times necessary to provide this protection. **Stewardship,** a subset of the anthropocentric worldview, is the careful and responsible management and care for Earth and its resources. The stewardship school of thought considers that not only does the natural world require protection but that it is our ethical responsibility to be good managers of Earth.

The **biocentric,** or life-centered, worldview holds that humans are just one of many species on Earth, all of which have equal intrinsic value. At the same time, this worldview considers that the ecosystems in which humans live have an instrumental value. There are various positions within the life-centered approach. While some consider that it is our obligation to protect a species, others consider that it is our obligation to protect every living creature.

The **ecocentric,** or Earth-centered, worldview places equal value on all living organisms and the ecosystems in which they live, and it demands that we consider nature free of any associations with our own existence. This worldview takes various forms. The environmental wisdom school, for example, believes that since resources on Earth are limited, we should adapt our needs to nature rather than adapt nature to our needs. The deep ecology school, meanwhile, insists that humans have no right to interfere with nature and its diversity. Our worldviews determine the decisions we make about our lives, our work, and the way we treat the planet.

Environmental worldviews can play a significant role in the policies a nation considers and how it implements them. For example, a nation that holds an anthropocentric worldview might proceed with a series of policies and regulations allowing economic development with no concern for the effects it will have on the natural environment. A country with an ecocentric worldview might have policies and regulations that carefully protect ecosystems and the species within them. In practice, the policies and regulations of most nations represent a variety of worldviews depending on the particular nation and the specific resource or region of the biosphere that is being affected. There is one more important concept that influences the policies and regulations that a particular nation may implement, the precautionary principle.

## The Precautionary Principle

In Chapter 17 we discussed the precautionary principle, which states that when the results of an action are uncertain, such as the effects caused by the introduction of a compound or chemical, it is better to choose an alternative known to be harmless. In many situations, scientific uncertainty complicates the estimation of the comparative risks of different actions. This is an important part of environmental decision making. In the United States, environmental law and policy has at times treated scientific uncertainty as a reason to discount or downplay scientific evidence of problems in the environment. Industrial and business groups have also used scientific uncertainty as a reason to avoid implementing expensive measures that would mitigate any future environmental harms. Those who favor using the precautionary principle argue that if we wait for widespread scientific consensus about the adverse effects of a particular compound or action, we run the risk of creating an environmentally unsustainable and inequitable future.

As mentioned in Chapter 17, critics of the precautionary principle maintain that "progress" and economic well-being will be hindered if we wait to use something until we verify that it is safe for the environment. There may also be an additional economic cost to waiting, or for choosing alternative means of achieving a goal.

In 1994, the International Union for Conservation of Nature, an organization composed of over 800 government and nongovernmental wildlife organizations, strengthened the 1992 Convention on Biological Diversity by publishing guidelines that included using the precautionary principle as a recommended tool in reaching decisions about the sustainable use of plant and animal species. The guidelines emphasize using the "best science available" in deciding whether to list a species as endangered and whether to ban any activity that could jeopardize that species.

The 1987 Montreal Protocol (see Science Applied "Were We Successful in Halting the Growth of the Ozone Hole?" on page 52) is an example of the precautionary principle being applied to global change. Despite the fact that there was still some scientific uncertainty at that time regarding evidence that CFCs, the chemicals used for refrigeration and other commercial applications, were a cause of ozone depletion, the protocol recommended eliminating the use of CFCs. In this case, economic and political factors were balanced with the scientific findings to reach an agreement that CFCs should be phased out over a period of decades rather than immediately. Part of the success of the Montreal Protocol has been credited to the availability of an affordable and fairly effective replacement to CFCs. As we have noted a number of times in this textbook, it has proven much more difficult to find affordable and effective replacements for the fossil fuels we currently use. Therefore, it is less likely that a similar scenario will unfold with respect to a reduction of greenhouse gases.

The precautionary principle is a relatively new and important part of environmental policy. It does not recommend or require any specific actions, but it does provide a reminder to environmental policy makers and managers that, in many cases, absolute scientific certainty may come too late when dealing with potentially serious environmental harms.

By considering the variety of worldviews presented earlier, and to what extent a particular country or agency subscribes to the precautionary principle, we can begin to understand more about the decision-making process that influences the various world agencies that have jurisdiction over global environmental issues.

## World Agencies

Global, national, or personal situations may prompt key beneficial decisions out of a sense of necessity and urgency. After World War II (1939–1945), leaders of the allied nations agreed to found the **United Nations (UN),** an institution dedicated to promoting dialogue among countries with the goal of maintaining world peace. When its charter was ratified in 1945, the UN had 51 member countries. In 2009, it had grown to 192. Since its establishment, the UN has created many internal agencies and institutions. Four of the many important UN organizations relating to the environment are the United Nations Environment Programme, the World Bank, the World Health Organization, and the United Nations Development Programme.

**THE UNITED NATIONS ENVIRONMENT PROGRAMME** The **United Nations Environment Programme (UNEP),** headquartered in Nairobi, Kenya, was founded in 1972. UNEP has a number of functions that include gathering environmental information, conducting research, and assessing environmental problems. UNEP is also the international agency responsible for negotiating certain environmental treaties. In particular, the Convention on Biological Diversity and the Convention on International Trade in Endangered Species (CITES) as well as the Montreal Protocol on Substances That Deplete the Ozone Layer are three important international treaties UNEP negotiated. The Global Environment Outlook (GEO) reports are prepared under the auspices of UNEP.

**THE WORLD BANK** The **World Bank,** headquartered in Washington, D.C., originated in 1944 along with the International Monetary Fund (IMF) at a monetary and financial meeting of the United Nations in Bretton Woods, New Hampshire. It provides technical and financial assistance to developing countries with the objectives of reducing poverty and promoting growth, especially in the poorest countries. The World Bank cites four goals for economic development: (1) educating government officials and strengthening governments; (2) creating infrastructure; (3) developing financial systems from microcredit to much larger systems; and (4) combating corruption. Critics of the World Bank maintain that there is too little consideration of environmental and ecological impacts when projects are evaluated and approved.

**THE WORLD HEALTH ORGANIZATION** Headquartered in Geneva, Switzerland, the **World Health Organization (WHO)** was created in 1945 to improve human health by monitoring and assessing health trends and providing medical advice to countries (**FIGURE 20.9**). It is the group within the UN responsible for human health, including combating the spread of infectious diseases such as those that are exacerbated by

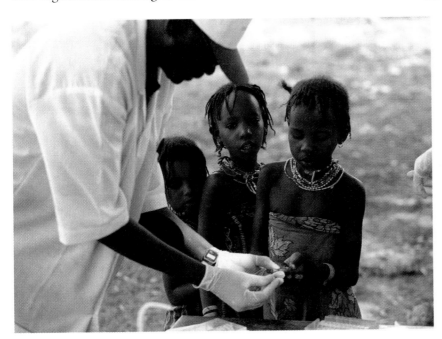

**FIGURE 20.9** World Health Organization workers. A WHO worker in Chad draws blood from children for disease testing.

global climatic changes. This organization is also responsible for health issues in crises and emergencies created by storms and other natural disasters. The five key objectives of the WHO are: (1) promoting development, which should lead to improved health of individuals; (2) fostering health security to defend against outbreaks of emerging diseases; (3) strengthening health care systems; (4) coordinating and synthesizing health research, information, and evidence; and (5) enhancing partnerships with other organizations.

**THE UNITED NATIONS DEVELOPMENT PROGRAMME** The **United Nations Development Programme (UNDP),** headquartered in New York City, was founded in 1965. It works in 166 countries around the world to advocate change that will help people obtain a better life through development. UNDP has a primary mission of addressing and facilitating issues of democratic governance, poverty reduction, crisis prevention and recovery, environment and energy issues, and prevention of the spread of HIV/AIDS. UNDP prepares an annual Human Development Report (HDR) that is an extremely useful measurement tool for the status of the human population.

**OTHER AGENCIES** There are also a great number of nongovernmental organizations (NGOs) that work on worldwide environmental issues. These include Greenpeace, the International Union for Conservation of Nature, World Wide Fund for Nature (formerly World Wildlife Fund), and Friends of the Earth International.

## United States Agencies

In January 1969, offshore oil platforms 10 kilometers (6 miles) from Santa Barbara, California, began to leak oil. Roughly 11.4 million liters (3 million gallons) of oil

would spill over the next 11 days and the leak would continue throughout the year. This was not the first oil spill during the 1960s, nor the largest, but its proximity to the southern California coast resulted in something new—vast media attention. Daily television news reports of dead seabirds, fish, and marine mammals, as well as large stretches of oil-soaked beaches, shocked the American public and government officials. The Santa Barbara oil spill was a major cause of a shift in federal policy toward incorporating an awareness of how human society affects the environment.

The first Earth Day, April 22, 1970, was partially the result of public reaction to the Santa Barbara oil spill and to other environmental problems that surfaced during the 1960s, such as those documented by Rachel Carson in *Silent Spring* (FIGURE 20.10). Earth Day 1970 is the symbolic birthday of the modern environmental belief that the natural environment and human society are inextricably connected. It also signals the beginning of modern environmental policy. Before 1970, environmental policy focused primarily on biological and physical systems as sources of economic resources for an industrial society. After 1970, sound environmental policy expanded to include the idea that economic benefits must be balanced by environmental science, environmental equity, and intergenerational equity—the interests of future generations in a healthy environment.

Since the early 1970s, several important agencies have been created in the United States that are dedicated to monitoring the human impact on the environment and promoting environmental and human health.

**THE ENVIRONMENTAL PROTECTION AGENCY** In 1970, President Richard Nixon signed the bill authorizing the creation of the **Environmental Protection Agency (EPA),** headquartered in Washington, D.C. The EPA oversees all governmental efforts related to the environment including science, research, assessment, and education. It also writes and develops regulations and works with the Department of Justice and Department of State and U.S. Native American governments to enforce those regulations.

**THE OCCUPATIONAL SAFETY AND HEALTH ADMINISTRATION** Also in 1970, President Nixon signed the act creating the **Occupational Safety and Health Administration (OSHA),** an agency of the U.S. Department of Labor. OSHA is the main federal agency responsible for the enforcement of health and safety regulations. Its main mission is to prevent injuries, illnesses, and deaths in the workplace. OSHA conducts inspections, workshops, and education efforts to achieve its goals. Limiting exposure to chemicals and pollutants in the workplace is one way that OSHA is involved in environmental protection.

**THE DEPARTMENT OF ENERGY** In 1977, President Jimmy Carter signed an act creating the **Department of Energy (DOE).** The DOE's main goal is to advance the energy and economic security of the United States. It includes scientific discovery, innovation, and environmental responsibility among its top goals. Within the DOE, the Energy Information Agency gathers data on the use of energy in the United States and elsewhere.

---

**GAUGE YOUR PROGRESS**

✓ How might an environmental worldview influence environmental policy or regulations?

✓ How does the precautionary principle relate to scientific uncertainty?

✓ What are the major world agencies that are concerned with the environment? Describe their functions.

---

# There are several approaches to measuring and achieving sustainability

Just as there are agencies, laws, and regulations designed to initiate and enhance sustainability, there are also a number of lenses through which to view the world, and a number of measurements or indexes used to evaluate sustainability. This section introduces some of these measurements and views. Eventually, some or all of these indices may become more directly involved in the measurement and assessment of sustainability.

**THE HUMAN DEVELOPMENT INDEX** The **human development index (HDI)** combines three basic measures of human status: life expectancy; knowledge and education, as shown in adult literacy rate and educational attainment; and standard of living, as shown in per capita GDP and individual purchasing power. HDI was developed in 1990 by economists from Pakistan, England, and the United States, and it has been used since then by the UNDP in its annual HDR. As an index, HDI serves to rank countries in order of development and determine whether they are developed, developing, or underdeveloped. **FIGURE 20.11** shows the range of HDI values and the distribution among countries. As you might expect, most of the developed countries have the highest HDI values.

**THE HUMAN POVERTY INDEX** The **human poverty index (HPI)** is the counterpart of HDI and was developed by the United Nations to investigate the proportion of a population suffering from deprivation in a country with a high HDI. This index measures three things: longevity, as indicated by the percentage of the population not expected to live past 40; knowledge, as measured by

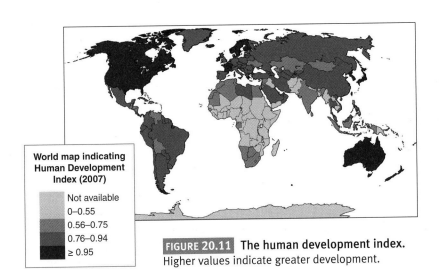

World map indicating Human Development Index (2007)

Not available
0–0.55
0.56–0.75
0.76–0.94
≥ 0.95

**FIGURE 20.11** The human development index. Higher values indicate greater development.

| Act | Abbreviation | Year enacted | Purpose | Prime example of a success |
|---|---|---|---|---|
| National Environmental Policy Act | NEPA | 1970 | Enhance environment; monitor with a tool: the Environmental Impact Assessment | Protection of coral formation and sea turtles has occurred. |
| Occupational Safety and Health Act | OSHA | 1970 | Prevent occupational injuries, illness, death from work-related exposure to physical and chemical harm | Worker training and knowledge of toxins has increased. |
| Endangered Species Act | ESA | 1973 | Protect animal and plant species from extinction | Bald eagle, peregrine falcon, and gray wolf populations have recovered. |
| Clean Air Act | CAA | 1970 | Promote clean air | Sulfur dioxide reductions from cap-and-trade have occurred. |
| Clean Water Act | CWA | 1972 | Promote clean water | Swimmable and fishable rivers across the United States have increased. |
| Resource Conservation and Recovery Act | RCRA | 1976 | Govern tracking and disposal of solid and hazardous waste | Numerous brownfields and contaminated lands have been cleaned up. |
| Comprehensive Environmental Response, Compensation, and Liability Act | CERCLA, also called Superfund | 1980 | Force and/or implement the cleanup of hazardous waste sites | Dozens of Superfund sites have have been cleaned up around the United States. |

**TABLE 20.1  Major U.S. legislation for promoting sustainability**

the adult illiteracy rate; and standard of living, as indicated by the proportion of the population without access to clean water and health services, as well as the percentage of children under 5 years of age who are underweight.

## U.S. Policies for Promoting Sustainability

Of the many regulations that have been established in the last 50 years or so in the United States, there are at least seven important pieces of legislation that may help move the United States toward sustainability. All of these regulations have been discussed in other chapters and are summarized in Table 20.1.

## The Policy Process in the United States

To be fair and effective, environmental policies should be based on scientific indicators that suggest a certain behavior or action will be best for the environment. When policy makers believe there is adequate understanding of the science, and there is a course of preferred action for states or individuals, they begin a process to develop a policy.

The five basic steps in a policy cycle are problem identification, policy formulation, policy adoption, policy implementation, and policy evaluation. **FIGURE 20.12** depicts this process as a circular or reiterative process. As a policy is evaluated, the need for amendment might arise. When an amendment is initiated, it follows

**FIGURE 20.12  The environmental policy cycle.** After an environmental problem is identified, environmental policy is formulated or modified. After a policy is adopted and implemented, it is evaluated and, if necessary, adjustments to the policy are made.

roughly the same steps. Many good environmental policies have had numerous amendments. For example, the Clean Air Act has been amended twice, and even the original Clean Air Act of 1970 was actually a modification of earlier clean air legislation.

## Deterrents and Incentives

United States governmental agencies have tried many ways to protect the environment, promote human safety and welfare and, in some cases, internalize externalities. Two prominent strategies are the **command-and-control approach,** which sets regulations for emissions, for example, and then controls them with fines or other punishments; and the **incentive-based approach,** which constructs financial and other incentives for lowering emissions based on profits and benefits. A combination of both approaches is likely to generate the maximum amount of desired changes.

Taxation is a major deterrent used to discourage companies from producing pollution and generating other negative impacts. A **green tax** is a tax placed on environmentally harmful activities or emissions in an attempt to internalize some of the externalities that may be involved in the life cycle of those activities or products. However, a tax alone may not be sufficient to achieve the desired results. Sometimes rebates and/or tax credits are given to individuals and businesses purchasing certain items such as energy-efficient appliances or building materials such as windows and doors. Cap-and-trade has been discussed in earlier chapters in relation to controlling compounds such as carbon dioxide and sulfur dioxide.

**THE TRIPLE BOTTOM LINE** In 1996, President Clinton's Council on Sustainable Development declared that "the essence of sustainable development is the recognition that the pursuit of one set of goals affects others and that we must pursue policies that integrate economic, environmental, and social goals." The **triple bottom line** concept states that we need to take into account three factors—economic, environmental, and social—when making decisions about business, the economy, and development (FIGURE 20.13). There are many organizations and businesses that place one of these three factors at the top of a priority list. Some businesses strive for economic well-being—a sound financial bottom line—to the exclusion of human welfare or the environment. They may be regarded as successful within certain communities, but the triple bottom line concept emphasizes that to be a true success, there must be adequate treatment of both humans and environment. Paul Hawken, the author of *Natural Capitalism,* states the objective as, "Leave the world better than you found it, take no more than you need, try not to harm life or the environment, and make amends if you do." The recent failures of several major U.S. automobile manufacturers

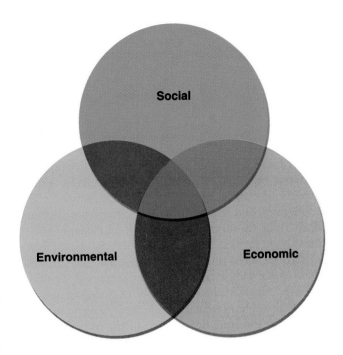

**FIGURE 20.13** **The triple bottom line.** Sustainability is believed to be achievable at the intersection of the three circles.

are considered to be prime examples of unsustainable business practices, certainly from an economic perspective and possibly from other perspectives as well.

### GAUGE YOUR PROGRESS

✓ What is the human development index (HDI)? How is it used?

✓ How do major environmental regulations help move the United States toward sustainability?

✓ How are deterrents and incentives used in environmental policy? Give examples of each.

# Two major challenges of our time are reducing poverty and stewarding the environment

The classic environmental dichotomy is "jobs versus the environment." Those primarily concerned with human well-being ask how we can make demands for environmental improvements when there is so much poverty and injustice in the world. Those primarily concerned with the environment ask how we can focus exclusively on human suffering when an impoverished environment cannot support human health and well-being.

## Poverty and Inequity

Approximately one-sixth of the human population—more than one billion people—lives in unsanitary conditions in informal settlements, slums, and shanty towns. Roughly one-sixth of the human population earns less than $1 a day, and half earns less than $2 a day. In the last 100 years, as developed countries have increased their GDPs and as many countries have modernized and developed their economies, the disparities between the rich and the poor have become greater. Poverty is simultaneously an issue of human rights, economics, and the environment. Every human has a basic right to survival, well-being, and happiness—all directly threatened by poverty. Indebted individuals and nations are often unable to pay what they owe. In 2005, the 8 major industrial countries of the world, known as the G8, cancelled the debt of the 18 poorest countries. From an environmental standpoint, poverty increases overuse of the land, degradation of the water, and incidence of disease.

In 2000, the United Nations offered an eight-point resolution listing what its member countries agreed were pressing issues that the world could no longer ignore. The member countries committed to reaching these Millennium Development Goals (MDGs), outlined in the United Nations' Millennium Declaration, by 2015:

- Eradicate extreme poverty and hunger
- Achieve universal primary education
- Promote gender equality and empower women
- Reduce child mortality
- Improve maternal health
- Combat HIV/AIDS, malaria, and other diseases
- Ensure environmental sustainability
- Develop a global partnership for development

As this book goes to press in 2011, just about 4 years from the agreed-upon deadline for achieving MDGs, some countries are well on their way to meeting these goals while others lag far behind. As with environmental laws and their implementation, the distance from resolutions to results is immense, and not all developed countries have committed as much as they had promised.

One proponent of the UN MDGs is Nobel Peace Prize Laureate Dr. Wangari Maathai from Kenya (**FIGURE 20.14**). Dr. Maathai is the founder of the Green Belt Movement, a Kenyan and international environmental organization that empowers women by paying them to plant trees, some of which can be harvested for firewood in a few years. The Green Belt Movement is

**FIGURE 20.14** **Wangari Maathai.** Dr. Maathai is the founder of the Green Belt Movement in Kenya.

credited with replanting large expanses of land in East Africa, thereby reducing erosion and improving soil conditions and moisture retention. In addition, the trees that have been replanted, provided that they are not overharvested, offer a renewable source of fuel for cooking. The Green Belt Movement is considered a global sustainability success story promoting both individual human and environmental well-being. Dr. Maathai has also been involved in environmental activism to achieve her goals, which has sometimes caused difficulties with certain governmental organizations.

## Environmental Justice

The typical North American uses many more resources than the average person in many other parts of the world. This situation is not equitable. We have seen how increased resource use usually increases harm to the environment. The subject of fair distribution of Earth resources, termed *environmental equity*, has received increasing international attention in recent years. Besides moral objections to inequity, there are concerns about sustainability. As more and more people develop a legitimate desire for better living conditions, the resources of Earth may not be able to support continued consumption at such high levels. Closely related to the equity of resource allocation are questions of the inequitable distribution of pollution and of environmental degradation with their adverse effects on humans and ecosystems. All of these topics fall under the subject of environmental equity.

As we discussed in Chapter 16 and elsewhere, African Americans and other minorities in the United States are more likely than Caucasians to live in a county or city with solid waste incinerators, chemical production plants, and other "dirty" industries. In a number of studies in the 1980s and later, investigators used the distribution of minority residents by postal zip code, corresponding to a specific town or division within a town, to relate race and class to location of hazardous sites. In Atlanta, 83 percent of the African American population lived in the same zip code area as the 94 uncontrolled toxic waste sites, while 60 percent of the whites lived in those areas. In Los Angeles, roughly 60 percent of Hispanics lived in the same areas as the toxic waste sites, while only 35 percent of the white population lived in those areas. One study in five southern states compared the size of specific landfill facilities with the percentage of minorities in the zip code area in which the landfill was located. The study concluded that the largest landfills are located in areas that have the greatest percentage of minorities. An important issue that has not been resolved, and can vary from case to case, is determining which came first to a given area—the affected population or the hazardous facility. By knowing which came first, people and organizations attempting to remedy the situation will have a better idea of how to modify existing legislation and regulations to reduce the number of people who live in degraded environments.

More recently, it has become clear that the subjects of disproportionate exposure to environmental hazards were not only African Americans, but those, both white and nonwhite, who were in lower income brackets. Moreover, the problem was not limited to the United States. The concept was broadened and renamed *environmental justice*, which is a social movement and a field of study that examines whether there is equal enforcement of environmental laws and elimination of disparities—intended or unintended—in the exposure to pollutants and other environmental harms affecting different ethnic and socioeconomic groups within a society. Delegates to the First National People of Color Environmental Leadership Summit in 1991 established 17 principles of environmental justice. Dr. Robert Bullard of Clark Atlanta University has published books and papers in the academic area of environmental justice and has been involved in the social movement as well. He is probably best known for his 1990 book *Dumping in Dixie,* which demonstrated that minority and lower socioeconomic groups were often the recipients of pollution related to dumping of MSW and hazardous wastes. More recently, Professors Paul Mohai and

Robin Saha of the University of Michigan reassessed the unequal distribution of hazardous waste dumping in the United States and found that the situation is actually worse than previously reported. In particular, they believe they have resolved the issue of whether the hazardous waste facility or the lower socioeconomic and minority population came first to an area. The authors maintain that the minority community in many cases was present first and that the hazardous waste facility was specifically targeted to that community.

## Individual and Community Action

There are a fair number of people who believe that, whether or not governments and private agencies are able to achieve their goals, individuals can and must act to further their own goals of sustaining human existence on the planet. These individuals have begun to make attempts to live a sustainable existence without government incentives, taxes, or other measures. They have begun activities such as calculating their own ecological footprint, carbon footprint, energy footprint, or other metrics to determine how much of an impact they are making on Earth. From this starting point, they have begun to make changes in their consumption, behavior, and lifestyle to reduce that impact. Some people act on their own while others act through groups and organizations. They have adopted a philosophy represented by the saying, "If the people lead, the leaders will follow." Some individuals have joined together to organize communities centered around philosophies of sustainability.

Van Jones, a graduate of Yale Law School, was a community organizer in San Francisco working on civil rights and human justice issues when he decided to combine concerns about the environment and global climate change with the need for creating jobs in cities (FIGURE 20.15). He founded an organization called Green

**FIGURE 20.15** A gathering of Green For All supporters in Oakland, California.

For All and in 2008 published a book titled *The Green Collar Economy: How One Solution Can Fix Our Two Biggest Problems*. The two problems, as he sees it, are global warming and urban poverty. He believes that creating green jobs, such as insulating buildings, constructing wind turbines and solar collectors, and building and operating mass transit systems, will improve the living conditions of some of the poorest people in the nation and reduce our impact on the environment. For Van Jones, this is a win-win solution—people will be employed and our emissions of global greenhouse gases will decrease.

Van Jones was a proponent of the Energy Independence and Security Act, a 2007 bill that authorizes expenditures for training and creation of green jobs around the country. The organization Green For All has now launched five pilot programs, in Seattle,

Pittsburgh, Philadelphia, Newark, and Atlanta. From March to September 2009, Van Jones was special advisor for green jobs at the White House Council on Environmental Quality.

**GAUGE YOUR PROGRESS**

✓ What is the connection between poverty and inequity?

✓ What is environmental justice? Give an example.

✓ What are some of the potential conflicts between human well-being and environmental protection?

# WORKING TOWARD SUSTAINABILITY

In the 1990s, athletic shoe manufacturer Nike, based in Beaverton, Oregon, drew a great deal of negative publicity for the conditions of its factories and treatment of its workers overseas. The prevalence of child labor, unsafe working conditions, and inadequate wages highlighted the social costs that often accompany commercial success. Environmental degradation was the most pressing unseen effect of a very profitable shoe manufacturing company. However, in subsequent years, Nike imposed a system of factory standards and inspections that, along with recently passed labor laws in countries like China, have considerably improved conditions for workers.

Unlike heavy industry and energy generation, the shoe industry is not often part of the environmental discussion in the United States. However, in Asian and South American countries where the factories are located (one-third of Nike shoes are produced in China), the environmental impacts of shoe manufacturing are far more visible. Emissions from energy use and transportation of materials and products, solid waste from all levels of production, and the effects of resource extraction prior to manufacturing reduce the quality of air, water, and soil both near to and far from production plants. Another threat to workers and the environment is the use of toxic solvents, adhesives, and rubber in shoe fabrication. The evaporation of these substances contributes to the concentration of volatile organic compounds (VOCs) in the atmosphere. VOCs also cause respiratory impairment in factory employees and contribute to hazardous waste. As shoe companies like Nike continue to increase production and profitability, the environmental effects are compounded. However, Nike is making significant efforts to lessen the environmental cost of a successful business.

## Reuse-A-Sneaker

In order to improve the sustainability of its industry, Nike has developed a comprehensive cradle-to-grave program that addresses environmental impacts in every stage of raw material extraction, product fabrication, sale, and disposal. The "Nike Considered" program, as it is called, offers more sustainable products and provides an index that evaluates each product and assigns it a score based on its environmental impact. Scores are assigned based on efficiency of design, solvent use, and waste creation. Essentially, Nike has created a life-cycle analysis of materials including extraction processes, energy and water use, manufacturing practices, and recyclability. Products labeled "Considered" must meet standards of sustainability significantly higher than the average Nike product. These products are made using water-based adhesives instead of solvent-based adhesives, thereby reducing VOC evaporation. Soles of the shoes consist of recycled and less toxic rubber. The design of the shoe as well as the manufacturing process must demonstrate efficiency. Also, the company is a member of the Organic Exchange, which supports the organic cotton industry, and currently offers a line of 100 percent organic cotton clothing. Nike plans for all shoes to meet "Considered" standards by 2011, clothing by 2015, and other gear by 2020, at which point waste will decrease by 17 percent and use of environmentally preferable materials will increase by 20 percent.

In addition to designing and fabricating more sustainable shoes and supporting organic agriculture, Nike reduces waste and positively affects communities through its Reuse-A-Shoe program. The company encourages the public to recycle used athletic shoes so that they can be broken down into a substance called Nike Grind. Nike Grind is the raw material that results from the recycling of athletic shoes collected through Nike's Reuse-A-Shoe program and from the recycling of scrap materials left over from the manufacture of Nike footwear. By recycling old shoes and manufacturing scrap, and incorporating this recycled material into sports surfaces and new Nike products, the program reduces landfill waste and reduces the need for extraction of raw materials for athletic facility surfaces. Additionally, Nike athletic shoes are now designed for easier breakdown, which decreases the energy needed to create Nike Grind. Nike has partnered with market-leading sports surfacing companies that use Nike Grind in the manufacture of tennis and basketball courts, running tracks, synthetic turf, children's playgrounds, and fitness room flooring, doubly benefiting the environment and the local population (FIGURE 20.16).

In April of 2009, volunteers used Nike Grind to build the first community playground in the Poncey-Highland neighborhood of Atlanta, Georgia. The rubber surface consisted of 92 percent recycled materials, and equipment was constructed from recycled milk jugs and other reused products. By using cradle-to-cradle practices to reduce environmental damage and support social programs, Nike is working to achieve a more ethical and sustainable business model. Presumably, over time, more companies and industries will develop innovative sustainable business practices. Sustainable practices are an essential part of being a consumer, but they should also be part of business and industry.

### References

Locke, R. M., et al. 2009. Nike Considered: Getting Traction on Sustainability. *MIT Sloan Teaching Innovation Resources (MSTIR)*. https://mitsloan.mit.edu/MSTIR/sustainability/NikeConsidered/Pages/default.aspx.

www.NikeConsidered.com.

**FIGURE 20.16  Nike Grind.** Making this athletic playing surface from recycled sneakers and sneaker manufacturing waste saved resources and energy.

## REVISIT THE KEY IDEAS

- **Discuss sustainability in a variety of environmental contexts including human well-being.**

Sustainable environmental systems must allow for maintaining air, water, land, and biosphere systems and must also maintain human well-being, the status of being healthy, happy, and prosperous. Sustainability will not be achieved if certain groups are exposed to a disproportionate share of dirty jobs or waste material in the home or workplace.

- **Evaluate ways in which traditional economic analysis can do a better job of including the costs of economic activities on the environment and on people.**

Sustainable systems must include a consideration of externalities. Gross domestic product (GDP) is the value of all products and services produced in a year in a given country. Genuine progress indicator (GPI) includes measures of personal consumption, income distribution, levels of higher education, resource depletion, pollution, and health of the population.

- **Understand that economic systems are based on three forms of capital—natural, human, and manufactured.**

Economic assets, or capital, can come from the natural systems on Earth, from humans, or from the manufactured products made by humans. Valuing all three kinds of capital is essential to systems that are sustainable.

- **Explain the role of laws and regulations in attempting to protect our natural and human capital.**

Once a society believes it has enough scientific information to act with the intent of protecting or reducing harm to the environment, it must determine the rules and regulations it wishes to enact. A group of government agencies in the United States handles the areas that offer protection to the environment and humans. Policies are enacted through passage and modification of laws.

- **Define and discuss the relationship among sustainability, poverty, personal action, and stewardship.**

One-sixth of the world population has inadequate housing and inadequate income. People will need access to food, housing, clean water, and adequate medical care before they can be concerned about environmental sustainability. The UN Millennium Goals have established objectives for improving the status of people and the sustainability of the environment. The human-centered worldview maintains that humans have intrinsic value and nature provides for our needs. The life-centered worldview holds that humans are one of many species on Earth, all of which have value. The Earth-centered worldview places equal value on both all living organisms and ecosystems. Individual and community action can lead to sustainable actions occurring at a greater level worldwide.

## CHECK YOUR UNDERSTANDING

1. Based on the supply and demand curve below, which of the following can be inferred?

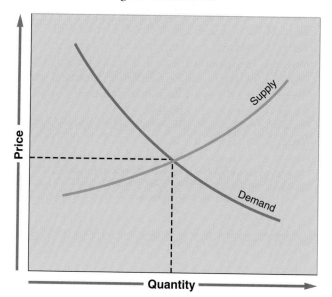

I   A lower price results in a greater demand.
II  A higher price results in a greater supply.
III Price changes as supply and demand fluctuate.

(a) I only             (d) I and II
(b) II only            (e) I, II, and III
(c) III only

2. All of the following are examples of negative externalities *except*

(a) global climate change as a result of greenhouse gas emissions from the burning of coal, oil, and gasoline.
(b) increased pollination rates of surrounding crop plants as a result of local beekeeping.
(c) a pulp mill that produces paper and pollutes the surrounding water and air.
(d) runoff of pesticides and fertilizers from a farm into a nearby river.
(e) acid deposition in the Adirondacks as a result of coal-burning power plants in the Midwest.

3. The genuine progress indicator is a more representative measure of the wealth and well-being of a country than the gross domestic product because the GPI

(a) measures productivity and consumption without taking externalities into account.
(b) has risen in the United States while the GDP has remained fairly constant since 1970.
(c) includes resource depletion and pollution and health of the population in its calculation.
(d) can be increased by higher health care costs and a greater incidence of illnesses.
(e) does not reflect personal consumption, income distribution, or levels of higher education.

4. Economic assets are the sum total of which of the following?

I   Natural capital
II  Human capital
III Manufactured capital

(a) II only            (d) I and III
(b) III only           (e) I, II, and III
(c) I and II

5. Valuation, according to environmental and ecological economics, would include all of the following *except*

(a) the revenue generated from tourists visiting a national park.
(b) the cost of wastewater treatment provided by a natural wetland.
(c) the benefits derived from medicinal plants found in tropical rainforests.
(d) the profits realized from hiring more employees to increase production.
(e) the cost of converting animal wastes into reusable organic matter by detritivores.

6. Full cost pricing by the internalization of externalities could result in which of the following?

(a) Higher prices and a reduction in the consumption of items with high negative impacts

(b) Lower prices and an increase in the consumption of items with high negative impacts

(c) Greater consumer demand for products with high negative impacts

(d) Lower production costs due to diminishing natural resources

(e) Reduced production of environmentally friendly goods and services

7. Cradle-to-cradle and cradle-to-grave analyses of manufactured goods can best be described as the study of the

(a) changes in the use of a product from one generation to the next.

(b) life cycle of a product from its production to use to ultimate disposal.

(c) use of resource extraction over the use of ecosystem services.

(d) options for the disposal of solid waste generated by the product.

(e) natural and human resources required for production.

8. Recently the Los Angeles Unified School District adopted a new policy on the use of pesticides in schools. This policy assumes that the use of pesticides constitutes a risk to the health of children and the environment. Pesticides will be employed only after nonchemical methods have been explored. The pest control measure that is the least harmful will be implemented. This is an example of

(a) the precautionary principle.

(b) ecosystem services.

(c) a market-driven approach.

(d) sustainable use.

(e) full cost pricing.

9. United Nations organizations that relate to the environment include which of the following?

(a) World Resources Institute (WRI)

(b) Occupational Safety and Health Administration (OSHA)

(c) Department of Energy (DOE)

(d) World Health Organization (WHO)

(e) Environmental Protection Agency (EPA)

10. Which of the following U.S. laws contributes to sustainability by governing the tracking and disposal of solid and hazardous waste?

(a) National Environmental Policy Act (NEPA)

(b) Resource Conservation and Recovery Act (RCRA)

(c) Clean Water Act (CWA)

(d) Comprehensive Environmental Response, Compensation, and Liability Act (CERCLA)

(e) Occupational Safety and Health Act (OSHA)

11. Strategies to implement environmental laws and regulations include all of the following *except*

(a) standards for emission levels with fines when these levels are exceeded.

(b) green taxes on environmentally harmful activities or emissions.

(c) buying and selling of pollution permits.

(d) an incentive-based approach based on profits.

(e) banning the cap-and-trade practice.

12. Harmful effects of poverty include which of the following?

(a) Decrease in unsanitary conditions

(b) Greater access to clean drinking water

(c) Increased overuse of the land

(d) Decreased malnutrition

(e) Lower infant mortality rates

13. The United Nations Millennium Declaration proposes to meet which of the following goals by 2015?

 I  Reduce environmental sustainability through economic development

 II  Eliminate extreme poverty and hunger and reduce child mortality

 III  Empower women and improve maternal health

(a) I only

(b) II only

(c) IIII only

(d) I and III

(e) II and III

14. The following is a summary report for the Distribution of Environmental Burdens for Allegheny County in Pennsylvania.

| Population categories | Number of facilities emitting criteria air pollutants per square mile |
|---|---|
| Minorities | 11 |
| Whites | 4.5 |
| Low-income families | 8 |
| High-income families | 3.9 |
| Families below poverty | 8.9 |
| Families above poverty | 4.1 |
| Non–high school graduates | 6.9 |
| High school graduates | 4.7 |

*Source:* http://www.scorecard.org/community/ej-summary.tcl?fips _county_code=42003&lang=eng#map>.

The information in this table is an example of

(a) an environmental equity issue.

(b) an anthropocentric worldview.

(c) a biocentric worldview.

(d) an ecocentric worldview.

(e) a stewardship school issue.

1. Use the following information to answer the questions that follow.

Gasoline consumption in the United States:

- In 2008 the United States consumed approximately 138 billion gallons of gasoline.
- The current federal tax on gasoline is 18.4 cents per gallon.
- The national average cost of a gallon of regular unleaded gasoline in June 2008 was $4.00 per gallon.
- 80 percent of the federal gasoline tax is used to subsidize road construction.

(a) Calculate the total amount of money spent in the United States on the purchase of gasoline in 2008 (when gasoline cost $4.00 per gallon).

(b) i. What percent of the cost per gallon is the gasoline tax?
  ii. How much revenue was generated by the gasoline tax in 2008?
  iii. How much was used in 2008 to subsidize road construction?

(c) Does the federal tax on gasoline qualify as a *green tax?* Explain your answer.

(d) Advocates of raising the gasoline tax suggest that the tax be increased to 80 cents per gallon. Identify *two* economic effects and *two* environmental effects of raising this tax.

2. In 1997, the ecological economist Robert Costanza and his associates published a report titled *The Value of the World's Ecosystem Services and Natural Capital.* They estimated that if all the ecosystem services provided worldwide had to be paid for, the cost would average $33 trillion per year with a range from $16 trillion to $54 trillion. In that same year the global gross national product (GNP) was $18 trillion.

(a) What is meant by ecosystem or ecological services? Give three specific examples and identify which United Nations organization might oversee these services.

(b) Define the term *valuation.* What would the worldwide consequences be if the world actually had to pay for ecosystem services and natural capital?

(c) Explain how this report could be used to develop a sustainable economic system.

(d) Which environmental worldview is most consistent with environmental economics? Explain.

## MEASURE YOUR IMPACT

**GDP and Footprints** The World Wide Fund for Nature (WWF) released the *Report on Ecological Footprint in China* in June 2008. Answer the following questions using the data below, which is based on this report.

| | Ecological footprint (global hectares/person) | | | GDP per capita (thousands of U.S. dollars) | | |
|---|---|---|---|---|---|---|
| | 1970 | 2000 | % change | 1970 | 2000 | % change |
| China | 0.7 | 1.5 | | 0.1 | 1.2 | |
| India | 0.7 | 0.7 | | 0.22 | 0.48 | |
| Japan | 2.5 | 4.3 | | 15 | 37 | |
| United States | 6.1 | 9.3 | | 18 | 34 | |

(a) Calculate the percentage change in ecological footprint and GDP for each country listed and complete the table.

(b) In the 30-year period cited in the data, what is the relationship between ecological footprint and GDP?

(c) Discuss environmental equity using India as an example of a developing country and the United States as an example of a developed country by comparing the ecological footprints and GDPs for the year 2000.

(d) Based on your answer in (c), identify *two* goals of the United Nations Millennium Declaration that would assist India in making its environmental inequities more equitable?

## ENGAGE YOUR ENVIRONMENT

How can you help your community move toward sustainability? This Engage Your Environment exercise asks you to assess the needs of your community in terms of sustainability, and then to inventory your talents and interests.

Through this exercise you will determine which projects in your community will be a good fit for your involvement. *See Engage Your Environment Activity 20 at the end of this book.*

# science applied

## Can We Solve the Carbon Crisis Using Cap-and-Trade?

In Chapter 19, we saw that the nations of the world have become concerned about the increasing amount of $CO_2$ in the atmosphere of Earth. While the Kyoto Protocol calls for developed nations to reduce total $CO_2$ production to 5.2 percent below 1990 levels by 2012, there has been a great deal of debate regarding the most effective way to achieve this reduction.

### What are our options for controlling $CO_2$ emissions?

Although many different options for controlling $CO_2$ emissions have been considered, most approaches can be categorized as either *command-and-control* or *cap-and-trade*. As discussed in Chapter 20, with a command-and-control approach, a government regulates the amount of pollution that can be emitted by different industries. For example, in 2010 the U.S. federal government announced substantially higher fuel-efficiency standards for cars and trucks (FIGURE SA8.1). These mandates would reduce the consumption of gasoline and thereby lower the output of $CO_2$ emitted into the atmosphere. The new standards, which ultimately received support from the major automobile manufacturers, require new technology that would make the vehicles more expensive for consumers, but this higher initial expense is more than offset by reduced fuel costs over the life of each vehicle.

While a command-and-control approach means the government sets a single mandatory standard to which all companies must adhere, **cap-and-trade** uses a different philosophy. As the name suggests, there are two elements to this approach. First, there is a cap, or upper limit, that is placed on the amount of a pollutant that

**FIGURE SA8.1** **The command-and-control approach.** When the federal government sets minimum standards for fuel efficiency, it is using a command-and-control approach to pollution.

can be emitted. Instead of placing a limit on individual companies or industries, the limit is placed on the total amount of the pollutant produced in a region, such as a state, country, or group of cooperating countries. In the case of $CO_2$, for example, a cap could be placed on the total amount of $CO_2$ that a state or country could emit. To meet the goals of the Kyoto Protocol, for example, the United States agreed to a 7 percent reduction in $CO_2$ from 1990 levels. This value could serve as a cap for the United States.

The *trade* portion of the cap-and-trade approach should be familiar from Chapter 16. It allows companies to buy and sell permits for a given amount of pollution. A government can either require all companies that emit pollutants to buy permits or it can distribute a large number of free permits and then slowly reduce the number of permits available to be traded. Allowing companies to buy and sell pollution permits provides an incentive for companies to invest in pollution reduction, since they would gain income by selling permits they no longer need to companies that continue to pollute. Companies still generating large amounts of pollution will also be motivated to reduce their emissions since they would have to spend more to buy the permits, assuming that the cost of the permit is higher than the cost of not polluting (FIGURE SA8.2).

Using this strategy, over time, the government would gradually reduce the number of pollution permits available. By the laws of supply and demand, reducing the supply of permits (lowering the cap) causes an increased demand for the remaining permits and results in higher prices for the permits. This price increase further motivates polluting companies to reduce their emissions, and the total amount of pollution declines. The cap-and-trade approach avoids the complex regulations that occur with the command-and-control approach and gives companies the freedom to choose from a wide range of possible solutions the approach that best suits them. At the same time, it offers an economic incentive to reduce pollution. The rate at which the pollution declines would depend on how quickly the government reduces the number of pollution permits.

## What are the concerns about controlling carbon using cap-and-trade?

There is currently a great deal of debate concerning the use of cap-and-trade as a means of controlling $CO_2$ emissions. One of the major issues is that most cap-and-trade proposals for $CO_2$ include an additional aspect known as *carbon offsets*. **Carbon offsets** are methods of promoting global $CO_2$ reduction that do not involve a direct reduction in the amount of $CO_2$ actually emitted by a company. For example, a company could pay for reforesting an area that could absorb $CO_2$ or pay a landowner not to log a forest, thereby keeping the carbon locked up in the biomass of the trees. By paying these costs, a company could avoid paying a potentially higher cost of reducing its own $CO_2$ emissions. Opponents

**FIGURE SA8.2** **Carbon cap-and-trade approach.** Many different industries emit large amounts of $CO_2$, including coal-fired electricity-generating power plants, automobiles, airplanes, and steel manufacturers. Under cap-and-trade, all of these industries could be involved in buying and selling permits for $CO_2$ emission.

contend that while these efforts would allow a particular piece of land to sequester carbon, it does not decrease the total amount of land needed to grow lumber or crops. As a result, the deforestation simply occurs in another location and the carbon offset does nothing to reduce the global production of $CO_2$. Furthermore, even if a section of land is allowed to sequester carbon, changes in government regulations or political control, or even a natural event such as a forest fire, could rapidly convert sequestered carbon to atmospheric carbon dioxide.

Exemption of certain companies or facilities is another concern about the cap-and-trade approach. Most experts agree that such exemptions are likely to occur. As we saw in Chapter 15, the cap-and-trade approach was used for the control of sulfur dioxide emissions from coal-powered plants in 1990 to reduce acid precipitation. Many of the older power plants were exempt from the program because at the time it was believed that these plants would soon be retired. However, nearly two-thirds of the coal-powered plants in operation today were constructed before 1975. For cap-and-trade to be a fair system, all polluting companies will need to pay for polluting permits.

## Has cap-and-trade ever worked successfully?

The concept of cap-and-trade, developed by economists in the 1970s, has experienced limited use in the decades since. As noted above, as part of the U.S. Clean Air Act of 1990, it was first implemented to reduce sulfur dioxide emissions that caused acid precipitation. Following a suggestion from the Environmental Defense Fund, the federal government developed a cap-and-trade program for coal-powered plants. Motivated by the new cost of buying pollution permits, the companies owning these plants had to retrofit existing facilities, build new and less polluting facilities, or buy pollution permits from other companies that had already reduced their emissions. Even though some of the older plants were exempt from certain provisions of the act, from 1990 to 2009 there has been a 64 percent reduction in sulfur emissions. The cap-and-trade system motivated investment in new technologies, which is one of the underlying reasons for the success of this program. For example, General Electric developed a system that converted sulfur into gypsum. Because gypsum is a valuable product used in construction, the economic incentive to reduce sulfur emissions motivated new research and development that produced a profitable solution.

In 2005, a carbon cap-and-trade system was instituted in the European Union. Critics of the European system point out that the national governments in the European Union have overestimated the baseline amount of $CO_2$ that their industries were producing and have given away most of the permits to the polluters for free. As a result, these companies have surplus $CO_2$ permits that they can sell at a profit. In rebuttal, proponents of the system point out that this strategy was necessary at the beginning to gain the support of companies. Starting the system with a surplus of permits and then gradually reducing the number of permits available allowed for a gradual adjustment by the companies that emitted $CO_2$. In 2010, the European Union started to lower the cap and more companies, including electricity-generating plants and airlines, will begin paying for the permits. Some industries, such as steel and cement manufacturers, however, have argued that they will need to continue having free permits if they are to remain competitive with companies that operate outside the European Union.

At this stage, it is too early to tell if the cap-and-trade system will be a successful strategy to reduce $CO_2$ emissions. Cap-and-trade was successful in reducing sulfur emissions, but in that case there were no offsets awarded that could be used as credit and allow continued pollution by a company. If carbon cap-and-trade is used in the United States, a major focus will be placed on energy companies, including those that produce petroleum products and those that generate electricity. A recent study by Point Carbon, a market research company, found that the major oil companies, including ExxonMobil and Chevron, would likely pay hundreds of millions of dollars in $CO_2$ permits while electricity-generating companies that have diversified to include hydroelectric and nuclear energy will pay substantially less. Ultimately, if cap-and-trade is successful in the European Union, it could serve as an effective model for other nations and help reduce the growing problem of $CO_2$ and the global changes that come with it.

## References

Hansen, J. 2009. Cap and fade. *New York Times,* December 7. http://www.nytimes.com/2009/12/07/opinion/07hansen.html?_r=1&scp=1&sq=cap%20and%20fade%20hansen&st=cse.

Krugman, P. 2009. Unhelpful Hansen. *New York Times,* December 7. http://krugman.blogs.nytimes.com/2009/12/07/unhelpful-hansen/.

# APPENDIX

# Fundamentals of Graphing

Many types of graphs are included throughout this text. If you are unfamiliar with how to read or interpret graphs or would like to review the basics, this appendix is for you.

## Scientists use graphs to present data and ideas

A graph is a tool that allows scientists to visualize data or ideas. Organizing information in the form of a graph can help us understand relationships more clearly. Throughout your study of environmental science you will encounter many different types of graphs. In the sections that follow we will look at the basics of how a graph is constructed and then at specific examples of graphs that environmental scientists use.

### Graphing Basics

Although many of the graphs in this book may look different from each other, they all follow the same basic principles. Let's begin by looking at the different parts of a graph. FIGURE A.1 presents data from a study that examined the number of bird species found on islands of different sizes in Lake Velence, Hungary. As in most scientific studies, the researchers looked at two variables, which are quantities that can take on more than one value. In this case, the variables were the area of land and the number of bird species.

Once the researchers gathered their data, they plotted the data on a graph. The points that you see represent different combinations of the two variables observed. For each observation a certain amount of land—shown on the horizontal, or $x$, axis—is associated with a particular number of bird

species—shown on the vertical, or $y$, axis. For example, on an island with an area of 50 ha, the researchers found 12 species of birds. Notice that on this graph, the $x$ axis is labeled in increments of 50 hectares (ha) from 0 to 200, and the $y$ axis is labeled in increments of 5 species, from 0 to 25. The difference in quantities reflects the different characteristics in the variables being measured; overall there are fewer species of birds than hectares of land.

When two variables are plotted, we can draw a line, described as a curve, through the points. This curve allows us to visualize a general trend: as the size of the island grows larger, the number of species also increases. This is known as a positive relationship between the two variables because they move together in the same direction.

## Scientists choose from many different types of graphs

After researchers have gathered a set of data and have determined that it would be useful to see the results in a visual display, they can choose from many possible formats. This section looks at some standard types of graphs.

### Scatter Plots

A scatter plot graph allows a researcher to observe whether or not the data shows a general trend, such as one variable increasing and the other decreasing. For the graph shown in FIGURE A.2, researchers who were investigating a possible relationship between per capita income and total fertility rate plotted data points for 11 countries. By convention, the place where the two axes converge in the bottom left corner, called the origin, represents a value of 0 for each variable. The units of measurement tend to get larger as we move from left to right on the $x$ axis, and from bottom to top on the $y$ axis.

The pattern of points shows a negative relationship between income and fertility rate. That is, as income increases, the fertility rate decreases. Although the points do not connect to form a smooth line, the graph includes a line of approximation calculated to fit the general relationship of the variables. Although this figure shows only 11 data points, scatter plot graphs may include hundreds of data points.

### Line Graphs

A line graph is among the most familiar types of graphs; it is used most often to display data that occur as a sequence of measurements over time or distance. Figures A.1 and A.2 are both line graphs. There are several types of line graphs including linear, nonlinear, and time series plots.

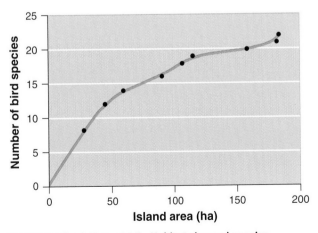

**FIGURE A.1 (book Figure 6.26)** Habitat size and species richness.

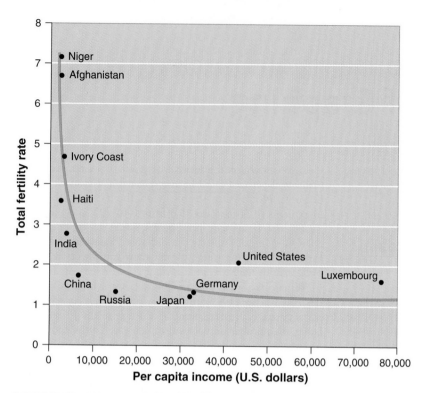

**FIGURE A.2 (book Figure 7.10)** Total fertility rate versus per capita income.

**LINEAR GRAPHS** FIGURE A.3 is an example of a line graph that illustrates an idea rather than a specific set of data. In this figure from Chapter 20, the law of supply and demand is expressed as two linear relationships. In the line representing supply (S) we see that as price goes up the quantity supplied also goes up. The smooth, straight curve indicates that this is a linear relationship; at every increase of price there is an equivalent rise in the amount of the item supplied. Conversely, for the line that represents the demand (D), at every increase of price the quantity demanded goes down. The point where the two lines meet is known as the equilibrium point, where the amount supplied and the quantity demanded are the same.

**NONLINEAR GRAPHS** In contrast to Figure A.3, FIGURE A.4 shows a nonlinear graph. The *x* axis represents the amount of disturbance in an ecosystem from infrequent to frequent. The *y* axis represents the number of species found in an ecosystem from low to high. The line climbs upward to a peak and then falls again. This shape reflects the intermediate disturbance hypothesis, which states that species diversity is highest—the peak of the arc—at intermediate levels of disturbance. Where disturbance is infrequent, the best competitors outcompete other species and species richness suffers. At the other end of the spectrum, where disturbance is high, only those species that have evolved to live under such conditions are able to survive.

Figure A.1 is also an example of a nonlinear graph. Looking at the data, we can see why: the rate of growth in species richness does not remain constant for every increment of additional land area. When land area increases from 0 ha to 50 ha, the number of species increases from 0 to 12. But when we add another 50 ha, moving from 50 ha to 100 ha, the number of species goes from 12 to 16, increasing by only 4. Throughout your study of environmental science you will become very familiar with nonlinear relationships, in which a variable changes more quickly at first and then slows down, as in Figures A.1 and A.4.

**TIME SERIES PLOTS** One form of graph that appears often in this book is a linear graph known as a time series plot. A time series graph plots the change of a variable over a period of time. Time is shown on the *x* axis and the variable

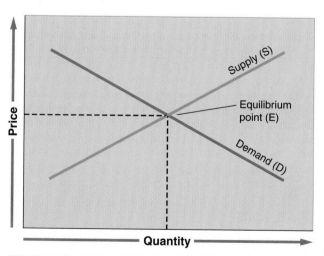

**FIGURE A.3 (book Figure 20.1)** Supply and demand.

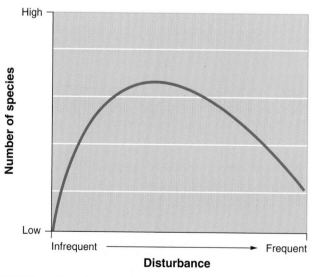

**FIGURE A.4 (book Figure 3.19)** Intermediate disturbance hypothesis.

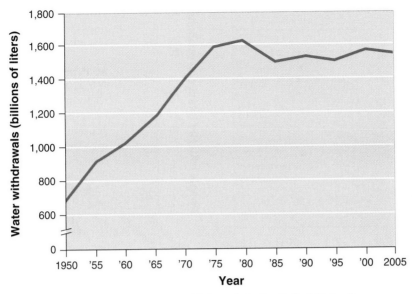

**FIGURE A.5 (book Figure 9.24)** Water withdrawals in the United States from 1950 to 2005.

being studied is on the *y* axis. **FIGURE A.5** shows a simple time series plot. The variable on the *y* axis is the amount of water withdrawals in the United States. The time period under consideration is 1950 to 2005. As you can see, a clear picture emerges: water use peaked in 1980 and has since leveled off.

Note the double hatch mark on the *y* axis between 0 and 600. This indicates a break in the scale. As you can see, from 600 to 1,800 the *y* axis increases in increments of 200. But from 0 to 600, the *y* axis increases by an increment of 600. The double hatch mark in that part of the graph allows us to condense part of the *y* axis, especially when we wish to focus on how the data change after the double hatch mark. In Figure A.5, there are no data points less than 600, so we compress the *y* axis between 0 and 600 so that we can better view the change in water use between 600 and 1,800.

A more complicated time series plot can illustrate several different variables against a period of time. **FIGURE A.6** presents data on changes in populations of two different animals on Isle Royale in the years 1955 to 2005. The left *y* axis represents the population growth of wolves. The right *y* axis plots the population changes of their prey, moose, during the same period of time. You can see that the changes in population sizes seem to mirror each other: as the wolf population falls, the moose population rises.

It is important to realize that an apparent relationship between two variables, such as the moose population increasing in number at the same time that the wolf population declines in number, may not reflect a relationship of cause and effect.

In other words, although it may appear that one event causes the other, this may not be the case. For example if we were to construct a time series plot that showed the months of the year on the *x* axis, the sales of sunscreen on one *y* axis, and the sales of hot chocolate on the other *y* axis, we might find that when sales of sunscreen fall to their lowest point, say in January, sales of hot chocolate reach their highest point. But this would not mean that drinking a lot of hot chocolate reduces the need for sunscreen. Rather a third variable, temperature, can account for both results. In the case of Figure

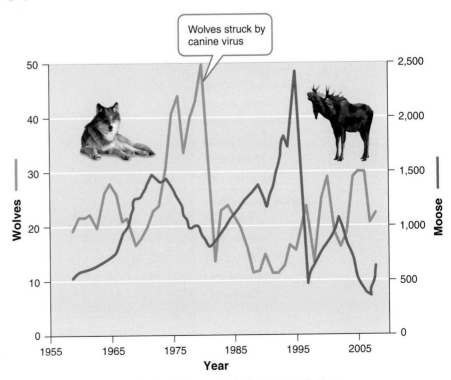

**FIGURE A.6 (book Figure 6.11)** Predator control of prey populations.

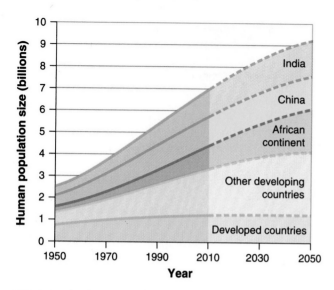

FIGURE A.7 (book Figure 7.14) Population growth past and future.

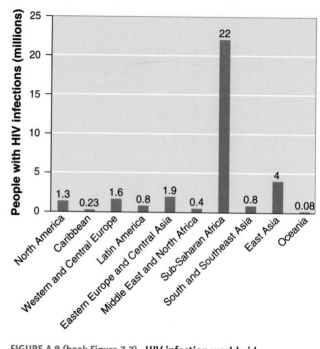

FIGURE A.8 (book Figure 7.7) HIV infection worldwide.

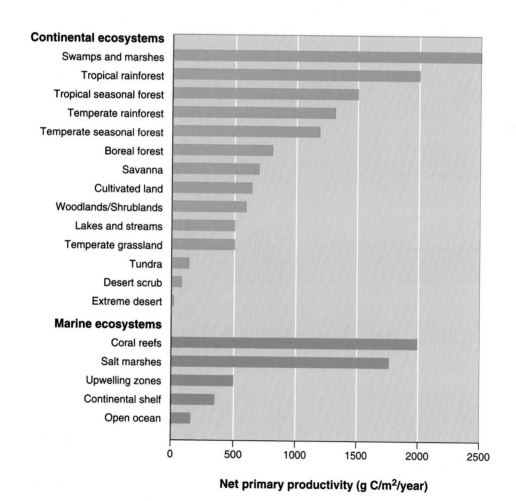

FIGURE A.9 (book Figure 3.8) Net primary productivity varies among ecosystems.

A.6, there is a plausible cause and effect relationship. During the time period under consideration, the wolf population on Isle Royale became ill with a canine virus. As you might expect, when the population of wolves plunged, there was a corresponding rise in the population of their prey. So, in fact, the change in one variable caused the change in the other. Scientists must always be very careful when suggesting cause and effect; the data must clearly support the conclusion.

**PROJECTIONS** Sometimes scientists want to show not only data they have gathered in their research but also projections, or extrapolations, of future trends based on that research. For example, the graph in FIGURE A.7 presents population growth statistics for developed and developing countries. Notice that the solid lines representing the five data sets change to dashed lines at the year 2010. The dashed lines indicate projections, in this case projections of population growth beyond 2010, the last year for which actual data are available.

## Bar Graphs

A bar graph, or bar chart, shows a relationship between a group of classifications or categories and a numerical scale. FIGURE A.8 is a good example. In this figure, the *x* axis contains categories—regions of the world. The *y* axis represents a numerical value—the number of people infected with HIV. The visual impact of the different bar heights provides a dramatic comparison of the incidence of HIV in different regions.

A bar graph is a very flexible tool and can be altered in several ways to accommodate data sets of different sizes or even several data sets that a researcher wishes to compare. In FIGURE A.9, the categories—various ecosystems—are on the *y* axis and the scale—net primary productivity—is on the *x* axis. This orientation makes it easier to accommodate the relatively large amount of text needed to name each ecosystem.

FIGURE A.10 shows an example of a bar graph that presents two data sets for comparison. The left *y* axis represents total annual energy consumption of each country. The right *y* axis plots annual energy consumption per capita (per person). Note that we saw this type of graph in the example shown in Figure A.6. Notice how much information we can gather from the way these data are presented; the graph allows us to compare total annual energy consumption versus per capita annual energy consumption within countries and energy consumption between countries.

## Pie Charts

A pie chart is a graph represented by a circle with slices of various sizes representing categories within the whole; it provides a certain amount of flexibility to further subdivide these categories. Each slice is sized according to the proportion, or percentage, of the pie that it represents. For example, in FIGURE A.11, the entire pie represents leading causes of death in the world. The pie is divided into seven slices, each representing one of the leading causes. As you can see, the slice of the pie given to cardiovascular diseases is significantly larger than cancers, but not much larger than infectious diseases. In this chart, the category of infectious diseases is further defined by its own pie chart, which reveals that respiratory infections are the leading cause of death by infectious disease.

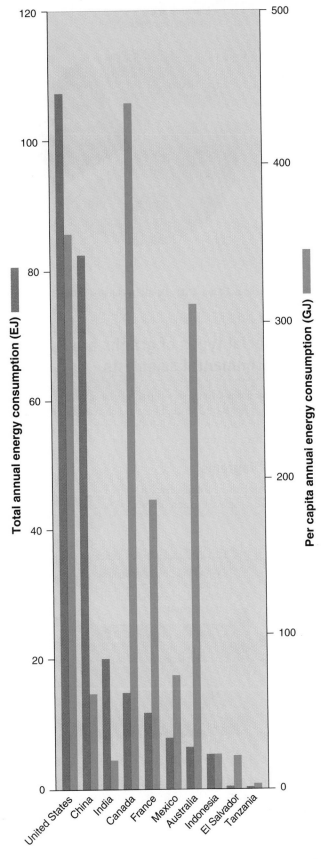

FIGURE A.10 (book Figure 12.2)  Global variation in total annual energy consumption and per capita energy consumption.

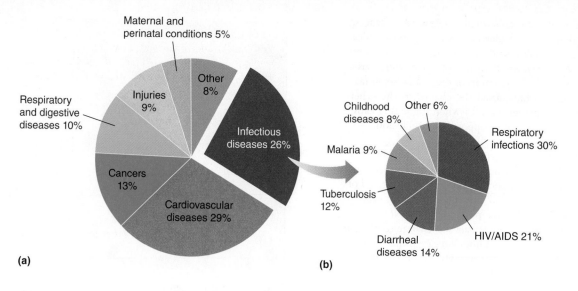

FIGURE A.11 (book Figure 17.1)  **Leading causes of death in the world.**

# Two special types of graphs are used by environmental scientists

This text includes two types of graphs that are not common in most other fields of science. These are climate diagrams and age structure diagrams. Although these graphs are discussed within the text, we provide them here for review.

## Climate Diagrams

To understand the productivity of a biome, environmental scientists use a special graph known as a climate diagram.

The examples shown in **FIGURE A.12** present two hypothetical biomes. By graphing the average monthly temperature and precipitation of a biome, we can see how conditions in a biome vary during a typical year. We can also observe the specific time period when the temperature is warm enough for plants to grow. In the biome illustrated in panel (a), the growing season is mid-March through mid-October. In panel (b) it is mid-April through mid-September.

In addition to allowing the reader to identify the growing season, climate diagrams can show the relationships among precipitation, temperature, and plant growth. In panel (a), the precipitation line is above the temperature line in every

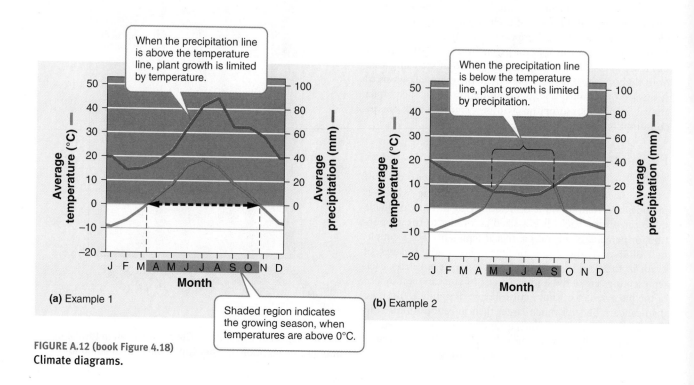

FIGURE A.12 (book Figure 4.18)
**Climate diagrams.**

month. This means that water supply exceeds demand, so plant growth is more constrained by temperature than by precipitation. In panel (b), the precipitation line intersects the temperature line. At this point, the amount of precipitation available to plants equals the amount of water lost by plants through evapotranspiration. When the precipitation line falls below the temperature line, water demand exceeds supply and plant growth will be constrained more by precipitation than by temperature.

## Age Structure Diagrams

Age structure diagrams are visual representations of age distribution for both males and females in a country. FIGURE A.13 presents four examples. Each horizontal bar of the diagram represents a 5-year age group. The total area of all the bars in the diagram equals the size of the whole population.

Every nation has a unique age structure, but we can group countries very broadly into three categories. Panel (a)

(a) India

(b) United States

(c) Germany

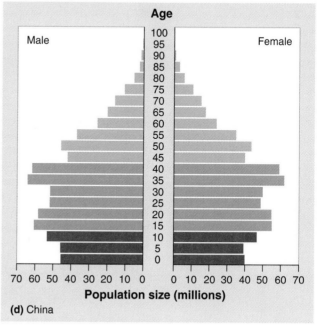

(d) China

FIGURE A.13 (book Figure 7.8)  Age structure diagrams.

shows a country with many more young people than older people. The age structure diagram of a country with this population will be in the shape of a pyramid, with its widest part at the bottom, moving toward the smallest at the top. Because of this, an age structure diagram with this shape is known as a "population pyramid." Population pyramids are typical of countries in the developing world, such as Venezuela and India.

A country with less difference between the number of individuals in the younger and older age groups has an age structure diagram that begins to look more like a column. With fewer individuals in the younger age groups, we can deduce that the country has slow population growth or is approaching no growth at all. The United States, Canada, Australia, Sweden, and many other developed countries have this type of age structure diagram, seen in panel (b).

Panels (c) and (d) show countries with a proportionally larger number of older people. This age structure diagram resembles an inverted pyramid. Such a country has a decreasing number of males and females within each younger age range and that number will continue to shrink. Italy, Germany, Russia, and a few other developed countries display this pattern. In recent years China has also begun to show this pattern.

# BIBLIOGRAPHY

## Chapter 1

Meadows, D. H., J. Randers, and D. L. Meadows. 2004. *Limits to Growth: The 30-Year Update*. Chelsea Green. See also http://www.sustainer.org/.

United Nations Development Programme. 2006. *Human Development Report 2006*. Oxford University Press.

*Vital Signs, 2009*. Worldwatch Institute.

Wackernagel, M., et al. 2004. *Ecological Footprint and Biocapacity Accounts 2004: The Underlying Calculation Method*. Global Footprint Network. www .footprintnetwork.org.

## Chapter 2

Hart, J. 1996. *Storm Over Mono: The Mono Lake Battle and the California Water Future*. University of California Press.

Kendall, K. E., and J. E. Kendall. 2008. *Systems Analysis and Design*. 7th ed. Prentice Hall.

Meadows, D. 2008. *Thinking in Systems: A Primer*. Chelsea Green.

Moore, J. T. E. 2005. *Chemistry Made Simple*. Broadway Press.

## Chapter 3

Bush, M. B. 2000. *Ecology of a Changing Planet*. 2nd ed. Prentice Hall.

Crowder, L. B., et al. 2008. The impacts of fisheries on marine ecosystems and the transition to ecosystem-based management. *Annual Review of Ecology, Evolution, and Systematics* 39: 259–278.

Fiedler, A. K., et al. 2008. Maximizing ecosystems services from conservation biological control: The role of habitat management. *Biological Control* 45: 254–271.

Letourneau, D. K., et al. 2008. Comparison of organic and conventional farms: Challenging ecologists to make biodiversity functional. *Frontiers in Ecology and the Environment* 6: 430–438.

Odum, E. P. 1997. *Ecology: A Bridge Between Science and Society*. 3rd ed. Sinauer.

Ricklefs, R. E. 2008. *The Economy of Nature*. 6th ed. W. H. Freeman.

## Chapter 4

Marimi, M. A., et al. 2009. Predicted climate-driven bird distribution changes and forecasted conservation conflicts in a neotropical savanna. *Conservation Biology* 23: 1558–1567.

National Science Teachers Association. World biomes.com. http://www.worldbiomes.com/.

U.S. Environmental Protection Agency. Wetlands. http://www.epa.gov/owow/wetlands/.

## Chapter 5

Coyne, J. A. 2009. *Why Evolution Is True*. Penguin.

Evolution: A Journey into Where We're From and Where We're Going. http://www.pbs.org/wgbh/evolution/.

Futuyma, D. J. 2009. *Evolution*. 2nd ed. Sinauer.

Magurran, A. E. 2003. *Measuring Biological Diversity*. Wiley-Blackwell.

University of California Museum of Paleontology. *Understanding Evolution: Your One-Stop Source for Information on Evolution*. http://evolution.berkeley.edu/.

## Chapter 6

Hanski, I., and O. E. Gaggiotti. 2004. *Ecology, Genetics, and Evolution of Metapopulations*. Elsevier.

Ricklefs, R. 2008. *The Economy of Nature*. 6th ed. W. H. Freeman.

Stein, B. A., L. S. Kutner, and J. S. Adams. 2000. *Precious Heritage: The Status of Biodiversity in the United States*. Oxford University Press. http://www.natureserve.org/publications/preciousHeritage.jsp.

U.S. Department of the Interior, National Park Service. Exploring Nature: Preserving Biodiversity. http://www.nature.nps.gov/biodiversity/index.cfm.

## Chapter 7

Cohen, J. E. 2003. Human population: The next half-century. *Science* 304: 1172–1175.

Dodds, W. K. 2008. *Humanity's Footprint*. Columbia University Press.

Global Footprint Network. http://www.footprintnetwork .org.

Myrskyla, M., et al. 2009. Advances in development reverse fertility declines. *Nature* 460: 741–743.

Weeks, J. R. 2007. *Population: An Introduction to Concepts and Issues*. 10th ed. Wadsworth.

## Chapter 8

Christopherson, R. W. 2008. *Geosystems: An Introduction to Physical Geography*. 7th ed. Prentice Hall.

Grotzinger, J., and T. H. Jordan. 2010. *Understanding Earth*. 6th ed. W. H. Freeman.

McPhee, J. 1982. *Basin and Range*. Farrar, Straus and Giroux.

U.S. Geological Survey. http://www.usgs.gov/.

## Chapter 9

Gleick, P. H. 2009. *The World's Water 2008–2009*. Island Press.

Postel, S., and B. Richter. 2004. *Rivers for Life: Managing Water for People and Nature*. Island Press.

## Chapter 10

Farr, D. 2007. *Sustainable Urbanism: Urban Design with Nature*. Wiley.

Flint, A. 2006. *This Land: The Battle Over Sprawl and the Future of America*. Johns Hopkins University Press.

Prescott, S. T. 2003. *Federal Land Management: Current Issues and Background*. Nova Science Publishers.

U.S. Environmental Protection Agency. Smart Growth. http://www.epa.gov/dced/index.htm.

## Chapter 11

Butterly, J. R., and J. Shepherd. 2010. *Hunger: The Biology and Politics of Starvation*. Dartmouth College Press.

Ikerd, J. 2008. *Crisis and Opportunity: Sustainability in American Agriculture*. Bison Books.

Pollan, Michael. 2007. *The Omnivore's Dilemma*. Penguin.

Seafood Watch, Monterey Bay Aquarium. http://www.montereybayaquarium.org/cr/seafoodwatch.aspx.

*The State of Food Insecurity in the World*. 2009. United Nations Food and Agriculture Organization.

U.S. Department of Agriculture. http://www.usda.gov/wps/portal/usdahome.

U.S. Department of Agriculture, National Agricultural Statistics Service. http://www.nass.usda.gov.

## Chapter 12

Nersesian, R. L. 2010. *Energy for the 21st Century*. 2nd ed. M. E. Sharpe.

Ristinen, R. A., and J. J. Kraushaar. 2006. *Energy and the Environment*. 2nd ed. Wiley.

Toossi, R. 2008. *Energy and the Environment*. 2nd ed. Verve Publishers.

U.S. Department of Energy, Energy Information Administration. http://www.eia.doe.gov.

Wolfson, R. 2008. *Energy, Environment, and Climate*. W. W. Norton.

## Chapter 13

Friedland, A. J., and K. T. Gillingham. 2010. Carbon accounting a tricky business (letter). *Science* 327: 410–411.

International Energy Agency. http://www.iea.org.

Kruger, P. 2006. *Alternative Energy Resources*. Wiley.

Randolph, J., and G. M. Masters. 2008. *Energy for Sustainability*. Island Press.

U.S. Department of Energy, National Renewable Energy Laboratory. http://www.nrel.gov.

U.S. Energy Information Administration, Independent Statistics and Analysis. http://www.eia.doe.gov.

Wolfson, R. 2011. *Energy, Environment, and Climate*. 2nd ed. W. W. Norton.

## Chapter 14

Evers, D. C., et al. 2007. Biological mercury hotspots in the northeastern United States and southeastern Canada. *BioScience* 57: 29–43.

Payne, J. R., et al. 2008. Long-term monitoring for oil in the *Exxon Valdez* spill region. *Marine Pollution Bulletin* 56: 2067–2081.

Revkin, A. C. 2009. Dredging of pollutants begins in Hudson. *New York Times*, May 16.

## Chapter 15

Baird, C., and M. Cann. 2008. *Environmental Chemistry*. 4th ed. W. H. Freeman.

Manahan, S. E. 2009. *Environmental Chemistry*. 9th ed. CRC Press.

Ramanathan, V. M., et al. 2008. *Atmospheric Brown Clouds: Regional Assessment Report with Focus on Asia*. United Nations Environment Programme. U.S. Environmental Protection Agency. Section on Air: http://www.epa.gov/ebtpages/air.html.

## Chapter 16

Appelhof, M. 1997. *Worms Eat My Garbage: How to Set Up and Maintain a Worm Composting System*. Flower Press.

McDonough, W., and M. Braungart. 2002. *Cradle to Cradle: Remaking the Way We Make Things*. North Point Press.

UN-HABITAT. 2010. *Solid Waste Management in the World's Cities: Water and Sanitation in the World's Cities*. Earthscan.

U.S. Environmental Protection Agency. 2008. *Municipal Solid Waste Generation, Recycling and Disposal in the United States: Facts and Figures for 2008*. http://www.epa.gov/osw/nonhaz/municipal/pubs/msw2008rpt.pdf.

## Chapter 17

Newman, M. C., and M. A. Unger. 2009. *Fundamentals of Ecotoxicology*. 3rd ed. CRC Press.

World Health Organization. 2009. *Global health risks: Mortality and burden of disease attributable to selected major risks*. WHO Press. http://www.who.int/entity/healthinfo/global_burden_disease/GlobalHealthRisks_report_full.pdf.

## Chapter 18

Convention on Biological Diversity. 2010. *Global Biodiversity Outlook 3*. www.cbd.int/gbo/gbo3/doc/GBO3-Summary-final-en.pdf.

Rosenthal, E. 2005. Food for thought: Crop diversity is dying. *New York Times*, August 18. http://www.nytimes.com/2005/08/17/world/europe/17iht-food. html?_r=1&scp=1&sq=Food%20for%20thought:%20Crop%20diversity%20is%20dying&st=cse.

## Chapter 19

Houghton, J. 2004. *Global Warming: The Complete Briefing*. 3rd ed. Cambridge University Press.

Intergovernmental Panel on Climate Change. *Climate Change 2007: Synthesis Report. Fourth Assessment Report of the Intergovernmental Panel on Climate Change*. http://www.ipcc.ch/pdf/assessment-report/ar4/syr/ar4_syr.pdf.

National Research Council. 2002. *Abrupt Climate Change: Inevitable Surprises*. National Academies Press.

U.S. Global Change Research Program. 2009. *Global Climate Change Impacts in the United States*. http://downloads.globalchange.gov/usimpacts/pdfs/climate-impacts-report.pdf.

## Chapter 20

Bullard, R. 2009. *Race, Place, and Environmental Justice After Hurricane Katrina*. Westview Press.

Goodstein, E. 2007. *Economics and the Environment*. 5th ed. Wiley.

Hawken, Paul. 1994. *The Ecology of Commerce*. Collins.

Jones, Van. 2009. *The Green Collar Economy: How One Solution Can Fix Our Two Biggest Problems*. HarperOne.

# GLOSSARY

**A horizon** Frequently the top layer of soil, characterized by mixing of organic material and mineral material.

**abiotic** Nonliving.

**accuracy** How close a measured value is to the actual or true value.

**acid** A substance that contributes hydrogen ions to a solution.

**acid deposition** Acids deposited on Earth as rain and snow or as gases and particles that attach to the surfaces of plants, soil, and water.

**acid precipitation** Precipitation high in sulfuric acid and nitric acid from reactions between sulfur dioxide and water vapor and nitrogen oxides and water vapor in the atmosphere; also known as *acid rain*.

**acid rain** See *acid precipitation*.

**acquired immune deficiency syndrome (AIDS)** An infectious disease caused by the human immunodeficiency virus (HIV).

**active solar energy** Energy captured from sunlight with intermediate technologies.

**acute disease** A disease that rapidly impairs the functioning of an organism.

**acute study** An experiment that exposes organisms to an environmental hazard for a short duration.

**adaptation** A trait that improves an individual's fitness.

**adaptive management plan** A plan that provides flexibility so that managers can modify it as changes occur.

**adiabatic cooling** The cooling effect of reduced pressure on air as it rises higher in the atmosphere and expands.

**adiabatic heating** The heating effect of increased pressure on air as it sinks toward the surface of Earth and decreases in volume.

**affluence** The state of having plentiful wealth; the possession of money, goods, or property.

**age structure** A description of how many individuals fit into particular age categories.

**age structure diagram** A diagram that shows the numbers of individuals within each age category, typically expressed for males and females separately.

**agribusiness** See *industrial agriculture*.

**agroforestry** An agricultural technique in which trees and vegetables are intercropped.

**albedo** The percentage of incoming sunlight reflected from a surface.

**algal bloom** A rapid increase or accumulation in the algal population of a waterway.

**alien species** A species living outside its historical range.

**allergen** A chemical that causes allergic reactions.

**allopatric speciation** The process of speciation that occurs with geographic isolation.

**anemia** A deficiency of iron.

**annual plant** A plant that lives only one season.

**anthropocentric worldview** A worldview that focuses on human welfare and well-being.

**anthropogenic** Derived from human activities.

**aphotic zone** The layer of ocean water that lacks sufficient sunlight for photosynthesis.

**aquaculture** Farming aquatic organisms such as fish, shellfish, and seaweeds.

**aqueduct** A canal or ditch used to carry water from one location to another.

**aquifer** A permeable layer of rock and sediment that contains groundwater.

**area of seismic activity** See *fault zone*.

**artesian well** A well created by drilling a hole into a confined aquifer.

**asbestos** A long, thin, fibrous silicate mineral with insulating properties, which can cause cancer when inhaled.

**ash** The residual nonorganic material that does not combust during incineration.

**asthenosphere** The layer of Earth located in the outer part of the mantle, composed of semi-molten rock.

**atom** The smallest particle that can contain the chemical properties of an element.

**atomic number** The number of protons in the nucleus of a particular element.

**autotroph** See *producer*.

**B horizon** Frequently the second major soil horizon, composed primarily of mineral material with very little organic matter.

**background extinction rate** The average rate at which species become extinct over the long term.

**base** A substance that contributes hydroxide ions to a solution.

**base saturation** The proportion of soil bases to soil acids, expressed as a percentage.

**becquerel (Bq)** Unit that measures the rate at which a sample of radioactive material decays; 1 Bq = decay of 1 atom or nucleus per second.

**benthic zone** The muddy bottom of a lake, pond, or ocean.

**bioaccumulation** An increased concentration of a chemical within an organism over time.

**biocentric worldview** A worldview that considers human beings to be just one of many species on Earth, all of which have equal intrinsic value.

**biochemical oxygen demand (BOD)** The amount of oxygen a quantity of water uses over a period of time at specific temperatures.

**biodiesel** A diesel substitute produced by extracting and chemically altering oil from plants.

**biodiversity** The diversity of life forms in an environment.

**biofuels** Liquid fuels created from processed or refined biomass.

**biogeochemical cycles** The movements of matter within and between ecosystems.

**biomass** The total mass of all living matter in a specific area.

**biome** A geographic region categorized by a particular combination of average annual temperature, annual precipitation, and distinctive plant growth forms on land, and a particular combination of salinity, depth, and water flow in water.

**biophilia** An appreciation for life.

**biosphere** The region of our planet where life resides, the combination of all ecosystems on Earth.

**biosphere reserve** A protected area consisting of zones that vary in the amount of permissible human impact.

**biotic** Living.

**bitumen** A degraded petroleum that forms when petroleum migrates to the surface of Earth and is modified by bacteria; also called tar or pitch.

**boreal forest** A forest made up primarily of coniferous evergreen trees that can tolerate cold winters and short growing seasons.

**bottleneck effect** A reduction in the genetic diversity of a population caused by a reduction in its size.

**bottom ash** Residue collected at the bottom of the combustion chamber in a furnace.

**broad-spectrum pesticide** A pesticide that kills many different types of pests.

**bycatch** The unintentional catch of nontarget species while fishing.

**C horizon** The least-weathered soil horizon, which always occurs beneath the B horizon and is similar to the parent material.

**capacity** In reference to an electricity-generating plant, the maximum electrical output.

**capacity factor** The fraction of time a power plant operates in a year.

**cap-and-trade** An approach to controlling $CO_2$ emissions, where a *cap* places an upper limit on the amount of pollutant that can be emitted and *trade* allows companies to buy and sell allowances for a given amount of pollution.

**capillary action** A property of water that occurs when adhesion of water molecules to a surface is stronger than cohesion between the molecules.

**carbohydrate** A compound composed of carbon, hydrogen, and oxygen atoms.

**carbon neutral** An activity that does not change atmospheric $CO_2$ concentrations.

**carbon offsets** Methods of promoting global $CO_2$ reduction that do not involve a direct reduction in the amount of $CO_2$ actually emitted by a company.

**carbon sequestration** An approach to stabilizing greenhouse gases by removing $CO_2$ from the atmosphere.

**carcinogens** Chemicals that cause cancer.

**carrying capacity (K)** The limit of how many individuals in a population the food supply can sustain.

**cation exchange capacity (CEC)** The ability of a particular soil to absorb and release cations.

**cell** A highly organized living entity that consists of the four types of macromolecules and other substances in a watery solution, surrounded by a membrane.

**cellular respiration** The process by which cells convert glucose and oxygen into energy, carbon dioxide, and water.

**chemical energy** Potential energy stored in chemical bonds.

**chemical reaction** A reaction that occurs when atoms separate from molecules or recombine with other molecules.

**chemical weathering** The breakdown of rocks and minerals by chemical reactions, the dissolving of chemical elements from rocks, or both.

**chemosynthesis** A process used by some bacteria in the ocean to generate energy with methane and hydrogen sulfide.

**child mortality rate** The number of deaths of children under age 5 per 1,000 live births.

**chlorofluorocarbons (CFCs)** A family of organic compounds whose properties make them ideal for use in refrigeration and air-conditioning.

**chronic disease** A disease that slowly impairs functioning of an organism.

**chronic study** An experiment that exposes organisms to an environmental hazard for a long duration.

**CITES** See *Convention on International Trade in Endangered Species of Wild Fauna and Flora (CITES)*.

**clear-cutting** A method of harvesting trees that involves removing all or almost all of the trees within an area.

**climate** The average weather that occurs in a given region over a long period of time.

**closed-loop recycling** Recycling a product into the same product.

**closed system** A system in which matter and energy exchanges do not occur across boundaries.

**coal** Solid fuel formed primarily from the remains of trees, ferns, and other plant materials preserved 280 million to 360 million years ago.

**cogeneration** The use of a single fuel to generate electricity and to produce heat.

**combined cycle** A power plant that uses both exhaust gases and steam turbines to generate electricity.

**command-and-control approach** A strategy for pollution control that involves regulations and enforcement mechanisms.

**commensalism** A relationship between species in which one species benefits and the other species is neither harmed nor helped.

**commercial energy source** An energy source that is bought and sold.

**community** All of the populations of organisms within a given area.

**community ecology** The study of interactions between species.

**competition** The struggle of individuals to obtain a limiting resource.

**competitive exclusion principle** The principle stating that two species competing for the same limiting resource cannot coexist.

**compound** A molecule containing more than one element.

**concentrated agriculture feeding operation (CAFO)** A large indoor or outdoor structure used to raise animals at very high densities.

**cone of depression** An area from which the groundwater has been rapidly withdrawn.

**confined aquifer** An aquifer surrounded by a layer of impermeable rock or clay that impedes water flow.

**consumer** An organism that must obtain its energy by consuming other organisms.

**contour plowing** An agricultural technique in which plowing and harvesting are done parallel to the topographic contours of the land.

**control group** In a scientific investigation, a group that experiences exactly the same conditions as the experimental group, except for the single variable under study.

**control rod** A cylindrical device inserted between the fuel rods in a nuclear reactor to absorb excess neutrons and slow or stop the fission reaction.

**Convention on Biological Diversity** A 1992 international treaty formed to help protect biodiversity.

**Convention on International Trade in Endangered Species of Wild Fauna and Flora (CITES)** A 1973 treaty formed to control the international trade of threatened plants and animals.

**conventional agriculture** See *industrial agriculture*.

**convergent plate boundary** An area where plates move toward one another and collide.

**coral bleaching** A phenomenon in which algae inside corals die, causing the corals to turn white.

**coral reef** The most diverse marine biome on Earth, found in warm, shallow waters beyond the shoreline.

**core** In reference to Earth, the innermost layer.

**Coriolis effect** The deflection of an object's path due to the rotation of Earth.

**corridor** A strip of natural habitat that connects separated populations.

**covalent bond** The bond formed when elements share electrons.

**cradle-to-grave analysis** A systems tool that looks at the materials used and released throughout the manufacturing, use, and disposal of a product; also known as *life-cycle analysis*.

**critical thinking** The process of questioning the source of information, considering the methods used to obtain the information, and drawing conclusions; essential to all scientific endeavor.

**crop rotation** An agricultural technique in which crop species in a field are alternated from season to season.

**crude birth rate (CBR)** The number of births per 1,000 individuals per year.

**crude death rate (CDR)** The number of deaths per 1,000 individuals per year.

**crude oil** Liquid petroleum removed from the ground.

**crust** In geology, the chemically distinct outermost layer of the lithosphere.

**crustal abundance** The average concentration of an element in Earth's crust.

**CTL (coal to liquid)** The process of converting solid coal into liquid fuel.

**cultural eutrophication** An increase in fertility in a body of water, the result of anthropogenic inputs of nutrients.

**curie** A unit of measure for radiation; 1 curie = 37 billion decays per second.

**dam** A barrier that runs across a river or stream to control the flow of water.

**dead zone** In a body of water, an area with extremely low oxygen concentration and very little life.

**decomposers** Fungi or bacteria that recycle nutrients from dead tissues and wastes back into an ecosystem.

**deductive reasoning** The process of applying a general statement to specific facts or situations.

**demographer** A scientist in the field of demography.

**demography** The study of human populations and population trends.

**density-dependent factor** A factor that influences an individual's probability of survival and reproduction in a manner that depends on the size of the population.

**density-independent factor** A factor that has the same effect on an individual's probability of survival and the amount of reproduction at any population size.

**deoxyribonucleic acid** See *DNA (deoxyribonucleic acid)*.

**Department of Energy (DOE)** A U.S. government agency created in 1977 with the goal of advancing the energy and economic security of the United States.

**deposition** The accumulation or depositing of eroded material such as sediment, rock fragments, or soil.

**desalination** The process of removing the salt from salt water.

**desertification** The transformation of arable, productive land to desert or unproductive land due to climate change or destructive land use.

**detritivore** An organism that specializes in breaking down dead tissues and waste products into smaller particles.

**developed country** A country with relatively high levels of industrialization and income.

**developing country** A country with relatively low levels of industrialization and income.

**development** Improvement in human well-being through economic advancement.

**die-off** A rapid decline in a population due to death.

**dike** A structure built to prevent ocean waters from flooding adjacent land.

**disease** Any impaired function of the body with a characteristic set of symptoms.

**distribution** Areas of the world in which a species lives.

**disturbance** An event, caused by physical, chemical, or biological agents, resulting in changes in population size or community composition.

**divergent plate boundary** An area beneath the ocean where tectonic plates move away from each other.

**DNA (deoxyribonucleic acid)** A nucleic acid, the genetic material that contains the code for reproducing the components of the next generation, and which organisms pass on to their offspring.

**dose-response study** A study that exposes organisms to different amounts of a chemical and then observes a variety of possible responses, including mortality or changes in behavior or reproduction.

**doubling time** The number of years it takes a population to double.

**E horizon** The zone of leaching that forms under the O horizon or, less often, the A horizon.

**earthquake** The sudden movement of Earth's crust caused by a release of potential energy along a geologic fault and usually causing a vibration or trembling at Earth's surface.

**Ebola hemorrhagic fever** An infectious disease with high death rates, caused by the Ebola virus.

**ecocentric worldview** A worldview that places equal value on all living organisms and the ecosystems in which they live.

**ecological economics** The study of economics as a component of ecological systems.

**ecological efficiency** The proportion of consumed energy that can be passed from one trophic level to another.

**ecological footprint** A measure of how much an individual consumes, expressed in area of land.

**ecological succession** The replacement of one group of species by another group of species over time.

**ecologically sustainable forestry** An approach to removing trees from forests in ways that do not unduly affect the viability of other trees.

**economies of scale** The observation that average costs of production fall as output increases.

**ecosystem** A particular location on Earth distinguished by its mix of interacting biotic and abiotic components.

**ecosystem diversity** The variety of ecosystems within a given region.

**ecosystem engineer** A keystone species that creates or maintains habitat for other species.

**ecosystem service** The process by which natural environments provide life-supporting resources.

**ED50** An abbreviation for the effective dose of a chemical that causes 50 percent of the individuals in a dose-response study to display a harmful, but nonlethal, effect.

**edge habitat** A habitat that occurs where two different communities come together, typically forming an abrupt transition.

**El Niño–Southern Oscillation (ENSO)** The periodic changes in winds and ocean currents, causing cooler and wetter conditions in the southeastern United States and unusually dry weather in southern Africa and Southeast Asia.

**electrical grid** A network of interconnected transmission lines that joins power plants together and links them with end users of electricity.

**electrolysis** The application of an electric current to water molecules to split them into hydrogen and oxygen.

**electromagnetic radiation** A form of energy emitted by the Sun that includes, but is not limited to, visible light, ultraviolet light, and infrared energy.

**element** A substance composed of atoms that cannot be broken down into smaller, simpler components.

**emergent infectious disease** An infectious disease that has not been previously described or has not been common for at least 20 years.

**emigration** The movement of people out of a country or region, to settle in another country or region.

**eminent domain** A principle that grants government the power to acquire a property at fair market value even if the owner does not wish to sell it.

**endangered** At serious risk of extinction.

**Endangered Species Act** A 1973 U.S. act that implements CITES, designed to protect species from extinction.

**energy** The ability to do work or transfer heat.

**energy carrier** Something that can move and deliver energy in a convenient, usable form to end users.

**energy conservation** The implementation of methods to use less energy.

**energy efficiency** The ratio of the amount of work done to the total amount of energy introduced to the system.

**energy intensity** The energy use per unit of gross domestic product.

**energy quality** The ease with which an energy source can be used for work.

**energy subsidy** The energy input per calorie of food produced.

**entropy** Randomness in a system.

**environment** The sum of all the conditions surrounding us that influence life.

**environmental economics** A subfield of economics that examines costs and benefits of various policies and regulations related to environmental degradation.

**environmental hazard** Anything in the environment that can potentially cause harm.

**environmental impact statement (EIS)** A document outlining the scope and purpose of a development project, describing the environmental context, suggesting alternative approaches to the project, and analyzing the environmental impact of each alternative.

**environmental indicator** An indicator that describes the current state of an environmental system.

**environmental justice** A social movement and field of study that focuses on equal enforcement of environmental laws and eliminating disparities in the exposure of environmental harms to different ethnic and socioeconomic groups within a society.

**environmental mitigation plan** A plan that outlines how a developer will address concerns raised by a project's impact on the environment.

**Environmental Protection Agency (EPA)** A U.S. government agency that creates federal policy and oversees enforcement of regulations related to the environment, including science, research, assessment, and education.

**environmental science** The field of study that looks at interactions among human systems and those found in nature.

**environmental studies** The field of study that includes environmental science, environmental policy, economics, literature, and ethics, among others.

**environmental worldview** A worldview that encompasses how people think the world works, how they view their role in it, and what they believe to be proper behavior regarding the environment.

**environmentalist** A person who participates in environmentalism, a social movement that seeks to protect the environment through lobbying, activism, and education.

**epicenter** The exact point on the surface of Earth directly above the location where rock ruptures during an earthquake.

**epidemic** A situation in which a pathogen causes a rapid increase in disease.

**epidemiology** The study of the causes of illness and disease in the populations of humans and other organisms.

**erosion** The physical removal of rock fragments from a landscape or ecosystem.

**ethanol** Alcohol made by converting starches and sugars from plant material into alcohol and $CO_2$.

**eutrophic lake** A lake with a high level of productivity.

**eutrophication** A phenomenon in which a body of water becomes rich in nutrients.

**evapotranspiration** The combined amount of evaporation and transpiration.

**evolution** A change in the genetic composition of a population over time.

**evolution by artificial selection** A change in the genetic composition of a population over time as a result of humans selecting which individuals breed, typically with a preconceived set of traits in mind.

**evolution by natural selection** A change in the genetic composition of a population over time as a result of the environment determining which individuals are most likely to survive and reproduce.

**exotic species** See *alien species.*

**exponential growth model** ($N_t = N_0 e^{rt}$) A growth model that estimates a population's future size ($N_t$) after a period of time ($t$), based on the intrinsic growth rate ($r$) and the number of reproducing individuals currently in the population ($N_0$).

**externality** The cost or benefit of a good or service that is not included in the purchase price of that good or service.

**extinction** The death of the last member of a species.

**extrusive igneous rock** Rock that forms when magma cools above the surface of Earth.

**exurban** An area similar to a suburb, but unconnected to any central city or densely populated area.

**family planning** The practice of regulating the number or spacing of offspring through the use of birth control.

**famine** The condition in which food insecurity is so extreme that large numbers of deaths occur in a given area over a relatively short period.

**fault** A fracture in rock caused by a movement of Earth's crust.

**fault zone** A large expanse of rock where a fault has occurred.

**fecal coliform bacteria** A group of microorganisms in the human intestines that can serve as an indicator species for potentially harmful microorganisms associated with contamination by sewage.

**feedback** An adjustment in input or output rates caused by changes to a system.

**first law of thermodynamics** A law of nature stating that energy can neither be created nor destroyed.

**fish ladder** A stair-like structure that allows migrating fish to get around a dam.

**fishery** A commercially harvestable population of fish within a particular ecological region.

**fishery collapse** The decline of a fish population by 90 percent or more.

**fission** A nuclear reaction in which a neutron strikes a relatively large atomic nucleus, which then splits into two or more parts, releasing additional neutrons and energy in the form of heat.

**fitness** An individual's ability to survive and reproduce.

**flex-fuel vehicle** A vehicle that runs on either gasoline or ethanol.

**floodplain** The land adjacent to a river.

**fly ash** The residue collected from the chimney or exhaust pipe of a furnace.

**food chain** The sequence of consumption from producers through tertiary consumers.

**food insecurity** A condition in which people do not have adequate access to food.

**food security** A condition in which people have access to sufficient, safe, and nutritious food that meets their dietary needs for an active and healthy life.

**food web** A complex model of how energy and matter move between trophic levels.

**forest** Land dominated by trees and other woody vegetation and sometimes used for commercial logging.

**fossil** The remains of an organism that has been preserved in rock.

**fossil carbon** Carbon in fossil fuels.

**fossil fuel** A fuel derived from biological material that became fossilized millions of years ago.

**founder effect** A change in a population descended from a small number of colonizing individuals.

**fracture** In geology, a crack that occurs in rock as it cools.

**freshwater wetland** An aquatic biome that is submerged or saturated by water for at least part of each year, but shallow enough to support emergent vegetation.

**fuel cell** An electrical-chemical device that converts fuel, such as hydrogen, into an electrical current.

**fuel rod** A cylindrical tube that encloses nuclear fuel within a nuclear reactor.

**fundamental niche** The suite of ideal environmental conditions for a species.

**gene** A physical location on the chromosomes within each cell of an organism.

**genetic diversity** The variety of genes within a given species.

**genetic drift** A change in the genetic composition of a population over time as a result of random mating.

**genetically modified organism (GMO)** An organism produced by copying genes from a species with a desirable trait and inserting them into another species.

**genotype** The complete set of genes in an individual.

**genuine progress indicator (GPI)** A measurement of the economy that considers personal consumption, income distribution, levels of higher education, resource depletion, pollution, and the health of the population.

**geographic isolation** Physical separation of a group of individuals from others of the same species.

**geothermal energy** Heat energy that comes from the natural radioactive decay of elements deep within Earth.

**global change** Change that occurs in the chemical, biological, and physical properties of the planet.

**global climate change** Changes in the climate of Earth; an aspect of global change.

**global warming** The warming of the oceans, landmasses, and atmosphere of Earth; an aspect of global climate change.

**gray water** Wastewater from baths, showers, bathrooms, and washing machines.

**Green Revolution** A shift in agricultural practices in the twentieth century that included new management techniques, mechanization, fertilization, irrigation, and improved crop varieties, and resulted in increased food output.

**green tax** A tax placed on environmentally harmful activities or emissions.

**greenhouse gas** A gas in Earth's atmosphere that traps heat near the surface.

**greenhouse warming potential** An estimate of how much a molecule of any compound can contribute to global warming over a period of 100 years relative to a molecule of $CO_2$.

**gross domestic product (GDP)** A measure of the value of all products and services produced in a country in a year.

**gross primary productivity (GPP)** The total amount of solar energy that producers in an ecosystem capture via photosynthesis over a given amount of time.

**ground source heat pump** A technology that transfers heat from the ground to a building.

**groundwater recharge** A process by which water percolates through the soil and works its way into an aquifer.

**growth rate** The number of offspring an individual can produce in a given time period, minus the death of the individual or any of its offspring during the same period.

**gyre** A large-scale pattern of water circulation that moves clockwise in the Northern Hemisphere and counterclockwise in the Southern Hemisphere.

**Hadley cell** A convection current in the atmosphere that cycles between the equator and 30° N and 30° S.

**half-life** The time it takes for one-half of an original radioactive parent atom to decay.

**hazardous waste** Waste material that is dangerous or potentially harmful to humans or ecosystems.

**haze** Reduced visibility.

**herbicide** A pesticide that targets plant species that compete with crops.

**herbivore** A predator that consumes plants as prey.

**heterotrophs** See *consumers*.

**Highway Trust Fund** A U.S. federal fund that pays for the construction and maintenance of roads and highways.

**hot spot** In geology, a place where molten material from Earth's mantle reaches the lithosphere.

**Hubbert curve** A bell-shaped curve representing oil use and projecting both when world oil production will reach a maximum and when we will run out of oil.

**human capital** Human knowledge, potential, and abilities.

**human development index (HDI)** A measure of economic well-being that combines life expectancy, knowledge, education, and standard of living as shown in GDP per capita and purchasing power.

**human immunodeficiency virus (HIV)** A virus that causes acquired immune deficiency syndrome.

**hydrogen bond** A weak chemical bond that forms when hydrogen atoms that are covalently bonded to one atom are attracted to another atom on another molecule.

**hydrologic cycle** The movement of water through the biosphere.

**hydroponic agriculture** The cultivation of plants in greenhouse conditions by immersing roots in a nutrient-rich solution.

**hypothesis** A testable theory or supposition about how something works.

**igneous rock** Rock formed directly from magma.

**immigration** The movement of people into a country or region, having come from another country or region.

**impermeable surfaces** Pavement or buildings that do not allow water penetration.

**inbreeding depression** A genetic phenomenon in which individuals with similar genotypes breed with each other and produce offspring with an impaired ability to survive and reproduce.

**incentive-based approach** A program that constructs financial and other incentives for lowering emissions, based on profits and benefits.

**incineration** The process of burning waste materials to reduce volume and mass, sometimes to generate electricity or heat.

**indicator species** A species that indicates whether or not disease-causing pathogens are likely to be present.

**individual transferable quota (ITQ)** A fishery management program in which individual fishers are given a total allowable catch of fish in a season that they can either catch or sell.

**induced demand** The phenomenon in which increase in the supply of a good causes demand to grow.

**inductive reasoning** The process of making general statements from specific facts or examples.

**industrial agriculture** Agriculture that applies the techniques of mechanization and standardization.

**infant mortality rate** The number of deaths of children under 1 year of age per 1,000 live births.

**infectious disease** A disease caused by a pathogen.

**infill** Development that fills in vacant lots within existing communities.

**innocent-until-proven-guilty principle** A principle based on the philosophy that a potential hazard should not be considered an actual hazard until the scientific data definitively demonstrate that it actually causes harm.

**inorganic compound** A compound that does not contain the element carbon or contains carbon bound to elements other than hydrogen.

**inorganic fertilizer** See *synthetic fertilizer*.

**input** An addition to a system.

**insecticide** A pesticide that targets species of insects and other invertebrates.

**instrumental value** Something that has worth as an instrument or a tool that can be used to accomplish a goal.

**integrated pest management (IPM)** An agricultural practice that uses a variety of techniques designed to minimize pesticide inputs.

**integrated waste management** A waste management technique that employs several waste reduction, management, and disposal strategies to reduce the environmental impact of municipal solid waste (MSW).

**intercropping** An agricultural method in which two or more crop species are planted in the same field at the same time to promote a synergistic interaction.

**intermediate disturbance hypothesis** The hypothesis that ecosystems experiencing intermediate levels of disturbance are more diverse than those with high or low disturbance levels.

**intertidal zone** The narrow band of coastline between the levels of high tide and low tide.

**intertropical convergence zone (ITCZ)** An area of Earth that receives the most intense sunlight; where the ascending branches of the two Hadley cells converge.

**intrinsic growth rate (r)** The maximum potential for growth of a population under ideal conditions with unlimited resources.

**intrusive igneous rock** Igneous rock that forms when magma rises up and cools in place underground.

**invasive species** A species that spreads rapidly across large areas.

**inversion layer** The layer of warm air that traps emissions in a thermal inversion.

**ionic bond** A chemical bond between two oppositely charged ions.

**IPAT equation** Impact = population × affluence × technology.

**isotopes** Atoms of the same element with different numbers of neutrons.

**J-shaped curve** The curve of the exponential growth model when graphed.

**joule** The amount of energy used when a one-watt electrical device is turned on for one second.

**K-selected species** A species with a low intrinsic growth rate that causes the population to increase slowly until it reaches carrying capacity.

**keystone species** A species that is far more important in its community than its relative abundance might suggest.

**kinetic energy** The energy of motion.

**Kyoto Protocol** An international agreement to reduce global emissions of greenhouse gases from all industrialized countries to 5.2 percent below their 1990 levels by 2012.

**Lacey Act** A U.S. act that prohibits interstate shipping of all illegally harvested plants and animals.

**latent heat release** The release of energy when water vapor in the atmosphere condenses into liquid water.

**law of conservation of matter** A law of nature stating that matter cannot be created or destroyed.

**LD50** The lethal dose of a chemical that kills 50 percent of the individuals in a dose-response study.

**leach field** A component of a septic system, made up of underground pipes laid out below the surface of the ground.

**leachate** Liquid that contains elevated levels of pollutants as a result of having passed through municipal solid waste (MSW) or contaminated soil.

**leaching** The transportation of dissolved molecules through the soil via groundwater.

**leapfrogging** The situation in which less developed countries use newer technology without first using the precursor technology.

**levee** An enlarged bank built up on each side of a river to prevent flooding.

**life-cycle analysis** See *cradle-to-grave analysis.*

**life expectancy** The average number of years that an infant born in a particular year in a particular country can be expected to live, given the current average life span and death rate in that country.

**limiting nutrient** A nutrient required for the growth of an organism but available in a lower quantity than other nutrients.

**limiting resource** A resource that a population cannot live without and that occurs in quantities lower than the population would require to increase in size.

**limnetic zone** A zone of open water in lakes and ponds.

**lipids** Smaller organic biological molecules that do not mix with water.

**lithosphere** The outermost layer of Earth, including the mantle and crust.

**littoral zone** The shallow zone of soil and water in lakes and ponds where most algae and emergent plants grow.

**logistic growth model** A growth model that describes a population whose growth is initially exponential, but slows as the population approaches the carrying capacity of the environment.

**macroevolution** Evolution that gives rise to new species, genera, families, classes, or phyla.

**macronutrients** The six key elements that organisms need in relatively large amounts: nitrogen, phosphorus, potassium, calcium, magnesium, and sulfur.

**mad cow disease** A disease in which prions mutate into deadly pathogens and slowly damage a cow's nervous system.

**magma** Molten rock.

**malaria** An infectious disease caused by one of several species of protists in the genus *Plasmodium.*

**malnourished** Having a diet that lacks the correct balance of proteins, carbohydrates, vitamins, and minerals.

**mangrove swamp** A swamp that occurs along tropical and subtropical coasts, and contains salt-tolerant trees with roots submerged in water.

**mantle** The layer of Earth above the core, containing magma.

**manufactured capital** All goods and services that humans produce.

**manure lagoons** Human-made ponds lined with rubber built to handle large quantities of manure produced by livestock.

**Marine Mammal Protection Act** A 1972 U.S. act to protect declining populations of marine mammals.

**market failure** The economic situation that results when the economic system does not appropriately account for all costs.

**mass** A measurement of the amount of matter an object contains.

**mass extinction** A large extinction of species in a relatively short period of time.

**mass number** A measurement of the total number of protons and neutrons in an element.

**matter** Anything that occupies space and has mass.

**maximum contaminant level (MCL)** The standard for safe drinking water established by the EPA under the Safe Drinking Water Act.

**maximum sustainable yield (MSY)** The maximum amount of a renewable resource that can be harvested without compromising the future availability of that resource.

**meat** Livestock or poultry consumed as food.

**mesotrophic lake** A lake with a moderate level of productivity.

**metal** An element with properties that allows it to conduct electricity and heat energy, and perform other important functions.

**metamorphic rock** Rock that forms when sedimentary rock, igneous rock, or other metamorphic rock is subjected to high temperature and pressure.

**metapopulation** A group of spatially distinct populations that are connected by occasional movements of individuals between them.

**microevolution** Evolution occurring below the species level.

**microlending** The practice of loaning small amounts of money to help people in less developed countries start small businesses.

**mineral** A solid chemical substance with a uniform, often crystalline, structure that forms under specific temperatures and pressures.

**mining spoils** Unwanted waste material created during mining.

**modern carbon** Carbon in biomass that was recently in the atmosphere.

**molecule** A particle containing more than one atom.

**monocropping** An agricultural method that utilizes large plantings of a single species or variety.

**mountaintop removal** A mining technique in which the entire top of a mountain is removed with explosives.

**multiple-use lands** A U.S. classification used to designate lands that may be used for recreation, grazing, timber harvesting, and mineral extraction.

**multi-use zoning** A zoning classification that allows retail and high-density residential development to coexist in the same area.

**municipal solid waste (MSW)** Refuse collected by municipalities from households, small businesses, and institutions.

**mutagens** Carcinogens that cause damage to the genetic material of a cell.

**mutation** A random change in the genetic code produced by a mistake in the copying process.

**mutualism** An interaction between species that increases the chances of survival or reproduction for both species.

**National Environmental Policy Act (NEPA)** A 1969 U.S. federal act that mandates an environmental assessment of all projects involving federal money or federal permits.

**national wilderness area** An area set aside with the intent of preserving a large tract of intact ecosystem or a landscape.

**national wildlife refuge** A federal public land managed for the primary purpose of protecting wildlife.

**native species** A species that lives in a historical range, typically where it has lived for thousands or millions of years.

**natural capital** The natural resources of Earth, such as air, water, and minerals.

**natural experiment** A natural event that acts as an experimental treatment in an ecosystem.

**natural law** A theory for which there is no known exception and that has withstood rigorous testing.

**negative feedback loop** A feedback loop in which a system responds to a change by returning to its original state, or by decreasing the rate at which the change is occurring.

**net migration rate** The difference between immigration and emigration in a given year per 1,000 people in a country.

**net primary productivity (NPP)** The energy captured by producers in an ecosystem minus the energy producers respire.

**net removal** The process of removing more than is replaced by growth, typically used when referring to carbon.

**neurotoxin** A chemical that disrupts the nervous systems of animals.

**niche generalist** A species that can live under a wide range of abiotic or biotic conditions.

**niche specialist** A species that is specialized to live in a specific habitat or to feed on a small group of species.

**nitrogen fixation** A process by which some organisms can convert nitrogen gas molecules directly into ammonia.

**nomadic grazing** Feeding herds of animals by moving them to seasonally productive feeding grounds, often over long distances.

**nondepletable energy source** An energy source that cannot be used up.

**nonpersistent pesticide** A pesticide that breaks down rapidly, usually in weeks or months.

**nonpoint source** A diffuse area that produces pollution.

**nonrenewable energy source** An energy source with a finite supply, primarily the fossil fuels and nuclear fuels.

**no-till agriculture** An agricultural method in which farmers do not turn the soil between seasons, used as a means of reducing erosion.

**nuclear fuel** Fuel derived from radioactive materials that give off energy.

**nuclear fusion** A reaction that occurs when lighter nuclei are forced together to produce heavier nuclei.

**nucleic acids** Organic compounds found in all living cells, which form in long chains to make DNA and RNA.

**null hypothesis** A statement or idea that can be falsified, or proved wrong.

**O horizon** The organic horizon at the surface of many soils, composed of organic detritus in various stages of decomposition.

**Occupational Safety and Health Administration (OSHA)** A U.S. federal agency responsible for the enforcement of health and safety regulations in the workplace.

**oil sands** Slow-flowing, viscous deposits of bitumen mixed with sand, water, and clay.

**oligotrophic lake** A lake with a low level of productivity as a result of low amounts of nutrients in the water.

**open-loop recycling** Recycling one product into a different product.

**open-pit mining** A mining technique that uses a large pit or hole in the ground, visible from the surface of Earth.

**open system** A system in which exchanges of matter or energy occur across system boundaries.

**ore** A concentrated accumulation of minerals from which economically valuable materials can be extracted.

**organic agriculture** Production of crops with the goal of improving the soil each year without the use of synthetic pesticides or fertilizers.

**organic compound** A compound that contains carbon-carbon and carbon-hydrogen bonds.

**organic fertilizer** Fertilizer composed of organic matter from plants and animals.

**output** A loss from a system.

**overnutrition** Ingestion of too many calories and improper foods.

**overshoot** When a population becomes larger than the environment's carrying capacity.

**oxygen-demanding waste** Organic matter that enters a body of water and feeds microbes that are decomposers.

**pandemic** An epidemic that occurs over a large geographic region.

**parasite** A predator that lives on or in the organism it consumes.

**parasitoid** An organism that lays eggs inside other organisms.

**parent material** Rock underlying soil; the material from which the inorganic components of a soil are derived.

**particles** See *particulate matter (PM)*.

**particulate matter (PM)** Solid or liquid particles suspended in air; also known as *particles* and *particulates*.

**particulates** See *particulate matter (PM)*.

**passive solar design** Construction designed to take advantage of solar radiation without active technology.

**pathogen** An illness-causing bacterium, virus, or parasite.

**PBDE (polybrominated diphenyl ether)** An environmentally harmful compound commonly found in flame retardants and other materials.

**PCBs (polychlorinated biphenyls)** A group of industrial compounds formerly used to manufacture plastics and insulate electrical transformers, and responsible for many environmental problems.

**peak demand** The greatest quantity of energy used at any one time.

**peak oil** The point at which half the total known oil supply is used up.

**perennial plant** A plant that lives for multiple years.

**periodic table** A chart of all chemical elements currently known, organized by their properties.

**permafrost** An impermeable, permanently frozen layer of soil.

**persistence** The length of time a chemical remains in the environment.

**persistent pesticide** A pesticide that remains in the environment for a long time.

**pesticide** A substance, either natural or synthetic, that kills or controls organisms that people consider pests.

**pesticide resistant** An individual that survives a pesticide application.

**pesticide treadmill** A cycle of pesticide development, followed by pest resistance, followed by new pesticide development.

**petroleum** A fossil fuel that occurs in underground deposits, composed of a liquid mixture of hydrocarbons, water, and sulfur.

**pH** The number indicating the strength of acids and bases on a scale of 0 to 14, where 7 is neutral, a value below 7 is acidic, and a value above 7 is basic (alkaline).

**phenotype** A set of traits expressed by an individual.

**photic zone** The upper layer of water in the ocean that receives enough sunlight for photosynthesis.

**photochemical oxidants** A class of air pollutants formed as a result of sunlight acting on compounds such as nitrogen oxides.

**photon** A massless packet of energy that carries electromagnetic radiation at the speed of light.

**photosynthesis** The process by which producers use solar energy to convert carbon dioxide and water into glucose.

**photovoltaic solar cells** A system of capturing energy from sunlight and converting it directly into electricity.

**phylogenies** The branching patterns of evolutionary relationships.

**physical weathering** The mechanical breakdown of rocks and minerals.

**phytoplankton** Floating algae.

**pioneer species** A species that can colonize new areas rapidly.

**placer mining** A mining technique in which metals and precious stones are sought in river sediments.

**plague** An infectious disease caused by the bacterium *Yersinia pestis,* carried by fleas.

**plate tectonics** The theory that the lithosphere of Earth is divided into plates, most of which are in constant motion.

**point source** A distinct location from which pollution is directly produced.

**polar cell** A convection cell in the atmosphere, formed by air that rises at 60° N and 60° S and sinks at the poles, 90° N and 90° S.

**polar molecule** A molecule in which one side is more positive and the other side is more negative.

**population** The individuals that belong to the same species and live in a given area at a given time.

**population density** The number of individuals per unit area at a given time.

**population distribution** A description of how individuals are distributed with respect to one another.

**population ecology** The study of factors that cause populations to increase or decrease.

**population momentum** Continued population growth that does not slow in response to growth reduction measures.

**population pyramid** An age structure diagram that is widest at the bottom and smallest at the top, typical of developing countries.

**population size (N)** The total number of individuals within a defined area at a given time.

**positive feedback loop** A feedback loop in which change in a system is amplified.

**potential energy** Stored energy that has not been released.

**potentially renewable** An energy source that can be regenerated indefinitely as long as it is not overharvested.

**power** The rate at which work is done.

**precautionary principle** A principle based on the philosophy that action should be taken against a plausible environmental hazard.

**precision** How close the repeated measurements of a sample are to one another.

**predation** The use of one species as a resource by another species.

**predator-mediated competition** Competition in which a predator is instrumental in reducing the abundance of a superior competitor, allowing inferior competitors to persist.

**prescribed burn** A fire deliberately set under controlled conditions in order to reduce the accumulation of dead biomass on a forest floor.

**primary consumer** An individual incapable of photosynthesis; must obtain energy by consuming other organisms.

**primary pollutant** A polluting compound that comes directly out of the smokestack, exhaust pipe, or natural emission source.

**primary succession** Ecological succession occurring on surfaces that are initially devoid of soil.

**prion** A small, beneficial protein that occasionally mutates into a pathogen.

**producer** An organism that uses the energy of the Sun to produce usable forms of energy.

**profundal zone** A region of water where sunlight does not reach, below the limnetic zone in very deep lakes.

**prospective study** A study that monitors people who might become exposed to harmful chemicals in the future.

**protein** A long chain of nitrogen-containing organic molecules known as amino acids, critical to living organisms for structural support, energy storage, internal transport, and defense against foreign substances.

**provision** A good that humans can use directly.

**r-selected species** A species that has a high intrinsic growth rate, which often leads to population overshoots and die-offs.

**radioactive decay** The spontaneous release of material from the nucleus of radioactive isotopes.

**radioactive waste** Nuclear fuel that can no longer produce enough heat to be useful in a power plant but continues to emit radioactivity.

**rain shadow** A region with dry conditions found on the leeward side of a mountain range as a result of humid winds from the ocean causing precipitation on the windward side.

**range of tolerance** The limits to the abiotic conditions that a species can tolerate.

**rangeland** A dry, open grassland.

**REACH (registration, evaluation, authorisation, and restriction of chemicals)** A 2007 agreement among the nations of the European Union about regulation of chemicals.

**realized niche** The range of abiotic and biotic conditions under which a species actually lives.

**recombination** The genetic process by which one chromosome breaks off and attaches to another chromosome during reproductive cell division.

**recycling** The process by which materials destined to become municipal solid waste (MSW) are collected and converted into raw material that is then used to produce new objects.

**Red List** A list of worldwide threatened species maintained by the International Union for Conservation of Nature.

**Reduce, Reuse, Recycle (three Rs)** A popular phrase promoting the idea of diverting materials from the waste stream.

**renewable** In energy management, an energy source that is either potentially renewable or nondepletable.

**replacement-level fertility** The total fertility rate required to offset the average number of deaths in a population in order to maintain the current population size.

**replication** The data collection procedure of taking repeated measurements.

**reproductive isolation** The result of two populations within a species evolving separately so they can no longer interbreed and produce viable offspring.

**reserve** In resource management, the known quantity of a resource that can be economically recovered.

**reservoir** A body of water created by blocking the natural flow of a waterway.

**resilience** The rate at which an ecosystem returns to its original state after a disturbance.

**resistance** A measure of how much a disturbance can affect flows of energy and matter in an ecosystem.

**resource conservation ethic** The belief that people should maximize use of resources, based on the greatest good for everyone.

**resource partitioning** A situation in which two species divide a resource, based on differences in their behavior or morphology.

**restoration ecology** The study and implementation of restoring damaged ecosystems.

**retrospective study** A study that monitors people who have been exposed to an environmental hazard at some time in the past.

**reuse** Using a product or material that was intended to be discarded.

**ribonucleic acid** See *RNA (ribonucleic acid)*.

**Richter scale** A scale that measures the largest ground movement that occurs during an earthquake.

**RNA (ribonucleic acid)** A nucleic acid that translates the code stored in DNA and allows for the synthesis of proteins.

**rock cycle** The continuous formation and destruction of rock on and below the surface of Earth.

**route of exposure** The way in which an individual might come into contact with an environmental hazard.

**runoff** Water that moves across the land surface and into streams and rivers.

**run-of-the-river** Hydroelectricity generation in which water is retained behind a low dam or no dam.

**S-shaped curve** The shape of the logistic growth model when graphed.

**salinization** A form of soil degradation that occurs when the small amount of salts in irrigation water becomes highly concentrated on the soil surface through evaporation.

**salt marsh** A marsh containing nonwoody emergent vegetation, found along the coast in temperate climates.

**saltwater intrusion** An infiltration of salt water in an area where groundwater pressure has been reduced from extensive drilling of wells.

**sample size (n)** The number of times a measurement is replicated in the data collection process.

**sanitary landfill** An engineered ground facility designed to hold municipal solid waste (MSW) with as little contamination of the surrounding environment as possible.

**saturation point** The maximum amount of water vapor in the air at a given temperature.

**scavenger** A carnivore that consumes dead animals.

**scientific method** An objective method to explore the natural world, draw inferences from it, and predict the outcome of certain events, processes, or changes.

**seafloor spreading** The formation of new ocean crust as a result of magma pushing upward and outward from Earth's mantle to the surface.

**second law of thermodynamics** The law stating that when energy is transformed, the quantity of energy remains the same, but its ability to do work diminishes.

**secondary consumer** A carnivore that eats primary consumers.

**secondary pollutant** A primary pollutant that has undergone transformation in the presence of sunlight, water, oxygen, or other compounds.

**secondary succession** The succession of plant life that occurs in areas that have been disturbed but have not lost their soil.

**sedimentary rock** Rock that forms when sediments such as muds, sands, or gravels are compressed by overlying sediments.

**seismic activity** The frequency and intensity of earthquakes.

**selective cutting** The method of harvesting trees that involves the removal of single trees or a relatively small number of trees from among many in a forest.

**selective pesticide** A pesticide that targets a narrower range of organisms.

**sense of place** The feeling that an area has distinct and meaningful character.

**septage** A layer of fairly clear water found in the middle of a septic tank.

**septic system** A relatively small and simple sewage treatment system, made up of a septic tank and a leach field, often used for homes in rural areas.

**septic tank** A large container that receives wastewater from a house as part of a septic system.

**sex ratio** The ratio of males to females.

**shifting agriculture** An agricultural method in which land is cleared and used for a few years until the soil is depleted of nutrients.

**sick building syndrome** A buildup of toxic compounds and pollutants in an airtight space; seen in newer buildings with good insulation and tight seals against air leaks.

**siltation** The accumulation of sediments, primarily silt, on the bottom of a reservoir.

**siting** The designation of a landfill location, typically through a regulatory process involving studies, written reports, and public hearings.

**SLOSS (single large or several small)** The debate over whether it is better for biodiversity to preserve a single large area or several different smaller areas.

**sludge** Solid waste material from wastewater.

**smart grid** An efficient, self-regulating electricity distribution network that accepts any source of electricity and distributes it effectively to end users.

**smart growth** A set of principles for community planning that focuses on strategies to encourage the development of sustainable, healthy communities.

**soil** A mix of geologic and organic components that forms a dynamic membrane covering much of Earth's surface.

**soil degradation** The loss of some or all of a soil's ability to support plant growth.

**solubility** How well a chemical dissolves in a liquid.

**source reduction** The reduction of waste through minimizing the use of materials destined to become municipal solid waste (MSW) from the early stages of design and manufacture.

**speciation** The evolution of new species.

**species** A group of organisms that is distinct from other groups in its morphology, behavior, or biochemical properties.

**species diversity** The variety of species within a given ecosystem.

**species evenness** The relative proportion of different species in a given area.

**species richness** The number of species in a given area.

**spring** A natural source of water formed when water from an aquifer percolates up to the ground surface.

**stakeholder** A person or organization with an interest in a particular place or issue.

**standing crop** The amount of biomass present in an ecosystem at a particular time.

**steady state** A state in which inputs equal outputs, so that the system is not changing over time.

**stewardship** The careful and responsible management of Earth and its resources.

**Stockholm Convention** A 2001 agreement among 127 nations concerning 12 chemicals to be banned, phased out, or reduced.

**stratosphere** The layer of the atmosphere above the troposphere, extending roughly 16 to 50 km (10–31 miles) above the surface of Earth.

**strip mining** The removal of strips of soil and rock to expose ore.

**subduction** The process of one crustal plate passing under another.

**sublethal effects** The effects of an environmental hazard that are not lethal, but which may impair an organism's behavior, physiology, or reproduction.

**subsistence energy sources** Energy sources gathered by individuals for their own immediate needs.

**subsurface mining** Mining techniques used when the desired resource is more than 100 m (328 feet) below the surface of Earth.

**subtropical desert** A biome prevailing at approximately 30° N and 30° S, with hot temperatures, extremely dry conditions, and sparse vegetation.

**suburban** An area surrounding a metropolitan center, with a comparatively low population density.

**Superfund** The Comprehensive Environmental Response, Compensation, and Liability Act (CERCLA), a 1980 U.S. federal act that imposes a tax on the chemical and petroleum industries, funds the cleanup of abandoned and nonoperating hazardous waste sites, and authorizes the federal government to respond directly to the release or threatened release of substances that may pose a threat to human health or the environment.

**surface tension** A property of water that results from the cohesion of water molecules at the surface of a body of water and creates a sort of skin on the water's surface.

**survivorship curve** A graph that represents the distinct patterns of species survival as a function of age.

**sustainability** Living on Earth in a way that allows humans to use its resources without depriving future generations of those resources.

**sustainable agriculture** Agriculture that fulfills the need for food and fiber while enhancing the quality of the soil, minimizing the use of nonrenewable resources, and allowing economic viability for the farmer.

**sustainable development** Development that balances current human well-being and economic advancement with resource management for the benefit of future generations.

**symbiotic** A relationship of two species that live in close association with each other.

**sympatric speciation** The evolution of one species into two, without geographic isolation.

**synergistic interactions** Risks that cause more harm together than expected based on separate individual risks.

**synthetic fertilizer** Fertilizer produced commercially, normally with the use of fossil fuels.

**system** Any set of interacting components that influence one another by exchanging energy or materials.

**systems analysis** An analysis to determine inputs, outputs, and changes in a system under various conditions.

**tailings** See *mining spoils*.

**technology transfer** The phenomenon of less developed countries adopting technological innovations that originated in wealthy countries.

**tectonic cycle** The cycle of processes that build up and break down the lithosphere.

**temperate grassland/cold desert** A biome characterized by cold, harsh winters, and hot, dry summers.

**temperate rainforest** A coastal biome typified by moderate temperatures and high precipitation.

**temperate seasonal forest** A biome with warmer summers and colder winters than temperate rainforests and dominated by deciduous trees.

**temperature** The measure of the average kinetic energy of a substance.

**teratogens** Chemicals that interfere with the normal development of embryos or fetuses.

**tertiary consumer** A carnivore that eats secondary consumers.

**texture** The property of soil determined by relative proportions of sand, silt, and clay.

**theory** A hypothesis that has been repeatedly tested and confirmed by multiple groups of researchers and has reached wide acceptance.

**theory of demographic transition** The theory that as a country moves from a subsistence economy to industrialization and increased affluence it undergoes a predictable shift in population growth.

**theory of island biogeography** A theory that demonstrates the dual importance of habitat size and distance in determining species richness.

**thermal inertia** The ability of a material to maintain its temperature.

**thermal inversion** A situation in which a relatively warm layer of air at mid-altitude covers a layer of cold, dense air below.

**thermal pollution** Nonchemical water pollution that occurs when human activities cause a substantial change in the temperature of water.

**thermal shock** A dramatic change in water temperature that can kill organisms.

**thermohaline circulation** An oceanic circulation pattern that drives the mixing of surface water and deep water.

**three Rs** See *Reduce, Reuse, Recycle (three Rs)*.

**tidal energy** Energy that comes from the movement of water driven by the gravitational pull of the Moon.

**tiered rate system** A billing system used by some electric companies in which customers pay higher rates as their use goes up.

**tipping fee** A fee charged for disposing of material in a landfill or incinerator.

**topsoil** See *A horizon*.

**total fertility rate (TFR)** An estimate of the average number of children that each woman in a population will bear throughout her childbearing years.

**tragedy of the commons** The tendency of a shared, limited resource to become depleted because people act from self-interest for short-term gain.

**transform fault boundary** An area where tectonic plates move sideways past each other.

**transit-oriented development (TOD)** Development that attempts to focus dense residential and retail development around stops for public transportation, a component of smart growth.

**transpiration** The release of water from leaves during photosynthesis.

**tree plantation** A large area typically planted with a single rapidly growing tree species.

**triple bottom line** An approach to sustainability that advocates consideration of economic, environmental, and social factors in decisions about business, the economy, the environment, and development.

**trophic levels** Levels in the feeding structure of organisms. Higher trophic levels consume organisms from lower levels.

**trophic pyramid** A representation of the distribution of biomass, numbers, or energy among trophic levels.

**tropical rainforest** A warm and wet biome found between 20° N and 20° S of the equator, with little seasonal temperature variation and high precipitation.

**tropical seasonal forest/savanna** A biome marked by warm temperatures and distinct wet and dry seasons.

**troposphere** A layer of the atmosphere closest to the surface of Earth, extending up to approximately 16 km (10 miles) and containing most of the atmosphere's nitrogen, oxygen, and water vapor.

**true predator** A predator that typically kills its prey and consumes most of what it kills.

**tuberculosis** A highly contagious disease caused by the bacterium *Mycobacterium tuberculosis* that primarily infects the lungs.

**tundra** A cold and treeless biome with low-growing vegetation.

**turbine** A device with blades that can be turned by water, wind, steam, or exhaust gas from combustion that turns a generator in an electricity-producing plant.

**uncertainty** An estimate of how much a measured or calculated value differs from a true value.

**unconfined aquifer** An aquifer made of porous rock covered by soil, which water can easily flow into and out of.

**undernutrition** The condition in which not enough calories are ingested to maintain health.

**United Nations (UN)** An institution dedicated to promoting dialogue among countries with the goal of maintaining world peace.

**United Nations Development Programme (UNDP)** A program of the United Nations that works to improve living conditions through economic development.

**United Nations Environment Programme (UNEP)** A program of the United Nations responsible for gathering environmental information and conducting research and assessing environmental problems.

**upwelling** The upward movement of ocean water toward the surface as a result of diverging currents.

**urban area** An area that contains more than 385 people per square kilometer (1,000 people per square mile).

**urban blight** The degradation of the built and social environments of the city that often accompanies and accelerates migration to the suburbs.

**urban growth boundary** A restriction on development outside a designated area.

**urban sprawl** Urbanized areas that spread into rural areas, removing clear boundaries between the two.

**valuation** The practice of assigning monetary value to seemingly intangible benefits and natural capital.

**volatile organic compound (VOC)** An organic compound that evaporates at typical atmospheric temperatures.

**volcano** A vent in the surface of Earth that emits ash, gases, or molten lava.

**waste** Material outputs from a system that are not useful or consumed.

**waste stream** The flow of solid waste that is recycled, incinerated, placed in a solid waste landfill, or disposed of in another way.

**waste-to-energy** A system in which heat generated by incineration is used as an energy source rather than released into the atmosphere.

**wastewater** Water produced by human activities including human sewage from toilets and gray water from bathing and washing of clothes and dishes.

**water impoundment** The storage of water in a reservoir behind a dam.

**water pollution** The contamination of streams, rivers, lakes, oceans, or groundwater with substances produced through human activities.

**water table** The uppermost level at which the water in a given area fully saturates rock or soil.

**waterlogging** A form of soil degradation that occurs when soil remains under water for prolonged periods.

**watershed** All land in a given landscape that drains into a particular stream, river, lake, or wetland.

**well-being** The status of being healthy, happy, and prosperous.

**wind energy** Energy generated from the kinetic energy of moving air.

**wind turbine** A turbine that converts wind energy into electricity.

**woodland/shrubland** A biome characterized by hot, dry summers and mild, rainy winters.

**World Bank** An international organization that provides technical and financial assistance to help reduce poverty and promote growth, especially in the world's poorest countries.

**World Health Organization (WHO)** A group within the United Nations responsible for human health, including combating the spread of infectious diseases and health issues related to natural disasters.

**zoning** A planning tool used to separate industry and business from residential neighborhoods.

# PHOTO CREDITS

(b) Peter Menzel, www.menzelphoto.com Material World. **p. 194:** *Figure 7.17* Publiphoto/Photo Researchers. **p. 196:** *Figure 7.19* Reuters/Mariana Bazo/Landov. **p. 197:** *Figure 7.20* Courtesy of Fatima Mata National College, Kerala, India/Dr. Sr. Soosamma Kavumpurath.

**Science Applied p. 203:** *Figure SA3.1* Gary Griffen/Earth Scenes/Animals Animals; *Figure SA3.2* Mark Newman/FLPA/Minden Pictures. **p. 204:** *Figure SA3.4* Gialuigi Guercia/AFP/Getty Images.

**Chapter 8 p. 206:** Robin Hammond. **p. 207:** izmostock/Alamy. **p. 214:** *Figure 8.9(b)* European Space Agency; *Figure 8.10 (inset)* age fotostock/SuperStock. **p. 216:** *Figure 8.12* Dominic Nahr/Getty Images. **p. 217:** *Figure 8.13* Mario Cipollini/Aurora Photos; *Figure 8.14(a), (b)* John Grotzinger/Ramon Rivera-Moret/Harvard Mineralogical Museum, (c) The Natural History Museum/Alamy. **p. 220:** *Figure 8.17* Mauritius/SuperStock; *Figure 8.18* Carr Clifton/Minden Pictures. **p. 226:** *Figure 8.25* Tim McCabe/USDA Natural Resources Conservation Service. **p. 230:** *Figure 8.28(a), (b)* Courtesy of Seneca Coal Company.

**Chapter 9 p. 234:** John W. Wall/Alamy. **p. 235:** Patrick McCully/Flickr. **p. 237:** *Figure 9.3* Jerry Moorman/iStockphoto.com. **p. 241:** *Figure 9.8* NOAA/Photo Researchers. **p. 242:** *Figure 9.9* Anthony Souffle/MCT/Landov. **p. 243:** *Figure 9.10* AP Photo/Xinhua, Du Huaju; *Figure 9.11* Dr. John D. Cunningham/Visuals Unlimited. **p. 244:** *Figure 9.12* Mark Hanauer/Corbis. **p. 247:** *Figure 9.17(a)* Jenny E. Ross/Corbis, (b) Jeff Vanuga/NRCS/USDA, (c) Hans Reinhard/Corbis, (d) Lynn Betts/NRCS/USDA. **p. 248:** *Figure 9.18* Kerry Sherck/Aurora Photos. **p. 249:** *Figure 9.19* Hans F. Meier/iStockphoto.com. **p. 250:** *Figure 9.22* David Pickford/Robert Harding Travel/Photolibrary. **p. 251:** *Figure 9.23* Jan Paul Schrage/iStockphoto.com. **p. 253:** *Figure 9.25* Sevin Coskun.

**Science Applied p. 257:** *Figure SA4.1 (inset)* Jim West/The Image Works. **p. 258:** *Figure SA4.2* Ed Young/age fotostock.

**Chapter 10 p. 260:** Paul Rogers/MCT/Landov. **p. 261:** John Storey/Time Life Pictures/Getty Images. **p. 263:** *Figure 10.1(a)* David McNew/Getty Images, (b) Ryan Pyle/Corbis, (c) Gerry Ellis/Minden Pictures, (d) Richard Cummins/SuperStock. **p. 268:** *Figure 10.7* Sal Maimone/Grant Heilman Photography. **p. 270:** *Figure 10.9* Peter Menzel/Photo Researchers; *Figure 10.10* Michael Thompson/Earth Scenes/Animals Animals. **p. 271:** *Figure 10.11* Ted Wood/Getty Images. **p. 272:** *Figure 10.12* Transtock/SuperStock. **p. 276:** *Figure 10.16* age fotostock/SuperStock. **p. 277:** AP Photo/Steve C. Wilson. **p. 278:** *Figure 10.18:* Boston Globe/George Rizer/Landov. **p. 281:** Orah Moore/Shelburne Farms.

**Chapter 11 p. 282:** Michael Wickes/The Image Works. **p. 283:** Michael Wickes. **p. 289:** *Figure 11.5* Doug Wilson/ARS/USDA. **p. 291:** *Figure 11.7* Lynn Betts/USDA Natural Resources Conservation Service;

*Figure 11.8* Inga Spence Visuals Unlimited. **p. 294:** *Figure 11.10* International Rice Institute. **p. 295:** Pete Oxford/DanitaDelimont.com. **p. 297:** *Figure 11.13(a)* Nigel Cattlin/Visuals Unlimited, (b) Chris R. Sharp/Photo Researchers, (c) Tim McCabe/USDA Natural Resources Conservation Service. **p. 298:** *Figure 11.15* Lynn Betts/USDA Natural Resources Conservation Service. **p. 299:** *Figure 11.16* Albert Mans/Foto Natura/Minden Pictures. **p. 300:** *Figure 11.18* Stephen Matera/DanitaDelimont.com. **p. 301:** *Figure 11.19* Grant Heilman/Grant Heilman Photography. **p. 303:** *Figure 11.23* Kevin Schafer/Minden Pictures. **p. 304:** *Figure 11.24* From J. D. Glover et al., *Science* 328 (2010). Courtesy of Jerry Glover, The Land Institute, Salina, Kansas.

**Science Applied p. 311:** *Figure SA5.1* Jerry Mason/Photo Researchers; *Figure SA5.2* U.S. Department of Agriculture. **p. 312:** *Figure SA5.3* Fagundes Brothers Organic Dairy.

**Chapter 12 p. 314:** Eberhard Streichan/Corbis. **p. 315:** Robert Ellis/iStockphoto.com. **p. 323:** *Figure 12.7 (left)* Siemens/Wikipedia, (right) Pegasus/Visuals Unlimited. **p. 326:** *Figure 12.10* J. Miles Cary/Knoxville News Sentinel. **p. 328:** *Figure 12.12* Sean Gardner/Reuters/Landov; *Figure 12.13 (inset)* Michio Hoshino/Minden Pictures. **p. 329:** *Figure 12.14* Joel Sartore/National Geographic/Getty Images. **p. 333:** *Figure 12.18 (inset)* U.S. Department of Energy/Photo Researchers. **p. 335:** *Figure 12.19* AP Photo/Holtec International. **p. 338:** *(inset)* Energy, Inc.

**Chapter 13 p. 342:** Chris Walsh/Getty Images. **p. 343:** Courtesy of Verdant Power, Inc. **p. 344:** *Figure 13.1 (top row)* Ian Hamilton/iStockphoto.com, Sandra Nicol/iStockphoto.com, Michael Utech/iStockphoto.com, Hans F. Meier/iStockphoto.com; *(middle row)* Bernd Lang/iStockphoto.com, Inga Spence/Photo Researchers; *(bottom row)* acilo/iStockphoto.com, Solar Technologies, Santa Cruz, CA, Chris McCooey/Getty Images, Rhoberazzi/iStockphoto.com. **p. 346:** *Figure 13.4 (top)* David Papazian/Uppercut Images/Getty Images, *(middle)* Lawrence Manning/Corbis/Fotosearch, *(bottom)* Courtesy of Apple. **p. 349:** *Figure 13.7* Michael Layefsky. **p. 352:** *Figure 13.9(a)* Jay Directo/AFP/Getty Images, (b) Eduardo Martino/Panos Pictures. **p. 353:** *Figure 13.10* Andrew Lank/iStockphoto.com. **p. 354:** *Figure 13.11* Kawakahi Kaeo Amina. **p. 356:** *Figure 13.13* Sal Maimone/SuperStock. **p. 357:** *Figure 13.14(a), (b)* Courtesy of Portland General Electric. **p. 359:** *Figure 13.16* Tim Dirven/Panos Pictures. **p. 360:** *Figure 13.18(b)* Julia Elman. **p. 361:** *Figure 13.19* Sandia National Laboratories/DOE. **p. 364:** Jorgen Schytte/Peter Arnold. **p. 366:** *Figure 13.25(b)* Courtesy of American Honda Motor Co., Inc. **p. 371:** *Figure 13.27* Martin Bond/Photo Researchers.

**Science Applied p. 377:** *Figure SA6.1 (inset)* Lee Wilcox. **p. 378:** *Figure SA6.2* Jonathan Kingston/Aurora Photos/Alamy. **p. 379:** *Figure SA6.3* S. Long, University of Illinois.

**Chapter 14 p. 380:** Peter Essick/Getty Images. **p. 383:** *Figure 14.1(a)* age fotostock/SuperStock, (b) The Irish Image Collection/SuperStock; *Figure 14.2* Bamboosil/age fotostock. **p. 384:** *Figure 14.3(a)* NASA/Goddard Space

# INDEX

Page numbers in **boldface** indicate a definition. Page numbers in *italics* indicates a figure. Page numbers followed by *t* indicate a table.

hybrid electric, *207, 207–208*
and urban sprawl, 274, *275*
Autotrophs, **60**

B horizon, **223,** *223*
Background extinction rate, **6**
Bacteria
denitrifying, 71
drug-resistant, 467
genetically modified, 134, 294, 396
and infectious diseases, 466, 467
nitrogen-fixing, 69
as pathogens, 163, 385
Baghouse filters, 422, *422*
Barnacles, goose, *111*
Basalt, 218
Base saturation, **225**
Bases, 33–34
Bass, largemouth, 381
Bats, 166
Bauxite, 226–227
*Beagle,* HMS [ship], 127
Bears
grizzly, 59
as omnivores, 62
polar, 137, 495, 507, *516,* 517–518
Beavers, *167,* 167–168
Becquerel (Bq), **335**
Beech trees, American, 149
Beetles
dung, 438, *439*
leaf, 149, 166
skeletonizing leaf, 134–135, *135*
Benthic zones, **108,** *109*
oceanic, 112, *112*
toxic chemicals in, 478
Bhopal (India), chemical spill in, 476,
*477*
Big Bend National Park, 511, *511*
Bioaccumulation, **292,** 414, 478,
518
Biocentric worldviews, **558**
Biochemical oxygen demand (BOD),
**384**
Biodiesel, **352,** 353–354, *354*
Biodiversity, 5–8, 120–124
causes for declines in, 500–507
conservation of, ecosystem,
509–511
conservation of, single-species,
507–509
declines in, 496–500
as determined by speciation and
extinction, 130–134
ecosystem, 6–8, 509–511
and ecosystem resilience, 75, 79, *79*
evolution as mechanism for,
124–130
genetic, 5–6
and genetically modified organisms,
294
and intermediate disturbance
hypothesis, *76,* 76–77

and island biogeography, 172,
509–510
single-species, 507–509
species, 6, 7
and species distribution, 134–139
Biodiversity hotspots, **144**–146, *145,*
146*t,* 500
Biofuels, **350**–351, 368*t*–369*t*
Biogeochemical cycles, **66**
Biogeography, island. *See* Island
biogeography
BioLite cookstove, 430–431, *431*
Biological molecules, 35–36
Bioluminescence, 112
Biomagnification, *478,* **478**–479
Biomass, **64,** 350–354, *352,*
368*t*–369*t. See also* Energy
Biomes, 99–101, *100,* **100,** 151
aquatic, 108–112
terrestrial, 101–107
Biophilia, **13**
Biosphere, **65**–73, *150,* 151
Biosphere reserves, **510**–511, *511*
Biotic components, **2**
Bird flu, **470,** *470*
Birds
biodiversity hotspots for, 146*t*
condors, 151, 152
conures, 125, *125*
cowbirds, *502,* 502–503
crows, 471
decline in species diversity of, 499,
*499*
dodos, 504, *505*
in dose-response studies, 474–475
eagles, 6, 62, 392, 508, *508*
falcons, 6, *8*
finches, 131, *132,* 163
flycatchers, 538, *538*
gannets, *152*
geese, 203
grouse, 230
jays, 471
murrelets, 271
owls, 163, *263,* 500, 508
parakeets, 202–203
pigeons, 505
robins, 471
Birth rates, 182
and demographic transitions, *188,*
188–189
Bison, American
conservation of, 6, *8*
overharvesting of, 505
Bisphenol A, 477
Bitumen, **330**
Bituminous coal, 325, *325*
Black Death, 466–467, *467*
Blackouts, 346
BLM. *See* Bureau of Land
Management (BLM)
Blue Fields Oyster Company, 140
Blue jays, 471

BOD. *See* Biochemical oxygen
demand (BOD)
Bogs, 109, *109*
Boiling, of water, 32–33
Bonds. *See* Chemical bonds
Boreal forests, *102,* **102**–103
Borers
emerald ash, 151
European corn, 294
Borlaug, Norman, 288
Boston (MA), and urban renewal, 278,
*278*
Bottleneck effect, *129,* **130**
Bottom ash, **450,** *451*
Bovine spongiform encephalopathy
(BSE), 469–470, *470*
BP Deepwater Horizon well, 315,
327, *328,* 394
Braungart, Michael, 456
Breeding, 126, *126, 127,* 131, *132*
Broad-spectrum pesticides, **292**
Brown, Lester, 378
Brown smog, **413**
Brownfields, **454**
Brownouts, 346
BSE. *See* Bovine spongiform
encephalopathy (BSE)
*Bubo virginianus. See under* Owls
Bubonic plague, 466–467, *467*
Bullard, Robert, 565
Bullfrogs, American, 155
Bureau of Land Management (BLM),
*267,* 267–268
Bureau of Reclamation, U.S., 235
Burkholder, JoAnn, 1, 15
Bush, George W., 352, 495, 542, 543
Bycatch, **302**

C horizon, **223,** *223*
Cacti, 99–100
*Leuchtenbergia,* 128
organ pipe, *99*
trade in, 506
CAFO. *See* Concentrated animal
feeding operations (CAFO)
Calcium, in ecosystems, 73
Calcium carbonate, and the carbon
cycle, 67, *68*
Calcium hydroxide, as base, 34
California Academy of Sciences, *349,*
349–350
Cancer. *See* Human Health
Capacity factors of power plants, **324**
Capacity of power plants, **324**
Capillary action, **32,** *32*
Capital, types of, 555
Carbohydrates, **35**
Carbon, modern vs. fossil, 351
Carbon cycle, 66–69, *68*
Carbon dating, 30–31
Carbon dioxide. *See also* Global
change; Greenhouse gases
and air pollution, 411–412, 412*t*

# Engage Your Environment Activities

# Engage Your Environment: Environmental Challenges

## What are the most critical environmental issues today?

### Learning Objectives

You will:

1. Understand the scope of current environmental issues.

2. Identify local environmental problems.

## Introduction

As our human population and demand for resources grow, we are confronted with many environmental challenges. Scientific study of environmental systems and challenges is inherently difficult because of their complexity and because they exist on both local and global scales.

Environmental scientists work to understand these challenges, determine which challenges are most critical, and propose ways to solve them.

## Background Questions

1. What factors make environmental systems difficult for scientists to study, understand, and predict?

2. How do human population growth and our demand for resources affect global environmental challenges?

## Methods

Using the following categories, list what you think are the most pressing environmental issues today. Also provide an example of a local component to each problem.

## Results

## Environmental issue categories

1. Food supply and food production:

2. Resource availability (water, energy, etc.):

3. Weather and climate:

4. Human health:

5. Species diversity and open space:

## Follow-up

1. Are any of the problems you listed above more pressing than others for your local community? Explain.

2. Do local programs address environmental issues in your community? If so, what are they?

3. On the global scale, does one problem stand out as more pressing or critical than the others? If so, what is that problem and why is it more critical than the others?

4. Should we choose to tackle environmental challenges by concentrating on one or two problems, or by spreading our resources among all problems?

5. Environmental scientists often refer to a current "ecological crisis." Now that you have assembled a list of environmental challenges, do you feel this statement is an exaggeration or a fair description of the status of Earth?

6. Choose one of your examples of today's most pressing environmental issues from the list above and propose an action to address this problem in your local community.

## CHAPTER 2

# Engage Your Environment: Automobile Efficiency

## How efficient are best-selling automobile models?

### Learning Objectives

You will:

1. Compare the efficiency of various automobiles.
2. Calculate the cost of operating an automobile.

### Resource

http://www.fueleconomy.gov/feg/findacar.htm (optional)

## Introduction

- - - - - - - - - - - - - - - - - -

Because of increasing gasoline prices, many people shop for automobiles based on the fuel efficiency of the model. We will use liters of gasoline consumed per 100 kilometers of travel in our comparisons.

Hybrid electric vehicles (HEVs) are becoming more common in the United States. Although HEVs work in a variety of ways, most are propelled both by a gasoline engine and by electric motors that receive electricity from batteries. The batteries are charged when the driver applies the brakes and the kinetic energy of the motion of the vehicle is converted to energy stored in the battery. The hybrid car switches between gasoline and battery power as necessary.

### Background Question

What factors might influence a person's decision to purchase a particular automobile make and model?

### Methods

Study the data in Results Table 1. Each automobile in the table is one of the top 10 best-selling cars in the United States for the model year shown, with the exception of the Toyota Prius, which is the best-selling hybrid car for the year 2010 but not among the top 10.

## Results

| TABLE 1 | Automobile efficiency, fuel cost, and oil usage* | | |
|---|---|---|---|
| Automobile make and model | Combined fuel efficiency liters/100 km† | Annual fuel cost | Petroleum use (barrels/year)‡ |
| 2000 Ford F-150. 2-wheel drive, automatic transmission | 14.7 | $2,897 | 21.4 |
| 2004 Toyota Camry. 2-wheel drive, automatic transmission | 9.8 | $1,933 | 14.3 |
| 2006 Chevrolet Impala. 2-wheel drive, automatic transmission | 11.8 | $2,318 | 17.1 |
| 2009 Honda Civic. 2-wheel drive, automatic transmission | 8.1 | $1,599 | 11.8 |
| 2010 Toyota Prius hybrid. 2-wheel drive, automatic transmission | 4.7 | $927 | 6.9 |

*Data from http://www.fueleconomy.gov/feg/findacar.htm as of 1/13/11. †Based on 45% highway and 55% city driving.
‡1 barrel of petroleum = 42 gallons of gasoline.

## Follow-up

1. Which nonhybrid automobile is the most efficient?

2. What characteristics of this automobile make it the most efficient?

3. The Ford F-series truck has been the best-selling automobile in the United States for 28 years. Over a 5-year period of ownership, how much gas money would you save if you drove a Toyota Camry instead of an F-150? How much would you save if you drove a Toyota Prius instead of an F-150?

4. What are some advantages of refining less petroleum to produce gasoline for automobiles?

5. Hybrid automobiles often cost more than conventional models. Calculate the time it would take to earn back your additional initial investment in purchasing a hybrid over a nonhybrid car. (Note that the cost of a base model 2011 Honda Civic Hybrid is $23,950, according to http://automobiles.honda.com/civic=hybrid/.)

## Reference

http://www.fueleconomy.gov/feg/findacar.htm

## CHAPTER 3

# Engage Your Environment: Your Campus as Wildlife Habitat

## How does your campus rate as a habitat for wildlife?

### Learning Objectives

You will:

1. Evaluate your campus as a wildlife habitat.
2. Propose and prioritize three improvements to this habitat.

### Resource

Wildlife habitat checklist (see Results)

## Introduction

Providing food, water, shelter, and space for wildlife will allow a surprisingly large number of insects, birds, and other small animals to coexist with humans.

### Background Questions

1. What evidence is there of insects, birds, and other small animals on your campus?

2. What is the predominant type of land use surrounding your campus?

## Methods

Take a walk around your campus. Use the checklist in the Results section to evaluate how well your campus provides habitat for wildlife. Your task is to:

1. Assess how well your campus provides habitat for wildlife. Rate each section on a scale of 1 (lowest) to 5 (highest).
2. Make notes on ways to improve this habitat.
3. Keep a list of places on campus where you think the habitat can be improved.

## Results

### Wildlife habitat checklist

| FOOD | Rating (lowest) (highest) | Ways to improve |
|---|---|---|
| nuts | 1 2 3 4 5 | _____ |
| seeds | 1 2 3 4 5 | _____ |
| berries | 1 2 3 4 5 | _____ |
| nectar | 1 2 3 4 5 | _____ |
| insects | 1 2 3 4 5 | _____ |
| **WATER** | | |
| a wet area present all year | 1 2 3 4 5 | _____ |
| a wet area present part of the year | 1 2 3 4 5 | _____ |
| **SHELTER** | | |
| thick brush and brambles | 1 2 3 4 5 | _____ |
| tall grassy fields | 1 2 3 4 5 | _____ |
| woodlands with many layers of plants | 1 2 3 4 5 | _____ |
| streams with forested buffers | 1 2 3 4 5 | _____ |
| dead standing trees | 1 2 3 4 5 | _____ |
| rotting logs on the ground | 1 2 3 4 5 | _____ |
| brush piles | 1 2 3 4 5 | _____ |
| **SPACE (natural habitats)** | | |
| meadows | 1 2 3 4 5 | _____ |
| forests | 1 2 3 4 5 | _____ |
| scrub/shrub | 1 2 3 4 5 | _____ |
| wetlands | 1 2 3 4 5 | _____ |
| streams | 1 2 3 4 5 | _____ |

## Follow-up

1. Based on your assessment, what improvements to the habitat are most needed?

2. Which improvements to the habitat will be relatively easy?

3. Which improvements to the habitat will be relatively low in cost?

## Reference

U.S. Fish and Wildlife Service, Schoolyard Habitat Project Guide, online at http://www.fws.gov/chesapeakebay/pdf/habitatguide.pdf

## How might water shortages in your community affect you personally?

### Learning Objectives

You will:

1. Examine the climatic characteristics of your local area, particularly the pattern of precipitation throughout the year.

2. Estimate how vulnerable your community is to water shortages.

3. Identify your personal pattern of water use.

4. Consider both your personal and your community's responses to water shortages.

### Resources

Figure 4.17 and biome descriptions in Chapter 4

## Introduction

Climate diagrams show the annual month-by-month patterns of variation in temperature and precipitation, averaged over a 30-year period. Climate affects the type of vegetation that can survive in an area with grasses more adapted to survive a lack of moisture than trees. Desert plants can actually grow under a wide range of conditions and prevail in areas too dry for grasses or trees.

Droughts—prolonged periods of lower-than-average precipitation—are part of cyclic changes in weather and can affect natural ecosystems and community water supplies. A community's water supply is also affected by its source of water, how it is "recharged" by precipitation, and local use patterns. Water shortages have led to lingering environmental damage as well as conflicts among farmers, conservationists, industries, and city dwellers.

In 2003, the Government Accounting Office surveyed water managers in 47 states and found that managers of 36 states expected some portion of their state to experience water shortages during the next 10 years, even under average water conditions.

### Background Questions

1. Is the place where you live formerly forest land, grassland, desert, or some other biome?

2. How much of the native vegetation has been converted to farmland or developed for urban/suburban uses?

**3.** What is the source of your community's water supply?

**4.** Do people in your neighborhood water their lawns and gardens in the summer or do they depend solely on rainfall?

**5.** Are you aware of any water shortages that have affected your community in the past?

## Methods

**1.** Use Figure 4.17 on page 100 to determine the biome type(s) where you live. Examine the climate diagram for your local biome and read its description; note your biome's annual pattern of precipitation.

**2.** Complete Results Table 1.

**3.** Complete Results Table 2. (For each pair of alternatives, choose the one most similar to your own habits.)

**4.** Tally your water use patterns score. The higher your score, the less you will be affected by a water shortage.

## Results

### TABLE 1 Biome types

| Major U.S. biome types | Driest months of the year (based on a 30-year average) |
|---|---|
| Desert | _____ |
| Grassland | _____ |
| Deciduous forest | _____ |
| Temperate rainforest | _____ |

### TABLE 2 Personal water use inventory (For each pair of choices, circle one.)

| Lower water use alternative (10 points each) | Higher water use alternative (5 points each) |
|---|---|
| Short, low-flow showers | Tub baths or long, high-flow showers |
| Going to the carwash | Washing your own car |
| Drip irrigation or using water from a rain barrel or no watering | Regular use of sprinklers for lawn and garden |
| Drought-tolerant landscape plants | Grassy lawn |
| Total points: _____ | Total points: _____ |

## Follow-up

1. How does your local climate affect your awareness of water issues and use of water?

2. Most states have plans in place to deal with water shortages. What are three feasible ways your community could respond to a water shortage?

3. Given likely changes in population patterns for your state or community, how is your community's vulnerability to water shortages likely to change over the next decade?

## References

Climographs for over 100 U.S. cities are available online at http://www.drought.unl.edu/whatis/ climographsdomesticenglish.htm

Drought-Ready Communities: A Guide to Community Drought Preparedness: http://www .drought.unl.edu/plan/DRC_Guide.pdf

GAO Report to Congress in 2003: http://www.gao.gov/new.items/d03514.pdf

State Drought Response Plans: http://drought.unl.edu/plan/stateplans.htm

## CHAPTER 5

# Engage Your Environment: Campus Species Richness

## Does species richness differ in areas of human disturbance?

### Learning Objectives

You will:

1. Understand the concept of species richness.
2. Compare species richness in both natural and disturbed habitats.

### Resources

Two sampling habitats, one natural and one disturbed, on your campus

## Introduction

The number of species present in a defined area is known as species richness. It varies greatly among habitats and can be influenced by many factors, including altitude, latitude, the availability of water, and the degree of disturbance. Environmental scientists use species richness as one indicator of biological diversity that gauges ecosystem health and function.

Determining species richness in two habitats on your campus will allow you to evaluate the variety of plant species present and the level to which disturbance influences the number of species.

### Background Questions

1. Why is species richness a valuable indicator of ecosystem health and function?

2. How might new species arrive in a habitat?

3. Why might a species "decide" to colonize one particular habitat over another?

4. How do humans influence the colonization of species into new habitats?

## Methods

Determine the number of plant species present in two habitats on your campus. Choose two habitats of approximately the same size but differing in the levels of human disturbance. The "natural" habitat should be as undisturbed by humans as possible. This habitat may be a wooded area, a park, or open space. The "disturbed" habitat should exhibit some level of human disturbance, and could be located at the edge of a parking lot, near a building, or beside a cleared lot.

In the appropriate column of Results Table 1 list the names of all plant species present in the natural and disturbed habitats. If you do not know the name of a species, name it yourself (for example, Plant species 1, Tree species 1) and make note of key characteristics (for example, size, leaf shape, flower color, the number of leaves, or type of stem, such as woody, green stem, or vine). Be consistent in naming and describing your species to avoid counting the same species multiple times.

## Results

| TABLE 1 | Plant species richness |
|---|---|
| **Species Name and Description** ||
| **Natural habitat** | **Disturbed habitat** |
| _____ | _____ |
| _____ | _____ |
| _____ | _____ |
| _____ | _____ |
| Total number of species: \_\_\_\_\_ | Total number of species: \_\_\_\_\_ |

## Follow-up

1. Describe the size and basic features of both the natural habitat and the disturbed habitat.

2. How many species were present in the natural habitat? How many were present in the disturbed habitat? What factors do you think account for this difference?

3. What differences, if any, were present in the types of species found in disturbed and undisturbed areas?

4. Propose a hypothesis to explain how disturbance affected species richness in these particular habitats.

## CHAPTER 6

# Engage Your Environment: Population Size, Distribution, and Density Habitat

## How would you describe species abundance and distribution in a local ecosystem?

### Learning Objectives

You will:

1. Understand species richness.
2. Determine population density in a sample plot of land.
3. Map species distribution in a sample plot of land.

## Introduction

To understand population structure, ecologists examine population size, distribution, and density. These measures allow ecologists to characterize the ecosystems they study.

## Background Questions

1. What factors influence population size?

2. How would the distribution of one species influence that of another species?

3. How do environmental factors influence species distribution?

4. How might the population density of a species affect resource availability for other species?

## Methods

1. Choose a small, vegetated area in your yard or on campus.

2. Within that area, randomly choose three plots, each 1 m² (about 40 inches by 40 inches).

3. Record the names of the plant species and the number of individuals you find in each plot in Results Table 1. If you don't know the name of a species, assign a descriptive name and number to the species (for example, Tree 1, Green plant 1, Flower 1), and make note of its identifying characteristics (for example, leaf shape, leaf number, color, flower, stem type). Be sure to keep the names consistent.

4. Pick one species from each sample plot and draw its distribution in Results Figure 1.

## Results

| TABLE 1 | Species name and abundance | | |
|---|---|---|---|
| Species name and/or species description | Plot number | Number of individuals | Population density (number of individual plants/m²) |
| | | | |
| | | | |
| | | | |
| | | | |
| | | | |
| | | | |
| | | | |
| | | | |

FIGURE 1 Species distribution.

## Follow-up

1. How many species did you find in each sample plot?

2. Which species had the highest density?

3. Were the species you recorded similar in size and shape?

4. How would you characterize the distribution of species in the sample plots: random, clumped, or even?

5. How was each species distributed within the sample plots: random, clumped, or even?

6. What local environmental factors might have influenced the distribution of species in the sample plots?

# How can census data be used to estimate future needs in your state or community?

## Learning Objectives

You will:

1. Use data from the U.S. Census Bureau to project changes in numbers of schoolchildren over the next 10 years in selected states.

2. Estimate requirements in these states for more schools and teachers in the next decade.

## Resources

1. Census data (in Results chart)

2. Calculator

## Introduction

The U.S. census is taken every 10 years. Different questions are asked for different reasons. For example, information from veterans is used to estimate ("project") needs for more veterans hospitals.

An important use of age distribution data is the projection of future needs for schools and teachers. (In some areas, projections indicate that fewer schools and fewer teachers will be needed in the future!)

Census data is also used to ensure that adequate community services are provided for the increased numbers of senior citizens expected as the "baby boomers" reach retirement age.

## Background Questions

1. Why would states need to estimate how many new teachers will be required in 10 years?

2. Once a community decides to build a new school, what are some of the other decisions they will need to make?

## Methods

1. Using the census data in the first three columns of the Results chart, estimate how the number of schoolchildren will change over the next 10 years in the United States as a whole as well as in four specific states. Enter your findings in the fourth column of the chart.

2. Assuming one teacher for every 30 students, calculate how the number of teachers will change. Enter your findings in the fifth column of the chart.

3. Assuming one school per 1,000 students, calculate how the number of schools will change. Enter your findings in the fifth column of the chart.

## Results

| TABLE 1 | Census data | | | |
|---|---|---|---|---|
| Data on 5–17 age group | 2010 estimate | 2020 estimate | Increase or decrease in number of students | Changes in needed number of teachers and number of schools (+/–)* |
| Florida | 2,890,955 | 3,512,166 | | _____ Teachers<br>_____ Schools |
| Texas | 4,708,080 | 5,676,403 | | _____ Teachers<br>_____ Schools |
| New York | 3,170,009 | 3,107,262 | | _____ Teachers<br>_____ Schools |
| New Jersey | 1,501,004 | 1,526,729 | | _____ Teachers<br>_____ Schools |
| Entire U.S. | 61,810,000 | 65,955,000 | | _____ Teachers<br>_____ Schools |

*Assume one teacher per 30 students and one school per 1,000 students.

## Follow-up

1. How might the accuracy of your calculations be affected by changes in immigration patterns or birthrates in different regions of the United States?

2. In light of your calculations, is the demand for teachers increasing or decreasing; how might this affect job availabilities for current college students? In some cities, many young children were not counted in past censuses. Why is it important to count even young children?

3. According to the U.S. Census Bureau, 74 percent of households in the United States filled out and mailed back their 2010 census questionnaires. The rest were visited by census takers. Do you think getting an accurate census is worth the effort? Why, or why not?

## References

Information on uses of census questions: http://www.census.gov/dmd/www/content.htm

Source of state age distribution data: http://www.census.gov/population/www/projections/statepyramid.html

U.S. Census Bureau population projections: http://www.census.gov/population/www/projections/index.html

# CHAPTER 8
# Engage Your Environment: Viewing Change from Afar

## How does remote sensing help us view environmental change?

### Learning Objectives

You will:

1. Understand how remote sensing is used as an environmental assessment tool.

2. Understand how remote sensing data can be used to document environmental change.

### Resource

U.S. Geological Survey Earthshots: Satellite Images of Environmental Change: http://earthshots.usgs.gov/tableofcontents

## Introduction

Remote sensing monitors Earth from afar using a variety of mostly electronic devices that include aircraft, watercraft, spacecraft, and satellites. The data obtained from these remote sensing devices can be used to collect baseline data, detect change over time, and open previously inaccessible locations to study.

## Background Questions

1. What advantage does remote sensing offer to environmental scientists in evaluating environmental change?

2. Propose a question that could be evaluated by an environmental scientist using remote sensing data.

## Methods

Using the U.S. Geological Survey Earthshots Web site provided, navigate to the following links:

Las Vegas, Nevada

Iraq and Kuwait

Orlando, Florida

Chernobyl, Ukraine

Mount St. Helens, Washington

Muskingum mines, Ohio

## Follow-up

For each Earthshot, answer the following questions:

1. What environmental change is being addressed?

2. What types of data are being collected to address this change?

3. How are remote sensing data being used to address or identify this change?

4. To what larger environmental issue is the change linked?

5. Was a conclusion reached based on the remote sensing data collected?

6. Suggest how other data might be combined with remote sensing data to further address the environmental problem.

## Reference

U.S. Geological Survey Earthshots: http://earthshots.usgs.gov/tableofcontents

## CHAPTER 9

# Engage Your Environment: The Trade-offs of Bottled Water

## How can you enjoy the convenience of bottled water while minimizing cost and plastic waste?

### Learning Objectives

You will:

1. Consider ways of avoiding or minimizing the impacts of using bottled water while retaining the convenience of a "portable water source."

2. Compare possible alternative courses of action.

3. Propose ways to reduce the impact of the consumption of bottled water on campus.

### Resource

Campus data (from your own observations or from talking with other students)

## Introduction

According to the Container Recycling Institute, people in the United States buy more than 30 billion single-serving bottles of water each year. That averages out to about 100 bottles per person. Leading reasons for buying bottled water are concerns about the safety of tap water and the convenience of the portability of bottled water.

Scientific studies do not support the claim that bottled water is safer to drink than tap water. In fact, nearly half of bottled water is actually tap water, though the price of bottled water is generally hundreds of times greater than that of the same amount of tap water.

Four out of five empty water bottles are not recycled, adding to the challenge of managing municipal waste. Energy costs to make and fill the bottles are also much higher than the costs of delivering tap water to a faucet.

On college campuses, students want the convenience of carrying drinking water around during the day. To minimize cost and waste, some campuses have installed "refilling stations" on water fountains that allow students to refill water bottles. In addition, many campuses provide extra recycle bins for discarded water bottles, especially at events like football games that may generate large numbers of empty bottles.

### Background Questions

1. How many bottles of water did you buy in the last year? If the average cost of bottled water is $1, how much did you spend?

2. Why did you buy bottled water?

3. How did you dispose of the empty bottles?

## Methods

Complete the campus survey column in the Results table based on your own observations around campus or on data collected from talking with other students.

## Results

| TABLE 1 | Bottled water on campus | |
|---|---|---|
| Issue | Ways to avoid or minimize impacts | Survey of your campus |
| Cost of bottled water | Drink from a fountain<br><br>Refill refillable bottles | Ample number of water fountains? (Y/N)<br><br>Refilling stations? (Y/N) |
| Waste (empty plastic bottles) | Buy and use refillable bottles<br><br>Recycle waste bottles | Refillable bottles for sale? (Y/N)<br><br>Ample number of recycle bins? (Y/N) |

## Follow-up

1. What potential changes on campus could provide ample drinking water for students while avoiding cost and waste?

   Consider the following in your response:

   Number, convenience, and functioning of water fountains
   Number and location of refilling stations
   Availability and cost of refillable bottles
   Number and location of recycling bins for plastic bottles

2. In light of current options for obtaining drinking water as well as any potential barriers to refilling stations and recycling, what top-priority improvements do you propose for implementation on your campus?

## Reference

Water, Water, Everywhere! (2007 report from the Container Recycling Institute): http://www.container-recycling.org/assets/pdfs/reports/2007-waterwater.pdf

## Do you live in a smart community?

### Learning Objectives

You will:

1. Understand the concept of smart growth.
2. Recognize the core principles of the U.S. Environmental Protection Agency (EPA) smart growth program.
3. Evaluate your community using two smart growth principles.

### Resource

List of the 10 basic principles of smart growth (pp. 276–277).

## Introduction

Urban sprawl, the spread of urban development into suburban and rural areas, continues to engulf land at a rapid rate. The EPA smart growth program suggests a range of strategies to promote the growth of environmentally sound and economically vigorous communities.

## Background Questions

1. What could prevent a city or town from integrating all 10 smart growth principles into its growth plan?

2. How could the age of a city affect the ability of planners to follow smart growth principles?

3. How could these principles taken together lead to a more sustainable community?

## Methods

Determine how "smart" your community is by evaluating two principles of smart growth: preservation of open space and transportation options. Limit the sample area to your neighborhood, which should be an area where you spend most of your time. Fill out Results Table 1, and answer the questions that follow. You can tour the area on foot and make note of open space and of transportation options, or you can consult an online aerial map such as Google satellite images.

## Results

| TABLE 1 | Smart growth principles | | |
|---|---|---|---|
| **Open space** | **Assign a percentage for each category** | **Transportation variety** | **Add a checkmark if available** |
| Working farmland | _____ | Subway | _____ |
| National parks/protected lands | _____ | Light rail | _____ |
| Wetlands | _____ | Buses | _____ |
| Local parks/ballfields | _____ | Bicycle lanes/paths | _____ |
| Meadows, community gardens | _____ | Car-sharing program | _____ |
| Other (describe) | | Carpool lanes | _____ |
| | _____ | Sidewalks | _____ |
| | 100% | Other | _____ |

## Follow-up

1. Refer to the results listed in Table 1, and then describe how well your community supports open space, including both strong and weak points. In the same way, evaluate the transportation modes that are available.

2. Do you support existing public transportation options? If so, how often do you take public transportation? If there is no public transportation available, what other transportation options would work best in your community?

3. What type(s) of open spaces in your community do you use most frequently?

4. How would you suggest your city or town address any shortage of open space?

5. What challenges do you foresee your city or town encountering as it addresses its smart growth weaknesses?

## Reference

U. S. Environmental Protection Agency: www.epa.gov/smartgrowth/pdf/whtiss4v2.pdf. April 2001. Web. 18 Sept. 2010.

## CHAPTER 11

# Engage Your Environment: Food Miles

# How can you use "food miles" to choose which fresh fruits and vegetables to buy?

## Learning Objectives

You will:

1. Consider trade-offs involved in buying produce that is grown locally versus that grown in other regions or countries.

2. Compare the "food miles" for four choices of fresh fruits and vegetables.

3. Identify and justify your likely choices of what to buy.

## Resources

1. Local grocery ads (optional)

2. Map atlas or distance calculator (optional)

## Introduction

"Food miles" refers to the distance food travels from the field where it was grown to your plate. Since transporting food contributes to carbon dioxide emissions and air pollution, many people believe that food miles should be considered in deciding what produce to buy at local groceries. Other factors that may be important in making purchase decisions are product availability, taste, and freshness, as well as how much energy and pesticide were used in growing it. Following are some additional benefits of reducing your food miles:

- Obtaining fresher produce

- Supporting local farmers

- Keeping your money in the community

- Knowing where your produce comes from

Calculating the food miles for particular food items can be difficult. The energy used in transporting food is not the only contributor to energy use related to food production and consumption. Another important consideration is the energy used in growing food, particularly if it is being grown out of season in a greenhouse. In addition, the food miles total should include your travel from where you buy the food to your home. And according to Cornell University ecologist David Pimentel, a typical American may use over 300 calories of fossil fuel energy in driving home after purchasing a 375-calorie can of corn!

About half of the produce sold in grocery stores in the United States is grown in California; the rest comes from other states and from other countries. Grocery stores may specify the country (or state) of origin in labels, stickers, ads, and signage. A simple, albeit incomplete, way to compare food miles is to consider the sources of fresh fruits and vegetables. One strategy to reduce food miles is to avoid "out-of-season" fruits and vegetables shipped from distant locations.

## Background Questions

1. What specific types of fresh fruits and vegetables have you purchased in the last year?

2. What factors were most important to you when you selected produce?

3. Were you aware of where your produce was grown? How did you know?

4. How far do you live from the orange-producing state of Florida? From the produce basket of California? (From the East Coast to the West Coast is about 3,000 miles; from the Canadian border to the Mexican border is about 1,000 miles.)

## Methods

Using the food miles examples in Results Table 1 as a guide, complete Results Table 2. Read grocery advertisements, visit a local grocery or convenience store, or look at your local supermarket's Web site for produce price, source, and availability information.

## Results

| TABLE 1 Food miles examples | | |
|---|---|---|
| Fresh produce item | Country of origin | "Food miles" (shipping distance to California ports) |
| Tomatoes | Mexico | 1,000 miles |
| Blueberries | Chile | 6,000 miles |
| Asparagus | Peru | 6,000 miles |
| Blueberries | Argentina | 6,000 miles |
| Tomatoes | Netherlands | 6,000 miles |
| Blueberries | New Zealand | 9,000 miles |
| Oranges | Australia | 9,000 miles |

| TABLE 2 | Your food miles | | |
|---------|-----------------|-----|-----|
| Fresh produce item | Origin choices | "Food miles" (shipping distance to your local store) | Your likely choices |
| Tomatoes | California | _____ miles | |
| | Locally grown | _____ miles | |
| | Mexico | _____ miles | |
| Blueberries | Chile | _____ miles | |
| | Argentina | _____ miles | |
| | New Zealand | _____ miles | |
| | Locally grown | _____ miles | |
| Grapes | Locally grown | _____ miles | |
| | California | _____ miles | |
| | Chile | _____ miles | |
| Oranges | California | _____ miles | |
| | Florida | _____ miles | |
| | Australia | _____miles | |

## Follow-up

1. Consulting Results Table 2, list the factors you would consider in making a decision to purchase produce. What is the most important factor on your list? Why?

2. How could you keep your food miles low if a favorite fruit were out of season locally?

3. Can you come up with a few "green shopping tips" to reduce your food miles at the grocery?

## References

Distance Calculator: http://www.daftlogic.com/projects-google-maps-distance-calculator.htm
Do Food Miles Matter? http://www.organicconsumers.org/articles/article_11662.cfm
Food Miles: http://www.nrdc.org/health/effects/camiles/foodmiles.pdf
Food Miles: Background and Marketing: http://attra.ncat.org/attra-pub/foodmiles.html
Food Miles Calculator: http://www.organiclinker.com/food-miles.cfm
Increased U.S. Imports of Fruits and Vegetables: http://www.ers.usda.gov/Publications/ fts/2007/08Aug/fts32801/fts32801.pdf
USDA Fruit and Vegetable Backgrounder article: http://www.ers.usda.gov/publications/vgs/apr06/ vgs31301/vgs31301.pdf

## CHAPTER 12

# Engage Your Environment: Preventing Energy Waste

## Do you unintentionally waste electric energy?

### Learning Objectives

You will:

1. Inventory all the electrical devices plugged into outlets in your dorm room or your bedroom at home and determine energy waste.
2. Inventory your use of incandescent light bulbs.

### Resource

Electrical device inventory chart (Table 1) and Incandescent light bulb inventory chart (Table 2) in the Results section

## Introduction

Every day, energy is wasted by inefficient incandescent light bulbs and other devices. Televisions, cell phone chargers, kitchen appliances, computers, and stereos continue to draw energy for the remote control, for "standby" status, or for the timer, even when these devices are turned off. This wasted energy accounts for 5 to 10 percent of the average total electric bill for a typical family.

## Background Questions

1. How many electrical outlets are there in your dorm room or bedroom (including ones that are part of a power strip or extension cord)?

2. How many devices, and what types, do you keep plugged in or are plugged in for long periods of time?

3. How many ordinary incandescent bulbs do you use in your room, including desk lamps, table lamps, and light fixtures? What wattages are these bulbs?

## Methods

1. In Results Table 1 (Electrical device inventory chart), list all items you plug into electrical outlets.

2. Estimate how often each of these is plugged in (all the time, sometimes, only when needed).

3. Consider which devices use energy continually. Determine which of your devices (such as desk lamps and hair dryers) do not use energy continually.

4. Complete Results Table 2 (Incandescent light bulb inventory chart).

## Results

| TABLE 1 | Electrical device inventory chart | |
|---|---|---|
| Plugged-in device | How often is the device plugged in? | Does the device draw energy continually? |
| | | |
| | | |
| | | |
| | | |
| | | |
| | | |
| | | |
| | | |

| TABLE 2 | Incandescent light bulb inventory chart |
|---|---|
| Wattage of bulb | Use (table lamp, desk lamp, etc.) |
| | |
| | |
| | |
| | |
| | |
| | |

## Follow-up

1. Of the devices that continually use electricity, which ones could you unplug without causing much inconvenience?

2. Of the devices that continually use electricity, which ones could you unplug when you will be away for several days?

3. A power strip can be turned off so that it does not draw energy when it is plugged into an electrical outlet. How many power strips would you need to turn off all devices that use energy continually?

4. How many compact fluorescent bulbs would you need to replace all the incandescent bulbs currently in use?

5. If your numbers are typical, estimate how many power strips and compact fluorescent bulbs would be necessary if all the students on your campus used them to prevent energy waste.

## References

Energy Efficiency Benefits: http://www.p2pays.org/ref/12/11365.pdf

Energy Star College Student Guide to Improving Energy Performance on Campus:
   http://www.energystar.gov/ia/business/higher_ed/ENERGYSTAR_Student_Activity_Guide.pdf

Energy Star Products: http://www.energystar.gov/

Standby Power and Energy Vampires: http://www.energystar.gov/index.cfm?c=about.vampires

## CHAPTER 13

# Engage Your Environment: Energy Audit

## How energy efficient is your home?

### Learning Objectives

You will:

1. Identify the major construction, maintenance, and design features that make a building energy efficient.

2. Determine energy-saving steps.

### Resource

Create a draft detector by taping one edge of a piece of tissue paper or plastic wrap to a pencil. If made correctly, the draft detector should look like a flag.

## Introduction

An energy-efficient building is more comfortable than a building that wastes energy. It requires less fuel for heat and less electricity for cooling. A building that is badly designed and maintained wastes money because the heat and air conditioning escape to the outdoors.

### Background Questions

1. What are the places of greatest energy loss in most homes?

2. What are some reasons that homes and buildings are not as energy efficient as they might be?

### Methods

1. Using Table 1, complete the observations on your house, apartment, or a building suggested by the instructor. Use the draft detector to help locate air infiltration.

2. Develop a set of recommendations for improving energy conservation in the house, apartment, or building that was studied.

## Results

| TABLE 1 | Interior and exterior building survey |
| --- | --- |

| Interior | Yes | No | Does not apply (or comments) |
| --- | --- | --- | --- |
| How much of the floor is covered with rugs, padding, and carpeting (percent or fraction)? | | | |
| Are heating and cooling equipment (e.g., ducts, grills, radiators) blocked by furniture, rugs, drapes, or other items? | | | |
| Do the colors of the walls and ceilings reflect light well? | | | |
| Are insulating drapes or other tight window treatments such as framed shades in place? | | | |
| If there is a fireplace, does it have a damper and glass doors? | | | |
| Is the damper closed when the fireplace is not in use? | | | |
| Are the glass doors kept closed when the fireplace is in use so that warm air from the room does not escape up the chimney? | | | |
| Does the draft detector move when placed along edges of doors and windows on both calm and windy days? | | | |
| On windy days does the draft detector show air currents coming in through electrical outlets that are positioned on the outside walls? | | | |
| Are cracks in walls and foundations sealed and holes plugged? | | | |
| Check other openings for drafts, and make a list. Examples are where pipes enter walls, exhaust fans in kitchen or bathroom, and where chimneys exit through walls or ceilings. | | | |
| Are hot water faucets free from drips? | | | |
| Have flow restrictors been placed in pipes connected to showerheads? | | | |
| Is the thermostat automatically set to lower temperatures at bedtime? | | | |

| Exterior | Yes | No | Does not apply (or comments) |
| --- | --- | --- | --- |
| Is there weather stripping around both inner and outer doors? Around the windows? | | | |
| Are cracks in walls and foundations sealed and holes plugged? | | | |
| Is there an air lock entry hall, double door, or insulated storm door at each outside entrance? | | | |
| Are evergreen shrubs and trees planted as windbreaks around the north and west sides of the building? | | | |
| Are deciduous (leaf-shedding) trees planted on the south side for summer shade and winter sun? | | | |
| Are basement doors insulated and tightly sealed? | | | |
| Are cracks and joints around windows, doors, stairways, pipes, and electrical wires caulked? | | | |
| Are there fewer windows on the north side of the building than elsewhere? | | | |
| Are the north windows smaller than those on the other sides? | | | |
| Are storm windows in place and tightly sealed? (If moisture condenses on the inside of windows and freezes on cold days, then the storm windows are not working properly.) | | | |

## Follow-up

1. How would you characterize the energy efficiency of the building you examined?

2. Which energy "fixes" would be inexpensive to implement?

3. Have good energy conservation principles been used in the landscaping around the building?

4. Do you think the windows were positioned for energy efficiency? Why or why not?

5. For what purposes is hot water really needed in a home? Why do many people use more hot water than they need?

6. In your building what are the three areas most urgently in need of energy repair? Will the fixes be expensive and time consuming to implement? Explain.

## Reference

Alliance to Save Energy: http://ase.org/; http://www.energysavers.gov/

## CHAPTER 14

# Engage Your Environment: From Runoff to Rain Gardens

## How can your community minimize its storm-water runoff?

### Learning Objectives

You will:

1. Explore ways to reduce the volume and velocity of polluted runoff reaching local waterways.

2. Compare different development strategies and their impacts on streams and the people living in affected areas.

### Resource

Low-impact development strategies comparison chart

## Introduction

The volume and velocity of storm water that runs off to nearby streams will likely increase whenever a new housing development is built. Storm-water runoff is a problem when it picks up garbage, debris, chemicals, fertilizers, and other pollutants from parking lots, yards, streets, and roofs. Many communities apply low-impact development strategies to reduce the impact of storm water on streams. These strategies can be used to plan new development or to modify existing development.

## Background Questions

1. What happens to rain that falls in the area in which you live? Does the water stay on top of the ground or does it soak into the ground? How much runs off into storm drains?

2. Where does the water in these storm drains go?

## Methods

1. Imagine that you are serving on a committee charged with granting approval for new subdivisions in your community.

2. Examine the eight low-impact development strategies given in the Results table. Which of these strategies are already being applied in subdivisions in your area and in construction projects on your campus?

3. Rate the eight strategies from 1 to 8, where 1 is the strategy you feel offers the greatest potential for protecting streams while minimizing inconveniences for people in the subdivision.

## Results

| TABLE 1 | Comparing eight "low-impact development" strategies | | |
|---|---|---|---|
| Strategy | Effect on runoff | Possible concerns | Ranking |
| Use pervious paving materials. | Allows more rainwater to soak in. | Initial installation costs may be higher than impervious surfaces. | |
| Build sidewalks on just one side of street. | Reduces area of impervious surfaces. | Some walkers must cross the street to access the sidewalk. | |
| Make streets narrower. | Reduces area of impervious surfaces. | Harder for school buses to pick up and let off students. | |
| Use vegetated channels along roads instead of curbs and gutters. | Allows water to flow to and soak into adjacent ground. | Curbs and gutters get rid of runoff faster but unfiltered runoff harms streams. | |
| Make narrower turnarounds (cul-de-sacs) with a vegetated island. | Reduces area of impervious surfaces. | Fewer parking spaces for guests. | |
| Plant rain gardens. | Absorbs runoff and pollutants, slowly releases runoff. | Some maintenance required, but much less than for ordinary gardens. | |
| Minimize clearing and grading. | Promotes absorption by avoiding soil compaction. | Some home buyers want a flat grassy lawn. | |
| Conserve trees during construction. | Promotes absorption by avoiding soil compaction. | Some contractors and homeowners prefer to landscape from scratch. | |
| Establish a streamside buffer zone. | Absorbs pollutants, slows down runoff. | Some people want streamside property. | |

## Follow-up

1. For which, if any, of these strategies do you think the concerns outweigh the benefits?

2. Which of these strategies would you recommend that your community apply in planning for new construction projects?

3. Which strategies would help your community better manage its existing storm-water problems?

## Reference

EPA Low Impact Development Web site: http://www.epa.gov/owow/NPS/lid/

# CHAPTER 15

# Engage Your Environment: The Air Quality Index

## How can you use air quality data to reduce your exposure to air pollution?

### Learning Objectives

You will:

1. Consider how to alter your schedule of activities to reduce your exposure to air pollution.

2. Consider how you can reduce your contribution to air pollution.

### Resources

Air quality index color code chart (Introduction Table 1)
Feasibility analysis chart (Results Table 2)

## Introduction

The Environmental Protection Agency has established national air quality standards to protect public health. The EPA calculates the air quality index for five criteria air pollutants regulated by the Clean Air Act: ground-level ozone, particulate matter, carbon monoxide, sulfur dioxide, and nitrogen dioxide. The first two of these pollutants pose the greatest threat to human health in the United States. The air quality index scale runs from 0 to 500. An index below 100 is considered "acceptable"; an index above 100 indicates progressively worsening air quality. A system of color codes makes the air quality index easy to understand.

| TABLE 1 | Air quality index color code chart | |
|---|---|---|
| Air quality index | Color | Air quality conditions |
| 0–50 | Green | Good |
| 51–100 | Yellow | Low levels of pollution |
| 101–150 | Orange | Unhealthy for sensitive groups |
| 151 or above | Red, purple, or maroon | Unhealthy for everyone |

Sensitive groups include children, older adults, and people with lung or heart disease. Healthy people who are active outdoors (including runners, hikers, and athletes) are also considered to be "sensitive" to air pollution, and they should avoid outdoor activities during "Code Orange" conditions.

## Background Questions

1. Do you or does anyone in your family have asthma?

2. How much time do you spend outdoors in the summer in sporting activities, running, or hiking?

3. How could "Code Red" or "Code Orange" air quality affect summer activities for children in your community?

4. How could "Code Red" or "Code Orange" air quality affect summer training for athletes on your campus?

## Methods

1. Consider your usual summer activities and their schedules.
2. Determine how you could change your behavior on days forecast to have dangerous levels of ground-level ozone ("Code Red" or "Code Orange").
3. Decide if it would be easy for you to reduce your contribution to air pollution.
4. Complete Results Table 2, the feasibility analysis chart.

## Results

| TABLE 2 | Feasibility analysis chart |
|---|---|

| Actions to reduce exposure to air pollution | Feasibility (10 = very feasible; 1 = not feasible) |
|---|---|
| When air quality is poor, stay indoors during the hottest portion of the day. | |
| Schedule outdoor exercise for early morning. | |
| Alert other people in your community about poor air quality days. | |

| Actions to reduce contribution to air pollution | Feasibility (10 = very feasible; 1 = not feasible) |
|---|---|
| If your electricity comes from coal or other fossil fuels, conserve electricity (e.g., turn off lights when leaving the room; take showers instead of baths). | |
| Set your air conditioner to a higher temperature. | |
| Bicycle or walk instead of driving. | |
| Share a ride or use public transit. | |
| Limit engine idling when driving. | |
| Refuel your car after dusk. | |
| Combine errands and reduce trips. | |

## Follow-up

1. Which behavior changes are you and your friends most likely to implement on "Code Red" or "Code Orange" days?

2. How could air quality data influence your travel plans?

3. How could local air quality information help you make informed, health-protective decisions about moving to a new community?

4. How does the air quality in your community compare with that in other communities? What time of year has the best air quality? Has the air quality in your community been improving or declining in recent years?

## References

Air Compare Web site: http://www.epa.gov/aircompare/index.htm

Air Quality Index: A Guide to Air Quality and Your Health: http://www.epa.gov/airnow/aqi_brochure_08-09.pdf

Local Air Quality Conditions and Forecasts: http://www.airnow.gov/

Ozone Pollution Information: http://www.epa.gov/air/ozonepollution/

## How does your production of municipal solid waste compare to that of the average person in the United States?

### Learning Objectives

You will:

1. Understand the current trends in U.S. waste production.

2. Analyze your personal daily waste production.

3. Compare your personal waste production to that of the average person in the United States.

4. Identify ways to reduce the amount of material that enters the municipal solid waste stream.

### Resources

1. Figure 16.3

2. Garbage bag

3. Scale (optional)

## Introduction

Modern consumers buy and consume many things throughout the day, and they produce waste, or municipal solid waste (MSW), which includes, but is not limited to, product packaging, food waste, appliances, and furniture. The U.S. Environmental Protection Agency (EPA) estimates that the average person in the United States produces approximately 2 kg (4.5 pounds) of waste each day. Measuring your waste production is one way to gauge your impact on the environment.

### Background Questions

1. Using the data presented in Figure 16.3 on page 439, describe patterns of MSW production for each decade in the United States from 1960 to 2008.

2. Why did the total MSW production rise from 1970 to 2008 while the per capita MSW production remained fairly constant?

## Methods

1. Choose one day to tally your garbage.

2. On the tally day, place all MSW that you will discard in a garbage bag. Include paper, glass, plastics, metals, rubber, leather, textiles, and wood.

3. Weigh (or estimate the weight of) anything that is too messy (such as food waste) or too heavy to place in your garbage bag. Record the weight in Results Table 1. If you have no scale, sample weights are provided to help you estimate the weight of your discards.

4. At the end of the day, separate your MSW into the categories described in the table below. Weigh (or estimate the weight of) your MSW for each material type and enter the data in the table.

## Results

| TABLE 1 | Garbage tally | |
|---|---|---|
| Material type | Sample weights for each material type | Discarded item and its estimated or actual weight |
| Paper and paper products | 8–10 pieces of 8.5 × 11 paper weigh 2 oz. | |
| Glass | An empty beer bottle weighs 6 oz. | |
| Metals | An empty 28 oz. aluminum can weighs 3 oz. | |
| Plastics | A plastic food container weighs 1 oz. | |
| Rubber and leather | A man's shoe weighs 14 oz. | |
| Textiles | A sock weighs 1 oz.; a shirt weighs 7 oz. | |
| Wood | A pencil weighs .5 oz. | |
| Other | An apple core weighs 1 oz. | |

## Follow-up

1. In total pounds, how much MSW did you produce during your one-day tally?

2. Did you collect more or less MSW during your one-day tally than you expected?

3. In which category did you produce the most MSW?

**4.** How does your daily production of MSW compare to the U.S. average in Figure 16.3?

**5.** In what way does the kind of work you do at business or school contribute to the type or quantity of MSW you produce?

**6.** How could you reduce your MSW production?

**7.** If you implemented some of the waste reduction strategies in question 6, by how much would you reduce the weight of your one-day tally?

## Reference

U.S. Environmental Protection Agency: www.epa.gov/osw/nonhaz/municipal/pubs/msw2008rpt .pdf. Feb. 2009. Web. 20 Sept. 2010.

## CHAPTER 17

# Engage Your Environment: Making Sunwise Decisions

## How can you reduce risks associated with sun exposure?

### Learning Objectives

You will:

1. Consider the behaviors currently putting you at risk for the effects of UV radiation.

2. Prioritize alternative behaviors that would reduce your risks.

### Resource

EPA SunWise Action Steps inventory

## Introduction

In the 1970s, the stratospheric ozone layer began to shrink due to the release of chlorine compounds into the atmosphere. Because of subsequent bans on these compounds the ozone layer is expected to gradually recover over the coming decades although, until it does recover, exposure levels for UV radiation will be high. UV radiation can cause premature aging of the skin, skin cancer, cataracts, and suppression of the immune system.

The Environmental Protection Agency recommends a number of specific steps to prevent overexposure to this radiation. While most of these steps are intended to prevent overexposure to UV light outdoors, even tanning beds can expose you to UV radiation and should be avoided

## Methods

1. Complete the EPA SunWise Action Steps inventory in the Results section.

2. Add up your scores to determine your total.

## Results

| TABLE 1 | EPA SunWise Action Steps inventory |
|---|---|
| **Your score** | **EPA SunWise Action Steps survey** |
| | How often do you get a sunburn? (Frequently= 10; Seldom = 5; Never = 0) |
| | How often do you deliberately "get a tan"? (Frequently = 10; Seldom = 5; Never = 0) |
| | How often do you generously apply sunscreen before getting in the sun? (Never = 10; Seldom = 5; Always = 0) |
| | How often do you wear protective clothing in the sun? (Never = 10; Seldom = 5; Always = 0) |
| | How often do you wear sunglasses in the sun? (Never = 10; Seldom = 5; Always = 0) |
| | How often do you seek shade between 10 A.M. and 4 P.M. when outside? (Never = 10; Seldom = 5; Always = 0) |
| | How often do you use extra caution in the sun near water, snow, and sand? (Never = 10; Seldom = 5; Always = 0) |
| | How often do you check the UV Index before spending time in the sun? (Never = 10; Seldom = 5; Always = 0) |
| | How do you get your vitamin D? (By deliberately spending time in the sun = 10; By taking vitamin D supplements = 0) |
| **Total:** | 70 or more: You need to change your behaviors, and the sooner the better! |
| | 25–65: You are at elevated risk for effects of UV exposure! |
| | 20 or less: You are very SunWise! |

## Follow-up

1. For which behaviors did you receive a high score?

2. Which of your high-point behaviors would be easiest for you to change in order to better protect yourself from the effects of UV exposure?

3. Which of your high-point behaviors would be the hardest for you to change and why?

4. Given the risks of UV exposure, do you feel that the benefits of changing your behavior would be worth it? Why, or why not?

## References

Daily UV Index: www.epa.gov/sunwise/uvindex.html
EPA SunWise Program: http://epa.gov/sunwise/index.html
Safeguarding the Ozone Layer: http://www.ipcc.ch/pdf/special-reports/sroc/sroc_spm.pdf

## CHAPTER 18

# Engage Your Environment: Personal Conservation Strategies

## What are ways to conserve biodiversity?

### Learning Objectives

You will:

1. Evaluate actions that could potentially benefit plant and animal life.
2. Identify conservation activities in which you can participate.

### Resources

1. Pages 496–500
2. Personal Conservation Strategies chart in Results section

## Introduction

The Millennium Ecosystem Assessment shows that human actions often lead to irreversible losses in terms of the diversity of life on Earth. These losses have been more rapid in the past 50 years than ever before in human history.

### Background Questions

1. List three reasons why loss of biodiversity should concern humans.

2. What are two species in your state that are threatened or endangered? If you are unaware of local species that are endangered or threatened, a list is available online: http://www.fws.gov/endangered/.

## Methods

Complete the table in the Results section. If the conservation action does not apply to you, write "does not apply" in column 3 (e.g., if you don't live near a beach or a body of water).

## Results

| TABLE 1 | Personal conservation strategies | | |
|---|---|---|---|
| Conservation action | | Difficulty of task where 10 = most difficult and 1 = least difficult | Will you complete conservation action? (no, maybe, yes) |
| **General** | | | |
| Educate yourself about biodiversity issues by reading books such as *The Diversity of Life* by E. O. Wilson. | | | |
| Study environmental science and conservation biology in school and pursue it as a career. | | | |
| Vacation in ecotourism destinations that provide funds for conservation activities. | | | |
| Vote for candidates supporting wildlife conservation and contact your representatives to notify them of your concerns for wildlife issues. | | | |
| Join a conservation organization and volunteer for conservation projects. | | | |
| Teach children to respect nature and the environment. Take them on hikes or camping. Be a good example and role model. | | | |
| **Aquatic/marine protection** | | | |
| If you must keep a saltwater tank, buy only fish certified by the Marine Aquarium Council as sustainably caught or reared in captivity. | | | |
| Never return aquarium fish to the ocean or other body of water. This practice has introduced non-native species to many areas, disrupting the balance of marine ecosystems. | | | |
| Avoid eating unsustainably caught seafood. Carry a sustainable seafood wallet guide available from the Monterey Bay Aquarium: http://www.montereybayaquarium.org/ | | | |
| Keep beaches clean. Participate in beach cleanups if you live in a coastal area. | | | |
| Don't walk on sand dunes. Dunes provide a barrier to prevent beach erosion from wind and water and often contain native plants vital to local ecosystems. | | | |
| Practice safe and clean boating. Obey no-wake zones, and watch out for marine life. | | | |
| Don't dispose of trash or toilet waste in the ocean. | | | |
| **Terrestrial protection** | | | |
| Familiarize yourself with invasive species and remove them from your home, or volunteer to remove them from local parks. | | | |
| Don't buy wood that isn't certified by the Forest Stewardship Council. | | | |
| Opt for an alternative to a grass lawn (which uses a lot of water and fertilizer and doesn't provide shelter for wildlife). With a nontraditional yard, you can liven up your home and create habitats for animals by planting a variety of native plants. | | | |
| Avoid using pesticides; use natural predators (e.g., the praying mantis) and insect-deterring plants (e.g., onions) to deter pests in the garden. | | | |
| Pull weeds instead of using herbicides, or better yet let them grow. | | | |
| Avoid use of chemical fertilizers. Instead, make your own mulch and use organic fertilizers only when needed. | | | |
| Put up birdfeeders, birdhouses, and birdbaths. | | | |
| Keep outside trash cans closed. Use lids that snap shut to prevent wildlife from eating hazardous materials and becoming a nuisance. | | | |
| Plant native trees and plants. | | | |

## Follow-up

1. How much disruption will your conservation actions cause you in your daily life?

2. What are the primary barriers to your participation in conservation activities?

3. Besides the activities included in the Results table, list two more actions you could take to conserve biodiversity.

## References

Conservation International: http://www.conservation.org
Invasive species information: http://Invasivespeciesinfo.gov
IUCN Red List of Threatened Species: http://www.iucnredlist.org/
Natural Resources Defense Council: http://www.nrdc.org
World Wildlife Fund: http://www.worldwildlife.org

## CHAPTER 19

# Engage Your Environment: Greenhouse Gas Emissions

## How could you reduce your carbon dioxide emissions?

### Learning Objectives

You will:

1. Understand the contribution greenhouse gases make to climate change.

2. Determine how you can reduce your emissions of carbon dioxide.

### Resource

EPA Household Emissions Calculator: http://epa.gov/climatechange/emissions /ind_calculator.html (optional)

## Introduction

Global climate change refers to any change in climate such as changes in surface temperature, precipitation, and weather patterns. Modern global climate change has been linked to the greenhouse effect, where energy absorbed by atmospheric gases is redirected to Earth's surface, thereby warming it. One gas implicated in intensifying this process is $CO_2$. Although there are many natural sources of $CO_2$, limiting anthropogenic sources of this gas has attracted great interest.

### Background Questions

1. How has the greenhouse effect been linked to increased global surface temperatures?

2. What are the anthropogenic sources of the following greenhouse gases: $CO_2$, methane, ozone, nitrous oxide, and CFCs?

## Methods

Review the table in the Results section and answer the questions that follow.

## Results

| TABLE 1 | Practical steps to reduce $CO_2$ emissions | | |
| --- | --- | --- | --- |
| Actions | | Dollar savings per year | Reduction in $CO_2$ production per year (pounds) |
| Drive 5 fewer miles each week. | | $42 | 212 |
| Perform routine auto maintenance. | | $124 | 892 |
| Drive a car that gets 10 more miles per gallon. | | $591 | 4,248 |
| Enable sleep feature on home computer. | | $10 | 90 |
| Wash 2 loads of laundry each week in cold water instead of warm water. | | $12 | 103 |
| Dry 50% of weekly laundry on a clothesline. | | $40 | 346 |
| Replace five 75-watt incandescent bulbs with 20-watt Energy Star bulbs. | | $35 | 369 |
| Total | | $854 | 6,260 |

For comparison, a household of one person produces about 20,750 pounds of $CO_2$ in one year.

## Follow-up

1. If you could implement all of the changes shown in the Results table, by what percentage would your annual $CO_2$ production decrease?

2. Which of these actions have you already implemented? Which others might you implement?

3. What are some limitations in implementing actions to reduce $CO_2$ emissions?

**4.** In addition to the actions listed in the Results table, what other ways could you reduce your contribution to greenhouse gases?

## Reference

EPA Household Emissions Calculator: http://epa.gov/climatechange/emissions/ind_calculator.html

# CHAPTER 20
# Engage Your Environment: Reducing Your Community's Footprint

## How can you help your community move toward sustainability?

### Learning Objectives

You will:

1. Evaluate the sustainability needs of your community.

2. Identify ways you can apply your talents to help address these needs.

### Resource

Community Issues Chart

## Introction

According to Chapter 1, one way to assess sustainability is by measuring our impact on world resources, including energy, food, and natural ecosystems. Using up resources more quickly than they can be produced or producing wastes more quickly than they can be processed are evidences of unsustainable living.

By now you have probably completed several Engage Your Environment activities and have considered ways of reducing your personal ecological footprint. This exercise broadens the scope and examines how your talents and interests can help you move your campus or hometown toward sustainability

## Background Questions

1. What ongoing programs does your campus or hometown have that relate to sustainability? (Examples would be cleanup campaigns, greenway projects, recycling programs, storm-water management, farmers' markets, and home weatherizing campaigns.)

2. What talents could you bring to community efforts to increase sustainability? (Examples would be communication skills, organization skills, "people skills," or expertise in particular activities such as gardening, home weatherizing, landscaping, or graphic design.)

## Methods

1. If you are a full-time college student, you may wish to consider how you could help your campus move toward sustainability.

2. Alternatively, you may wish to consider how you could help your hometown move toward sustainability.

3. Complete the chart in the Results section by rating the need in your community and considering ways you could contribute to solving a community issue.

## Results

| Potential community issue | Need (10 = very great; 1 = not great at all) | What you would be interested in doing (based on your talents) |
|---|---|---|
| Solid waste (reducing waste generation, recycling, litter cleanup, etc.) | | |
| Storm water (rain barrels, rain gardens, pervious surfaces, etc.) | | |
| Wildlife habitat and greenways (native plants, tree plantings, parks, etc.) | | |
| Energy (conservation, weatherizing, landscaping, transportation, etc.) | | |
| Food sustainability (community gardens, farmers' markets, farm-to-schools, etc.) | | |

## Follow-up

1. Based on your community's needs and your talents, is there one specific existing program in which you intend to participate? If so, what is it?

2. Based on community needs and your talents, is there one new program or project you would be interested in helping to organize? If so, what is it?

## References

College Sustainability Report Card: http://www.greenreportcard.org/
Sustainable Communities Online: http://www.sustainable.org/

# International System of Units (Metric System) and Common U.S. Unit Conversions

| Measurement | Unit Abbreviation | Metric Equivalent | Metric to U.S. | U.S. to Metric |
|---|---|---|---|---|
| Area | 1 square meter (m²) | = 10,000 square centimeters | 1 m² = 1.1960 square yards | 1 square yard = 0.8361 m² |
| | | | 1 m² = 10.764 square feet | 1 square foot = 0.0929 m² |
| | 1 square centimeter (cm²) | = 100 square millimeters | 1 cm² = 0.155 square inch | 1 square inch = 6.4516 cm² |
| Length | 1 kilometer (km)<br>1 meter (m) | = 1,000 (10³) meters<br>= 100 (10²) centimeters<br>= 1,000 millimeters | 1 km = 0.62 mile<br>1 m = 1.09 yards<br>1 m = 3.28 feet<br>1 m = 39.37 inches | 1 mile = 1.61 km<br>1 yard = 0.914 m<br>1 foot = 0.305 m |
| | 1 centimeter (cm)<br>1 millimeter (mm) | = 0.01 (10⁻²) meter<br>= 0.001 (10⁻³) meter | 1 cm = 0.394 inch<br>1 mm = 0.039 inch | 1 inch = 2.54 cm |
| Mass | 1 metric tonne (t) | = 1,000 kilograms | 1 t = 1.103 ton (referred to in text as "U.S. ton") | 1 ton = 0.907 t |
| | 1 kilogram (kg)<br>1 gram (g)<br>1 milligram (mg) | = 1,000 grams<br>= 1,000 milligrams<br>= 0.001 gram | 1 kg = 2.205 pounds<br>1 g = 0.0353 ounce | 1 pound = 0.4536 kg<br>1 ounce = 28.35 g |
| Volume–solids | 1 cubic meter (m³) | = 1,000,000 cubic centimeters | 1 m³ = 1.3080 cubic yards | 1 cubic yard = 0.7646 m³ |
| | 1 cubic centimeter (cm³ or cc)<br>1 cubic millimeter (mm³) | = 0.000001 cubic meter<br>= 1 milliliter<br>= 0.000000001 cubic meter | 1 m³ = 35.315 cubic feet<br>1 cm³ = 0.0610 cubic inch | 1 cubic foot = 0.0283 m³<br>1 cubic inch = 16.387 cm³ |
| Volume–liquids and gases | 1 liter (L)<br>1 kiloliter (kL) | = 1,000 milliliters<br>= 1,000 liters | 1 L = 0.264 gallons<br>1 kL = 264.17 gallons<br>1 L = 1.057 quarts | 1 quart = 0.946 L<br>1 gallon = 3.785 L |
| | 1 milliliter (ml) | = 0.001 liter<br>= 1 cubic centimeter | 1 ml = 0.034 fluid ounce<br>1 ml = approximately 1/5 teaspoon | 1 quart = 946 ml<br>1 pint = 473 ml<br>1 fluid ounce = 29.57 ml<br>1 teaspoon = approx. 5 ml |
| Time | 1 millisecond (ms) | = 0.001 second | | |
| Temperature | Degrees Celsius (°C) | | °C = 5/9 (°F − 32) | °F = 9/5 °C + 32 |
| Energy and Power | 1 kilowatt-hour | = 3,413 BTUs<br>= 860,421 calories | | |
| | 1 watt | = 3.413 BTU/hr<br>= 14.34 calorie/min | | |
| | 1 calorie | = amount of energy needed to raise the temperature of 1 gram (1 cm³) of water by 1 degree Celsius | | |
| | 1 horsepower<br>1 joule | = 7.457 × 10² watts<br>= 9.481 × 10⁻⁴ BTU<br>= 0.239 cal<br>= 2.778 × 10⁻⁷ kilowatt-hour | | |